CRC SERIES IN CHROMATOGRAPHY

Gunter Zweig and Joseph Sherma
Editors-in-Chief

GENERAL DATA AND PRINCIPLES
Editors
Gunter Zweig, Ph.D.
U.S. Environmental Protection Agency
Washington, D.C.
Joseph Sherma, Ph.D.
Lafayette College
Easton, Pennsylvania

CARBOHYDRATES
Editor
Shirley C. Churms, Ph.D.
Research Associate
C.S.I.R. Carbohydrate Chemistry Research Unit
Department of Organic Chemistry
University of Cape Town, South Africa

DRUGS
Editor
Ram Gupta, Ph.D.
Department of Laboratory Medicine
St. Joseph's Hospital
Hamilton, Ontario, Canada

TERPENOIDS
Editor
Carmine J. Coscia, Ph.D.
Professor of Biochemistry
St. Louis University School of Medicine
St. Louis, Missouri

STEROIDS
Editor
Joseph C. Touchstone, B.S., M.S., Ph.D.
Professor
School of Medicine
University of Pennsylvania
Philadelphia, Pennsylvania

PESTICIDES AND RELATED ORGANIC CHEMICALS
Editors
Joseph Sherma, Ph.D.
Joanne M. Follweiler, Ph.D.
Lafayette College
Easton, Pennsylvania

LIPIDS
Editor
H. K. Mangold, Dr. rer. nat.
Executive Director and Professor
Federal Center for Lipid Research
Munster, Federal Republic of Germany

HYDROCARBONS
Editors
Walter L. Zielinski, Jr., Ph.D.
Air Program Manager
National Bureau of Standards
Washington, D.C.

INORGANICS
Editor
M. Qureshi, Ph.D.
Professor, Chemistry Section
Zakir Husain College of Engineering and
  Technology
Aligarh Muslim University
Aligarh, India

PHENOLS AND ORGANIC ACIDS
Editor
Toshihiko Hanai, Ph.D.
University of Montreal
Quebec, Canada

AMINO ACIDS AND AMINES
Editor
Dr. S. Blackburn
Leeds, England

POLYMERS
Editor
Charles G. Smith
Dow Chemical, USA
Midland, Michigan

PLANT PIGMENTS
Editor
Dr. Hans-Peter Köst
Botanisches Institüt der Universität München
München, Federal Republic of Germany

CRC
Handbook
of
Chromatography

Lipids

Volume I

Editor

Helmut K. Mangold, Dr. rer. nat.

Executive Director and Professor
Federal Center for Lipid Research
Institute for Biochemistry and Technology
H. P. Kaufmann-Institute
Münster, Federal Republic of Germany

Editors-in-Chief

Gunter Zweig, Ph.D.

School of Public Health
University of California
Berkeley, California

Joseph Sherma, Ph.D.

Professor of Chemistry
Lafayette College
Easton, Pennsylvania

CRC Press, Inc.
Boca Raton, Florida

Library of Congress Cataloging in Publication Data
Main entry under title:

Lipids

 (CRC handbook of chromatography)
 Includes bibliographies and index.
 1. Lipids—Analysis. 2. Chromatographic analysis.
I. Mangold, H. K. (Helmut K.) II. Zweig, Gunter.
III. Sherma, Joseph. IV. Series. [DMLM: 1. Lipids—
Analysis—Handbooks. 2. Chromatography—Handbooks.
QU 85 L76525]
QD305.F2L45   1984   547.7'7046   82-25510
ISBN 0-8493-3037-8 (v. 1)
ISBN 0-8493-3038-6 (v. 2)

Direct all inquiries to CRC Press, Inc., 2000 Corporate Blvd., N.W., Boca Raton, Florida, 33431.

© 1984 by CRC Press Inc.
International Standard Book Number 0-8493-3037-8 (v. 1)
International Standard Book Number 0-8493-3038-6 (v. 2)

Library of Congress Card Number 82-25510
Printed in the United States

SERIES PREFACE

The present two volumes on the chromatography of lipids is a monumental task undertaken by Professor H. K. Mangold of the H. P. Kaufmann Institut in Muenster, Germany and a group of experts who contributed chapters in their specialties. Thus, we believe that this two-volume treatise represents one of the most comprehensive work not only on the chromatography of lipids and derivatives but also on their chemistry and other properties. The volume editor and the contributors are to be commended for having put together such a detailed book.

The Handbook of Chromatography — Lipids is another one in the continuing series of handbooks which was originally begun by us in 1972 and is still available as *Handbook of Chromatography,* Volumes I and II, of this series. These two volumes represented the state-of-the-art of chromatography of most organic and inorganic compounds compiled from the literature known up till then. When we contemplated revising the original volumes ten years later, it became apparent that the field of chromatography had grown to such proportions that we decided to invite volume editors for most important classes of organic and inorganic compounds and ions, respectively, and to edit separate volumes devoted to specific topics. Thus, we have edited volumes on Drugs (2); Carbohydrates; Polymers; Phenols and Organic Acids. Other volumes in this series will cover the fields of pesticides; pigments; amines and amino acids; hydrocarbons; terpenoids; peptides; and inorganics.

As in the past, we are asking our devoted readers and users of these handbooks to point out errors or omissions and to suggest other topics, and possibly editors, which should be covered in this series.

Gunter Zweig, Ph.D.
Joseph Sherma, Ph.D.
Editors-in-Chief

THE EDITORS-IN-CHIEF

Gunter Zweig, Ph.D., received his undergraduate and graduate training at the University of Maryland, College Park, where he was awarded the Ph.D. in biochemistry in 1952. Two years following his graduation, Dr. Zweig was affiliated with the late R. J. Block, pioneer in paper chromatography of amino acids. Zweig, Block, and Le Strange wrote one of the first books on paper chromatography which was published in 1952 by Academic Press and went into three editions, the last one authored by Gunter Zweig and Dr. Joe Sherma, the co-Editor-in-Chief of this series. *Paper Chromatography* (1952) was also translated into Russian.

From 1953 till 1957, Dr. Zweig was research biochemist at the C. F. Kettering Foundation, Antioch College, Yellow Springs, Ohio, where he pursued research on the path of carbon and sulfur in plants, using the then newly developed techniques of autoradiography and paper chromatography. From 1957 till 1965, Dr. Zweig served as lecturer and chemist, University of California, Davis and worked on analytical methods for pesticide residues, mainly by chromatographic techniques. In 1965, Dr. Zweig became Director of Life Sciences, Syracuse University Research Corporation, New York (research on environmental pollution), and in 1973 he became Chief, Environmental Fate Branch, Environmental Protection Agency (EPA) in Washington, D.C.

During his government career, Dr. Zweig continued his scientific writing and editing. Among his works are (many in collaboration with Dr. Sherma) the now 11-volume series on *Analytical Methods for Pesticides and Plant Growth Regulators* (published by Academic Press); the pesticide book series for CRC Press; co-editor of *Journal of Toxicology and Environmental Health*; co-author of basic review on paper and thin-layer chromatography for *Analytical Chemistry* from 1968 to 1980; co-author of applied chromatography review on pesticide analysis for *Analytical Chemistry*, beginning in 1981.

Among the scientific honors awarded to Dr. Zweig during his distinguished career are the Wiley Award in 1977, Rothschild Fellowship to the Weizmann Institute in 1963/64; the Bronze Medal by the EPA in 1980.

Dr. Zweig has authored or co-authored over 75 scientific papers on diverse subjects in chromatography and biochemistry, besides being the holder of three U.S. patents.

At the present time (1980/84), Dr. Zweig is Visiting Research Chemist in the School of Public Health, University of California, Richmond, where he is doing research on farmworker safety as related to pesticide exposure.

Joseph Sherma, Ph.D., received a B.S. in Chemistry from Upsala College, East Orange, N.J., in 1955 and a Ph.D. in Analytical Chemistry from Rutgers University in 1958. His thesis research in ion exchange chromatography was under the direction of the late William Rieman, III. Dr. Sherma joined the faculty of Lafayette College in September 1958, and is presently Charles A. Dana Professor of Chemistry in charge of two courses in analytical chemistry. At. Lafayette he has continued research in chromatography and had additionally worked a total of 12 summers in the field with Harold Strain at the Argonne National Laboratory, James Fritz at Iowa State University, Gunter Zweig at Syracuse University Research Corporation, Joseph Touchstone at the Hospital of the University of Pennsylvania, Brian Bidlingmeyer at Waters Associates, and Thomas Beesley at Whatman, Inc., Clifton, N.J.

Dr. Sherma and Dr. Zweig (who is now with the University of California-Richmond) co-authored or co-edited the original Volumes I and II of the *CRC Handbook of Chromatography*, a book on paper chromatography, seven volumes of the series *Analytical Methods for Pesticides and Plant Growth Regulators*, and the Handbooks of Chromatography of drugs, carbohydrates, polymers, and phenols and organic acids. Other books in the pesticide series and further volumes of the *CRC Handbook of Chromatography* are being edited with Dr. Zweig, and Dr. Sherma has co-authored the handbook on pesticide chromatography. A book on quantitative TLC was edited jointly with Dr. Touchstone, and a general book on TLC was co-authored with Dr. B. Fried. Dr. Sherma has been co-author of eight biennial reviews of column liquid and thin layer chromatography (1968—1982) and the 1981 and 1983 reviews of pesticide analysis for the journal *Analytical Chemistry*.

Dr. Sherma has written major invited chapters and review papers on chromatography and pesticides in *Chromatographic Reviews* (analysis of fungicides), *Advances in Chromatography* (analysis of nonpesticide pollutants), Heftmann's *Chromatography* (chromatography of pesticides), Race's *Laboratory Medicine* (chromatography in clinical analysis), *Food Analysis: Principles and Techniques* (TLC for food analysis), *Treatise on Analytical Chemistry* (paper and thin layer chromatography), *CRC Critical Reviews in Analytical Chemistry* (pesticide residue analysis), *Comprehensive Biochemistry* (flat bed techniques), *Inorganic Chromatographic Analysis* (thin layer chromatography), and *Journal of Liquid Chromatography* (advances in quantitative pesticide TLC). He is editor for residues and elements for the JAOAC.

Dr. Sherma spent 6 months in 1972 on sabbatical leave at the EPA Perrine Primate Laboratory, Perrine, Fla., with Dr. T. M. Shafik, and two additional summers (1975, 1976) at the USDA in Beltsville, Md., with Melvin Getz doing research on pesticide residue analysis methods development. He spent 3 months in 1979 on sabbatical leave with Dr. Touchstone developing clinical analytical methods. A total of more than 230 papers, books, book chapters, and oral presentations concerned with column, paper, and thin layer chromatography of metal ions, plant pigments, and other organic and biological compounds; the chromatographic analysis of pesticides; and the history of chromatography have been authored by Dr. Sherma, many in collaboration with various co-workers and students. His major research area at Lafayette is currently quantitative TLC (densitometry), applied mainly to clinical analysis and pesticide residue and food additive determinations.

Dr. Sherma has written an analytical quality control manual for pesticide analysis under contract with the U.S. EPA and has revised this and the EPA Pesticide Analytical Methods Manual under a 4-year contract jointly with Dr. M. Beroza of the AOAC. Dr. Sherma has also written an instrumental analysis quality assurance manual and other analytical reports for the U.S. Consumer Product Safety Commission, and is currently preparing two manuals on the analysis of food additives for the U.S. FDA, both of these projects also in collaboration with Dr. Beroza of the AOAC.

Dr. Sherma taught the first prototype short course on pesticide analysis with Henry Enos of the EPA for the Center for Professional Advancement. He was editor of the Kontes TLC quarterly newsletter for 6 years and also has taught short courses on TLC for Kontes and the Center for Professional Advancement. He is a consultant for numerous industrial companies and federal agencies on chemical analysis and chromatography and regularly referees papers for analytical journals and research proposals for government agencies. At Lafayette, Dr. Sherma, in addition to analytical chemistry, teaches general chemistry and a course in thin layer chromatography.

Dr. Sherma has received two awards for superior teaching at Lafayette College and the 1979 Distinguished Alumnus Award from Upsala College for outstanding achievements as an educator, researcher, author, and editor. He is a member of the ACS, Sigma Xi, Phi Lambda Upsilon, SAS, and AIC.

PREFACE

Chromatographic techniques have become indispensible tools for the analysis and char-aterization of lipophilic compounds. The present volume deals with methods for the analysis of naturally occurring fats and other lipids including less common compounds, such as waxes, diol lipids, and alkoxylipids, and rather unusual lipids, such as sterylglycolipids. The constituents of the surface lipids of plants as well as animals are also considered.

Special emphasis is on human tissue lipids. Methods for the analysis of lipids that are of importance as integral components of serum lipoproteins or as constituents of central and peripheral nervous tissues are dealt with in detail. the analysis of physiologically active compounds, such as the prostaglandins, is also treated. Thus, this volume should be of help to all those engaged in biomedical research and clinical analysis.

Methods for the characterization of intermediates in lipid synthesis are described for the benefit of organic chemists. Heretofore, this subject has not been covered in any treatise.

The systematic analysis and characterization of detergents, emulsifiers, plasticizers as well as oligomers and polymers of technological importance are considered because these substances are ever-present in daily life. Previous reviews on lipid chromatography have not included these groups of substances.

The members of the Advisory Board, R. G. Ackman, A. Kuksis, L. D. Metcalfe, K. D. Mukherjee, and F. Spener and all the authors of this volume have contributed substantially to the development of chromatographic methods of lipid analysis. They have prepared their chapters in such a way that the newcomer as well as the experienced worker can profit from their expertise. They have provided numerous hints from personal experience and have included a great deal of unpublished results. Thus this team of authors has assembled an up-to-date laboratory handbook on the analysis of naturally occurring lipids and their synthetic derivatives.

I thank all contributors for their help and cooperation.

H. K. Mangold
Münster, September 1981.

THE EDITOR

Helmut K. Mangold, Dr. rer. nat., Executive Director and Professor, Federal Center for Lipid Research, Münster, Federal Republic of Germany, was born June 19, 1924 in Heilbronn, Germany. He studied chemistry at the universities of Tübingen, Würzburg, and Heidelberg. He received his doctorate degree in 1952. He is married to Anne Mangold, nee Wenzel, and has three children, Barbara, Michael, and Ulrike.

After having worked for a chemical company in Germany, Dr. Mangold emigrated to the United States, where he was affiliated with the University of Minnesota, St. Paul, from 1954 through 1969. During this time, he was promoted to Assistant Professor (1957), Associate Professor (1962), and Professor of Biochemistry (1966). From 1959 through 1969 he headed a research group at The Hormel Institute of the University of Minnesota in Austin, Minnesota. He assumed his present position in 1970.

Dr. Mangold has been engaged in the development of chromatographic and thermoanalytical methods of lipid research, the synthesis and biosynthesis of unusual lipids, especially ether lipids, and the lipids of plant cell cultures. More recently, he has become involved in various biotechnological areas of research and development, such as the isolation and production of plant proteins and lipoproteins. He has published about 120 original papers and over 30 chapters in books. In addition, he holds 3 patents.

Throughout the years, Dr. Mangold has been active as co-editor or member of the editorial board of the journals, *Journal of Lipid Research, Separation Science and Technology, Journal of Chromatographic Science,* and *Annals of Nutrition Metabolism.* He was one of the editors of *Chemistry and Physics of Lipids* at its inception in 1966, and in 1981 he became managing editor of that periodical.

In 1960, Dr. Mangold won the Bond Award of the American Oil Chemists' Society and in 1978 the coveted Heinrich Wieland Prize. He has been a Consultant to the Oak Ridge Associated Universities, Oak Ridge, Tennessee, a Guest Professor at the universities of Alexandria and Cairo, Egypt, and recipient of a travel award by the Japan Society for the Promotion of Science. He has presented lectures at meetings and symposia in the United States, Canada and Peru, in various European countries, also in Japan as well as Egypt and South Africa.

Dr. Mangold's hobbies, besides the chemistry and biochemistry of natural products, are nautical sports, water coloring, literature, and classical music.

For Anne,
Barbara, Michael, and Ulrike

ADVISORY BOARD

CONTRIBUTORS

R. G. Ackman, Ph.D.
Professor
Faculty of Engineering
Technical University of Nova Scotia
Halifax, Nova Scotia, Canada

Gerd Assmann, Dr. med.
Professor
Zentrallaboratorium
Universitat Münster
Münster, Federal Republic of Germany

D. R. Body
Scientist
Applied Biochemistry Division
Department of Scientific and Industrial
　Research
Palmerston North, New Zealand

William W. Christie, B.Sc., Ph.D., D.Sc.
Head
Department of Lipid Biochemistry and
　Enzymology
Hannah Research Institute
Ayr, Scotland, United Kingdom

R.M.C. Dawson, Ph.D., D.Sc., F.R.S.
Deputy Director
Head of Biochemistry Department
Institute of Animal Physiology
Cambridge, England, United Kingdom

W. Fischer, Dr. med.
Professor
Institute of Physiological Chemistry
University of Erlangen-Nürnberg
Erlangen, Federal Republic of Germany

F. A. Fitzpatrick, Ph.D.
Senior Research Scientist
Drug Metabolism Research Unit
The Upjohn Company
Kalamazoo, Michigan

P. J. Holloway, B.Pharm., Ph.D.
Principal Scientific Officer
Long Ashton Research Station
University of Bristol
Bristol, England, United Kingdom

Richard T. C. Huang, Ph.D.
Professor
Institute for Molecular Biology and
　Biochemistry
Free University of Berlin
Berlin, Federal Republic of Germany

Jaroslav Janák, D.Sc.
Senior Research Scientist
Institute of Analytical Chemistry
Czechoslovak Academy of Sciences
Brno, Czechoslovakia

Rolf Georg Kladetsky, M.D.
Department of Cardiology
University of Dusseldorf
Dusseldorf, Federal Republic of Germany

Arnis Kuksis, Ph.D.
Professor
Banting and Best Department of Medical
　Research
University of Toronto
Toronto, Ontario, Canada

Hartmut K. Lichtenthaler, Dr. rer. nat.
Professor of Plant Physiology and Phar-
　maceutical Biology
University of Karlsruhe
Karlsruhe, Federal Republic of Germany

M.S.F. Lie Ken Jie, Ph.D., F.R.S.C.
Lecturer
Department of Chemistry
University of Hong Kong
Hong Kong

John M. Lowenstein, Ph.D.
Professor of Biochemistry
Graduate Department of Biochemistry
Brandeis University
Waltham, Massachusetts

Walter O. Lundberg
Emeritus Professor of Biochemistry
University of Minnesota
Federal Center for Lipid Research
Institute for Biochemistry and Technology
H. P. Kaufmann Institute
Münster, Federal Republic of Germany

Vaidyanath Mahadevan, Ph.D.
Lipid Research Laboratory
Temple University School of Medicine
Philadelphia, Pennsylvania

Lincoln D. Metcalfe
Assistant Director of Research
AKZO Chemie-America
McCook, Illinois

H. R. Morris
Department of Biochemistry
Imperial College of Science and Technology
London, England, United Kingdom

Kumar D. Mukherjee, Dr. rer. nat.
Scientist
Federal Center for Lipid Research
Münster, Federal Republic of Germany

Toshio Muramatsu, Ph.D.
Professor of Chemisty
Tokyo Medical and Dental University
Ichikawa-Shi
Chiba, Japan

Christense Nickell
The Hormel Institute
University of Minnesota
Austin, Minnesota

George M. Patton
Veterans Administration Hospital
Boston, Massachusetts

Jan Pokorny, Ph.D., D.Sc.
Professor of Food Science
Department of Food Chemistry
Prague Institute of Chemical Technology
Prague, Czechoslovakia

Orville S. Privett, Ph.D.
Professor
The Hormel Institute
University of Minnesota
Austin, Minnesota

S. S. Radwan, Ph.D.
Professor of Botany
Faculty of Science
Ain Shams University
Cairo, Egypt

Arthur A. Schmitz
Analytical Research Chemist (Retired)
AKZO Chemie-America
McCook, Illinois

Friedrich Spener, Ph.D.
Professor
Department of Biochemistry
University of Münster
Münster, Federal Republic of Germany

Milan Streibl, Ph.D.
Head Research Scientist
Department of Natural Substances
Institute of Organic Chemistry and
  Biochemistry
Czechoslovak Academy of Sciences
Prague, Czechoslovakia

Akemi Suzuki
Section Chief
Metabolism Section
Tokyo Metropolitan Institute of Medical
  Science
Tokyo, Japan

Gerald Szajer
Analytical Section Head
AKZO Chemie-America
McCook, Illinois

Graham W. Taylor
Lecturer in Chemistry
Department of Clinical Pharmacology
Royal Postgraduate Medical School
Hammersmith Hospital
London, England, United Kingdom

Eduardo Vioque
Research Professor
Department of Oilseeds and Proteins
Instituto de la Grasa y sus Derivados
Sevilla, Spain

Chu-nan Wang
Group Leader
Analytical Section
AKZO Chemie-America
McCook, Illinois

Tamio Yamakawa
Director
Tokyo Metropolitan Institute of Medical
  Science
Tokyo, Japan

Linda Madeline Yodual
Chemist
Analytical Section
AKZO Chemie-America
McCook, Illinois

TABLE OF CONTENTS

Volume I

Chapter 1
Introduction to Lipidology... 1

Chapter 2
Extraction and Hydrolysis of Lipids and Some Reactions of
  Their Fatty Acid Components...33

Chapter 3
Fractionation of Complex Lipid Mixtures 47

Chapter 4
Hydrocarbons.. 57

Chapter 5
Alcohols, Aldehydes, and Ketones 73

Chapter 6
Straight-Chain Fatty Acids 95

Chapter 7
Branched-Chain Fatty Acids...241

Chapter 8
Carbocyclic and Furanoid Fatty Acids..277

Chapter 9
Hydroxy-, Epoxy-, and Keto-Acids 295

Chapter 10
Cutins and Suberins, The Polymeric Plant Lipids 321

Chapter 11
Surface Lipids of Plants and Animals 347

Chapter 12
Acylglycerols (Glycerides) 381

Chapter 13
Glycerophospholipids...481

Chapter 14
Ether Lipids (Alkoxylipids) 509

Chapter 15
Products of Partial Degradation of Phospholipids....................................545

Chapter 16
Glycoglycerolipids..555

Index 589

Volume II

Chapter 17
Sphingoglycolipids and Sphingophospholipids... 1

Chapter 18
Sphingolipid Bases 65

Chapters 19
Icosanoids 75
(A) Prostaglandins and Thromboxanes 77
(B) Leukotrienes 101

Chapter 20
Prenyllipids, Including Chlorophylls, Carotenoids, Prenylquinones, and Fat-Soluble
   Vitamins 115

Chapter 21
Ozonides and Products of the Ozonolysis of Lipids 171

Chapter 22
Nonvolatile Lipid Oxidation Products 205

Chapter 23
Antioxidants..219

Chapter 24
Fatty Acids Labeled with Deuterium 225

Chapter 25
Radioactively Labeled Lipids...233

Chapter 26
Plasma Lipoproteins 241

Chapter 27
Nitrogen Derivatives of Fatty Acids..267

Chapter 28
Spray Reagents 309

Chapter 29
Preparation of Reference Compounds 319

Commercial Suppliers 331

Index 335

Chapter 1

LIPIDOLOGY

W. O. Lundberg*

I. INTRODUCTION

This chapter is designed solely for those readers of this volume who have some interest in a general background for the science and technology of lipids. At the outset of a comprehensive and authoritative treatise such as this on naturally occurring and synthetic lipids and some of their important derivatives, an introductory chapter seems appropriate to provide some background about lipid chemistry and technology that includes developments in the definition and classification of lipid substances and other orientational information.

It will be readily apparent to readers of this introductory chapter that a vast number of important discoveries in the area of lipid research are not mentioned. Many contributions have been made that are perhaps more important than some of those cited in this chapter, by acknowledged and highly honored experts whose names are not mentioned. The fact remains that the contributions of many or most of these experts are so highly specialized that they simply cannot be adequately recognized in a brief introductory chapter.

In this chapter an effort is made to point out a few somewhat representative milestones to provide a very general portrayal of the history and course of developments in the chemistry and technology of lipids during the 20th century. This necessarily requires that the author be allowed to make some highly arbitrary selections, good or bad, of what to include in a chapter that will, in his opinion, serve this purpose.

This volume should be useful to scientists who have an interest in one or more of the many subjects covered, in the opinion of this author. The reader will find this volume particularly useful because it is not just a collection of review articles, but it provides detailed experimental procedures.

II. HISTORICAL ASPECTS OF PROGRESS IN LIPID RESEARCH

For decades, progress in lipid chemistry lagged well behind that in other major biological substances, most notably carbohydrates and proteins. There appear to be two principal reasons for this.

First, as has been pointed out elsewhere,[1] it was more simple and easy to devise laboratory methods for handling carbohydrates and proteins to yield well defined and mostly crystalline compounds that could be readily analyzed and characterized. On the other hand, fats and substances closely associated with them were generally amorphous, and they were distressingly difficult to separate into individual components by methods commonly employed in organic chemistry in the early part of this century.

Second, the principal lipid substances, mainly triacylglycerols (triglycerides) in fats and oils, were considered to be relatively inert biologically. Fats and oils were generally considered to serve primarily for storage of energy that could be drawn upon as needed. Thus, in general, there was little interest in research to discover the properties, structures, biosynthetic pathways, biological utilization, and functions of triacylglycerols, and of quantitatively less abundant or even unusual lipids. Such lack of research interest was preva-

* This chapter was written while Dr. Lundberg was a Senior U.S. Scientist Awardee of the Alexander von Humboldt Foundation working at the H. P. Kaufmann-Institute, Münster, West Germany.

lent even in large scale technological applications, such as the use of fats and oils in the manufacture of soaps and protective coatings.

To be sure, there were isolated exceptions such as the chaulmoogra oils that had been found to be helpful in the treatment of leprosy; various isoprenoid lipids that were highly active biologically as vitamins and hormones; and a few other types of lipid compounds. Also, the highly important discovery that oils could be "hardened" by hydrogenation had been made in 1903 by Normann,[2] which led to the development of margarines and vegetable shortenings. Although it is well known that unnatural acyl components that contain *trans* double bonds and conjugated double bond systems are formed by catalytic hydrogenation, and although hydrogenated fats have been consumed by humans on a large scale for most of the present century without clear-cut evidence of damaging effects, there continues to be waxing and waning controversy concerning possible biologically harmful effects of such dietary fats. This controversy has provided a stimulus for various aspects of lipid research.

A noticeable and gradually increasing interest in lipids began sometime in the first quarter of the 20th century. Rather dramatic developments have occurred in the last 50 years, partly because it was realized that lipids are highly important in many previously unsuspected ways, and partly because, on the basis of this realization, many new preparative and analytical techniques were developed to expedite further discoveries pertaining to their biological and technological importance. Some of these will be summarized briefly in Sections IV, V, and VI below to provide a broad overview and orientation for the more detailed chapters that follow. The pace of progress in the lipids field now at least equals that in the carbohydrate field and is comparable to the exciting pace of progress in the areas of protein and nucleic acid research.

A high proportion of the references quoted in this chapter are to publications that appeared before the last decade began. Because it is impossible to provide an adequate summary of all the more recent developments, a few treatises and periodic review publications concerning lipids, as well as some of the better known journals devoted primarily to fats, oils, and other lipids, are listed later in this chapter.

III. DEVELOPMENT OF DEFINITIONS AND CLASSIFICATION OF LIPIDS

Because very little was known until after the turn of the 20th century about the vast number of different classes of lipid compounds, no serious attempts had been made to define and classify lipids. Perhaps the first of such attempts to receive general acceptance was that of Bloor in 1920,[3] with further modifications in 1925.[4] In essence, Bloor described "lipides" (later changed to "lipids") as a major class of biological substances including fatty acids, their naturally occurring compounds, and other substances related to them chemically, or found naturally in association with them.

Bloor further characterized lipids as follows:

- They are insoluble in water but soluble in fat solvents such as diethyl ether, chloroform, benzene, and boiling ethyl alcohol, which were their most distinguishing properties in contrast with the other main groups of biological substances. These properties were recognized as being not absolute, i.e., some lipids such as lecithins form dispersions in water that approach true solutions. On the other hand, some lipids were known to be relatively insoluble in some fat solvents, e.g., lecithins in acetone, cephalins in alcohol, and sphingomyelins and other glycolipids in diethyl ether.
- They are related to fatty acids as esters, either actually or potentially.
- They are utilized by living organisms.

The latter two characterizations were introduced to exclude organic compounds that have no biochemical relationship to fats and fatty acids but which, on the basis of solubilities alone, would be classifiable as lipids.

Bloor's classification, which was widely accepted without appreciable modification for many years, was essentially as follows:

- **Simple lipids.** Esters of fatty acids with alcohols. Fats — esters of fatty acids with glycerol. Waxes — esters of fatty acids with alcohols other than glycerol.
- **Compound lipids.** Esters of fatty acids containing additional functional groups or moieties. Phospholipids — substituted fats containing phosphoric acid and nitrogenous substituents, e.g., lecithins, cephalins, sphingomyelins. Glycolipids — compounds containing both constituent fatty acids and a carbohydrate moiety, e.g., cerebrosides. Aminolipids, sulfolipids, etc. — groups that could not be sufficiently well characterized for detailed classification.
- **Derived lipids.** Substances derived from the above groups by hydrolysis. Fatty acids including hydroxy fatty acids, etc. Sterols and aliphatic alcohols that are found in nature combined with fatty acids, and which are soluble in fat solvents, e.g., cholesterol and myricyl alcohol, etc.

It was also carefully pointed out that in addition to naturally occurring lipids, synthetic compounds should be regarded as lipids if they fulfill the three lipid characteristics mentioned earlier. Because hydrogenated fats and oils were discussed,[4] such products even though they contain unnatural fatty acids, presumably met the above criteria and apparently were regarded as lipids. However, there appears to have been some uncertainty about the lipid character of the fat-soluble vitamins that were known at that time, A, D, and E, because although they were discussed, and their close association with fats was emphasized, they are not specifically included in Bloor's classification. Also, sterols are included, but steroids, although then known, were not considered to be lipids until a later time (although now sterols are also considered to be steroids). On the basis of developments referred to in Section I, a vast number of new lipids and lipid complexes have been discovered in the meantime. Many new dipolar lipids, i.e., compounds containing hydrophilic and hydrophobic portions, have been found that appear to be highly important biologically in membrane structures. Other lipids, e.g., gangliosides, that are highly water soluble have been found, although these are also soluble to some extent in certain organic solvent mixtures. Proteolipids and lipoproteins, and other lipids and lipid complexes have come to the forefront. Although a surprising number of these can be accommodated in Bloor's classification and characterization of lipids, it became inevitable that broader schemes of classification would be needed to accommodate many of the new compounds that for one reason or another should logically be regarded as lipids.

In a recent volume,[1] lipids are classified, but only biochemically, according to three schemes, based respectively on chemical structure, solubilities, and functional biological roles. Only the first of these will be considered in this chapter. Compounds of industrial importance although many of them appear to have all of the commonly accepted characteristics of lipids, are usually still referred to as derivatives of lipids.

The first of the three classifications relating primarily to lipids of mammalians, especially humans, is summarized as follows, with some modifications to adapt it to the purposes of this volume:

Aliphatic *hydrocarbons*
    Straight-chain and branched-chain, such as octadecane and squalene
Aliphatic *alcohols*
    Straight-chain and branched-chain, such as octadecanol and phytol
Aliphatic *aldehydes*
    Straight-chain and branched-chain, such as octodecanal and geranylgeranial

Aliphatic *ketones* and *quinones*
  Straight-chain and branched-chain, such as stearone and ubiquinones
Aliphatic *acids*
  Straight-chain and branched-chain
    Saturated fatty acids
      Such as stearic and pristanic acids
    Unsaturated fatty acids
      Such as oleic and geranoic acids
    Hydroxy fatty acids
      Such as cerebronic and ricinoleic acids
    Epoxy fatty acids
      Such as vernolic acid
    Keto fatty acids
      Such as licanic acid
    Cyclopropane fatty acids
      Such as lactobacillic acid
    Cyclopropene fatty acids
      Such as sterculic acid
    Cyclopentenyl fatty acids
      Such as chaulmoogric acid
    Prostaglandins
    Furane fatty acids
Waxes
  Such as octadecyl stearate
Diol lipids
  Such as diacyl, alkylacyl, and alk-1-enylacylethanediols
Neutral glycerolipids
  Neutral acyl lipids (ester lipids)
    Such as acyl-, diacyl-, and triacylglycerols (glycerides)
  Neutral alkoxylipids (ether lipids)
    Such as alkyldiacylglycerols and alk-1-enyldiacylglycerols
  Glyceroglycolipids
    Such as monogalactosyldiacylglycerols and digalactosyldiacylglycerols
Glycerophospholipids (glycerophosphatides, phosphoglycerides)
  Phosphatidic acids
  Phosphatidyl glycerols
  Diphosphatidyl glycerols (cardiolipins)
  Phosphatidyl ethanolamines (cephalins)
  Phosphatidyl cholines (lecithins)
  Phosphatidyl serines
  Phosphatidyl inositols (mono-, di-, and triphosphoinositides)
    In addition to the diacyl (phosphatidyl) glycerophospholipids (ionic ester lipids), such as diacylglycerophos-
      phocholines, there occur alkylacyl and alk-1-enylacylglycerophospholipids (ionic alkoxylipids, ether lipids)
      such as alkylacylglycerophosphocholines and alk-1-enylacylglycerophosphocholines (choline plasmalogens)
  Phosphone analogs of the various glycerophosphatides are known
Sulfolipids
  Such as sulfogalactolipids and in the plant kingdom, sulfoquinovosyldiacylglycerols
Sphingolipids
  Sphingophospholipids
    Such as sphingomyelins
  Sphingoglycolipids
    Such as cerebrosides and gangliosides
Steroids
  Sterols and sterol esters
    Such as cholesterol and cholesterylstearate
  Sterylglycosides and acylsterylglycosides
  Sterol sulfates
  Bile acids and their conjugates
Fat-soluble vitamins (A,D,E,K)

Plant lipids are not comprehensively included in the foregoing sections of this classification, which is a somewhat modified version of a biochemical classification designed in

relation primarily to lipids produced or utilized by humans.[1] No effort to include plant lipids in a more comprehensive modern classification appears to have been made elsewhere and will not be attempted here. In general, most lipids of plants and lower forms of living matter appear to be generally classifiable within the foregoing framework; but obviously, knowledge about the lipids of most lower forms of living matter, including microfauna and microflora, is still quite meager. A modernized classification of lipids should include newer plant lipids that are described elsewhere in this volume, such as polymeric lipids (cutins and suberins) and others. However, all efforts to provide any complete classification are futile, because new lipids are being discovered at a rapid rate.

IV. DEVELOPMENT OF METHODS OF ISOLATION AND ANALYSIS

The characteristics of natural lipids and lipid mixtures make them less susceptible than carbohydrates and proteins to separation from their source materials and to further characterization by the more simple conventional chemical procedures. In the 1920s, two general procedures were used to remove lipids from tissue that are still used today, but with many modifications.

The first and foremost method is extraction by various procedures using individual or mixed lipid solvents in one or more steps that are carefully designed to provide maximal recovery with minimal alteration of the desired lipids. The more simple early methods, and recent more sophisticated methods, have been well described.[5,6]

A second method, now used less frequently because it usually involves destruction of the *in situ* character of the desired lipids, is hydrolysis (or saponification with strong alkali), followed by sequential extraction of the unsaponifiable and saponifiable fractions with appropriate organic solvents, before and after acidification of the saponified mixture, respectively.[7,8]

The methods that were available for further separation and characterization of lipid mixtures following their isolation by either of the foregoing methods were quite limited. The separation of the major lipid components, i.e., triacylglycerols and fatty acids, and also of unsaponifiable components, depended largely on distillation, crystallization (usually from selected solvents or solvent mixtures), or partitioning between immiscible solvents. Such methods, often tedious, were used to produce concentrates of individual lipids or lipid classes but seldom, and only with painstaking repetitions could be used to prepare essentially pure products.

Nevertheless, even with these older procedures, much valuable information was obtained. It was established that isomerism in all of its various forms was a prevalent feature of most of the important lipids, e.g., configurational isomers, positional isomers, and conjugated and nonconjugated isomers of polyunsaturated compounds. Also, polymorphism in triacylglycerols, although it was a confusing and controversial aspect for many years, became very important in the processing of food fats such as margarines and shortenings.[9-11] It was recognized that most lipids are optically active, although in many instances, e.g., in triacylglycerols, the activity is difficult to detect.

Although methods pertaining to physical properties of triacylglycerols and other lipids, such as melting points, refractive indices, solubilities, colors, etc. were important in the practical utilization of lipids, especially fats, methodology that provided information about chemical structure, biosynthesis, metabolism, and technological utilization was rudimentary. As regards chemical characterization, it had been found that under carefully selected conditions, halogens could be added almost quantitatively to unsaturated fatty acids and their esters without any complicating substitution reactions. Thus the iodine number, (or iodine value) defined in terms of the number of grams of iodine absorbed under standardized conditions by 100 g of fat, became a universal measure of unsaturation. Other common routine chemical analyses on fats included determination of free fatty acid content, saponification number, and other values, usually to characterize these products for commercial purposes.

Very limited information about other lipids was also obtained by rudimentary analyses. Phospholipids were commonly characterized by nothing more than a determination of phosphorus content. The determination of cholesterol was often based on a measurement of the color developed in a solution of the sample in chloroform by addition of acetic anhydride and sulfuric acid (the Liebermann-Burchard reaction), or, for more precise quantification, by a gravimetric method that involved precipitation of a digitonin derivative.

Many other examples of test tube chemistry could be cited that although they were useful for monitoring, provided very little real information about structure and pathways of biosynthesis.

Noteworthy progress in preparative and analytical lipidology began to take place in the second quarter of this century, involving, at first, primarily chemical and physicochemical techniques. A prodigious example concerning the triacylglycerol composition of natural fats was the work of Hilditch and his co-workers,[12] which involved tedious and not completely quantitative separations of such compounds, largely by solvent segregation. The main discovery in this tremendous effort was that, in the biosynthesis of triacylglycerols, there is a degree of specificity in the positioning of various fatty acids on the glycerol moiety.

On the subject of unsaturation in the acyl moieties of triacylglycerols, it was well recognized that the iodine values alone were unsatisfactory indicators of the nature of the double bonds. A temporarily important contribution in meeting this problem was made by Kaufmann when he discovered that unsaturated fatty acids react with thiocyanogen in a manner that differs from their addition of halogens.[13,14] Whereas, under suitable conditions of reaction, 1 mol of oleic, linoleic, or linolenic acids (or their monoesters) will add approximately 1, 2, or 3 mol of halogen, respectively, they will add approximately 1, 1, and somewhat less than 2 mol of thiocyanogen, respectively. This made it possible to determine quantitatively the proportions of fatty acids having one, two, and three double bonds in fats and oils that contained no fatty acids that were more unsaturated than linolenic acid. Because the addition of halogen and of thiocyanogen is not strictly quantitative even under the best of reaction conditions, this method of analysis was later improved by using empirical iodine and thiocyanogen constants that were determined for pure unsaturated fatty acids.[15]

Beginning in the 1930s, problems associated with the autoxidation of food fats, i.e., rancidity and flavor reversion, began to receive much attention. Analytical methods for detecting the early products of autoxidation such as peroxide values,[16] the active oxygen (AOM) test,[17] the TBA test which measured the reaction of thiobarbituric acid with secondary oxidation products, notably malonyl dialdehyde,[18,19] and others were developed. The autoxidation reactions have been studied extensively[20-22] and also methods for combatting autoxidation in fat-containing foods. In the field of protective coatings, on the other hand, the drying of products derived from triacylglycerols requires the autoxidation of acyl moieties coupled with polymerization,[23] using controlling prooxidants and polymerization agents known as ''driers''.[24]

Methodology has recently been developed for studying a special kind of in vivo autoxidation that is related to autoxidation, but is usually termed ''peroxidation'', which has become prominent because of possible involvement in the development of cardiovascular problems.[25] It may well be that in vivo peroxidations of lipids play an important role in aging that therefore may by one means or another be retarded.[26]

Also in the 1930s, the coupling of physical with chemical methodology began to develop in earnest. The analysis of fatty acid mixtures on the basis of iodine and thiocyanogen values was supplanted by alkali isomerization coupled with measurements of spectral absorption in the ultraviolet region of the spectrum.[27] Ultraviolet (UV) spectrophotometry that in those days required a sizeable room full of equipment was replaced in the 1940s by UV spectrophotometers that could be accommodated on a small table. The development of compact infrared (IR) spectrophotometers quickly followed, which were useful in lipid chemistry to detect and measure *trans* double bonds and also substituents such as hydroxy and keto components.

Extremely important advances in lipid chemistry are certainly based on techniques of chromatography. The discovery of this technique is usually attributed to the botanist Tswett who first reported a visual separation of various colored plant pigments on an adsorbent column in 1906,[28] and hence the term "chromatography". According to Deuel,[29] others had preceded Tswett in the use of this technique as early as 1897, but most historical authorities attribute the invention of chromatography as practiced today to Tswett.

In his pioneering work, Trappe[30] showed as early as 1940 that complex lipid mixtures could be resolved into classes of compounds by adsorption chromatography on columns of alumina or silicic acid. He described the eluotropic series of solvents, in which solvents are arranged in the order of increasing polarity and eluting power.[31] Schroeder[32] found that instead of a number of individual solvents, various mixtures of petroleum hydrocarbons and diethyl ether could be used for the fractionation of neutral lipids. In the 1950s, several authors described detailed procedures for the chromatography of lipids, especially serum lipids, on silicic acid columns.[33-35] A few years later, Marinetti[36] worked out procedures for the analysis of complex lipid mixtures including phospholipids on silicic acid-impregnated filter paper.

Boldingh[37] as well as Howard and Martin[38] found that mixtures of fatty acids could be resolved by reversed-phase partition chromatography using a polar solvent for their fractional elution from a nonpolar stationary phase. Shortly thereafter, reversed-phase partition chromatography on impregnated filter paper became available for the analysis of mixtures of fatty acids[39,40] as well as other lipid classes such as triacylglycerols.[40]

In 1952, James and Martin[41] announced the discovery that gas-liquid chromatography (GLC) could be used to separate fatty acid mixtures. This immediately proved to be a remarkable tool as an adjunct to methods that had enjoyed only a modicum of success in separating and analyzing lipid materials. This versatile technique has been vastly improved by the development of new detection devices, new capillary columns that require only minute amounts of material for complete analyses, equipment for automatic recording and interpretation of results, and methods for coupling with mass spectrometric analysis and with computerized systems for recording and interpreting the data.[42,43]

A method for chromatographic analysis on thin layers of adsorbent was described as early as 1938.[44] Sporadic descriptions of surface chromatography methods appeared without extensive application until Kirchner et al. used adsorbent coated glass strips in a noteworthy series of experiments to separate and analyze terpenes.[45] However, surface chromatography remained quite obscure until Stahl published several papers, the first in 1956,[46] describing new equipment and procedures for the preparation of chromatoplates and demonstrating that thin-layer chromatography (TLC) could be used to separate a wide variety of compounds. Inexplicably, in spite of these developments, especially the results of the systematic studies of terpenes, the potential of TLC in the field of lipids was apparently never really appreciated. In 1959, Mangold began a systematic study in his laboratory at the Hormel Institute that initiated the development of TLC into one of the most powerful and widely applicable tools for preparative and analytical purposes in the lipid field yet discovered.[47] Among more recent advances in this field is the development of a technique in which glass tubes of small diameter are coated with adsorbent on their interior surfaces, on which lipid mixtures are separated by selected solvents. The surface is instrumentally scanned automatically to provide continuously recorded information about the identities and quantities of the various components in the sample.[48]

A major advance in the analysis of lipids was the development of argentation chromatography, which permits the fractionation of unsaturated compounds according to the number and configuration of double bonds.[49]

A more recent development in column chromatography is high performance liquid chromatography (HPLC), which might become useful in lipid chemistry to produce appreciable quantities of lipid materials in highly purified form with a minimum expenditure of time and effort. Because many new sorbent (stationary phase) materials and also new eluting solvents (mobile phase) have been developed, reversed-phase HPLC and high performance reversed-phase TLC are now being used.

The development of physical methods and equipment for the study of lipid chemistry and functions has moved on apace. In addition to techniques for measuring optical rotary dispersion and circular dichroism, equipment has been devised for measuring fluorimetric phenomena, Raman spectra, nuclear magnetic resonance (NMR), electron spin resonance (ESR), mass spectra, X-ray diffraction, and various thermodynamic and other physical and chemical properties.[50]

The special methods that have been employed have included a vast variety of techniques and have involved the use of intact organisms, isolated tissues, cells and cell cultures, subcellular fractions, and separated individual chemical components.

V. BIOLOGICAL AND MEDICAL ASPECTS OF LIPIDS

Historically, because fatty acids were the principal components of the predominant lipids, biochemists were greatly interested in how fatty acids were utilized nutritionally. It had long been known that fatty acids (or fats) were major sources of energy, and that the energy was obtained by their ultimate conversion, primarily to carbon dioxide and water.

The first really important step in gaining an insight into the mechanisms involved was provided by Knoop,[51] who fed various phenylated short-chain fatty acids and obtained quite strong evidence that they were degraded in animals by a stepwise removal of units of two carbons at a time from the carboxyl end of the fatty acid chain. Subsequent clarification of the details of the β-oxidation mechanisms progressed hand-in-hand with the growth of knowledge about the roles of enzymes in biological processes. Coenzyme A and various other cofactors were found to be essential in the metabolism of fatty acids. Today, the oxidation mechanisms appear to be almost completely understood, and many reviews have been published on this subject, only one of which is cited.[52]

As was to be expected, progress in understanding the mechanisms of the oxidative metabolism of fatty acids was followed by a similar widespread interest in the mechanisms of their biosynthesis, because it had long been known that at least some of the common fatty acids could be synthesized in vivo from dietary carbohydrates and proteins. At an early stage of these investigations, it was thought that fatty acids were synthesized from units of two carbon atoms that were formed by degradation of carbohydrates and proteins, and that the mechanisms simply involved a reversal of the two-carbon steps that were found in oxidative degradation of fatty acids. However, it has been found that palmitic acid is formed in humans and other mammals primarily and directly by a condensation of seven units of malonyl CoA with one unit of acetate CoA. One of the many reviews that covers this and other aspects of fatty acid synthesis is cited.[53]

A third area of widespread interest and activity in the biosynthesis and metabolism of certain unsaturated fatty acids will be discussed briefly. Fifty years ago, Burr and Burr[54,55] reported that fats are not so inert biologically as had been supposed previously. Rats, fed a fat-free but otherwise adequate diet, developed a deficiency disease characterized by a formidable array of overt symptoms, which could be prevented or cured by one or another of several polyunsaturated dietary fatty acids. Thus, either linoleic or arachidonic acids (or their esters) were effective.

Shortly after the turn of the century, von Euler had discovered that some highly active biological substances (effective at concentrations too low to be measurable) were present in human semen, which stimulated smooth muscle in tissues of rabbits and lowered blood pressure.[56-59] These unidentified substances were named prostaglandins.

In 1963, Bergström and co-workers, after having spent about 15 years collecting hundreds of kilograms of dehydrated sheep prostate glands as a known source of prostaglandins, and after having prepared therefrom a few milligrams of several pure prostaglandins that were painstakingly characterized, announced their structure. This proved to be a monumental discovery.[60,61] Bergström surmised from their structure that they were derived from one or more of the highly unsaturated essential fatty acids. Independently, he and his co-workers,[62] and van Dorp and co-workers,[63] using enzyme preparations obtained from vesicular glands, proved that arachidonic acid and γ-dihomolinolenic acid were precursors of prostaglandins PGE_2 and PGE_1. Many other prostaglandins, other intermediate substances, and related substances such as thromboxanes have been found that stem from these same precursors,[64] Most of them are highly active biologically, having effects on smooth muscle, blood pressure, thrombocyte aggregation, and other physiological effects that are involved in essential fatty acid deficiency. Although little is known about the mechanisms by which these compounds exert their effects, it has been found that they sometimes counteract each other in their activities, and it has therefore been postulated that in normal situations they are produced as needed to maintain balanced control of their end functions. Although it must be emphasized that, in most cases, little if anything is known about the functions or mechanisms of functioning of lipid metabolites; to varying degrees the biosynthesis and metabolism of many lipids have been at least partly elucidated, as in the cases of various complex glycerolipids including some triacylglycerols and phospholipids, sphingolipids, and sterols.

The remainder of this section, except for brief consideration of one major medical problem in which lipids are involved, will include some general references to the dietary and biological importance of lipids. The role of fats in relation to cholesterol metabolism and cardiovascular disease, e.g., atherosclerosis, coronary thrombosis, and cerebral hemorrhage, has a long and controversial history. Many questions still remain about the roles that fats play in these phenomena in relation to the structures of dietary fatty acids. There have been many recent developments relating to complexes where lipids and proteins are held together by forces of the van der Waal type, as in lipoproteins. It now appears that there is a good correlation between the incidence of cardiovascular disease and the blood levels of high density lipoproteins (HDL), or the ratio of high density and low density lipoproteins (HDL:LDL). The relationship is inverse.

The Food and Agricultural Organization of the United Nations (FAO) and the World Health Organization (WHO) have recently[65] listed five of the most important functions of dietary fats as follows:

1.   As a source of energy
2.   For cell structure and membrane functions
3.   As a source of essential fatty acids for cell structures and prostaglandin synthesis
4.   As a vehicle for oil-soluble vitamins
5.   For control of blood lipids

Secondary functions include improvement of food palatability and use in food processing, including use for heat transfer in cooking.

Beyond this, however, dietary and tissue lipids contain many components other than triacylglycerols that are usually present in relatively small amounts, many of which undoubtedly have biological activities and functions not yet known.

VI. DEVELOPMENTS IN THE CHEMICAL SYNTHESIS OF LIPIDS

Two former major industrial uses of fats and oils, i.e., as sources of soaps and detergents and as protective coatings, have suffered major incursions by petroleum-based derivatives. Soaps are still a major product based on fats and oils, but have not undergone any dramatic chemical developments.

Derivatives of fatty acids and other lipids have been used in significant amounts in many other applications.[66] One of the more interesting technological developments is the production of synthetic pheromones.[67] These are lipids that occur naturally, primarily in insects, but also in other species, that are released by a member of one sex to attract a member of the opposite sex of the same species. Pheromones usually are unsaturated lipids that are highly species-specific. The synthesized products are useful in attracting many harmful insects to permit their destruction.

In recent years, routes of lipid synthesis have been devised that lead to pure products in high yields. These new preparative methods can often be utilized by investigators who need pure lipid compounds but cannot devote time to working out suitable procedures for their preparation. Progress in the synthesis of lipids has been described in a series of comprehensive review articles.[67-80]

Technological applications of materials derived from lipids are constantly being developed, and some of the other potential developments are pointed out in this volume.

VII. PRESENT STATUS OF NOMENCLATURE

Rapid developments in the lipid field have inevitably required attention to problems of nomenclature. The generally accepted authoritative sources of biochemical nomenclature are the International Union of Pure and Applied Chemistry and the International Union of Biochemistry which have formed a Commission of Biochemical Nomenclature. An article entitled Nomenclature of Lipids: Recommendations (1976) is at the end of this chapter. Information concerning trivial names of lipid substances, names recommended for biochemical usage, systematic names (as for organic compounds), and abbreviations and shorthand names are provided.

VIII. GLANCES INTO THE FUTURE

Where do we go from here? Lipids and their technical derivatives have obviously become highly important in many ways. On the basis of past observations, it appears to be extremely hazardous to predict future developments in lipidology more than a very few years in advance, especially in the face of rapidly occurring new developments.

Sophisticated methods of lipid and lipoprotein analysis are being applied in clinical diagnosis.[81,82] and in medical research.[83] Thus the threat of respiratory distress syndrome in newborn babies can be assessed by prenatal analysis of the lecithin/sphingomyelin ratio in amniotic fluid.[84] Similarly, enzyme defects in heritable lipidoses can be detected by chromatographic lipid analyses.[85]

One of the most rapidly growing areas of research is concerned with the role of lipids in biological membranes.[86-92] It can be expected that changes in the properties of such membranes, which can be induced by modifications of their lipid composition, may be utilized in the therapy of disease states.

Considering the rapid pace of developments in the field of lipids in the past decade, many important new biological and technological discoveries may be anticipated before this century ends.

REFERENCES

1. **Burton, R. M.,** in *Fundamentals of Lipid Chemistry,* Burton, R. M. and Guerra, F. C., Eds., BI Science Publications Division, Webster Grove, Mo., 1974, 1.
2. **Normann, W.,** British Patent, 1,515, 1903.
3. **Bloor, W. R.,** *Proc. Soc. Exp. Biol. Med.,* 17, 138, 1920.
4. **Bloor, W. R.,** *Chem. Rev.,* 2, 243, 1925.
5. **Bloor, W. R.,** *J. Biol. Chem.,* 59, 543, 1924.
6. **Burton, R. M.,** in *Fundamentals of Lipid Chemistry,* Burton, R. M. and Guerra, F. C., Eds., BI Science Publications Division, Webster Grove, Mo., 1974, 11.
7. **Kumagawa, M. and Suto, K.,** *Biochem. Z.,* 8, 212, 1908.
8. **Lemeland, P.,** *Bull. Soc. Chim. Biol.,* 5, 110, 1923.
9. **Malkin, T. and Meara, J.,** *J. Chem. Soc.,* 1141, 1939.
10. **Bailey, A. E., Jefferson, M. E., Kreeger, F. B., and Bauer, S. T.,** *Oil Soap,* 22, 10, 1945.
11. **Lutton, E. S.,** *J. Am. Chem. Soc.,* 70, 248, 1948.
12. **Hilditch, T. P.,** *The Chemical Constitution of Natural Fats,* 3rd ed., Chapman & Hall, London, 1956.
13. **Kaufmann, H. P.,** *Z. Unters. Lebensm.,* 51, 15, 1926.
14. **Kaufmann, H. P.,** *Analyst,* 51, 264, 1926.
15. **Kass, J. P., Lundberg, W. O., and Burr, G. O.,** *J. Am. Oil Chem. Soc.,* 17, 50, 1940.
16. **Wheeler, D. H.,** *J. Am. Oil Chem. Soc.,* 9, 89, 1932.
17. **King, A. E., Roschen, H. L., and Irwin, W. H.,** *J. Am. Oil Chem. Soc.,* 10, 105, 1933.
18. **Schmidt, H.,** *Fette Seifen Anstrichm.,* 61, 127, 1959.
19. **Sinnhuber, R. O. and Yu, T. C.,** *Yukagaku,* 26, 259, 1977.
20. **Lundberg, W. O.,** A Survey of Present Knowledge, Researches and Practices in the United States Concerning the Stabilization of Fats, Hormel Institute Publ. No. 20, University of Minnesota, Minneapolis, 1947.
21. **Lundberg, W. O., Ed.,** *Autoxidation and Antioxidants,* Vols. 1 and 2, John Wiley & Sons, New York, 1961 and 1962.
22. **Scott, G.,** *Atmospheric Oxidation and Antioxidants,* Elsevier, New York, 1965.
23. **Sims, R. P. A. and Hoffman, W. H.,** in *Autoxidation and Antioxidants,* Vol. 2, Lundberg, W. O., Ed., John Wiley & Sons, New York, 1962, 629.
24. **Stewart, W. J.,** in *Autoxidation and Antioxidants,* Vol. 2, Lundberg, W. O., Ed., John Wiley & Sons, New York, 1962, 683.
25. **Wilson, R. B.,** *Critical Reviews of Food Science and Nutrition,* CRC Press, Boca Raton, Fla., 1976, 325.
26. **Wolman, M.,** *Isr. J. Med. Sci.,* Suppl. 2, 40, 1975.
27. **Mitchell, J. H., Kraybill, H. R., and Zscheile, F. P.,** *Ind. Eng. Chem. Anal. Ed.,* 15, 1, 1943.
28. **Tswett, M.,** *Ber. Dtsch. Bot. Ges.,* 24, 316, 1906.
29. **Deuel, H. J., Jr.,** *Lipids,* Vol. 1, Interscience, New York, 1951, 652.
30. **Trappe, W.** *Biochem. Z.,* 305, 150, 1940.
31. **Trappe, W.,** *Biochem. Z.,* 306, 316, 1940.
32. **Schroeder, W. A.,** *Ann. N.Y. Acad. Sci.,* 49, 204, 1948.
33. **Borgström, B.,** *Acta Physiol. Scand.,* 25, 111, 1952.
34. **Fillerup, D. L. and Mead, J. F.,** *Proc. Soc. Exp. Biol. Med.,* 38, 574, 1953.
35. **Hirsch, J. and Ahrens, E. H., Jr.,** *J. Biol. Chem.,* 233, 311, 1958.
36. **Marinetti, G. V.,** *J. Lipid Res.,* 3, 1, 1962.
37. **Boldingh, J.,** *Rec. Trav. Chim. Pays Bas,* 69, 247, 1950.
38. **Howard, G. A. and Martin, A. J. P.,** *Biochem. J.,* 46, 532, 1950.
39. **Kaufmann, H. P. and Nitsch, W. H.,** *Fette Seifen Anstrichm.,* 56, 154, 1954.
40. **Mangold, H. K., Lamp, B. G., and Schlenk, H.,** *J. Am. Chem. Soc.,* 77, 6070, 1955.
41. **James, A. T. and Martin, A. J. P.,** *Biochem. J.,* 50, 679, 1952.
42. **Supina, W. R.,** in *Fundamentals of Lipid Chemistry,* Burton, R. M. and Guerra, F. C., Eds., BI Science Publications Division, Webster Grove, Mo., 1974, 89.
43. **Sweeley, C. C.,** in *Fundamentals of Lipid Chemistry,* Burton, R. M. and Guerra, F. C., Eds., BI Science Publications Division, Webster Grove, Mo., 1974, 119.
44. **Izmailov, N. A. and Shraiber, M. S.,** *Farmatsiya,* 3, 1, 1938.
45. **Kirchner, J. G., Miller, J. M., and Keller, G. E.,** *Anal. Chem.,* 23, 420, 1951.
46. **Stahl, E.,** *Pharmazie,* 11, 633, 1956.
47. **Mangold, H. K.,** *Fette Seifen Anstrichm.,* 61, 877, 1959.
48. **Mukherjee, K. D., Spaans, H., and Haahti, E.,** *J. Chromatogr. Sci.,* 10, 193, 1972.
49. **Morris, L. J.,** *J. Lipid Res.,* 7, 717, 1966.
50. **Chapman, D.,** *The Structure of Lipids,* Methuen, London, 1965.
51. **Knoop, F.,** *Beitr. Chem. Physiol. Pathol.,* 6, 150, 1905.

52. **Wakil, S. J.,** in *Lipid Metabolism,* Wakil, S. J., Ed., Academic Press, New York, 1970, 1.
53. **Stoops, J. K., Arslanian, M. J., Chalmers, J. H., Jr., Joshi, V. C., and Wakil, S. J.,** in *Bioorganic Chemistry,* Vol. 1, van Tamelen, E. E., Ed., Academic Press, New York, 1977.
54. **Burr, G. O. and Burr, M.,** *J. Biol. Chem.,* 82, 345, 1929.
55. **Burr, G. O. and Burr, M.,** *J. Biol. Chem.,* 86, 587, 1930.
56. **von Euler, U. S.** *Arch. Exp. Pathol. Pharmak.,* 175, 78, 1934.
57. **von Euler, U. S.,** *Klin. Wochenschr.,* 15, 1182, 1935.
58. **von Euler, U. S.,** *J. Physiol.,* 88, 213, 1936.
59. **von Euler, U. S.,** *Skand. Arch. Physiol.,* 81, 65, 1939.
60. **Bergström, S., Ryhage, R., Samuelsson, B., and Sjövall, J.,** *Acta Chem. Scand.,* 16, 501, 1962.
61. **Bergström, S., Dressler, G., Ryhage, R., Samuelsson, B., and Sjovall, J.,** *Ark. Kemi,* 19, 563, 1962.
62. **Bergström, S., Danielsson, H., and Samuelsson, B.,** *Biochim. Biophys. Acta,* 90, 207, 1964.
63. **van Dorp, D. A., Beerthuis, R. K., Nugteren, D. H., and Vonkema, H.,** *Biochim. Biophys. Acta,* 90, 204, 1964.
64. **Samuelsson, B.,** *Org. Chem. (New York),* 1977, 36.
65. **Anonymous,** *Dietary Fats and Oils in Human Nutrition,* Food and Agriculture Organization of the United Nations, Rome, 1977.
66. **Markley, K. S., Ed.,** *Fatty Acids, Their Chemistry, Properties, Production, and Uses,* Vols. 1 to 5. Interscience, New York, 1960—1968.
67. **Bestmann, H. J. and Vostrowsky, O.,** *Chem. Phys. Lipids,* 24, 335, 1979.
68. **Spener, F.,** *Chem. Phys. Lipids,* 11, 229, 1973.
69. **Mangold, H. K.,** *Chem. Phys. Lipids,* 11, 244, 1973.
70. **Kunau, W.-H.,** *Chem. Phys. Lipids,* 11, 254, 1973.
71. **Paltauf, F.,** *Chem. Phys. Lipids,* 11, 270, 1973.
72. **Slotboom, A. J., Verheij, H. M., and de Haas, G. H.,** *Chem. Phys. Lipids,* 11, 295, 1973.
73. **Stoffel, W.,** *Chem. Phys. Lipids,* 11, 318, 1973.
74. **Schäfer, H. J.,** *Chem. Phys. Lipids,* 24, 321, 1979.
75. **Verkuijlen, E. and Boelhouwer, C.,** *Chem. Phys. Lipids,* 24, 305, 1979.
76. **Lie Ken Jie, M. S. F.,** *Chem. Phys. Lipids,* 24, 407, 1979.
77. **Tulloch, A. P.,** *Chem. Phys. Lipids,* 24, 391, 1979.
78. **Spener, F.,** *Chem. Phys. Lipids,* 24, 431, 1979.
79. **Gigg, R.,** *Chem. Phys. Lipids,* 25, 287, 1979.
80. **Eibl, H.,** *Chem. Phys. Lipids,* 25, 405, 1979.
81. **Mangold, H. K. and Bezzegh, T.,** in *Clinical Biochemistry, Principles and Methods,* Curtius, H. Ch. and Roth, M., Eds., de Gruyter, New York, 1974, 1009.
82. **Mangold, H. K.,** in *Clinical Biochemistry, Principles and Methods,* Curtius, H. Ch. and Roth, M., Eds., de Gruyter, New York, 1974, 1042.
83. **Peeters, H., Ed.,** *The Lipoprotein Molecule,* Plenum Press, New York, 1978.
84. **Gluck, L.,** *Clin. Chem. (Winston-Salem, N.C.),* 23, 1107, 1977.
85. **Brady, R. O.,** in *Clinical Biochemistry, Principles and Methods,* Curtius, H. Ch. and Roth, M., Eds., de Gruyter, New York, 1974, 1277.
86. **Chapman, D., Ed.,** *Biological Membranes,* Vols. 1 to 3, Academic Press, New York, 1968—1976.
87. **Rothfield, L. I.,** *Structure and Function of Biological Membranes,* Academic Press, New York, 1971.
88. **Saier, M. H., Jr. and Stiles, C. D.,** *Molecular Dynamics in Biological Membranes,* Springer-Verlag, New York, 1975.
89. **Giebisch, G., Tosteson, D. C., and Ussing, H. H., Eds.,,** *Membrane Transport in Biology,* Vols. 1 to 3, Springer-Verlag, New York, 1978.
90. **Wallach, D. F. H. and Winzler, R. J.,** *Evolving Strategies and Tactics in Membrane Research,* Springer-Verlag, New York, 1974.
91. **Fleischer, S. and Packer, L., Eds.,** *Biomembranes,* in *Methods in Enzymology,* Vols. 31, 32, 52, 53, 54, 55, 56, Academic Press, New York, 1974—1982.
92. **Packer, L., Ed.,** *Biomembranes,* in *Methods in Enzymology,* Vol. 81, Academic Press, New York, 1983.

REVIEW PUBLICATIONS

Readers interested in new developments in lipid research may wish to refer to the following review publications, which, as a rule, are published annually.

Advances in Lipid Research, Paoletti, R. and Kritchevsky, D., Eds., Academic Press, New York, 1963-present.

Handbook of Lipid Research, Hanahan, D. J., Ed., Plenum Press, New York, 1978-present.

Monographs in Lipid Research, Kritchevsky, D., Ed., Plenum Press, New York, 1977—present.

Progress in Lipid Research, Holman, R. T., Ed. (formerly, *Progress in the Chemistry of Fats and Other Lipids*), Pergamon Press, Oxford, 1952—present.

JOURNALS

The following journals are devoted primarily to lipid chemistry, biochemistry, and technology. Numerous publications in the fields of biochemistry and medicine appear in more general biological and medical journals.

Biochimica et Biophysica Acta, (Lipids and Lipid Metabolism), Elsevier/North-Holland Biomedical Press, Amsterdam.

Chemistry and Physics of Lipids, Elsevier/North-Holland, Limerick, Ireland.

Fette Seifen Anstrichmittel, Deutsche Gesellschaft für Fettwissenschaft, Münster, Germany.

Grasas y Aceites, Instituto de la Grasa y sus Derivados, Sevilla, Spain.

Journal of the American Oil Chemists' Society, Champaign, Ill., U.S.

Journal of Lipid Research, Federation of American Societies for Experimental Biology, Bethesda, Md., U.S.

Lipids, American Oil Chemists' Society, Champaign, Ill., U.S.

Oleagineux, Institut de Recherches pour les Huiles et Oléagineux, Paris, France.

La Rivista Italiana delle Sostanze Grasse, Stazione Sperimentale per le Industrie degli Oli e dei Grassi, Milan, Italy.

Yukagaku (Journal of the Japanese Oil Chemists' Society), Tokyo, Japan.

APPENDIX

IUPAC-IUB COMMISSION ON BIOCHEMICAL NOMENCLATURE (CBN)*

THE NOMENCLATURE OF LIPIDS

RECOMMENDATIONS, 1976

INTRODUCTION

In 1967, a "Document for Discussion" on lipid nomenclature[1] was issued by CBN. It included a special system for the designation of configuration in glycerol derivatives that deviated considerably from standard stereochemical nomenclature. This system is based upon a fixed numbering ("stereo-specific numbering") for glycerol, regardless of substituents. It was hoped[1] that "discussion will lead shortly to the formulation" of recommendations acceptable to chemists in the field of lipids.

In subsequent years, there has been little discussion about this principle of stereospecific numbering; it has been well accepted within the field of glycerol derivatives for which it has been especially useful.** and is widely used. However, during this same period, many new and complex lipids and glycolipids have been isolated. Moreover, the Commissions on the Nomenclature of Organic Chemistry (CNOC) and Inorganic Chemistry (CNIC) issued, in 1973, *Nomenclature of Organic Chemistry, Section D,*[2] which includes a section on the nomenclature of phosphorus-containing organic compounds and necessitates a reconsideration of the earlier nomenclature[1] in this area.

The present "Recommendations 1976" are based on reports of working groups in lipids*** and glycolipids[†]. The main features are (a) the system of stereospecific numbering is retained; (b) semisystematic nomenclature is extended to the plasmalogens; (c) a semisystematic nomenclature for higher glycosphingolipids, based on trivial names for specific tri- and tetrasaccharides, is proposed.

FATTY ACIDS, NEUTRAL FATS, LONG-CHAIN ALCOHOLS AND LONG-CHAIN BASES

Generic Terms

Lip-1.1

The term 'fatty acid' designates any one of the aliphatic monocarboxylic acids that can

---

*   Document of the IUPAC-IUB Commission on Biochemical Nomenclature (CBN) approved by IUPAC and IUB in 1976. These Recommendations are a revision and extension of 'The Nomenclature of Lipids' which appeared in 1967 and was amended in 1970.[1] Comments and suggestions for future revisions may be sent to any member of CBN: O. Hoffmann-Ostenhof (chairman), W. E. Cohn (secretary), A. E. Braunstein, H. B. F. Dixon, B. L. Horecker, W. B. Jakoby, P. Karlson, W. Klyne, C. Liébecq, and E. C. Webb.

** CBN does not wish to imply that the idea of stereospecific numbering should be applied to other groups of compounds. It is the symmetry of glycerol itself, but the asymmetry of its derivatives carrying different substituents as 0-1 and 0-3, as well as the unique place of these compounds in lipid metabolism, that makes this special treatment desirable.

***Members of the Working Group on Lipid Nomenclature; H. Hirschmann (U.S.A.), P. Karlson (Federal Republic of Germany; convenor), W. Stoffel (Federal Republic of Germany), F. Snyder (U.S.A.), S. Veibel (Denmark), F. Vögtle (Federal Republic of Germany).

[†] Members of the Working Group on Glycolipid Nomenclature: S. Basu (U.S.A.), R. O. Brady (U.S.A.), R. M. Burton (U.S.A.), R. Caputto (Argentina), S. Gatt (Israel), S. I. Hakomori (U.S.A.), M. Philippart (U.S.A.), L. Svennerholm (Sweden), D. Shapiro (Israel), C. C. Sweeley (U.S.A.), H. Weigandt (Federal Republic of Germany; convenor).

be liberated by hydrolysis from naturally occurring fats and oils. In the terms "free fatty acids" or "nonesterified fatty acids", now widely in use, "free" and "nonesterified" are actually redundant and should be omitted (see Lip-1.14). [The designation "aliphatic carboxylate (C_{10}-C_{26}, nonesterified)" used by the Commission on Quantities and Units in Clinical Chemistry is correct, but rather cumbersome.] Whenever the sum of fatty acids and their esters is determined by an analytical method, this should be explicitly stated (see also Lip-1.6).

Lip-1.2

"Neutral fats" are mono-, di, or triesters of glycerol with fatty acids, and are therefore termed monoacylglycerol, diacylglycerol, or triacylglycerol, as appropriate. "Acylglycerols" includes mixtures of any or all of these.

Comments: (a) The term "acyl" is used in Organic Nomenclature[3] to denote the radical formed by loss of the OH group from the acid function of any acid (cf. Lip-1.6). We are concerned here with acyl radicals of aliphatic carboxylic acids with four or more carbon atoms, the larger members of which ($> C_{10}$) are also known as 'higher fatty acids'; (b) The old terms monoglyceride, diglyceride, and triglyceride are discouraged and should progressively be abandoned, not only for consistency, but mainly because strict interpretation does not convey the intended meaning. "Triglyceride", taken literally, indicates three glycerol residues (e.g., cardiolipin), diglyceride two (e.g., phosphatidylglycerol), and a monoglyceride is a monoacylglycerol.

Lip-1.3

The generic term 'long-chain alcohol' or 'fatty alcohol' refers to an aliphatic compound with a chain-length greater than C_{10} that possesses a terminal CH_2OH group. Such alcohols should be named according to systematic nomenclature[3] (see Lip-1.7).

Lip-1.4

The term 'sphingoid' or "sphingoid base" refers to sphinganine (cf. Lip-1.8), [D-erythro-2-amino-1,3-octadecanediol (I)], to its homologs and stereo-isomers (II,III), and to the hydroxy and unsaturated derivatives of these compounds (IV-VI). The term "long-chain base" may be used in a wider sense to indicate any base containing a long-chain aliphatic radical.

Lip-1.5

The following generic terms are used for the following groups of compounds: (a) sphingolipid, for any lipid containing a sphingoid; (b) ceramide, for an *N*-acylated sphingoid; (c) sphingomyelin, for a ceramide-1-phosphocholine; (see Reference 2 for this use of "phospho", also Lip-2.11); (d) glycosphingolipid, for any lipid containing a sphingoid and one or more sugars. (See below for other generic terms.)

Individual Compounds
Fatty Acids and Alcohols
Lip-1.6

Fatty acids (cf. Lip-1.1) and their acyl radicals (cf. Lip-1.2. comment [a]) are named according to the IUPAC Rules for the Nomenclature of Organic Chemistry (Reference 3 Rule C-4). (A list of trivial names is given in Appendix A.) Fatty acids are numbered with the carbon atom of the carboxyl group as C-1. By standard biochemical convention, the ending "-ate" in, e.g., palmitate denotes any mixture of the free acid and its ionized form in which the cations are not specified. The ending "-ate" is also used to designate esters, e.g., cholesteryl palmitate, ethylidene dilaurate, etc. (cf. Lip-1.12). Structural isomers of polyunsaturated acids, hitherto distinguished by Greek letters (e.g., α- and γ-linolenic acids), are better distinguished by the locants of the unsaturated linkages [e.g., (9, 12, 15)- and (6, 9, 12)-linolenic acids, respectively] (see Lip-1.15). However, the Greek letter prefixes may be useful in (defined) abbreviations (see Appendix A).

Lip-1.7

Long-chain alcohols (fatty alcohols) and the radicals derived from them should be designated by their systematic names (Reference 3 Rules C-201 and A-1 et seq.), but not by trivial names that are derived from those of fatty acids.

Examples: (a) 1-hexadecanol and hexadecyl-, not palmityl alcohol and palmityl-; (b) 1-dodecanol and dodecyl-, not lauryl alcohol and lauryl-.

Sphinganine and Derivatives
Lip-1.8

The compound previously known as dihydrosphingosine [2*D*-amino-1,3*D*-octadecanediol or D-*erythro*-2-amino-1,3-octadecanediol or (2*S*, 3*R*)-2-amino-1,3-octadecanediol] is called sphinganine (I).

Lip-1.9

Trivial names of higher or lower homologs of sphinganine may be derived by adding a prefix (Reference 3, Rule A-1) denoting the total number of carbon atoms in the main chain of the homolog, e.g., icosasphinganine* for the C_{20} compound (III), hexadecasphinganine for the C_{16} compound.

Lip-1.10

Affixes denoting substitution of sphinganine (hydroxy, oxo, methyl, etc.) are used as usual, according to existing rules.[3] The configurations of additional substituents should be specified by the prefixes "*D*" or "*L*" (italic capitals, cf. Reference 4), following the locant of substitution. These prefixes refer to the orientation of the functional groups to the right or left, respectively, of the carbon chain when written vertically in a Fischer projection with C-1 at the top (cf. Formulae I-VI). If the configuration is unknown, the prefix "*X*" may be used. In the case of a racemic mixture, "*rac*" should be used as a prefix to the name.

```
        CH₂OH                        CH₂OH
          |                            |
   H—C—NH₂                      H₂N—C—H
          |                            |
   H—C—OH                       H—C—OH
          |                            |
      (CH₂)₁₄                      (CH₂)₁₄
          |                            |
        CH₃                          CH₃

          I                           II

     Sphinganine                (2R,3R)- (or D-threo-
   [D-erythro or 2S,3R           -2-Amino-1,3-octadecanediol
   configuration implied]
```

Sphingoids differing from sphinganine in their configurations at C-2 and/or C-3 should be named not as derivatives of sphinganine, but with fully systematic names,[3] using the prefixes D-*threo*, L-*erythro*, as appropriate, e.g., D-*threo*-2-amino-1,3-octadecanediol, or (2*R*, 3*R*)-2-amino-1,3-octadecanediol, for II (cf. Rule Carb-8 in Reference 5) (cf. Lip-1.11, example [d]).

---

* See footnote d in Appendix A re "icosa" for "eicosa".

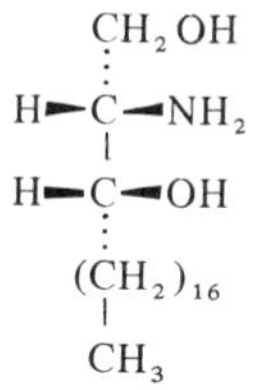

CH_2OH
$H-C-NH_2$
$H-C-OH$
$(CH_2)_{16}$
CH_3

III

Icosasphinganine (formerly eicosasphinganine, see footnote[d] in Appendix A)

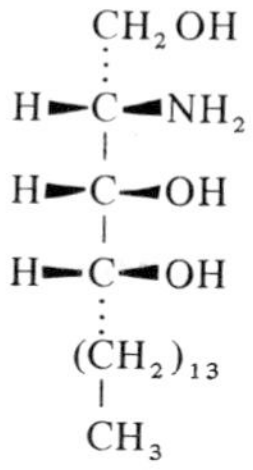

CH_2OH
$H-C-NH_2$
$H-C-OH$
$H-C-OH$
$(CH_2)_{13}$
CH_3

IV

4D-Hydroxysphinganine;
(2S,3S,4R)-2-amino-1,3,4-octadecanetriol;
(phytosphingosine).

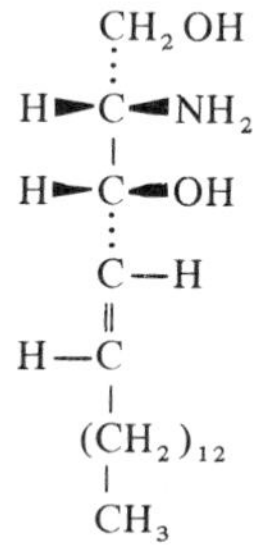

CH_2OH
$H-C-NH_2$
$H-C-OH$
$C-H$
$\|$
$H-C$
$(CH_2)_{12}$
CH_3

V

Sphingosine; (4E)-sphingenine;
trans-4-sphingenine;
(2,S,3R,4E)-2-amino-4-octadecene-1,3-diol;

CH_2OH
$H-C-NH_2$
$H-C-OH$
$C-H$
$\|$
$C-H$
$(CH_2)_{12}$
CH_3

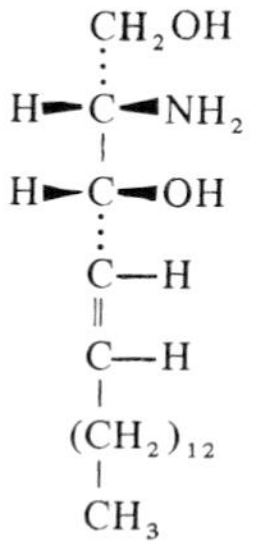

VI

cis-4-Sphingenine; (4Z)-sphingenine.

Comments: (a) The semisystematic names for the sphingoids are significantly shorter than the fully systematic names only if the terms chosen imply not only substituents but also configurations. Therefore, the name "sphinganine" specifies the D-*erythro* configuration, and it is logical that the names of stereoisomers of sphinganine differing in configuration at C-2 and/or C-3 should not include "sphinganine" as a root. This recommendation differs from that in the previous document;[1] (b) The configurations usually encountered have identical configuration prefixes only if a D/L but not if the R/S system[6] is used; e.g., C-3 is D and R in icosasphinganine (III) and D and S in 4D-hydroxysphinganine (IV). Whenever it is desirable to use the R/S system, the fully systematic name should be used with the specification of configuration at every center (and, when applicable, of the configuration at the double bond).

Examples: (a) (2R,3R)-2-amino-1,3-octadecanediol, for II; (b) (2S,3S,4R)-2-amino-1,3,4-octadecanetriol for IV; (c) (2S,3R,4R)-2-amino-4-octadecene-1,3-diol for sphingosine (V) (see also Lip-1.11).

Lip-1.11

Names for unsaturated compounds are derived from the names of the corresponding saturated compounds by the appropriate infixes, namely ene, diene, yne, etc.[3] If the geometry of the double bond is known, it should be indicated by the more modern E-Z system (cf. Reference 6, Rule E-2.2), e.g., (4E)-sphingenine for sphingosine (V).

Comment: The trivial name "sphingosine" (V) is retained. If trivial names other than sphingosine are used, they should be defined in each paper in terms of this nomenclature, or of the general nomenclature of organic chemistry.[3]

Other names for compounds described in Lip-1.10 and 1.11: (a) 4D-hydroxysphinganine for IV, formerly known as phytosphingosine; (b) (4*E*)-sphingenine for sphingosine (V); (c) (4*Z*)-sphingenine for the geometric isomer of sphingosine (VI); (d) D-*threo*-2-amino-1,3-octadecanediol for the C-2 epimer of sphinganine (II); cf. Lip-1.10, example (a).

Glycerol Derivatives

Lip-1.12

Esters, ethers and other *O*-derivatives of glycerol are designated according to Carb-15 of the *Rules of Carbohydrate Nomenclature*,[5] i.e., by a prefix, denoting the substituent, preceded by a locant. If the substitution is on a carbon atom, the compound is designated by its systematic name and not as a derivative of glycerol (e.g., 1,2,3-nonadecanetriol for $C_{16}H_{33}$CHOH-CHOH-CH_2OH, which could be considered as 1-*C*-hexadecylglycerol). It is permissible to omit the locant "*O*" if the substitution is on the oxygen atoms of glycerol.

Examples: (a) tristearoylglycerol or tri-*O*-stearoylglycerol or glycerol tristearate, or glyceryl tristearate; (b) 1,3-benzylideneglycerol or 1,3-*O*-benzylideneglycerol; (c) glycerol 2-phosphate (a permissible alternative to this term is 2-phosphoglycerol).[10]

Comment: The alternative system set forth in Carb-16 of the Rules on Carbohydrate Nomenclature,[5] i.e., the use of the suffix "-ate", is less suitable for glycerol esters, with the exception of the phosphates (see Examples). However, this system may be used to designate esters of monofunctional alcohols, e.g., cholesteryl palmitate (cf. Lip-1.6).

Lip-1.13

In order to designate the configuration of glycerol derivatives, the carbon atoms of glycerol are numbered stereospecifically. The carbon atom that appears on top in that Fischer projection that shows a vertical carbon chain with the hydroxyl group at carbon-2 to the left is designated as C-1. To differentiate such numbering from conventional numbering conveying no steric information, the prefix "*sn*" (for stereospecifically numbered) is used. This term is printed in lower-case italics, even at the beginning of a sentence, immediately preceding the glycerol term, from which it is separated by hyphen. The prefix "*rac-*" (for racemo) precedes the full name if the product is an equal mixture of both antipodes; the prefix "*X*"- may be used when the configuration of the compound is either unknown or unspecified (cf. Lip-1.10).

Examples: (a) *sn*-glycerol 3-phosphate for the stereoisomer (VII ≡ VIII), previously known as either L-α-glycerophosphate or as D-glycerol 1-phosphate; (b) *rac*-1-hexadecylglycerol; (c) 1,2-dipalmitoyl-3-stearoyl-*X*-glycerol.

$$\begin{array}{ccc}
CH_2OH & & CH_2OPO_3H_2 \\
HO \blacktriangleright \overset{\vdots}{C} \blacktriangleleft H & & H \blacktriangleright \overset{\vdots}{C} \blacktriangleleft OH \\
CH_2OPO_3H_2 & & CH_2OH \\
\\
VII & \equiv & VIII
\end{array}$$

sn-Glycerol 3-phosphate

[L-(glycerol 3-phosphate) ≡ D-(glycerol 1-phosphate)]

Comments: (a) The problem of distinguishing between stereoisomers was discussed *in extenso* in the 1967 document.[1] In brief, difficulties arise because glycerol is a prochiral compound. The parent substance of many phospholipids, natural glycerol phosphate, has been named both as L-α-glycerol phosphate[7] (VII) and, according to standard rules of nomenclature, D-glycerol 1-phosphate[8] (VIII). When the *R/S* system (sequence rule) is applied, substitution of one of the primary hydroxyl groups often leads to changes in the configurational prefix, thus obscuring chemical and biogenetical relationships; it is generally inapplicable to the steric description of such mixtures as occur in triacylglycerols isolated from natural sources. The stereospecific numbering of glycerol and its derivatives as proposed by Hirschmann,[9] described above and in Reference 1, avoids these difficulties; it has proved useful and is widely accepted. (b) The enantiomer of *sn*-glycerol 3-phosphate (VII) is *sn*-glycerol 1-phosphate (IX), as is evident from the structures.

$$CH_2OPO_3H_2$$
$$HO \!-\! C \!\blacktriangleleft\! H$$
$$CH_2OH$$

IX

sn-Glycerol 1-phosphate
[L-(glycerol 1-phosphate) ≡ D-(glycerol 3-phosphate)]

Symbols and Abbreviations
Lip-1.14
The term 'fatty acids' (cf. Lip-1.1) should not be abbreviated. The use of abbreviations like "FFA" for "free fatty acids" or "NEFA" for "non-esterified fatty acids" is strongly discouraged.
Comment: The words "acids" and "esters" serve to distinguish the 'free' (nonesterified) and 'bound' (esterified) fatty acids and are as short or shorter than the abbreviations themselves.

Lip-1.15
In tables and discussions where various fatty acids are involved, the notation giving the number of carbon atoms and of double bonds (separated by a colon) is acceptable, e.g., 16:0 for palmitic acid, 18:1 for oleic acid. Also, terms such as "(18:0)acyl" may be used to symbolize radicals of fatty acids (see Appendix A).
Comment: This system is already widely used. It should, however, be kept in mind that it sometimes does not completely specify the fatty acid. For example, α-linolenic acid and γ-linolenic acid are both 18:3 acids; the designation 18:3 is therefore ambiguous. In such a case, the position of double bonds should be indicated, e.g., 18:3(9,12,15) for (9,12,15)-linolenic acid, formerly known as α-linolenic acid.

Lip-1.16
It is sometimes desirable (for example, in discussing the biosynthesis of lipids) to indicate the position of each double bond with reference not to the carboxyl group (always C-1), but to the end of the chain remote from the carboxyl. If *n* is the number of carbon atoms in the chain (i.e., the locant of the terminal methyl group) and *x* is the (lower) locant of the double bond, the position of the double bond may be defined as (*n* minus *x*). Thus, the common position of the double bond in oleic and nervonic acids may be given as 18-9 and 24-9, respectively. This structural regulatory should not be expressed as ω 9.

Lip-1.17
The system described in Lip-1.15 may also be used to denote alcohols and aldehydes related to fatty acids, provided that the nature of the residue is clearly indicated either by the appropriate name of the compound(s) (e.g., 18:1 alcohol) or in the heading of the table. The 1-ene of alk-1-en-1-yl (i.e., 1-alkenyl) compounds is not counted in this system (see Lip-2.10 comment).

Lip-1.18
For many complex lipids, a representation of the structure using symbols rather than

structural formulae may be useful. Symbols proposed for various constituents are given in Appendix B (see also Reference 10), and, for glycolipids, in Lip-3.13. They are constructed in analogy to those in use for amino acids,[11] nucleosides[12] and saccharides.[13]

PHOSPHOLIPIDS

Generic Terms

The Rules of the *Nomenclature of Organic Phosphorus Compounds*, also known as D-Rules,[2] recognize, for biochemical usage, the prefix "phospho-" as an alternate to "*O*-phosphono-" (or *N*-phosphono-"). By a similar convention,[10] "-phospho-" may be used as an infix to designate the phosphodiester bridge present in such compounds as glycero-phosphono-" (or "*N*-phosphono-"). By a similar convention,[10] "-phospho-" may be used as an infix to designate the phosphodiester bridge present in such compounds as glycero-

Lip-2.1

"Phospholipid" may be used for any lipid containing phosphoric acid as mono- or diester. Likewise, lipids containing *C*-phosphono groups (e.g., compound X) may be termed "phosphonolipids".

Lip-2.2

"Glycerophospholipid" signifies any derivative of glycerophosphoric acid that contains at least one *O*-acyl, or *O*-alkyl, or *O*-(1-alkenyl) group attached to the glycerol residue. Generic names for other classes of phospholipids may be coined according to this scheme, e.g., sphingophospholipid, inositolphospholipid.

Comment: The old terms, "phosphatide", "phosphoglyceride", and "phosphoinositide" are no longer recommended because they do not convey the intended meaning (see also Lip-1.2).

Lip-2.3

"Phosphatidic acid" signifies a derivative of a glycerol phosphate (glycerophosphate) in which both remaining hydroxyl groups of glycerol are esterified with fatty acids. The position of the phosphate group may be emphasized by stereospecific numbering.

Comment: For the most common (3-*sn*) phosphatidic acid and its derivatives, the locants are often omitted. However, "phosphatidyl" without locants can lead to ambiguities. It is therefore preferable to use the proper locants, for example, 2-phosphatidic acid for compound XI, and 3-*sn*-phosphatidylserine for XIIa.

Lip-2.4

The common glycerophospholipids are named as derivatives of phosphatidic acid, e.g., 3-*sn*-phosphatidylcholine (this term is preferred to the trivial name, lecithin; the systematic name is 1,2-diacyl-*sn*-glycero-3-phosphocholine); 3-*sn*-phosphatidylserine; 1-phosphatidyl-inositol (see comment [b] below); 1,3-bis(3-*sn*-phosphatidyl)glycerol.

Comments: (a) It is understood that, in combination with compounds like ethanolamine (properly, 2-aminoethanol) or serine, which bear both hydroxyl and amino groups, substitution by phosphorus is at the hydroxyl group of the ethanolamine or serine. Substitution at any other position, or where confusion may arise, requires a locant. (b) The phosphorylated derivatives of 1-(3-*sn*-phosphatidyl)inositol should be called 1-phosphatidylinositol 4-phosphate and 1-phosphatidylinositol 3,4-bisphosphate, respectively. The use of "diphospho-inositide" and "triphosphoinositide" for these is discouraged, as these names do not convey the chemical structures of the compounds and can be misleading (cf. also Table 4 in Reference 10).

```
        CH₂O—CO—R'                                    CH₂O—CO—R'
R"CO—O►C◄H              NH₂          H₂O₃P—O►C◄H
        CH₂O—PO(OH)—CH₂CH—CO₂H               CH₂O—CO—R"
```

X   XI

A phosphonolipid   2-Phosphatidic acid

```
        CH₂O—CO—R'                                    CH₂O—CO—R
R"CO—O►C◄H              NH₂          HO►C◄H                    NH₂
        CH₂O—PO(OH)—O—CH₂CH—CO₂H            CH₂O—PO(OH)—OCH₂CH—CO₂H
```

XIIa   XIIb

Phosphatidylserine   A 2-lysophospholipid
(3-sn-phosphatidylserine)

Lip-2.5

As an alternative, generic names may be coined according to Lip-1.13, i.e. using glycerol phosphate (glycerophosphate) as the stem. In this case, the stereospecific numbering of glycerol should be used to indicate the position of the phosphoric residue as well as the other substituents (acyl-, alkyl-, 1-alkenyl). If the nature of these substituents cannot be specified, the prefix 'radyl' may be used.

Lip-2.6

Derivatives of phosphatidic acids resulting from hydrolytic removal of one of the two acyl groups may be designated by the old prefix "lyso", e.g., lysophosphatidylethanolamine for compound XIIb. A locant may be added to designate the site of (hydro)-lysis, 2-lyso designating hydrolysis at position 2, leaving a free hydroxyl group at this carbon atom.

Comment: The "lyso" term originated from the fact that these compounds are hemolytic. It is here redefined to indicate a limited hydrolysis of the phosphatidyl derivative (i.e., "deacyl").

Lip-2.7

The term "plasmalogen" may be used as a generic term for glycerophospholipids in which the glycerol moiety bears an 1-alkenyl ether group.

Lip-2.8

The term "plasmenic acid" signifies a derivative of *sn*-glycero-3-phosphate in which carbon-1 bears an *O*-(1-alkenyl) residue, and position 2 is esterified with a fatty acid (XIII). This term can also be used to name derivatives, e.g., plasmenylethanolamine.

Comments: (a) The use of 'phosphatidyl' as a name for the acyl radical of phosphatidic acid has facilitated the nomenclature of its various compounds (see Lip-2.4). Therefore, it seems logical to offer a similar short term for XIII, i.e., "plasmenic acid', as an alternative to the more systematic name, 2-ayl-1-alkenyl-*sn*-glycero 3-phosphate, which, of course, may be used if desired. "Plasmenic" is a contraction of 'plasmalogenic' and may be especially useful in naming derivatives, e.g., plasmenylserine. (b) Isomers like those bearing the phosphate residue in position 2 (e.g., compound XIV) should not be named in this way but as derivatives of the corresponding glycerophosphate, using stereospecific numbering.

Lip-2.9

The term 'lysoplasmenic acid' may be used for a derivative of *sn*-glycero-3-phosphate that has an *O*-(1-alkenyl) residue on carbon-1, the hydroxyl group in position 2 being unsubstituted (XVa). This name may also be used in combinations like "lysoplasmenyl-ethanolamine" (XVb).

Lip-2.10

For compounds of type XVI, bearing a saturated ether group in position 1 and an acyl group in position 2 of *sn*-glycero-3-phosphate, the term "plasmanic acid" is proposed. Compounds deacylated in position 2, or with a substituent on the phosphoric residue, can be treated as are the plasmenic acids (Lip-2.9).

Comment: The proposed names will be especially useful for naming phosphoric diesters (phosphodiesters), e.g., plasmanylethanolamine, instead of 2-acyl-1-alkyl-*sn*-glycero-3-phosphoethanolamine. The terms "plasmanic acid" and "plasmanyl" may also be applied to ethers with an alkyl group bearing a double bond within the chain, e.g., a 9-hexadecenyl residue (derived from palmitoleic acid). In such cases, the proper term "alkenyl" if used without the "ene" locant(s), would be misleading (see Lip-1.17).

Individual Compounds

Lip-2.11

Individual glycerophospholipids in which the substituents can be specified are named according to existing Rules[2,3,5,6] using the infix "-phospho-"[2,10] to indicate the phospho-diester bridge.

Example: 1-palmitoyl-2-stearoyl-*sn*-glycero-3-phosphoethanolamine for compound XVII.

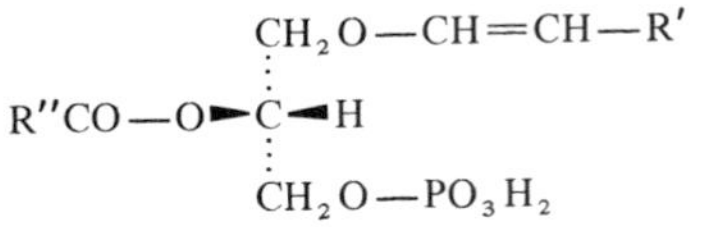

$$CH_2O—CH=CH—R'$$
$$R''CO—O►\overset{|}{C}◄H$$
$$CH_2O—PO_3H_2$$

XIII

A plasmenic acid

$$OH \quad CH_2O—CH=CH—R'$$
$$HO—\overset{|}{\underset{\|}{P}}—O►\overset{|}{C}◄H$$
$$O \quad CH_2O—CO—R''$$

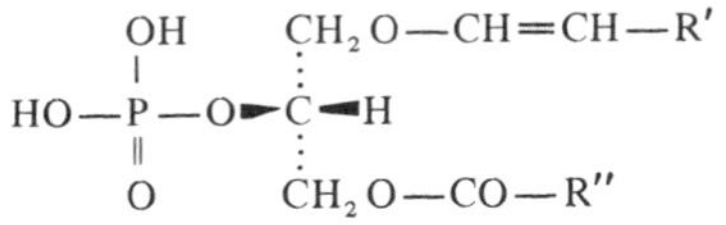

XIV

3-Acyl-1-(1-alkenyl)-
sn-glycerol 2-phosphate

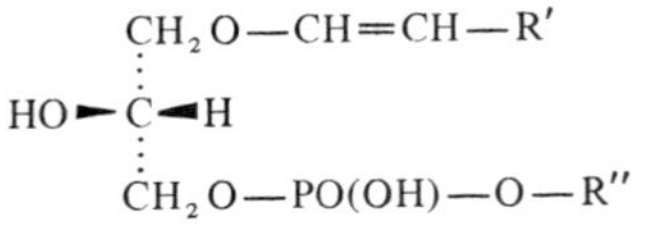

$$CH_2O—CH=CH—R'$$
$$HO►\overset{|}{C}◄H$$
$$CH_2O—PO(OH)—O—R''$$

XV

$$CH_2O—CH_2CH_2R'$$
$$R''CO—O►\overset{|}{C}◄H$$
$$CH_2O—PO_3H_2$$

XVI

A plasmanic acid

XVa: (R'' = H):
   a lysoplasmenic acid
XVb: (R'' = $CH_2CH_2NH_2$)
   a lysoplasmenylethanolamine

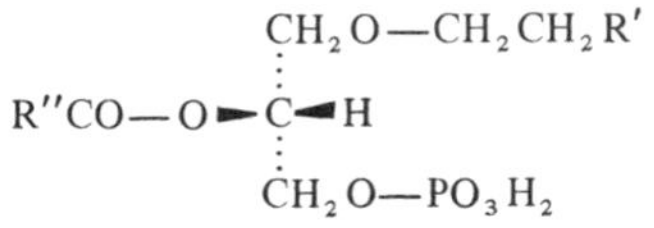

$$CH_2O—CO—C_{15}H_{31}$$
$$C_{17}H_{35}—CO—O►\overset{|}{C}◄H$$
$$CH_2O—PO(OH)—OCH_2CH_2NH_2$$

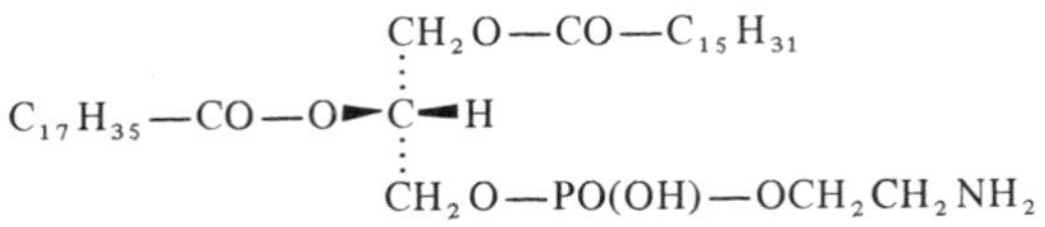

XVII

1-Palmitoyl-2-stearoyl-*sn*-glycero-3-phosphoethanolamine

Lip-2.12

The ketone derived from glycerol, 1,3-dihydroxy-2-propanone, also known as dihydroxyacetone, may be termed "glycerone", if desired. The name is a contraction of "glycero-ketone" and may be useful to emphasize the relationship with glycerol, glyceraldehyde (glyceral), and glycerate. It also permits a simple symbolism (Appendix B) and the naming of derived lipids, e.g., 1-palmitoyl-3-phosphoglycerone.

GLYCOLIPIDS

General Considerations

Glycolipids (a contraction of glycosyllipids) are generally named as glycosyl derivatives of the corresponding lipid, e.g., diacylgalactosylglycerol, glucosylceramide. Because of the heterogeneity of the fatty acids and long-chain bases encountered in most cases, a generic name for the lipid moiety is needed, i.e., ceramide. With higher glycosphingolipids, especially the gangliosides, naming problems arise from the complexity of the carbohydrate moiety of these compounds. The systematic names of the oligosaccharides linked to ceramide are so cumbersome that they are of the same practical value as, e.g., the systematic name for a peptide hormone such as insulin. It was felt that this difficulty could be overcome only by creating suitable trivial names for some parent oligosaccharides. In constructing these names (see Table 1) the following principles were observed:

a. The number of monosaccharide units in an oligosaccharide is indicated by the suffixes "-biose", "-triaose", "-tetraose", etc. This follows the well-established practice in the carbohydrate field (cf. cellobiose, cellotetraose, maltotetraose, etc.), with the exception that the suffix "-triose", as used in maltotriose, has been changed to "-triaose" to avoid confusion with the monosaccharides called trioses.

b. The oligosaccharides are grouped in series according to their structure and biogenetic relationship.

c. Differences in linkage (e.g., 1→4 vs. 1→3) in otherwise identical sequences are indicated by "iso-" or "neo-", used as a prefix. On the basis of these names, the semisystematic nomenclature for neutral glycosphingolipids and gangliosides described below is recommended. A set of symbols has been devised that allows a simple representation of complex neutral and acidic glycosphingolipids (Table 1).

Generic Terms

Lip-3.1

The term "glycolipid" designates any compound containing one or more monosaccharide residues linked by a glycosyl linkage to a lipid part [e.g., a mono- or diacylglycerol, a long-chain base (sphingoid) like sphingosine, or a ceramide].

Lip-3.2

The term "glycoglycerolipid" may be used to designate glycolipids containing one or more glycerol residues.

Lip-3.3

The term "glycosphingolipid", as hitherto, includes all compounds containing at least one monosaccharide and a sphingoid. The glycosphingolipids can be subdivided as follows:

Neutral glycosphingolipids: monoglycosyl- and oligoglycosylsphingoids; monoglycosyl- and oligoglycosylceramides.

Acidic glycosphingolipids: sialosylglycosylsphingolipids (gangliosides); sulfoglycosylsphingolipids (formerly "sulfatides", which is not recommended) (cf. Lip-3.11).

Lip-3.4

"Psychosine" may be used as a generic name for 1-monoglycosylsphingoids, although

Table 1
NAMES AND ABBREVIATIONS OF SIMPLE GLYCOLIPIDS

Structure[a]	Trivial name of oligosaccharide[b]	Symbols[c]	Short symbol[d]
Gal(α1-4)Gal(β1-4)GlcCer	Globotriaose	GbOse$_3$	Gb$_3$
GalNAc(β1-3)Gal(α1-4)Gal(β1-4)GlcCer	Globotetraose	GbOse$_4$	Gb$_4$
Gal(α1-3)Gal(β1-4)GlcCer	Isoglobotriaose	iGbOse$_3$	iGb$_3$
GalNAc(β1-3)Gal(α1-3)Gal(β1-4)GlcCer	Isoglobotetraose	iGbOse$_4$	iGb$_4$
Gal(β1-4)Gal(β1-4)GlcCer	Mucotriaose	McOse$_3$	Mc$_3$
Gal(β1-3)Gal(β1-4)Gal(β1-4)GlcCer	Mucotetraose	McOse$_4$	Mc$_4$
GlcNAc(β1-3)Gal(β1-4)GlcCer	Lactotriaose	LcOse$_3$	Lc$_3$
Gal(β1-3)GlcNAc(β1-3)Gal(β1-4)GlcCer	Lactotetraose	LcOse$_4$	Lc$_4$
Gal(β1-4)GlcNAc(β1-3)Gal(β1-4)GlcCer	Neolactotetraose	nLcOse$_4$	nLc
GalNAc(β1-4)Gal(β1-4)Gal(β1-4)GlcCer	Gangliotriaose	GgOse$_3$	Gg$_3$
Gal(β1-3)GalNAc(β1-4)Gal(β1-4)GlcCer	Gangliotetraose	GgOse$_4$	Gg$_4$
Gal(α1-4)GalCer	Galabiose	GaOse$_2$	Ga$_2$
Gal(1-4)Gal(α1-4)GalCer	Galatriaose	GaOse$_3$	Ga$_3$
GalNAc(1-3)Gal(1-4)Gal(α1-4)GalCer	*N*-Acetylgalactosaminylgalatriaose	GalNAc1-3GaOse$_3$	—

[a] Symbols and arrangement are discussed in Lip-3.13. Hyphens replace left-to-right arrows (see Section 3.4 of Reference 13).

[b] Name of glycolipid is formed by converting ending "-ose" to "-osyl", followed by "-ceramide", without space; e.g., globotriaosylceramide.

[c] Should be followed by Cer for the glycolipid, without space; e.g., McOse$_3$Cer, Mc$_4$Cer (see Lip-3.13).

[d] The short form should be used only in situations of limited space or in case of frequent repetition.

the latter is preferred. The nature of the monosaccharide and the sphingoid is not specified in this name.

Lip-3.5

The term "fucolipid" may be used to designate fucose-containing neutral or acid glycolipids.

Individual* Compounds

Lip-3.6

Glycoglycerolipids may be named either as glycosyl compounds according to Rule Carb-24 or as glycosides according to Rule Carb-23.[5]

Example: the compound XVIII may be named either 1,2-diacyl-3-β-D-galactosyl-*sn*-glycerol or 1,2-diacyl-*sn*-glycerol 3-β-D-galactoside.

Comment: The first form is preferred, as the glycosphingolipids are also named this way.

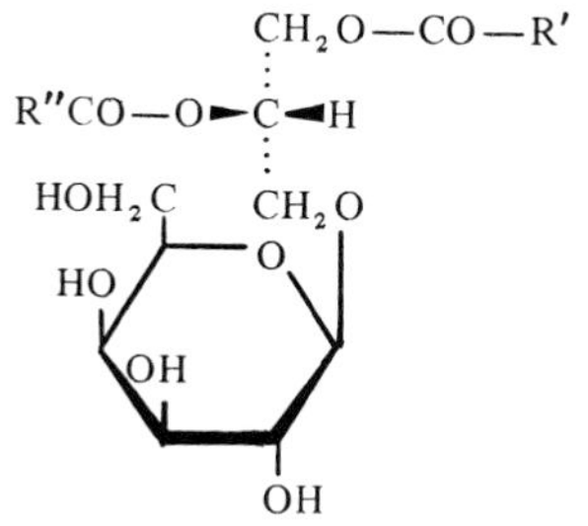

XVIII

Lip-3.7

A glycosphingolipid is generally named as a "glycosylsphingoid" or a "glycosylceramide", using the appropriate trivial name of the mono- or oligosaccharide residue for "glycosyl". It is understood that the sugar residue is attached to the C-1 hydroxyl group of the ceramide. For glycosphingolipids carrying two to four saccharide residues, the trivial names listed in Table 1 are recommended.

Comment: It is strongly recommended that the name of the oligosaccharide be defined in each publication by means of the standard symbols for sugars (as in Table 1, column 1) rather than by the full name, which is often so long as to be confusing.

Lip-3.8

The trivial name "cerebroside" designates 1-β-glycosylceramide (the natures of the sphingoid and of the fatty acid are not specified in this name).

Lip-3.9

Glycosphingolipids carrying fucose either as a branch or at the end of an oligohexosyl-ceramide are named as "fucosyl(X)osylceramide" where (X) stands for the root name of the oligosaccharide. The location of the fucosyl residue is indicated by a Roman numeral designating the position of the monosaccharide residue in the parent oligosaccharide (counting from the ceramide end) to which the fucose residue is attached, with an Arabic numeral superscript indicating the position within that residue to which the fucose is attached. If necessary, the anomeric symbol can be used as usual, i.e., preceding "fucosyl-".

Examples for Lip-3.7 and Lip-3.9 (structures given in the symbols of Lip-3.13). ** (a)

---

*   "Individual" in this section refers to the carbohydrate moiety only.

**   D is omitted by convention in the abbreviated formulas, but D (or L) may be inserted when desirable. Hyphens may replace left-to-right arrows (see Section 3.4 of Reference 13).

lactosylceramide for Gal(β 1→4)GlcCer;(b) mucotriaosylceramide for Gal(β 1→4)Gal(β 1→4)GlcCer; (c) III²-α-fucosylisoglobotriaosylceramide for Fuc(α1→2)Gal(α1→3)Gal(β 1→4)Glc(β 1→1)Cer.

Lip-3.10

Sialoglycosphingolipids (synonym: gangliosides) are glycosphingolipids carrying one or more sialic residues. Sialic acid is the generic term for *N*-acetyl- or *N*-glycoloylneuraminic acid (cf. section 3 in Reference 1). Gangliosides are named as *N*-acetyl-(or *N*-glycololy-)-neuraminosyl-(X)osylceramide, where (X) stands for the root name of the neutral oligosaccharide to which the sialosyl residue is attached (cf. Table 1). The position of the sialosyl residue is indicated in the same way as in the case of fucolipids (see Lip-3.9).

Example: II³-*N*-acetylneuraminosyllactosylceramide for AcNeu(α2→3)Gal(β 1→4)Glc(β 1→1)Cer.

Lip-3.11

Glycosphingolipids carrying a sulfuric ester (sulfate) group, formerly called sulfatides, are are preferably named as sulfates of the parent neutral glycosphingolipid. The location of the sulfate group may be indicated as in Lip-3.9.

Example: Lactosylceramide II³-sulfate.

Lip-3.12

Phosphoglycosphingolipids with phosphodiester structures are named according to the recommendation for the phospholipids (see above).

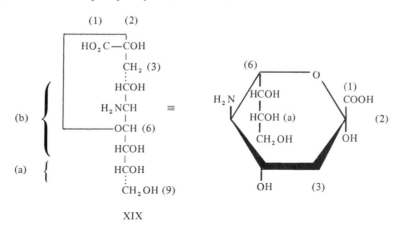

XIX

Neuraminic acid

5-Amino-3,5-dideoxy-D-*glycero*-D-*galacto*-nonulopyranosonic acid

(a)   (b)

Symbols and Abbreviations

Lip-3.13

Simple or complex glycosphingolipids can be represented according to existing rules, using the symbols Cer, Sph, AcNeu, etc. (Appendix B), together with the recommended[13] symbols for the hexoses (Glc, Gal, etc.). Examples are given above, and in Table 1 and Appendix C. However, due to the complexity of the higher glycosphingolipids, this often results in very long and cumbersome series that are not easy to comprehend. It is therefore recommended that the oligosaccharides listed in Table 1 be represented by specific symbols in which the number of monosaccharide units (-oses) is indicated by Ose$_n$, preceded by two letters representing the trivial name of the oligosaccharide (column 3). For a short form, which may be required in the case of limited space or frequent repetition, Ose can be omitted (column 4); however, the long form is preferred as being more evocative.

Examples: (a) McOse$_3$Cer for mucotriaosylceramide, Gal(β1-4)Galβ1-4)Glc(1-1)Cer; (b) II^{3}AcNeu-GgOse$_4$Cer for II^{3}N-acetylneuraminosylgangliotetraosylceramide. Galβ1→-3GalNAcβ1→4Gal(3←2αNeuAc)β1→4Glcβ1→1Cer (see Lip-3.14 for this mode of representing a branched chain).

Abbreviations for the more important gangliosides are given in Appendix C.

Lip-3.14

When it is desirable to represent a branched oligosaccharide on a single line, as in running text or a table, the parentheses surrounding the locants in the main chain may be omitted and used instead to enclose the symbols for the branched portion(s) of the molecule. The branches follow, in parentheses and with appropriate arrows, the residues to which they are attached.

Examples: (a) NeuGcα2→3Galβ1→3GalNAcβ1→4Gal(3←2αNeuGc)β1→4Glcβ1-→1Cer; (b) NeuAcα2→3Galβ1→3GalNAcβ1→4Gal(3←2αNeuAc8←2αNeuAc)β1-→4Glcβ1→1Cer; (c) GalNAcα1→3Gal(2←1αFuc)β1→4GlcNAc(3←1αFuc)β1→3Galβ1-→4Glcβ1→1Cer ≡ III^3,IV^{2}α,α-difucosyl-IV^3-α-2-acetamido-2-deoxygalactosylneolactotetraosylceramide ≡ III^3,IV^2(Fucα$_2$,IV^{3}GalNacα-nLcOse$_4$Cer.

NEURAMINIC ACID

Lip-4.1

The compound 5-amino-3,5-dideoxy-D-glycero-D-galacto-nonulosonic acid is neuraminic acid (XIX), with the symbol Neu.[11]

Lip-4.2

The term "sialic acid" signifies the N-acylneuraminic acids and their esters and other derivatives of the alcoholic hydroxyl groups.

Lip-4.3

The radicals resulting from the removal of a hydroxyl group of neuraminic acid or sialic acid are designated as neuraminoyl or sialoyl, respectively, if the hydroxyl is removed from the carboxyl group, and as neuraminosyl and sialosyl, respectively, if the hydroxyl group is removed from the anomeric carbon atom of the cyclic structure.

Appendix A
NAMES OF AND SYMBOLS FOR HIGHER FATTY ACIDS

	Numerical symbol	Structure H$_3$C—(R)—CO$_2$H	Systematic names[a]	Trivial names[b]	"Name" symbol
1	10:0	—(CH$_2$)$_8$—	Decano-	Capr.[c]	Dec
2	12:0	—(CH$_2$)$_{10}$—	Dodecano-	Laur-	Lau
3	14:0	—(CH$_2$)$_{12}$—	Tetradecano-	Myrist-	Myr
4	16:0	—(CH$_2$)$_{14}$—	Hexadecano-	Palmit-	Pam
5	16:1	—(CH$_2$)$_5$CH=CH(CH$_2$)$_7$—	9-Hexadeceno-	Palmitole-	ΔPam
6	18:0	—(CH$_2$)$_{16}$—	Octadecano-	Stear-	Ste
7	18:1(9)	—(CH$_2$)$_7$CH=CH(CH$_2$)$_7$—	*cis*-9-Octadeceno-	Ole-	Ole
8	18:1(11)	—(CH$_2$)$_5$CH=CH(CH$_2$)$_9$—	11-Octadeceno-	Vaccen-	Vac
9	18:2(9,12)	—(CH$_2$)$_3$(CH$_2$CH=CH$_2$(CH$_2$)$_7$—	*cis, cis*-9,12-Octadecadieno-	Linole-	Lin
10	18:3(9,12,15)	—(CH$_2$CH=CH)$_3$(CH$_2$)$_7$—	9,12,15-Octadecatrieno-	(9,12,15)-Linolen-	αLnm
11	18:3(6,9,12)	—(CH$_2$)$_3$(CH$_2$CH=CH)$_3$(CH$_2$)$_4$—	6,9,12-Octadecatrieno-	(6,9,12)-Linolen-	γLnm
12	18:3(9,11,13)	—(CH$_2$)$_3$(CH=CH)$_3$(CH$_2$)$_7$—	9,11,13-Octadecatrieno-	Eleostear-	eSte
13	20:0	—(CH$_2$)$_{18}$—	Icosano-[d]	Arachid-	Ach
14	20:2(8,11)	—(CH$_2$)$_6$(CH$_2$CH=CH)$_2$(CH$_2$)$_6$—	8,11-Icosadieno-[d]		Δ$_2$Ach
15	20:3(5,8,11)	—(CH$_2$)$_6$(CH$_2$CH=CH)$_3$(CH$_2$)$_3$—	5,8,11-Icosatrieno-[d]		[d]Δ$_3$Ach
16	20:4(5,8,11,14)	—(CH$_2$)$_3$(CH$_2$CH=CH)$_4$(CH$_2$)$_3$—	5,8,11,14-Icosatetraeno-[d]	Arachidon-	[d]Δ$_4$Ach
17	22:0	—(CH$_2$)$_{20}$—	Docosano-	Behen-	Beh
18	24:0	—(CH$_2$)$_{22}$—	Tetracosano-	Lignocer-	Lig
19	24:1	—(CH$_2$)$_7$CH=CH(CH$_2$)$_{13}$—	*cis*-15-Tetracoseno-	Nervon-	Ner
20	26:0	—(CH$_2$)$_{24}$—	Hexacosano-	Cerot-	Crt
21	28:0	—(CH$_2$)$_{26}$—	Octacosano-	Montan-	Mon

[a]   Ending in "-ic", "-ate", "-yl", for acid, salt or ester, acyl radical, respectively.

[b]   Ending in "-ic", "-ate", "-oyl" for acid, salt or ester, or acyl radical, respectively.

[c]   Not recommended because of confusion with caproic (hexanoic) and caprylic (octanoic) acids. Decanoic is preferred.

[d]   Formerly "eicosa-" (Changed by IUPAC Commission on Nomenclature of Organic Chemistry, 1975).

Appendix B
SYMBOLS RECOMMENDED FOR VARIOUS CONSTITUENTS OF LIPIDS

Name	Symbol[a]
For alkyl radicals[b]	R
Methyl, ethyl . . . dodecyl	Me, Et, Pr, Bu, Pe, Hx, Hp, Oc, Nn, Dec, Und, Dod
For aliphatic carboxylic acids[b]	Acyl (not abbreviated), RCO-
Formyl, acetyl, glycoloyl, propionyl	Fo (or HCO), Ac, Gc, Pp
Butyryl, valeryl	Br, Vl
Hexanoyl, heptanoyl, octanoyl	Hxo, Hpo, Oco
Nonanoyl, decanoyl, undecanoyl	Nno, Dco, Udo
Lauroyl, myristoyl, palmitoyl	Lau, Myr, Pam
Stearoyl, eleostearoyl, linoleoyl, arachidonoyl	Ste, eSte, Lin, $\triangle_4$Ach
For glycerol and its oxidation products[c]	
Glycerol, glyceraldehyde, glycerone, glyceric acid	Gro, Gra, Grm, Gri
For "glycosyl"	Ose
Glucose, galactose, fucose	Glc[d], Gal, Fuc
Gluconic acid, glucuronic acid	GlcA, GlcU[e]
Glucosamine[f], N-acetylglucosamine	GlcN, GlcNAc
Neuraminic, sialic, muramic acids	Neu, Sia, Mur
N-Acetylneuraminic acid, N-glycoloylneuraminic acid	NeuAc[g], NeuGe
Deoxy	d
Miscellaneous	
Ceramide, choline, ethanolamine	Cer, Cho, Etn[h]
Inositol, serine	Ins, Ser
Phosphatidyl, sphingosine, sphingoid, phosphoric residue	Ptd, Sph, Spd, P

[a]   These symbols are constructed in analogy to those already in use for amino acids and saccharides[11,13]; they may assist the abbreviated representation of more complex lipids in a way similar to the peptides and polysaccharides. Prefixes such as "iso-", "tert-", "cyclo-" are specified in the symbols by lower-case superscripts (Pr[i], Bu[t], Hx[c]) or lower-case prefixes (iPr, tBu, cHx), unsaturation by, e.g., $\triangle^3$ for a 3,4 double bond, $\triangle^3$ for a 3,4 triple bond (cf. Proteins, Vol. I. pp. 96—108, in *Handbook of Biochemistry*, 3rd ed., edited by G. Fasman, CRC Press, Cleveland, Ohio, 1976). Many of these symbols are drawn from previously published Recommendations.[11,12] See also Appendix A.

[b]   Systematic and recommended trivial names of unbranched, acyclic compounds only (cf. Appendix A). Other forms are created by prefixes (e.g., "iso-", "tert-", "cyclo-"). See Appendix A.

Appendix B (continued)
SYMBOLS RECOMMENDED FOR VARIOUS CONSTITUENTS OF LIPIDS

c   These symbols form a self-consistent series for a group of closely related compounds. It is recognized that other abbreviations (but no symbols) are currently in use, (see Lip-2.12).

d   Not Glu (glutamic acid) or G (nonspecific).

e   Recommended in place of GlcUA, the "A" begin unnecessary.

f   Approved trivial name for 2-amino-2-deoxyglucose; similarly for galactose (GalNAc), etc.

g   AcNeu was recommended earlier.[11] When it is necessary to differentiate between *N*-acetyl and *O*-acetyl derivatives. NeuN Ac and NeuO Ac (italicized locants, in contradistinction to GalNAc, etc.) may be employed.

h   May take the form OEtN< if substitution on the nitrogen atom is to be indicated.

Appendix C
ABBREVIATED REPRESENTATION OF GANGLIOSIDES

Lipid document[a]	Designation according to Wiegandt[b]	Svennerholm[c]
1. I^{3}NeuAc-GalCer	G$_{Gal}$ 1 NeuAc	—
2. II^{3}NeuAc-LacCer	G$_{Lac}$1 NeuAc	
3. II^{3}NeuGc-LacCer	G$_{Lac}$1 NeuNGl	G$_{M3}$
4. II^3(NeuAc)$_2$-LacCer	G$_{Lac}$2 NeuAc	—
5. II^{3}NeuAc/NeuGc-LacCer	G$_{Lac}$2 NeuAc/NeuNGl	G$_{D3}$
6. II^{3}NeuGc-LacCer	G$_{Lac}$2 NeuNGl	—
7. II^{3}NeuAc-GgOse$_3$Cer	G$_{Gtri}$1 NeuAc	G$_{M2}$
8. II^{3}NeuAc-GgOse$_4$Cer	G$_{Gtet}$1 NeuAc	G$_{M1}$
9. IV^{3}NeuAc-nLcOse$_4$Cer	G$_{Lntet}$1a NeuAc	G$_{M1-GlcNAc}$
10. IV^{6}NeuAc-nLcOse$_4$Cer	G$_{Lntet}$1b NeuAc	
11. IV^{2}Fuc,II^3 NeuAc-GgOse$_4$Cer	G$_{Gfpt}$1 NeuAc	—
12. IV^{3}NeuAc-nLcOse$_4$Cer	—	—
13. II^3(NeuAc)$_2$-GgOse$_4$Cer	G$_{Gtet}$2b NeuAc	G$_{D1b}$
14. IV^{3}NeuAc, II^{3}NeuAc-GgOse$_4$Cer	G$_{Gtet}$2a NeuAc	G$_{D1a}$
15. II^3(NeuAc)$_3$-GgOse$_4$Cer	G$_{Gtet}$3b NeuAc	—
16. IV^{3}NeuAc,II^3(NeuAc)$_2$-GgOse$_4$Cer	G$_{Gtet}$3a NeuAc	G$_{T1}$
17. IV^{3}NeuAc,II^3(NeuAc)$_3$-GgOse$_4$Cer	G$_{Gtet}$4b NeuAc	—
18. IV^3(NeuAc)$_2$II^3(NeuAc)$_3$-GgOse$_4$Cer	G$_{Gtet}$5 NeuAc	—
19. IV^{3}NeuAc.II^{3}NeuAc-GgOse$_5$Cer	G$_{Gpt}$2a NeuAc	—

[a] To indicate linkage points and anomeric form: Fuc should be written ($\leftarrow$1αFuc); NeuAc should be written ($\leftarrow$2αNeuAc); (NeuAc)$_2$ should be written ($\leftarrow$2αNeuAc8)$_2$; etc. If these features are assumed or defined, the short form used in this column is more convenient for use in texts and tables.

[b] The subscripts to G (for ganglioside), from 7 on, have the meanings; Gtri = gangliotriose, Gtet = gangliotetraose, Litet = lactoisotetraose, Gpt = gangliopentaose, Gfpt = gangliofucopentaose, Wiegand, H., *Hoppe-Seyler's Z. Physiol. Chem.*, 354, 1049, 1973.

[c] G = ganglioside, M = monosialo, D = disialo, T = trisialo. Arabic numerals indicate sequence of migration in thin-layer chromatograms. (Svennerholm, L., *J. Neurochem.*, 10, 613, 1963.)

EDITORIAL COMMENT

These Recommendations are being published as a service to the biochemical community. Publication does not imply endorsement of the recommendations by the journals but rather presents this information for consideration.

ACKNOWLEDGMENT

We are indebted to the editor of *Lipids* for permission to reproduce photographically the chemical structures that appeared in their publication of these Recommendations in *Lipids*, 12, 455—468, 1977.

REFERENCES

1. IUPAC-IUB Commission on Biochemical Nomenclature (1967), *Eur. J. Biochem.*, 2, 127, 1967; 12, 1, 1970.
2. International Union of Pure and Applied Chemistry, *Information Bulletin*, Appendix, 31, 1973.
3. International Union of Biochemistry, *Nomenclature of Organic Chemistry (Sections A, B, and C)*, 2nd ed., Butterworths, London, 1966.
4. **Mills, J. A. and Klyne, W.**, *Progr. Stereochem. 1*, 181, 1954.
5. IUPAC Commission on the Nomenclature of Organic Chemistry and IUPAC-IUB Commission on Biochemical Nomenclature, *Eur. J. Biochem.*, 21, 455, 1971; also 25, 4, 1972.

6. International Union of Pure and Applied Chemistry, *J. Org. Chem.*, 35, 2849, 1970; also *Eur. J. Biochem.*, 18, 151, 1971.
7. **Baer, E. and Fischer, H. O. L.,** *J. Biol. Chem.*, 128, 475, 1939.
8. **Baddiley, J., Buchanan, J. G. and Carss, B.,** *J. Chem. Soc.*, 1869, 1957.
9. **Hirschmann,H.,** *J. Biol. Chem.*, 235, 2762, 1960.
10. IUPAC-IUB Commission on Biochemical Nomenclature (1977), *Proc. Natl. Acad. Sci. U.S.A.*, 74, 2222, 1977; *Hoppe-Seyler's Z. Physiol. Chem.*, 358, 599, 1977 also *Eur. J. Biochem.*, 79, 1, 1977.
11. IUPAC-IUB Commission on Biochemical Nomenclature (1972), *Eur. J. Biochem.*, 27, 201, 1972.
12. IUPAC-IUB Commission on Biochemical Nomenclature (1970), *Eur. J. Biochem.*, 15, 203, 1970.
13. IUPAC-IUB Combined Commission on Biochemical Nomenclature (1967), *Eur. J. Biochem.*, 1, 259, 1967.

Chapter 2

EXTRACTION AND HYDROLYSIS OF LIPIDS AND SOME REACTIONS OF THEIR FATTY ACID COMPONENTS

W. W. Christie

I. INTRODUCTION

Lipids exist in tissues in a variety of physical forms. In the adipose tissue of animals or the oil bodies of seeds, for example, lipids are found as discrete droplets in separate compartments from most of the other cellular constituents. On the other hand in other tissues, lipids occur in intimate association with the other components of the cells (especially proteins) and indeed may be strongly bound to them by hydrophobic forces or by hydrogen bonding. Before these lipids can be analyzed, they must be isolated quantitatively from the tissues by extraction with suitable organic solvents, and the extracts must be freed of nonlipid contaminants by an appropriate washing procedure. The solvents selected for the purpose must be sufficiently polar to disrupt the association between the lipids and the other tissue constituents, but should not react chemically with the lipids in any way.

The lipid extracts thus obtained may be separated into simpler lipid classes by the chromatographic procedures described in other chapters of this volume, but for complete analysis it may be necessary to break down these lipids, or often the complete lipid extracts, into simpler component parts by chemical means. For example, the fatty acid components of lipids are obtained by hydrolysis or by transesterification before their composition is determined. The methods adopted for this should permit complete recovery of the desired compounds but should not attack other functional groups that may be present.

At all stages, care must be taken to prevent the introduction of artifacts or other extraneous substances into the lipid extracts or into fractions derived from the lipids. In particular, precautions must be taken to prevent loss of unsaturated fatty acids that can autoxidize rapidly if left unprotected. Natural antioxidants, for example tocopherols, are generally present in tissues and will afford some protection to lipids, but it is generally advisable to add an additional antioxidant such as 2,6-di-*tert*-butyl-*p*-cresol (BHT or butylated hydroxy toluene) at a level of approximately 50 mg/ℓ to the solvents.[1] This need not interfere with any subsequent analyses as it can be removed if need be by volatilization under vacuum, or in a stream of nitrogen, or by chromatography. In addition, lipid extraction and all chemical procedures should be carried out wherever possible in an atmosphere of nitrogen, and samples should be stored at low temperatures (down to $-20°C$) under nitrogen at all times. Excessively high temperatures should be avoided when extracting or otherwise treating lipids. Many of the procedures described in this chapter were adapted from those published elsewhere by the author.[2]

II. EXTRACTION OF LIPIDS FROM TISSUES

A. Storage of Tissues Prior to Extraction

Under optimum conditions, tissues from animal, plant, or other sources should be extracted as soon as possible after separation from the living entity to minimize the opportunities for changes to occur in the nature of the lipid components. If immediate extraction is not possible, the samples should be frozen rapidly and stored in sealed glass (not plastic) containers at $-20°C$ with an atmosphere of nitrogen. Holman[3] has advised that the samples should also be stored under chloroform rather than in saline solution as is often recommended. Storage in the frozen state with subsequent thawing can irreversibly damage tissues by disrupting the cell membranes, and the lipids can come into contact with enzymes from which they are normally excluded. For example, lipases may be released that can effect hydrolysis of lipids so that appreciable amounts of unesterified (free) fatty acids are ultimately found in the extracts. In plant tissues, phospholipase D may attack lipids, especially in the presence of some organic solvents, so that there is an accumulation of phosphatidic acid. These unwanted side effects can be minimized by homogenizing and extracting the samples in suitable solvents without allowing them to thaw first.

B. General Principles of Solvent Extraction

A variety of solvents can be used to dissolve purified single lipid classes. Nonpolar solvents such as hexane, benzene, or cyclohexane will readily dissolve neutral lipids such as tri-acylglycerols (triglycerides) or cholesterol esters, but are less suitable for polar complex lipids such as phospholipids or glycolipids. Somewhat more polar solvents such as chloroform or diethyl ether are equally good solvents for both neutral and polar lipids, while highly polar solvents such as methanol will dissolve polar complex lipids but are less suitable for many neutral lipids, especially triacylglycerols. In general, the greater the number and chain lengths of the aliphatic moieties in a molecule of a given lipid relative to the polar functional groups that may be present, the more soluble is the lipid likely to be in nonpolar solvents. Phospholipids are not at all soluble in acetone, which is often used to precipitate a fraction enriched in these compounds from solution.[4] The gangliosides are not soluble in many common organic solvents and are in fact more soluble in water or water-methanol/ethanol mixtures.

In selecting a solvent for extracting lipids from tissues, the absolute solubility of particular components in that solvent is only one factor that must be considered. The solvent also must be sufficiently polar to overcome the binding forces between the lipids and the other cellular constituents such as proteins and lipoproteins; in some circumstances, the solvent can also be used to inhibit any enzymatic hydrolysis of lipids that could occur during extraction. It should not react chemically with the lipids or otherwise stimulate side reactions, nor should it dissolve appreciable amounts of nonlipid components of the tissue at the same time.

The nature of the tissue to be extracted may also be important. For example, oil seeds or adipose tissue, which contain high proportions of triacylglycerols, can often be extracted with nonpolar solvents such as hexane or diethyl ether since these lipids aid the solubilization of the small amounts of complex lipids that may be present. More polar solvents are required in most circumstances, however. It should also be noted that most tissues contain a high proportion of water, and this can interact with other solvents to promote solubilization of particular components.

Single pure solvents are rarely used for extraction purposes although ethanol has been employed under reflux for 5 min to extract lipids quantitatively from liver tissue.[5] Also, isopropanol is frequently used in a preliminary extraction of plant tissues, as it is a potent inhibitor of the action of tissue phospholipase D,[6-8] an enzyme that may in fact be stimulated by diethyl ether or chloroform. Ethanol-diethyl ether mixtures (3:1 or 9:1 v/v) have been used often to extract lipids from lipoprotein fractions, but by far the most commonly used solvent mixture for extraction of tissues in general is chloroform-methanol (2:1 v/v), although the water present in a given tissue should perhaps be considered as a third component of the mixture.[9] This is suitable, with occasional modification, for animal, plant, and bacterial tissues and gives essentially quantitative recoveries of most of the common neutral and polar lipids. Among the important exceptions are the polyphosphoinositides, which require the addition of acid[10] or inorganic salts[11,12] to the solvents for quantitative recovery and lyso-phosphatides. These are best extracted with butanol saturated with water.[13] In extracting with chloroform-methanol, the solvents, (20 to 30 volumes) may be added together to the tissue prior to homogenization, or better results may be obtained if the tissue is first homogenized with methanol (10 volumes) before chloroform (20 mℓ) is added and the blending is repeated. For routine analyses of large numbers of samples, a single extraction may be sufficient. For optimum recovery of lipids, two or three extractions may be required; a five-stage extraction in which both acidic and basic solvent mixtures are used has been advised when essentially complete recovery of all lipids is required.[14] Other solvent mixtures have been recommended for lipid extractions, e.g., butanol-water,[13] hexane-isopropanol,[15] or toluene-ethanol,[16] but these have yet to be widely tested.

Most polar solvents will extract significant amounts of nonlipid contaminants such as carbohydrates, amino acids, and salts from tissues in addition to the lipids, and these impurities must be removed before the lipid extracts can be analyzed further. The simplest and most widely used methods consist of washing the chloroform-methanol (2:1 v/v) extracts with one quarter of their total volume of a dilute salt solution (e.g., 0.88% potassium chloride).[17] A two phase system is formed consisting of a lower layer containing chloroform-methanol-water in the approximate ratio (86:14:1 v/v) in which most of the lipids remain, and an upper phase in which the same solvents are present in proportions (3:48:47 v/v) and which contains most of the nonlipid contaminants. The lower phase, which will comprise about 60% of the total volume, can be further washed if necessary with chloroform-methanol (1:1 v/v) to remove any traces of impurities remaining. A related procedure has been developed that is particularly suitable for large samples[18] in that it uses smaller volumes of solvents while the water already present in the tissue is allowed for when further water is added during the washing step. In any washing procedure, it is very important that the proportions of chloroform, methanol, and water in the combined phases should be close to the optimum otherwise losses of lipid may occur. Water-saturated butanol extracts of tissues are washed with butanol-saturated water to remove nonlipid contaminants.[13]

Any gangliosides present in the tissues together with some of the other glycosphingolipids will tend to partition into the upper aqueous phase, and they can be recovered from this phase by dialysis and lyophilization or by a variety of other methods.[19]

An alternative more exhaustive, if more laborious, method for removing nonlipid contaminants with minimal losses of lipid consists of performing the washing procedure by means of liquid-liquid partition chromatography on columns.[20-22] As a preliminary step, chloroform, methanol, and water (8:4:3 v/v) are mixed and allowed to partition. The upper aqueous phase is then immobilized on a column of a dextran gel such as Sephadex G-25, and the lipid extract is washed through the column in some of the lower organic phase; the lipids elute rapidly, while nonlipid contaminants remain on the column. The modified procedure described by Wuthier[21] is probably that most suitable for rapid purification of large numbers of samples, while that of Siakotos and Rouser[22] also permits the isolation of a separate pure ganglioside fraction. A procedure for the separation of lipids from nonlipids is described in Chapter 3.

C. Some Recommended Extraction Procedures

If a rapid procedure is required for large amounts of sample and less than complete recovery of lipids is acceptable, the procedure described below has been modified somewhat to minimize losses of acidic lipids such as phosphatidylinositol.[23] Recoveries of most lipids are likely to be of the order of 95%.

Procedure for the Extraction of Lipids[18,23]
The fresh tissue is assumed to contain 80% water. The tissue, 100 g, is homogenized with chloroform (100 mℓ) and methanol (200 mℓ) by means of a Waring blender for about 4 min. (If two liquid phases form, more solvent should be added). After filtration, the residue is homogenized once more with chloroform (100 mℓ) and refiltered. Potassium chloride solution (0.88% 100 mℓ) is added to the combined filtrates and the whole is shaken vigorously before being allowed to settle. The upper layer together

with any interfacial material is removed by aspiration while the lower phase, which contains the lipids, is filtered and the solvent is removed under vacuum on a rotary evaporator. The lipid so obtained is stored in a small volume of chloroform at $-20°C$ until it is required for further analysis.

The ratio of chloroform-methanol-water in the biphasic system should be near 10:10:9 v/v. Diethyl ether can be used as a preliminary extractant for tissues rich in triacylglycerols such as adipose tissue or oil seeds, as this solvent does not coextract significant amounts of nonlipid contaminants before the above procedure is applied to the residue.

The extraction procedure of Folch et al.[17] is that most often used when more exhaustive recoveries of lipids are required and the modified procedure[24] described below gives recoveries of the order of 95 to 99% with most animal tissues.

Procedure for the Extraction of Lipids[17,24]
The tissue (1 g) is homogenized in methanol (10 ml) for 1 min after which chloroform (20 ml) is added and homogenization is continued for a further 2 min. The mixture is filtered and potassium chloride solution (0.88%, 7.5 ml) is added to the filtrate, and the whole is shaken vigorously before being allowed to settle. The upper layer is removed by aspiration and water-methanol (1:1 v/v) amounting to one quarter of the volume of the lower layer is added and the process is repeated. The lipid is recovered from the solvent in the lower layer in the manner described in the previous method.

If the volume of the solvent used is at least 30 times that of the sample, a single extraction should suffice, but if it is thought necessary for particular samples, the solid residue can be reextracted to improve the recovery provided that the ratio of chloroform-methanol-water in the biphasic system remains in these proportions (8:4:3 v/v). In some circumstances even more thorough extraction procedures may be required.[14]

Procedure for the Extraction of Lipids from Plant Tissues[7,8]
Plant tissues should be extracted first with hot isopropanol (100 parts by weight), then with chloroform-isopropanol (1:1 v/v) (200 parts by weight) to inhibit the phospholipases.[7,8] After filtering, the solvent is largely removed and the lipids are taken up in chloroform-methanol (2:1 v/v) and purified by the washing procedure in the last method described above.

The dextran gel purification procedures, in particular that described by Wuthier,[21] can also be incorporated into the above methods if it is necessary to replace the washing steps. (See Chapter 3.)

D. Some Practical Considerations

Precautions necessary to minimize autoxidation of samples during handling were described in the introduction to this chapter. In addition, large volumes of solvents should be evaporated under vacuum in rotary evaporators at a temperature below 40°C; small amounts of solvents can be evaporated by directing a stream of nitrogen onto the surface (in a ventilated area). The tissues to be extracted should be homogenized in a Waring Blender or in a machine of similar construction, i.e., in which there is no contact between the solvent and any washers or greased bearings. It may be necessary to rehydrate lyophilized tissues prior to extraction to improve the recoveries of lipids. Lipid extracts should not be left in the dry state but are better dissolved in a small volume of chloroform and stored under nitrogen at $-20°C$ until required for further analysis.

Potentially troublesome contaminants can be found from time to time in all solvents that should preferably be distilled before use. Any equipment or containers made of plastics other than Teflon (fluon) should be strictly avoided, as plasticizers (e.g., dibutylphthalate) can be leached out surprisingly easily, and may appear in subsequent analyses as spurious peaks on chromatograms.

If tissues containing small amounts of carbonates or bicarbonates are held in the presence of alcoholic solvents for long periods, transesterification of the lipids can occur and appreciable amounts of fatty acid esters of the alcohol are found in the extracts.[25] A number of related reports have appeared, and it is possible that both acidic and basic constituents of tissues can catalyze the reaction so that it may be necessary to adjust the pH of the reaction mixture to prevent this. Methyl and ethyl esters have been found in trace amounts as natural constituents of some tissues, however, so it cannot always be assumed that they are artifacts of the extraction procedure. Where there is doubt, this can be checked by extracting a separate piece of the tissue with nonalcoholic solvents such as diethyl ether,[26] hexane,[27] or chloroform-acetone mixtures.[28]

Contact with solvents can affect other lipid bonds. For example, plasmalogens stored in methanol for long periods have been shown to undergo rearrangement,[29] while acetone can promote dephosphorylation of polyphosphoinositides in brain lipids[10,30] and form an imine-derivative of phosphatidylethanolamine in lyophilized tissue.[31,32]

Care should always be taken in handling chloroform-methanol as it is a powerful skin irritant and can produce some pain and temporary hypo- or hyperpigmentation if spilled on the skin.

The weight percentage of lipid in the wet tissue should always be recorded after an extraction, together, if possible, with the weight percentage of lipid relative to the amount of dry matter in the tissue. In the latter instance, it may be necessary to determine the dry matter content of a piece of tissue separately.

III. HYDROLYSIS OF LIPIDS

A. Alkaline Hydrolysis of Ester Bonds

Hydrolysis or saponification of the fatty acid ester bonds in most natural lipids can be achieved readily by treatment with ethanolic or methanolic alkali. Generally, the reaction is carried out by heating the lipids under reflux with an excess of dilute aqueous ethanolic potassium hydroxide for a short period. The fatty acids and any diethyl ether-soluble non-saponifiable products are recovered for further analysis as in the following procedure.

Alkaline Hydrolysis ("Saponification") of Lipids
The lipid sample (100 mg) is dissolved in a solution of potassium hydroxide (1 N) in 95% ethanol (2 mℓ) and is heated under reflux for 1 hr. Water (5 mℓ) is then added and the mixture is extracted thoroughly with diethyl ether (3 × 5 mℓ) with centrifugation, if necessary, to break any emulsions. The ether layer is washed with water and dried over anhydrous sodium sulfate before the solvent is removed under vacuum to yield the nonsaponifiable materials. The aqueous layer, together with the water washings, are acidified with 6 N hydrochloric acid and extracted thoroughly with diethyl ether in a manner analogous to that above to recover the unesterified fatty acids.

Any sterols, hydrocarbons, or long-chain alcohols that were in the lipid sample in free or esterified form are recovered in the nonsaponifiable fraction together with the deacylated forms of any neutral ether lipids. If the lipids contain short-chain fatty acids (less than C_{12}), it is necessary to be very thorough in the extraction step to optimize the recovery, but it is virtually impossible to obtain quantitative recovery of hexanoic and shorter-chain acids. Sterol esters are hydrolyzed comparatively slowly by this procedure, and if these are major components of the lipid sample, it is necessary to reflux the mixture for a longer period (about 2 hr). The free fatty acid and nonsaponifiable fractions obtained from a hydrolytic reaction of this kind can also be extracted together, then separated by means of column chromatography on alkali-impregnated silicic acid[33] or on ion-exchange resins.[34] Procedures are available for the isolation and analysis of the water-soluble products of alkaline hydrolysis of complex lipids.[35] (See Chapter 15, "Products of Partial Degradation of Phospholipids".) If the mild hydrolytic procedure described above is used and the conventional precautions are taken against autoxidation, polyunsaturated fatty acids are not affected by the reaction.[36] Prolonged reaction times or excessive concentrations of alkali can cause some isomerization of double bonds, however.

B. Hydrogenolysis of Ester Bonds

1-Alkylglycerols and 1-alk-1'-enylglycerols can be obtained in good yield from alkyldiacylglycerols or alk-1'-enyldiacylglycerols (neutral plasmalogens), respectively, or from the analogous phospholipids by means of hydrogenolysis with lithium aluminum hydride.[37,38] The fatty acids are reduced to aliphatic alcohols during the reaction, but their double bond structure is retained.

A procedure for the hydrogenolysis of lipids is described in Chapter 14, "Ether Lipids (Alkoxylipids)". The same chapter includes methods for the further analysis of the products formed, such as a procedure for the acid-catalyzed cleavage of alk-1-enyl moieties, and procedures for the preparation of isopropylidene derivatives, as well as TMS-derivatives of alkylglycerols.

C. Alkaline Hydrolysis of Amide Bonds

Fatty acids occur naturally linked to long-chain bases by amide bonds in sphingomyelin and glycosphingolipids. Such bonds are comparatively resistant to alkaline hydrolysis, but the procedure described above is effective if the reflux time is prolonged to about 18 hr,[39,40] and if 2 N potassium hydroxide in methanol is used as the hydrolyzing reagent.[41]

Procedures for the hydrolysis of sphingolipids and the further analysis of the products formed are described in Chapters 17 and 18.

D. Enzymatic Hydrolysis of Lipids

Lipolytic enzymes are widely distributed in nature and have highly specific functions in animal and plant tissues; they have also been much used by lipid chemists and biochemists to hydrolyze specific ester bonds in lipids. The properties of such enzymes have been reviewed by Brockerhoff and Jensen[42] and their use in lipid structural studies has been described by Christie.[2]

One of the enzymes most used is pancreatic lipase (EC. 3.1.1.3), which hydrolyzes fatty acids from the primary positions of triacylglycerols, leaving 2-monoacylglycerols that can be analyzed to determine the composition of the fatty acids originally present in position *sn*-2. Although triacylglycerols such as fish oils that contain longer-chain polyunsaturated fatty acids are hydrolyzed comparatively slowly by the enzyme and those that contain shorter-chain fatty acids are hydrolyzed comparatively rapidly, little fatty acid specificity is observed with triacylglycerols that contain a more conventional range of fatty acid components. The following method is recommended:

Procedure for the Hydrolysis of Triacylglycerols with Pancreatic Lipase[43]
A mixture of tris(hydroxymethyl)methylamine ("tris") buffer (1 M, pH 8.0, 1 mℓ), calcium chloride solution (2.2%, 0.1 mℓ), and bile salt solution (0.05%, 0.25 mℓ) are added to the triacylglycerols in a stoppered test tube and the solution is equilibrated to a tempeature of 40°C when the pancreatic lipase preparation (1 mg) is added. A mechanical shaker is used to shake this mixture vigorously for 2 to 4 min at 40°C until 50 to 60% hydrolysis is achieved. The reaction is then curtailed by adding ethanol (1 mℓ) followed by hydrochloric acid (6 N, 1 mℓ). The mixture is extracted thoroughly

with diethyl ether (3 × 10 mℓ), and the ether layer is washed with water (2 × 5 mℓ) and dried over anhydrous sodium sulfate before the solvent is removed under vacuum. The monoacylglycerols are obtained for analysis by preparative TLC on silica gel G layers using hexane-diethyl ether-formic acid (80:20:2 v/v) as developing solvent; the required component is found just above the origin, and it is recovered from the adsorbent by elution with chloroform-methanol (19:1 v/v) for further analysis.

It is essential that the triacylglycerols fully disperse for the reaction to occur, and with comparatively high melting samples containing high proportions of saturated fatty acids, it may be necessary to add a small amount of hexane[44] or methyl oleate[45] to assist the process. Pancreatic lipase will also hydrolyze the primary ester bonds in phospholipids[46,47] and in glycosyldiacylglycerols[48,49] and has been used in structural analyses of such lipids.

Phospholipase A_2 (EC. 3.1.1.4) hydrolyzes ester bonds in position *sn-2* of glycerophosphatides, yielding free fatty acids from this position and a lysophosphatide with fatty acids in position *sn-1*. Both these products can be isolated to determine the fatty acid compositions of each position. Snake venoms are the most important sources of the enzyme and while that from *Crotalus adamanteus* or *C. atrox* has probably been used most often, that from *Ophiophagus hannah*[50] is now considered more suitable. Although the enzyme has been used most frequently in studies of the structure of phosphatidylcholine, it has also been used with cardiolipin,[51] phosphatidylethanolamine,[52] phosphatidylserine,[53] and many other phospholipids. The following procedure has been used in a number of laboratories for the analysis of phosphatidylcholine.

Procedure for the Hydrolysis of Phosphatidylcholines and other Phospholipids with Phospholipase A_2[54]
A stock solution of O. hannah snake venom (2.5 mg) in tris buffer (0.5 mℓ, pH 7.5) containing calcium chloride (4 mM) is made up, and 100 μℓ of this is added to phosphatidylcholine (up to 10 mg) in diethyl ether (2 mℓ). The mixture is shaken vigorously for 1 hr, and then is washed into a flask with methanol (10 mℓ) and chloroform (20 mℓ). The whole is dried over sodium sulfate. The solution is filtered, the solvent is removed under vacuum, and the lipids and the products are separated by preparative TLC on silica gel G; the plate is developed half way in a solvent consisting of chloroform-methanol-acetic acid-water (25:15:4:2 v/v); then the plate is dried and redeveloped to the top in hexane-diethyl ether (4:1 v/v). The free fatty acids are the only component in the top half of the plate, and can be recovered from the adsorbent by elution with diethyl ether, while lysophosphatidylcholine should be the only component in the bottom half of the plate and can be recovered by elution with chloroform-methanol-water (5:5:1 v/v) for further analysis.

IV. ESTERIFICATION AND TRANSESTERIFICATION OF LIPIDS

The hydrolytic procedures described above provide a convenient method for obtaining the fatty acid components of a lipid in free (unesterified) form. However, for many analytical purposes, it is preferable that the fatty acids be converted to the less polar and more volatile methyl ester derivatives (other ester derivatives may also be useful for specific purposes). It is not necessary to hydrolyze lipids prior to esterification as they can readily be transesterified directly in the presence of a large excess of the required alcohol by means of an appropriate catalyst; the choice of reagent will depend on the circumstances and is governed by the nature of the lipid and of the fatty acid constituents. A number of reviews of methods for the preparation of alkyl esters have appeared.[2,55,56]

In all the following procedures, it is assumed that the precautions against autoxidation described above will be taken. Although methods of preparing methyl esters only are described below, other esters can be prepared by substituting the appropriate alcohol in most procedures. Methyl esters can be purified, if necessary, by preparative TLC on layers of silica gel G; hexane-diethyl ether (9:1 v/v) is a suitable developing solvent.

A. Base-Catalyzed Transesterification of Lipids

Lipids containing *O*-acyl moieties are transesterified rapidly in solution in anhydrous methanol containing a basic catalyst. Free fatty acids, on the other hand, simply form salts and are not esterified by the reaction, and indeed, if care is not taken to exclude water, some hydrolysis can occur with formation of free fatty acids. The most widely used reagent is 0.5 N sodium methoxide in methanol, which is prepared by dissolving clean sodium metal in anhydrous methanol. This is stable for several months at room temperature, especially if oxygen is excluded. A homogeneous solution of the lipid is essential for the reaction to work properly, and it may be necessary to add benzene to the reaction medium to effect solution. Chloroform is also used occasionally for this purpose, but it is possible that dichlorocarbene could be generated and react with double bonds. The following procedure can be recommended.

> *Procedure for the Base-Catalyzed Transesterification of Lipids*
> *Benzene (1 mℓ) and 0.5 N sodium methoxide in anhydrous methanol (2 mℓ) are added to the lipids (2 to 50 mg) in a test tube and the whole is heated at 50°C for 10 min. Glacial acetic acid (0.1 mℓ) and water (5 mℓ) are added and the esters are extracted with hexane (2 × 5 mℓ) using a Pasteur pipet to separate the layers. After drying the hexane layer over anhydrous sodium sulfate the solvent is removed under vacuum or in a stream of nitrogen and the required esters are dissolved in a little fresh hexane for storage.*

The procedure works well with triacylglycerols, phospholipids, and other glycerolipids, but cholesterol esters are transesterified comparatively slowly and require a reaction time of 20 to 30 min. No degradation of the cholesterol occurs, however, and this can be recovered for further analysis or quantification if necessary. If the reaction is carried out as described above, no alteration of the double bond structure of the fatty acids occurs. Aldehydes are not liberated from plasmalogens by the procedure, and the amide-linked fatty acids of sphingolipids are not transesterified, a property that is occasionally used in the purification of such lipids.[40]

Triacylglycerols containing short-chain fatty acids such as those in milk fat can be esterified by the above and related procedures, but the methyl esters are highly volatile and some loss can occur during the working-up process. The following method, which is similar to that described by Christopherson and Glass,[57] has no aqueous extraction or solvent removal steps and is carried out at room temperature; it has been used frequently in the preparation of methyl esters from milk fats for gas-liquid chromatography (GLC) analysis.

> *Procedure for the Base-Catalyzed Transesterification of Lipids with Short-Chain Acyl Moieties*
> *Hexane (0.3 mℓ) and 2 N sodium methoxide in dry methanol (0.1 mℓ) are added to the triacylglycerols (5 to 20 mg) in a stoppered test tube. The solution is shaken gently at room temperature for 5 min when further hexane (2.5 mℓ) is added followed by powdered anhydrous calcium chloride (about 1 g). After 1 hr, the tube is centrifuged at approximately 2000 g for 2 min to compact the drying agent. An aliquot of the supernatant solution can be taken for GLC analysis.*

If unesterified fatty acids are also present in the sample, they can be esterified with diazomethane (see below) before the glycerolipids are transesterified by the above method. Although the latter is in use in a number of laboratories, some doubts have recently been cast on its efficacy, and an alternative procedure for the preparation of butyl esters has been recommended.[58]

B. Acid-Catalyzed Esterification and Transesterification of Lipids

O-Acyl lipids can be transesterified and free fatty acids esterified by reaction with a large excess of anhydrous methanol in the presence of an acidic catalyst. The most popular reagent is 5% hydrogen chloride in methanol, prepared by bubbling hydrogen chloride gas from a cylinder or prepared *in situ*[59] into anhydrous methanol; alternatively, acetyl chloride can be added to anhydrous methanol to prepare the reagent.[60]

The reaction is performed by heating the lipid with the reagent under reflux for 2 hr or in a stoppered tube at 50°C overnight; benzene or chloroform should again be added to effect solution of nonpolar lipids. A solution of 1% concentrated sulfuric acid in anhydrous methanol can be used in the same manner and is more easily prepared, but it can cause some oxidation of unsaturated components if used carelessly. Boron trifluoride in methanol (12 to 14% w/v)[61] has also been used a great deal, especially for esterification of free fatty acids, but the reagent cannot be stored for long periods and artifacts may be formed with loss of unsaturated fatty acids if old or too concentrated solutions are used.[62,63] The following procedure is recommended.

Procedure for the Acid-Catalyzed Esterification and Transesterification of Lipids
Benzene (1 mℓ) and 5% hydrogen chloride in methanol (2 mℓ) are added to the lipid sample in a test tube and the mixture is refluxed for 2 hr. Aqueous sodium chloride solution (5 mℓ; 5% w/v) is then added, and the esters are extracted with hexane (2 × 5 mℓ). The hexane layer is washed with aqueous potassium bicarbonate solution (5 mℓ; 4%, w/v) and dried over anhydrous sodium sulfate, before the solvent is removed in vacuo. The esters are stored in a small volume of fresh hexane under nitrogen.

The same method is used to prepare dimethylacetals from aliphatic aldehydes or plasmalogens so that when it is used on lipid samples containing such compounds, acetals may contaminate the methyl esters.[2] If cholesterol esters are transesterified with this and other acidic reagents, some of the cholesterol released can react to form cholestadiene and cholesteryl methyl ethers that may interfere with subsequent analyses.[61,64-66] Also, cyclopropane, cyclopropene, and epoxy groups in fatty acids are disrupted by acidic reagents, and conjugated double bond systems can undergo some rearrangement.[2,55] Lipids containing these unusual fatty acids and cholesterol esters are best transesterified by alkaline reagents. Despite such potential drawbacks, the method has wide applicability to more conventional lipids.

Amide-linked fatty acids are not readily transesterified by acidic catalysts in anhydrous methanol, but reaction can be achieved by refluxing sphingolipids with methanol-concentrated hydrochloric acid (5:1 v/v) for 5 hr,[67] or by heating with the reagent at 50°C for 24 hr.[55] Alternative procedures should be adopted if the long-chain bases are also required for analysis.[2]

C. Esterification of Free Fatty Acids with Diazomethane

Diazomethane, which is usually prepared as a solution of diethyl ether, reacts almost instantaneously at room temperature with unesterified fatty acids in the presence of a small amount of methanol that acts as a catalyst to form methyl esters. The reagent is prepared by the action of alkali on a nitrosamide, e.g., *N*-methyl-*N*-nitroso-*p*-toluenesulfonamide (Diazald, Aldrich Chemical Co., Milwaukee, Wis.) in alcoholic solution. Diazomethane is a neutral reagent, is easy to use, and has few side reactions, so it is especially suited to the esterification of free fatty acids that contain acid-labile functional groups. On the other hand, nitrosamides are potentially carcinogenic, and diazomethane is highly toxic and potentially explosive, so that great care must be exercised in its preparation and use. Fortunately, two microprocedures have been described for the preparation of small amounts of diazomethane for immediate use;[68,69] with these methods, the risk to the careful user is negligible.

V. SOME REACTIONS OF THE DOUBLE BONDS OF FATTY ACIDS

A. Hydrogenation
1. Catalytic Hydrogenation
Catalytic hydrogenation can be a useful procedure in a number of aspects of lipid analysis. For example, it is used to hydrogenate unsaturated fatty acids to the corresponding saturated components that can then be identified unequivocally by GLC methods and the chain length of the original fatty acid determined. In the analysis of intact lipids such as triacylglycerols by means of GLC at high temperatures, better recoveries and resolutions may be achieved after hydrogenation. The following procedure,[2] which is much simpler than many that have been described, is recommended.

Procedure for the Catalytic Hydrogenation of Unsaturated Lipids
The lipid (2 to 5 mg) is dissolved in hexane (1 mℓ) in a test tube and platinum oxide (Adams' catalyst) (1 mg) is added. A two-way tap is used to connect the tube to either a reservoir of hydrogen at just above atmospheric pressure (e.g., in a gas sampling bag) or to a water pump. The tube is evacuated briefly and then flushed with hydrogen, and the process is repeated twice before the tube is left open to the hydrogen atmosphere and is shaken vigorously for 2 hr. At the end of this period, the tube is flushed with nitrogen, and the hydrogenated lipid is recovered from the catalyst by washing with chloroform.

Methyl esters may be hydrogenated in methanol solution. The amount of hydrogen taken up can be measured, if necessary, by using a suitably calibrated hydrogen reservoir (experimental details can be obtained in most textbooks of practical organic chemistry). In an alternative procedure for quantitative hydrogenation,[70] both the platinum catalyst and the hydrogen are generated *in situ* by treatment of platinum salts with a standard solution of sodium borohydride; an application of the reaction to the determination of unsaturation in fats and oils has been described.[71]

2. Hydrazine and Deuterohydrazine Reduction
If the double bonds in an unsaturated fatty acids are hydrogenated catalytically and the reaction is not allowed to go to completion, considerable rearrangement of the hydrogen atoms and of the positions and configurations of the double bonds are found in the partially hydrogenated products. Such rearrangements do not occur during hydrazine reduction, a reaction that is commonly used to prepare partially hydrogenated fatty acids in order to simplify the task of determining double bond position and configuration in polyunsaturated fatty acids.[72,73] Such procedures are described in detail in Chapter 21.

A related procedure is used to add deuterium without rearrangement across double bonds to aid in the location of the latter by means of mass spectrometry.[74]

Procedure for the Deuterohydrazine Reduction of Unsaturated Lipids
Deuterohydrazine (5 mmol) in twice its volume of deuterium oxide is added to the unsaturated ester (0.5 mmol) in anhydrous dioxane (5 mℓ). The solution is stirred at 55 to 60°C in air but with rigid exclusion of atmospheric water for 8 hr, after which the reagents are evaporated in vacuo. The product can be purified by preparative TLC

on layers of silica gel G impregnated with silver nitrate (described in detail in other chapters) with hexane-diethyl ether (9:1 v/v) as developing solvent.

B. Oxidation

1. Hydroxylation

A number of different reagents can be used to oxidize double bonds with formation of vicinal diols, but those most commonly used are alkaline potassium permanganate solution and osmium tetroxide. *cis*-Addition occurs with both reagents yielding diols of the *erythro*-configuration from *cis*-double bonds and of the *threo*-configuration from *trans*-double bonds. The following procedure has been recommended for mono- and dienoic fatty acids in particular.

Procedure for the Hydroxylation of Unsaturated Fatty Acids[75]
The unsaturated fatty acid (20 μmol) is dissolved in sodium hydroxide solution (0.25 M; 0.2 mℓ) and diluted with water (1 mℓ) and the whole cooled in an ice bath. Potassium permanganate solution (0.05 M; 0.2 mℓ) is added and the mixture is stirred for 5 min, after which the reaction is stopped by bubbling in sulfur dioxide. The required products are found in the lower layer formed when chloroform-methanol (2:1 v/v; 7 mℓ) is added to the solution, and they are recovered after removal of the solvent.

Osmium tetroxide is reported to give better yields of fully hydroxylated derivatives from the corresponding polyunsaturated fatty acids.[75] Such hydroxy-fatty acids in the form of the more volatile methoxy,[75] isopropylidene,[76] or trimethylsilyl ether[77,78] derivatives are useful for locating the positions of the original double bonds by means of mass spectrometry.

2. Epoxidation

Peracetic, perbenzoic, and *m*-chloroperbenzoic acids react with double bonds to form epoxides, *cis*-epoxides being formed from *cis*-olefins and *trans*-epoxides from *trans*-olefins. The following procedure is relatively simple to perform.

Procedure for the Epoxidation of Unsaturated Fatty Acids[79]
m-Chloroperbenzoic acid (8 mg) in chloroform (1 mℓ) is added to the methyl ester of a monoenoic fatty acid (10 mg) at room temperature. After 4 hr, a solution of potassium bicarbonate (5%; 2 mℓ) is added and the product is extracted thoroughly with diethyl ether. The ether layer is dried over anhydrous sodium sulfate, the solvent is removed in vacuo and the pure epoxy ester is obtained by preparative TLC on silica gel G layers; hexane-diethyl ether (4:1 v/v) is a suitable developing solvent.

Epoxy derivatives have been used as an aid in the location of double bonds in fatty acids by mass spectrometry[80,81] and as a means of estimating the relative proportions of *cis*- and *trans*-isomers by gas chromatography.[82,83]

3. Oxidative Cleavage

The positions of double bonds in fatty acids can be ascertained by oxidative fission at the double bond followed by identification of the products by gas chromatography. Ideally, the method used should give a single product from each half of the aliphatic chain, i.e., with no overoxidation. These conditions can best be achieved by using ozonolysis procedures,[72,73] although slightly more vigorous methods such as permanganate-periodate[2,84] or chromic acid-celite[85] oxidation or periodate fission of epoxides[86] are frequently used because of their convenience in practice. Such procedures are described in detail in Chapter 21.

C. Stereomutation of Double Bonds

A variety of reagents has been used to convert the *cis*-double bonds of fatty acids to the *trans*-configuration, including selenium,[87,88] diphenyl sulfide with UV irradiation,[88,89] and sodium nitrite and nitric acid (NO$_2$ radical).[90-92] All yield an equilibrium mixture in which the thermodynamically more stable *trans*-isomer is present in the highest proportion (70 to 80%). The vigorous conditions required for isomerization with selenium can lead to some migration of the double bond, but the author[99] has found that the following method in which nitrous acid is used gives excellent conversion of *cis*-double bonds to the corresponding *trans*-isomers with no detectable double bond migration.

> *Procedure for the Conversion of cis- to trans-Double Bonds in Methyl Esters of Monounsaturated Fatty Acids*
> *The cis-monoenoic ester (10 mg) is dissolved in diethyl ether (1 mℓ) and sodium nitrite (2 mg) and nitric acid (3 N; 10 μℓ) are added. The mixture is stirred in an atmosphere of nitrogen at room temperature for 1 hr, then water (2 mℓ) and hexane (10 mℓ) are added. The hexane layer is recovered and dried over sodium sulfate before the solvent is removed in vacuo. The required trans-isomer is obtained by preparative TLC on layers of silica gel impregnated with silver nitrate; hexane-diethyl ether (9:1 v/v) is a suitable developing solvent.*

Some polar byproduct formation occurs, but this can be kept to the minimum by using sufficiently dilute nitric acid and by not prolonging the reaction time unduly. Yields of 65% of the required *trans*-isomer are readily attainable.

D. Mercuric Acetate Derivatives

Mercuric acetate in methanol solution reacts quantitatively with the double bonds of the methyl esters of unsaturated fatty acids to form polar acetoxy-mercuriamethyoxy derivatives that have been used to separate components according to degree of unsaturation by means of adsorption chromatography[93] and in the bulk preparation of pure unsaturated fatty acids.[94,95] These derivatives can be reduced by sodium borohydride to methoxy fatty acid esters[96] or converted by reaction with halogens to methoxymercurihalogenoesters[97] that have distinctive mass spectrometric fragmentation patterns of value in locating the positions of the original double bonds. The following method is recommended.

> *Procedure for the Preparation of Mercuric Acetate Adducts of Methyl Esters of Unsaturated Fatty Acids[94]*
> *The methyl ester of the unsaturated fatty acid is heated under reflux with a 20% excess of the theoretical amount of mercuric acetate in methanol (5% solution) for 1 hr. The solution is cooled, diethyl ether (2.6 times the volume of the methanol) is added to precipitate the excess mercury salts; the solution is filtered, and the required derivatives are obtained on removal of the solvent.*

The original double bonds can be regenerated if necessary without double bond migration or stereomutation by refluxing with concentrated hydrochloric acid in methanol.[94] Mercuric acetate (2 mol) adds to triple bonds in a similar manner, but the latter cannot be regenerated and on acidic hydrolysis yield keto-groups. This reaction has been utilized to produce derivatives suitable for mass spectrometric location of the original triple bonds.[98]

REFERENCES

1. **Wren, J. J. and Szczepanowska, A. D.**, *J. Chromatogr.*, 14, 405, 1964.
2. **Christie, W. W.**, *Lipid Analysis*, 2nd ed., Pergamon Press, Oxford, 1982.
3. **Holman, R. T.**, in *Progress in the Chemistry of Fats and Other Lipids*, Vol. 9, Holman, R. T., Ed., Pergamon Press, Oxford, 1968, 1.
4. **Hanahan, D. J., Turner, M. B., and Jayko, M. E.**, *J. Biol. Chem.*, 192, 623, 1951.
5. **Lucas, C. C. and Ridout, J. H.**, in *Progress in the Chemistry of Fats and Other Lipids*, Vol. 10, Holman, R. T., Ed., Pergamon Press, Oxford, 1970, 1.
6. **Kates, M.**, *Can J. Biochem. Physiol.*, 34, 967, 1956.
7. **Nichols, B. W.**, *Biochim. Biophys. Acta*, 70, 417, 1963.
8. **Nichols, B. W.**, in *New Biochemical Separations*, James, A. T. and Morris, L. J., Eds., Van Nostrand, New York, 1964, 321.
9. **Schmid, P.**, *Physiol. Chem. Phys.*, 5, 141, 1973.
10. **Wells, M. A. and Dittmer, J. C.**, *Biochemistry*, 4, 2459, 1965.
11. **Michell, R. H., Hawthorne, J. N., Coleman, R., and Karnovsky, M. L.**, *Biochim. Biophys. Acta*, 210, 86, 1970.
12. **Hauser, G. and Eichberg, J.**, *Biochim. Biophys. Acta*, 326, 201, 1973.
13. **Bjerve, K. S., Daae, L. N. W., and Bremer, J.**, *Anal. Biochem.*, 58, 238, 1974.
14. **Rouser, G., Kritchevsky, G., and Yamamoto, A.**, in *Lipid Chromatographic Analysis*, Vol. 1, Marinetti, G. V., Ed., Edward Arnold, London, 1967, 99.
15. **Hara, A. and Radin, N. S.**, *Anal. Biochem.*, 90, 420, 1978.
16. **Schmid, P., Calvert, J., and Steiner, R.**, *Physiol. Chem. Phys.*, 5, 157, 1973.
17. **Folch, J., Lees, M., and Stanley, G. H. S.**, *J. Biol. Chem.*, 226, 497, 1957.
18. **Bligh, E. G. and Dyer, W. J.**, *Can. J. Biochem. Physiol.*, 37, 911, 1957.
19. **Carter, T. P. and Kanfer, J.**, *Lipids*, 8, 537, 1973.
20. **Wells, M. A. and Dittmer, J. C.**, *Biochemistry*, 2, 1259, 1963.
21. **Wuthier, R. E.**, *J. Lipid Res.*, 7, 558, 1966.
22. **Siakotos, A. N. and Rouser, G.**, *J. Am. Oil Chem. Soc.*, 42, 913, 1965.
23. **Palmer, F. B. St. C.**, *Biochim. Biophys. Acta*, 231, 134, 1971.
24. **Ways, P. and Hanahan, D. J.**, *J. Lipid Res.*, 5, 318, 1964.
25. **Lough, A. K., Felinski, L., and Garton, G. A.**, *J. Lipid Res.*, 3, 478, 1962.
26. **Dhopeshwarkar, G. A. and Mead, J. F.**, *Proc. Soc. Exp. Biol. Med.*, 109, 425, 1962.
27. **Kaufmann, H. P. and Viswanathan, C. V.**, *Fette Seifen Anstrichm.*, 65, 925, 1963.
28. **Leikola, E., Nieminen, E., and Solomaa, E.**, *J. Lipid Res.*, 6, 490, 1965.
29. **Viswanathan, C. V., Hoevet, S. P., Lundberg, W. O., White, J. M., and Muccini, G. A.**, *J. Chromatogr.*, 40, 225, 1969.
30. **Dawson, R. M. C. and Eichberg, J.**, *Biochem. J.*, 96, 634, 1965.
31. **Ando, N., Ando, S., and Yamakawa, T.**, *J. Biochem. (Tokyo)*, 70, 341, 1971.
32. **Helmy, F. M. and Hack, M. H.**, *Lipids*, 1, 279, 1966.
33. **McCarthy, R. D. and Patton, S.**, *J. Lipid Res.*, 3, 117, 1962.
34. **Zinkel, D. F. and Rowe, J. W.**, *Anal. Chem.*, 36, 1160, 1964.
35. **Dawson, R. M. C.**, in *Lipid Chromatographic Analysis*, Vol. 2, 2nd ed., Marinetti, G. V., Ed., Edward Arnold, London, 1976, 663.
36. **Jamieson, G. R. and Reid, E. H.**, *J. Chromatogr.*, 20, 232, 1965.
37. **Thompson, G. A. and Lee, P.**, *Biochim. Biophys. Acta*, 98, 151, 1965.
38. **Wood, R. and Snyder, F.**, *Lipids*, 3, 129, 1968.
39. **Carter, H. E., Rothfus, J. A., and Gigg, R.**, *J. Lipid Res.*, 2, 228, 1961.
40. **Morrison, W. R.**, *Biochim. Biophys. Acta*, 176, 537, 1969.
41. **Morrison, W. R. and Hay, J. D.**, *Biochim. Biophys. Acta*, 202, 460, 1970.
42. **Brockerhoff, H. and Jensen, R. G.**, *Lipolytic Enzymes*, Academic Press, New York, 1974.
43. **Luddy, F. E., Bradford, R. A., Herb, S. F., Magidman, P., and Riemenschneider, F.**, *J. Am. Oil Chem. Soc.*, 41, 603, 1964.
44. **Brockerhoff, H.**, *Arch. Biochem. Biophys.*, 110, 586, 1965.
45. **Barford, R. A., Luddy, F. E., and Magidman, P.**, *Lipids*, 1, 287, 1966.
46. **Slotboom, A. J., De Haas, G. H., Bonson, P. P. M., Burbach-Westerhuis, G. J., and Van Deenen, L. L. M.**, *Chem. Phys. Lipids*, 4, 15, 1970.
47. **Slotboom, A. J., De Haas, G. H., Burbach-Westerhuis, G. J., and Van Deenen, L. L. M.**, *Chem. Phys. Lipids*, 4, 30, 1970.
48. **Noda, M. and Fujiwara, N.**, *Biochim. Biophys. Acta*, 137, 199, 1967.
49. **Arunga, R. O. and Morrison, W. R.**, *Lipids*, 6, 768, 1971.

50. **Nutter, L. J. and Privett, O. S.,** *Lipids,* 1, 258, 1966.
51. **Okuyama, H. and Noyima, S.,** *J. Biochem. (Tokyo),* 57, 529, 1965.
52. **Van Golde, L. M. G. and Van Deenen, L. L. M.,** *Chem. Phys. Lipids,* 1, 157, 1967.
53. **Yabuuchi, H. and O'Brien, J. S.,** *J. Lipid Res.,* 9, 65, 1968.
54. **Robertson, A. F. and Lands, W. E. M.,** *Biochemistry,* 1, 804, 1962.
55. **Christie, W. W.,** in *Topics in Lipid Chemistry,* Vol. 3, Gunstone, F . D., Ed., Paul Elek Books, London, 1972, 171.
56. **Sheppard, A. J. and Iverson, J. L.,** *J. Chromatogr. Sci.,* 13, 448, 1975.
57. **Christopherson, S. W. and Glass, R. L.,** *J. Dairy Sci.,* 52, 1289, 1969.
58. **Iverson, J. L. and Sheppard, A. J.,** *J. Assoc. Off. Anal. Chem.,* 60, 284, 1977.
59. **Vogel, A. I.,** *Practical Organic Chemistry,* 3rd ed., Longmans-Green, London, 1956.
60. Applied Science Laboratories, *Gas-Chrom Newsletter,* 11(4), 1970.
61. **Morrison, W. R. and Smith, L. M.,** *J. Lipid Res.,* 5, 600, 1964.
62. **Lough, A. K.,** *Biochem. J.,* 890, 44, 1964.
63. **Fulk, W. K. and Shorb, M. S.,** *J. Lipid Res.,* 11, 276, 1970.
64. **Shapiro, I. L. and Kritchevsky, D.,** *J. Chromatogr.,* 18, 599, 1965.
65. **Kawamura, N. and Taketomi, T.,** *Jpn. J. Exp. Med.,* 43, 157, 1973.
66. **Kramer, J. K. G. and Hulan, H. W.,** *J. Lipid Res.,* 17, 674, 1976.
67. **Sweeley, C. C. and Moscatelli, E. A.,** *J. Lipid Res.,* 1, 40, 1959.
68. **Schlenk, H. and Gellerman, J. L.,** *Anal. Chem.,* 32, 1412, 1960.
69. **Levitt, M. J.,** *Anal. Chem.,* 45, 618, 1973.
70. **Brown, H. C., Sivasankaran, K., and Brown, C. A.,** *J. Org. Chem.,* 28, 214, 1963.
71. **Miwa, T. K., Kwolek, W. F., and Wolff, I. A.,** *Lipids,* 1, 152, 1966.
72. **Privett, O. S.,** in *Progress in the Chemistry of Fats and Other Lipids,* Vol. 9, Holman, R. T., Ed., Pergamon Press, Oxford, 1966, 91.
73. **Privett, O. S. and Nickell, E. C.,** *Lipids,* 1, 98, 1966.
74. **Dinh-Nguyen, N., Ryhage, R., and Ställberg-Stenhagen, S.,** *Arkiv Kemi,* 15, 433, 1960.
75. **Niehaus, W. G. and Ryhage, R.,** *Anal. Chem.,* 40, 1840, 1968.
76. **McCloskey, J. A. and McClelland, M. J.,** *J. Am. Chem. Soc.,* 87, 5090, 1965.
77. **Argoudelis, C. J. and Perkins, E. G.,** *Lipids,* 3, 379, 1968.
78. **Capella, P. and Zorzut, C. M.,** *Anal. Chem.,* 40, 1458, 1968.
79. **Gunstone, F. D. and Jacobsberg, F. R.,** *Chem. Phys. Lipids,* 9, 26, 1972.
80. **Audier, H., Bory, S., Fetizon, M., Longevialle, P., and Toubiana, R.,** *Bull. Soc. Chim. France,* 1964, 3034.
81. **Kenner, G. W. and Stenhagen, E.,** *Acta Chem. Scand.,* 18, 1551, 1964.
82. **Emken, E. A.,** *Lipids,* 6, 686, 1971.
83. **Emken, E. A.,** *Lipids,* 7, 459, 1972.
84. **von Rudloff, E.,** *Can. J. Chem.,* 34, 1413, 1956.
85. **Schwarz, D. P.,** *Anal. Biochem.,* 74, 320, 1976.
86. **Emken, E. A. and Dutton, H. J.,** *Lipids,* 9, 272, 1974.
87. **Butterfield, R. O., Scholfield, C. R., and Dutton, H. J.,** *J. Am. Oil Chem. Soc.,* 41, 397, 1964.
88. **Gunstone, F. D. and Ismail, I. A.,** *Chem. Phys. Lipids,* 1, 264, 1967.
89. **Moussebois, C. and Dale, J.,** *J. Chem. Soc. (C),* 1966, 260.
90. **Harlow, R. D., Litchfield, C., and Reiser, R.,** *J. Am. Oil Chem. Soc.,* 40, 505, 1963.
91. **Litchfield, C., Lord, J. E., Isbell, A. F., and Reiser, R.,** *J. Am. Oil Chem. Soc.,* 40, 553, 1963.
92. **Litchfield, C., Harlow, R. D., Isbell, A. F., and Reiser, R.,** *J. Am. Oil Chem. Soc.,* 42, 73, 1965.
93. **Mangold, H. K.,** in *Thin Layer Chromatography. A Laboratory Handbook,* Stahl, E., Ed., Academic Press, New York, 1965, 137.
94. **Stearns, E. M., White, H. B., and Quackenbush, F. W.,** *J. Am. Oil Chem. Soc.,* 39, 61, 1962.
95. **White, H. B. and Quackenbush, F. W.,** *J. Am. Oil Chem. Soc.,* 39, 517, 1962.
96. **Minnikin, D. E., Abley, P., McQuillin, F. J., Kusamran, K., Maskens, K., and Polgar, N.,** *Lipids,* 9, 135, 1974.
97. **Minnikin, D. E.,** *Lipids,* 10, 55, 1975.
98. **Bu'Lock, J. D. and Smith, G. N.,** *J. Chem. Soc. (C),* 1967, 322,
99. **Christie, W. W.,** *J. Labelled Compd. Radiopharm.,* 16, 263, 1979.

Chapter 3

FRACTIONATION OF COMPLEX LIPID MIXTURES

H. K. Mangold

I. INTRODUCTION

The treatment of microorganisms and of plant and animal tissues with organic solvents such as chloroform-methanol invariably extracts not only lipids but also various proportions of nonlipid substances. The latter material can be removed, at least partially, by washing the extracts with an aqueous salt solution,[1,2] by dialysis through a semipermeable membrane,[3-5] or by passage through a column of cellulose[6,7] or Sephadex.[8-10] The procedures used most commonly for the extraction and purification of lipids are described in the preceding chapter.

Most natural lipid mixtures consist of a large number of classes of compounds. Each lipid class, — i.e., lipids having the same type and number of functional groups per molecule — is composed of a multiplicity of individual compounds (molecular species) having straight-chain and branched-chain saturated and unsaturated acyl moieties and, at least in animal tissues, alkyl and alk-1-enyl moieties of different chain lengths. The unsaturated chains differ with regard to the number, position and configuration of double bonds; occasionally, they contain triple bonds.

A complete resolution of the molecular species in such extracts cannot be accomplished by any single process. Therefore, complementary principles of separation must be employed consecutively to obtain initially a few well defined fractions and then to resolve each of these fractions by one or several more discriminating techniques.

Methods that are based on solvent fractionation, such as the separation of neutral lipids and phospholipids either by precipitation of the latter with acetone,[11,12] or by partitioning between two immiscible solvents[13-15] as well as procedures based on dialysis,[3-5] can be used on fairly large scales. The fractionations achieved by these methods, however, are rather crude. Chromatographic methods are useful for fractionating only fairly small amounts. Yet, the consecutive application of several chromatographic techniques that are based on complementary principles allows the complete resolution of complex lipid mixtures.

II. SAMPLE PREPARATION

After removal of water and nonlipid organic substances from the extract, the lipids should be dissolved in a fairly apolar solvent and stored under nitrogen or argon in a glass vessel (never a plastic vessel) at a temperature of $-10°C$ or less. Mixtures containing large proportions of triacylglycerols or other nonpolar lipids are best dissolved in hexane; more polar lipids such as phospholipids are dissolved in chloroform or chloroform-methanol. All solvents must be made free of peroxides, then redistilled and flushed with nitrogen before use. To retard the autoxidation of unsaturated lipids, an antioxidant such as butylated hydroxy toluene (BHT) may be added to the lipid solution at a level of 50 to 100 mg/ℓ.

The prolonged storage of solutions containing methanol may lead to the formation of methyl esters by transesterification of lipids, especially in the presence of carbonate or bicarbonate.[16] Acetone should not be used, as it may react with phosphatidylethanolamines[17] and may lead to the dephosphorylation of polyphosphoinosites.[18]

III. CHROMATOGRAPHY

Chromatographic methods may be used to fractionate a complex lipid mixture, to isolate its components, to characterize, and, to a limited extent, classify the original mixture as well as its constituent compounds. For the complete fractionation of such a mixture, various principles of chromatography are employed consecutively in the following sequence.

1. <u>Adsorption chromatography</u> on a column or a thin layer of silica gel usually is employed first to resolve a complex lipid mixture into classes of compounds. Subfractionations within one class of lipids, though often perceptible, do not interfere with class separations. Thus hydrocarbons, alcohols, aldehydes, and fatty acids may be fractionated into classes of compounds.

2. <u>Argentation chromatography</u> on a column or a layer of silica gel impregnated with silver nitrate is then used to fractionate each of the various lipid classes into groups of compounds having the same number and configuration of double bonds. Thus a series of alcohols may be separated into saturated, monounsaturated, diunsaturated, and triunsaturated compounds, *cis-* and *trans*-unsaturated isomers being resolved simultaneously.

3. <u>Reversed-phase partition chromatography</u> on cellulose or paper impregnated with paraffin or silicone can be used to resolve mixtures of compounds differing in chain lengths or in the number of double bonds. Relative to a saturated compound, the effect of *one* double bond is approximately equal to that found for a compound of a chain length shorter by *two* methylene groups (critical pairs). *cis-trans*-Unsaturated isomers are not resolved. Thus a series of saturated alcohols of different chain lengths, or a series of saturated, mono-, di-, and triunsaturated alcohols of the same chain length may be resolved, but not mixtures containing alcohols that differ both in chain length and number of double bonds.

4. <u>Gas-liquid partition chromatography</u> is capable of resolving the components of a class of lipids simultaneously, both according to chain length and number of double bonds per molecule. As a rule, straight-chain and branched-chain compounds are also separated, and even positional and configurational isomers of unsaturated substances may be resolved. Thus straight-chain and branched-chain saturated and mono-, di-, and triunsaturated alcohols including various isomers can be efficiently separated.

Each of these principles of chromatography effects a distinct pattern of fractionation, and the regularities of these separations are known, as indicated above.

The fractionation steps complement each other and follow a sequence that starts with the least discerning and concludes with the most discriminating method. Moreover, this sequence is practical with regard to the sample size needed for each method.

Not all studies require this systematic approach to lipid analysis. The depth of the inquiry will, of course, depend on the specific needs of the investigator. Thus it must be emphasized that a complex lipid mixture can be characterized by its *chromatographic pattern* without isolating and identifying individual molecular species. The pattern of fractionation, as seen on a thin-layer chromatogram, for example, is often sufficient for assessing the source of a fat, for distinguishing lipid extracts of normal and diseased tissues, and for detecting decomposition or adulteration of a preparation.

The following procedure is recommended for the removal of nonlipid contaminants from crude lipid extracts.

Procedure for the Purification of Lipids from Nonlipid Contaminants by Chromatography on Sephadex[10]
The mixture chloroform-methanol-water (8:4:3 v/v) is equilibrated by shaking, and the upper and lower phases formed are collected separately and stored at room temperature. A glass column (1 × 15 cm, equipped with a sintered glass bottom) is packed

Table 1
COLUMN CHROMATOGRAPHY OF NONPOLAR NEUTRAL LIPIDS ON SILICIC ACID

Adsorbent	Eluting solvent	Volume (mℓ)	Lipids eluted
Silicic acid	Hexane	45	Hydrocarbons
	Hexane-diethyl ether (99:1 v/v)	95	Wax esters, steryl esters
	Hexane-diethyl ether (95:5 v/v)	60	Triradylglycerols
	Hexane-diethyl ether (92:2 v/v)	75	Fatty acids
	Hexane-diethyl ether (85:15 v/v)	115	Diradylglycerols, sterols
	Diethyl ether	45	Monoradylglycerols

Adsorbent: "Unisil", 100-200 mesh (Clarkson Chemical Co., Inc., Williamsport, PA 17701).
Column: 5 × 2.5 cm.
Sample: 100 mg in chloroform.
Elution: 3 mℓ/min.

REFERENCE

Rouser, G., Kritchevsky, G., Simon, G., and Nelson, G. J., *Lipids*, 2, 37, 1967.

with a slurry of 25 g of Sephadex G-25, fine, bead form (Pharmacia Fine Chemicals, Uppsala, Sweden) that was soaked overnight in 100 mℓ of upper phase, then rinsed with 4 × 100 mℓ of upper phase. After settling under slight pressure, the column of Sephadex beads is covered with filter paper disks. The column is then rinsed with 10 mℓ of upper phase, followed by 10 mℓ of lower phase. A crude lipid extract that has been obtained by the procedure of Folch et al.[1] is concentrated in a stream of nitrogen and dried under high vacuum. Up to 200 mg of the dry crude lipid mixture is dissolved in 2 to 5 mℓ of lower phase. Precipitated protein is removed by filtration through a sintered glass funnel of medium porosity, and the filtrate is directly applied to the Sephadex column. Purified lipids are then recovered by eluting the column with 25 to 30 mℓ of lower phase at a flow rate of 1 mℓ/min. Gangliosides remain on the column. They are eluted together with nonlipids with at least 50 mℓ of upper phase. The Sephadex column can then be regenerated for further use by washing with 20 mℓ of lower phase. After repeated regeneration, flow rates decrease to a point where repacking of the column becomes necessary.

In routine work, nonpolar neutral lipids and phospholipids in blood are often separated by shaking plasma with chloroform in the presence of a synthetic zeolite,[19] or by chromatographing on a short column of Florisil.[20] Both the zeolite and Florisil retain the phospholipids. A very detailed procedure for the analysis of complex lipids by chromatography on a column of silicic acid is hardly used anymore.[21]

The following three procedures are applicable to the fractionation of neutral nonpolar lipids and their removal from ionic and other polar lipids by adsorption chromatography.

Procedure for the Fractionation of Neutral Lipids on a Column of Silicic Acid[22]
Silicic acid, 12 g, such as Unisil (100 to 200 mesh, Clarkson Chemical Co., Williamsport, Pa., U.S.) or Bio-Sil BH (Bio-Rad Laboratories, Richmond, Calif., U.S.) is heated at 120°C for 2 hr and then slurried with about 30 mℓ of chloroform. This slurry is poured in a column (1.2 × 24 cm) whose lower end is plugged with glass wool, and the adsorbent is allowed to settle. After covering the adsorbent with a filter paper disk, 120 to 180 mg of lipids dissolved in 2 to 3 mℓ of hexane is added. The various classes of nonpolar neutral lipids and fatty acids are eluted with hexane, mixtures of hexane with diethyl ether, and diethyl ether, as specified in Table 1.

Table 2
COLUMN CHROMATOGRAPHY OF NONPOLAR NEUTRAL LIPIDS ON FLORISIL[1,2]

Adsorbent	Eluting solvent	Volume (mℓ)	Lipids eluted
Florisil	Hexane	40	Hydrocarbons
containing	Hexane-diethyl ether (95:5 v/v)	90	Wax esters, steryl esters
7% water	Hexane-diethyl ether (85:15 v/v)	150	Triradylglycerols
	Hexane-diethyl ether (75:25 v/v)	100	Sterols
	Hexane-diethyl ether (50:50 v/v)	100	Diradylglycerols
	Diethyl ether-methanol (98:2 v/v)	135	Monoradylglycerols
	Diethyl ether-acetic acid (96:4 v/v)	50	Fatty acids

Adsorbent: 30 g "Florisil", 60—100 mesh (Floridin Co., 3 Penn Center, Pittsburgh, PA 15235).
Column: 17 × 2 cm.
Sample: 200—300 mg in hexane solution.
Elution: 10 mℓ fractions are collected.

REFERENCES

1. **Carroll, K. K.,** *J. Lipid Res.,* 2, 135, 1961.
2. **Carroll, K. K., Cutts, J. H., and Murray, G. D.,** *Can. J. Biochem.,* 46, 899, 1968.

Alternatively, all nonpolar neutral lipids can be eluted together with 180 mℓ of chloroform, glycolipids containing traces of phospholipids with 700 mℓ of acetone, and phospholipids with 180 mℓ of methanol.

Procedure for the Fractionation of Neutral Lipids on a Column of Hydrated Florisil[23,24]
Florisil (60 to 100 mesh, Fisher Scientific Co., Pittsburgh, Pa., U.S.) and water in a ratio of 100:7 are shaken overnight in a stoppered flask. A column (1,2 × 24 cm) is packed with 12 g of the resulting hydrated adsorbent, which is covered with a filter paper disk. Lipids (120 to 180 mg) are applied to the column and fractionated into classes of compounds using the solvents specified in Table 2. Column chromatography on hydrated Florisil has the advantage of allowing high flow rates and of strongly retaining fatty acids. It should be noted, however, that monoacylglycerols and diacylglycerols are isomerized on this adsorbent unless it is impregnated with 10% of boric acid.

In column chromatography on silicic acid or Florisil, the minor constituents of a lipid mixture are not satisfactorily resolved from the major compounds having similar physical properties. Better separations are achieved by chromatography on layers of the same adsorbents.

The following procedure is particularly useful for the isolation of lipids that occur in minor proportions, such as the alkyldiacylglycerols and alk-1-enyldiacylglycerols in human and animal depot fats. First, the bulk of the major constituents is removed by chromatography on thick layers of silica gel. Thus a concentrate is obtained in which the minor components are present in a ratio more favorable for efficient fractionation. This concentrate is resolved by chromatography on thin layers, and pure lipid classes are isolated.

Procedure for the Fractionation of Neutral Lipids on Layers of Silica Gel[25]
The sample (about 0.5 g) dissolved in 5 mℓ of hexane is applied in an even streak parallel to one edge of a 20 × 20 cm glass plate coated with a 2 mm thick layer of silica gel H (E. Merck, Darmstadt, Germany). Lipid classes less polar than triacylglycerols are fractionated by developing the plate twice or thrice with hexane-diethyl ether (95:5 v/v). The major fractions are, in most cases, visible without the use of an

Table 3
FRACTIONATION OF COMPLEX LIPID MIXTURES BY TWO-DIMENSIONAL THIN-LAYER CHROMATOGRAPHY

Sorbent	Mobile Phases	Ref.
Plant lipids		
Silica Gel H	Chloroform-methanol-7 *N* ammonium hydroxide (65:25:4 v/v) followed by Chloroform-methanol-acetic acid-water (170:25:25:4 v/v)	1
Silica Gel H containing 1% ammonium sulfate	Acetone-benzene-water (91:30:8 v/v) followed by Chloroform-methanol-acetic acid-water (85:10:10:2 v/v)	2
Animal lipids		
Silica Gel HR	Chloroform-methanol-water-28% aq. ammonia (130:70:8:0.5 v/v) followed by Chloroform-acetone-methanol-acetic acid-water (100:40:20:20:10 v/v)	3
Silicic acid containing 10% magnesium silicate	Chloroform-methanol-water (65:25:4 v/v) followed by 1-Butanol-acetic acid-water (60:20:20 v/v)	4

REFERENCES

1. **Radwan, S. S., Spener, F., Mangold, H. K., and Staba, E. J.,** *Chem. Phys. Lipids,* 14, 72, 1975.
2. **Radwan, S. S.,** *J. Chromatog. Sci.,* 16, 538, 1978.
3. **Parson, J. G. and Patton, S.,** *J. Lipid Res.,* 8, 696, 1967.
4. **Rouser, G., Galli, C., Lieber, E., Blank, M. L., and Privett, O. S.,** *J. Am. Oil Chem. Soc.,* 41, 836, 1964.

indicator, whereas the positions of the minor fractions are deduced relative to the major ones. For example, alkyldiacylglycerols and alk-1-enyldiacylglycerols are located in and above the leading edge of the triacylglycerol fraction. These neutral ether lipids are obtained by scraping off the adsorbent layer between the solvent front and a line 2 mm below the leading edge of the triacylglycerol fraction, elution with diethyl ether, and filtration through a jacketed sintered glass funnel at room temperature. After evaporation of the diethyl ether, each of the two classes of neutral ether lipids is isolated from the concentrate by repeated chromatography on layers of silica gel H, 0.5 mm thick, applying 50 to 75 mg of the mixture, dissolved in hexane, to a 20 × 20 cm plate. Chromatography and elution are carried out as described above. Each fraction must not only be pure as a class but must contain all constituents of that class in true proportions. To assure this, all fractions of the same lipid class must be combined.

Column chromatographic procedures that include the fractionation of phospholipids into classes of compounds are described in Chapter 13.

Table 3 lists experimental conditions for the two-dimensional fractionation of complex lipid mixtures on an analytical scale; additional information can be found in the chapter on phospholipids. Two-dimensional thin-layer chromatography of lipids in conjunction with gas chromatography of their constituent fatty acids, using an internal standard, is the method of choice for the quantitative analysis of complex mixtures; the fatty acid patterns of the individual lipid classes are obtained simultaneously.[26]

Some classes of lipids that are not resolved by chromatography on silicic acid can be separated on other adsorbents. For example, mixtures of 1-monoacylglycerols and 2-monoacylglycerols can be separated by thin-layer chromatography on hydroxyapatite,[27] and mixtures of wax esters and cholesteryl esters can be separated according to classes on magnesium oxide.[28] In some cases, changes in the composition of the developing solvent may influence the rate of migration of various lipid classes and facilitate their separation.[29]

The impregnation of silicic acid with complexing agents such as boric acid permits the separation of the *threo* and *erythro* isomers of vicinal dihydroxy compounds.[30] The addition of silver nitrate to silicic acid effects the fractionation of lipid mixtures on the basis of the number, configuration, and position of double bonds in their components.[30,31] Mixtures of compounds differing both with regard to the configuration of vicinal hydroxy groups and the number and configuration of double bonds can be completely resolved by chromatography on silicic acid containing both boric acid and silver nitrate.[30]

Table 4 lists solvent systems that can be used in argentation chromatography and reversed-phase partition chromatography for the resolution of pure lipid classes into simple groups of compounds or even individual molecular species.

In thin-layer chromatography it is possible to apply two principles of chromatography in succession. Thus a complex lipid mixture can be submitted to two-dimensional thin-layer chromatography on a layer of silicic acid, part of which is impregnated with silver nitrate.[32] The consecutive application of adsorption chromatography and argentation chromatography on a single plate offers great advantages as it permits separations of lipids both according to the type and number of functional groups and the type and number of double bonds of their constituent compounds.[32,33] Combinations of adsorption with reversed-phase partition[34] and of argentation chromatography with reversed-phase partition chromatography[35,36] are also possible. A review on the use of thin-layer chromatography in the lipid field has been published recently.[37]

Table 4
SOLVENTS FOR THE FRACTIONATION OF PURE COMPOUND CLASSES USING ADSORPTION, ARGENTATION, AND REVERSED PHASE PARTITION TLC

Compound class	Adsorption TLC[a] hRf	Solvents for argentation TLC (5% AgNO$_3$ in Silica Gel G)		Solvents for partition TLC in reversed phase (paraffin oil on Kieselguhr G)	
Hydrocarbons	98	Hexane-diethyl ether	(95:5 v/v)		
Alcohols	30	Hexane-diethyl ether	(50:50 v/v)	Acetic acid-water	75:25[c]
(as) Acetates	78	Hexane-diethyl ether	(85:15 v/v)	Acetic acid-water	85:15[c]
Aldehydes	73	Hexane-diethyl ether	(80:20 v/v)	Acetic acid-water	85:15[c]
(as) Dimethylacetals	73	Hexane-diethyl ether	(85:15 v/v)	Acetone-water	90:10[c]
Methyl ketones	63	Hexane-diethyl ether	(75:25 v/v)	Acetic acid-water	90:10[c]
Acids	41	Hexane-diethyl ether-acetic acid	(50:50:1 v/v)	Acetic acid-water	80:20[c]
(as) Methyl esters	77	Hexane-diethyl ether	(85:15 v/v)	Acetic acid-water	85:15[c]
Wax esters	88	Hexane-diethyl ether	(90:10 v/v)		
1-Alkylglycerols	3	Diethyl ether-methanol	(90:10 v/v)	Tetrahydrofuran-water	60:40[b]
(as) Dimethylketals	62	Hexane-diethyl ether	(75:25 v/v)		
(as) Acetates	34	Hexane-diethyl ether	(60:40 v/v)	Acetic acid-water	60:40[c]
(as) Alkoxy acetaldehydes	25	Hexane-diethyl ether	(15:85 v/v)	Acetic acid-water	85:15[c]
1,2-Dialkylglycerols	30	Hexane-diethyl ether	(50:50 v/v)	Chloroform-methanol-water	25:75:5[b]
(as) Acetates	63	Hexane-diethyl ether	(75:25 v/v)	Acetic acid-water	90:10[b]
Trialkylglycerols	85	Hexane-diethyl ether	(80:20 v/v)	Acetone-diethyl ether	50:50
Acylglycerols	2	Diethyl ether-methanol	(90:10 v/v)	Tetrahydrofuran-water	50:50[b]
(as) Dimethylketals	60	Hexane-diethyl ether	(70:30 v/v)		
(as) Acetates	30	Hexane-diethyl ether	(60:40 v/v)	Acetic acid-water	60:40[c]
(as) Alkoxy acetaldehydes	23	Hexane-diethyl ether	(15:85 v/v)	Acetic acid-water	85:15[c]
Diacylglycerols	21	Hexane-diethyl ether	(15:85 v/v)	Chloroform-methanol-water	25:75:15[b]
(as) Acetates	45	Hexane-diethyl ether	(70:30 v/v)	Acetic acid-water	90:10[c]
Triacylglycerols	62	Hexane-diethyl ether	(70:30 v/v)	Acetic acid-water	99.5:0.5[c]
				Acetone-acetonitrile	70:30
Alk-1-enyl diacylglycerols ("Neutral plasmalogens")	82	Hexane-diethyl ether	(75:25 v/v)	Acetone-diethyl ether	50:50[b]
Alkyldiacylglycerols	78	Hexane-diethyl ether	(75:25 v/v)	Acetone-diethyl ether	50:50[b]
Dialkylglycerols	88	Hexane-diethyl ether	(80:20 v/v)	Acetone-diethyl ether	50:50[b]

Table 4 (continued)

SOLVENTS FOR THE FRACTIONATION OF PURE COMPOUND CLASSES USING ADSORPTION, ARGENTATION, AND REVERSED PHASE PARTITION TLC

Compound class	Adsorption TLC[a] hRf	Solvents for argentation TLC (5% AgNO$_3$ in Silica Gel G)		Solvents for partition TLC in reversed phase (paraffin oil on Kieselguhr G)	
α-Hydroxyacids	11	Hexane-diethyl ether-acetic acid	(20:80:1 v/v)	Acetic acid-water	75:25
(as) Methyl esters		Hexane-diethyl ether	(40:60 v/v)	Acetic acid-water	80:20[c]
(as) Acetates		Diethyl ether-methanol	(90:10 v/v)	Acetic acid-water	80:20[c]
(as) Acetylated methyl esters		Hexane-diethyl ether	(60:40 v/v)	Acetic acid-water	80:20[c]
Hydroxyethylamides of fatty acids	0			Tetrahydrofuran-water	40:60[b]
(as) Acetates		Diethyl ether-methanol	(90:10 v/v)	Tetrahydrofuran-water	50:50[b]
Cholesterol	19	Hexane-diethyl ether	(10:90 v/v)		
Cholesteryl esters	94	Hexane-diethyl ether	(90:10 v/v)	Butanone-acetonitrile	70:30
Vitamin A		Hexane-diethyl ether	(10:90 v/v)		
Vitamin A esters		Hexane-diethyl ether	(90:10 v/v)		
Amines					
(as) Acetates	0			Tetrahydrofuran-water	60:40[b]
Amides	5	Diethyl ether-methanol	(90:10 v/v)	Acetic acid-water	70:30[c]
Nitriles	62	Hexane-diethyl ether	(70:30 v/v)	Acetic acid-water	75:25[c]

a   Layer: Silica Gel G: solvent: hexane-diethyl ether-acetic acid (80:20:1 v/v).

b   Solvent contains 5% paraffin oil.

c   Solvent saturated with paraffin oil.

REFERENCES

1. **Folch, J., Lees, M., and Sloane-Stanley, G. H.,** *J. Biol. Chem.,* 226, 497, 1957.
2. **Bligh, E. G. and Dyer, W. J.,** *Can. J. Biochem. Physiol.,* 37, 911, 1959.
3. **van Beers, G. J., de Iongh, H., and Boldingh, J.,** in *Essential Fatty Acids,* Sinclair, H. M., Ed., Butterworths, London, 1958, 43.
4. **Olmsted, P. S.,** *Biochim. Biophys. Acta,* 41, 158, 1960.
5. **Eberhagen, D. and Betzing, H.,** *J. Lipid Res.,* 3, 382, 1962.
6. **Lea, C. H. and Rhodes, D. N.,** *Biochem. J.,* 54, 467, 1953.
7. **Garcia, M. D., Lovern, J. A., and Olley, J.,** *Biochem. J.,* 62, 99, 1956.
8. **Wells, M. A. and Dittmer, J. C.,** *Biochemistry,* 2, 1259, 1963.
9. **Siakotos, A. N. and Rouser, G.,** *J. Am. Oil Chem. Soc.,* 42, 913, 1965.
10. **Wuthier, R. E.,** *J. Lipid Res.,* 7, 558, 1966.
11. **Sinclair, R. G. and Dolan, M.,** *J. Biol. Chem.,* 142, 659, 1942.
12. **Grossman, C. M.,** *Biochim. Biophys. Acta,* 36, 541, 1959.
13. **Ahrens, E. H., Jr. and Craig, L. C.,** *J. Biol. Chem.,* 195, 763, 1952.
14. **Blankenhorn, D. H. and Ahrens, E. H., Jr.,** *J. Biol. Chem.,* 212, 69, 1955.
15. **Galanos, D. S. and Kapoulas, V. M.,** *J. Lipid Res.,* 3, 134, 1962.
16. **Lough, A. K., Felinski, L., and Garton, G. A.,** *J. Lipid Res.,* 3, 478, 1962.
17. **Ando, N., Ando, S., and Yamakawa, T.,** *J. Biochem. (Tokyo),* 70, 341, 1971.
18. **Dawson, R. M. C. and Eichberg, J.,** *Biochem. J.,* 96, 634, 1965.
19. **Cheng, A. L. S. and Zilversmit, D. B.,** *J. Lipid Res.,* 1, 190, 1960.
20. **Blankenhorn, D. H., Rouser, G., and Weimer, T. C.,** *J. Lipid Res.,* 2, 281, 1961.
21. **Hirsch, J. and Ahrens, E. H., Jr.,** *J. Biol. Chem.,* 233, 311, 1958.
22. **Rouser, G., Kritchevsky, G., Simon, G., and Nelson, G. J.,** *Lipids,* 2, 37, 1967.
23. **Carroll, K. K.,** *J. Lipid Res.,* 2, 135, 1961.
24. **Carroll, K. K., Cutts, J. H., and Murray, G. D.,** *Can. J. Biochem.,* 46, 899, 1968.
25. **Schmid, H. H. O., Jones, L. L., and Mangold, H. K.,** *J. Lipid Res.,* 8, 692, 1967.
26. **Radwan, S. S.,** *J. Chromatogr. Sci.,* 16, 538, 1978.
27. **Hofmann, A. F.,** *J. Lipid Res.,* 3, 391, 1962.
28. **Kaufmann, H. P., Mangold, H. K., and Mukherjee, K. D.,** *J. Lipid Res.,* 12, 506, 1971.
29. **Schmid, H. H. O. and Mangold, H. K.,** *Biochim. Biophys. Acta,* 125, 182, 1966.
30. **Morris, L. J.,** *J. Chromatogr.,* 12, 321, 1963.
31. **Morris, L. J.,** *J. Lipid Res.,* 7, 717, 1966.
32. **Cubero, J. M. and Mangold, H. K.,** *Microchem. J.,* 9, 227, 1965.
33. **Schmid, H. H. O., Baumann, W. J., Cubero, J. M., and Mangold, H. K.,** *Biochim. Biophys. Acta,* 125, 189, 1966.
34. **Kaufmann, H. P. and Makus, Z.,** *Fette Seifen Anstrichm.,* 62, 1014, 1960.
35. **Bergelson, L. D., Dyatlovitskaya, E. V., and Voronkova, V. V.,** *J. Chromatogr.,* 15, 191, 1964.
36. **Totani, N. and Mangold, H. K.,** *J. Am. Oil Chem. Soc.,* in press.
37. **Mangold, H. K.,** in *Dietary Fats and Health,* Perkins, E. G. and Visek, W. J., Eds., American Oil Chemists' Society, Champaign, Ill., 1983, 85.

Chapter 4

HYDROCARBONS

M. Streibl and J. Janák

I. INTRODUCTION

This chapter pertains to hydrocarbons that are components of complex lipid mixtures in plants and animals. Crude oil hydrocarbons, including ozocerites, and mixtures of synthetic hydrocarbons are not dealt with, since they are not found together with other lipids in natural materials and their composition is much more complex[1,2] than that of the lipid hydrocarbons. (Chromatography of hydrocarbons is also covered in a separate volume of this series.)

The hydrocarbons occurring in natural lipid mixtures have simple chemical structures. Their carbon skeleton is usually straight; one or more side chains, mostly the methyl group only, occur infrequently. They may be either saturated or unsaturated. Characteristic features of the aliphatic lipid hydrocarbons are the typical length of their carbon chains and their occurrence in a homologous series,[3] ranging from approximately C_{10} to C_{35}.

Hydrocarbons can be found primarily as a part of extracellular epicuticular lipids (waxes, both recent and fossil) with higher plants,[4,5] insects,[6] and other lower[7] and higher animals.[8] In smaller amounts they are also a part of the intracellular lipids in bacteria,[9] fungi,[10] algae,[10] and also higher organisms[11] inclusive of the mammals.[12] Hydrocarbons are also components of soil lipids,[13] and are included in fats and oils.[14,15]

Figure 1 shows an outline of hydrocarbon types found in other lipids. As can be seen, there are also isoprenoid hydrocarbons, such as squalene, phytane, farnesene, and cholestene that are composed of subunits containing five carbon atoms arranged according to the "isoprene rule".[16] They are distinguished[17,18] from the nonisoprenoid lipid hydrocarbons by a unique chemical constitution and by the fact that they are not found in homologous series in lipids since they arise by means of another biosynthetic path.[19]

The simple chemical structure and a relatively low chemical activity of the majority of hydrocarbons enable us to carry out group separation and identification without particular difficulties.

II. SAMPLE PREPARATION

A. Extraction

The hydrocarbons are components of the total lipid extract, especially as far as the intracellular lipids from separate tissues are concerned. They are separated from the extracts either directly or after alkaline hydrolysis.[20] With the epicuticular surface waxes, particularly those of plants, the *n*-alkanes and the accompanying branched and unsaturated hydrocarbons composition is of interest for chemotaxonomical purposes.[21-23] In such cases it is best to extract the least polar lipid fractions by means of petroleum hydrocarbon, hexane, chlorinated hydrocarbons, or their mixtures. The extraction may be shorter or longer timewise, at ambient or elevated temperatures, according to the nature of the material. Mere dipping in a solvent,[24] or washing and spraying,[25,26] is used with the surface lipids. If a fresh, nondried material containing water, such as plants or insects, is extracted, water prevents the lipophilic solvent from penetrating, and as a result, an extract mainly from the surface of the material is obtained.[27]

On the other hand, if one wishes to get lipids from the interior of organs without admixture of surface lipids, it is necessary to extract the surface substances first, and only thereafter extract the intracellular lipids.[28,29] Extraction by means of nonpolar solvents dissolves the

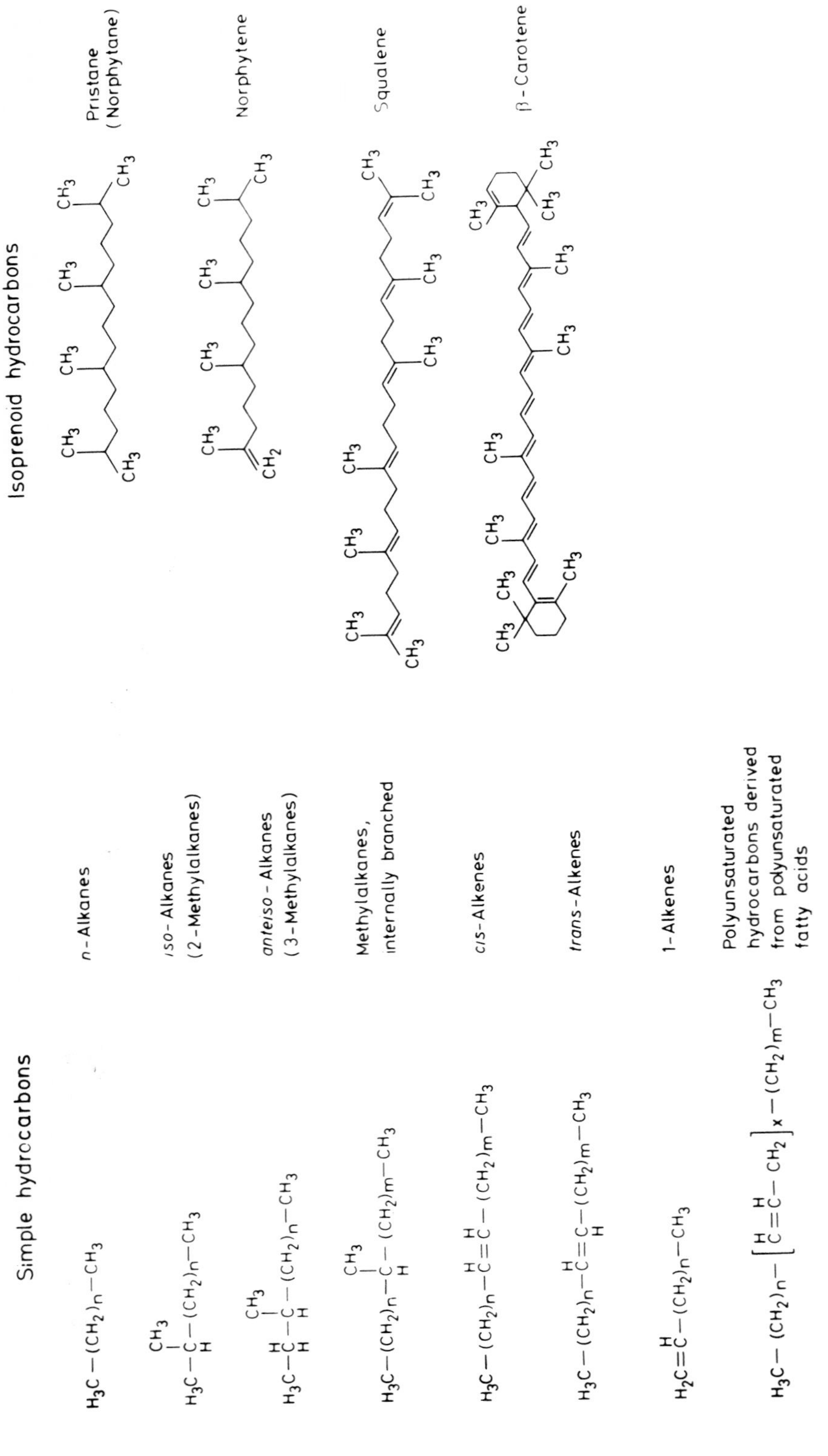

Simple hydrocarbons

Isoprenoid hydrocarbons

$H_3C-(CH_2)_n-CH_3$ *n*-Alkanes

$H_3C-\overset{CH_3}{\underset{H}{C}}-(CH_2)_n-CH_3$ *iso*-Alkanes (2-Methylalkanes)

$H_3C-\overset{CH_3}{\underset{H}{C}}-\overset{}{\underset{H}{C}}-(CH_2)_n-CH_3$ *anteiso*-Alkanes (3-Methylalkanes)

$H_3C-(CH_2)_n-\overset{CH_3}{\underset{H}{C}}-(CH_2)_m-CH_3$ Methylalkanes, internally branched

$H_3C-(CH_2)_n-\overset{H}{C}=\overset{H}{C}-(CH_2)_m-CH_3$ *cis*-Alkenes

$H_3C-(CH_2)_n-\overset{H}{\underset{}{C}}=\overset{}{\underset{H}{C}}-(CH_2)_m-CH_3$ *trans*-Alkenes

$H_2C=\overset{H}{C}-(CH_2)_n-CH_3$ 1-Alkenes

$H_3C-(CH_2)_n-\left[\overset{H}{C}=\overset{H}{C}-CH_2\right]_x-(CH_2)_m-CH_3$ Polyunsaturated hydrocarbons derived from polyunsaturated fatty acids

Pristane (Norphytane)

Norphytene

Squalene

β-Carotene

FIGURE 1.   An outline of hydrocarbon types in other lipids.

hydrocarbons without dissolving very much of the polar lipids. It should be pointed out that when extracting small quantities of hydrocarbons from a large amount of material, it is essential to avoid contamination from solvent impurities, grease, plasticizers, fingerprints, etc.[30] Recently, the omnipresence of hydrocarbons from air pollution[31-33] has been recognized. Also, losses of lower homologs, from C_{16} down, have been shown to occur during long storage, and this, as well as evaporation under reduced pressure,[34] can sometimes lead to wrong conclusions.

B. Derivatives

Linear, as well as branched alkanes are further analyzed in an unchanged form. Additional derivatization may be used for the separation,[35] partial identification,[36] and localization of double bonds[37] of the unsaturated hydrocarbons. The mutual separation of different types of hydrocarbons will be dealt with later.

III. CHROMATOGRAPHY

There are two main kinds of chromatographic methods used for the analyses of lipid hydrocarbons, i.e., liquid-solid adsorption, either in columns or on thin layers, or gas-liquid chromatography, sometimes in combination with mass spectrometry. Liquid-liquid and gel permeation chromatography are used less frequently.

A. Liquid-Solid Chromatography

Either column or thin-layer techniques are used for separation of hydrocarbons from the other lipid constituents. When separating larger amounts of sample and evaluating quantitatively, column chromatography is used;[38] for smaller amounts and an orientative separation, the thin layer is preferable. Because of their weak interaction with the sorbent, hydrocarbons are the most easily eluted of all lipids. They can be separated by means of petroleum hydrocarbon, hexane or heptane, from all more polar lipids, including simple wax esters,[39] which are eluted only after the hydrocarbons. When using a 50- to 100-fold amount of sorbent compared with the amount of the sample, it is possible, by column chromatography to separate some types of hydrocarbons from each other, e.g., *n*-alkanes from the unsaturated terpenoid hydrocarbons farnesene[40] and squalene.[41]

Suitable sorbents and mobile phases are selected according to the nature of the sample. As a rule, if a total lipid extract is analyzed, separation conditions are designed that are most suited for the separation of the various classes of ester lipids, and the separation of hydrocarbons is not optimal.[42] If hydrocarbons are the mainly desired fraction of the extracts, silica gel is used,[43] sometimes partly deactivated by water. Also alumina[44] and magnesium silicate (Florisil)[43] are often used as column sorbents. Alumina is suitable for the chromatography of stable substances, by stepwise change of single solvents or by means of gradually changing gradients, often automatically controlled.[45] We refer to typical examples of the separation of hydrocarbons and other substances in neutral lipids on silica gel[46] and Florisil.[47]

A sample dissolved in an appropriate solvent is introduced in the column of the sorbent that has been first washed with the same solvent. If the sample is not sufficiently soluble in a nonpolar solvent, it is adsorbed first from a more polar solvent, such as chloroform, on a deactivated silica gel and, after removal of the solvent, the gel with the sample is introduced on the top of the column.[48] In order to increase the solubility of the sample, chromatography is sometimes carried out at a slightly elevated temperature (up to 40°C) in columns with a thermostatic water jacket.

Retention of olefins on silica gel impregnated with silver nitrate[49-51] is used for separating saturated from unsaturated hydrocarbons. With suitable processing and using sorbents with sufficient capacity, it is also possible to separate different types of unsaturated hydrocarbons[52,53]

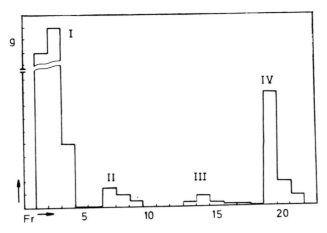

FIGURE 2.   Column chromatography of the hydrocarbons from sugar cane wax. Sorbent: silicic acid containing 15% silver nitrate. Solvents: hexane (1—10), hexane-chloroform (5:1 v/v) (11—18), hexane-chloroform-ethanol (10:2:1 v/v) (18—31). (I) Alkanes. (II) *cis*-Alkenes. (III) Alkenes with a trisubstituted double bond. (IV) 1-Alkenes. (From Streibl, M. and Stránský, K., unpublished.)

from each other. Thus hydrocarbons with different numbers of double bonds (Figure 2) can be resolved on the basis of varying affinities of π-complexes these double bonds form with silver ions.[54] If the level of impregnation of silica gel with silver nitrate is higher (more than 20 wt %), the impregnation agent[55] is sometimes washed out during the percolation of more polar solvents, such as CH_2Cl_2 — diethyl ether. Separation of the various hydrocarbon classes is not influenced, but it is necessary to remove silver nitrate from the fraction obtained by means of rechromatographing or water extraction.

Selective separations of hydrocarbons in chromatographic columns have been described; thus, in a celite column with urea, separation can be effected due to the formation of inclusion compounds.[56] In this system, straight-chain hydrocarbons are retained as inclusion compounds, whereas branched-chain and cyclic hydrocarbons[57] are eluted. As a rule, this separation is carried out not in a chromatographic column, but as a batch process.

The direct detection of hydrocarbons in the column effluent has not yet been solved satisfactorily, whereas lipids containing functional groups can well be monitored,[58] either continuously or in intervals. Only some of the methods that measure physical properties or pyrolysis products[59,60] continuously can be used for the detection of hydrocarbons.[43] Monitoring the eluted fractions with the help of thin-layer chromatography (TLC) or gas chromatography (GC) is still used as an easy method of detection; this process may be automated.[61] It is also possible to measure radioactivity[62,63] of eluted hydrocarbons if the substances are labeled with ^{14}C or ^{3}H. Currently used techniques are described elsewhere.[64]

The equipment and procedures employed in the column chromatography of hydrocarbons do not differ from the commonly used techniques.[65] Reference is made to an article in which factors responsible for high efficiency in column chromatography have been described.[66]

B. Thin-Layer Chromatography (TLC)

The other principal technique used currently for the separation of hydrocarbons is thin-layer chromatography, both in analytical and preparative versions.[67,68] Because it is facile and does not require expensive equipment, it can be used for rapidly finding suitable elution systems and optimal conditions for separation using column chromatography with the same sorbents. With TLC, as with column chromatography, knowledge of the approximate composition of the sample to be separated and the aim of the work are important for selecting the most suitable method of analysis.

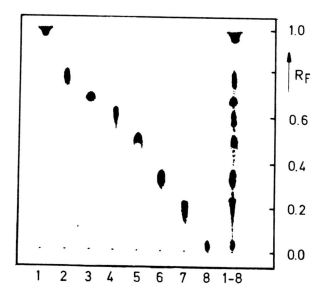

FIGURE 3. Thin-layer chromatogram of hydrocarbons from Czech montan wax. Sorbent: silica gel G containing 20% silver nitrate. Solvent: hexane, developed twice. (1) Alkanes. (2) *trans*-Alkenes. (3) *cis*-Alkenes. (4) Alkenes with a trisubstituted double bond. (5) 1-Alkenes. (6) 1-Octadecine. (7) Alkadienes. (8) Terpenoid polyenes. (From Streibl, M. and Stránský, K., unpublished.)

Separation of the hydrocarbon fraction from oxygen containing lipids takes place most satisfactorily with judicious selection of the nonpolar solvents.[69] Mainly silica gel and, to a lesser extent, Florisil,[70] both fixed by means of calcium sulfate, are used as sorbents for hydrocarbon separation. In the presence of polar lipids, calcium ions interfere with the separation, but it is possible to work without this binding agent. Detailed systems for the separation of lipid classes, including hydrocarbons, have been worked out.[71-76] Multiple development, either in one[77] or in two dimensions, is always of advantage.

A disadvantage of TLC compared to column chromatography is the limited possibility of quantitative analysis; exhaustive recovery of the separated spots is not sufficiently precise. Nevertheless, TLC has been used also for the quantitative determination of hydrocarbons.[78,79] *In situ* densitometry is also becoming widely used in TLC for quantitations. An advantage of separation by TLC compared to column chromatography is the possibility of utilizing various derivatization reagents[72] that can, after spotting on the start line and before development, change the chemical nature of certain sample components; the difference found after developing two identical samples, one of which had been derivatized, for example by acetyl chloride or diazomethane, can facilitate the determination of the structure of the substances separated. Another advantage of TLC is the possibility of treatment of the layer by various reagents to achieve specific separations. It is possible to impregnate a part of the layer with an ethanol solution of $AgNO_3$ and subsequently, by employing another principle of chromatography, to separate the same hydrocarbon mixture in two directions.[77] Argentation chromatography of unsaturated hydrocarbons[49,54] is presently used as a very effective method of separation; not only are saturated and unsaturated hydrocarbons separated, but also all types of mono- and polyolefins,[52,74] as shown in Figure 3.

C. Gas-Liquid Chromatography (GLC)

This is certainly the most efficient method of lipid analysis. While the liquid-solid chromatographic methods are suitable for the separation of individual lipid classes, including hydrocarbons, GC represents the main chromatographic method enabling separation of the individual members of homologous series; further, it provides very precise quantitative data on the amounts of the compounds separated. With the help of standards and knowledge of the dependencies of the retention data[80,81] on the structure of the separated substances,[82,83] it is sometimes possible to identify unknown peaks.

All variants of GC are applied for the analysis of the hydrocarbons, including analytical and preparative separation in packed columns and in capillaries. Separations may be performed isothermally or in gradient temperature systems, on polar and nonpolar stationary phases, and with multiple systems of columns and detectors. Practically all hydrocarbons occurring in lipids are sufficiently volatile and stable to permit analysis by GC.

The restricted temperature stability of most currently employed phases has been a problem. Recently, columns packed with nonpolar phases such as silicone greases, as well as polar phases such as carborane siloxane that are stable up to 350 to 400°C, have been employed. These permit not only the analysis of triterpene hydrocarbons with 30 C-atoms, but even allow chromatographing of hydrocarbons up to $C_{7.0}$.[83,84] In selecting suitable stationary phases for special purposes, Rohrschneider's criteria[85] can be used. The newly developed thermally stable nonchiral phase can also be utilized for the separation of hydrocarbons. The most frequently used nonpolar phases[86] are silicone polymers (SE-30, SE-52, OV-1, OV-17) and the hydrocarbon phase APIEZON-L; polar phases include borosilicone DEXSIL 300, fluorosilicone QF-1, and the polyesters PEGA and DEGS. If the column is not overloaded there is no tailing of hydrocarbons, and the peaks are symmetrical. Some decomposition of hydrocarbons in the injection port or in the column is usually observed.

GC of aliphatic hydrocarbons is also utilized for determination of the carbon skeleton of lipid substances of the other lipid classes.[52,84,87] Such lipids are converted by means of chemical reactions into structurally equivalent hydrocarbons whose structures have been established on the basis of relationships between structures and retention data of hydrocarbons; these are used to provide tentative identification of unknown peaks.[79,88,89] Comparison of retention times on GC columns of different polarities is useful in identification.[89]

GC offers the possibility of carrying out various chemical reactions in the gas phase, usually at the column inlet or outlet. These reactions techniques[90] include hydrogenation, dehydration, and hydrogenolysis. Hydrogen as the carrier gas saturates double bonds on the catalyst and removes many hetero atoms (S, O, N, etc.) of the molecules of various substances, thus producing saturated hydrocarbons. Analogously, molecular sieves for separating straight-chain alkanes,[87,91,92] and sulfuric acid for reaction with olefins are sometimes used.

Direct gas chromatography without preceding class separation has been used for characterization of some commercially produced natural waxes[93-95] in order to develop criteria for their rapid evaluation.

Tandem coupling of GC and TLC makes it possible to utilize, to a maximum, differences in volatility and chemical functionality of molecules.[96] This is done as follows: a stream of the carrier gas leaving the chromatographic column is directed to the start line of the chromatoplate, where vapor of high-boiling components in the carrier gas are adsorbed by the layer. When moving the chromatoplate stepwise or continuously, either linearly or according to a program, e.g., logarithmically,[61] the substances from the gas chromatograph can be fixed on the origin of the thin layer and thus the first dimension of the chromatographic development is determined by volatility. This, in a suitable arrangement, reflects the factor of the molecular size, and a thin-layer chromatogram may then be developed by means of a suitable solvent system. Separation in the second direction of the chromatogram is then representative of the chemical functionality of the compounds resolved. A logarithmic movement of the layer under the gas chromatographic column outlet permits easy identification of zones on the layer because the individual members of homologous series appear in equidistant intervals. This method has been successfully applied in lipid chemistry.[97]

D. Gas Chromatography — Mass Spectrometry

A very powerful tool in the analysis of lipid hydrocarbons is the tandem coupling of gas chromatography and mass spectrometry (GC/MS). Special advantages of this tandem for hydrocarbon analysis are that practically all hydrocarbons can be separated by gas chromatography; that it is possible by means of auxiliary methods to separate in advance the complex mixtures to obtain well resolved peaks; and finally, that mass spectra of hydrocarbons are easy to distinguish from those of the other lipids and are thus easier to interpret.[40,98] Combining highly efficient capillary GC with MS is now an outstanding tool for solving analytical problems in the field.[99-102]

Fragmentation patterns of the individual hydrocarbon types have been studied in detail, and thus the determination of the structure of an unknown hydrocarbon is largely a matter of routine and experience. *n*-Alkanes, for example, have a fragmentation characterized by a lack of M-5 ion,[103] and the parent molecular ion, which is useful in establishing the molecular weight, is usually small but readily observable. The most abundant peaks are the ions C_nH_{2n+1} and C_nH_{2n-1}. Mass spectra of 2-methylalkanes are characterized by M-43 and M-15 ions due to loss of isopropyl and methyl groups; *anteiso*-alkanes loose ethyl (M-29), and secondary butyl groups (M-57). Internally branched monomethyl- and polymethylalkanes have an abundancy of ions that arise in the cleavage of all C–C bonds at the point of branching.[43,51,80,101,104,105] In alkenes it was observed that the parent molecular ion is abundant, accompanying the fragments C_nH_{2n+1}. A precise MS determination of the double bond position, however, is not possible due to an extensive migration of the double bond along the chain.[43]

If a GC/MS interface is not available, we may use each method separately. In such cases, it is necessary to isolate separately[105-107] each GC fraction from the effluent, and to transfer the sample manually to the MS inlet; that, of course, increases the amount of sample needed. It is obvious that the mass spectrum of hydrocarbons contains typical ions for the given series, which occur in the spectra of all members of a homologous series as well. It is thus possible to determine the hydrocarbon type from such spectra, as well as the original number of the homologs, such as the chain-length range of the series and to a certain degree also their quantitative ratios in accordance with the intensities of the major peaks.

E. Other Chromatographic Methods

Because TLC and GC are very efficient, other chromatographic methods are not applied very frequently in hydrocarbon analysis. A basic study was made on the neutral and polar lipids[108] by gel permeation chromatography. Sephadex LH 20 was used in separating cyclic and acyclic hydrocarbons of similar molecular weights;[109] further applications of gel permeation chromatography for the analysis of hydrocarbons[110-112] have been published. In ion-exchange chromatography of amphiphilic lipids on modified celluloses, the group of hydrocarbons, if present, may be eluted within the portion of neutral lipids.

F. Separation of Hydrocarbon Types

Several auxiliary methods are used to complement the separation of hydrocarbons from complex mixtures. With their help certain types of hydrocarbons are separated by adduct formation, molecular sieving, thermal diffusion, etc. Infrared (IR), ultraviolet (UV), and nuclear magnetic resonance (NMR) spectroscopy can aid in identifying fractions. The chemical reactivity of the double bond is utilized to modify the hydrocarbon molecule.

For separation of straight-chain hydrocarbons the technique of urea inclusion is quite useful.[114,115] The rate of adduct formation and the stability of the adducts depend on the nature of the included compound. Urea will adduct normal, *iso*- and *anteiso*-alkanes and other compounds possessing an unbranched carbon chain of more than six carbon atoms at one end of the molecule. This method is also used for separating mixtures of hydrocarbons into fractions of aliphatic and cyclic plus isoprenoid compounds.[53,107] Similarly, thiourea accepts isoprenoids, both aliphatic and cyclic.[114] Selective segregation of *n*-alkanes is effected by inorganic "sieves"[115] with effective pore diameter of 5 Å usually in 2,2,4-trimethyl-pentane or benzene as solvent[39,99,101,107] (Figure 4). Thermal diffusion[116] was explored for separating hydrocarbons from human hair lipids, but the method seems to be time consuming, although many isomers may be separated according to their molecular shapes.

Spectroscopic methods give more or less information on the functional groups in the molecule or on certain groupings being present. IR spectroscopy is a powerful tool in discovering the nature of the double bond,[52,74,117] and it helps in the determination of the carbon-chain skeleton.[27,98] It is also a simple and suitable method for proving the identity of an isolated hydrocarbon with a standard. UV spectroscopy[118] reveals the presence of chromophores, notably in the case of hydrocarbons that contain systems of conjugated double bonds.[29] ^{1}H-Nuclear magnetic resonance spectrometry is suitable for determining structural environments of all protons in the hydrocarbon molecule, and thus specifies the patterns of unsaturation and branching; it may serve also for comparison of the isolated substances with authentic specimens.[119]

While the determination of the number of double bonds in the molecules of olefins by hydrogenation, bromine addition, or peracid titration represents no difficulties, the localization of double bonds in the chain requires rather sophisticated methods. Oxidation of alkenes to carboxylic acids followed by methylation and GC of esters yields adequate results.[29,51,52,117,120] By means of OsO_4, olefins can be converted to diols. The trimethylsilyl derivative of these diols can be well analyzed by means of MS since the specific cleavage takes place between the carbons of the original double bond.[39,120,121] Another fast and precise method for the localization of double bonds is based on the formation of acetoxymercuri-methoxy adducts, reduction of the adduct, and an analysis of the resulting product by means of MS.[122] Detailed descriptions of these methods for the structural analysis of fatty acids have been published.[37,125] Hydroboration along with oxidation provides the possibility of turning alkenes into alcohols that can be separated from the saturated hydrocarbons and analyzed by means of GC.[124]

Mixtures of hydrocarbons isolated from lipid extracts are often very complex and the choice of suitable methods of separation and the use of adequate methods of identification require much experience. Typical applications can be found in the literature.[29,50,63,74,80,107,117] Isoprenoid hydrocarbons require a special approach to isolation, since they do not occur in homologous series and frequently have a relatively high molecular weight. They are isolated by argentation TLC or preparative GC. The GC/MS technique is very suitable for their identification, and all other spectral methods are used in an auxiliary way. For illustration, we refer to some methods that have been published in the literature.[29,50,74,99,102,119]

It should be mentioned that fossil organic materials (geolipids) from sedimentary deposits frequently contain hydrocarbons as the dominant fraction. Most interesting from the paleo-biological viewpoint are the findings of isoprenoid hydrocarbons, such as terpanes and steranes, that are important as biological markers because, with regard to high structural specificity, they indicate the occurrence of ancient biological processes.[125] Methods of isolation and identification of geolipid hydrocarbons are not different from the procedures used for lipids of recent origin.

IV. FUTURE DEVELOPMENT

The analysis of lipid hydrocarbons, including isoprenoids, has no basic methodological or instrumental problems at present. In the authors' opinion, there are no revolutionary changes in methodology to be expected in the near future. Certainly, this field will continue to be influenced by progress in methodology of analyzing the petroleum hydrocarbons, as

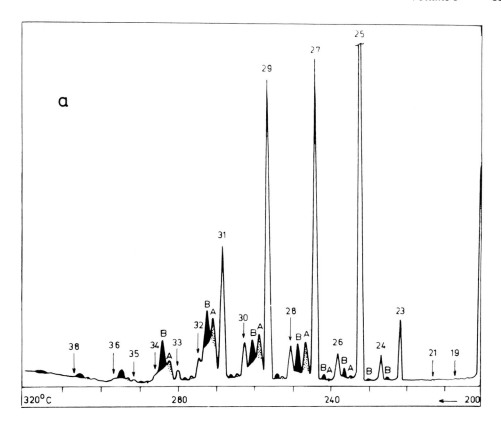

FIGURE 4. Gas chromatogram of the saturated hydrocarbons from blossoms of *Matricaria chamomilla* L. Instrument: Packard 427 gas chromatograph. Column: glass tubing, 185 cm × 0.24 cm I.D., packed with 3% SE–30 on Gas Chrom Q, 80 to 100 mesh. Temperature: programmed from 200 to 310°C, 3°C/min. Carrier gas: nitrogen, 50 mℓ/min. (a) Total hydrocarbon fraction, C_{19} through C_{38}. (b) Hydrocarbons obtained after sieving with Linde 5 Å sieve. A, B, and C are members of homologous series of methylalkanes and dimethylalkanes. (From Streibl, M. and Stránský, K., unpublished.)

it always had been; some methods, such as thermal diffusion,[116] can surely find more application in the analysis of hydrocarbons. Thermal field-flow fractionation,[126] especially of high molecular hydrocarbons, may become useful. Further development of GC, mainly of high efficiency columns and improvement of their operation,[127] may be anticipated. Also, high performance liquid chromatography, especially variants involving reversed phases, will broaden both analytical and preparative applications.[65]

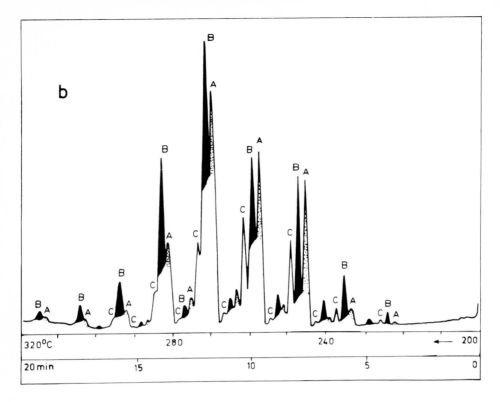

FIGURE 4b.

Table 1
HYDROCARBONS (POLAR PACKED COLUMNS, REPORTED AS RETENTION TIMES RELATIVE TO OCTADECANE) (GC)

Column packing	P1	P2	P3	P4	P5
Temperature (°C)	125	125	135	130	130
Gas:	Ar	Ar	Ar	N_2,	N_2,
Flow rate (mℓ/min)	na	na	na	30	30
Column					
Length (cm)	122	122	122	183	183
Diameter					
(cm I.D.)	na	na	na	na	na
(cm O.D.)	0.64	0.64	0.64	0.318	0.318
Form	Tube	Tube	Tube	Helix	Helix
Material	G	G	G	SS	SS
Detector	Argon	Argon	Argon	FID	FID
Reference	1	1	1	1	1

Compound	$r_{Octadecane}$				
Tetradecane	0.152	0.152	0.162	—	0.186
Tetradecene	0.205	0.205	0.209	—	0.239
Pentadecane	0.244	0.248	0.254	—	0.275
Pentadecene	0.312	0.311	0.334	—	0.347
Hexadecane	0.406	0.393	0.399	—	0.420
cis-9-Hexadecene	0.480	0.449	0.468	—	0.477
Heptadecane	0.628	0.632	0.640	—	0.647
Heptadecene	0.760	0.714	0.718	—	0.727
Octadecane	1.00	1.00	1.00	1.00	1.000
cis-9-Octadecene	1.18	1.13	1.12	1.12	1.13
cis,cis-9,12-Octadecadiene	1.56	1.41	1.42	1.32	1.37
cis,cis,cis-6,9,12-Octadecatriene	1.89	1.67	1.75	1.53	1.51
cis,cis,cis-9,12,15-Octadecatriene	2.28	1.89	1.92	1.70	1.82
cis,cis,cis,cis-6,9,12,15-Octadecatetraene	—	—	—	1.98	—
cis-11-Icosene	2.88	2.77	2.69	2.43	2.55
cis,cis-11,14-Icosadiene	—	—	—	2.86	—
cis,cis,cis-8,11,14-Icosatriene	—	—	—	3.35	—
cis,cis,cis,cis-5,8,11,14-Icosatetraene	—	—	—	3.84	—
cis,cis,cis-8,11,14,17-Icosatetraene	—	—	—	4.28	—
cis,cis,cis,cis-5,8,11,14,17-Icosapentaene	—	—	—	5.05	—
cis-13-Docosene	—	—	—	5.33	—
cis,cis-13,16-Docosadiene	—	—	—	6.32	—
cis,cis,cis-10,13,16-Docosatriene	—	—	—	7.51	—
cis,cis,cis,cis-7,10,13,16-Docosatetraene	—	—	—	8.42	—
cis,cis,cis,cis,cis-4,7,10,13,16-Docosapentaene	—	—	—	9.60	—
cis,cis,cis,cis,cis-7,10,13,16,19-Docosapentaene	—	—	—	10.8	—
cis,cis,cis,cis,cis,cis-4,7,10,13,16,19-Docosahexaene	—	—	—	13.7	—
cis-15-Tetracosene	—	—	—	11.6	—

Column packing:
P1 = 17% EGSS-X (ethylene glycol succinate polyester combined with methyl silicone) on Gas-Chrom P
P2 = 15.5% EGSS-Y (ethylene glycol succinate polyester combined with methyl silicone) on Gas-Chrom P
P3 = 10% Poly (ethylene glycol adipate) on Celite 545, AW (100-120 mesh)
P4 = 8% Poly (1,4-butanediol succinate) on Chromosorb W, HMDS (80-100 mesh)
P5 = 8% Poly (1,4-butanediol succinate) on Chromosorb W, HMDS (80-100 mesh)

REFERENCE

1. **Jamieson, G. R. and Reid, E. H.,** *J. Chromatogr.*,26, 8, 1967.

Table 2
HYDROCARBONS (POLAR PACKED COLUMNS, REPORTED AS EQUIVALENT CHAIN LENGTHS BASED ON SATURATED STRAIGHT-CHAIN HYDROCARBONS

	P1	P2	P3	P4
Column packing				
Temperature (°C)	125	125	135	130
Gas:	Ar	Ar	Ar	N_2
Flow rate (mℓ/min)	na	na	na	30
Column				
Length (cm)	122	122	122	183
Diameter	na	na	na	na
(cm I.D.)				
(cm O.D.)	0.64	0.64	0.64	0.318
Form	Tube	Tube	Tube	Helix
Material	G	G	G	SS
Detector	Argon	Argon	Argon	FID
Reference	1	1	1	1

Compound	ECL (based on *n*-alkanes)			
Tetradecane	14.00	14.00	14.00	14.00
Tetradecene	14.60	14.60	14.53	14.62
Pentadecane	15.00	15.00	15.00	15.00
Pentadecene	15.50	15.48	15.60	15.51
Hexadecane	16.00	16.00	16.00	16.00
cis-9-Hexadecene	16.42	16.29	16.34	16.27
Heptadecane	17.00	17.00	17.00	17.00
Heptadecene	17.40	17.25	17.26	17.27
Octadecane	18.00	18.00	18.00	18.00
cis-9-Octadecene	18.37	18.30	18.27	18.27
cis, cis-9, 12-Octadecadiene	18.96	18.75	18.77	18.75
cis, cis, cis-6, 9, 12-Octadecatriene	19.40	19.12	19.21	19.00
cis, cis, cis-9, 12, 15-Octadecatiene	19.78	19.36	19.43	19.45
Icosane	20.00	20.00	20.00	20.00
cis-11-Icosane	20.30	20.22	20.15	20.23

Column packing:
- P1 = 17% EGSS-X (ethylene glycol succinate polyester combined with methyl silicone) on Gas-Chrom P
- P2 = 15.5% EGSS-Y (ethylene glycol succinate polyester combined with methyl silicone) on Gas-Chrom P
- P3 = 10% Poly (ethylene glycol adipate) on Celite 545, AW, (100-120 mesh)
- P4 = 8% Poly (1,4-butanediol succinate) on Chromosorb W, HMDS, (80-100 mesh)

REFERENCE

1. **Jamieson, G. R. and Reid, E. H.**, *J. Chromatogr.*, 26, 8, 1967.

REFERENCES

1. **Mair, B. J.**, *Oil Gas J.*, 62, 130, 1964.
2. **Rossini, F. D., Mair, B. J., and Streiff, A. J.**, *J. Chem. Educ.*, 37, 554, 1960.
3. **Kolattukudy, P. E., Ed.**, *Chemistry and Biochemistry of Natural Waxes*, Elsevier, Amsterdam, 1976, 7.
4. **Tulloch, A. P.**, in *Chemistry and Biochemistry of Natural Waxes*, Kolattukudy, P. E., Ed., Elsevier, Amsterdam, 1976, 241.
5. **Hamilton, S. and Hamilton, R. J.**, *Top. Lipid Chem.*, 3, 226, 1972.

6. **Jackson, L. L. and Blomquist, G. J.**, in *Chemistry and Biochemistry of Natural Waxes*, Kolattukudy, P. E., Ed., Elsevier, Amsterdam, 1976, 211.
7. **Sargent, J. R., Lee, R. F., and Nevenzel, J. C.**, in *Chemistry and Biochemistry of Natural Waxes*, Kolattukudy, P. E., Ed., Elsevier, Amsterdam, 1976, 50.
8. **Jacob, J.**, in *Chemistry and Biochemistry of Natural Waxes*, Kolattukudy, P. E., Ed., Elsevier, Amsterdam, 1976, 94.
9. **Albro, P. W.**, in *Chemistry and Biochemistry of Natural Waxes*, Kolattukudy, P. E., Ed., Elsevier, Amsterdam, 1976, 422.
10. **Weete, J. D.**, in *Chemistry and Biochemistry of Natural Waxes*, Kolattukudy, P. E., Ed., Elsevier, Amsterdam, 1976, 350.
11. **Gülz, P. G.**, *Z. Pflanzenphysiol.*, 64, 462, 1971.
12. **Downing, D. T.**, in *Chemistry and Biochemistry of Natural Waxes*, Kolattukudy, P. E., Ed., Elsevier, Amsterdam, 1976, 18.
13. **Morrison, R. I.**, in *Organic Geochemistry*, Eglington, G. and Murphy, M. T. J., Eds., Springer-Verlag, Berlin, 1969, 566.
14. **Fedeli, E. and Jacini, G.**, *Ad. Lipid Res.*, 9, 359, 1971.
15. **Kuksis, A.**, *Biochemistry*, 3, 1086, 1964.
16. **Ruzicka, L.**, *Experientia*, 9, 357, 1953.
17. **Kolattukudy, P. E., Croteau, R., and Buckner, J. S.**, in *Chemistry and Biochemistry of Natural Waxes*, Kolattukudy, P. E., Ed., Elsevier, Amsterdam, 1976, 294.
18. **Kolattukudy, P. E.**, *Lipids*, 5, 259, 1970.
19. **Goodwin, T. W.**, *Biochem. J.*, 123, 293, 1971.
20. **Herbin, G. A.**, *Proc. E. Afr. Acad.*, 2, 11, 1964.
21. **Stránský, K. and Streibl, M.**, *Collect. Czech. Chem. Commun.*, p. 1103, 1969.
22. **Douglas, A. G. and Eglington, G.**, in *Comparative Phytochemistry*, Swain, T., Ed., Academic Press, London, 1966, 57.
23. **Herbin, G. A. and Robins, P. A.**, *Phytochemistry*, 7, 1325, 1968.
24. **Laseter, J. L., Weber, D. J., and Oró, J.**, *Phytochemistry*, 7, 1005, 1968.
25. **Holloway, P. J., Brown, G. A., Baker, E. A., and Macey, M. J. K.**, *Chem. Phys. Lipids*, 19, 114, 1977.
26. **Tulloch, A. P.**, *Lipids*, 9, 664, 1974.
27. **Kaneda, T.**, *Phytochemistry*, 8, 2039, 1969.
28. **Jackson, L. L.**, *Lipids*, 5, 38, 1970.
29. **Hamilton, R. J., Raie, N. Y., Weatherstone, I., Brooks, C. J. W., and Borthwick, H. J.**, *J. Chem. Soc. Perkin Trans. I*, 354, 1975.
30. **Mecklenburg, H. C.**, *Phytochemistry*, 5, 1201, 1966.
31. **Nicolaides, N.**, *J. Am. Oil Chem. Soc.*, 42, 691, 1965.
32. **Gelpi, E. and Oró, J.**, *J. Chromatogr. Sci.*, 8, 210, 1970.
33. **Gelpi, E., Nooner, D. W., and Oró, J.**, *Geochim. Cosmochim. Acta*, 34, 421, 1970.
34. **Streibl, M., Stránský, K., and Herout, V.**, *Collect. Czech. Chem. Commun.*, 43, 320, 1978.
35. **Spengler, G. and Hauf, G.**, *Fette Seifen Anstrichm.*, 59, 509, 1957.
36. **Davenport, J. B. and Fogerty, A. C.**, in *Biochemistry and Methodology of Lipids*, Johnson, A. R. and Davenport, J. R., Eds., Wiley-Interscience, New York, 1971, 285.
37. **Privett, O. S.**, *Prog. Chem. Fats Lipids*, 9, 91, 1966.
38. **Bianchi, G. and Corbellini, M.**, *Phytochemistry*, 16, 943, 1977.
39. **Stránský, K., Ubik, K., Holman, J., and Streibl, M.**, *Collect. Czech. Chem. Commun.*, 38, 770, 1973.
40. **Stránský, K., Trka, A., Kohoutová, J., and Streibl, M.**, *Collect. Czech. Chem. Commun.*, 41, 1799, 1976.
41. **Hirsch, J. and Ahrens, E. H., Jr.**, *J. Biol. Chem.*, 233, 311, 1958.
42. **Body, D. R.**, *Phytochemistry*, 13, 1527, 1974.
43. **Oró, J., Nooner, D. W., and Olson, R. J.**, in *Lipid Chromatographic Analysis*, Marinetti, G. V., Ed., Marcel Dekker, New York, 1969, 479.
44. **Eglington, G., Gonzales, A. G., Hamilton, R. J., and Raphael, R. A.**, *Phytochemistry*, 1, 89, 1962.
45. **Hanny, B. W. and Gueldner, R. C.**, *J. Agric. Food Chem.*, 24, 401, 1976.
46. **Vandenburg, L. E. and Wilder, S. C.**, *J. Am. Oil Chem. Soc.*, 47, 514, 1970.
47. **Macey, M. J. K. and Barber, H. N.**, *Phytochemistry*, 9, 13, 1970.
48. **Stránský, K. and Streibl, M.**, *Collect. Czech. Chem. Commun.*, 36, 2267, 1971.
49. **Janák, J., Jagarič, Z., and Dressler, M.**, *J. Chromatogr.*, 53, 525, 1970.
50. **Stojanova-Ivanova, B., Mladenova, K., and Malova, I.**, *Phytochemistry*, 10, 2525, 1971.
51. **Hutchins, R. F. N. and Martin, M. M.**, *Lipids*, 3, 250, 1968.
52. **Streibl, M. and Stránský, K.**, *Fette Seifen Anstrichm.*, 70, 343, 1968.

53. **Streibl, M. and Šorm, F.,** *Collect. Czech. Chem. Commun.,* 31, 1585, 1966.
54. **Guha, O. K. and Janák, J.,** *J. Chromatogr.,* 68, 325, 1972.
55. **Souček, J., Engelhardt, G., Stránský, K., and Schraml, J.,** *Collect. Czech. Chem. Commun.,* 41, 234, 1976.
56. **Zimmerschmied, W. J., Dinerstein, R. A., Weitkamp, A. W., and Marschner, R. F.,** *J. Am. Chem. Soc.,* 71, 2947, 1949.
57. **Coles, L.,** *J. Chromatogr.,* 32, 657, 1968.
58. **Davenport, J. B.,** in *Biochemistry and Methodology of Lipids,* Johnson, A. R. and Davenport, J. B., Eds., Wiley-Interscience, New York, 1971, 151.
59. **Haathi, E. and Nikkari, T.,** *Acta Chem. Scand.,* 17, 2565, 1965.
60. **Karmen, A., Walker, T., and Bowman, R. L.,** *J. Lipid Res.,* 4, 103, 1963.
61. **Janák, J.,** in *Progress in Thin-Layer Chromatography and Related Methods,* Vol. II, Niederwieser, A. and Pataki, G., Eds., Humphrey, Ann Arbor, 1971, 63.
62. **van der Horst, D. J. and Oudejans, R. C. H. M.,** *Comp. Biochem. Physiol.,* 41B, 823, 1972.
63. **Oudejans, R. C. H. M.,** *Comp. Biochem. Physiol.,* 42B, 15, 1972.
64. **Johnson, A. R.,** in *Biochemistry and Methodology of Lipids,* Johnson, A. R. and Davenport, J. B., Eds., Wiley-Interscience, New York, 1971, 330.
65. **Deyl, Z., Macek, K., and Janák, J., Eds.,** *Modern Column Liquid Chromatography,* Elsevier, Amsterdam, 1975.
66. **Snyder, L. R.,** *J. Chromatogr. Sci.,* 7, 352, 1969.
67. **Shenstone, F. S.,** in *Biochemistry and Methodology of Lipids,* Johnson, A. R. and Davenport, J. B., Eds., Wiley-Interscience, New York, 1971, 171.
68. **Oró, J., Nooner, D. W., and Olson, R. J.,** in *Lipid Chromatographic Analysis,* Marinetti, G. V., Ed., Marcel Dekker, New York, 1969, 479.
69. **Tulloch, A. P. and Hoffman, L. L.,** *J. Am. Oil Chem. Soc.,* 54, 587, 1977.
70. **Kaufmann, H. P., Mangold, H. K., and Mukherjee, K. D.,** *J. Lipid Res.,* 12, 506, 1971.
71. **Baker, E. A., Procopiou, J., and Hunt, G. M.,** *J. Sci. Food Agric.,* 26, 1093, 1975.
72. **Holloway, P. J. and Challen, S. B.,** *J. Chromatogr.,* 25, 236, 1966.
73. **Tulloch, A. P.,** *J. Chromatogr. Sci.,* 13, 403, 1975.
74. **Brieskorn, C. H. and Beck, K. H.,** *Phytochemistry,* 9, 1633, 1970.
75. **Kaufmann, H. P. and Das, B.,** *Fette Seifen Anstrichm.,* 65, 398, 1963.
76. **Barbará, N. H. and Cadenas, R. A.,** *Phytochemistry,* 13, 671, 1964.
77. **Arnold, M. T., Blomquist, G. J., and Jackson, L. L.,** *Comp. Biochem. Physiol.,* 31, 685, 1969.
78. **Cmelik, S. H. W.,** *Insect Biochem.,* 1, 439, 1971.
79. **Atkinson, P. W., Brown, W. V., and Gilby, S. R.,** *Insect Biochem.,* 3, 103, 1973.
80. **Nishimoto, S.,** *J. Sci. Hiroshima Univ. Ser. A,* 38, 165, 1974.
81. **Jarolímek, P., Wollrab, V., and Streibl, M.,** *Coll. Czech. Chem. Commun.,* 29, 2528, 1964.
82. **Soják, L., Hrivnák, J., Majer, P., and Janák, J.,** *Anal. Chem.,* 45, 293, 1973.
83. **Nelson, D. R. and Sukkestad, D. R.,** *J. Lipid Res.,* 12, 18, 1975.
84. **Wellburn, A. R. and Hemming, E. W.,** *J. Chromatogr.,* 23, 51, 1966.
85. **Rohrschneider, L.,** *J. Chromatogr.,* 22, 6, 1966.
86. **Riedo, R., Fritz, D., Tarján, G., and Kováts, E. S.,** *J. Chromatogr.,* 126, 63, 1976.
87. **Downing, D. T., Kranz, Z. H., and Murray, K. E.,** *Austr. J. Chem.,* 13, 80, 1966.
88. **Castello, G., Lunardelli, M., and Berg, M.,** *J. Chromatogr.,* 76, 31, 1973.
89. **Cramers, C. A., Rijks, J. A., Pacakova, J., and Ribeiro de Andrade, J.,** *J. Chromatogr.,* 51, 13, 1970.
90. **Beroza, M. and Coad, R. A.,** *J. Gas Chromatogr.,* 4, 199, 1966.
91. **Adlard, E. R. and Whitham, B. T.,** *Nature (London),* 192, 966, 1961.
92. **Netting, A. G., Macey, M. J. K., and Barber, H. N.,** *Phytochemistry,* 11, 579, 1972.
93. **Miltenberger, K. H.,** *Fette Seifen Anstrichm.,* 70, 736, 1968.
94. **Tulloch, A. P.,** *J. Am. Oil Chem. Soc.,* 49, 609, 1972.
95. **Tulloch, A. P.,** *J. Am. Oil Chem. Soc.,* 50, 367, 1973.
96. **Janák, J.,** *Nature (London),* 195, 696, 1962.
97. **Ruseva-Atanaseva, N. and Janák, J.,** *J. Chromatogr.,* 21, 207, 1966.
98. **Lockey, K. H.,** *Insect Biochem.,* 6, 457, 1976.
99. **Nagy, B., Modzeleski, V. E., and Scott, W. M.,** *Biochem. J.,* 114, 645, 1969.
100. **Oró, J., Nooner, D. W., Zlatkis, A., Wikstrom, S. A., and Barghoorn, E. S.,** *Nature (London),* 213, 1082, 1967.
101. **Bandursi, E. L. and Nagy, B.,** *Lipids,* 10, 67, 1975.
102. **Nooner, D. W., Oró, J., and Cerbulis, J.,** *Lipids,* 8, 489, 1973.
103. **Jackson, L. L. and Blomquist, G. J.,** in *Chemistry and Biochemistry of Natural Waxes,* Kolattukudy, P. E., Ed., Elsevier, Amsterdam, 1976, 207.

104. **Nelson, D. R., Sukkestad, D. R., and Zaylskie, R. G.,** *J. Lipid Res.,* 13, 413, 1972.
105. **Stránský, K., Streibl, M., and Sörm, F.,** *Collect. Czech. Chem. Commun.,* 31, 4694, 1966.
106. **Blomquist, G. J., Blailock, T. T., Scheetz, R. W., and Jackson, L. L.,** *Comp. Biochem. Physiol.,* 54B, 381, 1976.
107. **Stránský, K., Streibl, M., and Sörm, F.,** *Collect. Czech. Chem. Commun.,* 33, 416, 1968.
108. **Brooks, C. J. W. and Keates, R. A. B.,** *J. Chromatogr.,* 44, 509, 1969.
109. **Cooper, B. S.,** *J. Chromatogr.,* 46, 112, 1970.
110. **Sweeney, E. G., Thompson, R. E., and Ford, D. C.,** *J. Chromatogr. Sci.,* 8, 76, 1970.
111. **Pokorńy, S., Čoupek, J., Luan, N. T., and Pokorńy, J.,** *J. Chromatogr.,* 84, 319, 1973.
112. **Hillman, D. E.,** *Anal. Chem.,* 43, 1007, 1971.
113. **Baron, M.,** in *Physical Methods in Analytical Chemistry,* Vol. 4, Berl, W., Ed., Academic Press, New York, 1961, 226.
114. **Schliessler, R. W. and Flitter, D.,** *J. Am. Chem. Soc.,* 74, 1720, 1952.
115. **Thomas, T. and Mays, R.,** in *Physical Methods in Analytical Chemistry,* Vol. 4, Berl, W., Ed., Academic Press, New York, 1961, 45.
116. **Gershbein, L. L., Krotoszynski, B. K., and Singh, E. J.,** *J. Chromatogr.,* 27, 431, 1967.
117. **Wollrab, V.,** *Collect. Czech. Chem. Commun.,* 33, 1584, 1968.
118. **Jaffe, H. and Orchin, M.,** *Theory and Applications of Ultraviolet Spectroscopy,* John Wiley & Sons, New York, 1962.
119. **Enzell, C. R., Kimland, B., and Gunnardson, L. E.,** *Tetrahedron Lett.,* p. 1971, 1983.
120. **Jackson, L. L.,** *Comp. Biochem. Physiol.,* 41B, 331, 1972.
121. **Wolff, R. E., Wolff, G., and McCloskey, J. A.,** *Tetrahedron,* 22, 3093, 1966.
122. **Burlingame, A. L. and Schnoes, H. K.,** in *Organic Geochemistry,* Eglinton, G. and Murphy, M. T. J., Eds., Springer-Verlag, Berlin, 1969, 130.
123. **Plattner, R. D., Spencer, G. F., and Kleiman, R.,** *Lipids,* 11, 222, 1976.
124. **Ohtaki, T., Ozaki, H., Yata, N., Iton, M., and Suzuki, A.,** *Nippon Kagaku Kaishi,* 934, 939, 1972.
125. **Streibl, M. and Herout, V.,** in *Organic Geochemistry,* Eglinton, G. and Murphy, M. T. J., Eds., Springer-Verlag, Berlin, 1969, 416.
126. **Mayers, M. V., Caldwell, K. D., and Giddings, J. C.,** *Separation Sci.,* 9, 47, 1974.
127. **Rijks, J. A., van den Berg, J. H. M., and Diependaal, J. P.,** *J. Chromatogr.,* 91, 603, 1974.

Chapter 5

ALCOHOLS, ALDEHYDES, AND KETONES

V. Mahadevan and R. G. Ackman

I. INTRODUCTION

Many long-chain alcohols and alkyl acetates, also some aldehydes and a few ketones, are excreted by insects and serve as "chemical messengers" between individuals of the same species. The isolation of these pheromones, and the elucidation of their structures as well as methods for their analysis are described in recent reviews.[1,2]

The surface lipids of plants contain fairly large proportions of long-chain alcohols and their esters, the waxes, as well as aldehydes and ketones.[3] Methods for the isolation and analysis of such compounds are described in Chapter 11. In animal and human tissues there occur very small proportions of long-chain alcohols[3,4] and aldehydes.[3,5] Long-chain aldehydes are easily liberated by enzymatic or acid-catalyzed hydrolysis of lipids containing alk-1-enyl moieties. Such aldehydogenic lipids or plasmalogens are constituents of animal and human tissues, particularly of the central and peripheral nervous systems. Methods for the characterization of lipids containing alkyl or alk-1-enyl moieties including procedures for the analysis of the aldehydes derived from plasmalogens are described in Chapter 14.

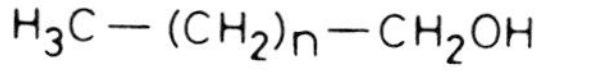

$$H_3C - (CH_2)_n - CH_2OH \qquad \textit{prim.} \text{ Alcohols}$$

$$H_3C - (CH_2)_n - \underset{\underset{H}{|}}{\overset{\overset{OH}{|}}{C}} - CH_3 \qquad \textit{sec.} \text{ Alcohols}$$

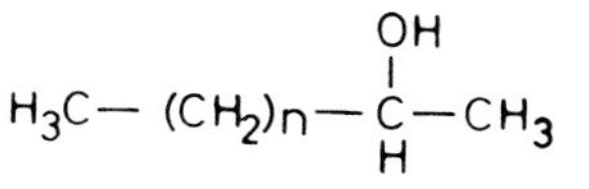

$$H_3C - (CH_2)_n - C \overset{H}{\underset{}{\diagup}}{=} O \qquad \text{Aldehydes}$$

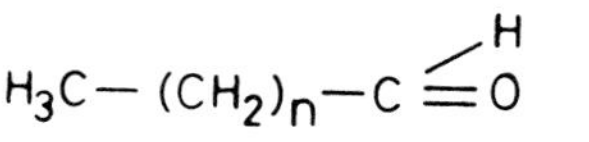

$$H_3C - (CH_2)_n - \underset{\underset{O}{\|}}{C} - CH_3 \qquad \text{Methyl Ketones}$$

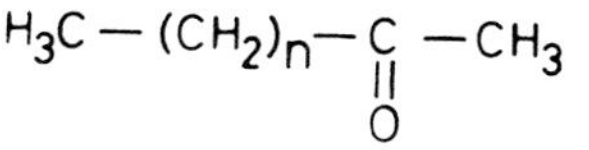

Reference is made to reviews on the chemistry and biochemistry of long-chain alcohols,[4] aldehydes,[5] and methyl ketones.[6]

This chapter has been written to complement the information presented in the previously mentioned contributions. Specifically, methods for the analysis of *primary* and *secondary* alcohols, aldehydes, and methyl ketones are described. The experimental procedures given are complemented by tables* of chromatographic data. Emphasis is placed on long-chain compounds, but aldehydes and methyl ketones of shorter chain lengths are also considered because they are encountered in nature, especially in plants and insects, and in deteriorated fats. Moreover, the analysis of short-chain aldehydes (and aldesters) is of interest in work aimed at determining the structure of unsaturated lipids by ozonolysis, as described in Chapter 21.

---

* The tables appear following the text.

II. PREPARATION OF SAMPLES

Detailed procedures for the isolation of surface lipids are described in Chapter 11, "Surface Lipids of Plants and Animals", and procedures for the extraction of mammalian tissue lipids can be found in Chapter 2.

The fractionation of complex lipid mixtures into classes of compounds is carried out by adsorption chromatography, preferably on thin layers of silica gel, as described in Chapter 3, and elsewhere in this treatise.

Gas chromatography (GC) is most suitable for the complete resolution of a lipid class and the quantitative determination of its constituents. Procedures for the preparation of derivatives of long-chain alcohols and aldehydes are described below.

Paper chromatography (PC) and thin-layer chromatography (TLC) are used occasionally for the fractionation of mixtures of the 3.5-dinitrobenzoates of short-chain alcohols or the 2.4-dinitrophenylhydrazones of the short-chain aldehydes and ketones. Procedures for the preparation of such derivatives can be found in textbooks of organic analysis.[7,8]

A. Alcohols

Mixtures of long-chain alcohols as well as mixtures of alkyl acetates can be resolved by GC. Procedures for the preparation of alkyl acetates are described below.

Procedure for the Acetylation of Alcohols[9]
The long-chain alcohols, 1 to 10 mg, are dissolved in 1 mℓ of acetic anhydride-anhydrous pyridine, (1:2 v/v) in a glass-stoppered tube and kept at 37°C with occasional shaking. After 15 min, 5 mℓ of water is added, the alkyl acetates are extracted with three 5 mℓ portions of hexane and the combined extracts are washed with water and dried over anhydrous sodium sulfate. After a few hours the solution is decanted, and most of the hexane is evaporated in a stream of nitrogen. The completeness of the reaction is monitored by TLC on silica gel with hexane-diethyl ether (9:1 v/v) as solvent.

Procedure for the Acetylation of Alcohols[10]
The long-chain alcohols, 1 to 10 mg, are added to a mixture of 1.6 mℓ acetonitrile and 0.4 mℓ acetic anhydride in a screw-capped 25 mℓ tube and stirred magnetically, at room temperature, in the presence of 20 mg of DOWEX 50W-X8 ion exchanger (previously washed with water, then with 1 N hydrochloric acid, and water, until neutral, and dried at 50°C overnight). After 15 min, 5 mℓ water is added to the reaction mixture, and the acetates formed are extracted with four 1 mℓ portions of hexane. The combined extracts are washed with water until neutral, and dried over sodium sulfate. The completeness of the reaction is monitored by TLC on silica gel with hexane-diethyl ether (9:1 v/v) as solvent.

B. Aldehydes

Mixtures of long-chain aldehydes can be resolved by GC, but, since the aldehydes tend to undergo condensation reactions, they are often chromatographed as dimethyl acetals.[12-14] The latter derivatives can form alk-1-enylmethyl ethers during chromatography, and these artifacts are then erroneously considered to be dimethyl acetals of branched-chain aldehydes,[11,13] although the latter do exist.[9,13-15] The dimethyl acetals themselves can mimic such natural structures.[16] In view of these problems with the possible inadvertent formation of dimethyl acetals as artifacts in aldehyde samples, and of vinyl ether artifacts from dimethyl acetals,[17,18] extreme caution should be exercised in dealing with the analysis of aldehydes. The aldehydes themselves are reported to be stable in carbon disulfide,[16] and the difficulties with acetals can be overcome by analyzing cyclic acetals of aldehydes with ethylene or propylene glycols.[14] Alternatively, the aldehydes may be reduced to alcohols, which are acetylated (see Procedures), and the resulting alkyl acetates are fractionated by gas chromatography.

Procedure for the Preparation of Dimethyl Acetals[9]
The sample, 1 to 10 mg, of aldehydes or lipids containing alk-1-enyl ether moieties is dissolved in 2 mℓ of a 5% solution of dry hydrogen chloride in absolute methanol. The solution is covered with nitrogen and heated to 80°C in a screw-capped 25 mℓ tube for 2 hr. After cooling to room temperature, 2 mℓ of water are added, and the lipids are extracted with three 5 mℓ portions of hexane. The combined extracts are washed with water and dried over anhydrous sodium sulfate. After an hour, the solution is decanted, and most of the hexane is evaporated in a stream of nitrogen. The dimethyl acetals are separated from methyl esters by TLC on silica gel with hexane-diethyl ether (95:5 v/v) or benzene as solvents. The plates are developed twice, and the dimethyl acetals, which migrate ahead of the methyl esters, are eluted from the absorbent with several portions of diethyl ether.

Procedure for the Reduction of Aldehydes to Alcohols[9,19]
The long-chain aldehydes, 1 to 10 mg, are dissolved in 1 mℓ absolute diethyl ether. The solution is cooled to −20°C, and 3 mℓ of a 3% solution of lithium aluminum hydride in absolute diethyl ether is added. After stirring at −20°C for 2 hr, methanol-water (9:1 v/v) is added dropwise to decompose excess lithium aluminum hydride. The suspension is centrifuged and the diethyl ether evaporated; traces of water are removed by adding a few drops of benzene and distilling off. The remaining long-chain alcohols are acetylated as described (see Procedures).

Procedure for the Cleavage of Acetals and Alk-1-enyl Ethers[20]
The lipid sample, 1 to 10 mg, or silica gel containing such an amount, 3 mℓ peroxide-free diethyl ether, and 1 mℓ concentrated hydrochloric acid are stirred under nitrogen in a 25 mℓ tube for 30 min. The tube is then placed in ice water, opened, and 5 mℓ of hexane is added followed by 10 mℓ water. The tube is closed, shaken, and if necessary, centrifuged to break emulsions. The lower layer is siphoned off with a syringe and the upper layer, which contains "free" aldehydes, is washed with several 5 mℓ portions of water, until neutral. To avoid the formation of secondary products, the solution must not be evaporated, and the aldehydes should be analyzed as soon as possible. The procedure is suitable also for the cleavage of the trialkyltrioxanes, which are sometimes found in lipid extracts (see below).

C. Ketones

Long-chain ketones are analyzed by GC without prior derivatization; the position of the keto group is best determined by a combination of GC with mass spectrometry (MS). Methyl ketones can be identified by their ability to undergo the haloform reaction, which leads to a fatty acid having one carbon less than the ketone.[21]

III. CHROMATOGRAPHY

Mixtures of alcohols, aldehydes, and ketones of comparable chain lengths can easily be separated into classes of compounds by chromatography on various adsorbents. In TLC on silica gel with the solvent hexane-diethyl ether-acetic acid (80:20:1 v/v) the aldehydes migrate slightly ahead of the methyl ketones, whereas the *secondary* alcohols, followed by the *primary* alcohols, stay far behind. With chloroform as solvent, however, the sequence is methyl ketones, aldehydes, *secondary* alcohols, *primary* alcohols. Classes of compounds that are not satisfactorily resolved by chromatography on silica gel may be well separated on layers of magnesium oxide or another adsorbent.

Aldehydes can be detected as violet fractions by spraying the chromatogram with Schiff's reagent (Chapter 28, Reagent No. 33); aldehydes and ketones appear as yellow fractions after spraying the chromatogram with 2.4-dinitrophenylhydrazine solution (Chapter 28, Reagent No. 12). If recovery of the different lipid classes is intended, the chromatogram is sprayed with 2',7'-dichlorofluorescein solution and viewed in UV light (Chapter 28, Reagent No. 8); the fractions are eluted with diethyl ether, and the eluates are filtered through sintered glass (see Chapter 3).

The aldehydes are much more reactive than the alcohols and ketones. Thus the aldehydes occurring in mammalian tissues tend to trimerize to 2,4,6-trialkyltrioxanes.[22] Upon chromatography on silica gel, these cyclic compounds migrate close to the hydrocarbons; they can be isolated and cleaved to aldehydes (see Procedures).

$$3 \ R\text{-CHO} \ \rightleftharpoons$$

Aldehyde   Trioxane

If lipid extracts containing aldehydes are stored, 2.3-dialkylacroleins may be formed by aldol condensation followed by dehydration; the condensation reaction is catalyzed by ethanolamine phospholipids.[23] In chromatography on silicic acid, the dialkylacroleins migrate ahead of the common long-chain aldehydes; they can be distinguished from the latter compounds by the deep orange color they yield with 2.4-dinitrophenylhydrazine solution (Chapter 28, Reagent No. 12).

The experimental conditions employed in the gas chromatographic analysis of alcohols and alkyl acetates, aldehydes and dimethyl acetals, and ketones are similar to those that have been developed for the fractionation of the methyl esters of fatty acids having comparable chain lengths (see Chapter 6).

IV. CONCLUSIONS

Both the *primary* and *secondary* alcohols occurring in plant surface lipids are exclusively saturated. The *primary* alcohols have an even number of carbon atoms, and their chain lengths range from C_{22} to C_{34}. The *secondary* alcohols have an odd number of carbon atoms, predominantly C_{29} or C_{31}. The hydroxy group is, as a rule, close to the center of the chain. The *primary* alcohols and the corresponding alkyl acetates that function as pheromones, are saturated or mono- or diunsaturated. The *primary* alcohols occurring in mammalian tissues, however, are saturated or monounsaturated, never diunsaturated. The insect pheromones as well as the alcohols of mammalian tissues have an even number of carbon atoms; their chain lengths, like those of the fatty acids, range from C_{14} to C_{18}.

The aldehydes occurring in plant surface lipids are exclusively saturated. Like the *primary* alcohols, they have an even number of carbon atoms, and their chain lengths range from C_{26} to C_{32}. The aldehydes functioning as pheromones as well as those occurring in mammalian tissues are saturated or monounsaturated, they have an even number of carbon atoms, and their chain lengths resemble those of the fatty acids bound in the common ester lipids.

The ketones in plant surface lipids are exclusively saturated. Like the *secondary* alcohols, they have an odd number of carbon atoms, predominantly C_{29} or C_{31}. The keto group is, as a rule, close to the center of the chain. The methyl ketones occurring in plants and insects

Table 1
LONG-CHAIN ALCOHOLS (GC)

Column packing	P1	P2	P3	P4	P5	P6
Temperature (°C)	163	163	185	200	211	200
Gas,	Ar	Ar	Ar	N_2	H_2	H_2
Flow rate (mℓ/min)	na	na	na	30	163	110
Column						
Length (cm)	122	122	122	183	400	200
Diameter						
(cm I.D.)	na	na	na	na	0.6	0.6
(cm O.D.)	0.64	0.64	0.64	0.318	na	na
Form	Tube	Tube	Tube	Helix	Tube	Tube
Material	G	G	G	SS	Cu	Cu
Detector	Argon	Argon	Argon	FID	TC	TC
Reference	1	1	1	1	2	3
Data form						
Compound			$r_{Octadecanol}$			
Tetradecanol	0.295	0.240	0.268	0.284	0.36	0.37
Tetradecenol	0.396	0.302	0.315	0.338	—	0.44
Pentadecanol	0.409	0.351	0.370	0.396	0.45	0.48
Pentadecenol	0.508	0.424	0.449	0.474	—	—
Hexadecanol	0.544	0.500	0.518	0.538	0.58	0.61
cis-9-Hexadecenol	0.678	0.576	0.594	0.601	0.70	0.73
Heptadecanol	0.735	0.712	0.716	0.742	—	0.81
Heptadecenol	0.895	0.812	0.823	0.807	—	—
Octadecanol	1.00	1.00	1.00	1.00	1.00	1.00
cis-9-Octadecenol	1.22	1.12	1.12	1.12	1.16	1.21
cis,cis-9,12-Octadecadienol	1.60	1.34	1.36	1.31	1.41	1.54
cis,cis,cis-6,9,12-Octadecatrienol	1.88	1.52	1.54	1.45	—	—
cis,cis,cis-9,12,15-Octadecatrienol	2.20	1.71	1.76	1.66	—	2.10
cis-11-Icosenol	2.20	2.19	2.15	2.04	1.71	2.02
cis-13-Docosenol						3.30

Column packing	P1	=	17% EGSS-X (ethylene glycol succinate polyester combined with methyl silicone) on Gas-Chrom P
	P2	=	15.5% EGSS-Y (ethylene glycol succinate polyester combined with methyl silicone) on Gas-Chrom P
	P3	=	10% Poly (ethylene glycol adipate) on Celite 545, AW (100-120 mesh)
	P4	=	8% Poly (1,4-butanediol succinate) on Chromosorb W, HMDS (80-100 mesh)
	P5	=	30% Poly (ethylene glycol α-ketopimelate) on Sterchamol, Merck (0.25-0.3 mm)
	P6	=	20% PAGN (cyanoethylated polyglycerol) on Sterchamol, Merck (0.25-0.3 mm)

REFERENCES

1. **Jamieson, G. R. and Reid, E. H.,** *J. Chromatogr.,* 26, 8, 1967.
2. **Falk, F.,** *J. Chromatogr.,* 17, 450, 1965.
3. **Falk, F. and Dietrich, P.,** *J. Chromatogr.,* 33, 422, 1968.

Note added in proof: Separations of geometrical and positional isomers of insect pheromone olefinic primary alcohols on the cyanosilicone liquid phases: SP-2300, SP-2310, SP-2330, SP-2340, and the polyglycol types: SP-1000, Carbowax-20M wall-coated in open tubular columns are illustrated and discussed by **Heath, R. R., Burnsed, G. E., Tumlinson, J. H., and Doolittle, R. E.,** *J. Chromatogr.,* 189, 199, 1980.

are saturated and monounsaturated, have an odd number of carbon atoms, and chain lengths up to C_{17}.

Table 2
ALKYL ACETATES (GC)

	P1	P2	P3	P4	P5	P6	P7	P8
Column packing								
Temperature (°C)	216	163	163	185	200	200	185	170
Gas	H_2	Ar	Ar	Ar	N_2	N_2	He	He
Flow rate (ml/min)	163	na	na	na	30	30	5	5
Column Length (cm)	400	122	122	122	183	183	4700	4700
Diameter (cm I.D.)	0.6	na	na	na	na	na	0.025	0.025
(cm O.D.)	na	0.64	0.64	0.64	0.318	0.318	na	na
Form	Tube	Tube	Tube	Tube	Helix	Helix	WCOT	WCOT
Material	Cu	G	G	G	SS	SS	SS	SS
Detector	TC	Argon	Argon	Argon	FID	FID	FID	FID
Reference	1	2	2	2	2	2	3	3
Data form					$r_{Octadecanoyl\ Acetate}$			

Compound	P1	P2	P3	P4	P5	P6	P7	P8
Tetradecanyl acetate	0.35	0.297	0.249	0.265	—	0.282	0.255	0.237
Tetradecenyl acetate	0.40	0.387	0.293	0.322	—	0.342	0.313	0.279
Pentadecanyl acetate	0.45	0.411	0.356	0.370	—	0.414	0.364	0.338
Pentadecenyl acetate	—	0.510	0.414	0.453	—	0.474	—	—
Hexadecanyl acetate	0.58	0.542	0.502	0.520	—	0.536	0.510	0.483
cis-9-Hexadecenyl acetate	0.67	0.671	0.578	0.598	—	0.599	0.591	0.542
Heptadecanyl acetate	—	0.736	0.733	0.714	—	0.739	0.710	—
Heptadecenyl acetate	—	0.895	0.797	0.823	—	0.809	—	—
Octadecanyl acetate	1.00	1.00	1.00	1.00	1.00	1.00	1.00	1.00
cis-9-Octadecenyl acetate	1.12	1.21	1.13	1.12	1.12	1.12	1.12	1.07
cis,cis-9,12-Octadecadienyl acetate	1.22	1.56	1.35	1.36	1.31	1.32	1.35	1.30
cis,cis-6,9,12-Octadecatrienyl acetate	—	1.57	1.52	1.54	1.43	1.43	—	—
cis,cis-9,12,15-Octadecatrienyl acetate	1.52	2.20	1.72	1.76	1.63	1.65	1.67	1.71
cis,cis,cis-6,9,12,15-Octadecatetraenyl acetate	—	—	—	—	1.82	—	—	1.87
cis-11-Icosenyl acetate	—	2.16	2.19	2.09	2.03	2.03	2.16	2.20
cis,cis-11,14-Icosadienyl acetate	—	—	—	—	2.41	—	2.60	2.64
cis,cis-8,11,14-Icosatrienyl acetate	—	—	—	—	2.61	—	2.97	2.92
cis,cis,cis-5,8,11,14-Icosatetraenyl acetate	—	—	—	—	2.90	—	3.24	3.27

cis,cis,cis,cis-8,11,14,17-Icosatetraenyl acetate	—	—	—	—	3.30	—	—	3.56	3.74
cis,cis,cis,cis,cis-5,8,11,14,17-Icosapentaenyl acetate	—	—	—	—	3.60	—	—	—	4.18
cis-11-Docosenyl acetate	—	—	—	—	3.63	—	4.08	4.32
cis,cis,cis,cis-7,10,13,16-Docosatetraenyl acetate	—	—	—	—	5.25	—	—	6.48
cis,cis,cis,cis,cis-4,7,10,13,16-Docosapentaenyl acetate	—	—	—	—	5.74	—	—	7.05
cis,cis,cis,cis,cis,cis-7,10,13,16,19-Docosapentaenyl acetate	—	—	—	—	6.49	—	—	8.22
cis,cis,cis,cis,cis,cis-4,7,10,13,16,19-Docosahexaenyl acetate	—	—	—	—	7.14	—	7.81	8.95
cis-15-Tetracosenyl acetate	—	—	—	—	6.50	—	7.95	9.84

Column packing P1 = 30% Poly (ethylene glycol α-ketopimelate) on Sterchamol, Merck (0.25—0.3 mm)

P2 = 17% EGSS-X (ethylene glycol succinate polyester combined with methyl silicone) on Gas-Chrom P

P3 = 15.5% EGSS-Y (ethylene glycol succinate polyester combined with methyl silicone) on Gas-Chrom P

P4 = 10% Poly (ethylene glycol adipate) on Celite 545, AW (100-120 mesh)

P5 = 8% Poly (1,4-butanediol succinate) on Chromosorb W, HMDS (80-100 mesh)

P6 = 8% Poly (1,4-butanediol succinate) on Chromosorb W, HMDS (80-100 mesh)

P7 = SILAR-5CP (50% cyanopropyl, 50% phenyl, siloxane) in form of wall coating in stainless steel tubing

P8 = Poly (1,4-butanediol succinate) in form of wall coating in stainless steel tubing

REFERENCES

1. **Falk, F.,** *J. Chromatogr.,* 17, 450, 1965.
2. **Jamieson, G. R. and Reid, E. H.,** *J. Chromatogr.,* 26, 8, 1967.
3. **Ratnayake, N. and Ackman, R. G.,** unpublished.

Table 3
ALKYL ACETATES (GC)

	P1	P2	P3	P4	P5	P6	P7	P8
Column packing	216	163	180	163	185	200	185	175
Temperature (°C)	H_2	Ar	N_2	Ar	Ar	N_2	He	He
Gas	163	na	5	na	na	30	5	5
Flow rate (mℓ/min)								
Column								
Length (cm)	400	122	5000	122	122	183	4700	4700
Diameter								
(cm I.D.)	0.6	na	0.05	na	na	na	0.025	0.025
(cm O.D.)	na	0.64	na	0.64	0.64	0.318	na	na
Form	Tube	Tube	WCOT	Tube	Tube	Helix	WCOT	WCOT
Material	Cu	G	SS	G	G	SS	SS	SS
Detector	TC	Argon	FID	Argon	Argon	FID	FID	FID
Reference	1	2	3	2	2	2	4	4
Data form								

Equivalent Chain Lengths Based on *n*-Alkanol Acetates

Compound	P1	P2	P3	P4	P5	P6	P7	P8
Tetradecanyl acetate	14.00	14.00	14.00	14.00	14.00	14.00	14.00	14.00
Tetradecenyl acetate	14.48	14.85	—	14.43	14.60	14.48	14.60	14.49
Pentadecanyl acetate	15.00	15.00	15.00	15.00	15.00	15.00	15.00	15.00
Pentadecenyl acetate	—	15.80	—	15.45	15.60	15.55	—	—
Hexadecanyl acetate	16.00	16.00	16.00	16.00	16.00	16.00	16.00	16.00
Hexadecenyl acetate	16.52	16.70	—	16.40	16.44	16.33	16.45	16.31
cis-9-Hexadedecenyl acetate	17.00	17.00	17.00	17.00	17.00	17.00	17.00	17.00
Heptadecanyl acetate	—	17.65	—	17.30	17.40	17.26	—	—
Heptadecenyl acetate								
Octadecanyl acetate	18.00	18.00	18.00	18.00	18.00	18.00	18.00	18.00
cis-9-Octadecenyl acetate	18.41	18.62	18.54	18.35	18.33	18.36	18.34	18.30
cis,cis-9,12-Octadecadienyl acetate	18.73	19.45	19.17	18.86	18.90	18.90	18.88	18.78
cis,cis,cis-6,9,12-Octadecatrienyl acetate	—	20.10	19.68	19.22	19.28	19.16	—	—
cis,cis,cis-9,12,15-Octadecatrienyl acetate	19.52	20.64	20.05	19.56	19.68	19.64	19.64	19.57
cis,cis,cis-6,9,12,15-Octadecatetraenyl acetate	—	—	20.60	—	—	—	—	19.88
Icosanyl acetate	20.00	20.00	20.00	20.00	20.00	20.00	20.00	20.00
cis-11-Icosenyl acetate	—	20.60	20.48	20.25	20.25	20.30	20.33	20.22
cis,cis-11,14-Icosadienyl acetate	—	—	21.12	—	—	—	20.91	20.72
cis,cis,cis-8,11,14-Icosatrienyl acetate	—	—	21.59	—	—	—	21.30	21.01

	P1	P2	P3	P4	P5	P6	P7	P8
cis,cis,cis,cis-5,8,11,14-Icosatetraenyl acetate	22.00	—	—	—	—	22.08	21.54	21.33
cis,cis,cis,cis-8,11,14,17-Icosatetraenyl acetate	—	—	—	—	—	22.48	21.85	21.70
cis,cis,cis,cis,cis-5,8,11,14,17-Icosapentaenyl acetate	—	22.00	22.00	—	—	23.01	—	22.02
Docosanyl acetate	22.00	22.00	22.00	22.00	22.00	22.00	22.00	22.00
cis-11-Docosenyl acetate	—	—	—	—	—	22.36	22.28	22.10
cis,cis,cis,cis-7,10,13,16-Docosatetraenyl acetate	—	—	—	—	—	—	—	23.20
cis,cis,cis,cis,cis-4,7,10,13,16-Docosapentaenyl acetate	—	—	—	—	—	—	—	23.47
cis,cis,cis,cis,cis-7,10,13,16,19-Docosapentaenyl acetate	—	—	—	—	—	24.81	—	23.96
cis,cis,cis,cis,cis,cis-4,7,10,13,16,19-Docosahexaenyl acetate	—	—	—	—	—	25.16	24.17	24.15
cis-15-Tetracosenyl acetate	—	—	—	—	—	—	24.22	24.43

Column packing P1 = 30% Poly (ethylene glycol α-ketopimelate) on Sterchamol, Merck (0.25-0.3 mm)

P2 = 17% EGSS-X (ethylene glycol succinate polyester combined with methyl silicone) on Gas-Chrom P

P3 = EGSS-X (ethylene glycol succinate combined with methyl silicone) in form of wall coating in stainless steel tubing

P4 = 15.5% EGSS-Y (ethylene glycol succinate polyester combined with methyl silicone) on Gas-Chrom P

P5 = 10% Poly (ethylene glycol adipate) on Celite 545, AW (100-120 mesh)

P6 = 8% Poly (1,4-butanediol succinate) on Chromosorb W, HMDS (80-100 mesh)

P7 = SILAR-5CP (50% cyanopropyl, 50% phenyl, siloxane) in form of wall coating in stainless steel tubing

P8 = Poly (1,4-butanediol succinate) in form of wall coating in stainless steel tubing

REFERENCES

1. **Falk, F.**, *J. Chromatogr.*, 17, 450, 1965.
2. **Jamieson, G. R. and Reid, E. H.**, *J. Chromatogr.*, 26, 8, 1967.
3. **Jamieson, G. R. and Reid, E. H.**, *J. Chromatogr.*, 40, 160, 1969.
4. **Ratnayake, N. and Ackman, R. G.**, unpublished.

Table 4
FATTY ALDEHYDES IN POLAR (POLYESTER TYPE) COLUMNS (GC)

	P1	P2	P3	P4	P5	P6	P7	P8	P9	P10	P11	P12
Column packing Temperature (°C)	115	90	100	100	154	173	185	197	204	80—200	129	192
Gas	N_2	N_2	H_2	H_2	Ar	Ar	Ar	Ar	Ar	na	Ar	He
Flow rate (mℓ/min)	75	75	60	60	50	50	40	30	30	na	23	98
Column Length (cm)	200	200	150	150	142	142	142	142	142	366	305	305
Diameter (cm I.D.)	na	na	0.44	0.44	0.4	0.4	0.4	0.4	0.4	na	na	na
(cm O.D.)	0.64	0.64	0.64	0.64	na	na	na	na	na	0.64	na	0.64
Form	Helix	Helix	Helix	Helix	Tube	Tube	Tube	Tube	Tube	Helix	Tube	Helix
Material	Cu	Cu	SS	SS	G	G	G	G	G	G	G	SS
Detector	FID	FID	TC	TC	Argon	Argon	Argon	Argon	Argon	FID	Argon	TC
Reference	1	1	2	2	3	3	3	3	3	4	5	6
Data form	$r_{10:0}$	$r_{10:0}$	Kovats	Kovats	$r_{18:0}$	$r_{18:0}$	$r_{18:0}$	$r_{18:0}$	$r_{18:0}$	ECL	$r_{12:0}$	$r_{18:0}$
Compound	**ALD**	**ALD**	**I**	**I**	**ALD**	**ALD**	**ALD**	**ALD**	**ALD**	**ME[a]**	**ALD**	**ALD**
Ethanal	—	—	—	742	—	—	—	—	—	—	—	—
Propanal	—	—	860	830	—	—	—	—	—	—	—	—
Butanal	—	—	948	920	—	—	—	—	—	—	—	—
Pentanal	—	—	1050	1015	—	—	—	—	—	—	—	—
Hexanal	—	0.11	1148	1110	—	—	—	—	—	5.2	0.07	—
Octanal	0.24	0.33	1340	1300	—	—	—	—	—	—	0.10	—
Nonanal	0.55	0.57	1432	—	—	—	—	—	—	8.7[b]	0.26	—
Decanal	1.00	1.00	1537	1390	—	—	—	—	—	—	0.40	—
Undecanal	—	—	—	—	0.066	0.086	0.102	0.132	—	—	0.64	—
Dodecanal	—	—	—	—	—	—	—	—	—	—	1.00[c]	0.17
Tetradecanal	—	—	—	—	0.212	0.245	0.265	0.309	0.309	—	—	0.33
Hexadecanal	—	—	—	—	0.460	0.495	0.521	0.549	0.550	—	—	0.57
Octadecanal	—	—	—	—	1.00[d]	1.00[e]	1.00[f]	1.00[g]	1.00[h]	—	—	1.00[i]
cis-9-Octadecenal	—	—	—	—	—	—	—	—	—	—	—	1.17
cis-9,12-Octadecadienal	—	—	—	—	—	—	—	—	—	—	—	1.46

a ME = methyl ester of fatty acid.
b One ethylenic bond.
c Retention time 2.60 relative to methyl nonanoate.
d Retention time 0.755 relative to methyl octadecanoate.
e Retention time 0.792 relative to methyl octadecanoate.
f Retention time 0.816 relative to methyl octadecanoate.
g Retention time 0.836 relative to methyl octadecanoate.
h Retention time 0.868 relative to methyl octadecanoate.
i Retention time 0.81 relative to methyl octadecanoate.

Column packing
P1 = 10% Ethylene glycol adipate, linked with pentaerythritol, on Embacel
P2 = 10% Ethylene glycol adipate, linked with pentaerythritol, on Embacel
P3 = 20% Reoplex 400 (polyethylene glycol adipate polyester) on Chromosorb W (60-80 mesh)
P4 = 20% Poly (diethylene glycol adipate) on Chromosorb W (60-80 mesh)
P5 = 18% Poly (ethylene glycol adipate) on Celite, AW, ALK-W (120-140 mesh)
P6 = 18% Poly (ethylene glycol adipate) on Celite, AW, ALK-W (120-140 mesh)
P7 = 18% Poly (ethylene glycol adipate) on Celite, AW, ALK-W (120-140 mesh)
P8 = 18% Poly (ethylene glycol adipate) on Celite, AW, ALK-W (120-140 mesh)
P9 = 18% Poly (ethylene glycol adipate) on Celite, AW, ALK-W (120-140 mesh)
P10 = 5% LAC-2R-446 (commercial poly diethylene glycol adipate polyester) on Chromosorb W-AW-DCMS
P11 = 17% Poly (ethylene glycol succinate) on Chromosorb-DMCS (80-100 mesh)
P12 = 10% Poly (ethylene glycol succinate) on Gas-Chrom P (60-80 mesh)

REFERENCES

1. **Fredericks, K. M. and Taylor, R.,** *Anal. Chem.,* 38, 1961, 1966.
2. **d'Abrigeon, C., Maume, B., and Baron, C.,** *Bull. Soc. Chim. France,* 2329, 1967.
3. **Farquhar, J.,** *J. Lipid Res.,* 3, 21, 1962.
4. **Kleiman, R., Spencer, G. F., Earle, F. R., and Wolff, I. A.,** *Lipids,* 4, 135, 1969.
5. **Stein, R. A. and Nicolaides, N.,** *J. Lipid Res.,* 3, 476, 1962.
6. **Rao, P. V., Ramachandran, S., and Cornwell, D. G.,** *J. Lipid Res.,* 8, 380, 1967.

Table 5
FATTY ALDEHYDES IN POLAR (MISCELLANEOUS LIQUID PHASES) COLUMNS (GC)

Column packing	P1	P2	P3	P4	P5
Temperature (°C)	100	75-228	80	130	150
Gas	H_2	na	H_2	He	He
Flow rate (mℓ/min)	60	80	75	20	30
Column					
Length (cm)	150	na	300	330	152
Diameter					
(cm I.D.)	0.44	na	0.4	0.3	0.3
(cm O.D.)	0.64	na	na	na	na
Form	Helix	na	Helix	Helix	Helix
Material	SS	na	G	G	G
Detector	TC	TC	TC	FID	FID
Reference	1	2	3	4	4
Data form	Kovats	Kovats	$r_{5:0}$	$r_{6:0}$	$r_{12:0}$
Compound	**I**	**I**	**ALD**	**ALD**	**ALD**
Ethanal	—	—	0.09	—	—
Propanal	762	—	0.20	0.02	—
Butanal	845	866	0.44	0.08	—
Pentanal	950	—	1.00	0.28	—
Hexanal	1045	1080	—	1.00	0.03
Heptanal	1137	1184	—	—	0.10
Nonanal	—	1386	—	—	0.20
Decanal	—	1498	—	—	0.33
Undecanal	—	1605	—	—	—
Dodecanal	—	1710	—	—	1.00

Column packing	P1	=	5% Carbowax 600 on Fluoropak
	P2	=	25% Carbowax 20M on Celite
	P3	=	Dioctylphthalate
	P4	=	0.4% Carbowax 1500 on Carbopack A
	P5	=	10% Apiezon M on Chromosorb W (100-120 mesh)

REFERENCES

1. **d'Abrigeon, C., Maume, B., and Baron, C.,** *Bull. Soc. Chim. France,* 2329, 1967.
2. **Van den Pool, H. and Dec. Kratz, P.,** *J. Chromatogr.,* 11, 463, 1963.
3. **Nishi, S.,** *Tokyo Industrial Research Institute Report,* 60(9), 345, 1965.
4. **Ke, P. J., Ackman, R. G., and Linke, B. A.,** *J. Am. Oil Chem. Soc.,* 52, 349, 1975.

Table 6
DIMETHYL ACETALS OF ALDEHYDES IN POLAR (ETHYLENE GLYCOL ADIPATE) COLUMNS

Column packing	P1	P2	P3	P4	P5	P6	P7
Temperature (°C)	154	174	185	197	204	150	150
Gas	Ar	Ar	Ar	Ar	Ar	N_2	N_2
Flow rate (mℓ/min)	50	50	40	30	30	50	50
Column							
Length (cm)	142	142	142	142	142	150	150
Diameter							
(cm I.D.)	0.4	0.4	0.4	0.4	0.4	na	na
(cm O.D.)	na	na	na	na	na	0.40	0.40
Form	Tube	Tube	Tube	Tube	Tube	Helix	Helix
Material	G	G	G	G	G	G	G
Detector	Argon	Argon	Argon	Argon	Argon	FID	FID
References	1	1	1	1	1	2	2
Data form	$r_{18:0}$	$r_{18:0}$	$r_{18:0}$	$r_{18:0}$	$r_{18:0}$	$r_{18:0}$	r_{ECL}
Compound base	**DMA**	**DMA**	**DMA**	**DMA**	**DMA**	**DMA**	**DMA**
Undecanal	0.067	0.087	0.096	0.126	—	—	—
Dodecanal	—	—	—	—	—	0.10	12.0
Tetradecanal	0.208	0.250	0.265	0.300	0.304	0.213	14.0
Hexadecanal	0.459	0.495	0.518	0.545	0.549	0.463	16.0
cis-9-Hexadecenal	—	—	—	—	—	0.523	16.4
Octadecanal	1.00[a]	1.00[b]	1.00[c]	1.00[d]	1.00[e]	1.00[f]	18.0
cis-9-Octadecenal	—	—	1.10	—	—	1.13	18.4
cis,cis-9,12-Octadecadienal	—	—	1.32	—	—	1.35	—
cis,cis,cis-9,12,15-Octadecatrienal	—	—	—	—	—	1.70	—

[a] Retention time 0.879 relative to methyl octadecanoate.
[b] Retention time 0.872 relative to methyl octadecanoate.
[c] Retention time 0.867 relative to methyl octadecanoate.
[d] Retention time 0.874 relative to methyl octadecanoate.
[e] Retention time 0.875 relative to methyl octadecanoate.
[f] Retention time 1.86 relative to methyl hexadecanoate.

Column packing		
P1	=	18% Poly (ethylene glycol adipate) on Celite, AW, ALK-W (120-140 mesh)
P2	=	18% Poly (ethylene glycol adipate) on Celite, AW, ALK-W (120-140 mesh)
P3	=	18% Poly (ethylene glycol adipate) on Celite, AW, ALK-W (120-140 mesh)
P4	=	18% Poly (ethylene glycol adipate) on Celite, AW, ALK-W (120-140 mesh)
P5	=	18% Poly (ethylene glycol adipate) on Celite, AW, ALK-W (120-140 mesh)
P6	=	10% Poly (ethylene glycol adipate) on Gas Chrom CLH (80-100 mesh)
P7	=	10% Poly (ethylene glycol adipate) on Gas Chrom CLH (80-100 mesh)

REFERENCES

1. **Farquhar, J.,** *J. Lipid Res.,* 3, 21, 1962.
2. **Gray, G. M.,** in *Lipid Chromatographic Analysis,* Vol. 1, Marinetti, G. V., Ed., Marcel Dekker, New York, 1967, 401.

Table 7
DIMETHYL ACETALS, AND 1,2- AND 1,3- CYCLIC ACETALS, OF ALDEHYDES IN POLAR (MISCELLANEOUS LIQUID PHASES) COLUMNS (GC)

	P1	P2	P3	P4	P5	P6	P7	P8	P9
Column packing									
Temperature (°C)	192	173	175	150	150	186	192	192	185
Gas	He	N_2	N_2	N_2	N_2	He	He	He	He
Flow rate (mℓ/min)	98	20	20	50	50	na	98	98	na
Column									
Length (cm)	305	183	150	150	150	4575	305	305	305
Diameter (cm I.D.)	na	na	0.2	na	na	na	na	na	na
(cm O.D.)	0.64	0.318	na	0.40	0.40	0.025	0.64	0.64	0.318
Form	Helix	Helix	na	Helix	Helix	WCOT	Helix	Helix	Helix
Material	SS	SS	G	G	G	SS	SS	SS	SS
Detector	TC	FID	FID	FID	FID	FID	TC	TC	FID
Reference	1	2	3	4	4	5	1	1	1
Data form	$r_{18:0}$	r_{ECL}	$r_{18:0}$	$r_{18:0}$	ECL	$r_{18:0}$	$r_{18:0}$	$r_{18:0}$	$r_{18:0}$
Compound base	DMA	DMA	DMA	DMA	DMA	DMA	1,2-CA	1,3-CA	1,3-CA
Dodecanal	0.18	—	—	0.12	12.0	—	0.18	0.19	0.19
Tetradecanal	0.29	—	0.31	0.25	14.0	—	—	0.32	0.33
Hexadecanal	0.55	16.0	0.55	0.49	16.0	0.489	0.57	0.58	0.58
cis-9-Hexadecenal	—	—	—	0.57	16.4	0.529	—	—	—
Octadecanal	1.00[a]	18.0	1.00	1.00[b]	18.0	1.00	1.00[c]	1.00[d]	1.00
cis-9-Octadecenal	—	18.5	1.11	1.16	18.4	1.05	1.13	1.15	1.16
cis-11-Octadecenal	—	—	—	—	—	1.08	—	—	—
cis,cis-9,12-Octadecadienal	—	19.1	1.31	—	—	—	1.40	1.39	—
cis,cis,cis-9,12,15-Octadecatrienal	—	—	—	—	—	—	—	—	—

[a] Retention time 0.84 relative to methyl octadecanoate.
[b] Retention time 1.61 relative to methyl hexadecanoate.
[c] Retention time 1.68 relative to methyl octadecanoate.
[d] Retention time 2.27 relative to methyl octadecanoate.

Column packing
P1 = 10% poly (ethylene glycol succinate) on Gas-Chrom P (60-80 mesh)
P2 = 20% Analabs C6 stabilized poly (diethylene glycol succinate) on Anakrom ABS (100-110 mesh)
P3 = 10% EGSS (ethylene glycol succinate poly ester combined with methyl silicone) on Gas-Chrom P (120-150 mesh)
P4 = 15% EGSS-X (ethylene glycol succinate polyester combined with methyl silicone) on Gas Chrom CLH (80-100 mesh)
P5 = 15% EGSS-X (ethylene glycol succinate polyester combined with methyl silicone) on Gas Chrom CLH (80-100 mesh)
P6 = Carbowax K20-M plus 1% V-930
P7 = 10% Poly (ethylene glycol succinate) on Gas-Chrom P (60-80 mesh)
P8 = 10% Poly (ethylene glycol succinate) on Gas-Chrom P (60-80 mesh)
P9 = 15% Poly (ethylene glycol succinate) on Gas-Chrom P (60-80 mesh)

REFERENCES

1. **Rao, P. V., Ramachandran, S., and Cornwell, D. G.,** *J. Lipid Res.,* 8, 380, 1967.
2. **Sun, G. Y. and Horrocks, L. A.,** *Lipids,* 3, 79, 1968.
3. **Reisch, J., Münnighoff, C., and Möllmann, H.,** *Arch. Pharmaz.,* 306, 173, 1973.
4. **Gray, G. M.,** in *Lipid Chromatographic Analysis,* Vol. 1, Marinetti, G. V., Ed., Marcel Dekker, New York, 1967, 401.
5. **Goldfine, H. and Panos, C.,** *J. Lipid Res.,* 12, 214, 1971.

Table 8
FATTY ALDEHYDES (P1, P2), DIMETHYL ACETALS (P3-P8), IN NONPOLAR COLUMNS

	P1	P2	P3	P4	P5	P6	P7	P8
Column packing	P1	P2	P3	P4	P5	P6	P7	P8
Temperature (°C)	75-228	80-200	174	197	185	190	190	197
Gas	na	na	Ar	Ar	Ar	Ar	Ar	Ar
Flow rate (ml/min)	na	na	na	112	na	100	100	112
Column Length (cm)	na	122	na	na	na	120	120	183
Diameter (cm I.D.)	na	na	na	na	na	na	na	na
(cm O.D.)	na	0.64	na	na	na	0.40	0.40	0.64
Form	na	Helix	Tube	Tube	Tube	Tube	Tube	Tube
Material	na	G	G	G	G	G	G	G
Detector	TC	FID	Argon	Argon	Argon	Argon	Argon	FID
Reference	1	2	3	3	3	4	4	5
Data form	Kovats	r_{ECL}	$r_{18:0}$	$r_{18:0}$	$r_{18:0}$	$r_{18:0}$	r_{ECL}	$r_{18:0}$

Compound or Compound Base	I	ME[a]	DMA	DMA	DMA	DMA	DMA	DMA
Hexanal	—	5.2	—	—	—	—	—	—
Heptanal	895	—	—	—	—	—	—	—
Octanal	—	6.6	—	—	—	—	—	—
Nonanal	1091	7.9[b]	—	—	—	—	—	—
Decanal	1193	8.7	—	—	—	—	—	0.046
Undecanal	1296	10.0	—	—	—	—	—	0.069
Dodecanal	1397	11.2	—	—	—	—	—	0.090
Tetradecanal	—	—	0.147	0.180	0.162	0.071	12.0	0.177
Hexadecanal	—	—	0.385	0.424	0.405	0.172	14.0	0.420
cis-9-Hexadecenal	—	—	—	0.361	—	0.418	16.0	0.362
Octadecanal	—	—	1.00[c]	1.00[c]	1.00[c]	1.00[d]	18.0	1.00
cis-9-Octadecenal	—	—	—	0.847	—	0.845	17.6	0.869
cis,cis-9,12-Octadecadienal	—	—	—	0.782	—	0.791	17.45	0.780
cis,cis,cis-9,12,15-Octadecatrienal	—	—	—	—	—	0.791	17.45	—

a   ME = methyl ester of fatty acid.
b   One ethylenic bond; P2 ECL = 7.8.
c   Retention time 1.28 relative to methyl octadecanoate.
d   Retention time 3.08 relative to methyl hexadecanoate.

Column packing   P1 = 25% SE-30 (methyl silicone) on Celite
                   P2 = 5% Apiezon-L on Chromosorb W, AW, DMCS
                   P3 = 14% Apiezon-M on Celite AW, ALK-W (140-170 mesh)
                   P4 = 14% Apiezon-M on Celite AW, ALK-W (140-170 mesh)
                   P5 = 14% Apiezon-M on Celite AW, ALK-W (140-170 mesh)
                   P6 = 12.5% Apiezon-L on Gas Chrom P, ALK-W (80-100 mesh)
                   P7 = 12.5% Apiezon-L on Gas Chrom P, ALK-W (80-100 mesh)
                   P8 = 16% Apiezon-M on Gas Chrom S, Alk-W, HMDS (60-80 mesh)

REFERENCES

1. **Van den Pool, H. and Dec. Kratz, P.**, *J. Chromatogr.*, 11, 463, 1963.
2. **Kleiman, R., Spencer, G. F., Earle, F. R., and Wolff, I. A.**, *Lipids*, 4, 135, 1969.
3. **Farquhar, J. W.**, *J. Lipid Res.*, 3, 21, 1962.
4. **Gray, G. M.**, in *Lipid Chromatographic Analysis*, Vol. 1, Marinetti, G. V., Ed., Marcel Dekker, New York, 1967, 401.
5. **Gilbertson, J. R., Ferrell, W. J., and Gelman, R. A.**, *J. Lipid Res.*, 8, 38, 1967.

Table 9
ALDEHYDE ESTERS (AE; P1-P3), ALKOXY ACETALDEHYDES (ALK-AC; P4, P5), DINITROPHENYLHYDRAZONES (DNPH; P6-P9); AND OXIMES (P10) AND DIALDEHYDES (P11, P12) ON VARIOUS LIQUID PHASES

	P1	P2	P3	P4	P5	P6	P7	P8	P9	P10	P11	P12
Column packing												
Temperature (°C)	183	80-200	80-200	160	190	205	245	205	245	88	80-200	80-200
Gas	Ar	na	na	Ar	Ar	Ar	Ar	Ar	Ar	He	na	na
Flow rate (ml/min)	28.3	na	na	80	80	10-45	10-45	10-14	10-14	35	na	na
Column												
Length (cm)	91	366	122	183	183	100	100	100	100	200	366	122
Diameter (cm I.D.)	na	na	na	na	na	0.2	0.2	0.2	0.2	na	na	na
Diameter (cm O.D.)	na	0.64	0.64	0.64	0.64	na	na	na	na	0.8	0.64	0.64
Form	Tube	Helix	Helix	Tube	Tube	Helix	Helix	Helix	Helix	na	Helix	Helix
Material	G	G	G	G	G	SS	SS	SS	SS	na	G	G
Detector	Argon	FID	FID	FID	FID	FID	FID	FID	FID	TC	FID	FID
Reference	1	2	2	3	3	4	4	4	4	5	2	2
Data form	$r_{14:0}$	ECL	ECL	$r_{18:0}$	$r_{18:0}$	Kovats	Kovats	Kovats	Kovats	$r_{4:0}$	ECL	ECL
Compound (or derivative, see above)	AE	ME[a]	ME[a]	ALK-AC	ALK-AC	I	I	I	I	Oxime	ME[a]	ME[a]
Ethanal	—	—	—	—	—	2198	2251	2302	2370	0.37	—	—
Propanal	—	—	—	—	—	2290	2336	2385	2438	0.56	8.6	—
Butanal	—	—	—	—	—	2378	2428	2472	2522	1.00	9.6	—
Pentanal	—	10.9	6.6	—	—	2472	2513	2569	2624	—	10.8	5.8
Hexanal	0.10	12.0	7.6	—	—	2568	2611	2666	2719	—	11.4	6.7
Octanal	0.18	13.8	9.5	—	—	2763	2802	2859	2910	—	13.9	8.8
Nonanal	0.24	15.2[b]	10.7[b]	—	—	2857	2897	2955	3004	—	—	—
Decanal	0.32	16.2	11.7	0.074	0.063	2951	2994	3050	3096	—	15.3	10.6
Undecanal	0.43	17.3	12.7	—	—	3052	3099	3154	3202	—	—	—
Dodecanal	0.57	—	—	0.129	0.102	—	3202	—	—	—	—	—
Tetradecanal	1.00	20.2	15.6	0.254	0.217	—	—	—	—	—	—	—
Hexadecanal	—	—	—	0.502	0.466	—	—	—	—	—	—	—
Octadecanal	—	—	—	1.00	1.00	—	—	—	—	—	—	—
cis-9-Octadecenal	—	—	—	1.150	0.880	—	—	—	—	—	—	—

[a]   ME = methyl ester of fatty acid.
[b]   One ethylenic bond; P2 ECL = 15.4; P3 ECL = 10.5.
[c]   10% Phenyl, 90% methyl, polysiloxane.
[d]   20% Phenyl, 80% methyl, polysiloxane.

Column packing

P1	=	12% Poly (ethylene glycol succinate) on Chromosorb-HMDS (60-100 mesh)
P2	=	5% LAC-2R-446 (commercial poly diethylene glycol adipate polyester) on Chromosorb W-AW-DCMS
P3	=	5% Apiezon-L on Chromosorb-W, AW, DMCS
P4	=	15% Poly (diethylene glycol succinate) on Anakrom-A (100-120 mesh)
P5	=	3.5% SE-30 (methyl silicone) on Gas-Chrom-S (80-100 mesh)
P6	=	1% OV-3[c] on Chromosorb G, AW-DMCS, H.P. (80-100 mesh)
P7	=	1% OV-3[c] on Chromosorb G, AW-DMCS, H.P. (80-100 mesh)
P8	=	1% OV-7[d] on Chromosorb G, AW-DMCS, H.P. (80-100 mesh)
P9	=	1% OV-7[d] on Chromosorb G, AW-DMCS, H.P. (80-100 mesh)
P10	=	3% Di-2-ethylhexyl-phthalate on Celite firebrick (30-60 mesh)
P11	=	5% LAC-2R-446 (commercial polydiethyleneglycoladipate polyester) on Chromosorb W-AW-DCMS
P12	=	5% Apiezon-L on Chromosorb W-AW-DCMS

REFERENCES

1. **Stein, R. A. and Nicolaides, N.,** *J. Lipid Res.,* 3, 476, 1962.
2. **Kleiman, R., Spencer, G. F., Earle, F. R., and Wolff, I. A.,** *Lipids,* 4, 135, 1969.
3. **Gelman, R. A. and Gilbertsen, J. R.,** *Anal. Biochem.,* 31, 463, 1969.
4. **Pias, J. B. and Gasco, L.,** *Chromatographia,* 8, 270, 1975.
5. **Cason, J. and Harris, E. R.,** *J. Org. Chem.,* 24, 676, 1959.

Table 10
METHYL KETONES

Column packing	P1	P2	P3	P4
Temperature (°C)	160	180	190	210
Gas	N_2	N_2	N_2	N_2
Flow rate (mℓ/min)	na	na	na	na
Column				
Length (cm)	183	183	183	183
Diameter	na	na	na	na
(cm I.D.)				
(cm O.D.)	0.318	0.318	0.318	0.318
Form	Helix	Helix	Helix	Helix
Material	SS	SS	SS	SS
Detector	FID	FID	FID	FID
Reference	1	1	1	1
Data form				

Compound	ECL (Based on 2-Alkanes)			
cis-Nonadec-10-en-2-one	18.20	18.49	18.60	12.58
cis,cis-Nonadec-10,13-dien-2-one	18.57	19.22	18.48	19.70
cis,cis,cis-Nonadec-7,10,13-trien-2-one	18.88	19.87	20.12	20.41
cis,cis,cis-Nonadec-10,13,16-trien-2-one	19.19	20.21	20.53	20.78
cis,cis,cis,cis-Nonadec-7,10,13,16-tetraen-2-one	19.51	20.85	21.24	21.51
cis,cis-Henicos-12,15-dien-2-one	20.57	21.21	21.41	21.59
cis,cis,cis-Henicos-9,12,15-trien-2-one	20.76	21.65	21.99	22.16
cis,cis,cis-Henicos-12,15,18-trien-2-one	21.18	22.18	22.49	22.65
cis,cis,cis,cis-Henicos-6,9,12,15-tetraen-2-one	20.93	22.18	22.60	22.88
cis,cis,cis,cis-Henicos-9,12,15,18-tetraen-2-one	21.36	22.71	23.10	23.41
cis,cis,cis,cis,cis-Henicos-6,9,12,15,18-pentaen-2-one	21.51	23.12	23.57	23.98
cis,cis,cis,cis-Tricos-8,11,14,17-tetraen-2-one	22.82	24.07	24.61	24.97
cis,cis,cis,cis,cis-Tricos-5,8,11,14,17-pentaen-2-one	22.99	24.48	24.94	25.34
cis,cis,cis,cis,cis-Tricos-8,11,14,17,20-pentaen-2-one	23.44	25.07	25.61	25.88
cis,cis,cis,cis,cis,cis-Tricos-5,8,11,14,17,20-hexaen-2-one	23.61	25.40	25.93	26.51
cis-Nonadec-10-en-7-yn-2-one	20.05	21.09	21.59	21.79
cis,cis-Nonadec-10,13-dien-7-yn-2-one	20.45	21.91	22.43	22.81
cis,cis,cis-Nonadec-10,13,16-trien-7-yn-2-one	21.06	22.96	23.51	23.96
cis,cis-Henicos-12,15-dien-9-yn-2-one	22.26	23.79	24.37	24.65
cis,cis,cis-Henicos-12,15,18-trien-9-yn-2-one	23.03	24.71	25.28	25.80

Column packing	P1	=	5% Poly (1,4-butanediol succinate) on Chromosorb W, HMDS
	P2	=	5% Poly (diethylene glycol succinate) on Chromosorb W, HMDS
	P3	=	10% EGSS-X (ethylene glycol succinate combined with methyl silicone) on Chromosorb W, HMDS
	P4	=	10% EGSS-X (ethylene glycol succinate combined with methyl silicone) on Chromosorb W, HMDS

REFERENCE

1. **Jamieson, G. R., McMinn, A. L., and Reid, E. H.,** *J. Chromatogr.*, 121, 327, 1978.

Note added in proof: For ECL data for very long chain (C_{37} - C_{39}) unsaturated methyl and ethyl ketones of interest to marine geologists and biochemists, on a dimethyl silicone (OV-1) wall-coated open-tabular column, see **Volkmann, J. K., Eglinton, G., Corner, E. D. S., and Forsberg, T.E.V.,** *Phytochemistry,* 19, 2619, 1980.

REFERENCES

1. **Birch, M. C., Ed.,** *Pheromones,* Elsevier, New York, 1974.
2. **Brand, J. M., Young, J. Chr., and Silverstein, R. M.,** in *Progress in the Chemistry of Organic Natural Products,* Vol. 37, Herz, W., Grisebach, H., and Kirby, G. W., Eds., Springer-Verlag, New York, 1979, 1.
3. **Kolattukudy, P. E., Ed.,** *Chemistry and Biochemistry of Natural Waxes,* Elsevier, Amsterdam, 1976.
4. **Mahadevan, V.,** in *Progress in the Chemistry of Fats and Other Lipids,* Vol. 15, Holman, R. T., Ed., Pergamon Press, Oxford, 1978, 255.
5. **Mahadevan, V.,** in *Progress in the Chemistry of Fats and Other Lipids,* Vol. 11, Holman, R. T., Ed., Pergamon Press, Oxford, 1970, 81.
6. **Forney, F. W. and Markowetz, A. J.,** *J. Lipid Res.,* 12, 383, 1971.
7. **Shriner, R. L., Fuson, R. C., and Curtin, D. Y.,** *Systematic Identification of Organic Compounds,* 5th ed., John Wiley & Sons, New York, 1964.
8. **Feigl, F.,** *Spot Tests in Organic Analysis,* 6th ed., Elsevier, Amsterdam, 1960.
9. **Farquhar, J. W.,** *J. Lipid Res.,* 3, 21, 1962.
10. **Totani, N. and Muramatsu, T.,** *Chem. Phys. Lipids,* 29, 121, 1981.
11. **Mahadevan, V., Viswanathan, C. V., and Phillips, F.,** *J. Lipid Res.,* 8, 2, 1967.
12. **Viswanathan, C. V.,** *Chromatogr. Rev.,* 10, 18, 1968.
13. **Gray, G. M.,** in *Lipid Chromatographic Analysis,* Vol. 1, Marinetti, G. V., Ed., Marcel Dekker, New York, 1967, 401; 2nd ed., Vol. 3, 1976, 897.
14. **Rao, P. V., Ramachandran, S., and Cornwell, D. G.,** *J. Lipid Res.,* 8, 380, 1967.
15. **Gray, G. M.,** *J. Chromatogr.,* 6, 236, 1961.
16. **Wood, R. and Harlow, R. D.,** *J. Lipid Res.,* 10, 463, 1969.
17. **Neudoerffer, T. S.,** *Chem. Phys. Lipids,* 1, 341, 1967.
18. **Evans, M. B. and Williamson, B.,** *Chromatographia,* 5, 264, 1972.
19. **Schmid, H. H. O. and Mangold, H. K.,** *Biochem. Z.,* 346, 13, 1966.
20. **Bandi, Z. L.,** *Chem. Phys. Lipids,* 3, 409, 1969.
21. **Kates, M.,** *Techniques of Lipidology,* Elsevier, New York, 1972, 522.
22. **Tuna, N. and Mangold, H. K.,** in *Evolution of the Atherosclerotic Plaque,* Jones, R. D., Ed., University of Chicago Press, Ill., 1963, 85.
23. **Schmid, H. H. O. and Takahashi, T.,** *Hoppe-Seyler's Z. Physiol. Chem.,* 349, 1673, 1968.

Chapter 6

STRAIGHT-CHAIN FATTY ACIDS

R. G. Ackman

I. INTRODUCTION

Gas chromatography (GC) is by far the most efficient technique for the identification and quantitation of fatty acids in complex mixtures. Methyl esters rather than the unesterified fatty acids are generally suitable for separation by GC.

The present chapter deals with the GC of methyl esters of a great variety of naturally occurring, as well as synthetic, fatty acids. The structures of the common naturally occurring unsaturated fatty acids are shown in the following scheme.

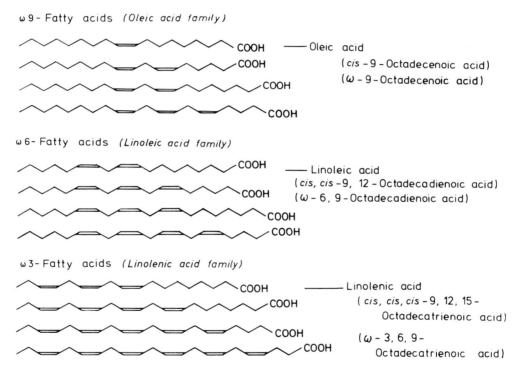

ω9- Fatty acids *(Oleic acid family)*

COOH —— Oleic acid
COOH (*cis* – 9 – Octadecenoic acid)
COOH (ω – 9 – Octadecenoic acid)
COOH

ω6- Fatty acids *(Linoleic acid family)*

COOH —— Linoleic acid
COOH (*cis, cis* –9, 12 – Octadecadienoic acid)
COOH (ω – 6, 9 – Octadecadienoic acid)
COOH

ω3- Fatty acids *(Linolenic acid family)*

COOH —— Linolenic acid
COOH (*cis, cis, cis* – 9, 12, 15 –
COOH Octadecatrienoic acid)
COOH (ω – 3, 6, 9 –
Octadecatrienoic acid)

As a rule, ω-7-fatty acids (palmitoleic acid family), such as palmitoleic acid (*cis*-9-hexadecenoic acid), occur in small proportions. Fatty acids having a *trans*-double bond or a triple bond are quite unusual. The same is true for polyunsaturated fatty acids whose double bonds are separated by more than one methylene group.

Comprehensive chromatographic data are presented on the separation of methyl esters on a large number of stationary phases in both packed and open tubular columns; the gas chromatography of some unesterified fatty acids is also considered.

Several terms, used commonly to express chromatographic data, are critically evaluated.

II. PREPARATION OF SAMPLES

Methods for the esterification of fatty acids and the transesterification of lipids are described in Chapter 2. It is mandatory to carry out, prior to GC, a preliminary fractionation of the methyl esters into classes having different functional groups, such as hydroxy, epoxy, and keto groups (see Chapter 9). In addition, fractionation into groups of vinylogs, either by argentation chromatography of the methyl esters or by adsorption chromatography of their mercuric acetate adducts, can be highly useful.

It should be emphasized that the gas chromatographic behavior of a methyl ester on one or several stationary phases is not the ultimate proof of its identity. Complementary techniques, such as ultraviolet (UV), infrared (IR), nuclear magnetic resonance (NMR), and mass spectrometry (MS) should, as a rule, be exploited to obtain additional evidence of identity. GC coupled with MS, though a highly expensive setup, should be the method of choice for the identification of methyl esters in complex mixtures, such as those derived from many naturally occurring lipids. Hydrogenation followed by GC is a useful and simple check.

Procedure for the Purification of Methyl Esters by Adsorption Thin-Layer Chromatography
Mixtures of crude methyl esters obtained by the esterification of fatty acids or by trans-esterification of ester lipids are dissolved in hexane at a ratio of 1:10. Up to 0.5 mℓ of this solution is applied as a band to a 0.5 mm thick silica gel layer, 20 × 20 cm. The chromatoplate is developed two or three times with hexane-diethyl ether (95:5 v/v) and the fractions visualized in UV light after spraying the layer with 2',7'-dichlorofluorescein solution (Chapter 28, Reagent No. 8). The methyl esters of the common fatty acids migrate further than those of oxygenated fatty acids, and further than other products, such as dimethylacetals and alkyl glycerols. Diethyl ether, saturated with water, is used to elute the methyl esters from the adsorbent. The solution is filtered through sintered glass, and the ether is evaporated. The purified methyl esters are dissolved in hexane and stored under nitrogen at −5°C or lower.

Procedure for the Fractionation of Methyl Esters into Groups of Vinylogs by Argentation Thin-Layer Chromatography (TLC)
Complex mixtures of purified methyl esters can be fractionated according to the number and configuration of double bonds by chromatography on layers of silica gel containing 5 to 10% silver nitrate. Up to 0.5 mℓ of a 10% solution of the methyl ester in hexane is applied as a band to a 0.5 mm thick silica gel layer, 20 × 20 cm. The chromatoplate is developed once or twice with hexane-diethyl ether (95:5 v/v) for the resolution of the methyl esters of saturated, monounsaturated, and diunsaturated fatty acids, or with diethyl ether-methanol (90:10 v/v) for the fractionation of the methyl esters of more highly unsaturated fatty acids. The fractions are visualized in UV light after spraying the layer with 2',7'-dichlorofluorescein solution (Chapter 28, Reagent No. 8). The methyl esters of saturated fatty acids migrate further than those of mono-, di-, tri-, and tetraunsaturated fatty acids. Diethyl ether, saturated with 2 N hydrochloric acid solution, is used to elute each of these groups of methyl esters. The solutions are filtered through sintered glass, washed with water, and dried over anhydrous sodium sulfate. After decantation, the ether is evaporated, and the various groups of methyl esters are dissolved in hexane and stored under nitrogen at −5°C or lower.

It must be emphasized that argentation chromatography does not only resolve the methyl esters of fatty acids differing in the number of double bonds, but also the esters of *cis-* and *trans*-isomeric fatty acids. In contrast, the following procedure is suitable for fractionating the methyl esters of fatty acids strictly according to the number of double bonds.

Procedure for the Fractionation of Methyl Esters as Mercuric Acetate Adducts
Mercuric acetate adducts are prepared as described in Chapter 2. Up to 0.2 mℓ of a 10% solution of mercuric acetate adducts in chloroform is applied as a band on a 0.3 mm thick

layer of silica gel, 20 × 20 cm. The chromatoplate is developed with hexane-dioxane
(3:2 v/v). The methyl esters of saturated fatty acids, which migrate close to the solvent
front, are detected in UV light after spraying the plates with 2',7'-dichlorofluorescein
solution (Chapter 28, Reagent No. 8), and the mercuric acetate adducts of methyl esters
of unsaturated fatty acids are visualized by spraying with s-diphenylcarbazone solution
(Chapter 28, Reagent No. 14). The methyl esters of saturated fatty acids are eluted with
water-saturated diethyl ether; whereas the various adduct fractions, on the adsorbent, are
shaken with 10 mℓ of methanol and 0.5 mℓ concentrated hydrochloric acid in an atmosphere of nitrogen. After standing for a few minutes, the solution is decanted and filtered
through sintered glass. The residue is treated a second time with methanol-hydrochloric
acid. The combined extracts are diluted with 25 mℓ of water and extracted once with 25
mℓ and then four times with successive 10 mℓ portions of diethyl ether. The etheral solution is dried over anhydrous sodium sulfate. After decantation, the solvent is evaporated,
and the residual methyl esters are dissolved in hexane and stored under nitrogen at −5°C
or lower.

III. CHROMATOGRAPHY

The most widely used standard methods of identification of fatty acids in gas-liquid
chromatography (GLC) are all oriented to comparisons of time of emergence of the peak to
standards. There are several ways of exploiting the time element.

Retention time (t_R) is the time from injection of the sample to the emergence of the peak
(Figure 1). Traditionally this time of emergence of the peak is measured at the peak apex.
Actual measurement on a recorder chart at the peak apex is especially convenient for sharp
peaks, but with electronic integrators the reversal of slope can be accurately measured and
is usually conveniently printed out on the chart near the peak. Early workers in gas chromatography (GC) tabulated absolute retention times for publication but it was soon realized
that the inaccuracies inherent in control of column temperature and carrier gas flow rate
made reproducibility even within an operating day very dubious. Column aging was also a
factor resulting in a progressive shortening of retention time. Currently, the combination of
modern proportional temperature controllers, higher quality gas pressure regulators and flow
system components, and thermostable liquid phases, with data processor or computer time
recording and printing, gives retention times with a higher degree of reproducibility. Nonetheless, to allow data processors to identify components, the time "window" must be
widened appreciably for later-eluting components to accommodate unpredictable variations
in retention times.

Adjusted retention time (t_R') is the retention time for component R measured from the
"air" peak. Thus the total dead volume of the system (originally volume V of carrier gas
was used as a parameter instead of time) is eliminated. For practical purposes, t_R' is usually
measured from the front of the solvent peak. The difference between t_R and t_R' is not critical
with packed columns; but with open-tubular columns, a time difference of 1, 2, or 3 min
may be observed. This can affect theoretical plate calculations (Figure 1). With a polar wall-
coated open-tubular column, 47 m in length and 0.25 mm I.D., operated at 175°C and 60
psig helium (ca. 5 mℓ/min), $t_R - t_R'$ was 0.78 min; for methyl hexadecanoate (methyl
palmitate), theoretical plates (n) based on t_R were 25,000 while "effective" theoretical plates
(N) based on t_R' were 18,000. Either time measurement (t_R or t_R') may be used for identification provided it is specified. Unfortunately, it is often impossible to tell from publications
which has been employed.

A more critical source of error can be the shape of the peak. All overloaded GC peaks
tend to be skewed, with steepening of the trailing edges.[1] The position of the peak top is
retarded accordingly. The slope of the leading edges also can change, but *the intercept of
the frontal tangent with the baseline is almost invariant* (Figure 2) as discussed elsewhere.[2,3]
It is therefore desirable, if possible, to measure retention times to the intercept of the frontal

tangent and the baseline. Alternatively, in lieu of the peak top, some data processors or computers can be programmed to register the time as the start of integration, which is nominally the same point. This reduces error due to skewing. A secondary problem, which can affect certain components of like structure in open-tubular column work, is the "load effect",[1,2] in which the minor, later eluting, component is retarded as shown for methyl *cis*-15-docosenoate (22:1ω7)* in Figure 2. This is not normally observed with packed columns. Nonetheless, it is desirable that standards be run at the same on-column loads as sample components.[3-5] It is also desirable to run these daily, or more often in close time proximity to analyses. Relative retention time (r_x) is simply the time for any component (r) divided by the time for a suitable reference component (x). This process tends to smooth out differences in retention time arising from variations in operating conditions occurring from one day to another, or within one day, or even within one analytical run. Tabulated results typical of this problem in the case of the $r_{18:0}$ values for methyl hexadecanoate have been published.[3] Not all problems reflect sample size, but newer gas-liquid chromatographic systems in good condition, with detent temperature settings and proportional temperature control can reduce such effects. It is still surprising to find reputable companies installing gas-flow controllers and pressure regulators in the vicinity of the oven without adequate isolation or insulation. Gradual warming will undoubtedly cause slight but important shifts in retention time.

In the analysis of methyl esters of fatty acids the reference methyl ester originally chosen was 16:0 (methyl hexadecanoate or palmitate), or sometimes 14:0 (methyl tetradecanoate or myristate). These eluted fairly early in most mixtures, especially those originating in animal biochemistry, which often included methyl docosahexaenoate (22:6ω3)*; however, the short retention time of 14:0 increased identification errors. Accordingly, 18:0 (methyl octadecanoate or stearate) became the most widely used reference methyl ester, although some workers used 18:1 (methyl *cis*-9-octadecenoate or oleate) when the 18:0 peak tended to be small or masked by other components.

Relative retention times are notoriously ineffective in correcting among compounds with *different* chain lengths for column aging, a process generally resulting in loss of the liquid phase. Aging of a polyester phase usually yields separations similar to those of a less polar column combination;[3,6,7] the polarity is obviously decreased (see discussion of column and operating variables under "Column Variables," p. 101).

A. Equivalent Chain Lengths (Carbon Numbers)

An inherent problem with relative retention times is that visualization of where component x actually falls in a strip-chart recording is easy only when x is close to the reference component. Tabulation of data or discussion of retention material also require one or more extra entries to specify the reference methyl ester. Independently, two groups[8-10] adapted the Kovats Index system,[11] based on four digits and hydrocarbons where the *n*-alkyl chains were 1200, 1300, etc. to methyl esters of fatty acids under the names "equivalent chain length" (ECL) or "carbon number" (CN). Thus the methyl esters of fatty acids are given two whole digits based on the actual chain length of the fatty acid in question followed by two decimal places in most instances, relating the peak position on the chart to the peaks for the methyl esters of any appropriate straight-chain saturated fatty acids (Figure 3). The system is open ended, but the six commonly occurring and readily available straight chain saturated fatty acids 14:0, 16:0, 18:0, 20:0, 22:0, and 24:0 cover most of the range of fatty acids found in animal and plant lipids. An advantage of this system is that at least three of

---

* A shorthand notation for chain length, number of ethylenic bonds, and position of the double bond closest to the terminal methyl group is often used in the literature for monoethylenic fatty acids and *cis*-methylene-interrupted fatty acids. In this chapter only the positions relative to the carboxyl group will be given for unsaturated acids, but the shorthand chain length:number of ethylenic bonds will be used in referring to methyl esters of saturated acids.

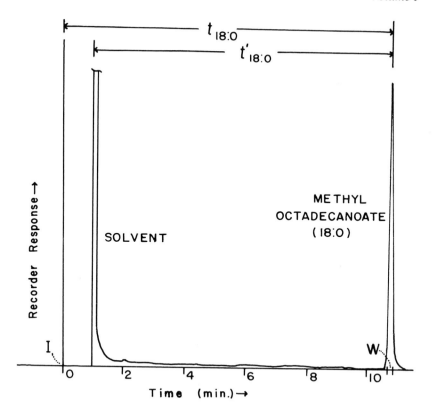

FIGURE 1. Reproduction of the gas-liquid chromatographic recording from an analysis of methyl octadecanoate (18:0) on a wall-coated open-tubular column coated with Apiezon-L, 4700 cm in length and 0.025 cm I.D., temperature 190°C, carrier gas helium at 50 psig (3.5 kg/cm²). The actual injection took place at point I (time 0), and retention time $t_{18:0}$ is 10.4 min (to peak apex). The "adjusted" retention time $t_{18:0}'$ is measured from the first emergence of solvent (hexane) and is 9.4 min. The time (W) between peak tangent intercepts with baseline is 0.172 min. Theoretical plates N are therefore

$$\left(\frac{10.4}{0.172}\right)^2 \times 16 = 58,500$$

and "effective" theoretical plates *N* are

$$\left(\frac{9.4}{0.172}\right)^2 \times 16 = 47,800$$

both being stipulated for methyl octadecanoate and the operating conditions specified.

14:0, 16:0, 18:0, and 20:0 are often found in natural mixtures of plant or animal origin. As shown in Figure 4, the method of calculation of equivalent chain lengths (to avoid confusion, carbon numbers[8] will not be listed in this presentation) follows from the long-established concept that the logarithms of the retention times of a series of saturated aliphatic chains plotted against the numbers of carbon atoms in the chains give a straight line. The degree of accuracy is such that two-cycle semilogarithmic graph paper allows the first decimal place to be accurately read and the second estimated to an accuracy of within ±0.02 ECL units. This concept is discussed in detail elsewhere,[3,6,7,12] but it must be noted that in-depth research now suggests that a straight line relationship is only a good approximation and that other mathematical relationships (e.g., quadratic curves[13]) may have to be used with sophisticated data processing. In practice, plotting can still be used in awkward cases of nonlinearity by inflecting straight lines after four carbons (e.g., 14:0 to 18:0, 18:0 to 22:0), or as often as

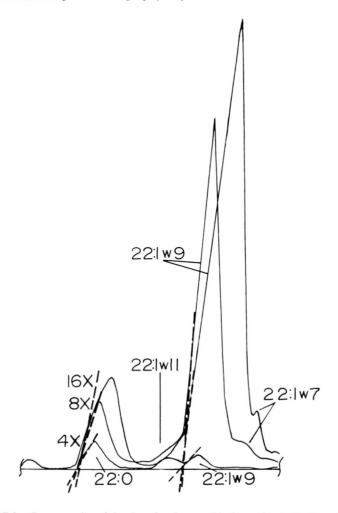

FIGURE 2.   Demonstration of skewing of peaks caused by increasing the load on a WCOT column coated with SILAR-5CP operated at 180°C and 3.5 kg/cm² (also observed on the same peak for BDS coatings). Note that 22:0 and methyl *cis*-13-docosenoate (22:1 ω 9) frontal tangent intercepts with baseline are not affected by skewing in the absence of a high proportion of a preceding similar isomer, but the methyl *cis*-15-docosenoate (22:1 ω 7) is retarded by the "load effect" discussed in the text. The smaller, earlier-eluting, methyl *cis*-11-docosenoate (22:1 ω 11) is not affected. (Reproduced by permission of *Fette Seifen Anstrichmittel.*)

necessary. This should be noted in data publication. It must be emphasized that ECL values are derived from *isothermal* operation and are not entitled to be designated ECL if from temperature-programmed operations.

A further development, the modified equivalent chain length (MECL), was proposed[3,14] because of improved relationships among fractional ECL values for polyunsaturated acid based on a plot line for methyl esters of the monoethylenic acids (i.e., 16:1, 18:1, 20:1, 22:1). Compared to the fractional ECL values based on the plot line for the corresponding saturated acids, the MECL reduced the influence of temperature. This objective has, however, been diminished in significance by the improved stability of liquid phases. Thus most analyses are now executed at 200°C on polyester or organosilicone polar liquid phases. The log plot lines for methyl esters of unsaturated fatty acids tend to be parallel to the log plot line for methyl esters of saturated acids, instead of convergent as formerly observed at low temperatures.[14,15]

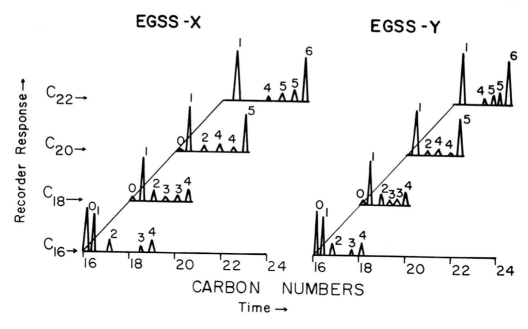

FIGURE 3. Schematic comparison of differing chain-length overlaps and distribution of peaks for methyl esters of cod liver oil fatty acids. Analyses on columns packed with a high polarity liquid phase (EGSS-X) or a medium polarity liquid phase (EGSS-Y). Numerals on peaks refer to number of ethylenic bonds, with ω3 isomer emerging after ω6 isomer where two isomers are shown.

Inexpensive and relatively stable standards for methyl esters of fatty acids are usually based on the saturated and monounsaturated fatty acids. Polyunsaturated fatty acids are more costly and less stable. In principle, the relationships of retention time among methyl esters of a number of common polyunsaturated fatty acids ranging from *cis,cis*-6,9-hexadecadienoate to *cis,cis,cis,cis,cis,cis*-4,7,10,13,16,19-docosahexaenoate, share a common susceptibility to the interacting factors making up column polarity, viz. temperature, liquid phase and percent of liquid phase, and support (see below). This is illustrated schematically in Figure 3 and elsewhere[6,16] and also in another format.[2] In addition, the principle has been extended by computer analysis of retention data to give a table of approximate ECL values.[12,17] Using these, an ECL value measured for a common and readily identifiable component such as methyl *cis,cis,cis*-9,12,15-octadecatrienoate provides the key to a whole set of ECL values that will suggest probable identifications for other polyunsaturated fatty acids. It must be emphasized that in publishing data based on speculative identifications, considerable care is necessary unless the fatty acid system is well defined by other detailed analyses.

B. Column Variables

The introduction of polyester liquid phases[18-20] was a major step in making the analyses of unsaturated fatty acids practical. An early systematic investigation of the interplay among type of polyester, percent of polyester, and type of support[21] clarified the polarity problem (cf. Figure 3). Systematic investigations have followed, and the effects of temperature on (relative) retention time and equivalent chain length for especially thermostable polar liquid phases are included in Tables 1 to 4.* The benefits of the expression of retention time as ECL units are clearly shown, as the *cis,cis,cis*-9,12,15-octadecatrienoate elutes later relative to the 20:0, that has an ECL of 20.00 by definition, as the temperature increases. On the other hand, the change in relative retention time ($r_{18:0}$) for 16:0 and 20:0 reveals that this

---

* All tables follow text.

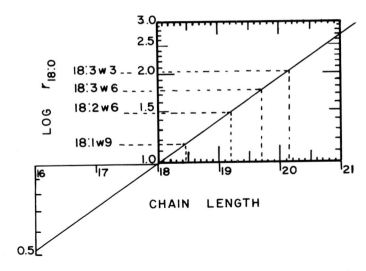

FIGURE 4.   Graphical conversion of relative retention times (16:0 = 0.52, 18:0 = 1.00, 18:1 ω 9 = 1.16, 18:2 ω 6 = 1.48, 18:3 ω 6 = 1.76, 18:3 ω 3 = 2.01, 20:0 = 1.90) to equivalent chain lengths (19:1 ω 9 = 18.46, 18:2 ω 6 = 19.20, 18:3 ω 6 = 19.70, 18:3 ω 3 = 20.15). Columns packed with 20% poly (diethylene-glycol succinate) on Anakrom ABS (70 to 80 mesh) and operated at 170°. (Redrawn after Ackman, R. G., *Lipids*, 2, 502, 1967. With permission of the American Oil Chemists' Society, Champaign, Illinois.)

change in ECL stems from the change in slope of the plot of log retention times against chain length (Table 1).

The fact that there is a difference in polarity due to differing types of liquid phase is emphasized by Table 2. The temperature, type of solid support, and percentage of liquid phase are fixed but it can be seen from the ECL values for methyl *cis,cis,cis*-9,12,15-octadecatrienoate that the succinate-ethylene glycol type of polyester tends to be the most polar liquid available (excluding certain newer cyanosilicone compounds which are not polyesters). The magnitudes of the relative retention times for this unsaturated acid are more or less inversely related to the polarity order based on equivalent chain length concepts.

The effect of column age on polarity with polyesters is due to loss of material and changes in the chemical nature of the polymer.[22] Table 3 shows that this is a progressive change overlapping the differences in new column packings from different commercial sources. This in turn may reflect subtle differences in polyester, support, or in the conditioning of the packing before sale or in the column before use.

C. Cyanosilicone Liquid Phases

Almost all of the theory and application in the gas-liquid chromatographic separation of the methyl esters of fatty acids was developed from the liquid phases Apiezon-L, the methyl silicone SE-30, and ethylene glycol adipate, supplemented by diethylene glycol succinate (DEGS), ethylene glycol succinate (EGS), and 1,4-butanediol succinate (BDS). The pure polyesters have been to some extent replaced by a variety of cyanosilicones of varying polarities. Retention data for methyl esters of fatty acids on these phases is tabulated in the appropriate tables. For details of composition the reader should refer to references on liquid phases compiled by J. K. Haken.[23-25] Two Tables (4 and 5) of contemporary data for the analysis of the common C_{18} fatty acids on cyanosilicone phases show how these have useful ranges of polarities, permitting selection of the right polarity for the analysis desired. At the same time, effects of the type of support and of operating temperature modify the polarity in the same manner as the polarity of the polyesters is modified (Tables 1 to 5). Temperature can often be exploited to achieve desired separations,[15] but in the long run is less satisfactory than the use of large[26] or small[27] shifts in polarity among polar liquid phases. The cyano-silicones greatly extend the variety of useful polar phases available for this purpose.

D. Tabulated Data

For the decade 1960—1970 almost all analyses of methyl esters of fatty acids were executed on polyesters, Carbowax, Apiezon-L, and methyl polysiloxanes.[12] In the last few years a variety of other polymeric chemicals have been introduced.[23] Some of these closely mimic the characteristics of the older materials, but are more stable. Others may interact more subtly with particular functional groups. To assist in the use for all classes of compounds the tabulations following are based on relative retention times, or equivalent chain lengths, mostly as originally published. Whenever possible the minimal parameters of liquid phase, support, percent of liquid phase, and temperature are tabulated,[7,28] although more information is desirable. Carrier gas is often specified, but the flow rate is less often available. The column dimensions are also given, with the type of detector.

Certain groups of liquid phases in packed columns are tabulated together, but broken down to some extent by the operating details or ranges of fatty acids described. Open-tubular data is mostly available for wall-coated (WCOT) columns. Support-coated (SCOT) columns will have more of the characteristics of packed columns. A limited number of special interest fatty acid analyses have been grouped together (e.g., naturally occurring acetylenic and ethylenic acid combinations). The ethyl esters can be perplexing contaminants in some preparations, and their basic retention relationship to methyl esters is also defined. Under-ivatized fatty acids can be analyzed for the range C_{12} to C_{18}, and the rules for relative retention times can be deduced from the data presented.

ACKNOWLEDGMENT

The preparation of this material was supported in part by funding from the Natural Sciences Research and Engineering Council of Canada.

Table 1
METHYL ESTERS OF FATTY ACIDS: POLARITY VARIATIONS OF IDENTICAL PACKED COLUMNS OPERATED AT DIFFERENT TEMPERATURES

	P1	P2	P3	P4	P5
Column packing					
Temperature (°C)	160	170	180	190	205
Gas	Ar	Ar	Ar	Ar	Ar
Flow rate (mℓ/min)	120	120	120	120	120
Column					
Length (cm)	275	275	275	275	275
Diameter					
(cm I.D.)	0.4	0.4	0.4	0.4	0.4
(cm O.D.)	na	na	na	na	na
Form	U-tube	U-tube	U-tube	U-tube	U-tube
Material	G	G	G	G	G
Detector	Argon	Argon	Argon	Argon	Argon
Reference	1	1	1	1	1

Compound	rMethyl Octadecanoate				
Methyl tetradecanoate	0.24	0.25	0.27	0.29	0.31
Methyl hexadecanoate	0.49	0.50	0.52	0.54	0.57
Methyl octadecanoate	1.00	1.00	1.00	1.00	1.00
Methyl *cis*-9-octadecenoate	1.14	1.14	1.14	1.15	1.15
Methyl *cis,cis*-9,12-octadecadienoate	1.45	1.44	1.43	1.44	1.44
Methyl *cis,cis,cis*-9,12,15-octadecatrienoate	1.99	1.96	1.92	1.91	1.90
Methyl icosanoate	2.06	2.00	1.95	1.84	1.77

	ECL				
Methyl *cis*-9-octadecenoate	18.37	18.38	18.38	18.47	18.49
Methyl *cis,cis*-9,12-octadecadienoate	19.03	19.05	19.06	19.20	19.27
Methyl *cis,cis,cis*-9,12,15-octadecatrienoate	19.91	19.95	19.94	20.12	20.24

Column packing P1 = 8% EGSS-X (ethylene glycol succinate polyester combined with methyl silicone) on Gas-Chrom P (100-120 mesh)

REFERENCE

1. **Supina, W. R.,** in *Biomedical Applications of Gas Chromatography,* Szymanski, H. A., Ed., Plenum Press, New York, 1964, 271.

Note added in proof: For additional RRT and ECL data recently tabulated for an OV-275 (15% on Gas Chrom RZ [100-120 mesh]) column operated at different temperatures see: **Itabashi, Y. and Takagi, T.,** *Yukagaku,* 27, 219, 1978.

Table 2

METHYL ESTERS OF FATTY ACIDS: POLARITY VARIATION OF PACKED COLUMNS, WITH A FIXED PROPORTION OF DIFFERENT LIQUID PHASES, ALL COATED ON THE SAME SOLID SUPPORT AND OPERATED AT THE SAME TEMPERATURE

Column packing	P1	P2	P3	P4	P5	P6	P7	P8
Temperature (°C)	175	175	175	175	175	175	175	175
Gas	Ar	Ar	Ar	Ar	Ar	Ar	Ar	Ar
Flow rate (mℓ/min)	54	54	54	54	54	54	54	54
Column								
Length (cm)	183	183	183	183	183	183	183	183
Diameter								
(cm I.D.)	4	4	4	4	4	4	4	4
(cm O.D.)	na	na	na	na	na	na	na	na
Form	U-tube	U-tube	U-tube	U-tube	U-tube	U-tube	U-tube	U-tube
Material	G	G	G	G	G	G	G	G
Detector	Argon	Argon	Argon	Argon	Argon	Argon	Argon	Argon
Reference	1	1	1	1	1	1	1	1

Compound	\(r_{\text{Methyl Octadecanoate}}\)							
Methyl tetradecanoate	0.32	0.33	0.32	0.27	0.27	0.27	0.26	0.29
Methyl pentadecanoate	0.43	0.45	0.43	0.37	0.36	—	—	0.40
Methyl hexadecanoate	0.56	0.57	0.57	0.53	0.52	0.51	0.52	0.53
Methyl heptadecanoate	0.76	0.76	0.75	0.72	0.69	—	—	0.74
Methyl octadecanoate	1.00	1.00	1.00	1.00	1.00	1.00	1.00	1.00
Methyl *cis*-9-octadecenoate	1.17	1.17	1.17	1.12	1.11	1.12	1.10	1.13
Methyl *cis,cis*-9,12-octadecadienoate	1.48	1.50	1.49	1.34	1.33	1.35	1.40	1.37
Methyl *cis,cis,cis*-9,12,15-octadecatrienoate	1.98	2.00	1.98	1.72	1.71	1.73	1.65	1.76
Methyl nonadecanoate	1.34	1.34	1.33	1.36	1.31	—	1.41	1.35
Methyl icosanoate	1.78	1.77	1.77	1.89	1.88	1.90	2.00	1.90

	ECL							
Methyl *cis*-9-octadecenoate	18.54	18.55	18.55	18.35	18.33	18.35	18.27	18.37
Methyl *cis,cis*-9,12-octadecadienoate	19.37	19.42	19.40	18.92	18.90	18.93	18.97	18.95
Methyl *cis,cis,cis*-9,12,15-octadecatrienoate	20.27	20.43	20.39	19.70	19.70	19.65	19.45	19.70

Column packing P1 = 15% Poly (ethylene glycol succinate) on Gas-Chrom P (80-100 mesh)
P2 = 15% Poly (diethylene glycol succinate) on Gas-Chrom P (80-100 mesh)
P3 = 15% EGSS-X (ethylene glycol succinate polyester combined with methyl silicone) on Gas-Chrom P (80-100 mesh)
P4 = 15% Poly (ethylene glycol adipate) on Gas-Chrom P (80-100 mesh)
P5 = 15% Poly (diethylene glycol adipate) on Gas-Chrom P (80-100 mesh)
P6 = 15% Poly (1,4-butanediol succinate) on Gas-Chrom P (80-100 mesh)
P7 = 15% Reoplex 400 (polyethylene glycol adipate polyester) on Gas-Chrom P (80-100 mesh)
P8 = 15% Poly (ethylene glycol phthalate) on Gas-Chrom P (80-100 mesh)

REFERENCE

1. **Supina, W. R.,** in *Biomedical Applications of Gas Chromatography,* Szymanski, H. A., Ed., Plenum Press, New York, 1964, 271.
(Equivalent chain lengths: **Jamieson, G. R.,** in *Topics in Lipid Chemistry,* Vol. 1, Gunstone, F. D., Ed., Logos Press, London, 1970, 107.)

Table 3
METHYL ESTERS OF FATTY ACIDS: POLARITY VARIATIONS OF PACKED COLUMNS WITH AGE (P1, P2, P3) OR SUPPLIER (P1, P4, P5). WITH FOUR COLUMNS ECL VALUES IN PARENTHESES ARE PREDICTED FROM MEASURED ECL FOR METHYL *CIS*-9, *CIS*-12, *CIS*-15-OCTADECATRIENOATE[2,3]

Column packing	P1	P2	P3	P4	P5
Temperature (°C)	200	200	200	200	200
Gas	N_2	N_2	N_2	N_2	N_2
Flow rate (mℓ/min)	na	na	na	na	na
Column					
Length (cm)	183	183	183	183	183
Diameter					
(cm I.D.)	na	na	na	na	na
(cm O.D.)	0.318	0.318	0.318	0.318	0.318
Form	Helix	Helix	Helix	Helix	Helix
Material	SS	SS	SS	SS	SS
Detector	FID	FID	FID	FID	FID
Reference	1	1	1	1	1

Compound	ECL				
Methyl *cis*-9-octadecenoate	18.38	18.31	18.24	18.41	18.31
Methyl *cis,cis*-9,12-octadecadienoate	18.87	18.71	18.52	18.91	18.89
Methyl *cis,cis,cis*-6,9,12-octadecatrienoate	19.22(19.25)	19.18(19.17)	19.12	19.29(19.29)	19.25(19.28)
Methyl *cis,cis,cis*-9,12,15-octadecatrienoate	19.58	19.50	19.42	19.63	19.61
Methyl *cis,cis,cis,cis*-6,9,12,15-octadecatetraenoate	19.98(19.98)	19.87(19.88)	19.75	20.05(20.05)	19.96(20.02)
Methyl *cis,cis,cis,cis,cis*-5,8,11,14,17-icosapentaenoate	22.24(22.11)	22.06(21.99)	21.84	22.09(22.19)	22.16(22.16)
Methyl *cis,cis,cis,cis,cis,cis*-4,7,10,13,16,19-docosahexaenoate	24.53(24.32)	24.27(24.18)	23.97	24.37(24.41)	24.39(24.38)

Column packing P1 = 8% Poly (1,4-butanediol succinate) on Chromosorb W, HMDS (80-100 mesh); in use 7 days; Supplier A

P2 = 8% Poly (1,4-butanediol succinate) on Chromosorb W, HMDS (80-100 mesh); in use 3 months

P3 = 8% Poly (1,4-butanediol succinate) on Chromosorb W, HMDS (80-100 mesh); in use 6 months

P4 = 8% Poly (1,4-butanediol succinate) on Chromosorb W, HMDS; Supplier B

P5 = 8% Poly (1,4-butanediol succinate) on Chromosorb W, HMDS; Supplier C

REFERENCES

1. **Jamieson, G. R.**, in *Topics in Lipid Chemistry*, Vol. I, Gunstone, F. D., Ed., Logos Press, London, 1970, 107.
2. **Jamieson, G. R. and Reid, E. H.**, *J. Chromatog.*, 42, 304, 1969.
3. **Jamieson, G. R.**, *J. Chromatog. Sci.*, 13, 491, 1975.

Table 4
METHYL ESTERS OF FATTY ACIDS ON POLAR COLUMNS PACKED WITH A CYANOSILICONE COATING ON EITHER AN ACID-WASHED SUPPORT (P1) OR ONE THAT IS ACID-WASHED AND SILANIZED (P2)

Column packing	P1	P1	P1	P2	P2	P2
Temperature (°C)	190	200	210	190	200	210
Gas	N_2	N_2	N_2	N_2	N_2	N_2
Flow rate (mℓ/min)	20	20	20	20	20	20
Column						
Length (cm)	183	183	183	183	183	183
Diameter						
(cm I.D.)	0.2	0.2	0.2	0.2	0.2	0.2
(cm O.D.)	na	na	na	na	na	na
Form	na	na	na	na	na	na
Material	G	G	G	G	G	G
Detector	FID	FID	FID	FID	FID	FID
Reference	1	1	1	1	1	1

Compound	$r_{\text{Methyl Octadecanoate}}$					
Methyl octadecanoate	1.00	1.00	1.00	1.00	1.00	1.00
Methyl *cis*-9-octadecenoate	1.20	1.21	1.21	1.22	1.21	1.23
Methyl *cis,cis*-9,12-octadecadienoate	1.58	1.58	1.56	1.63	1.62	1.61
Methyl *cis,cis,cis*-9,12,15-octadecatrienoate	2.12	2.10	2.07	2.27	2.22	2.19
Methyl icosanoate	1.66	1.60	1.52	1.52	1.52	1.42

	ECL					
Methyl octadecanoate	18.00	18.00	18.00	18.00	18.00	18.00
Methyl *cis*-9-octadecenoate	18.67	18.82	18.91	18.93	18.91	19.20
Methyl *cis,cis*-9,12-octadecadienoate	19.70	19.95	20.12	20.35	20.30	20.74
Methyl *cis,cis,cis*-9,12,15-octadecatrienoate	20.80	21.17	21.46	21.90	21.80	22.45
Methyl icosanoate	20.00	20.00	20.00	20.00	20.00	20.00

Column packing   P1 = 20% OV-275 (100% cyanoethyl siloxane) on Chromosorb W, AW (100-120 mesh)

                P2 = 20% OV-275 (100% cyanoethyl siloxane) on Supelcoport (AW, DMCS-treated, 100-120 mesh)

REFERENCE

1. Supelco, Inc., *Chromatography, Lipids,* VIII, (6), 2, 1974.

Table 5

METHYL ESTERS OF FATTY ACIDS ON POLAR COLUMNS PACKED WITH CYANOSILICONE PHASES OF DIFFERING POLARITIES AND ORIGINS AND ON TWO TYPES OF SUPPORT

Column packing	P1	P2	P3	P4	P5	P6	P7	P8
Temperature (°C)	200	200	200	200	180	180	180	180
Gas	N_2	N_2	N_2	N_2	N_2	N_2	N_2	N_2
Flow rate (mℓ/min)	20	20	20	20	40	40	40	40
Column								
Length (cm)	183	183	183	183	183	183	183	183
Diameter								
(cm I.D.)	0.2	0.2	0.2	0.2	0.4	0.4	0.4	0.4
(cm O.D.)	na	na	na	na	na	na	na	na
Form	na	na	na	na	U-tube	U-tube	U-tube	U-tube
Material	G	G	G	G	G	G	G	G
Detector	FID	FID	FID	FID	FID	FID	FID	FID
Reference	1	1	1	1	2	2	2	2

Compound	$r_{Methyl\ Octadecanoate}$							
Methyl octadecanoate	1.00	1.00	1.00	1.00	1.00	1.00	1.00	1.00
Methyl cis-9-octadecenoate	1.11	1.12	1.15	1.18	1.13	1.17	1.19	1.22
Methyl cis,cis-9,12-octadecadienoate	1.32	1.34	1.44	1.49	1.35	1.46	1.52	1.60
Methyl cis,cis,cis-9,12,15-octadecatrienoate	1.62	1.64	1.82	1.91	1.68	1.88	1.99	2.14
Methyl icosanoate	1.90	1.82	1.71	1.66	1.93	1.80	1.74	1.67

Compound	**ECL**							
Methyl octadecanoate	18.00	18.00	18.00	18.00	18.00	18.00	18.00	18.00
Methyl cis-9-octadecenoate	18.32	18.38	18.51	18.68	18.39	18.52	18.61	18.75
Methyl cis,cis-9,12-octadecadienoate	18.87	18.98	19.35	19.57	18.91	19.29	19.51	19.82
Methyl cis,cis,cis-9,12,15-octadecatrienoate	19.50	19.14	20.22	20.55	19.51	20.15	20.49	20.96
Methyl icosanoate	20.00	20.00	20.00	20.00	20.00	20.00	20.00	20.00

Column packing P1 = 10% SP-2300 (12.4% nitrile, = 50% cyanopropyl, 50% phenyl, siloxane) on Chromosorb W, AW (100-120 mesh)

P2 = 10% SP-2310 (20% nitrile, z 75% cyanopropyl, 25% phenyl,siloxane) on Chromosorb W, AW (100-120 mesh)

P3 = 10% SP-2330 (27% nitrile, = 90% cyanopropyl, 10% phenyl,siloxane) on Chromosorb W, AW (100-120 mesh)

P4 = 10% SP-2340 (28.4% nitrile, = 100% cyanopropyl, siloxane) on Chromosorb W, AW (100-120 mesh)

P5 = 10% SILAR-5CP (50% cyanopropyl, 50% phenyl, siloxane) on GAS-CHROM Q (100-120 mesh)

P6 = 10% SILAR-7CP (70% cyanopropyl, 30% phenyl, siloxane) on GAS-CHROM Q (100-120 mesh)

P7 = 10% SILAR-9CP (90% cyanopropyl, 10% phenyl, siloxane) on GAS-CHROM Q (100-120 mesh)

P8 = 10% SILAR-10CP (100% cyanopropyl, siloxane) on GAS-CHROM Q (100-120 mesh)

REFERENCES

1. Supelco, Inc., *Chromatography, Lipids,* VIII (5), 2, 1974.
2. Applied Science Laboratories, Inc. Newsletter, 2, 1975.

Note added in proof: For additional RRT data recently tabulated for SP-2300 cyanosilicone liquid phase on different supports or at different loadings see: **Shibahara, A., Yoshida, H., and Kajimoto, G.,** *Yukagaku,* 27, 233, 1978.

Table 6
METHYL ESTERS OF FATTY ACIDS: *CIS* AND *TRANS*, SHOWING INFLUENCE OF PROPORTION OF CYANO GROUPS IN A LIQUID PHASE WITHOUT SUPPORT EFFECT (I.E., IN A WALL-COATED OPEN-TUBULAR COLUMN) ON SEPARATIONS BY EITHER OR BOTH OF NUMBER OF ETHYLENIC BONDS AND THEIR GEOMETRY

Column packing	P1	P2	P3
Temperature (°C)	200	200	200
Gas	Ar	Ar	Ar
Flow rate (mℓ/min)	0.5	0.5	0.5
Column			
Length (cm)	6100	6100	6100
Diameter			
(cm I.D.)	0.025	0.025	0.025
(cm O.D.)	na	na	na
Form	WCOT	WCOT	WCOT
Material	SS	SS	SS
Detector	Argon	Argon	Argon
Reference	1	1	1

Compound	ECL		
Methyl *cis*-9-octadecenoate	18.06	18.33	18.48
Methyl *trans*-9-octadecenoate	18.00	18.13	18.26
Methyl *cis,cis*-9,12-octadecadienoate	18.33	18.77	19.21
Methyl *cis,trans*-9,12-octadecadienoate	18.24	18.62	18.99
Methyl *trans,cis*-9,12-octadecadienoate	18.30	18.68	19.02
Methyl *trans,trans*-9,12-octadecadienoate	18.07	18.39	18.64
Methyl *cis,cis,cis*-9,12,15-octadecatrienoate	18.70	19.20	20.04
Methyl *trans,trans,trans*-9,12,15-octadecatrienoate	18.54	18.97	19.66

Column packing P1 = XE-60 (25% cyanoethyl, with methyl, silicone) in form of wall coating in stainless steel tubing
P2 = Half XE-60, half GE 238-149-99 (= 50% cyanoethyl, with methyl, silicone) in form of wall coating in stainless steel tubing
P3 = GE 238-149-99 (50% cyanoethyl, with methyl, silicone) in form of wall coating in stainless steel tubing

REFERENCE

1. **Litchfield, C., Reiser, R., and Isbell, A. F.,** *J. Am. Oil Chem. Soc.,* 41, 52, 1964.

Table 7A
METHYL ESTERS OF FATTY ACIDS: ON POLAR (POLYESTER TYPE) PACKED COLUMNS OF LOW POLARITY, REPORTED AS RELATIVE RETENTION TIMES

Column packing	P1	P2	P3	P4	P5	P6
Temperature (°C)	na	180	180	220	200	216
Gas	—	Ar	N_2	He	N_2	H_2
Flow rate (mℓ/min)	na	na	na	90	30	na
Column						
Length (cm)	na	122	180	305	183	400
Diameter						
(cm I.D.)	na	na	0.4	na	na	0.6
(cm O.D.)	na	0.4	na	0.635	0.318	na
Form	na	na	na	Helix	Helix	na
Material	na	G	na	Cu	SS	Cu
Detector	na	Argon	FID	TC	FID	TC
Reference	1	2	3	4	5	6

Compound	$r_{Methyl\ Octadecanoate}$					
Methyl tetradecanoate	0.26	0.26	0.25	0.39	—	0.33
Methyl tetradecenoate	—	—	0.30	0.44	—	—
Methyl pentadecanoate	—	0.36	—	0.49	—	0.45
Methyl pentadecenoate	—	—	—	0.57	—	—
Methyl hexadecanoate	0.51	0.51	0.50	0.63	—	0.58
Methyl cis-9-hexadecenoate	0.58	0.59	0.57	0.73	—	—
Methyl cis,cis-6,9-hexadecadienoate	0.69	—	—	—	—	—
Methyl cis,cis-9,12-hexadecadienoate	—	—	—	0.89	—	—
Methyl cis,cis,cis-6,9,12-hexadecatrienoate	0.86	—	—	1.02	—	—
Methyl heptadecanoate	—	0.71	—	0.78	—	0.75
Methyl heptadecenoate	—	—	—	0.91	—	—
Methyl octadecanoate	1.00	1.00	1.00	1.00	1.00	1.00
Methyl cis-9-octadecenoate	1.12	1.12	1.09	1.14	1.12	1.08
Methyl cis,cis-9,12-octadecadienoate	1.35	1.37	1.28	1.36	1.32	1.27
Methyl cis,cis,cis-6,9,12-octadecatrienoate	—	—	1.45[a]	1.58	1.43	—
Methyl cis,cis,cis-9,12,15-octadecatrienoate	1.71	1.78	1.66	1.72	1.65	1.70
Methyl cis,cis,cis,cis-6,9,12,15-octadecatetraenoate	—	—	—	1.99	1.84	—
Methyl nonadecanoate	—	—	—	—	—	1.34
Methyl nonadecenoate	—	—	—	1.43	—	—
Methyl icosanoate	2.02	1.95	1.86[a]	1.59	—	1.79
Methyl cis-11-icosenoate	—	—	—	1.78	2.04	—
Methyl cis,cis-11,14-icosadienoate	2.67	—	—	2.14	2.42	—
Methyl cis,cis,cis-8,11,14-icosatrienoate	—	—	2.87	—	2.63	—
Methyl cis,cis,cis-11,14,17-icosatrienoate	3.32	—	—	—	—	—
Methyl cis,cis,cis,cis-5,8,11,14-icosatetraenoate	—	3.33	3.13[a]	2.73	2.94	—
Methyl cis,cis,cis,cis-8,11,14,17-icosatetraenoate	—	—	3.62[a]	3.11	3.31	—
Methyl cis,cis,cis,cis,cis-5,8,11,14,17-icosapentaenoate	—	4.33	3.90	3.43	3.61	—
Methyl cis,cis,cis,cis,cis-6,9,12,15,18-henicosapentaenoate	—	—	—	4.32	—	—
Methyl docosanoate	3.86	—	—	—	—	—
Methyl cis-11-docosenoate (or cis-13-docosenoate)	—	4.15	—	2.84	3.66	—
Methyl cis,cis-13,16-docosadienoate	5.10	—	—	—	—	—
Methyl cis,cis,cis,cis-7,10,13,16-docosatetraenoate	—	—	—	4.32	5.30	—

Table 7A (continued)
METHYL ESTERS OF FATTY ACIDS: ON POLAR (POLYESTER TYPE) PACKED COLUMNS OF LOW POLARITY, REPORTED AS RELATIVE RETENTION TIMES

Compound					$r_{Methyl\ Octadecanoate}$	
Methyl *cis,cis,cis,cis,cis*-4,7,10,13,16-docosapentaenoate	—	—	—	—	5.80	—
Methyl *cis,cis,cis,cis,cis*-7,10,13,16,19-docosapentaenoate	—	—	—	5.39	6.61	—
Methyl *cis,cis,cis,cis,cis,cis*-4,7,10,13,16,19-docosahexaenoate	—	9.25	8.51	6.07	7.34	—
Methyl tetracosanoate	—	—	—	—	6.70	—

[a] Tentative identifications assigned by R. G. Ackman.

Column packing P1 = Polyester, probably poly (ethylene glycol adipate)
　　　　　　　　　P2 = 20% Poly (ethylene glycol adipate) on Celite (100-210 mesh)
　　　　　　　　　P3 = 10% Poly (1,4-butanediol succinate) on Gas-Chrom W
　　　　　　　　　　　　P4 = 25% Poly (diethylene glycol succinate) on GC-22 Super-Support (60-80 mesh)
　　　　　　　　　P5 = 8% Poly (1,4-butanediol succinate) on Chromosorb W
　　　　　　　　　　　　P6 = 23% Poly (ethylene glycol - 4-oxopimelate) on Sterchamol (Merck)

REFERENCES

1. **Ishakov, N. I. and Vereschagin, A. G.,** *Biokhimiya,* 29, 487, 1964; Engl. translation 29, 420, 1964. (See also **Vereschagin, A. G.,** in *Gas-Chromatographie,* 551, 1965.
2. **James, A. T.,** *J. Chromatogr.,* 2, 552, 1959.
3. **Anderson, R., Livermore, B. P., Kates, M., and Volcani, B. E.,** *Biochim. Biophys. Acta,* 528, 77, 1978.
4. **Ackman, R. G. and Burgher, R. D.,** *Can. J. Biochem. Physiol.,* 41, 2501, 1963.
5. **Jamieson, G. R. and Reid, E. H.,** *J. Chromatogr.,* 26, 8, 1967.
6. **Falk, F.,** *J. Chromatogr.,* 17, 450, 1965.

Table 7B
METHYL ESTERS OF FATTY ACIDS: ON POLAR (POLYESTER TYPE) PACKED COLUMNS OF LOW POLARITY, REPORTED AS RELATIVE RETENTION TIMES

Column packing	P1	P2	P3	P4	P5	P6
Temperature (°C)	230	230	197	197.0	184.5	173.5
Gas	He	He	Ar	Ar	Ar	Ar
Flow rate (mℓ/min)	150	150	40	30	40	50
Column						
Length (cm)	305	305	142	142	142	142
Diameter						
(cm I.D.)	—	—	0.4	0.4	0.4	0.4
(cm O.D.)	0.635	0.635	na	na	na	na
Form	Helix	Helix	Tube	Tube	Tube	Tube
Material	Cu	Cu	G	G	G	G
Detector	TC	TC	Argon	Argon	Argon	Argon
References	1	2	3	3	3	3

Compound	$r_{\text{Methyl Octadecanoate}}$					
	P1	P2	P3	P4	P5	P6
Methyl tetradecanoate	0.33	0.34	0.32	0.30	0.27	0.25
Methyl tetradecenoate	0.37	0.38	0.36	0.34	0.32	0.30
Methyl pentadecanoate	0.45	0.45	0.43	0.40	0.37	0.35
Methyl pentadecenoate	0.49	—	—	—	—	—
Methyl hexadecanoate	0.59	0.57	0.57	0.55	0.52	0.50
Methyl *cis*-9-hexadecenoate	0.64	0.64	0.64	0.63	0.59	0.58
Methyl *cis,cis*-6,9-hexadecadienoate	—	—	0.79	0.78	0.75	0.73
Methyl *cis,cis*-9,12-hexadecadienoate	0.73	0.73	—	0.81	—	—
Methyl *cis,cis,cis*-6,9,12-hexadecatrienoate	0.80	—	0.90	0.90	0.89	0.86
Methyl *cis,cis,cis,cis*-4,7,10,13-hexadecatetraenoate	—	—	—	1.08	—	—
Methyl *cis,cis,cis,cis*-6,9,12,15-octadecatetraenoate	0.91	—	—	1.11	—	—
Methyl heptadecanoate	0.77	0.76	0.74	0.73	0.72	0.71
Methyl heptadecenoate	0.82	0.82	—	—	—	—
Methyl octadecanoate	1.00	1.00	1.00	1.00	1.00	1.00
Methyl *cis*-9-octadecenoate	1.08	1.06	1.11	1.12	1.12	1.12
Methyl *cis,cis*-9,12-octadecadienoate	1.21	1.20	1.32	1.34	1.35	1.35
Methyl *cis,cis,cis*-6,9,12-octadecatrienoate	1.30	—	—	1.54		
Methyl *cis,cis,cis*-9,12,15-octadecatrienoate	1.41	1.38	1.66	1.72	1.73	1.76
Methyl *cis,cis,cis,cis*-6,9,12,15-octadecatetraenoate	1.53	1.50	1.91	1.97	1.99	2.04
Methyl nonadecanoate	1.29	—	1.32	1.35	1.38	1.41
Methyl nonadecenoate	1.40	1.38	—	—	—	—
Methyl icosanoate	1.70	—	1.78	1.82	1.89	1.99
Methyl *cis*-11-icosenoate	1.81	1.76	1.96	2.02	2.09	2.16
Methyl *cis,cis*-11,14-icosadienoate	2.05	—	2.45	2.45	2.52	2.66
Methyl *cis,cis,cis*-8,11,14-icosatrienoate	—	—	2.68	2.76	2.88	3.02
Methyl *cis,cis,cis,cis*-5,8,11,14-icosatetraenoate	2.31	2.23	2.90	3.04	3.17	3.32
Methyl *cis,cis,cis,cis*-8,11,14,17-icosatetraenoate	2.53	2.52	3.33	3.51	3.70	3.91
Methyl *cis,cis,cis,cis,cis*-5,8,11,14,17-icosapentaenoate	2.78	2.61	3.67	3.85	4.08	4.33
Methyl *cis,cis,cis,cis,cis*-6,9,12,15,18-henicosapentaenoate	3.61	3.48	—	—	—	—
Methyl docosanoate	—	—	—	3.30	—	—
Methyl *cis*-11-docosenoate (or *cis*-13-docosenoate)	3.05	2.96	—	3.68	—	—
Methyl *cis,cis*-13,16-docosadienoate	—	—	—	4.38		
Methyl *cis,cis,cis,cis*-7,10,13,16-docosatetraenoate	—	3.69	—	5.50	5.99	6.40
Methyl *cis,cis,cis,cis,cis*-4,7,10,13,16-docosapentaenoate	—	—	5.80	6.09	6.75	7.43
Methyl *cis,cis,cis,cis,cis*-7,10,13,16,19-docosapentaenoate	4.63	4.31	6.67	7.00	7.69	8.40
Methyl *cis,cis,cis,cis,cis,cis*-4,7,10,13,16,19-docosahexaenoate	4.88	4.59	7.40	7.75	8.59	9.55

Table 7B (continued)
METHYL ESTERS OF FATTY ACIDS: ON POLAR (POLYESTER TYPE) PACKED COLUMNS OF LOW POLARITY, REPORTED AS RELATIVE RETENTION TIMES

Column packing
P1 = 25% Poly (neopentyl glycol succinate) on GC-22 "Super-Support" (60-80 mesh)

P2 = 25% Poly (neopentyl glycol succinate) on GC-22 "Super-Support" (60-80 mesh)

P3 = 25% Reoplex-400 (commercial propylene glycol adipate polyester) on Celite, AW, ALK-W (120-140 mesh)

P4 = 18% Poly (ethylene glycol adipate) on Celite, AW, ALK-W (120-140 mesh)

P5 = 18% Poly (ethylene glycol adipate) on Celite, AW, ALK-W (120-140 mesh)

P6 = 18% Poly (ethylene glycol adipate) on Celite, AW, ALK-W (120-140 mesh)

REFERENCES

1. **Ackman, R. G., Burgher, R. D., and Jangaard, P. M.,** *Can. J. Biochem. Physiol.,* 41, 1627, 1963.
2. **Ackman, R. G. and Burgher, R. D.,** *Can. J. Biochem. Physiol.,* 41, 2501, 1963.
3. **Farquhar, J. W., Insull, W., Jr., Rosen, P., Stoffel, W., and Ahrens, E. H., Jr.,** *Nutr. Rev. (Suppl.),* 17, 1, 1959.

Table 7C
METHYL ESTERS OF FATTY ACIDS: ON POLAR PACKED (POLYESTER TYPE) COLUMNS OF LOW POLARITY, REPORTED AS RELATIVE RETENTION TIMES

	P1	P2	P3	P4	P5	P6
Column packing						
Temperature (°C)	180	200	220	200	235	235
Gas	N$_2$	N$_2$	N$_2$	He	He	He
Flow rate (ml/min)	28	28	28	na	54	54
Column Length (cm)	100	100	100	183	200	200
Diameter (cm I.D.)	0.6	0.6	0.6	na	na	na
(cm O.D.)	na	na	na	0.318	0.5	0.5
Form	na	na	na	Helix	Helix	Helix
Material	na	na	na	G	Cu	Cu
Detector	FID	FID	FID	FID	TC	TC
Reference	1	1	1	2	3	3
Compound			$r_{\text{Methyl Octadecanoate}}$			
Methyl tetradecanoate	0.28	0.31	0.35	—	—	—
Methyl hexadecanoate	0.53	0.56	0.60	—	0.56	0.57
Methyl cis-9-hexadecenoate	0.60	—	—	—	0.60	0.61
Methyl octadecanoate	1.00	1.00	1.00	1.00	1.00	1.00
Methyl cis-9-octadecenoate	1.11	1.11	1.11	1.09	1.06	1.07
Methyl cis,cis-9,12-octadecadienoate	1.36	1.33	1.33	1.30	1.20	1.20
Methyl cis,cis,cis-6,9,12-octadecatrienoate	—	—	—	1.45	—	—
Methyl cis,cis,cis-9,12,15-octadecatrienoate	—	—	—	1.62	1.43	1.44
Methyl cis,cis,cis,cis-6,9,12,15-octadecatetraenoate	—	—	—	1.81	—	—
Methyl icosanoate	1.86	1.69	1.63	—	1.77	1.78
Methyl cis-11-icosenoate	2.07	1.92	1.84	—	1.87	1.90
Methyl docosanoate	3.49	3.08	2.78	—	—	—
Methyl cis-11-docosenoate (or cis-13-docosenoate)	3.87	3.41	3.05	—	3.15	3.52

Column packing   P1 = 20% Poly (ethylene glycol adipate) on Celite (0.15-0.2 mm)
                  P2 = 20% Poly (ethylene glycol adipate) on Celite (0.15-0.2 mm)
                  P3 = 20% Poly (ethylene glycol adipate) on Celite (0.15-0.2 mm)
                  P4 = 8% Poly (1,4-butanediol succinate)
                  P5 = 20% Poly (α-butyl-α^1-hydroxy-ethyl glyceryl diether adipate) on Daichrome A (60-80 mesh)
                  P6 = 20% Poly (α-butyl-α^1-hydroxy-ethyl glyceryl diether succinate) on Daichrome A (60-80 mesh)

REFERENCES

1. **Zeman, I.,** *J. Gas Chromatogr.,* 3, 18, 1965.
2. **Jamieson, G. R. and Reid, E. H.,** *J. Sci. Food Agric.,* 19, 628, 1968.
3. **Watanabe, S., Sato, Y., and Kitamura, K.,** *Yukagaku,* 16, 570, 1967.

Table 7D
METHYL ESTERS OF FATTY ACIDS: ON POLAR PACKED COLUMNS OF MODERATE POLARITY, REPORTED AS RELATIVE RETENTION TIMES

	P1	P2	P3	P4	P5
Column packing					
Temperature (°C)	180	180	200	190	206
Gas	N_2	N_2	H_2	N_2	N_2
Flow rate (mℓ/min)	50	na	110	30	20
Column					
Length (cm)	200	305	300	300	220
Diameter					
(cm I.D.)	0.4	na	0.4	0.3	na
(cm O.D.)	na	0.318	na	na	na
Form	na	Helix	na	na	na
Material	SS	SS	na	SS	G
Detector	FID	FID	TC	FID	FID
Reference	1	2	3	4	5

Compound	$r_{Methyl\ Octadecanoate}$				
Methyl tetradecanoate	0.26	0.27	—	0.32	0.39
Methyl tetradecenoate	0.31	—	0.32	0.38	—
Methyl pentadecanoate	0.36	0.33	0.39	0.43	—
Methyl pentadecenoate	0.40	—	—	—	—
Methyl hexadecanoate	0.52	0.52	0.55	0.56	0.62
Methyl *cis*-9-hexadecenoate	0.59	0.59	0.61[a]	0.66	0.73
Methyl *cis,cis*-6,9-hexadecadienoate	—	0.72[a]	0.74[a]	0.80	—
Methyl *cis,cis*-9,12-hexadecadienoate	—	—	—	0.84	—
Methyl *cis,cis,cis*-6,9,12-hexadecatrienoate	—	0.86[a]	—	1.04	—
Methyl *cis,cis,cis,cis*-4,7,10,13-hexadecatetraenoate	—	1.03	—	1.19	—
Methyl *cis,cis,cis,cis*-6,9,12,15-hexadecatetraenoate	—	—	—	1.27	—
Methyl heptadecanoate	0.72	0.73	—	0.75	—
Methyl heptadecenoate	0.80	0.81	—	0.85	—
Methyl octadecanoate	1.00	1.00	1.00	1.00	1.00
Methyl *cis*-9-octadecenoate	1.12	1.11	1.11	1.15	1.17
Methyl *cis,cis*-9,12-octadecadienoate	1.34	1.34	1.33	1.40	1.45
Methyl *cis,cis,cis*-6,9,12-octadecatrienoate	1.52	—	1.61	1.64	1.72
Methyl *cis,cis,cis*-9,12,15-octadecatrienoate	1.71	1.66	1.76	1.83	1.92
Methyl *cis,cis,cis,cis*-6,9,12,15-octadecatetraenoate	—	—	1.96	2.12	—
Methyl nonadecanoate	1.34		—	1.32	—
Methyl nonadecenoate	1.52	1.54	—	1.48	—
Methyl icosanoate	1.94	2.02	—	1.78	1.64
Methyl *cis*-11-icosenoate	2.09	2.03	2.07	1.99	1.90
Methyl *cis,cis*-11,14-icosadienoate	2.56	2.47	2.50	2.46	2.39
Methyl *cis,cis,cis*-5,8,11-icosatrienoate	—	—	—	2.63	2.58
Methyl *cis,cis,cis*-8,11,14-icosatrienoate	2.89	2.86	2.79	2.82	2.83
Methyl *cis,cis,cis*-11,14,17-icosatrienoate	—	—	—	3.20	—
Methyl *cis,cis,cis,cis*-5,8,11,14-icosatetraenoate	3.15	3.44	3.14	—	3.20
Methyl *cis,cis,cis,cis*-8,11,14,17-icosatetraenoate	—	3.64	—	3.66	—
Methyl *cis,cis,cis,cis,cis*-5,8,11,14,17-icosapentaenoate	4.05	4.09	—	4.16	4.27
Methyl henicosanoate	—	—	—	2.35	—
Methyl *cis,cis,cis,cis,cis*-6,9,12,15,18-heniocosapentaenoate	—	5.58	—	5.51	—
Methyl docosanoate	3.69	—	—	—	2.70
Methyl *cis*-11-docosenoate (or *cis*-13-docosenoate)	4.91	—	—	3.48	3.13
Methyl *cis,cis*-13,16-docosadienoate	—	—	—	4.28	—
Methyl *cis,cis,cis,cis*-7,10,13,16-docosatetraenoate	6.60	6.60	—	—	5.42
Methyl *cis,cis,cis,cis,cis*-4,7,10,13,16-docosapentaenoate	—	—	—	—	6.26
Methyl *cis,cis,cis,cis,cis*-7,10,13,16,19-docosapentaenoate	7.75	7.52	—	7.18	7.16
Methyl *cis,cis,cis,cis,cis,cis*-4,7,10,13,16,19-docosahexaenoate	8.69	8.49	—	8.15	8.26
Methyl tetracosanoate	—	—	—	—	4.59
Methyl *cis*-15-tetracosenoate	—	—	—	6.25	5.22

Table 7D (continued)
METHYL ESTERS OF FATTY ACIDS: ON POLAR PACKED COLUMNS OF MODERATE POLARITY, REPORTED AS RELATIVE RETENTION TIMES

^a   Isomer may be other than listed.

Column packing   P1 = 20% Poly (ethylene glycol adipate) on Chromosorb W (60-80 mesh)
P2 = 20% Poly (diethylene glycol succinate) on Chromosorb W (80-100 mesh)
P3 = 20% Poly (ethylene glycol adipate) on INZ-100 brick (0.18-0.25 mm)
P4 = 10% Poly (diethylene glycol succinate) on Shimalite W (60-80 mesh)
P5 = 20% Poly (diethylene glycol succinate) on Celite

REFERENCES

1. **Kluytmans, J. H. F. M.**, Ph.D. Thesis, University of Utrecht, The Netherlands, 1971.
2. **Graille, J., Huq, M. S., and Naudet, M.**, *Rev. Franc. Corps Gras,* 23, 151, 1976.
3. **Bardyshev, I. I., Papanov, G. Ya., Ivanov, S. A., Kryuk, S. I., and Pertsovskii, A. L.,** *Nauchni trudove Vissh. Pedagog. Inst., Plovdiv, Mat., Fiz., Khim., Biol.,* 9, 95, 1971.
4. **Kochi, M.,** *Nihon Suisan, Gakkai-Shi,* 42, 219, 1976.
5. **Takayasu, K., Okuda, K., and Yoshikawa, I.,** *Lipids,* 5, 743, 1970.

Table 8A
METHYL ESTERS OF FATTY ACIDS: ON POLAR (POLYESTER TYPE) PACKED COLUMNS OF LOW TO MODERATE POLARITY, REPORTED AS RELATIVE RETENTION TIMES

	P1	P2	P3	P4	P5	P6
Column packing						
Temperature (°C)	205	180	200	197	190	190
Gas	He	Ar	H₂	N₂	Ar	—
Flow rate (mℓ/min)	90	40	100	100	na	na
Column						
Length (cm)	305	122	137	244	183	na
Diameter						
(cm I.D.)	na	0.4	na	na	0.4	na
(cm O.D.)	0.635	na	0.4	0.48	na	na
Form	Helix	Tube	U-tube	Helix	U-tube	na
Material	Cu	G	Cu	SS	G	na
Detector	TC	Argon	TC	FID	Argon	na
Reference	1	2	3	4	5	6

Compound	\(r_{\text{Methyl Octadecanoate}} \)					
	P1	P2	P3	P4	P5	P6
Methyl tetradecanoate	0.36	0.25	0.32	0.36	—	0.28
Methyl tetradecenoate	0.44	0.30	—	0.42	—	0.34
Methyl pentadecanoate	0.45	0.35	0.42	0.47	—	0.38
Methyl pentadecenoate	0.52	—	—	0.50	—	0.46
Methyl hexadecanoate	0.54	0.50	0.57	0.60	0.56	0.53
Methyl cis-9-hexadecenoate	0.71	0.58	0.67	0.71	0.66	0.62
Methyl cis,cis-6,9-hexadecadienoate	0.85	—	—	0.84	—	—
Methyl cis,cis-9,12-hexadecadienoate	0.90	—	—	—	—	—
Methyl cis,cis,cis-6,9,12-hexadecatrienoate	1.08	—	—	1.06	—	0.99
Methyl cis,cis,cis,cis-4,7,10,13-hexadecatetraenoate	1.29	—	—	—	—	1.21
Methyl cis,cis,cis,cis-6,9,12,15-hexadecatetraenoate	1.39	—	—	1.33	—	1.26
Methyl heptadecanoate	0.79	0.70	0.74	0.76	—	0.72
Methyl heptadecenoate	0.89	—	—	0.91	—	0.81
Methyl octadecanoate	1.00	1.00	1.00	1.00	1.00	1.00
Methyl cis-9-octadecenoate	1.16	1.12	1.15	1.16	1.15	1.14
Methyl cis,cis-9,12-octadecadienoate	1.41	1.34	—	1.37	1.45	1.44
Methyl cis,cis,cis-6,9,12-octadecatrienoate	1.64	—	—	1.64	1.67	1.70
Methyl cis,cis,cis-9,12,15-octadecatrienoate	1.82	1.84	1.81	1.81	1.93	1.93

	P1	P2	P3	P4	P5	P6
Methyl cis,cis,cis,cis-6,9,12,15-octadecatetraenoate	2.12	—	—	2.14	2.24	2.20
Methyl nonadecanoate	1.30	1.41	1.30	1.29	—	1.37
Methyl nonadecenoate	1.49	—	—	1.61	—	—
Methyl icosanoate	1.71	2.00	1.78	1.71	1.75	1.87
Methyl cis-11-icosenoate	1.91	2.20	—	1.99	2.01	1.09
Methyl cis,cis-11,14-icosadienoate	2.29	—	—	2.22	2.53	2.60
Methyl cis,cis-5,8,11-icosatrienoate	—	3.00	—	—	—	2.81
Methyl cis,cis-8,11,14-icosatrienoate	2.70	—	—	—	2.97	3.03
Methyl cis,cis-11,14,17-icosatrienoate	2.96	—	—	—	3.45	3.45
Methyl cis,cis,cis-5,8,11,14-icosatetraenoate	3.00	3.55	3.11	3.08	4.00	3.45
Methyl cis,cis,cis-8,11,14,17-icosatetraenoate	3.55	—	—	—	4.60	3.90
Methyl cis,cis,cis,cis-5,8,11,14,17-icosapentaenoate	3.89	4.21	—	4.06	—	4.40
Methyl henicosanoate	—	—	2.36	—	3.13	—
Methyl cis,cis,cis,cis,cis-6,9,12,15,18-henicosapentaenoate	5.00	2.82	—	2.95	3.52	—
Methyl docosanoate	2.97	4.08	3.18	3.65	—	—
Methyl cis-11-docosenoate (or cis-13-docosenoate)	3.17	4.68	3.64	—	5.99	3.88
Methyl cis-13,16-docosadienoate	—	—	—	5.50	6.94	—
Methyl cis,cis,cis-7,10,13,16-docosatetraenoate	5.00	7.39	—	6.93	7.93	—
Methyl cis,cis,cis,cis-4,7,10,13,16-docosapentaenoate	5.40	—	—	7.80	9.22	—
Methyl cis,cis,cis,cis-7,10,13,16,19-docosapentaenoate	6.38	—	—	—	—	—
Methyl cis,cis,cis,cis,cis-4,7,10,13,16,19-docosahexaenoate	7.22	9.61	—	—	—	—
Methyl tetracosanoate	—	5.66	—	—	—	—

Column packing   P1 = 25% Poly (ethylene glycol succinate) on GC-22 Super-Support (60-80 mesh)
P2 = 5% Poly (vinyl acetate) on Celite, AW, ALK-W (100-120 mesh)
P3 = 20% Poly (diethylene glycol succinate) on Celite 545, AW, ALK-W (48-85 mesh)
P4 = 15% Poly (diethylene glycol succinate) on Chromosorb W (45-60 mesh)
P5 = 20% Poly (diethylene glycol succinate) on GC-22 Super-Support (60-80 mesh)
P6 = ?% Poly (diethylene glycol succinate)

REFERENCES

1. **Ackman, R. G., Burgher, R. D., and Jangaard, P. M.,** Can. J. Biochem. Physiol., 41, 1627, 1963.
2. **Kingsbury, K. J., Morgan, D. M., and Heyes, T. D.,** Biochem. J., 90, 140, 1964.
3. **Hawke, J. C., Hansen, R. P., and Shorland, F. B.,** J. Chromatogr., 2, 547, 1959.
4. **Gedam, P. H., Subbaram, M. R., and Aggarwal, J. S.,** Fette Seifen Anstrichm., 73, 748, 1971.
5. **Ackman, R. G.,** J. Gas Chromatogr., 1(4), 11, 1963.
6. **Kochi, M.,** J. Shimonoseki Univ. Fish., 22, 47, 1973.

Table 8B

METHYL ESTERS OF FATTY ACIDS: ON POLAR (POLYESTER TYPE) PACKED COLUMNS OF MODERATE TO HIGH POLARITY, REPORTED AS RELATIVE RETENTION TIMES

Column packing	P1	P2	P3	P4	P5	P6
Temperature (°C)	175	190	190	190	205	183
Gas	He	N₂	N₂	N₂	N₂	He
Flow rate (ml/min)	40	60	60	30	20	72
Column						
Length (cm)	183	300	300	150	200	183
Diameter						
(cm I.D.)	na	0.4	0.4	0.2	0.2	na
(cm O.D.)	0.318	na	na	na	na	0.318
Form	Helix	na	na	Helix	na	na
Material	na	SS	SS	G	na	SS
Detector	FID	FID	FID	FID	FID	FID
Reference	1	2	2	3	4	5
Compound	$r_{\text{Methyl Octadecanoate}}$					
Methyl tetradecanoate	0.29	0.28	0.32	0.33	0.32	0.30
Methyl tetradecenoate	—	0.34	0.38	—	0.38	0.35
Methyl pentadecanoate	0.38	0.38	0.43	—	0.44	0.41
Methyl pentadecenoate	—	0.46	—	—	—	0.47
Methyl hexadecanoate	0.54	0.53	0.56	0.58	0.58	0.56
Methyl cis-9-hexadecenoate	0.63	0.63	0.66	0.69	0.69	0.66
Methyl cis,cis-6,9-hexadecadienoate	—	0.74	0.80	—	0.72*	—
Methyl cis,cis-9,12-hexadecadienoate	—	—	—	—	—	0.87
Methyl cis,cis,cis-6,9,12-hexadecatrienoate	—	0.99	1.04	—	1.16*	1.08
Methyl cis,cis,cis,cis-4,7,10,13-hexadecatetraenoate	—	1.21	1.19	—	1.31	—
Methyl cis,cis,cis,cis-6,9,12,15-hexadecatetraenoate	—	1.26	1.27	—	1.35	1.39
Methyl heptadecanoate	0.71	0.72	0.75	0.75	0.76	0.73
Methyl heptadecenoate	0.85	0.81	0.85	—	—	0.87
Methyl octadecanoate	1.00	1.00	1.00	1.00	1.00	1.00
Methyl cis-9-octadecenoate	1.15	1.14	1.15	1.18	1.16	1.20
Methyl cis,cis-9,12-octadecadienoate	1.47	1.44	1.40	1.51	1.42	1.45
Methyl cis,cis,cis-6,9,12-octadecatrienoate	—	1.70	1.64	—	1.74	1.74
Methyl cis,cis,cis-9,12,15-octadecatrienoate	1.94	1.93	1.83	—	1.87	1.96

Compound	P1	P2	P3	P4	P5	P6
Methyl cis,cis,cis,cis-6,9,12,15-octadecatetraenoate	—	2.20	2.12	—	2.20	2.34
Methyl nonadecanoate	—	1.37	1.32	—	—	1.34
Methyl nonadecenoate	1.86	1.52	1.48	—	1.50	1.54
Methyl icosanoate	2.1x	1.87	1.78	—	1.70	1.78
Methyl cis-11-icosenoate	—	2.09	1.99	—	1.99	2.03
Methyl cis-11,14-icosadienoate	—	2.60	2.46	—	2.42	2.57
Methyl cis,cis,cis-5,8,11-icosatrienoate	—	2.81	2.63	—	2.92	—
Methyl cis,cis,cis-8,11,14-icosatrienoate	—	3.03	2.82	3.10	—	3.03
Methyl cis,cis,cis-11,14,17-icosatrienoate	—	3.45	3.20	—	3.20	3.44
Methyl cis,cis,cis,cis-5,8,11,14-icosatetraenoate	—	3.45	3.20	3.56	3.81	3.49
Methyl cis,cis,cis,cis-8,11,14,17-icosatetraenoate	—	3.90	3.66	—	4.17	4.11
Methyl cis,cis,cis,cis,cis-5,8,11,14,17-icosapentaenoate	—	4.40	4.16	—	—	4.68
Methyl cis,cis,cis,cis,cis-6,9,12,15,18-henicosapentaenoate	3.50	6.10	—	—	2.70	6.32
Methyl docosanoate	3.8x	—	—	—	3.20	3.20
Methyl cis-11-docosenoate (or cis-13-docosenoate)	—	3.88	3.48	—	3.20	3.57
Methyl cis-13,16-docosadienoate	—	—	4.28	—	5.48	4.14
Methyl cis,cis,cis-7,10,13,16-docosatetraenoate	—	—	—	—	—	6.15
Methyl cis,cis,cis,cis-4,7,10,13,16-docosapentaenoate	—	—	—	—	7.10	7.18
Methyl cis,cis,cis,cis-7,10,13,16,19-docosapentaenoate	—	—	7.18	—	8.10	8.29
Methyl cis,cis,cis,cis,cis-4,7,10,13,16,19-docosahexaenoate	—	—	8.15	—	—	9.64
Methyl tetracosanoate	6.35	—	—	—	5.48	5.90
Methyl cis-15-tetracosenoate	6.7x	—	6.25	—	—	6.59

x Isomer may be other than listed.

Column packing   P1 = 15% Poly (diethylene glycol succinate) on Chromosorb W,AW (80-100 mesh)
P2 = 10% Poly (diethylene glycol succinate) on Shimalite W (60-80 mesh)
P3 = 23.610% Poly (diethylene glycol succinate) on Shimalite W (60-80 mesh)
P4 = 10% Poly (diethylene glycol succinate) on Chromosorb W,AW (80-100 mesh)
P5 = 20% Poly (ethylene glycol succinate) on Chromosorb W,AW (80-100 mesh)
P6 = 15% Poly (diethylene glycol succinate) on Chromosorb W (80-100 mesh)

REFERENCES

1. **Hølmer, G. and Aaes-Jørgensen, E.**, *Lipids*, 4, 507, 1969.
2. **Kochi, M.**, *J. Shimonoseki Univ. Fish.*, 25, 83, 1976.
3. **Grunert, A.**, *Z. Klin. Chem. Klin. Biochem.*, 13, 407, 1975.
4. **Calzolari, C., Cerma, E., and Stancher, B.**, *Riv. Ital. Sost. Grasse*, 48, 605, 1971.
5. **Hølmer, G.**, Fiskeriministeriets Forsøgslaboratorium Kobenhavn, 102, 115, 1967.

Table 9A
METHYL ESTERS OF FATTY ACIDS: ON POLAR (POLYESTER TYPE) PACKED COLUMNS OF MODERATE TO HIGH POLARITY, REPORTED AS RELATIVE RETENTION TIMES

	P1	P2	P3	P4	P5	P6
Column packing						
Temperature (°C)	150	170	170	200	170	185
Gas	Ar	Ar	Ar	N_2	N_2	N_2
Flow rate (ml/min)	na	na	na	20	20	20
Column						
Length (cm)	183	183	183	200	400	400
Diameter						
(cm I.D.)	0.4	0.4	0.4	na	na	na
(cm O.D.)	na	na	na	0.318	0.318	0.318
Form	U-tube	U-tube	U-tube	na	Helix	Helix
Material	G	G	G	SS	SS	SS
Detector	Argon	Argon	Argon	FID	FID	FID
Reference	1	1	2	3	4	4

Compound	\(r_{Methyl\ Octadecanoate}\)					
	P1	P2	P3	P4	P5	P6
Methyl tetradecanoate	—	—	0.27	0.32	0.32	0.36
Methyl pentadecanoate	—	—	0.37	0.43	0.42	0.49
Methyl hexadecanoate	0.46	0.51	0.52	0.57	0.56	0.64
Methyl cis-9-hexadecenoate	0.56	0.62	0.64	0.70	0.67	0.76
Methyl heptadecanoate	—	—	0.72	0.75	0.75	0.83
Methyl heptadecenoate	—	—	0.86	0.92	—	—
Methyl octadecanoate	1.00	1.00	1.00	1.00	1.00	1.00
Methyl cis-9-octadecenoate	1.15	1.16	1.16	1.21	1.20	1.28
Methyl cis,cis-9,12-octadecadienoate	1.52	1.51	1.48	1.51	1.55	1.64
Methyl cis,cis,cis-6,9,12-octadecatrienoate	1.84	1.78	1.76	1.80	—	—
Methyl cis,cis,cis-9,12,15-octadecatrienoate	2.13	2.08	2.01	2.10	2.14	2.21
Methyl cis,cis,cis,cis-6,9,12,15-octadecatetraenoate	—	—	2.40	2.44	—	—
Methyl nonadecanoate	—	—	1.58	1.62	—	—
Methyl icosanoate	2.17	1.98	1.90	1.80	1.80	1.84
Methyl cis-11-icosenoate	2.37	2.21	2.15	2.10	—	—
Methyl cis,cis-11,14-icosadienoate	3.15	2.87	2.76	2.63	2.72	2.74
Methyl cis,cis,cis-8,11,14-icosatrienoate	3.77	3.38	3.26	3.11	—	—
Methyl cis,cis,cis,cis-5,8,11,14-icosatetraenoate	4.18	3.84	3.71	3.44	—	—
Methyl cis,cis,cis-8,11,14,17-icosatetraenoate	5.24	4.67	4.47	4.17	—	—
Methyl cis,cis,cis,cis-5,8,11,14,17-icosapentaenoate	5.80	5.26	5.08	4.78	—	—
Methyl docosanoate	4.69	3.85	—	3.11	3.27	3.22
Methyl cis-11-docosenoate (or cis-13-docosenoate)	4.84	4.20	3.98	3.44	3.85	3.83
Methyl cis,cis-13,16-docosadienoate	—	—	—	—	4.81	4.75
Methyl cis,cis,cis,cis-7,10,13,16-docosatetraenoate	8.51	7.33	6.90	5.90	—	—
Methyl cis,cis,cis,cis,cis-4,7,10,13,16-docosapentaenoate	9.95	8.55	8.02	7.25	—	—
Methyl cis,cis,cis,cis,cis-7,10,13,16,19-docosapentaenoate	11.8	11.2	9.42	8.22	—	—
Methyl cis,cis,cis,cis,cis,cis-4,7,10,13,16,19-docosahexaenoate	13.9	11.7	11.0	9.58	—	—
Methyl tetracosanoate	—	—	—	—	5.75	5.47
Methyl cis-15-tetracosenoate	—	—	7.44	6.50	6.55	6.25

Column packing P1 = 20% Poly (diethylene glycol succinate) on GC-22 Super-Support (60-80 mesh)

P2 = 20% Poly (diethylene glycol succinate) on GC-22 Super-Support (60-80 mesh)

P3 = 20% Poly (diethylene glycol succinate) on Anakrom ABS (70-80 mesh)

P4 = 10% LAC-3-R-728 (commercial diethylene glycol succinate polyester) on Gas-Chrom P (80-100 mesh)

Table 9A (continued)

METHYL ESTERS OF FATTY ACIDS: ON POLAR (POLYESTER TYPE) PACKED COLUMNS OF MODERATE TO HIGH POLARITY, REPORTED AS RELATIVE RETENTION TIMES

P5 = 10% LAC-4-R-886 (commercial ethylene glycol succinate polyester) on Chromosorb W, AW, silanized

P6 = 10% LAC-4-R-886 (commercial ethylene glycol succinate polyester) on Chromosorb W, AW, silanized

REFERENCES

1. **Ackman, R. G.,** *J. Gas Chromatogr.,* 1(4), 11, 1963.
2. **Ackman, R. G. and Burgher, R. D.,** *J. Am. Oil Chem. Soc.,* 42, 38, 1965.
3. **Cerma, E. and Stancher, B.,** *Riv. Ital. Sost. Grasse,* 54, 425, 1977.
4. **Hadorn, H. and Zürcher, K.,** *Mitt. Geb. Lebensm. Hyg.,* 58, 209, 1967.

Table 9B
METHYL ESTERS OF FATTY ACIDS: ON POLAR (POLYESTER TYPE) PACKED COLUMNS OF HIGH POLARITY, REPORTED AS RELATIVE RETENTION TIMES

Column packing	P1	P2	P3	P4	P5
Temperature (°C)	185	175	180	165	208
Gas	N_2	N_2	N_2	N_2	He
Flow rate (mℓ/min)	55	25	25	40	60
Column					
Length (cm)	180	450	450	183	244
Diameter					
(cm I.D.)	na	na	na	na	na
(cm O.D.)	0.31	0.318	0.318	0.318	0.48
Form	Helix	Helix	Helix	Helix	Helix
Material	na	SS	SS	SS	Cu
Detector	FID	FID	FID	FID	TC
Reference	1	2	2	3	4

Compound	$r_{Methyl\ Octadecanoate}$				
Methyl tetradecanoate	0.27	0.33	0.34	0.29	—
Methyl tetradecenoate	—	—	—	0.34	—
Methyl pentadecanoate	0.37	0.44	0.45	0.39	—
Methyl pentadecenoate	—	—	—	0.47	—
Methyl hexadecanoate	0.52	0.58	0.58	0.54	0.57
Methyl cis-9-hexadecenoate	0.64	0.71	—	0.65	—
Methyl cis,cis-6,9-hexadecadienoate	—	0.92[a]	0.89[a]	0.71	—
Methyl cis,cis-9,12-hexadecadienoate	—	—	—	0.82	—
Methyl cis,cis,cis,cis-4,7,10,13-hexadecatetraenoate	—	—	—	1.53	—
Methyl heptadecanoate	0.72	0.78	0.77	0.74	—
Methyl heptadecenoate	—	—	—	0.86	—
Methyl octadecanoate	1.00	1.00	1.00	1.00	1.00
Methyl cis-9-octadecenoate	1.16	1.22	1.17	1.18	1.18
Methyl cis,cis-9,12-octadecadienoate	1.48	1.58	1.51	1.53	1.49
Methyl cis,cis,cis-6,9,12-octadecatrienoate	1.76	—	—	1.76	1.79
Methyl cis,cis,cis-9,12,15-octadecatrienoate	2.05	2.15	2.03	2.06	2.02
Methyl cis,cis,cis,cis-6,9,12,15-octadecatetraenoate	—	—	—	2.47	2.35
Methyl nonadecanoate	1.37	—	—	1.35	—
Methyl nonadecenoate	—	—	—	1.53	—
Methyl icosanoate	1.90	1.86	1.75	1.82	—
Methyl cis-11-icosenoate	2.15	2.15	2.03	2.06	—
Methyl cis,cis-11,14-icosadienoate	2.76	—	2.58	—	—
Methyl cis,cis,cis-5,8,11-icosatrienoate	2.93	—	—	—	—
Methyl cis,cis,cis-8,11,14-icosatrienoate	3.26	—	—	3.24	—
Methyl cis,cis,cis,cis-5,8,11,14-icosatetraenoate	3.71	—	—	4.29	—
Methyl cis,cis,cis,cis-8,11,14,17-icosatetraenoate	4.47	—	—	4.82	—
Methyl cis,cis,cis,cis,cis-5,8,11,14,17-icosapentaenoate	5.08	—	—	6.23	—
Methyl henicosanoate	—	—	—	2.35	—
Methyl docosanoate	3.62	3.33	3.06	3.29	—
Methyl cis-11-docosenoate (or cis-13-docosenoate)	3.98	3.81	3.36	3.06[b]	—
Methyl cis-13,16-docosadienoate	—	—	4.42	—	—
Methyl cis,cis,cis,cis-7,10,13,16-docosatetraenoate	6.90	—	—	7.65	—
Methyl cis,cis,cis,cis,cis-4,7,10,13,16-docosapentaenoate	8.02	—	—	9.65	—
Methyl cis,cis,cis,cis,cis-7,10,13,16,19-docosapentaenoate	9.42	—	—	11.2	—
Methyl cis,cis,cis,cis,cis,cis-4,7,10,13,16,19-docosahexaenoate	11.00	—	—	12.4	—
Methyl tetracosanoate	—	—	—	6.11	—
Methyl cis-15-tetracosenoate	7.44	—	6.04	6.82	—

[a] Probably 16:2ω6.
[b] Changed from 3.60 by R. G. Ackman.

Table 9B (continued)
METHYL ESTERS OF FATTY ACIDS: ON POLAR (POLYESTER TYPE) PACKED COLUMNS OF HIGH POLARITY, REPORTED AS RELATIVE RETENTION TIMES

Column packing P1 = 15% Poly (diethylene glycol succinate) on Gas-Chrom P (100-120 mesh)

P2 = 10% LAC-4-R-886 (commercial polyethylene glycol succinate) on Chromosorb W, AW, HMDS (100-120 mesh)

P3 = 10% LAC-4-R-886 (commercial polyethylene glycol succinate) on Chromosorb W, AW, HMDS (100-120 mesh)

P4 = 15% Poly (diethylene glycol succinate) on Gas-Chrom P (100-200 mesh)

P5 = 15% Poly (ethylene glycol succinate) on Chromosorb W (60-80 mesh)

REFERENCES

1. **Kuksis, A.**, *Fette Seifen Anstrichm.*, 73, 130, 1971.
2. **Hadorn, H. and Zürcher, K.**, *Mitt. Geb. Lebensm. Hyg.*, 59, 78, 1968.
3. **Ghosh, A., Ghosh, M., Hoque, M., and Dutta, J.**, *J. Sci. Food Agric.*, 27, 159, 1976.
4. **Bhatty, M. and Craig, B. M.**, *Can. J. Biochem.*, 44, 311, 1966.

Note added in proof: For additional RRT data recently tabulated for a polyester liquid phase (10% Poly[diethylene glycol succinate] on Chromosorb-W, AW, HMDS [80-100 mesh]) column see: **Ashes, J. R., Haken, J. K., and Mills, S. C.**, *J. Chromatogr.*, 187, 297, 1980.

Table 10
METHYL ESTERS OF FATTY ACIDS: ON POLAR COLUMNS PACKED WITH ORGANOSILICONE (MODIFIED POLYESTER) LIQUID PHASES, REPORTED AS RETENTION TIMES RELATIVE TO OCTADECANOATE

	P1	P2	P3	P4	P5	P6	P7	P8	P9	P10
Column packing										
Temperature (°C)	180	200	200	200	170	200	185	na	178	194
Gas	Ar	Ar	Ar	Ar	Ar	N_2	N_2	na	N_2	N_2
Flow rate (mℓ/min)	na	na	na	na	na	18	55	na	50	50
Column Length (cm)	183	183	183	183	183	200	180	na	214	214
Diameter (cm I.D.)	0.3	0.3	0.3	0.3	0.4	na	na	na	na	na
(cm O.D.)	na	na	na	na	na	0.318	0.318	na	0.64	0.64
Form	U-tube	U-tube	U-tube	U-tube	U-tube	na	Helix	na	na	na
Material	G	G	G	G	G	SS	na	na	G	G
Detector	FID	FID	Argon	Argon	Argon	FID	FID	FID	FID	FID
Reference	1	1	2	2	3	4	5	6	7	7
Compound						$r_{\text{Methyl Octadecenoate}}$				
Methyl tetradecanoate	0.201	0.242	0.32	0.28	0.24	0.32	0.29	—	—	—
Methyl tetradecenoate	0.252	0.300	0.40	0.33	0.31	—	—	—	—	—
Methyl pentadecanoate	0.305	0.348	0.44	0.39	0.34	0.42	0.42	—	—	—
Methyl pentadecenoate	—	—	0.52	0.46	—	—	—	—	—	—
Methyl hexadecanoate	0.447	0.492	0.57	0.53	0.48	0.56	0.54	—	0.58	0.57
Methyl cis-9-hexadecenoate	0.522	0.568	0.70	0.61	0.61	0.68	0.67	—	0.69	0.68
Methyl cis,cis-6,9-hexadecadienoate	—	—	0.85	—	0.70	—	—	—	—	—
Methyl cis,cis-9,12-hexadecadienoate	0.658	0.723	0.94	0.76	0.80	—	—	—	0.90[a]	0.85[a]
Methyl cis,cis,cis-6,9,12-hexadecatrienoate	0.751	0.815	1.12	0.87	—	—	—	—	—	—
Methyl cis,cis,cis,cis-6,9,12,15-hexadecatetraenoate	0.905	0.935	—	—	—	—	—	—	—	—
Methyl heptadecanoate	0.670	0.713	0.75	0.74	0.70	0.75	0.75	—	—	—
Methyl heptadecenoate	0.760	0.795	0.93	0.83	0.84	0.91	—	—	—	—
Methyl octadecanoate	1.00	1.00	1.00	1.00	1.00	1.00	1.00	1.0	1.00	1.00
Methyl cis-9-octadecenoate	1.09	1.12	1.19	1.12	1.12	1.19	1.15	—	1.19	1.16
Methyl cis,cis-9,12-octadecadienoate	1.33	1.32	1.51	1.36	1.43	1.48	1.44	—	1.52	1.41

	P1	P2	P3	P4	P5	P6	P7	P8	P9	P10
Methyl *cis,cis,cis*-6,9,12-octadecatrienoate	1.48	1.48	1.84	1.54	1.64	1.77	—	—	1.80	1.60
Methyl *cis,cis,cis*-9,12,15-octadecatrienoate	1.68	1.64	2.00	1.72	1.89	2.05	2.04	—	2.02	1.78
Methyl *cis,cis,cis,cis*-6,9,12,15-octadecatetraenoate	1.90	1.85	2.43	1.94	2.19	2.36	—	—	2.40	2.04
Methyl nonadecanoate	1.50	1.43	—	—	1.43	—	1.40	—	—	—
Methyl nonadecenoate	1.63	1.58	1.61	1.54	1.57	1.56	—	—	—	—
Methyl icosanoate	2.26	2.06	1.79	1.87	2.03	1.77	1.84	—	1.77	1.76
Methyl *cis*-11-icosenoate	2.38	2.26	2.06	2.11	2.19	2.05	2.04	—	2.08	2.01
Methyl *cis*-11,14-icosadienoate	2.89	2.68	2.64	2.56	2.78	2.61	2.66	—	2.67	2.44
Methyl *cis,cis,cis*-5,8,11-icosatrienoate	—	—	2.83	—	—	—	2.92	—	2.87	2.57
Methyl *cis,cis,cis*-8,11,14-icosatrienoate	—	2.96	3.14	2.84	3.18	3.09	3.25	3.0	2.99	2.72
Methyl *cis,cis,cis*-11,14,17-icosatrienoate	—	—	—	—	—	—	—	—	3.13	2.78
Methyl *cis,cis,cis,cis*-5,8,11,14-icosatetraenoate	3.45	3.13	3.56	3.08	3.55	3.50	3.67	—	3.59	3.10
Methyl *cis,cis,cis,cis*-8,11,14,17-icosatetraenoate	4.13	3.70	4.15	3.54	4.22	4.03	4.43	—	4.18	3.54
Methyl *cis,cis,cis,cis,cis*-5,8,11,14,17-icosapentaenoate	4.35	3.92	4.75	3.88	4.72	4.61	4.95	—	4.85	3.91
Methyl *cis,cis,cis,cis,cis*-6,9,12,15,18-henicosapentaenoate	6.70	5.72	6.21	5.32	—	—	—	—	—	—
Methyl docosanoate	—	—	—	—	—	3.09	3.34	—	3.14	3.12
Methyl *cis*-11-docosenoate (or *cis*-13-docosenoate)	5.13	4.40	3.57	3.93	4.22	3.50	3.75	—	3.59	3.44
Methyl *cis,cis,cis,cis*-7,10,13,16-docosatetraenoate	7.61	6.30	6.21	5.76	6.97	6.25	6.66	6.0	6.34	5.48
Methyl *cis,cis,cis,cis,cis*-4,7,10,13,16-docosapentaenoate	8.22	6.80	7.26	6.34	8.16	7.02	7.50	6.9	6.55	5.27
Methyl *cis,cis,cis,cis,cis*-7,10,13,16,19-docosapentaenoate	9.67	7.85	8.32	7.19	9.35	8.14	8.75	8.0	8.60	6.92
Methyl *cis,cis,cis,cis,cis,cis*-4,7,10,13,16,19-docosahexaenoate	10.54	8.43	9.57	7.89	10.75	9.40	10.15	9.2	9.10	7.74
Methyl tetracosanoate	—	—	—	—	—	—	5.91	—	5.59	5.90
Methyl tetracosenoate	11.37	8.80	—	—	—	6.79	6.75	—	—	—

a   Methyl *cis,cis*-7,10-hexadecadienoate.

Column packing P1 = 3% EGSP-Z (ethylene glycol succinate polyester combined with phenyl silicone) on Gas-Chrom Q (100-120 mesh)

P2 = 3% EGSP-Z (ethylene glycol succinate polyester combined with phenyl silicone) on Gas-Chrom Q (100-120 mesh)

P3 = 10% EGSS-X (ethylene glycol succinate polyester combined with methyl silicone) on Gas-Chrom P (100-120 mesh)

P4 = 12% EGSS-Y (ethylene glycol succinate polyester combined with methyl silicone) on Gas-Chrom P (100-120 mesh)

P5 = 3% EGSS-X (ethylene glycol succinate polyester combined with methyl silicone) on Gas-Chrom P (100-120 mesh)

P6 = 10% EGSS-X (ethylene glycol succinate polyester combined with methyl silicone) on Gas-Chrom P (80-100 mesh)

P7 = 10% EGSS-X (ethylene glycol succinate polyester combined with methyl silicone) on Gas-Chrom P (100-120 mesh)

P8 = 15% EGSS-X (ethylene glycol succinate polyester combined with methyl silicone) on Gas-Chrom P (100-120 mesh)

P9 = 15% EGSS-X (ethylene glycol succinate polyester combined with methyl silicone) on Gas-Chrom P

P10 = 15% EGSS-X (ethylene glycol succinate polyester combined with methyl silicone) on Chromosorb W, AW, DMCS (100-120 mesh)

P10 = 15% EGSS-Y (ethylene glycol succinate polyester combined with methyl silicone) on Chromosorb W, AW, DMCS (100-120 mesh)

Table 10 (continued)

METHYL ESTERS OF FATTY ACIDS: ON POLAR COLUMNS PACKED WITH ORGANOSILICONE (MODIFIED POLYESTER) LIQUID PHASES, REPORTED AS RETENTION TIMES RELATIVE TO OCTADECANOATE

REFERENCES

1. **Ackman, R. G.,** *J. Gas Chromatogr.*, 4, 256, 1966.
2. **Ackman, R. G. and Burgher, R. D.,** *Can. J. Biochem. Physiol.*, 41, 2501, 1963.
3. **Ackman, R. G. and Burgher, R. D.,** *J. Am. Oil Chem. Soc.*, 42, 38, 1965.
4. **Cerma, E. and Stancher, B.,** *Riv. Ital. Sost. Grasse*, 54, 425, 1977.
5. **Grogan, W. M., Jr., Coniglio, J. G., and Rhany, R. K.,** *Lipids*, 8, 480, 1973.
6. **Kuksis, A.,** *Fette Seifen Anstrichm.*, 73, 130, 1971.
7. **Christie, W. W.,** *Lipid Analysis*, Pergamon Press, Oxford, 1973.

Note added in proof: For additional RRT data recently tabulated for an organosilicone liquid phase (15% EGSS-X [ethylene glycol-succinate polyester combined with methyl silicone] on Gas-Chrom P, [100-120 mesh]) column, see: **Walker, B. L.,** *Lipids*, 16, 468, 1981.

Table 11
METHYL ESTERS OF FATTY ACIDS: ON POLAR COLUMNS PACKED WITH SUPPORT COATED WITH CYANOSILICONE LIQUID PHASES, REPORTED RELATIVE TO METHYL OCTADECANOATE

Column packing	P1	P2	P3	P4	P5	P6
Temperature (°C)	200	180	170	195	200	200
Gas	N_2	N_2	N_2	N_2	N_2	N_2
Flow rate (mℓ/min)	23	20	10	20	20	20
Column						
Length (cm)	200	200	400	305	183	183
Diameter						
(cm I.D.)	na	na	0.2	na	0.2	0.2
(cm O.D.)	0.318	0.318	na	0.318	na	na
Form	na	na	Helix	Helix	na	na
Material	SS	SS	G	SS	G	G
Detector	FID	FID	FID	FID	FID	FID
Reference	1	1	2	3	4	4

Compound	$r_{Methyl\ Octadecanoate}$					
Methyl tetradecanoate	0.36	0.38	0.37	0.32	—	—
Methyl tetradecenoate	—	—	0.48	0.40	—	—
Methyl pentadecanoate	0.47	0.49	0.48	0.44	—	—
Methyl hexadecanoate	0.61	0.62	0.61	0.57	—	—
Methyl *cis*-9-hexadecenoate	0.76	0.81	0.76	0.68	—	—
Methyl heptadecanoate	0.78	0.79	0.78	0.76	—	—
Methyl heptadecenoate	0.95	1.01	—	—	—	—
Methyl octadecanoate	1.00	1.00	1.00	1.00	1.00	1.00
Methyl *cis*-9-octadecenoate	1.20	1.25	1.21	1.16	1.21	1.21
Methyl *cis*,*cis*-9,12-octadecadienoate	1.57	1.66	1.53	1.44	1.58	1.62
Methyl *cis*,*cis*,*cis*-6,9,12-octadecatrienoate	1.83	2.02	1.86	—	—	—
Methyl *cis*,*cis*,*cis*-9,12,15-octadecatrienoate	2.05	2.32	2.08	1.87	2.10	2.22
Methyl *cis*,*cis*,*cis*,*cis*-6,9,12,15-octadecatetraenoate	2.50	2.88	—	—	—	—
Methyl nonadecanoate	—	—	1.28	1.32	—	—
Methyl nonadecenoate	1.50	1.58	1.55	—	—	—
Methyl icosanoate	1.63	1.60	1.64	1.75	1.60	1.52
Methyl *cis*-11-icosenoate	1.97	2.02	1.98	2.02	—	—
Methyl *cis*,*cis*-11,14-icosadienoate	2.53	2.60	2.51	—	—	—
Methyl *cis*,*cis*,*cis*-8,11,14-icosatrienoate	3.10	3.30	3.04	—	—	—
Methyl *cis*,*cis*,*cis*,*cis*-5,8,11,14-icosatetraenoate	3.53	3.82	3.37	3.19	—	—
Methyl *cis*,*cis*,*cis*,*cis*-8,11,14,17-icosatetraenoate	4.02	4.44	—	—	—	—
Methyl *cis*,*cis*,*cis*,*cis*,*cis*-5,8,11,14,17-icosapentaenoate	4.61	5.29	4.48	4.13	—	—
Methyl henicosanoate	—	—	2.10	2.32	—	—
Methyl docosanoate	2.75	2.60	2.69	3.09	—	—
Methyl *cis*-11-docosenoate (or *cis*-13-docosenoate)	3.10	3.12	3.25	3.52	—	—
Methyl *cis*,*cis*-13,16-docosadienoate	—	—	4.11	—	—	—
Methyl *cis*,*cis*,*cis*,*cis*-7,10,13,16-docosatetraenoate	5.99	6.20	5.52	—	—	—
Methyl *cis*,*cis*,*cis*,*cis*,*cis*-4,7,10,13,16-docosapentaenoate	6.50	6.81	—	—	—	—
Methyl *cis*,*cis*,*cis*,*cis*,*cis*-7,10,13,16,19-docosapentaenoate	7.54	8.40	—	—	—	—
Methyl *cis*,*cis*,*cis*,*cis*,*cis*,*cis*-4,7,10,13,16,19-docosahexaenoate	8.53	9.57	8.00	8.34	—	—
Methyl tetracosanoate	—	—	4.40	5.44	—	—
Methyl *cis*-15-tetracosenoate	5.41	5.63	—	6.05	—	—

Table 11 (continued)
METHYL ESTERS OF FATTY ACIDS: ON POLAR COLUMNS PACKED WITH SUPPORT COATED WITH CYANOSILICONE LIQUID PHASES, REPORTED RELATIVE TO METHYL OCTADECANOATE

Column packing P1 = 10% SILAR-10C (APOLAR-10C; 100% cyanopropyl siloxane) on GAS-CHROM P (80-100 mesh)

P2 = 10% OV-275 (100% cyanoethyl siloxane) on GAS-CHROM P (80-100 mesh)

P3 = 12% SILAR-10C (100% cyanopropyl siloxane) on Chromosorb W, HP (80-100 mesh)

P4 = 10% ECNSS-S (ethylene glycol succinate combined with cyanoethyl silicone) on silanized GAS-CHROM P (100-120 mesh)

P5 = 20% OV-275 (100% cyanoethyl siloxane) on Chromosorb W, AW (100-120 mesh)

P6 = 20% OV-275 (100% cyanoethyl siloxane) on Supelcoport (AW, DMCS treated) (100-120 mesh)

REFERENCES

1. **Cerma, E. and Stancher, B.,** *Riv. Ital. Sost. Grasse,* 54, 425, 1977.
2. **Heckers, H., Dittmar, K., Melcher, F. W., and Kalinowski, H. O.,** *J. Chromatogr.,* 135, 93, 1977.
3. **Feldman, G. L.,** *J. Gas Chromatogr.,* 1(9), 26, 1963.
4. **Supelco, Inc.,** *Chromatography,* Lipids VIII (6), 1, 1974.

Note added in proof: For additional RRT data recently tabulated for the cyanosilicone liquid phases: (A) 15% OV-275 on Chromosorb-P, AW, HMDS; (B) 10% SILAR-5CP on GAS-CHROM P (80-100) mesh; (C) 15% OV-17 on Chromosorb-W, AW, HMDS (80-100 mesh); (D) 15% OV-275 on GAS-CHROM RZ (100-120 mesh); (E) 10% SILAR-10C on GAS-CHROM Q (100-120 mesh); (F) 15% OV-275 on Chromosorb P, AW, HMDS (100-120 mesh); in packed columns, see:(A) **Cantafora, A., David, E., and Di Biase, A.,** *Riv. Soc. Ital. Sci. Aliment.,* 8, 27, 1979; (B) **Homberg, E. and Bielefeld, B.,** *Fette Seifen Anstrichm.,* 81, 377, 1979; (C) **Ashes, J. R., Haken, J. K., and Mills, S. C.,** *J. Chromatogr.,* 187, 297, 1980; (D) **Itabashi, Y. and Takagi, T.,** *Yukugaku,* 27, 219, 1978; (E) **Court, W. A. and Hendel, J. G.,** *J. Chromatogr. Sci.,* 16, 314, 1978, and (F) **Walker, B. L.,** *J. Chromatogr.,* 16, 468, 1981.

Table 12A
METHYL ESTERS OF FATTY ACIDS: ON POLAR (POLYESTER TYPE WITH ONE VINYL ACETATE) PACKED COLUMNS OF LOW POLARITY, REPORTED AS EQUIVALENT CHAIN LENGTHS

Column packing	P1	P2	P3	P4	P5	P6
Temperature (°C)	240	180	200	200	180	180
Gas	He	Ar	—	Ar	Ar	N_2
Flow rate (mℓ/min)	na	na	na	40	na	na
Column						
Length (cm)	360	122	183	152	152	305
Diameter						
(cm I.D.)	na	na	na	na	na	na
(cm O.D.)	0.635	na	0.318	na	0.63	0.318
Form	Helix	Tube	Helix	na	Helix	Helix
Material	na	G	SS	na	SS	SS
Detector	TC	Argon	FID	FID	TC	FID
Reference	1	2	3	4	5	6

Compound			ECL			
Methyl tetradecanoate	14.00	14.00	14.00	14.00	14.00	14.0
Methyl tetradecenoate	—	14.50	—	14.55	—	—
Methyl pentadecanoate	15.00	15.00	15.00	15.00	15.00	15.0
Methyl hexadecanoate	16.00	16.00	16.00	16.00	16.00	16.0
Methyl cis-9-hexadecenoate	16.55	16.45	—	16.49	16.40	16.4
Methyl cis,cis-6,9-hexadecadienoate	—	—	—	—	17.00[a]	16.9[a]
Methyl cis,cis-9,12-hexadecadienoate	—	—	—	17.20	—	—
Methyl cis,cis,cis-6,9,12-hexadecatrienoate	—	—	—	17.70	17.80[a]	17.5[a]
Methyl cis,cis,cis,cis-4,7,10,13-hexadecatetraenoate	—	—	—	—	—	18.1[a]
Methyl heptadecanoate	17.00	17.00	17.00	17.00	17.00	17.0
Methyl heptadecenoate	—	—	—	17.35	17.45	17.4
Methyl octadecanoate	18.00	18.00	18.00	18.00	18.00	18.0
Methyl cis-9-octadecenoate	18.32	18.35	—	18.40	18.37	18.4
Methyl cis,cis-9,12-octadecadienoate	19.02	18.90	18.91	18.95	19.00	18.9
Methyl cis,cis,cis-6,9,12-octadecatrienoate	—	—	19.29	—	19.45	—
Methyl cis,cis,cis-9,12,15-octadecatrienoate	19.6x	19.75	19.64	19.80	19.80	19.5
Methyl cis,cis,cis,cis-6,9,12,15-octadecatetraenoate	—	—	10.05	20.20	20.26	—
Methyl nonadecanoate	19.00	19.00	19.00	19.00	19.00	19.0
Methyl nonadecenoate	—	—	—	19.65	—	19.4
Methyl icosanoate	20.00	20.00	20.00	20.00	20.00	20.1
Methyl cis-11-icosenoate	—	20.35	—	20.35	20.26	20.4
Methyl cis,cis-11,14-icosadienoate	20.65	—	—	20.90	20.90	20.9
Methyl cis,cis,cis-8,11,14-icosatrienoate	21.42	21.10	21.12	21.30	21.3	21.4
Methyl cis,cis,cis-11,14,17-icosatrienoate	—	—	—	—	—	—
Methyl cis,cis,cis,cis-5,8,11,14-icosatetraenoate	22.0x	21.60	21.44	21.75	21.60	21.9
Methyl cis,cis,cis,cis-8,11,14,17-icosatetraenoate	—	—	21.84	—	—	22.1
Methyl cis,cis,cis,cis,cis-5,8,11,14,17-icosapentaenoate	22.42	22.10	22.09	22.48	22.50	22.4
Methyl henicosanoate	—	21.00	21.00	—	20.90	21.0?
Methyl cis,cis,cis,cis,cis-6,9,12,15,18-henicosapentaenoate	—	—	—	—	—	23.4
Methyl docosanoate	22.0	22.00	22.00	22.00	21.90	22.0?
Methyl cis-11-docosenoate (or cis-13-docosenoate)	—	22.35	—	22.20	—	—
Methyl cis,cis,cis,cis-7,10,13,16-docosatetraenoate	23.35	23.60	23.29	23.55	—	24.1
Methyl cis,cis,cis,cis,cis-7,10,13,16,19-docosapentaenoate	24.0x	24.05	24.04	24.55	—	24.4
Methyl cis,cis,cis,cis,cis,cis-4,7,10,13,16,19-docosahexaenoate	24.32	24.40	24.38	24.85	—	24.7
Methyl tetracosanoate	24.00	24.00	24.00	24.00	23.80	24.0?
Methyl cis-15-tetracosenoate	24.32	—	—	—	24.10	—

Table 12A (continued)

METHYL ESTERS OF FATTY ACIDS: ON POLAR (POLYESTER TYPE WITH ONE VINYL ACETATE) PACKED COLUMNS OF LOW POLARITY, REPORTED AS EQUIVALENT CHAIN LENGTHS

ᵃ   Isomer other or uncertain.

Column packing
P1 = ?% Poly (diethylene glycol succinate)
P2 = 5% Poly (vinyl acetate); on Celite, AW, ALK-W (100-120 mesh)
P3 = 8% Poly (1,4-butananediol succinate) on Chromosorb W, HMDS (80-100 mesh)
P4 = 10% Poly (ethylene glycol adipate) on Celite (100-120 mesh)
P5 = Poly (ethylene glycol adipate)
P6 = 20% Poly (diethylene glycol succinate) on Chromosorb W (80-100 mesh)

REFERENCES

1. **Johnston, P. V. and Kummerow, F. A.,** *Proc. Soc. Exp. Biol. Med.,* 104, 201, 1960.
2. **Kingsbury, K. J., Morgan, D. N., and Heyes, T. D.,** *Biochem. J.,* 90, 140, 1964.
3. **Jamieson, G. R. and Reid, E.,** *J. Chromatogr.,* 39, 71, 1969.
4. **Neudoerffer, T. S. and Lea, C. H.,** *Br. J. Nutr.,* 20, 581, 1966.
5. **Haigh, W. G., Safford, R., and James, A. T.,** *Biochim. Biophys. Acta,* 176, 647, 1969.
6. **Graille, J., Huq, M. S., and Naudet, M.,** *Rev. Franc. Corps Gras,* 23, 151, 1976.

Note added in proof: For additional ECL data recently tabulated for the polyester liquid phases: (A) 15% Poly(diethylene glycol pentaerythritol adipate) on GAS-CHROM Q, AW, HMDS (60-80 mesh); (B) 10% Poly(diethylene glycol succinate) on Chromosorb W, AW, HMDS (80-100 mesh); in packed columns, see: (A) **Miyagawa, M., Miwa, T. K., and Spencer, G. F.,** *J. Am. Oil Chem. Soc.,* 56, 834, 1979; and (B) **Ashes, J. R., Haken, J. K., and Mills, S. C.,** *J. Chromatogr.,* 187, 297, 1980.

Table 12B
METHYL ESTERS OF FATTY ACIDS: ON POLAR (POLYESTER TYPE) PACKED COLUMNS OF LOW TO MODERATE POLARITY, REPORTED AS EQUIVALENT CHAIN LENGTHS

Column packing	P1	P2	P3	P4	P5	P6
Temperature (°C)	150	184	180	190	200	210
Gas	Ar	Ar	N_2	N_2	N_2	N_2
Flow rate (mℓ/min)	na	40	30	30	30	30
Column						
Length (cm)	183	142	183	183	183	183
Diameter						
(cm I.D.)	0.4	0.4	na	na	na	na
(cm O.D.)	na	na	0.318	0.318	0.318	0.318
Form	U-tube	Tube	Helix	Helix	Helix	Helix
Material	G	G	SS	SS	SS	SS
Detector	Argon	Argon	FID	FID	FID	FID
Reference	1	2	3	3	3	3

Compound	ECL					
Methyl hexadecanoate	16.00	16.00	16.00	16.00	16.00	16.00
Methyl *cis*-9-hexadecenoate	16.50	16.48	16.37	16.38	16.39	16.39
Methyl *cis,cis*-6,9-hexadecadienoate	—	16.99	—	—	—	—
Methyl *cis,cis*-9,12-hexadecadienoate	—	17.30	—	—	—	—
Methyl *cis,cis,cis*-6,9,12-hexadecatrienoate	—	17.66	—	—	—	—
Methyl *cis,cis,cis,cis*-4,7,10,13-hexadecatetraenoate	—	18.26	—	—	—	—
Methyl *cis,cis,cis,cis*-6,9,12,15-hexadecatetraenoate	—	18.36	—	—	—	—
Methyl heptadecanoate	17.00	17.00	17.00	17.00	17.00	17.00
Methyl octadecanoate	18.00	18.00	18.00	18.00	18.00	18.00
Methyl *cis*-9-octadecenoate	18.36	18.38	18.32	18.35	18.39	18.41
Methyl *cis,cis*-9,12-octadecadienoate	19.07	18.98	18.78	18.82	18.91	18.94
Methyl *cis,cis,cis*-6,9,12-octadecatrienoate	19.59	19.45	19.15	19.18	19.29	19.32
Methyl *cis,cis,cis*-9,12,15-octadecatrienoate	19.96	19.82	19.47	19.53	19.64	19.66
Methyl *cis,cis,cis,cis*-6,9,12,15-octadecatetraenoate	20.45	20.27	19.85	19.91	20.05	20.09
Methyl nonadecanoate	19.00	19.00	19.00	19.00	19.00	19.00
Methyl icosanoate	20.00	20.00	20.00	20.00	20.00	20.00
Methyl *cis*-11-icosenoate	20.24	20.35	20.27	20.31	20.33	20.38
Methyl *cis,cis*-11,14-icosadienoate	20.97	21.00	20.70	20.77	20.82	20.91
Methyl *cis,cis,cis*-5,8,11-icosatrienoate	—	—	—	—	20.79	—
Methyl *cis,cis,cis*-8,11,14-icosatrienoate	21.44	21.40	—	21.12	21.16	21.29
Methyl *cis,cis,cis*-11,14,17-icosatrienoate	—	—	—	21.50	21.59	—
Methyl *cis,cis,cis,cis*-5,8,11,14-icosatetraenoate	21.71	21.76	—	21.35	21.44	21.55
Methyl *cis,cis,cis,cis*-8,11,14,17-icosatetraenoate	22.30	22.21	—	21.82	21.88	22.00
Methyl *cis,cis,cis,cis,cis*-5,8,11,14,17-icosapentaenoate	22.61	22.52	—	22.12	22.24	22.34
Methyl henicosanoate	21.00	21.00	21.00	21.00	21.00	21.00
Methyl docosanoate	22.00	22.00	22.00	22.00	22.00	22.00
Methyl *cis*-11-docosenoate (or *cis*-13-docosenoate)	22.09	22.36	22.23	22.22	22.24	22.30
Methyl *cis,cis*-13,16-docosadienoate	—	22.95	—	—	—	—
Methyl *cis,cis,cis,cis*-7,10,13,16-docosatetraenoate	23.62	23.71	—	23.40	23.29	—
Methyl *cis,cis,cis,cis,cis*-4,7,10,13,16-docosapentaenoate	24.04	24.05	—	—	23.91	—
Methyl *cis,cis,cis,cis,cis*-7,10,13,16,19-docosapentaenoate	24.39	24.51	—	24.18	24.04	—
Methyl *cis,cis,cis,cis,cis,cis*-4,7,10,13,16,19-docosahexaenoate	24.81	24.86	—	24.57	24.38	24.50
Methyl tetracosanoate	24.00	24.00	24.00	24.00	24.00	24.00
Methyl *cis*-15-tetracosenoate	—	—	—	—	24.22	—

Table 12B (continued)
METHYL ESTERS OF FATTY ACIDS: ON POLAR (POLYESTER TYPE) PACKED COLUMNS OF LOW TO MODERATE POLARITY, REPORTED AS EQUIVALENT CHAIN LENGTHS

Column packing P1 = 20% Poly (diethylene glycol succinate) on GC-22 Super-Support (70-80 mesh)
P2 = 18% Poly (ethylene glycol adipate) on Celite, AW, ALK-W (120-140 mesh)
P3 = 8% Poly (1,4-butanediol succinate) on Chromosorb W, HMDS (80-100 mesh)
P4 = 8% Poly (1,4-butanediol succinate) on Chromosorb W, HMDS (80-100 mesh)
P5 = 8% Poly (1,4-butanediol succinate) on Chromosorb W, HMDS (80-100 mesh)
P6 = 8% Poly (1,4-butanediol succinate) on Chromosorb W, HMDS (80-100 mesh)

REFERENCES

1. **Ackman, R. G.,** *J. Gas Chromatogr.,* 1(4), 11, 1963.
2. **Farquhar, J. W., Insull, W., Jr., Rosen, P., Stoffel, W., and Ahrens, E. H., Jr.,** *Nutr. Rev. (Suppl.),* 17, 1, 1959. (See also **Jamieson, G. R.,** in *Topics in Lipid Chemistry,* Vol. I, Gunstone, F. D., Ed., Logos Press, London, 1970, 107; **West, C. E. and Rowbotham, T. R.,** *J. Chromatogr.,* 30, 62, 1967.)
3. **Jamieson, G. R.,** in *Topics in Lipid Chemistry,* Vol. I, Gunstone, F. D., Ed., Logos Press, London, 1970, 107.

Table 13A
METHYL ESTERS OF FATTY ACIDS: ON POLAR (POLYESTER TYPE) PACKED COLUMNS OF MODERATE TO HIGH POLARITY, REPORTED AS EQUIVALENT CHAIN LENGTHS

Column packing	P1	P2	P3	P4	P5	P6
Temperature (°C)	234	226	180	150	170	190
Gas	He	He	Ar	Ar	Ar	Ar
Flow rate (mℓ/min)	120	60	60	na	na	na
Column						
Length (cm)	360	200	200	183	183	183
Diameter						
(cm I.D.)	0.3	0.3	0.4	0.4	0.4	0.4
(cm O.D.)	na	na	na	na	na	na
Form	na	na	U-tube	U-tube	U-tube	U-tube
Material	na	na	G	G	G	G
Detector	TC	TC	Argon	Argon	Argon	Argon
Reference	1	1	1	2	2	2

Compound			ECL			
Methyl tetradecanoate	14.00	14.00	14.00	14.00	14.00	14.00
Methyl tetradecenoate	14.71	14.80	14.75	—	—	—
Methyl pentadecanoate	15.00	15.00	15.00	15.00	15.00	15.00
Methyl hexadecanoate	16.00	16.00	16.00	16.00	16.00	16.00
Methyl *cis*-9-hexadecenoate	16.55	16.55	16.56	16.50	16.55	16.51
Methyl *cis,cis*-6,9-hexadecadienoate	17.25	17.50	17.32	—	—	—
Methyl *cis,cis,cis*-6,9,12-hexadecatrienoate	18.10	18.52	18.33	—	—	—
Methyl heptadecanoate	17.00	17.00	17.00	17.00	17.00	17.00
Methyl heptadecenoate	17.56	17.60	17.55	—	—	—
Methyl octadecanoate	18.00	18.00	18.00	18.00	18.00	18.00
Methyl *cis*-9-octadecenoate	18.55	18.51	18.50	18.36	18.43	18.50
Methyl *cis,cis*-9,12-octadecadienoate	19.23	19.30	19.22	19.07	19.22	19.34
Methyl *cis,cis,cis*-6,9,12-octadecatrienoate	19.70	20.00	19.78	19.59	19.73	19.85
Methyl *cis,cis,cis*-9,12,15-octadecatrienoate	20.10	20.40	20.13	19.96	20.12	20.34
Methyl *cis,cis,cis,cis*-6,9,12,15-octadecatetraenoate	20.73	21.00	21.15	20.45	20.65	20.85
Methyl nonadecanoate	19.00	19.00	19.00	19.00	19.00	19.00
Methyl nonadecenoate	19.53	19.60	19.50	—	—	—
Methyl icosanoate	20.00	20.00	20.00	20.00	20.00	20.00
Methyl *cis*-11-icosenoate	20.32	20.44	20.38	20.24	20.36	20.47
Methyl *cis,cis*-11,14-icosadienoate	21.13	21.36	21.13	20.97	21.14	21.28
Methyl *cis,cis,cis*-5,8,11-icosatrienoate	21.53	21.65	21.57	—	—	—
Methyl *cis,cis,cis*-8,11,14-icosatrienoate	21.72	22.13	21.65	21.44	21.62	21.80
Methyl *cis,cis,cis,cis*-5,8,11,14-icosatetraenoate	22.15	22.43	22.25	21.71	22.00	22.38
Methyl *cis,cis,cis,cis*-8,11,14,17-icosatetraenoate	—	—	—	22.30	22.68	22.90
Methyl *cis,cis,cis,cis,cis*-5,8,11,14,17-icosapentaenoate	22.88	23.45	22.92	22.61	22.92	23.32
Methyl henicosanoate	21.00	21.00	21.00	21.00	21.00	21.00
Methyl docosanoate	22.00	22.00	22.00	22.00	22.00	22.00
Methyl *cis*-11-docosenoate (or *cis*-13-docosenoate)	22.27	22.28	22.30	22.09	22.30	22.45
Methyl *cis,cis,cis,cis*-7,10,13,16-docosatetraenoate	24.03	24.58	23.85	23.62	23.94	24.32
Methyl *cis,cis,cis,cis,cis*-4,7,10,13,16-docosapentaenoate	24.43	24.97	24.37	24.04	24.40	24.71
Methyl *cis,cis,cis,cis,cis*-7,10,13,16,19-docosapentaenoate	24.87	25.38	24.93	24.39	24.87	25.20
Methyl *cis,cis,cis,cis,cis,cis*-4,7,10,13,16,19-docosahexaenoate	25.35	26.03	25.40	24.81	25.26	25.72
Methyl tetracosanoate	24.00	24.00	24.00	24.00	24.00	24.00
Methyl *cis*-15-tetracosenoate	24.20	24.27	24.40	—	—	—

Table 13A (continued)
METHYL ESTERS OF FATTY ACIDS: ON POLAR (POLYESTER TYPE) PACKED COLUMNS OF MODERATE TO HIGH POLARITY, REPORTED AS EQUIVALENT CHAIN LENGTHS

Column packing P1 = 20% Poly β-cyclodextrin acetate on Gas-Chrom P (30-60 mesh)
  P2 = 20% Poly (diethylene glycol succinate) on Gas-Chrom P (80-100 mesh)
  P3 = 20% Poly (ethylene glycol succinate) on Gas-Chrom P (80-100 mesh)
  P4 = 20% Poly (diethylene glycol succinate) on GC-22 Super-Support (60-80 mesh)
  P5 = 20% Poly (diethylene glycol succinate) on GC-22 Super-Support (60-80 mesh)
  P6 = 20% Poly (diethylene glycol succinate) on GC-22 Super-Support (60-80 mesh)

REFERENCES

1. **Hofstetter, H. H., Sen, N., and Holman, R. T.,** *J. Am. Oil Chem. Soc.*, 42, 537, 1965.
2. **Ackman, R. G.,** *J. Gas Chromatogr.*, 1(4), 11, 1963.

Table 13B
METHYL ESTERS OF FATTY ACIDS: ON POLAR (POLYESTER TYPE) PACKED COLUMNS OF MODERATE TO HIGH POLARITY REPORTED AS EQUIVALENT CHAIN LENGTHS

Column packing	P1	P2	P3	P4	P5	P6	P7
Temperature (°C)	180	205	170	220	190	190	165
Gas	Ar	He	Ar	He	na	na	N_2
Flow rate (mℓ/min)	na	90	na	90	na	30	40
Column							
Length (cm)	244	305	183	305	90	152	183
Diameter							
(cm I.D.)	0.4	na	0.4	na	0.2	0.2	na
(cm O.D.)	na	0.635	na	0.635	na	na	0.318
Form	U-tube	Helix	U-tube	Helix	na	Helix	na
Material	G	Cu	G	Cu	G	SS	SS
Detector	Argon	TC	Argon	TC	FID	FID	FID
References	1	2	3	4	5	6	7

Compound	ECL						
Methyl tetradecanoate	14.00	14.00	14.00	14.00	14.00	14.0	14.0
Methyl tetradecenoate	—	14.80	—	—	—	—	14.5
Methyl pentadecanoate	15.00	15.00	15.00	15.00	15.00	15.0	15.0
Methyl pentadecenoate	—	15.67	—	—	—	—	15.6
Methyl hexadecanoate	16.00	16.00	16.00	16.00	16.00	16.0	16.0
Methyl cis-9-hexadecenoate	—	16.65	16.61	16.63	16.58	—	16.6
Methyl cis,cis-6,9-hexadecadienoate	—	17.36	—	—	—	—	16.9
Methyl cis,cis-9,12-hexadecadienoate	—	17.57	—	17.47	17.51[a]	—	17.4
Methyl cis,cis,cis-6,9,12-hexadecatrienoate	—	18.28	—	18.09	18.16[b]	—	17.8
Methyl cis,cis,cis,cis-4,7,10,13-hexadecatetraenoate	—	18.95	—	—	—	—	19.4
Methyl cis,cis,cis,cis-6,9,12,15-hexadecatetraenoate	—	19.28	—	—	—	—	—
Methyl heptadecanoate	17.00	17.00	17.00	17.00	17.00	17.0	17.0
Methyl heptadecenoate	—	17.57	17.53	—	—	—	17.4
Methyl octadecanoate	18.00	18.00	18.00	18.00	18.00	18.0	18.0
Methyl cis-9-octadecenoate	18.55	18.56	18.46	18.53	18.49	18.6	18.5
Methyl cis,cis-9,12-octadecadienoate	19.42	19.26	19.25	19.35	19.25	19.5	18.9
Methyl cis,cis,cis-6,9,12-octadecatrienoate	—	19.86	19.76	20.00	—	—	19.9
Methyl cis,cis,cis-9,12,15-octadecatrienoate	20.50	20.20	20.18	20.33	20.24	—	20.4
Methyl cis,cis,cis,cis-6,9,12,15-octadecatetraenoate	—	20.82	20.74	20.95	—	—	21.0
Methyl nonadecanoate	19.00	19.00	19.00	19.00	19.00	19.0	19.0
Methyl nonadecenoate	—	19.54	—	—	—	—	19.4
Methyl icosanoate	20.00	20.00	20.00	20.00	20.00	20.0	20.0
Methyl cis-11-icosenoate	—	20.43	20.39	20.50	20.44	—	20.4
Methyl cis,cis-11,14-icosadienoate	—	21.10	21.19	21.28	21.18	—	—
Methyl cis,cis,cis-5,8,11-icosatrienoate	—	—	—	—	—	21.9	—
Methyl cis,cis,cis-8,11,14-icosatrienoate	—	21.69	21.70	21.96	—	22.2	21.9
Methyl cis,cis,cis,cis-5,8,11,14-icosatetraenoate	22.48	22.10	22.01	22.35	—	22.8	22.8
Methyl cis,cis,cis,cis-8,11,14,17-icosatetraenoate	—	22.68	22.69	22.94	22.64[a]	—	23.5
Methyl cis,cis,cis,cis,cis-5,8,11,14,17-icosapentaenoate	23.55	23.05	23.06	23.32	23.19	23.7	24.0
Methyl henicosanoate	21.00	21.00	21.00	21.00	21.00	21.0	21.0
Methyl cis,cis,cis,cis,cis-6,9,12,15,18-henicosapentaenoate	—	23.94	—	—	—	—	—
Methyl docosanoate	22.00	22.00	22.00	22.00	22.00	22.0	22.0

Table 13B (continued)
METHYL ESTERS OF FATTY ACIDS: ON POLAR (POLYESTER TYPE) PACKED COLUMNS OF MODERATE TO HIGH POLARITY REPORTED AS EQUIVALENT CHAIN LENGTHS

Compound	ECL						
Methyl *cis*-11-docosenoate (or *cis*-13-docosenoate)	22.48	22.32	22.32	22.51	—	—	22.3
Methyl *cis,cis,cis,cis*-7,10,13,16-docosatetraenoate	—	23.94	24.00	24.33	—	24.8	24.7
Methyl *cis,cis,cis,cis,cis*-4,7,10,13,16-docosapentaenoate	—	24.18	24.49	24.83	—	—	25.5
Methyl *cis,cis,cis,cis,cis*-7,10,13,16,19-docosapentaenoate	—	24.83	24.97	25.25	25.14	—	26.0
Methyl *cis,cis,cis,cis,cis,cis*-4,7,10,13,16,19-docosahexaenoate	—	25.31	25.47	25.77	25.63	—	26.2
Methyl tetracosanoate	24.00	24.00	24.00	24.00	24.00	24.0	24.0

[a] Unspecified isomer.

[b] 16:3ω3 isomer.

Column packing
P1 = 10% Poly (diethylene glycol succinate) on Gas-Chrom P, AW, silanized (80/100 mesh)
P2 = 25% Poly (ethylene glycol succinate) on GC-22 Super-Support (60/80 mesh)
P3 = 20% Poly (diethylene glycol succinate) on Anakrom ABS (70-80 mesh)
P4 = 25% Poly (diethylene glycol succinate) on GC-22 Super-Support (60-80 mesh)
P5 = 15% Poly (diethylene glycol succinate) on Gas-Chrom P (100-120 mesh)
P6 = 15% Poly (ethylene glycol succinate) on Gas-Chrom P (80-100 mesh)
P7 = 15% Poly (diethylene glycol succinate) on Gas-Chrom P (100-120 mesh)

REFERENCES

1. **Nazir, D. J., Alcaraz, A. P., and Nair, P. P.,** *Lipids,* 1, 453, 1966.
2. **Ackman, R. G., Burgher, R. D., and Jangaard, P. M.,** *Can. J. Biochem. Physiol.,* 41, 1627, 1963.
3. **Ackman, R. G. and Burgher, R. D.,** *J. Am. Oil Chem. Soc.,* 42, 38, 1965.
4. **Ackman, R. G. and Burgher, R. D.,** *Can. J. Biochem. Physiol.,* 41, 2501, 1963.
5. **Withers, N. W. and Nevenzel, J. C.,** *Lipids,* 12, 989, 1977.
6. **Walker, B. L.,** *Arch. Biochem. Biophys.,* 114, 465, 1966.
7. **Ghosh, A., Ghosh, A., Hoque, M., and Datta, J.,** *J. Sci. Food Agric.,* 27, 159, 1976.

Table 14
METHYL ESTERS OF FATTY ACIDS: ON POLAR COLUMNS PACKED WITH ORGANOSILICONE (MODIFIED POLYESTER) LIQUID PHASES, REPORTED AS EQUIVALENT CHAIN LENGTHS

Column packing	P1	P2	P3	P4	P5	P6	P7	P8	P9
Temperature (°C)	180	200	200	175	175	160	190	178	194
Gas	He	Ar	Ar	He	He	He	N_2	N_2	N_2
Flow rate (mℓ/min)	40	na	na	na	na	na	na	50	50
Column Length (cm)	180	183	183	180	180	180	182	214	214
Diameter (cm I.D.)	0.3	0.3	0.3	0.2	0.2	0.2	0.24	na	na
(cm O.D.)	na	na	na	na	na	na	na	0.64	0.64
Form	na	U-tube	U-tube	na	na	na	Helix	na	na
Material	G	G	G	G	G	G	SS	G	G
Detector	FID	FID	FID	FID	FID	FID	FID	FID	FID
Reference	1	2	2	3	3	3	4	5	5

Compound					ECL				
Methyl tetradecanoate	14.00	14.00	14.00	14.00	14.00	14.00	14.00	14.00	14.00
Methyl tetradecenoate	—	14.80	14.47	—	—	—	14.96	—	—
Methyl pentadecanoate	15.00	15.00	15.00	15.00	15.00	15.00	15.00	15.00	15.00
Methyl pentadecenoate	—	15.69	15.56	—	—	—	—	—	—
Methyl hexadecanoate	16.00	16.00	16.00	16.00	16.00	16.00	16.00	16.00	16.00
Methyl cis-9-hexadecenoate	16.72	16.74	16.44	16.63	16.55	16.21	16.68	16.57	16.62
Methyl cis,cis-6,9-hexadecadienoate	—	17.39	16.82	—	—	—	—	17.65[a]	17.45[a]
Methyl cis,cis-9,12-hexadecadienoate	—	17.77	17.16	—	—	—	—	—	—
Methyl cis,cis,cis-6,9,12-hexadecatrienoate	—	18.42	17.50	—	—	—	—	—	—
Methyl cis,cis,cis,cis-6,9,12,15-hexadecatetraenoate	—	19.30	18.10	—	—	—	19.10	—	—
Methyl heptadecanoate	17.00	17.00	17.00	17.00	17.00	17.00	17.00	17.00	17.00
Methyl heptadecenoate	—	17.67	17.42	—	—	—	17.96	—	—
Methyl octadecanoate	18.00	18.00	18.00	18.00	18.00	18.00	18.00	18.00	18.00
Methyl cis-9-octadecenoate	18.64	18.62	18.38	18.58	18.50	18.25	18.65	18.53	18.52

Table 14 (continued)
METHYL ESTERS OF FATTY ACIDS: ON POLAR COLUMNS PACKED WITH ORGANOSILICONE (MODIFIED POLYESTER) LIQUID PHASES, REPORTED AS EQUIVALENT CHAIN LENGTHS

Compound	ECL								
Methyl cis,cis-9,12-octadecadienoate	19.42	19.45	18.95	19.44	19.14	18.72	19.44	19.42	19.20
Methyl cis,cis,cis-6,9,12-octadecatrienoate	—	20.15	19.38	—	—	—	20.02	20.00	19.67
Methyl cis,cis,cis-9,12,15-octadecatrienoate	20.62	20.50	19.76	20.50	20.00	19.38	20.44	20.40	20.02
Methyl cis,cis,cis-6,9,12,15-octadecatetraenoate	—	21.13	20.13	21.12	20.39	19.65	21.05	21.05	20.52
Methyl nonadecanoate	19.00	19.00	19.00	—	19.00	19.00	19.00	—	—
Methyl icosanoate	20.00	20.00	20.00	20.00	20.00	20.00	20.00	20.00	20.00
Methyl cis-11-icosenoate	20.65	20.45	20.25	—	—	—	20.60	20.50	20.45
Methyl cis,cis-11,14-icosadienoate	—	21.29	21.01	—	—	—	21.42	21.40	21.15
Methyl cis,cis,cis-5,8,11-icosatrienoate	—	—	—	—	—	—	—	21.63	21.33
Methyl cis,cis,cis-8,11,14-icosatrienoate	—	22.06	21.34	—	—	—	—	21.77	21.53
Methyl cis,cis,cis-11,14,17-icosatrienoate	—	22.40	21.72	—	—	—	22.41	21.95	21.60
Methyl cis,cis,cis-5,8,11,14-icosatetraenoate	22.68	22.50	21.60	—	—	—	22.42	22.43	22.00
Methyl cis,cis,cis-8,11,14,17-icosatetraenoate	—	23.08	22.08	—	—	—	23.03	23.00	22.47
Methyl cis,cis,cis,cis-5,8,11,14,17-icosapentaenoate	23.81	23.56	22.32	23.52	22.78	21.85	23.50	23.50	22.80
Methyl henicosanoate	21.00	21.00	21.00	21.00	21.00	21.00	21.00	21.00	21.00
Methyl cis,cis,cis,cis-6,9,12,15,18-henicosapentaenoate	—	24.56	23.34	—	—	—	24.54	—	—
Methyl docosanoate	22.00	22.00	22.00	22.00	22.00	22.00	22.00	22.00	22.00
Methyl cis-11-docosenoate (or cis-13-docosenoate)	22.55	22.47	22.38	—	—	—	22.64	22.43	22.35
Methyl cis,cis-13,16-docosadienoate	—	—	—	—	—	—	23.39	—	—
Methyl cis,cis,cis-7,10,13,16-docosatetraenoate	—	24.56	23.60	—	—	—	24.49	24.45	24.00
Methyl cis,cis,cis,cis-4,7,10,13,16-docosapentaenoate	25.23	24.95	23.90	—	—	—	25.00	24.57	23.85

	P1	P2	P3	P4	P5	P6	P7	P8	P9
Methyl *cis,cis,cis,cis,cis*-7,10,13,16,19-docosapentaenoate	25.79	24.42	24.30	—	—	—	25.52	25.53	24.80
Methyl *cis,cis,cis,cis,cis*-4,7,10,13,16,19-docosahexaenoate	26.40	25.91	24.59	26.08	25.12	24.05	26.01	26.18	25.20
Methyl tetracosanoate	24.00	24.00	24.00	24.00	24.00	24.00	24.00	24.00	24.00
Methyl *cis*-15-tetracosenoate	—	—	—	—	—	24.62	—	—	—

Column packing
P1 = 10% EGSS-X (ethylene glycol succinate polyester combined with methyl silicone) on Gas-Chrom P (100-120 mesh)

P2 = 10% EGSS-X (ethylene glycol succinate polyester combined with methyl silicone) on Gas-Chrom P (100-120 mesh)

P3 = 12% EGSS-Y (ethylene glycol succinate polyester combined with methyl silicone) on Gas-Chrom P (100-120 mesh)

P4 = 10% EGSS-X (ethylene glycol succinate polyester combined with methyl silicone) on Gas-Chrom Q (100-120 mesh)

P5 = 10% EGSS-Y (ethylene glycol succinate polyester combined with methyl silicone) on Gas-Chrom Q (100-120 mesh)

P6 = 3% EGSP-Z (ethylene glycol succinate polyester combined with phenyl-methyl silicone) on Gas-Chrom Q (100-120 mesh)

P7 = 10% EGSS-X (ethylene glycol succinate polyester combined with phenyl silicone) on Gas-Chrom P (100-120 mesh)

P8 = 15% EGSS-X (ethylene glycol succinate polyester combined with methyl silicone) on Chromosorb W, AW, DMCS (100-120 mesh)

P9 = 15% EGSS-Y (ethylene glycol succinate polyester combined with methyl silicone) on Chromosorb W, AW, DMCS (100-120 mesh)

REFERENCES

1. **Myher, J. J., Marai, L., and Kuksis, A.,** *Anal. Biochem.,* 62, 188, 1974.
2. **Ackman, R. G., and Burgher, R. D.,** *Can. J. Biochem. Physiol.,* 41, 250, 1963; Equivalent chain lengths: **Jamieson, G. R.,** in *Topics in Lipid Chemistry,* Vol. 1, Gunstone, F. D., Ed., Logos Press, London, 1970, 107.
3. **Joseph, J.,** *Lipids,* 10, 395, 1975.
4. **Morales, R. W. and Litchfield, C.,** *Biochim. Biophys. Acta,* 431, 206, 1976.
5. **Christie, W. W.,** *Lipid Analysis,* Pergamon Press, Oxford, 1973.

Table 15
METHYL ESTERS OF FATTY ACIDS: ON POLAR COLUMNS PACKED WITH CYANOSILICONE LIQUID PHASES, REPORTED AS EQUIVALENT CHAIN LENGTHS

	P1	P2	P3	P4	P5	P6	P7
Column packing							
Temperature (°C)	200	200	200	170	180	180	220
Gas	He	He	He	N₂	He	He	He
Flow rate (mℓ/min)	30+	30+	30+	10	40	40	30
Column Length (cm)	150	210	210	400	180	180	457
Diameter (cm I.D.)	0.4	0.4	0.4	0.2	0.2	0.3	na
(cm O.D.)	na	na	na	na	na	na	0.318
Form	na	na	na	Helix	na	Helix	Helix
Material	G	G	G	G	G	G	SS
Detector	FID	FID	FID	FID	FID	FID	FID
Reference	1	1	1	2	3	3	4
Compound				ECL			
Methyl tetradecanoate	14.00	14.00	14.00	14.00	14.00	14.00	14.00
Methyl tetradecenoate	14.23	14.42	14.91	15.04	—	—	—
Methyl pentadecanoate	15.00	15.00	15.00	15.00	15.00	15.00	15.00
Methyl pentadecenoate	—	—	—	—	—	—	—
Methyl hexadecanoate	16.00	16.00	16.00	16.00	16.00	16.00	16.00
Methyl *cis*-9-hexadecenoate	16.10	16.32	16.73	16.89	16.31	—	17.00
Methyl heptadecanoate	17.00	17.00	17.00	17.00	17.00	17.00	17.00
Methyl heptadecenoate	—	—	—	17.89	—	—	—
Methyl octadecanoate	18.00	18.00	18.00	18.00	18.00	18.00	18.00
Methyl *cis*-9-octadecenoate	18.08	18.24	18.62	18.77	18.25	18.51	18.90
Methyl *cis,cis*-9,12-octadecadienoate	18.39	18.69	19.47	19.71	18.73	19.31	20.00
Methyl *cis,cis,cis*-6,9,12-octadecatrienoate	18.55	18.93	20.08	20.51	19.28	20.33	21.40
Methyl *cis,cis,cis*-9,12,15-octadecatrienoate	18.80	19.21	20.44	20.91	19.63	—	—
Methyl *cis,cis,cis*-6,9,12,15-octadecatetraenoate	—	—	—	—	—	—	—
Methyl nonadecanoate	19.00	19.00	19.00	19.00	—	—	19.00
Methyl nonadecenoate	—	—	—	19.78	—	—	—

	P1	P2	P3	P4	P5	P6	P7
Methyl icosanoate	20.00	20.00	20.00	20.00	20.00	20.00	20.00
Methyl cis-11-icosenoate	20.05	20.22	20.55	20.77	20.23	—	20.90
Methyl cis,cis-11,14-icosadienoate	20.40	20.66	21.41	21.71	—	—	—
Methyl cis,cis,cis-8,11,14-icosatrienoate	20.54	20.90	22.00	22.51	—	—	—
Methyl cis,cis,cis-11,14,17-icosatrienoate	20.80	21.20	22.32	—	—	—	—
Methyl cis,cis,cis,cis-5,8,11,14-icosatetraenoate	20.51	21.02	22.41	22.92	21.15	22.16	—
Methyl cis,cis,cis,cis,cis-5,8,11,14,17-icosapentaenoate	20.91	21.55	23.36	24.08	21.80	23.11	—
Methyl henicosanoate	21.00	21.00	21.00	21.00	21.00	21.00	21.00
Methyl docosanoate	22.00	22.00	22.00	22.00	22.00	22.00	22.00
Methyl cis-11-docosenoate (or cis-13-docosenoate)	21.99	22.21	22.50	22.77	22.23	—	22.85
Methyl cis,cis-13,16-docosadienoate	—	—	—	23.71	—	—	—
Methyl cis,cis,cis-7,10,13,16-docosatetraenoate	22.51	23.05	24.38	24.92	—	—	—
Methyl cis,cis,cis,cis-4,7,10,13,16-docosapentaenoate	—	—	—	—	23.38	—	—
Methyl cis,cis,cis,cis,cis-7,10,13,16,19-docosapentaenoate	22.92	23.61	25.31	—	23.81	—	—
Methyl cis,cis,cis,cis,cis,cis-4,7,10,13,16,19-docosahexaenoate	—	23.72	25.66	27.16	23.96	25.36	—
Methyl tetracosanoate	22.00	24.00	24.00	24.00	24.00	24.00	24.00
Methyl cis-15-tetracosenoate	23.99	24.20	24.46	—	—	—	—

Column packing P1 = 5% OV-225 (50% phenyl methyl, 50% cyanopropyl methyl siloxane) on Chromosorb W-AW (80-100 mesh)

P2 = 3% SILAR-5CP (50% cyanopropyl, 50% phenyl siloxane) on Chromosorb W-AW (80/100 mesh)
P3 = 10% SILAR-10C (100% cyanopropyl siloxane) on Chromosorb W-AW (80/100 mesh)
P4 = 12% SILAR-10C (100% cyanopropyl siloxane) on Chromosorb W, HP 9 (80-100 mesh)
P5 = 3% SILAR-5CP (50% cyanopropyl, 50% phenyl siloxane) on Gas-Chrom Q
P6 = 3% SILAR-10C (100% cyanopropyl siloxane) on Gas-Chrom P
P7 = 10% SILAR-10C (100% cyanopropyl siloxane) on Gas-Chrom Q (100-120 mesh)

Table 15 (continued)

METHYL ESTERS OF FATTY ACIDS: ON POLAR COLUMNS PACKED WITH CYANOSILICONE LIQUID PHASES, REPORTED AS EQUIVALENT CHAIN LENGTHS

REFERENCES

1. **Golovnya, R. V., Uralets, V. P., and Kuz'menko, T. E.,** *Zh. Anal. Khim.,* 32, 340, 1977; Engl. trans., 32, 269, 1977.

2. **Heckers, H., Dittmar, K., Melcher, F. W., and Kalinowski, H. O.,** *J. Chromatogr.,* 135, 93, 1977.

3. **Myher, J. J., Marai, L., and Kuksis, A.,** *Anal. Biochem.,* 62, 188, 1974.

4. **Conacher, H. B. S., Iyengar, J. R., and Beare-Rogers, J. L.,** *J. Assoc. Offic. Anal. Chem.,* 60, 899, 1977.

Note added in proof: For additional ECL data recently tabulated for the cyanosilicone liquid phases: (A) 3% SILAR-5CP on GAS-CHROM Q, AW, HMDS (60-80 mesh); (B) 10% OV-17 on Chromosorb W, AW, HMDS (80-100 mesh); (C) 15% OV-275 on GAS-CHROM RZ (100-120 mesh); (D) 12% OV-275 on Chromosorb P, AW, DMCS (100-120 mesh); in packed columns, see: (A) **Miyagawa, M., Miwa, T. K., and Spencer, G. F.,** *J. Am. Oil Chem. Soc.,* 56, 834, 1979; (B) **Ashes, J. R., Haken, J. K., and Mills, S. C.,** *J. Chromatogr.,* 187, 297, 1980; (C) **Itabashi, Y. and Takagi, T.,** *Yukagaku,* 27, 219, 1978; and (D) **Özcimder, M. and Hammers, W. E.,** *J. Chromatogr.,* 187, 307, 1980.

Table 16A
METHYL ESTERS OF FATTY ACIDS: ON PACKED NONPOLAR COLUMNS, REPORTED AS RELATIVE RETENTION TIMES

Column packing	P1	P2	P3	P4	P5	P6
Temperature (°C)	193	188	197	207	197	200
Gas	Ar	Ar	Ar	Ar	Ar	—
Flow rate (mℓ/min)	30	30	30	na	na	na
Column						
Length (cm)	142	142	142	254	122	122
Diameter	0.4	0.4	0.4	na	0.4	0.4
(cm I.D.)	na	na	na	0.6	na	na
(cm O.D.)	Tube	Tube	Tube	Tube	Tube	na
Form	G	G	G	G	G	na
Material	Argon	Argon	Argon	Argon	Argon	na
Detector	1	1	1	2	3	4
Reference						

Compound	$r_{\text{Methyl Octadecanoate}}$					
Methyl tetradecanoate	0.144	0.162	0.174	0.203	0.182	0.182
Methyl tetradecenoate	0.125	0.146	0.159	—	0.167	—
Methyl pentadecanoate	0.232	0.257	0.272	0.308	0.280	0.278
Methyl hexadecanoate	0.380	0.405	0.422	0.456	0.425	0.434
Methyl *cis*-9-hexadecenoate	0.322	0.350	0.366	0.394	0.380	0.391
Methyl *cis,cis*-6,9-hexadecadienoate	—	—	0.360	—	—	—
Methyl *cis,cis*-9,12-hexadecadienoate	—	—	0.360	—	—	—
Methyl *cis,cis,cis*-6,9,12-hexadecatrienoate	0.278	0.308	0.324	—	—	—
Methyl *cis,cis,cis,cis*-4,7,10,13-hexadecatetraenoate	0.266	0.290	0.314	—	—	—
Methyl heptadecanoate	0.615	0.638	0.650	0.681	0.652	0.632
Methyl octadecanoate	1.00	1.00	1.00	1.00	1.00	1.00
Methyl *cis*-9-octadecenoate	0.815	0.844	0.860	0.879	0.864	20.878
Methyl *cis,cis*-9,12-octadecadienoate	0.754	0.780	0.792	0.839	0.827	0.86?
Methyl *cis,cis,cis*-9,12,15-octadecatrienoate	0.754	0.780	0.792	0.839	0.836	0.803
Methyl *cis,cis,cis,cis*-6,9,12,15-octadecatetraenoate	0.661	0.694	0.720	—	—	—
Methyl nonadecanoate	—	—	—	1.48	—	1.57
Methyl icosanoate	2.64	2.48	2.37	2.17	2.20	—
Methyl *cis*-11-icosenoate	—	—	1.99	—	—	—
Methyl *cis,cis*-11,14-icosadienoate	—	—	1.85	—	—	—
Methyl *cis,cis,cis*-8,11,14-icosatrienoate	—	1.64	1.64	—	—	—
Methyl *cis,cis,cis,cis*-5,8,11,14-icosatetraenoate	—	1.40	1.46	1.51	1.50	—
Methyl *cis,cis,cis,cis,cis*-5,8,11,14,17-icosapentaenoate	—	1.46	1.46	—	1.50	—
Methyl henicosanoate	—	—	—	3.21	—	—
Methyl docosanoate	—	—	—	4.58	4.78	—
Methyl *cis*-11-docosenoate (or *cis*-13-docosenoate)	—	—	—	4.28	4.65	—
Methyl *cis,cis,cis,cis,cis*-7,10,13,16,19-docosapentaenoate	—	—	3.30	—	—	—
Methyl *cis,cis,cis,cis,cis,cis*-4,7,10,13,16,19-docosatetraenoate	—	—	3.02	—	3.05	—
Methyl tetracosanoate	—	—	—	10.2	—	—

Column packing P1 = 14% Apiezon M on Celite AW, ALK-W (140-170 mesh)
P2 = 14% Apiezon M on Celite AW, ALK-W (140-170 mesh)
P3 = 14% Apiezon M on Celite, AW, ALK-W (140-170 mesh)
P4 = 4% Apiezon-L on Celite 545
P5 = 20% Apiezon-L on Celite ALK-W (120-210 mesh)
P6 = 20% Apiezon-M on Celite 545 AW, ALK-W (48-85 mesh)

Table 16A (continued)
METHYL ESTERS OF FATTY ACIDS: ON PACKED NONPOLAR COLUMNS, REPORTED AS RELATIVE RETENTION TIMES

REFERENCES

1. **Farquhar, J. W., Insull, W., Jr., Rosen, P., Stoffel, W., and Ahrens, E. H., Jr.,** *Nutr. Rev. (Suppl.),* 17, 1, 1959.
2. **Gerson, T.,** *J. Chromatogr.,* 6, 178, 1961.
3. **James, A. T.,** *J. Chromatogr.,* 2, 552, 1959.
4. **Hawke, J. C., Hansen, R. P., and Shorland, F. B.,** *J. Chromatogr.,* 2, 547, 1959.

Table 16B
METHYL ESTERS OF FATTY ACIDS: ON PACKED COLUMNS WITH NONPOLAR LIQUID PHASES, REPORTED AS RETENTION TIMES RELATIVE TO METHYL OCTADECANOATE

Column packing	P1	P2	P3	P4	P5
Temperature (°C)	195	220	240	260	200
Gas	Ar	N_2	N_2	N_2	N_2
Flow rate (mℓ/min)	40	44	44	44	na
Column					
Length (cm)	122	100	100	100	305
Diameter					
(cm I.D.)	0.4	0.6	0.6	0.6	na
(cm O.D.)	na	na	na	na	0.318
Form	Tube	na	na	na	Helix
Material	G	na	na	na	SS
Detector	Argon	FID	FID	FID	FID
Reference	1	2	2	2	3

Compound	$r_{\text{Methyl Octadecanoate}}$				
Methyl tetradecanoate	0.175	0.23	0.26	0.30	0.24
Methyl tetradecenoate	0.157	—	—	—	—
Methyl pentadecanoate	0.272	—	—	—	0.34
Methyl pentadecenoate	—	—	—	—	—
Methyl hexadecanoate	0.420	0.48	0.51	0.54	0.45
Methyl cis-9-hexadecenoate	0.362	0.43	—	—	0.49
Methyl hexadecadienoate	—	—	—	—	0.45
Methyl hexadecatrienoate	—	—	—	—	0.45
Methyl hexadecatetraenoate	—	—	—	—	0.44
Methyl heptadecanoate	0.649	—	—	—	0.69
Methyl heptadecenoate	—	—	—	—	0.63
Methyl octadecanoate	1.000	1.00	1.00	1.00	1.00
Methyl cis-9-octadecenoate	0.859	0.89	0.90	0.92	0.90
Methyl cis,cis-9,12-octadecadienoate	0.792	0.89	—	—	0.90
Methyl cis,cis,cis-9,12,15-octadecatrienoate	0.792	—	—	—	0.90
Methyl nonadecanoate	1.58	—	—	—	1.43
Methyl nonadecenoate	—	—	—	—	1.31
Methyl icosanoate	2.36	2.10	1.90	1.84	2.03
Methyl cis-11-icosenoate	2.00	1.85	1.72	1.65	1.88
Methyl cis,cis-11,14-icosadienoate	1.86	—	—	—	1.88
Methyl cis,cis,cis-8,11,14-icosatrienoate	1.65	—	—	—	1.88
Methyl cis,cis,cis-11,14,17-icosatrienoate	—	—	—	—	—
Methyl cis,cis,cis,cis-5,8,11,14-icosatetraenoate	1.47	—	—	—	1.66
Methyl cis,cis,cis,cis-8,11,14,17-icosatetraenoate	—	—	—	—	1.88
Methyl cis,cis,cis,cis,cis-5,8,11,14,17-icosapentaenoate	1.47	—	—	—	1.54
Methyl cis,cis,cis,cis,cis-6,9,12,15,18-henicosapentaenoate	—	—	—	—	2.18
Methyl docosanoate	5.50	4.31	3.48	3.11	—
Methyl cis-11-docosenoate (or cis-13-docosenoate)	4.74	3.87	3.24	2.97	—
Methyl cis,cis,cis,cis-7,10,13,16-docosatetraenoate	3.60	—	—	—	3.57
Methyl cis,cis,cis,cis,cis-7,10,13,16,19-docosapentaenoate	3.29	—	—	—	3.02
Methyl cis,cis,cis,cis,cis,cis-4,7,10,13,16,19-docosahexaenoate	3.02	—	—	—	2.85

Column packing
P1 = 5% Apiezon-L on Celite, AW, ALK-W (100-120 mesh)
P2 = 20% Apiezon-L on Celite (0.15—0.20 mm)
P3 = 20% Apiezon-L on Celite (0.15—0.20 mm)
P4 = 20% Apiezon-L on Celite (0.15—0.20 mm)
P5 = 10% SE-30 (methyl silicone) on C_{22} brick (80-100 mesh)

Table 16B (continued)
METHYL ESTERS OF FATTY ACIDS: ON PACKED COLUMNS WITH
NONPOLAR LIQUID PHASES, REPORTED AS RETENTION TIMES
RELATIVE TO METHYL OCTADECANOATE

REFERENCES

1. **Kingsbury, K. J., Morgan, D. M., and Heyes, T. D.,** *Biochem. J.,* 90, 140, 1964.
2. **Zeman, I.,** *J. Gas Chromatogr.,* 3, 18, 1965.
3. **Graille, J., Huq, M. S., and Naudet, M.,** *Rev. Franc. Corps Gras,* 23, 151, 1976.

Note added in proof: For additional RRT data recently tabulated for the non-polar liquid phases: (A) 10% Apiezon-M on Chromosorb-W, AW, HMDS (80-100 mesh); (B) 10% $C_{87} H_{176}$ on Chromosorb-W, AW, HMDS (80-100 mesh); in packed columns, see: (A) **Ashes, J. R., Haken, J. K., and Mills, S. C.,** *J. Chromatogr.,* 187, 297, 1980; and (B) **Ashes, J. R., Haken, J. K., and Mills, S. C.,** *J. Chromatogr.,* 187, 297, 1980.

Table 17
METHYL ESTERS OF FATTY ACIDS: ON PACKED NONPOLAR COLUMNS, REPORTED AS EQUIVALENT CHAIN LENGTHS

Column packing	P1	P2	P3	P4	P5	P6	P7
Temperature (°C)	240	165	200	195	200	172	200
Gas	He	Ar	He	Ar	na	He	N₂
Flow rate (mℓ/min)	60	na	30	40	na	—	na
Column							
Length (cm)	250	183	150	122	152	180	305
Diameter							
(cm I.D.)	0.3	0.5	0.4	0.4	0.2	0.2	na
(cm O.D.)	na	na	na	na	na	na	0.318
Form	Helix	U-tube	Helix	Tube	Helix	Helix	Helix
Material	na	G	G	G	SS	G	SS
Detector	TC	Argon	FID	Argon	FID	FID	FID
Reference	1	2	3	4	5	6	7
Compound				ECL			
Methyl tetradecanoate	14.00	14.00	14.00	14.00	14.0	14.00	14.0
Methyl tetradecenoate	13.83	—	13.82	13.70	—	—	—
Methyl pentadecanoate	15.00	15.00	15.00	15.00	15.0	15.00	15.0
Methyl hexadecanoate	16.00	16.00	16.00	16.00	16.0	16.00	16.0
Methyl cis-9-hexadecenoate	15.70	—	15.75	15.65	—	15.78	15.7
Methyl cis,cis-6,9-hexadecadienoate	15.47	—	—	—	—	—	15.7
Methyl cis,cis,cis-6,9,12-hexadecatrienoate	15.47	—	—	—	—	—	15.7
Methyl heptadecanoate	17.00	17.00	17.00	17.00	17.0	17.00	17.0
Methyl heptadecenoate	16.73	—	—	—	—	—	16.7
Methyl octadecanoate	18.00	18.00	18.00	18.00	18.0	18.00	18.0
Methyl cis-9-octadecenoate	17.71	17.70	17.68	17.60	17.6	—	17.7
Methyl cis,cis-9,12-octadecadienoate	17.53	17.66	17.61	17.45	17.5	17.60	17.7
Methyl cis,cis,cis-6,9,12-octadecatrienoate	17.30	—	17.44	—	—	—	—
Methyl cis,cis,cis-9,12,15-octadecatrienoate	17.51	17.66	17.66	17.45	—	—	17.7
Methyl cis,cis,cis-6,9,12,15-octadecatetraenoate	17.30	—	—	—	—	17.41	—
Methyl nonadecanoate	19.00	19.00	19.00	19.00	19.0	19.00	19.0

Table 17 (continued)
METHYL ESTERS OF FATTY ACIDS: ON PACKED NONPOLAR COLUMNS, REPORTED AS EQUIVALENT CHAIN LENGTHS

Compound	ECL					
Methyl nonadecenoate	18.60	—	—	—	—	19.7
Methyl icosanoate	20.00	20.00	20.00	20.00	20.0	20.0
Methyl cis-11-icosenoate	19.78	—	19.65	19.60	—	19.6
Methyl cis,cis-11,14-icosadienoate	19.48	—	19.65	19.50	—	19.6
Methyl cis,cis,cis-5,8,11-icosatrienoate	19.15	—	—	—	19.1	—
Methyl cis,cis,cis-8,11,14-icosatrienoate	19.23	—	19.38	19.20	19.1	19.6
Methyl cis,cis,cis-11,14,17-icosatrienoate	—	—	19.66	—	—	—
Methyl cis,cis,cis-5,8,11,14-icosatetraenoate	19.00	19.00	19.17	18.90	—	19.5
Methyl cis,cis,cis,cis,cis-5,8,11,14,17-icosapentaenoate	19.00	19.00	19.18	18.90	18.9	19.2
Methyl henicosanoate	21.00	21.00	21.00	21.00	21.0	21.0
Methyl docosanoate	22.00	22.00	22.00	22.00	22.0	22.0
Methyl cis-11-docosenoate (or cis-13-docosenoate)	21.57	21.61	21.66	21.65	—	—
Methyl cis,cis,cis-7,10,13,16-docosatetraenoate	20.93	—	21.13	21.00	20.7	21.7
Methyl cis,cis,cis,cis-4,7,10,13,16-docosapentaenoate	20.87	—	—	—	—	—
Methyl cis,cis,cis,cis,cis-7,10,13,16,19-docosapentaenoate	21.00	—	21.12	20.75	—	21.2
Methyl cis,cis,cis,cis,cis,cis-4,7,10,13,16,19-docosahexaenoate	20.73	—	20.95	20.55	—	20.9
Methyl tetracosanoate	24.00	24.00	24.00	24.00	24.0	24.0
Methyl cis-15-tetracosenoate	23.67	—	23.66	—	—	—

Column packing   P1 = 20% Apiezon-L on Gas-Chrom P (100-140 mesh)
                  P2 = 2% Apiezon-L on Gas-Chrom P, AW, Siliconized (80-100 mesh)
                  P3 = 3% SE-30 (methyl silicone) on Gas-Chrom Q (80-100 mesh)
                  P4 = 5% Apiezon-L on Celite, AW, ALK-W (100-120 mesh)
                  P5 = 15% Apiezon-L on Gas-Chrom P (80-100 mesh)
                  P6 = 3% SE-30 (methyl silicone) on Gas-Chrom Q (100-120 mesh)
                  P7 = 10% SE-30 (methyl silicone) on C_{22} Brick (80-100 mesh)

REFERENCES

1. **Hofstetter, H. H., Sen, N., and Holman, R. T.,** *J. Am. Oil Chem. Soc.,* 42, 537, 1965.
2. **Nazir, D. J., Alcaraz, A. P., and Nair, P. P.,** *Lipids,* 1, 453, 1966.
3. **Golovnya, R. V., Uralets, V. P., and Kuz'menko, T. E.,** *Zh. Anal. Khim.,* 32, 340, 1977; Engl. transl. 32, 269, 1977.
4. **Kingsbury, K. J., Morgan, D. M., and Hayes, T. D.,** *Biochem. J.,* 90, 140, 1964.
5. **Walker, B. L.,** *Arch. Biochem. Biophys.,* 114, 465, 1966.
6. **Joseph, J.,** *Lipids,* 10, 395, 1975.
7. **Graille, J., Hug, M. S., and Naudet, M.,** *Rev. Franc. Corps Gras,* 23, 151, 1976.

Note added in proof: For additional ECL data recently tabulated for the nonpolar liquid phases: (A) 5% Apiezon-L on GAS-CHROM Q, AW, HMDS (60-80 mesh); (B) 10% Apiezon-M on Chromosorb-W, AW, HMDS (80-100 mesh); (C) 10% C_{87} H_{176} on Chromosorb W, AW, HMDS (80-100 mesh); in packed columns, see: (A) **Miyagawa, M., Miwa, T. K., and Spencer, G. F.,** *J. Am. Oil Chem. Soc.,* 56, 834, 1979; (B) **Ashes, J. R., Haken, J. K., and Mills, S. C.,** *J. Chromatogr.,* 187, 297, 1980; and (C) **Ashes, J. R., Haken, J. K., and Mills, S. C.,** *J. Chromatogr.,* 187, 297, 1980.

Table 18A
METHYL ESTERS OF FATTY ACIDS: ON POLAR WALL-COATED OPEN-TUBULAR COLUMNS WITH POLYESTER LIQUID PHASES, REPORTED AS RELATIVE RETENTION TIMES

	P1	P2	P3	P4	P5	P6
Column packing						
Temperature (°C)	180	170	160	200	175	170
Gas	na	na	He	He	He	He
Flow rate (mℓ/min)	na	na	na	5	na	2
Column						
Length (cm)	10000	4500	5000	4500	4700	4700
Diameter						
(cm I.D.)	0.025	0.025	0.05	0.05	0.05	0.025
(cm O.D.)	na	na	na	na	na	na
Form	WCOT	WCOT	WCOT	WCOT	SCOT	WCOT
Material	SS	SS	SS	SS	SS	SS
Detector	FID	FID	FID	MS	FID	FID
Reference	1	1	2	3	4	5
Compound			$r_{\text{Methyl Octadecanoate}}$			
Methyl tetradecanoate	0.27	0.28	—	0.41	0.270	0.218
Methyl tetradecenoate	—	—	—	—	—	0.249
Methyl pentadecanoate	0.38	0.39	—	0.49	0.376	0.318
Methyl pentadecenoate	0.41	0.45	—	—	—	0.318
Methyl hexadecanoate	0.52	0.53	—	0.62	0.518	0.465
Methyl *cis*-9-hexadecenoate	0.60	0.65	—	0.67	0.615	0.524
Methyl *cis,cis*-9,12-hexadecadienoate	—	—	—	—	—	0.658
Methyl *cis,cis,cis*-6,9,12-hexadecatrienoate	—	—	—	—	—	—
Methyl *cis,cis,cis*-4,7,10,13-hexadecatetraenoate	—	—	—	—	—	0.877
Methyl heptadecanoate	0.72	0.73	—	0.77	0.720	0.686
Methyl heptadecenoate	0.81	0.87	—	0.85	0.830	0.746
Methyl octadecanoate	1.00	1.00	1.00	1.00	1.000	1.000
Methyl *cis*-9-octadecenoate	1.11	1.17	1.18	1.14	1.131	1.07
Methyl *cis,cis*-9,12-octadecadienoate	1.33	1.51	1.48	1.29	1.416	1.29
Methyl *cis,cis,cis*-6,9,12-octadecatrienoate	—	—	1.79	1.40	—	1.45
Methyl *cis,cis,cis*-9,12,15-octadecatrienoate	1.69	2.07	2.03	1.54	—	1.65
Methyl *cis,cis,cis*-6,9,12,15-octadecatetraenoate	—	—	2.46	—	—	1.86

	P1	P2	P3	P4	P5
Methyl nonadecanoate	1.38	—	—	—	—
Methyl nonadecenoate	—	1.37	—	—	1.59
Methyl icosanoate	1.90	1.90	1.65	1.94	—
Methyl cis-11-icosenoate	2.05	2.06	1.81	2.16	2.28
Methyl cis,cis-11,14-icosadienoate	—	—	2.14	2.66	2.69
Methyl cis,cis,cis-5,8,11-icosatrienoate	—	—	—	2.82	2.97
Methyl cis,cis,cis-8,11,14-icosatrienoate	—	—	2.43	3.09	3.42
Methyl cis,cis,cis-11,14,17-icosatrienoate	—	—	—	—	3.42
Methyl cis,cis,cis,cis-5,8,11,14-icosatetraenoate	—	—	2.65	3.49	3.21
Methyl cis,cis,cis,cis-8,11,14,17-icosatetraenoate	—	—	2.98	—	3.78
Methyl cis,cis,cis,cis,cis-5,8,11,14,17-icosapentaenoate	—	—	3.24	—	4.08
Methyl henicosanoate	2.61	2.62	—	—	—
Methyl docosanoate	3.59	3.60	—	3.80	—
Methyl cis-13-docosenoate	—	—	—	4.15	—
Methyl cis,cis-13,16-docosadienoate	—	—	—	5.04	—
Methyl cis,cis,cis-7,10,13,16-docosatetraenoate	—	—	—	6.59	6.67
Methyl cis,cis,cis,cis-4,7,10,13,16-docosapentaenoate	—	—	—	—	7.40
Methyl cis,cis,cis,cis,cis-7,10,13,16,19-docosapentaenoate	—	—	—	—	8.41
Methyl cis,cis,cis,cis,cis,cis-4,7,10,13,16,19-docosahexaenoate	—	—	—	9.98	8.94
Methyl tetracosanoate	—	—	—	7.51	—
Methyl cis-15-tetracosenoate	—	—	—	8.09	—

Column packing P1 = Poly (1,4-butanediol succinate) in form of wall coating in stainless steel tubing
P2 = Poly (diethylene glycol succinate) in form of wall coating in stainless steel tubing
P3 = Poly (diethylene glycol succinate) in form of wall coating in stainless steel tubing
P4 = Poly (1,4-butanediol succinate) in form of wall coating in stainless steel tubing
P5 = Poly (diethylene glycol succinate) in form of supporting coating in stainless steel tubing
P6 = Poly (1,4-butanediol succinate) in form of wall coating in stainless steel tubing

REFERENCES

1. **Hrivnak, J., Palo, V., and Krupcik, J.,** *Milchwissenschaft*, 28, 699, 1973.
2. **Jamieson, G. R. and Reid, E. H.,** *J. Sci. Food Agric.*, 19, 628, 1968.
3. **Masada, Y., Hashimoto, K., Inoue, T., Yoshida, H., Fukui, I., Masaki, K., and Takahara, J.,** *Yakugaku Zasshi*, 89, 579, 1969.
4. **Pullarkat, R. K. and Reha, H.,** Personal communication; see also **Pullarkat, R. K. and Reha, H.,** *J. Chromatogr. Sci.*, 14, 25, 1976.
5. **Perry, G. J.,** Ph.D. Thesis, University of Melbourne, Australia, 1977.

Table 18B
METHYL ESTERS OF FATTY ACIDS: (ON POLAR (POLYESTER TYPE) OPEN-TUBULAR COLUMNS, REPORTED AS RELATIVE RETENTION TIMES

	P1	P2	P3	P4	P5	P6
Column packing						
Temperature (°C)	170	170	175	180	185	180
Gas	na	na	na	N₂	N₂	N₂
Flow rate (ml/min)				1	1	1
Column						
Length (cm)	4700	4700	4700	5000	5000	4000
Diameter (cm I.D.)	0.025	0.025	0.05	0.05	0.05	0.03
(cm O.D.)	na	na	na	na	na	N2.1
Form	WCOT	WCOT	SCOT	WCOT	WCOT	WCOT
Material	SS	SS	SS	SS	SS	G
Detector	FID	FID	FID	FID	FID	FID
Reference	1	2	3	4	4	5

$r_{\text{Methyl Octadecanoate}}$

Compound	P1	P2	P3	P4	P5	P6
Methyl tetradecanoate	0.240	—	0.251	0.266	0.267	0.30
Methyl tetradecenoate	0.275	—	0.291	—	—	—
Methyl pentadecanoate	0.342	—	0.353	0.367	0.370	0.40
Methyl pentadecenoate	—	—	0.404	—	—	0.45
Methyl hexadecanoate	0.488	0.47	0.499	0.513	0.517	0.54
Methyl cis-9-hexadecenoate	0.555	0.53	0.573	0.597	0.587	0.64
Methyl cis,cis-6,9-hexadecadienoate	0.633	—	—	—	—	—
Methyl cis,cis-9,12-hexadecadienoate	0.680	—	—	—	—	—
Methyl cis,cis-6,9,12-hexadecatrienoate	0.810	—	—	—	—	—
Methyl cis,cis,cis-4,7,10,13-hexadecatetraenoate	0.948	—	—	—	—	—
Methyl cis,cis,cis-6,9,12,15-hexadecatetraenoate	0.948	—	—	—	—	—
Methyl heptadecanoate	0.700	—	0.705	0.706	0.714	0.74
Methyl heptadecenoate	0.770	—	0.782	—	—	0.86
Methyl octadecanoate	1.00	1.00	1.000	1.000	1.000	1.00
Methyl cis-9-octadecenoate	1.08	1.07	1.10	1.12	1.11	1.16
Methyl cis,cis-9,12-octadecadienoate	1.33	1.27	1.34	1.36	1.34	1.46
Methyl cis,cis,cis-6,9,12-octadecatrienoate	1.53	—	—	1.58	1.54	—

	P1	P2	P3	P4	P5	P6
Methyl *cis,cis,cis*-9,12,15-octadecatrienoate	1.73	1.63	—	1.78	1.71	1.91
Methyl *cis,cis,cis,cis*-6,9,12,15-octadecatetraenoate	1.96	—	—	—	—	—
Methyl nonadecanoate	1.44	—	1.42	1.39	1.38	—
Methyl nonadecenoate	1.57	—	1.56	—	—	—
Methyl icosanoate	2.06	2.10	2.01	1.97	1.92	1.85
Methyl *cis*-11-icosenoate	2.21	2.21	2.17	2.15	2.11	2.09
Methyl *cis,cis*-11,14-icosadienoate	2.70	2.63	2.61	2.58	2.53	2.60
Methyl *cis,cis,cis*-5,8,11-icosatrienoate	—	—	—	2.70	2.65	—
Methyl *cis,cis,cis*-8,11,14-icosatrienoate	3.03	—	—	2.91	2.86	—
Methyl *cis,cis,cis*-11,14,17-icosatrienoate	3.48	—	—	3.34	3.27	—
Methyl *cis,cis,cis,cis*-5,8,11,14-icosatetraenoate	3.29	—	—	3.23	3.14	—
Methyl *cis,cis,cis,cis*-8,11,14,17-icosatetraenoate	3.88	—	—	—	—	—
Methyl *cis,cis,cis,cis,cis*-5,8,11,14,17-icosapentaenoate	4.21	—	—	4.26	3.99	2.50
Methyl henicosanoate	2.92	—	2.87	2.70	2.65	—
Methyl *cis,cis,cis,cis,cis*-6,9,12,15,18-henicosapentaenoate	6.15	—	—	—	—	—
Methyl docosanoate	4.25	—	4.11	3.80	2.48	3.43
Methyl *cis*-11-docosenoate (or *cis*-13-docosenoate)	4.38	4.58	4.37	4.08	3.99	3.96
Methyl *cis,cis*-13,16-docosadienoate	—	5.46	—	—	—	4.73
Methyl *cis,cis,cis,cis*-7,10,13,16-docosatetraenoate	6.65	—	—	—	—	—
Methyl *cis,cis,cis,cis,cis*-4,7,10,13,16-docosapentaenoate	7.32	—	—	—	—	—
Methyl *cis,cis,cis,cis,cis*-7,10,13,16,19-docosapentaenoate	8.47	—	—	—	—	—
Methyl *cis,cis,cis,cis,cis,cis*-4,7,10,13,16,19-docosahexaenoate	9.37	—	—	9.77	8.39	—
Methyl tetracosanoate	8.80	—	8.51	7.17	6.86	6.35
Methyl *cis*-15-tetracosenoate	9.05	—	8.78	7.80	7.48	7.03

Column packing   P1 = Poly (1,4-butanediol succinate) in form of wall coating in stainless steel tubing
P2 = Poly (1,4-butanediol succinate) in form of wall coating in stainless steel tubing
P3 = Poly (diethylene glycol succinate) in form of support-coated open-tubular stainless steel column
P4 = Poly (diethylene glycol succinate) in form of wall coating in stainless steel tubing
P5 = Poly (ethylene glycol adipate) in form of wall coating in stainless steel tubing
P6 = Poly (diethylene glycol succinate) in form of wall coating in glass tubing

Table 18B (continued)
METHYL ESTERS OF FATTY ACIDS: (ON POLAR (POLYESTER TYPE) OPEN-TUBULAR COLUMNS, REPORTED AS RELATIVE RETENTION TIMES

REFERENCES

1. **Ackman, R. G., Sipos, J. C., and Jangaard, P. M.,** *Lipids,* 2, 251, 1967.
2. **Ackman, R. G. and Castell, J. D.,** *J. Gas Chromatogr.,* 5, 489, 1967.
3. **Nelson, G. J.,** *Lipids,* 9, 254, 1974.
4. **Kingsbury, K. J., Morgan, D. M. L., and Brett, C. G.,** *Biochem. Med.,* 18, 37, 1977.
5. **Pallotta, U., Losi, G., and Zorzut, C.,** *Riv. Ital. Sost. Grasse,* 42, 142, 1965.

Note added in proof: For additional RRT data recently tabulated for an open-tubular column wall-coated with the polyester liquid phase BDS, see: **Perry, G. J., Volkman, J. K., and Johns, R. B.,** *Geochim. Cosmochim. Acta,* 43, 1715, 1979.

Table 19
METHYL ESTERS OF FATTY ACIDS: (ON OPEN-TUBULAR COLUMNS COATED WITH POLAR LIQUID PHASES OF SPECIAL (NONPOLYESTER) COMPOSITION, REPORTED AS RETENTION TIMES RELATIVE TO METHYL OCTADECANOATE

Column packing	P1	P2	P3	P4
Temperature (°C)	180	200	180	180
Gas	N_2	N_2	He	He
Flow rate (mℓ/min)	1	1	1	1
Column				
Length (cm)	5000	5000	6000	6000
Diameter				
(cm I.D.)	0.05	0.05	0.025	0.025
(cm O.D.)	na	na	na	na
Form	WCOT	WCOT	WCOT	WCOT
Material	SS	SS	G	G
Detector	FID	FID	FID	FID
Reference	1	1	2	2

Compound	$r_{\text{Methyl Octadecanoate}}$			
Methyl tetradecanoate	0.279	0.263	0.530	0.529
Methyl pentadecanoate	0.380	0.368	—	0.604
Methyl hexadecanoate	0.525	0.517	0.704	0.704
Methyl *cis*-9-hexadecenoate	0.612	0.567	—	0.807
Methyl heptadecanoate	0.720	0.715	—	—
Methyl octadecanaote	1.00	1.00	1.00	1.00
Methyl *cis*-7-octadecenoate	—	—	—	1.133
Methyl *cis*-9-octadecenoate	1.12	1.08	1.152	1.151
Methyl *cis*-11-octadecenoate	—	—	1.170	1.171
Methyl *cis*-13-octadecenoate	—	—	—	1.205
Methyl *cis,cis*-9,12-octadecadienoate	1.38	1.24	1.416	1.410
Methyl *cis,cis,cis*-6,9,12-octadecatrienoate	—	1.34	—	—
Methyl *cis,cis,cis*-9,12,15-octadecatrienoate	1.78	1.52	1.809	1.804
Methyl *cis,cis,cis,cis*-6,9,12,15-octadecatetraenoate	—	—	—	2.123
Methyl nonadecanoate	1.80	1.27	—	—
Methyl icosanoate	1.92	1.92	1.505	—
Methyl *cis*-9-icosenoate	—	—	—	1.717
Methyl *cis*-11-icosenoate	2.11	2.05	1.745	1.744
Methyl *cis*-13-icosenoate	—	—	1.787	1.783
Methyl *cis,cis*-11,14-icosadienoate	2.54	2.35	2.175	2.175
Methyl *cis,cis,cis*-5,8,11-icosatrienoate	2.66	2.38	—	—
Methyl *cis,cis,cis*-8,11,14-icosatrienoate	2.88	2.55	—	—
Methyl *cis,cis,cis*-11,14,17-icosatrienoate	3.23	2.88	—	—
Methyl *cis,cis,cis,cis*-5,8,11,14-icosatetraenoate	3.23	2.73	—	2.889
Methyl *cis,cis,cis,cis*-8,11,14,17-icosatetraenoate	—	—	—	3.323
Methyl *cis,cis,cis,cis,cis*-5,8,11,14,17-icosapentaenoate	4.17	3.30	—	3.789
Methyl henicosanoate	2.66	2.63	—	—
Methyl docosanoate	3.71	3.66	2.365	—
Methyl *cis*-11-docosenoate	—	—	—	2.698
Methyl *cis*-13-docosenoate	3.99	3.83	2.783	2.744
Methyl *cis*-15-docosenoate	—	—	2.836	2.818
Methyl *cis,cis*-13,16-docosadienoate	—	—	3.458	—
Methyl *cis,cis,cis,cis*-7,10,13,16-docosatetraenoate	—	—	—	—

Table 19 (continued)
METHYL ESTERS OF FATTY ACIDS: (ON OPEN-TUBULAR COLUMNS COATED WITH POLAR LIQUID PHASES OF SPECIAL (NONPOLYESTER) COMPOSITION, REPORTED AS RETENTION TIMES RELATIVE TO METHYL OCTADECANOATE

Compound	$r_{\text{Methyl Octadecanoate}}$			
Methyl *cis,cis,cis,cis,cis*-7,10,13,16,19-docosapentaenoate	—	—	—	6.246

Column packing
P1 = EGSS-X (ethylene glycol succinate polyester combined with methyl silicone) in form of wall coating in stainless steel tubing
P2 = FFAP (Carbowax-20M terminated with 2-nitroterephthalic acid) in form of wall coating in stainless steel tubing
P3 = SP2340 (Supelco 100% cyanopropyl siloxane) in form of wall coating on glass tubing
P4 = SP2340 (Supelco 100% cyanopropyl siloxane) in form of wall coating on glass tubing

REFERENCES

1. **Kingsbury, K. J., Morgan, D. M. L., and Brett, C. G.,** *Biochem. Med.,* 18, 37, 1977.
2. **Ackman, R. G. and Eaton, C. A.,** *Fette Seifen Anstrichm.,* 80, 21, 1978; (from the analyses by H. T. Slover).

Note added proof: For additional RRT data recently tabulated for the cyanosilicone or polar liquid phases: (A) SILAR-5CP; (B) OV-351; (C) SP-2300; (D) Carbowax 20M; in wall-coated open-tubular columns, see: (A) **Itabashi, Y. and Takagi, T.,** *Yukagaku,* 29, 855, 1980; (B) **Korhonen, I. O. O.,** *Chromatographia,* 17, 70, 1983; (C) **Takagi, T., Hayashi, K., and Itabashi, Y.,** *Bull. Fac. Fish. Hokkaido Univ.,* 33, 255, 1982; and (D) **Korhonen, I. O. O.,** *J. Chromatogr.,* 209, 96, 1981.

Table 20
METHYL ESTERS OF FATTY ACIDS:
ON WALL-COATED OPEN-TUBULAR COLUMNS
COATED WITH C$_2$ GLYCOL TYPE POLYESTERS GIVING
LOW POLARITIES, REPORTED AS EQUIVALENT CHAIN
LENGTHS

Column packing	P1	P2	P3
Temperature (°C)	180	180	180
Gas	N$_2$	N$_2$	N$_2$
Flow rate (mℓ/min)	1	1	5
Column			
Length (cm)	5000	5000	1525
Diameter			
(cm I.D.)	0.05	0.05	0.05
(cm O.D.)	na	na	
Form	WCOT	WCOT	SCOT
Material	SS	SS	SS
Detector	FID	FID	FID
Reference	1	1	2

Compound	ECL		
Methyl tetradecanoate	14.00	14.00	14.00
Methyl pentadecanoate	15.00	15.00	15.00
Methyl hexadecanoate	16.00	16.00	16.00
Methyl *cis*-9-hexadecenoate	16.39	16.46	—
Methyl heptadecanoate	17.00	17.00	—
Methyl octadecanoate	18.00	18.00	18.00
Methyl *cis*-9-octadecenoate	18.36	18.38	—
Methyl *cis,cis*-9,12-octadecadienoate	18.86	18.97	—
Methyl *cis,cis,cis*-6,9,12-octadecatrienoate	19.18	19.42	18.65
Methyl *cis,cis,cis*-9,12,15-octadecatrienoate	19.66	19.78	18.94
Methyl *cis,cis,cis,cis*-6,9,12,15-octadecatetraenoate	—	—	19.15
Methyl nonadecanoate	19.00	19.00	19.00
Methyl icosanoate	20.00	20.00	20.00
Methyl *cis*-11-icosenoate	20.32	20.38	—
Methyl *cis,cis*-11,14-icosadienoate	20.84	20.90	20.35
Methyl *cis,cis,cis*-5,8,11-icosadienoate	21.00	21.04	—
Methyl *cis,cis,cis*-8,11,14-icosatrienoate	21.22	21.26	20.51
Methyl *cis,cis,cis*-11,14,17-icosatrienoate	21.66	21.90	20.77
Methyl *cis,cis,cis,cis*-5,8,11,14-icosatetraenoate	21.50	21.59	20.60
Methyl *cis,cis,cis,cis*-8,11,14,17-icosatetraenoate	—	—	21.00
Methyl *cis,cis,cis,cis,cis*-5,8,11,14,17-icosapentaenoate	22.22	22.43	21.14
Methyl henicosanoate	21.00	21.00	21.00
Methyl docosanoate	22.00	22.00	22.00
Methyl *cis*-13-docosenoate	22.22	22.32	—
Methyl *cis,cis,cis,cis,cis*-7,10,13,16,19-docosapentaenoate	—	—	22.62
Methyl *cis,cis,cis,cis,cis,cis*-4,7,10,13,16,19-docosahexaenoate	24.60	25.33	23.09
Methyl tetracosanoate	24.00	24.00	24.00
Methyl *cis*-15-tetracosenoate	24.24	24.32	—

Column packing　　P1　=　　Poly (ethylene glycol adipate) in form of wall coating in stainless steel tubing

Table 20 (continued)
METHYL ESTERS OF FATTY ACIDS:
ON WALL-COATED OPEN-TUBULAR COLUMNS
COATED WITH C$_2$ GLYCOL TYPE POLYESTERS GIVING
LOW POLARITIES, REPORTED AS EQUIVALENT CHAIN
LENGTHS

P2 =   Poly (diethylene glycol succinate) in form of wall
        coating in stainless steel tubing
P3 =   Poly (diethylene glycol succinate) in stainless steel
        tubing on support coating

REFERENCES

1. **Kingsbury, K. J., Morgan, D. M. L., and Brett, C. G.,** *Biochem. Med.,* 18, 37, 1977.
2. **Jamieson, C. R. and Reid, E. H.,** *J. Chromatogr.,* 42, 304, 1969.

Table 21
METHYL ESTERS OF FATTY ACIDS: ON WALL-COATED OPEN-TUBULAR COLUMNS COATED WITH LOW POLARITY 1,4-BUTANEDIOL SUCCINATE POLYESTER, REPORTED AS EQUIVALENT CHAIN LENGTHS

Column packing	P1	P2	P3	P4	P5	P6
Temperature (°C)	180	170	195	170	170	190
Gas	na	He	H₂	N₂	N₂	N₂
Flow rate (mℓ/min)	na	na	2	5	5	5
Column Length (cm)	10000	4600	3000	5000	5000	5000
Diameter (cm I.D.)	0.025	0.025	0.03	0.05	0.05	0.05
(cm O.D.)	na	na	na	na	na	na
Form	WCOT	WCOT	WCOT	WCOT	WCOT	WCOT
Material	SS	SS	G	SS	SS	SS
Detector	FID	FID	FID	FID	FID	FID
Reference	1	2	3	4	4	4
Compound			ECL			
Methyl tetradecanoate	14.00	14.00	14.00	14.00	14.00	14.00
Methyl tetradecenoate	—	—	—	14.38	—	—
Methyl pentadecanoate	15.00	15.00	15.00	15.00	15.00	15.00
Methyl pentadecenoate	15.27	—	—	—	—	—
Methyl hexadecanoate	16.00	16.00	16.00	16.00	16.00	16.00
Methyl *cis*-9-hexadecenoate	16.41	16.36	16.41	16.33	—	—
Methyl *cis,cis*-6,9-hexadecadienoate	—	—	—	16.73	—	—
Methyl *cis,cis*-9,12-hexadecadienoate	—	—	—	16.93	—	—
Methyl *cis,cis,cis*-6,9,12-hexadecatrienoate	—	—	—	17.42	—	—
Methyl *cis,cis,cis,cis*-4,7,10,13-hexadecatetraenoate	—	—	—	17.85	—	—
Methyl *cis,cis,cis,cis*-6,9,12,15-hexadecatetraenoate	—	—	—	18.05	—	—
Methyl heptadecanoate	17.00	17.00	17.00	17.00	17.00	17.00
Methyl heptadecenoate	17.35	—	—	17.27	—	—
Methyl octadecanoate	18.00	18.00	18.00	18.00	18.00	18.00
Methyl *cis*-9-octadecenoate	18.32	18.26	18.32	18.20	18.25	18.32

Table 21 (continued)
METHYL ESTERS OF FATTY ACIDS: ON WALL-COATED OPEN-TUBULAR COLUMNS COATED WITH LOW POLARITY 1,4-BUTANEDIOL SUCCINATE POLYESTER, REPORTED AS EQUIVALENT CHAIN LENGTHS

Compound			ECL			
Methyl cis,cis-9,12-octadecadienoate	18.90	18.84	18.90	18.71	18.90	18.96
Methyl cis,cis,cis-6,9,12-octadecatrienoate	—	—	—	19.15	19.32	19.39
Methyl cis,cis,cis-9,12,15-octadecatrienoate	19.62	19.58	19.64	19.50	19.65	19.74
Methyl cis,cis,cis-6,9,12,15-octadecatetraenoate	—	19.96	—	19.86	20.01	20.15
Methyl nonadecanoate	19.00	19.00	19.00	19.00	19.00	19.00
Methyl nonadecenoate	—	—	—	19.18	—	—
Methyl icosanoate	20.00	20.00	20.00	20.00	20.00	20.00
Methyl cis-11-icosenoate	20.23	—	20.29	20.19	20.23	20.29
Methyl cis,cis-11,14-icosadienoate	—	—	20.87	20.75	20.80	20.92
Methyl cis,cis,cis-5,8,11-icosatrienoate	—	—	—	20.70	—	—
Methyl cis,cis,cis-8,11,14-icosatrienoate	—	—	—	21.06	21.24	21.30
Methyl cis,cis,cis-11,14,17-icosatrienoate	—	—	—	21.45	21.63	21.68
Methyl cis,cis,cis-5,8,11,14-icosatetraenoate	—	—	—	21.29	21.48	21.62
Methyl cis,cis,cis-8,11,14,17-icosatetraenoate	—	—	—	21.75	21.91	22.00
Methyl cis,cis,cis,cis-5,8,11,14,17-icosapentaenoate	—	22.11	—	21.97	22.20	22.25
Methyl henicosanoate	21.00	21.00	21.00	21.00	21.00	21.00
Methyl docosanoate	22.00	22.00	22.00	22.00	22.00	22.00
Methyl cis-13-docosenoate	—	—	—	22.14	22.18	22.35
Methyl cis,cis-13,16-docosadienoate	—	—	—	22.60	—	—
Methyl cis,cis,cis-7,10,13,16-docosatetraenoate	—	—	—	23.23	23.48	23.54
Methyl cis,cis,cis,cis-4,7,10,13,16-docosapentaenoate	—	—	—	23.50	—	—
Methyl cis,cis,cis-7,10,13,16,19-docosapentaenoate	—	24.38	—	23.90	24.18	24.26
Methyl cis,cis,cis,cis-4,7,10,13,16,19-docosahexaenoate	—	—	—	24.18	24.43	24.60
Methyl tetracosanoate	24.00	—	24.00	24.00	24.00	24.00
Methyl cis-15-tetracosenoate	—	—	—	24.08	—	—

Column packing P1 = Poly (1,4-butanediol succinate) in form of wall coating in stainless steel tubing
P2 = Poly (1,4-butanediol succinate) in form of wall coating in stainless steel tubing
P3 = Poly (1,4-butanediol succinate) in form of wall coating in glass tubing
P4 = Poly (1,4-butanediol succinate) in form of wall coating in stainless steel tubing
P5 = Poly (1,4-butanediol succinate) in form of wall coating in stainless steel tubing
P6 = Poly (1,4-butanediol succinate) in form of wall coating in stainless steel tubing

REFERENCES

1. **Hrivnak, J., Palo, V., and Krupcik, J.,** *Milchwissenschaft,* 28, 699, 1973.
2. **Joseph, J.,** *Lipids,* 10, 395, 1975.
3. **Holmbom, B.,** *J. Am. Oil Chem. Soc.,* 54, 289, 1977.
4. **Jamieson, G. R.,** in *Topics in Lipid Chemistry,* Vol. I, Gunstone, F. D., Ed., Logos Press, London, 1970, 107.

Note added in proof: For additional ECL data recently tabulated for methyl-branched fatty acids on BDS liquid phase wall-coated in open-tubular columns see: **Ackman, R. G.,** in *Progress in the Chemistry of Fats and Other Lipids,* Vol.12, Holman, R. T., Ed., Pergamon Press, New York, 1972, 165; **Smith, A. and Duncan, W. R. H.,** *Lipids,* 14, 350, 1979.

Table 22
METHYL ESTERS OF FATTY ACIDS: ON WALL-COATED OPEN-TUBULAR COLUMNS COATED WITH HIGH-POLARITY POLYESTERS, REPORTED AS EQUIVALENT CHAIN LENGTHS

Column packing	P1	P2	P3	P4	P5
Temperature (°C)	160	188	180	170	190
Gas	N_2	Ar	N_2	na	N_2
Flow rate (mℓ/min)	na	2	1	na	na
Column					
Length (cm)	5000	4700	4000	4500	5000
Diameter					
(cm I.D.)	0.05	0.025	0.03	0.025	0.05
(cm O.D.)	na	na	na	na	na
Form	WCOT	WCOT	WCOT	WCOT	WCOT
Material	SS	SS	G	SS	SS
Detector	FID	Argon	FID	FID	FID
Reference	1	2	3	4	5

Compound	ECL				
Methyl tetradecanoate	14.00	14.00	14.00	14.00	14.00
Methyl pentadecanoate	15.00	15.00	15.00	15.00	15.00
Methyl pentadecenoate	—	—	15.42	15.46	—
Methyl hexadecanoate	16.00	16.00	16.00	16.00	16.00
Methyl *cis*-9-hexadecenoate	16.53	—	16.57	16.61	—
Methyl *cis,cis*-6,9,-hexadecadienoate	17.16	—	—	—	—
Methyl *cis,cis,cis,cis*-4,7,10,13-hexadecatetraenoate	18.68	—	—	—	—
Methyl heptadecanoate	17.00	17.00	17.00	17.00	17.00
Methyl heptadecenoate	—	—	17.50	17.54	—
Methyl octadecanoate	18.00	18.00	18.00	18.00	18.00
Methyl *cis*-9-octadecadienoate	18.49	18.55	18.48	18.50	18.45
Methyl *cis,cis*-9,12-octadecadienoate	19.14	19.36	19.25	19.31	19.04
Methyl *cis,cis,cis*-6,9,12-octadecatrienoate	19.69	—	—	—	19.68
Methyl *cis,cis,cis*-9,12,15-octadecatrienoate	20.06	20.44	20.12	20.32	20.06
Methyl *cis,cis,cis,cis*-6,9,12,15-octadecatetraenoate	20.62	—	—	—	20.60
Methyl nonadecanoate	19.00	19.00	19.00	19.00	19.00
Methyl icosanoate	20.00	20.00	20.00	20.00	20.00
Methyl *cis*-11-icosenoate	20.36	—	20.40	20.31	—
Methyl *cis,cis*-11,14-icosadienoate	21.06	—	20.22	—	—
Methyl *cis,cis,cis*-5,8,11-icosatrienoate	21.20	—	21.22	—	—
Methyl *cis,cis,cis*-8,11,14-icosatrienoate	21.54	—	—	—	—
Methyl *cis,cis,cis*-11,14,17-icosatrienoate	22.00	—	—	—	—
Methyl *cis,cis,cis,cis*-5,8,11,14-icosatetraenoate	21.94	—	—	—	—
Methyl *cis,cis,cis,cis*-8,11,14,17-icosatetraenoate	22.43	—	—	—	—
Methyl *cis,cis,cis,cis,cis*-5,8,11,14,17-icosapentaenoate	22.84	—	—	—	—
Methyl henicosanoate	21.00	21.00	21.00	21.00	21.00
Methyl docosanoate	22.00	22.00	22.00	22.00	22.00
Methyl *cis*-13-docosenoate	22.22	—	22.50	—	—
Methyl *cis,cis*-13,16-docosadienoate	—	—	23.20	—	—
Methyl *cis,cis,cis,cis*-7,10,13,16-docosatetraenoate	23.84	—	—	—	—
Methyl *cis,cis,cis,cis,cis*-4,7,10,13,16-docosapentaenoate	24.32	—	—	—	—
Methyl *cis,cis,cis,cis,cis*-7,10,13,16,19-docosapentaenoate	24.79	—	—	—	—

Table 22 (continued)
METHYL ESTERS OF FATTY ACIDS: ON WALL-COATED OPEN-TUBULAR COLUMNS COATED WITH HIGH-POLARITY POLYESTERS, REPORTED AS EQUIVALENT CHAIN LENGTHS

Compound	ECL				
Methyl *cis,cis,cis,cis,cis,cis*-4,7,10,13,16,19-docosahexaenoate	25.22	—	—	—	—
Methyl tetracosanoate	24.00	24.00	24.00	—	24.00
Methyl *cis*-15-tetracosenoate	—	—	—	—	—

Column packing		
P1	=	Poly (diethylene glycol succinate) in form of wall coating in stainless steel tubing
P2	=	Poly (diethylene glycol succinate) in form of wall coating in stainless steel tubing
P3	=	Poly (diethylene glycol succinate) in form of wall coating in stainless steel tubing
P4	=	Poly (diethylene glycol succinate) in form of wall coating in stainless steel tubing
P5	=	Poly (diethylene glycol succinate) in form of wall coating in stainless steel tubing

REFERENCES

1. **Jamieson, G. R.**, in *Topics in Lipid Chemistry*, Vol. I, Gunstone, F. D., Ed., Logos Press, London, 1970, 107.
2. **Litchfield, C., Reiser, R., and Isbell, A. P.**, *J. Am. Oil Chem. Soc.*, 40, 302, 1963.
3. **Pallotta, V., Losi, G., and Zorzut, C.**, *Riv. Ital. Sost. Grasse*, 42, 142, 1965.
4. **Hrivnak, J., Palo, V., and Krupcik, J.**, *Milchwissenschaft*, 28, 699, 1973.
5. **Barve, J. A., Gunstone, F. D., Jacobsberg, F. R., and Winlow, P.**, *Chem. Phys. Lipids*, 8, 117, 1972.

Table 23
METHYL ESTERS OF FATTY ACIDS: ON WALL-COATED POLAR COLUMNS WITH SPECIAL (NONPOLYESTER) LIQUID PHASES, REPORTED AS EQUIVALENT CHAIN LENGTHS

	P1	P2	P3	P4	P5	P6
Column packing						
Temperature (°C)	200	190	190	180	180	200
Gas	N₂	N₂	H₂	N₂	na	N₂
Flow rate (ml/min)	1	2	na	1	—	5
Column						
Length (cm)	5000	10000	5000	5000	5000	5000
Diameter						
(cm I.D.)	0.05	0.05	0.04	0.05	0.05	0.05
(cm O.D.)	na	na	na	na	na	na
Form	WCOT	WCOT	WCOT	WCOT	WCOT	WCOT
Material	SS	G	G	SS	SS	SS
Detector	FID	FID	FID	FID	FID	FID
References	1	2	3	1	4	5

Compound	ECL					
	P1	P2	P3	P4	P5	P6
Methyl tetradecanoate	14.00	14.00	14.00	14.00	14.00	14.00
Methyl tetradecenoate	—	—	14.39	—	—	—
Methyl pentadecanoate	15.00	15.00	15.00	15.00	15.00	15.00
Methyl hexadecanoate	16.00	16.00	16.00	16.00	16.00	16.00
Methyl *cis*-9-hexadecenoate	16.35	16.27	16.24	16.50	—	—
Methyl heptadecanoate	17.00	17.00	17.00	17.00	17.00	17.00
Methyl heptadecenoate	—	17.23	17.20	—	—	—
Methyl octadecanoate	18.00	18.00	18.00	18.00	18.00	18.00
Methyl *cis*-9-octadecenoate	18.28	18.20	18.13	18.34	18.00	18.65
Methyl *cis,cis*-9,12-octadecadienoate	18.68	18.63	18.58	18.97	—	19.50
Methyl *cis,cis,cis*-6,9,12-octadecatrienoate	18.94	18.91	18.90	—	19.90	19.95
Methyl *cis,cis,cis*-9,12,15-octadecatrienoate	19.34	19.29	19.23	19.75	20.31	20.55
Methyl *cis,cis,cis,cis*-6,9,12,15-octadecatetraenoate	—	19.53	—	—	20.88	20.96
Methyl nonadecanoate	19.00	19.00	19.00	19.00	19.00	19.00
Methyl nonadecenoate	—	19.17	—	—	—	—
Methyl icosanoate	20.00	20.00	20.00	20.00	20.00	20.00
Methyl *cis*-11-icosenoate	20.24	20.17	20.15	20.34	—	20.36

	P1	P2	P3	P4	P5	P6
Methyl *cis,cis*-11,14-icosadienoate	20.67	20.61	20.54	20.83	21.25	21.39
Methyl *cis,cis,cis*-6,8,11-icosatrienoate	20.74	20.65	—	21.00	21.58	21.70
Methyl *cis,cis,cis*-8,11,14-icosatrienoate	20.93	20.86	—	21.26	21.81	21.90
Methyl *cis,cis,cis*-11,14,17-icosatrienoate	21.27	21.22	—	21.52	22.25	22.32
Methyl *cis,cis,cis,cis*-5,8,11,14-icosatetraenoate	21.10	21.06	—	21.52	22.25	22.32
Methyl *cis,cis,cis,cis*-8,11,14,17-icosatetraenoate	—	21.47	—	—	22.78	22.95
Methyl *cis,cis,cis,cis,cis*-5,8,11,14,17-icosapentaenoate	21.74	21.67	—	22.35	23.23	23.50
Methyl docosanoate	22.00	22.00	22.00	22.00	22.00	22.00
Methyl *cis*-11-docosenoate (or *cis*-13-docosenoate)	22.18[a]	22.08	22.06	22.20[a]	—	22.52[a]
Methyl *cis,cis,cis*-7,10,13,16-docosatetraenoate	—	23.02	—	—	24.17	—
Methyl *cis,cis,cis,cis*-4,7,10,13,16-docosapentaenoate	—	23.28	—	—	24.69	—
Methyl *cis,cis,cis,cis*-7,10,13,16,19-docosapentaenoate	—	23.63	—	—	25.14	—
Methyl *cis,cis,cis,cis,cis*-4,7,10,13,16,19-docosahexaenoate	24.04	23.90	—	—	25.68	—
Methyl tetracosanoate	24.00	24.00	24.00	24.00	24.00	24.00
Methyl *cis*-15-tetracosenoate	24.14	24.14	24.12	—	—	—

[a] Isomer is *cis*-13-docosenoate.

Column packing
P1 = FFAP (Carbowax 20M terminated with 2-nitroterephthalic acid) in form of wall coating in stainless steel tubing
P2 = Carbowax 20M-AT (copolymer with terephthalic acid) in form of wall coating in stainless steel tubing
P3 = Carbowax 20M in form of wall coating in glass tubing
P4 = EGSS-X (ethylene glycol-succinate polyester combined with methyl silicone) in form of wall coating in stainless steel tubing
P5 = EGSS-X (ethylene glycol-succinate polyester combined with methyl silicone) in form of wall coating in stainless steel tubing
P6 = EGSS-X (ethylene glycol-succinate polyester combined with methyl silicone) in form of wall coating in stainless steel tubing

Table 23 (continued)

METHYL ESTERS OF FATTY ACIDS: ON WALL-COATED POLAR COLUMNS WITH SPECIAL (NONPOLYESTER) LIQUID PHASES, REPORTED AS EQUIVALENT CHAIN LENGTHS

REFERENCES

1. Kingsbury, K. J., Morgan, D. M. L., and Brett, C. G., *Biochem. Med.*, 18, 37, 1977.
2. Flanzy, J., Boudon, M., Leger, C., and Pihet, J., *J. Chromatogr. Sci.*, 14, 17, 1976.
3. Mordret, F., Prevot, A., Le Barbanchon, N., and Barbati, C., *Rev. Franc. Corps Gras*, 24, 467, 1977.
4. Jamieson, G. R. and Reid, E. H., *J. Chromatogr.*, 39, 71, 1969.
5. Jamieson, G. R., in *Topics in Lipid Chemistry*, Vol. 1, Gunstone, F. D., Ed., Logos Press, London, 1970, 107.

Note added in proof: For additional ECL data recently tabulated for Carbowax 20M type liquid phases in wall-coated open-tubular columns, see: **Ackman, R. G.,** *Fette. Seifen. Anstrichm.*, 82, 351, 1980. **Ackman, R. G.,** *Chem. Ind. (London),* 715, 1981. **Ackman, R. G.,** *J. Am. Oil Chem. Soc.,* 57, 821A, 1980. **Ackman, R. G. and Sipos, J. C.,** *Progr. Lipid Res.,* 20, 773, 1980. **Dorris, G. M., Douek, M., and Allen, L. H.,** *J. Am. Oil Chem. Soc.,* 59, 494, 1982 (for 9,12-18:2 correct to 18.65; for 5,9-18:2 correct to 18.77: for 11-18:1 correct to 18.27; add 18.39 for 13-18:1). **Kaitaranta, J. and Linko, R. R.,** *J. Sci. Food Agric.,* 30, 921, 1979. **Ramananarivo, R., Artaud, J., Estienne, J., Peiffer, G., and Gaydon, E. M.,** *J. Am. Oil Chem. Soc.,* 58, 1038, 1981. **Nichols, P. D., Klumpp, D. W., and Johns, R. B.,** *Phytochemistry,* 21, 1613, 1982.

For additional ECL data recently tabulated for the organosilicone liquid phase EGSS-X (ethylene glycol-succinate polyester combined with methyl silicone) wall-coated in an open-tubular column, see: **Patton, G. M., Cann, S., Brunengraber, H., and Lowenstein, J. M.,** *Methods in Enzymology,* Vol. 72D, Lowenstein, J. M., Ed., Academic Press, New York, 1981, 8.

Table 24
METHYL ESTERS OF FATTY ACIDS: ON POLAR (CYANOSILICONE TYPE) OPEN-TUBULAR COLUMNS, REPORTED AS EQUIVALENT CHAIN LENGTHS

	P1	P2	P3	P4	P5	P6
Column packing						
Temperature (°C)	180	180	170	170	160	175
Gas	He	He	He	He	N_2	N_2
Flow rate (mℓ/min)	5	5	5	5	0.5	0.5
Column						
Length (cm)	4700	4700	4700	4700	5000	5000
Diameter (cm I.D.)	0.025	0.025	0.025	0.025	0.03	0.03
(cm O.D.)	na	na	na	na	na	na
Form	WCOT	WCOT	WCOT	WCOT	WCOT	WCOT
Material	SS	SS	SS	SS	G	G
Detector	FID	FID	FID	FID	FID	FID
References	1	1	1	1	2	2
Compound			ECL			
Methyl hexadecanoate	16.00	16.00	16.00	16.00	16.00	16.00
Methy cis-9-hexadecenoate	16.43	16.42	16.40	16.55	16.70	16.76
Methyl cis,cis-9,12-hexadecadienoate	17.10	—	—	11.35	—	—
Methyl cis,cis,cis-6,9,12-hexadecatrienoate	—	—	—	11.82	—	—
Methyl cis,cis,cis,cis-4,7,10,13-hexadecatetraenoate	17.79	—	17.82	—	—	—
Methyl cis,cis,cis,cis-6,9,12,15-hexadecatetraenoate	17.97	—	—	—	—	—
Methyl heptadecanoate	17.00	17.00	17.00	17.00	17.00	17.00
Methyl heptadecenoate	17.38	17.38	—	17.52	—	—
Methyl octadecanoate	18.00	18.00	18.00	18.00	18.00	18.00
Methyl cis-9-octadecenoate	18.35	18.32	18.32	18.48	18.58	18.66
Methyl cis,cis-9,12-octadecadienoate	18.89	18.81	18.82	19.17	—	—
Methyl cis,cis,cis-6,9,12-octadecatrienoate	19.20	19.12	—	19.60	—	—
Methyl cis,cis,cis-9,12,15-octadecatrienoate	19.52	19.42	19.44	19.98	—	—
Methyl cis,cis,cis,cis-6,9,12,15-octadecatetraenoate	19.85	19.72	19.72	—	—	—
Methyl nonadecanoate	19.00	19.00	19.00	19.00	—	—
Methyl icosanoate	20.00	20.00	20.00	20.00	20.00	20.00

Table 24 (continued)

METHYL ESTERS OF FATTY ACIDS: ON POLAR (CYANOSILICONE TYPE) OPEN-TUBULAR COLUMNS, REPORTED AS EQUIVALENT CHAIN LENGTHS

Compound	ECL					
Methyl cis-11-icosenoate	20.31	20.25	20.25	20.46	20.53	20.64
Methyl cis,cis-11,14-icosadienoate	20.85	20.80	20.76	21.12	—	—
Methyl cis,cis,cis-8,11,14-icosatrienoate	21.16	21.10	—	21.55	—	—
Methyl cis,cis,cis-11,14,17-icosatrienoate	21.50	21.41	—	21.90	—	—
Methyl cis,cis,cis,cis-5,8,11,14-icosatetraenoate	21.35	21.26	21.18	21.82	—	—
Methyl cis,cis,cis,cis-8,11,14,17-icosatetraenoate	21.80	21.70	21.64	—	—	—
Methyl cis,cis,cis,cis,cis-5,8,11,14,17-icosapentaenoate	21.99	21.85	21.81	22.61	—	—
Methyl henicosanoate	21.00	21.00	21.00	21.00	21.00	21.00
Methyl cis,cis,cis,cis,cis-6,9,12,15,18-henicosapentaenoate	23.07	—	—	23.81	—	—
Methyl docosanoate	22.00	22.00	22.00	22.00	22.00	22.00
Methyl cis-11-docosenoate	22.12	22.23	22.16	22.35	—	22.49
Methyl cis-13-docosenoate	22.25	22.28	22.21	22.45	—	22.57
Methyl cis,cis,cis,cis-7,10,13,16-docosatetraenoate	23.37	—	—	23.85	—	—
Methyl cis,cis,cis,cis,cis-4,7,10,13,16-docosapentaenoate	23.53	—	—	24.11	—	—
Methyl cis,cis,cis,cis,cis-7,10,13,16,19-docosapentaenoate	24.01	—	—	24.61	—	—
Methyl cis,cis,cis,cis,cis,cis-4,7,10,13,16,19-docosahexaenoate	24.18	—	—	24.88	—	—
Methyl tetracosanoate	24.00	24.00	24.00	24.00	24.00	24.00
Methyl cis-15-tetracosenoate	—	—	24.20	24.36	—	—

Column packing P1 = SILAR-5CP (50% cyanopropyl, 50% phenyl, siloxane) in form of wall coating in stainless steel tubing
P2 = SILAR-5CP (50% cyanopropyl, 50% phenyl, siloxane) in form of wall coating in stainless steel tubing
P3 = SILAR-5CP (50% cyanopropyl, 50% phenyl, siloxane) in form of wall coating in stainless steel tubing
P4 = SILAR-7CP (70% cyanopropyl, 30% phenyl, siloxane) in form of wall coating in stainless steel tubing
P5 = SILAR-10C (100% cyanopropyl siloxane) in form of wall coating in glass tubing

REFERENCES

1. **Ackman, R. G. and Eaton, C. A.**, *Fette Seifen Anstrichm.*, 80, 21, 1978.
2. **Ojanpera, S. H.**, *J. Am. Oil Chem. Soc.*, 55, 290, 1977.

Note added in proof: For additional ECL data recently tabulated for the polar cyanosilicone liquid phases: (A) SILAR-5CP; (B) SILAR-7CP; (C) SILAR-10C; (D) SIL-47-CNP; wall-coated in open-tubular columns, see: (A) **Itabashi, Y. and Takagi, Y.,** *Yukagaku,* 29, 855, 1980; (A) **Nagao, A. and Yamazaki, M.,** *Yukagaku,* 32, 207, 1983; (A) **Sebedio, J-L. and Ackman, R. G.,** *J. Chromatogr. Sci.,* 20, 231, 1982; (A) **Takagi, T. and Itabashi, Y.,** *Lipids,* 17, 716, 1982; (A) **Patton, G. M., Cann, S., Brunengraber, H., and Lowenstein, J. N.,** *Methods in Enzymology,* Vol 72D, Lowenstein, J. M., Ed., Academic Press, New York, 1981, 8; (B) **Sebedio, J-L. and Ackman, R. G.,** *J. Chromatogr. Sci.,* 20, 231, 1982; (C) **Kai, Y. and Pryde, E. H.,** *J. Am. Oil Chem. Soc.,* 59, 300, 1982; (D) **Wannigama, G. P., Volkman, J. R., Gillan, F. T., Nichols, P. D., and Johns, R. B.,** *Phytochemistry,* 20, 659, 1981; (D) **Johns, R. B., Nichols, P. D., Gillan, F. T., Perry, G. J., and Volkman, J. K.,** *Comp. Biochem. Physiol.,* 69B, 843, 1981; and (D) **Nichols, P. D., Volkman, J. K., and Johns, R. B.,** *Phytochemistry,* 22, in press, 1983.

Table 25
METHYL ESTERS OF FATTY ACIDS: ON WALL-COATED OPEN-TUBULAR COLUMNS WITH NONPOLAR LIQUID PHASES, REPORTED AS RETENTIVE TIMES RELATIVE TO METHYL OCTADECANOATE

Column packing	P1	P2	P3	P4
Temperature (°C)	207	220	200	190
Gas	Ar	N_2	na	N_2
Flow rate (m𝓁/min)	na	1	na	1
Column				
Length (cm)	65600	5000	4500	9000
Diameter				
(cm I.D.)	0.025	0.05	0.025	0.025
(cm O.D.)	na	na	na	na
Form	WCOT	WCOT	WCOT	WCOT
Material	SS	SS	SS	SS
Detector	Argon	FID	FID	FID
Reference	1	2	3	4

Compound	$r_{Methyl\ Octadecanoate}$			
Methyl tetradecanoate	0.22	0.227	0.178	0.29
Methyl tetradecenoate	—	—	—	0.28
Methyl pentadecanoate	0.32	0.328	0.277	0.37
Methyl pentadecenoate	—	—	0.254	—
Methyl hexadecanoate	0.47	0.477	0.426	0.50
Methyl *cis*-9-hexadecenoate	0.42	0.439	0.380	0.45
Methyl *cis,cis*-6,9-hexadecadienoate	0.89	—	—	—
Methyl heptadecanoate	0.68	0.685	0.655	0.68
Methyl heptadecenoate	—	—	0.573	—
Methyl octadecanoate	1.00	1.000	1.000	1.00
Methyl *cis*-9-octadecenoate	—	0.917	0.831	0.88
Methyl *cis,cis*-9,12-octadecadienoate	—	0.899	0.815	0.82
Methyl *cis,cis,cis*-6,9,12-octadecatrienoate	—	0.814	—	—
Methyl *cis,cis,cis*-9,12,15-octadecatrienoate	0.85	0.899	0.815	0.82
Methyl nonadecanoate	—	1.43	—	—
Methyl icosanoate	2.13	2.07	—	—
Methyl *cis*-11-icosenoate	—	1.86	—	—
Methyl *cis,cis*-11,14-icosadienoate	—	1.81	—	—
Methyl *cis,cis,cis*-5,8,11-icosatrienoate	—	1.56	—	—
Methyl *cis,cis,cis*-8,11,14-icosatrienoate	—	1.62	—	—
Methyl *cis,cis,cis*-11,14,17-icosatrienoate	—	1.86	—	—
Methyl *cis,cis,cis,cis*-5,8,11,14-icosatetraenoate	1.47	1.54	—	—
Methyl *cis,cis,cis,cis,cis*-5,8,11,14,17-icosapentaenoate		1.56	—	—
Methyl henicosanoate	—	2.92	—	—
Methyl docosanoate	—	4.17	—	—
Methyl *cis*-13-docosenoate	—	3.70	—	—
Methyl *cis,cis,cis,cis,cis,cis*-4,7,10,13,16,19-docosahexaenoate	—	2.94	—	—
Methyl tetracosanoate	—	8.38	—	—
Methyl *cis*-15-tetracosenoate	—	7.40	—	—

Column packing	P1	=	Apiezon-L in form of wall coating in stainless steel tubing
	P2	=	E301 (methyl silicone) in form of wall coating in stainless steel tubing
	P3	=	Apiezon-L in form of wall coating in stainless steel tubing
	P4	=	Apiezon-L in form of wall coating in stainless steel tubing

Table 25 (continued)
METHYL ESTERS OF FATTY ACIDS: ON WALL-COATED OPEN-TUBULAR COLUMNS WITH NONPOLAR LIQUID PHASES, REPORTED AS RETENTIVE TIMES RELATIVE TO METHYL OCTADECANOATE

REFERENCES

1. **Lipsky, S. R., Landowne, R. A., and Lovelock, J. E.,** *Anal. Chem.,* 31, 852, 1959.
2. **Kingsbury, K. J., Morgan, D. M. L., and Brett, C. G.,** *Biochem. Med.,* 18, 37, 1977.
3. **Hrivnak, J., Palo, V., and Krupcik, J.,** *Milchwissenschaft,* 28, 699, 1973.
4. **Immamura, M., Niiya, I., Takagi, K., and Matsumoto, T.,** *Yukagaku,* 18, 72, 1969.

Note added in proof: For additional RRT data recently tabulated for the nonpolar liquid phase SE-30 wall-coated in an open-tubular column, see: **Korhonen, I.O.O.,** *Chromatographia,* 17, 70, 1983.

Table 26
METHYL ESTERS OF FATTY ACIDS: ON WALL-COATED OPEN-TUBULAR COLUMNS WITH NONPOLAR LIQUID PHASES, REPORTED AS EQUIVALENT CHAIN LENGTHS

Column packing	P1	P2	P3	P4	P5
Temperature (°C)	200	220	200	200	210
Gas	He	N_2	Ar	na	N_2
Flow rate (mℓ/min)	na	1	1	na	na
Column					
Length (cm)	3-5000	5000	4700	4500	5000
Diameter					
(cm I.D.)	0.03	0.05	0.025	0.025	0.05
(cm O.D.)	na	na	na	na	na
Form	WCOT	WCOT	WCOT	WCOT	WCOT
Material	G	SS	SS	SS	SS
Detector	FID	FID	Argon	FID	FID
Reference	1	2	3	4	5

Compound			ECL		
Methyl pentadecanoate	15.00	15.00	15.00	15.00	15.00
Methyl pentadecenoate	—	—	—	14.80	—
Methyl hexadecanoate	16.00	16.00	16.00	16.00	16.00
Methyl *cis*-9-hexadecenoate	15.77	15.80	15.67	15.73	—
Methyl heptadecanoate	17.00	17.00	17.00	17.00	17.00
Methyl heptadecenoate	—	—	—	16.70	—
Methyl octadecanoate	18.00	18.00	18.00	18.00	18.00
Methyl *cis*-9-octadecenoate	17.71	17.82	17.62	17.64	17.63
Methyl *cis,cis*-9,12-octadecadienoate	17.64	17.76	17.48	17.52	17.50
Methyl -*cis,cis,cis*-6,9,12-octadecatrienoate	—	17.50	—	—	17.27
Methyl *cis,cis,cis*-9,12,15-octadecatrienoate	17.61	17.78	17.48	17.52	17.49
Methyl *cis,cis,cis,cis*-6,9,12,15-octadecatetraenoate	—	—	—	—	17.26
Methyl icosanoate	20.00	20.00	20.00	20.00	20.00
Methyl *cis*-11-icosenoate	19.69	19.80	—	—	—
Methyl *cis,cis*-11,14-icosadienoate	19.61	19.74	—	—	—
Methyl *cis,cis,cis*-5,8,11-icosatrienoate	—	19.36	—	—	—
Methyl *cis,cis,cis*-8,11,14-icosatrienoate	—	19.40	—	—	—
Methyl *cis,cis,cis*-11,14,17-icosatrienoate	—	19.80	—	—	—
Methyl *cis,cis,cis,cis*-5,8,11,14-icosatetraenoate	—	19.24	—	—	—
Methyl *cis,cis,cis,cis,cis*-5,8,11,14,17-icosapentaenoate	—	19.34	—	—	—
Methyl henicosanoate	21.00	21.00	21.00	21.00	21.00
Methyl *cis,cis,cis,cis,cis*-6,9,12,15,18-henicosapentaenoate	—	—	—	19.91[a]	—
Methyl docosanoate	22.00	22.00	22.00	22.00	22.00
Methyl *cis*-13-docosenoate	—	21.74	—	—	—
Methyl *cis,cis,cis,cis,cis,cis*-4,7,10,13,16,19-docosahexanenoate	—	20.94	—	—	—
Methyl tetracosanoate	24.00	24.00	24.00	24.00	24.00
Methyl *cis*-15-tetracosenoate	—	23.80	—	—	—

[a] Contributed from Mayzaud, R. G. and Ackman, R. G., The 6,9,12,15,18-henicosapentaenoic acid of seal oil, *Lipids* 13, 24-28, 1978.

Column packing   P1 = SE-30 (methyl silicone) in form of wall coating in glass tubing.
  P2 = E301 (methyl silicone) in form of wall coating in stainless steel tubing
  P3 = Apiezon-L in form of wall coating in stainless steel tubing
  P4 = Apiezon-L in form of wall coating in stainless steel tubing
  P5 = Apiezon-L in form of wall coating in stainless steel tubing

Table 26 (continued)
METHYL ESTERS OF FATTY ACIDS: ON WALL-COATED OPEN-TUBULAR COLUMNS WITH NONPOLAR LIQUID PHASES, REPORTED AS EQUIVALENT CHAIN LENGTHS

REFERENCES

1. **Holmbom, B.,** *J. Am. Oil Chem. Soc.,* 54, 289, 1977.
2. **Kingsbury, K. J., Morgan, D. M. L., and Brett, C. G.,** *Biochem. Med.,* 18, 37, 1977.
3. **Litchfield, C., Reiser, R., and Isbell, A. F.,** *J. Am. Oil Chem. Soc.,* 40, 302, 1963.
4. **Hrivnak, J., Palo, V., and Krupcik, J.,** *Milchwissenschaft,* 28, 699, 1973.
5. **Barve, J. A., Gunstone, F. D., Jacobsberg, F. R., and Winlow, P.,** *Chem. Phys. Lipids,* 8, 117, 1972.

Note added in Proof: For additional ECL data recently tabulated for the dimethyl silicone liquid phases: (A) SE-30; (B) OV-1; (C) SP-2100; wall-coated in open-tubular columns, see: (A) **Dorris, G. M., Douek, M., and Allen, L. H.,** *J. Am. Oil Chem. Soc.,* 59, 494, 1982; (B) **Pörschmann, J., Welsch, T., Herzchuh, R., Engewald, W., and Müller, K.,** *J. Chromatogr.,* 241, 73, 1982; (C) **Gillan, F. T.,** *J. Chromatogr. Sci.,* 21, 293, 1982; and (C) **Dorris, G. M., Douek, M., and Allen, L. H.,** *J. Am. Oil Chem. Soc.,* 59, 494, 1982. For additional ECL data recently tabulated for SE-54 low polarity cyanosilicone (5% phenyl, 1% vinyl) liquid phase wall-coated in open-tubular columns, see: **Pörschmann, J., Welsch, J., Herzschuh, R., Engewald, W., and Müller, K.,** *J. Chromatogr.,* 241, 73, 1982; and **Ayanoglu, E., Walkup, R. D., Sica, D., and Djerassi, C.,** *Lipids,* 17, 617, 1982. For ECL data tabulated for methyl-branched fatty acids on Apiezon-L liquid phase wall-coated in an open-tubular column see: **Ackman, R. G.,** in *Progress in the Chemistry of Fats and Other Lipids,* Vol. 12, Holman, R. T., Ed., Pergamon Press, New York, 1972, 165.

Table 27
METHYL ESTERS OF FATTY ACIDS: WITH CHAIN LENGTHS OF C_{24} OF MORE, ON PACKED COLUMNS WITH POLAR OR NONPOLAR LIQUID BASES, REPORTED AS RETENTION TIMES RELATIVE TO METHYL OCTADECANOATE OR METHYL cis-9-OCTADECENOATE, OR AS EQUIVALENT CHAIN LENGTHS

	P1	P2	P2	P1	P2	P2	P2	P3	P4	P5	P6	P7	P8	P9	P10	P11	P12
Column packing																	
Temperature (°C)	185	210	210	185	210	210	210	185	185	200	200	200	180	207	190	200	200
Gas	N_2	na	na	N_2	na	na	na	N_2	N_2	N_2	N_2	N_2	N_2	na	na	na	na
Flow rate (mℓ/min)	30	na	na	30	na	na	na	55	55	18	20	23	20	na	na	na	na
Column Length (cm)	183	na	na	183	na	na	na	180	180	200	200	200	200	182	182	180	180
Diameter (cm I.D.)	0.318	na	na	0.318	na	na	na	0.31	0.31	0.318	0.318	0.318	0.318	0.24	0.24	0.24	0.24
(cm O.D.)	na	na	na	na	na	na	na	na	na	na	na	na	na	na	na	na	na
Form	Helix	U-tube	U-tube	Helix	U-tube	U-tube	U-tube	Helix	Helix	na	na	na	na	na	na	na	na
Material	G	G	G	G	G	G	G	G	G	SS	SS	SS	SS	SS	SS	SS	SS
Detector	FID	na	na	FID	na	na	na	FID	FID	FID	FID	FID	FID	FID	FID	FID	FID
References	1	1	1	1	1	1	1	2	2	3	3	3	3	4	5	6	7,8
Compound	ECL	ECL	ECL	$r_{18:1}$	$r_{18:1}$	$r_{18:1}$	$r_{18:1}$	$r_{18:0}$	$r_{18:0}$	$r_{18:0}$	$r_{18:0}$	$r_{18:0}$	$r_{18:0}$	ECL	ECL	ECL	ECL
Methyl octadecanoate	—	—	—	1.00	1.00	1.00	1.00	1.00	1.00	1.00	1.00	1.00	1.00	18.00	18.00	18.00	18.00
Methyl cis-9-octadecenoate	—	—	—	1.00	1.00	1.00	1.00	1.16	1.15	1.19	1.21	1.20	1.25	—	18.65	—	—
Methyl tetracosanoate	24.00	24.00	24.00	4.80	—	7.40	—	5.91	—	—	6.50	5.41	5.63	24.00	24.00	24.00	—
Methyl cis-15-tetracosenoate	24.78	23.63	23.69	5.45	6.60	6.70	7.44	6.75	6.79	—	—	—	—	—	24.62	—	—
Methyl cis,cis-5,9-tetracosadienoate	—	—	—	—	—	—	—	8.01	—	—	—	—	—	—	24.87	—	—
Methyl cis,cis-12,15-tetracosadienoate	—	—	—	—	—	—	—	—	—	—	—	—	—	—	—	—	—
Methyl cis,cis-15,18-tetracosadienoate	—	—	—	—	—	—	—	—	8.62	8.14	8.22	7.02	6.81	—	25.36	—	—
Methyl cis,cis,cis-15,18,21-tetracosatrienoate	—	—	—	—	—	—	—	—	—	—	—	—	—	—	26.32	—	—
Methyl cis,cis,cis,cis-12,15,18,21-tetracosatetraenoate	26.85	23.38	23.36	10.0	6.03	6.03	—	—	—	13.06	13.23	11.30	11.75	—	—	—	—
Methyl cis,cis,cis,cis,cis-6,9,12,15,18-tetracosapentaenoate	—	—	—	—	—	—	—	—	—	12.40	12.53	10.07	10.76	—	—	—	—
Methyl cis,cis,cis,cis,cis-9,12,15,18,21-tetracosapentaenoate	27.46	23.12	23.08	11.8	5.60	5.50	—	—	—	—	—	—	—	—	—	—	—

Methyl ester	P1	P2	P3	P4	P5	P6	P7	P8	P9	P10	P11	P12
Methyl *cis,cis,cis,cis,cis,cis*-6,9,12,15,18,21-tetracosahexaenoate	28.00	—	—	5.05	20.35	—	—	—	—	—	—	25.84
Methyl *cis,cis*-5,9-pentacosadienoate	—	22.80	13.6	—	—	—	—	—	—	—	—	26.35
Methyl *cis,cis*-16,19-pentacosadienoate	—	—	—	—	—	—	—	—	—	—	—	26.00
Methyl hexacosanoate	26.00	26.00	26.00	—	—	—	26.00	26.00	26.00	—	—	26.00
Methyl *cis*-9-hexacosenoate	—	—	—	—	10.35	—	—	—	—	26.39	—	—
Methyl *cis*-17-hexacosenoate	27.06	25.69	10.4	12.7	11.05	11.80	11.31	9.30	8.70	26.62	26.70	—
Methyl *cis,cis*-5,9-hexacosadienoate	—	—	—	—	—	—	—	—	—	27.08	26.80	27.00
Methyl *cis,cis*-17,20-hexacosadienoate	—	—	—	—	—	—	—	—	—	—	27.34	—
Methyl *cis,cis,cis*-5,9,19-hexacosatrienoate	—	—	—	—	—	—	—	—	—	27.74	27.43	—
Methyl *cis,cis,cis*-17,20,23-hexacosatrienoate	—	—	—	—	—	—	—	—	—	—	28.24	—
Methyl *cis,cis,cis,cis*-11,14,17,20,23-hexacosapentaenoate	29.78	25.20	25.07	22.6	10.6	10.3	—	—	—	—	—	—
Methyl *cis,cis,cis,cis,cis,cis*-8,11,14,17,20,23-hexacosahexaenoate	30.90	24.97	24.79	24.0	9.90	9.40	—	—	—	—	—	—
Methyl *cis,cis,cis*-5,9,20-heptacosatrienoate	—	—	—	—	—	—	—	—	—	—	—	28.38
Methyl *cis,cis,cis*-5,9,19-octacosatrienoate	—	—	—	—	—	—	—	—	—	—	29.52	—
Methyl *cis,cis,cis,cis*-4,7,10,13,16,19,22-octacosaheptaenoate	32.73	26.43	51.0	15.8	15.6	—	—	—	—	—	—	—
Methyl *cis,cis,cis*-5,9,23-triacontatrienoate	—	—	—	—	—	—	—	—	—	—	—	32.04
Methyl *cis,cis,cis,cis*-15,18,21,24-triacontatetraenoate	—	—	—	—	—	—	—	—	—	—	—	33.28
Methyl *cis,cis,cis,cis,cis*-15,18,21,24,27-triacontapentaenoate	—	—	—	—	—	—	—	—	—	—	—	34.47

Column packing   P1 = 3% EGSS-X (ethylene glycol succinate polyester combined with methyl silicone) on Gas-Chrom P (100-200 mesh)
P2 = 1% SE-30 (methyl silicone) on Gas-Chrom P (100-200 mesh)
P3 = 15% Poly (diethylene glycol succinate) on Gas-Chrom P (100-200 mesh)
P4 = 10% EGSS-X (ethylene glycol succinate polyester combined with methyl silicone) on Gas-Chrom P (100-120 mesh)
P5 = 10% EGSS-X (ethylene glycol succinate polyester combined with methyl silicone) on Gas-Chrom P (100-120 mesh)
P6 = 10% LAC-3-R-728 (commercial diethylene glycol succinate polyester) on Gas-Chrom P (80-100 mesh)
P7 = SILAR-10C (APOLAR 10C; 100% cyanopropyl siloxane) on Gas-Chrom P (80-100 mesh)
P8 = 10% OV-275 (100% cyanoethyl siloxane) on Gas-Chrom P (80-100 mesh)
P9 = 10% EGSS-X (ethylene glycol succinate polyester combined with methyl silicone) on Gas-Chrom P (100-120 mesh)
P10 = 10% EGSS-X (ethylene glycol succinate polyester combined with methyl silicone) on Gas-Chrom P (100-120 mesh)
P11 = 10% EGSS-X (ethylene glycol succinate polyester combined with methyl silicone) on Gas-Chrom P (100-120 mesh)
P12 = 10% SILAR-10C (APOLAR 10C; 100% cyanopropyl siloxane) on Gas-Chrom Q (100-120 mesh)

Table 27 (continued)
METHYL ESTERS OF FATTY ACIDS: WITH CHAIN LENGTHS OF C$_{24}$ OF MORE, ON PACKED COLUMNS WITH POLAR OR NONPOLAR LIQUID BASES, REPORTED AS RETENTION TIMES RELATIVE TO METHYL OCTADECANOATE OR METHYL *cis*-9-OCTADECENOATE, OR AS EQUIVALENT CHAIN LENGTHS

P8 = 10% OV-275 (100% cyanoethyl siloxane) on Gas-Chrom P (80-100 mesh)
P9 = 10% EGSS-X (ethylene glycol succinate polyester combined with methyl silicone) on Gas-Chrom P (100-120 mesh)
P10 = 10% EGSS-X (ethylene glycol succinate polyester combined with methyl silicone) on Gas-Chrom P (100-120 mesh)
P11 = 10% EGSS-X (ethylene glycol succinate polyester combined with methyl silicone) on Gas-Chrom P (100-120 mesh)
P12 = 10% SILAR-10C (APOLAR 10C; 100% cyanopropyl siloxane) on Gas-Chrom Q (100-120 mesh)

REFERENCES

1. **Linko, R. R. and Karinkanta, H.,** *J. Am. Oil Chem. Soc.*, 47, 42, 1970.
2. **Kuksis, A.,** *Fette Seifen Anstrichm.*, 73, 130, 1971.
3. **Cerma, E. and Stancher, B.,** *Riv. Ital. Sost. Grasse*, 54, 425, 1977.
4. **Jefferts, E., Morales, R. W., and Litchfield, C.,** *Lipids*, 9, 244, 1974.
5. **Morales, R. W. and Litchfield, C.,** *Biochim. Biophys. Acta*, 431, 206, 1976.
6. **Litchfield, C. and Marcantonio, E. E.,** *Lipids*, 13, 199, 1978.
7. **Litchfield, C., Tyszkiewicz, J., Marcantonio, E. E., and Noto, G.,** *Lipids*, 14, 619, 1979.
8. **Litchfield, C., Tyszkiewicz, J., and Dato, V.,** *Lipids*, 15, 200, 1980.

Note added in proof: For additional ECL data, especially for methyl-branched polyunsaturated very long-chain fatty acids, recently tabulated for SE-54 low polarity cyanosilicone (5% phenyl, 1% vinyl) wall-coated in an open-tubular column see: **Ayanoglu, E., Walkup, R. D., Sica, D., and Djerassi, C.,** *Lipids*, 17, 617, 1982.

Table 28
METHYL ESTERS OF FATTY ACIDS: OF LONGER-CHAIN N-9 SERIES, IN PACKED COLUMNS, REPORTED IN EQUIVALENT CHAIN LENGTHS

	P1	P2	P3	P4	P5	P6
Column packing						
Temperature (°C)	234	226	180	190	210	240
Gas	He	He	Ar	na	na	He
Flow rate (mℓ/min)	120	60	60	na	na	60
Column						
Length (cm)	360	200	200	150	150	250
Diameter (cm I.D.)	0.3	0.3	0.4	0.2	0.2	0.3
(cm O.D.)	na	na	na	na	na	na
Form	na	na	U-tube	na	na	Helix
Material	na	na	G	SS	SS	na
Detector	TC	TC	Argon	FID	FID	TC
References	1	1	1	2	2	1
Compound			**ECL**			
Methyl *cis,cis,cis*-5,8,11-icosatrienoate	21.53	21.65	21.57	21.22	19.13	19.15
Methyl *cis,cis,cis*-7,10,13-docosatrienoate	23.30	23.73	23.17	23.37	21.23	21.15
Methyl *cis,cis,cis*-9,12,15-tetracosatrienoate	—	—	—	25.21	22.90	—

Column packing P1 = 20% Poly (β-cyclodextrin acetate) on Gas-Chrom P (30-60 mesh)
P2 = 20% Poly (diethylene glycol succinate) on Gas-Chrom P (80-100 mesh)
P3 = 20% Poly (ethylene glycol succinate) on Gas-Chrom P (80-100 mesh)
P4 = 15% Poly (ethylene glycol succinate) on Chromosorb-W
P5 = 15% Apiezon-L on Chromosorb-W
P6 = 20% Apiezon-L on Gas-Chrom P (100-140 mesh)

REFERENCES

1. **Hofstetter, H. H., Sen, N., and Holman, R. T.,** *J. Am. Oil Chem. Soc.,* 42, 537, 1965.
2. **Carney, J. A. and Walker, B. L.,** *Nutr. Rept. Int.,* 3, 62, 1971.

Table 29
METHYL ESTERS OF FATTY ACIDS: WITH Δ5 ETHYLENIC UNSATURATION AND RELEVENT OTHER NATURALLY OCCURRING STRUCTURES, ON PACKED COLUMNS, REPORTED AS EQUIVALENT CHAIN LENGTHS

	P1	P2	P3	P4	P5	P6	P7	P8	P9	P10
Column packing										
Temperature (°C)	210	250	200	190	190	145	na	200	177	196
Gas	Ar	na	na	He	He	He	Ar	na	Ar	Ar
Flow rate (ml/min)	na	na	na	100	100	20	na	na	na	na
Column										
Length (cm)	153	na	367	200	200	183	240	na	153	153
Diameter										
(cm I.D.)	na	na	na	0.6	0.6	na	0.65	na	na	na
(cm O.D.)	na	na	0.6	na	na	0.318	na	na	na	na
Form	na	na	na	na	na	na	Helix	na	na	na
Material	na	na	na	na	na	SS	Al	na	na	na
Detector	Argon	na	na	na	na	FID	Argon	na	Argon	Argon
References	1	2	3	4	5	6	7	2	1	1

ECL

Compound	P1	P2	P3	P4	P5	P6	P7	P8	P9	P10
Methyl tetradecanoate	14.00	14.00	14.00	14.00	14.00	14.00	14.00	14.00	14.00	14.00
Methyl cis-5-tetradecenoate	—	13.62	—	—	—	—	—	14.34	—	—
Methyl hexadecanoate	16.00	16.00	16.00	16.00	16.00	16.00	16.00	16.00	16.00	16.00
Methyl cis-5-hexadecenoate	15.89	15.67	—	—	—	—	—	16.29	—	16.42
Methyl cis-9-hexadecenoate	15.58	15.48	—	—	—	16.52	—	16.90	16.60	16.70
Methyl cis,cis-5,9-hexadecadienoate	—	—	—	—	—	17.03	—	—	17.08	—
Methyl octadecanoate	18.00	18.00	18.0	18.00	18.00	18.00	18.00	18.00	18.00	18.00
Methyl cis-5-octadecenoate	17.60	17.66	18.2	18.33	18.14	18.60	18.40	18.27	18.40	18.31
Methyl cis-9-octadecenoate	17.34	17.70	18.3	18.6x	18.14	18.98	—	18.40	19.00	18.61
Methyl cis,cis-5,9-octadecadienoate	17.46	17.46	18.6	18.91	—	19.35	19.02	18.84	19.20	18.90
Methyl cis,cis-9,12-octadecadienoate	—	17.53	18.8	19.18	18.80	19.77	19.20	19.04	—	—
Methyl cis,cis,cis-5,9,12-octadecatrienoate	—	—	—	—	—	20.35	19.91	—	—	—
Methyl cis,cis,cis-9,12,15-octadecatrienoate	—	—	19.8	19.68	19.58	20.80	—	—	—	—
Methyl cis,cis,cis,cis-5,9,12,15-octadecatetraenoate	—	—	—	—	—	—	20.42	—	—	—
Methyl cis,cis,cis,cis-6,9,12,15-octadecatetraenoate	—	—	—	—	—	—	—	—	—	—

	P1	P2	P3	P4	P5	P6	P7	P8	P9	P10
Methyl icosanoate	20.00	20.00	20.00	20.00	20.00	20.00	20.00	20.00	20.00	20.00
Methyl *cis*-5-icosenoate	—	19.67	20.1	20.24	20.24	—	—	20.25	—	—
Methyl *cis*-11-icosenoate	—	19.67	20.3	20.24	20.69	—	—	20.25	—	—
Methyl *cis,cis*-5,11-icosadienoate	—	19.34	20.6	20.46	20.68	—	—	20.68	—	—
Methyl *cis,cis*-11,14-icosadienoate	—	—	20.8	—	—	21.05	—	—	—	—
Methyl *cis,cis,cis*-5,11,14-icosatrienoate	—	—	21.2	20.96	—	21.25	—	—	—	—
Methyl *cis,cis,cis,cis*-5,8,11,14-icosatetraenoate	—	—	—	—	22.38	—	—	—	—	—
Methyl *cis,cis,cis,cis*-5,11,14,17-icosatetraenoate	—	—	—	21.80	—	—	—	—	—	—
Methyl docosanoate	22.00	22.00	22.00	22.00	22.00	22.00	22.00	22.00	22.00	22.00
Methyl *cis*-5-docosenoate	21.60	—	—	—	—	—	—	22.23	—	—
Methyl *cis*-13-docosenoate	21.66	—	—	—	—	—	—	22.36	—	—
Methyl *cis,cis*-5,13-docosadienoate	21.24	—	—	—	—	—	—	22.60	—	—
Methyl *cis,cis*-13,16-docosadienoate	21.60	—	—	—	—	—	—	23.03	—	—
Methyl *cis,cis,cis*-13,16,9-docosatrienoate	21.60	—	—	—	—	—	—	23.80	—	—
Methyl *cis,cis,cis,cis*-5,13,16,19-docosatetraenoate	21.10	—	—	—	—	—	—	24.08	—	—
Methyl tetracosanoate	24.00	24.00	—	—	—	—	—	24.00	—	—
Methyl *cis*-5-tetracosenoate	23.60	—	—	—	—	—	—	24.24	—	—
Methyl *cis,cis*-5,15-tetracosadienoate	23.22	—	—	—	—	—	—	24.58	—	—

Column packing P1 = 10% Apiezon M
P2 = 5% Apiezon L on Gas-Chrom Q
P3 = 5% LAC-2-R-446 (commercial poly diethylene glycol adipate polyester) on Celite 545 (100-150 mesh)
P4 = 20% LAC-2S-446 (commercial poly diethylene glycol adipate polyester) on Celite 545 (100-150 mesh)
P5 = 20% LAC-2-R 446 (commercial poly diethylene glycol adipate polyester) on Celite 545 (100-150 mesh)
P6 = 10% Poly (diethylene glycol succinate, plus H_3PO_4) on Supelcoport (80-100 mesh)
P7 = 20% Poly (ethylene glycol adipate) + (10% Episcote Resin, Shell Chem. Co.) on Celite (30-80 mesh)
P8 = 5% LAC-2-R (Resoflex)-446 (commercial poly ethylene glycol adipate) on Gas-Chrom Q
P9 = 16.7% Poly (ethylene glycol succinate)
P10 = 20% Poly (ethylene glycol adipate)

REFERENCES

1. **Davidoff, F. and Korn, E. D.**, *J. Biol. Chem.*, 238, 3199, 1963.
2. **Miwa, T. K.**, Unpublished results.
3. **Madrigal, R. V. and Smith, C. R., Jr.**, *Lipids*, 10, 502, 1975.
4. **Smith, C. R., Jr., Freidinger, R. M., Hagemann, J. W., Spencer, G. F., and Wolff, I. A.**, *Lipids*, 4, 462, 1969.
5. **Smith, C. R., Jr., Kleiman, R., and Wolff, I. A.**, *Lipids*, 3, 37, 1968.
6. **Weeks, G.**, *Biochim. Biophys. Acta*, 450, 21, 1976.
7. **Hansen, R. P. and Boderick, D. F.**, *Tappi*, 51, 48, 1968.

Table 30
METHYL ESTERS OF FATTY ACIDS: WITH Δ5 ETHYLENIC UNSATURATION AND RELEVANT OTHER NATURALLY OCCURRING STRUCTURES, ON WALL-COATED OPEN-TUBULAR COLUMNS, REPORTED AS EQUIVALENT CHAIN LENGTHS

	P1	P2	P3	P4	P5	P6	P7	P8	P9	P10	P11
Column packing											
Temperature (°C)	180	180	180	180	195	185	200	170	170	180	180
Gas	N$_2$	N$_2$	N$_2$		He	He	He	He	He	He	He
Flow rate (mℓ/min)	na	na	na		na	5	na	5	5	5	5
Column Length (cm)	5000	5000	5000	5000	3000	4700	35000	4700	4700	4700	4700
Diameter (cm I.D.)	0.05	0.05	0.05	0.05	0.03	0.025	0.03	0.025	0.025	0.025	0.025
(cm O.D.)	na	na	na	na	na	na	na	na	na	na	na
Form	WCOT	WCOT	SCOT	SCOT	WCOT	WCOT	WCOT	WCOT	WCOT	WCOT	WCOT
Material	SS	SS	SS	SS	G	SS	G	SS	SS	SS	SS
Detector	FID	FID	FID	FID	FID	FID	FID	FID	FID	FID	FID
References	1	1	1	1	2	3	2	4	4	4	3
Compound						ECL					
Methyl tetradecanoate	14.00	14.00	14.00	14.00	14.00	14.00	14.00	14.00	14.00	14.00	14.00
Methyl *cis*-5-tetradecenoate	—	—	—	—	—	13.67	—	—	—	—	14.22
Methyl pentadecanoate	15.00	15.00	15.00	15.00	15.00	15.00	15.00	15.00	15.00	15.00	15.00
Methyl *cis*-5-pentadecenoate	—	—	—	—	—	14.64	—	—	—	—	15.21
Methyl hexadecanoate	16.00	16.00	16.00	16.00	16.00	16.00	16.00	16.00	16.00	16.00	16.00
Methyl *cis*-5-hexadecenoate	—	—	—	—	—	—	—	—	16.15	16.18	—
Methyl *cis*-9-hexadecenoate	—	—	—	—	16.41	15.70	15.77	16.39	16.34	16.39	16.40
Methyl octadecanoate	18.00	18.00	18.00	18.00	18.00	18.00	18.00	18.00	18.00	18.00	18.00
Methyl *cis*-5-octadecenoate	—	—	—	—	—	—	—	—	18.20	18.20	—
Methyl *cis*-9-octadecenoate	18.49	18.44	18.24	18.22	18.32	17.62	17.71	18.34	18.29	18.33	18.28
Methyl *cis,cis*-5,9-octadecadienoate	18.93	18.79	18.48	18.32	18.58	—	17.51	18.90	—	—	18.78
Methyl *cis,cis,cis*-9,12-octadecadienoate	19.25	19.16	18.79	18.63	18.90	17.45	17.64	—	18.80	18.82	—
Methyl *cis,cis,cis*-5,9,12-octadecatrienoate	19.74	19.57	19.02	18.83	19.16	—	17.43	19.16	—	—	—
Methyl *cis,cis,cis*-6,9,12-octadecatrienoate	19.96	19.74	19.18	18.95	—	—	—	—	—	—	—
Methyl *cis,cis,cis*-9,12,15-octadecatrienoate	20.33	20.12	19.51	19.26	19.64	17.42	17.61	19.50	19.51	19.47	19.38

Compound	P1	P2	P3	P4	P5	P6	P7	P8	P9	P10	P11
Methyl cis,cis,cis,cis-5,9,12,15-octadecatetraenoate	20.78	20.46	19.71	19.40	—	—	—	—	—	—	—
Methyl cis,cis,cis,cis-6,9,12,15-octadecatetraenoate	20.99	20.69	19.89	19.56	—	—	19.80	19.89	19.72	—	—
Methyl icosanoate	20.00	20.00	20.00	20.00	20.00	20.00	20.00	20.00	20.00	20.00	20.00
Methyl cis-5-icosenoate	—	—	—	—	—	—	—	20.14	20.17	—	—
Methyl cis-11-icosenoate	—	—	—	—	20.29	19.69	—	20.25	—	—	—
Methyl cis,cis-11,14-icosadienoate	—	—	—	—	20.87	19.61	20.84	20.78	—	—	—
Methyl cis,cis,cis-5,11,14-icosatrienoate	21.66	21.45	20.89	20.56	21.10	19.34	21.12	—	—	—	—
Methyl cis,cis,cis-8,11,14-icosatrienoate	21.77	21.57	21.04	20.77	—	—	21.48	—	—	—	—
Methyl cis,cis,cis-11,14,17-icosatrienoate	—	—	—	—	—	—	21.30	—	—	—	—
Methyl cis,cis,cis,cis-5,8,11,14-icosatetraenoate	22.59	22.30	21.60	21.18	—	—	21.61	22.01	21.87	—	—
Methyl cis,cis,cis,cis-5,11,14,17-icosatetraenoate	22.87	22.50	21.71	21.37	—	—	21.75	—	—	—	—
Methyl cis,cis,cis,cis-8,11,14,17-icosatetraenoate	—	—	—	—	—	—	21.90	—	—	—	—
Methyl cis,cis,cis,cis,cis-5,8,11,14,17-icosapentaenoate	—	—	—	—	—	—	—	—	—	—	—
Methyl docosanoate	22.00	22.00	22.00	22.00	22.00	22.00	22.00	22.00	22.00	22.00	22.00
Methyl cis-5-docosenoate	—	—	—	—	—	—	—	22.09	22.12	—	—
Methyl cis-13-docosenoate	—	—	—	—	—	—	—	22.17	22.28	—	—
Methyl cis,cis-5,13-docosadienoate	—	—	—	—	—	—	—	22.30	22.41	—	—

Column packing P1 = EGSS-X (ethylene glycol succinate combined with methyl silicone) in form of wall coating in stainless steel tubing

P2 = EGSS-X (ethylene glycol succinate combined with methyl silicone) in form of wall coating in stainless steel tubing

P3 = EGSS-X (ethylene glycol succinate combined with methyl silicone) in form of support coated open-tubular stainless steel column

P4 = EGSS-X (ethylene glycol succinate combined with methyl silicone) in form of support coated open-tubular stainless steel column

P5 = Poly (1,4-butanediol succinate) in form of wall coating in glass tubing

P6 = Apiezon-L in form of wall coating in stainless steel tubing

P7 = SE-30 (methyl silicone) in form of wall coating in glass tubing

P8 = SILAR-5CP (50% cyanopropyl, 50% phenyl, siloxane) in form of wall coating in stainless steel tubing

P9 = Poly (1,4-butanediol succinate) in form of wall coating in stainless steel tubing

P10 = SILAR-5CP (50% cyanopropyl, 50% phenyl, siloxane) in form of wall coating in stainless steel tubing

P11 = SILAR-5CP(50% cyanopropyl, 50% phenyl, siloxane); in form of wall coating in stainless steel tubing

Table 30 (continued)

METHYL ESTERS OF FATTY ACIDS: WITH Δ5 ETHYLENIC UNSATURATION AND RELEVANT OTHER NATURALLY OCCURRING STRUCTURES, ON WALL-COATED OPEN-TUBULAR COLUMNS, REPORTED AS EQUIVALENT CHAIN LENGTHS

REFERENCES

1. **Jamieson, G. R. and Reid, E. H.,** *J. Chromatogr.*, 61, 346, 1971.
2. **Holmbom, B.,** *J. Am. Oil Chem. Soc.*, 54, 289, 1977.
3. **Sebedio, J. L. and Ackman, R. G.,** *J. Am. Oil Chem. Soc.*, 56, 15, 1979.
4. **Ackman, R. G.,** unpublished results.

Note added in proof: For additional ECL data for fatty acids of these types recently tabulated for a variety of liquid phases: (A) SE-30; (B) SP-2100; (C) SE-54; (D) Carbowax 20M; (E) BDS; (F) SILAR-5CP; wall-coated in open-tubular columns, see: (A) **Dorris, G. M., Doeuk, M., and Allen, L. H.,** *J. Am. Oil Chem. Soc.*, 59, 494, 1982; (B) **Dorris, G. M., Doeuk, M., and Allen, L. H.,** *J. Am. Oil Chem. Soc.*, 59, 494, 1982; (C) **Ayanoglu, E., Walkup, R. D., Sica, D., and Djerassi, C.,** *Lipids,* 17, 617, 1982; (D) **Dorris, G. M., Doeuk, M., and Allen, L. H.,** *J. Am. Oil Chem. Soc.*, 59, 494, 1982; (E) **Ekman, R.,** *Phytochemistry,* 19, 147, 1980; **Yildirim, H. and Holmbom, B.,** *Acta Academiae Aboensis (B),* 37[5], 1978; and (F) **Takagi, J. and Itabashi, Y.,** *Lipids,* 17, 716, 1982.

Table 31
METHYL ESTERS OF FATTY ACIDS OF PARTICULAR INTEREST IN BIOSYNTHESIS AND INTERCONVERSION OF MODERATE CHAIN LENGTH FATTY ACIDS THAT ARE PRECURSORS OF LONGER-CHAIN FATTY ACIDS, ON VARIOUS COLUMNS, REPORTED AS RELATIVE RETENTION TIMES; VALUES IN PARENTHESES ARE CALCULATED FROM GRAPHIC DATA (1)

	P1	P2	P3	P4	P5	P6
Column packing						
Temperature (°C)	180	180	180	120	175	150
Gas						
Flow rate (mℓ/min)	na	na	na	na	AR	na
	na	na	na	na	na	na
Column						
Length (cm)	183	183	183	na	183	na
Diameter						
(cm I.D.)	na	na	na	na	na	na
(cm O.D.)	0.318	0.318	0.318	na	na	na
Form	na	na	na	na	U-tube	na
Material	na	na	na	na	na	na
Detector	FID	FID	FID	na	Argon	na
References	1	1	1	2	3	2

Relative Retention Times

Compound	P1	P2	P3	P4	P5	P6
Methyl decanoate	0.088	1.00	—	1.00	—	1.00
Methyl *cis*-2-decenoate	(0.09)	—	—	1.06	—	1.00
Methyl *trans*-2-decenoate	(0.15)	1.70	—	1.83	—	1.37
Methyl *cis/trans*-3-decenoate	(0.11)	1.28	—	1.38	—	0.93
Methyl dodecanoate	0.161	—	—	—	—	—
Methyl *trans*-2-dodecenoate	(0.27)	—	—	—	—	—
Methyl *cis/trans*-3-dodecenoate	(0.20)	—	—	—	—	—
Methyl tetradecanoate	0.296	—	—	—	—	—
Methyl *trans*-2-tetradecenoate	0.485	—	0.890	—	0.874	—
Methyl *cis/trans*-3-tetradecenoate	0.367	—	0.675	—	0.67	—
Methyl *cis*-7-tetradecenoate	0.347	—	0.637	—	—	—
Methyl hexadecanoate	0.545	—	1.00	—	1.00	—
Methyl *trans*-2-hexadecenoate	(0.89)	—	—	—	—	—

Table 31 (continued)
METHYL ESTERS OF FATTY ACIDS OF PARTICULAR INTEREST IN BIOSYNTHESIS AND INTERCONVERSION OF MODERATE CHAIN LENGTH FATTY ACIDS THAT ARE PRECURSORS OF LONGER-CHAIN FATTY ACIDS, ON VARIOUS COLUMNS, REPORTED AS RELATIVE RETENTION TIMES; VALUES IN PARENTHESES ARE CALCULATED FROM GRAPHIC DATA (1)

Compound	Relative Retention Times						
Methyl *cis/trans*-3-hexadecenoate	(0.67)	—	—	—	—	—	—
Methyl hexadecenoate	0.641	—	—	—	—	—	—
Methyl octadecanoate	1.000	—	—	—	—	—	—
Methyl octadecenoate	1.181	—	—	—	—	—	—

Column packing P1 = 10% EGSS-X (ethylene glycol succinate combined with methyl silicone), on Gas-Chrom P (100-120 mesh)
P2 = 10% EGSS-X (ethylene glycol succinate combined with methyl silicone), on Gas-Chrom P (100-120 mesh)
P3 = 10% EGSS-X (ethylene glycol succinate combined with methyl silicone), on Gas-Chrom P (100-120)
P4 = 10% Poly (diethylene glycol succinate)
P5 = 10% Poly (diethylene glycol succinate)
P6 = 10% Apiezon-L

REFERENCES

1. **Rooney, S. A., Goldfine, H., and Sweeley, C. C.,** *Biochim. Biophys. Acta,* 270, 289, 1972.
2. **Lennarz, W. J., Light, R. J., and Bloch, K.,** *Proc. Natl. Acad. Sci. U.S.A.,* 48, 840, 1962.
3. Nesbitt, J. A., III, and Lennarz, W. J., *J. Bacteriol.,* 89, 1020, 1965.

Table 32
METHYL ESTERS OF FATTY ACIDS: REFERENCE AND NATURAL MIXTURES EMPHASIZING DIVERSE ISOMERIC MONOETHYLENIC FATTY ACIDS, ON POLAR OR NONPOLAR LIQUID PHASES, REPORTED AS RELATIVE RETENTION TIMES

	P1	P2	P3	P4	P5	P6	P7	P8	P9	P10	P11	P12	P13	P14
Column packing														
Temperature (°)	180	200	180	180	170	180	170	180	180	185	180	220	230	240
Gas	N_2	N_2	He	He	He	He	He	He	He	He	He	N_2	N_2	N_2
Flow rate (mℓ/min)	1	1	1	1	5	5	5	5	5	5	5	8	8	8
Column														
Length (cm)	5000	5000	6000	6000	4700	4700	4700	4700	4700	4700	4700	600	600	600
Diameter (cm I.D.)	0.05	0.05	0.025	0.025	0.025	0.025	0.025	0.025	0.025	0.025	0.025	0.3	0.3	0.3
(cm O.D.)	na	na	na	na	na	na	na	na	na	na	na	na	na	na
Form	WCOT	WCOT	WCOT	WCOT	WCOT	WCOT	WCOT	WCOT	WCOT	WCOT	WCOT	na	na	na
Material	SS	SS	G	G	SS	SS	SS	SS	SS	SS	SS	SS	SS	SS
Detector	FID	FID	FID	FID	FID	FID	FID	FID	FID	FID	FID	FID	FID	FID
References	1	1	2	2	3	4	4	5	6	6	3	7	7	7
Compound								[b]Methyl octadecanoate						
Methyl tetradecanoate	0.279	0.263	0.530	0.529	—	0.256	0.244	0.229	0.233	0.156	0.153	0.590	0.647	0.686
Methyl *cis*-5-tetradecenoate	—	—	—	—	—	—	—	—	—	0.133	—	—	—	—
Methyl *cis*-7-tetradecenoate	—	—	—	—	—	—	—	—	0.269	0.135	—	—	—	—
Methyl *cis*-9-tetradecenoate	—	—	—	—	—	0.308	0.295	0.278	0.284	0.141	0.141	—	—	—
Methyl pentadecanoate	0.380	0.368	—	0.604	—	—	0.350	0.331	0.338	0.251	0.246	—	—	—
Methyl *cis*-5-pentadecenoate	—	—	—	—	—	—	—	—	0.366	0.211	—	—	—	—
Methyl *trans*-5-pentadecenoate	—	—	—	—	—	—	—	—	0.366	0.224	—	—	—	—
Methyl hexadecanoate	0.525	0.517	0.704	0.704	—	0.508	0.497	0.478	0.486	0.399	0.394	0.756	0.792	0.817
Methyl *cis*-7-hexadecenoate	—	—	—	—	0.533	—	0.560	0.536	0.546	0.336	0.335	—	—	—
Methyl *cis*-9-hexadecenoate	0.612	0.567	—	—	0.543	0.585	0.579	0.551	0.562	0.344	0.341	—	—	—
Methyl *cis*-11-hexadecenoate	—	—	—	—	0.570	—	—	0.566	0.585	0.359	—	—	—	—
Methyl *trans*-3-hexadecenoate	—	—	—	—	0.585	—	—	0.597	—	—	0.373	—	—	—
Methyl *trans*-9-hexadecenoate	—	—	—	—	—	—	—	—	—	0.354	—	—	—	—
Methyl heptadecanoate	0.720	0.715	—	—	—	—	0.701	0.695	0.651	0.634	0.629	—	—	—
Methyl octadecanoate	1.00	1.00	1.000	1.000	1.000	1.000	1.000	1.000	1.000	1.000	1.000	1.000	1.000	1.000
Methyl *cis*-7-octadecenoate	—	—	—	—	1.068	—	—	—	—	—	—	—	—	—
Methyl *cis*-9-octadecenoate	1.12	1.08	1.152	1.151	1.092	1.116	1.12	1.112	1.118	0.836	0.839	1.155	1.144	1.134
Methyl *cis*-11-octadecenoate	—	—	1.170	1.171	1.121	—	1.15	1.146	1.148	0.858	0.856	—	—	—
Methyl *cis*-13-octadecenoate	—	—	—	1.205	1.182	—	1.18	—	—	—	—	—	—	—
Methyl *trans*-9-octadecenoate	—	—	—	—	—	—	—	—	—	—	—	—	—	—
Methyl *cis,cis*-9,12-octadecadienoate	1.38	1.24	1.416	1.410	—	—	1.35	1.337	1.337	0.776	0.785	1.094	1.086	1.079
Methyl *cis,cis,cis*-6,9,12-octadecatrienoate	—	1.34	—	—	—	—	1.50	1.497	—	—	—	1.401	1.365	1.339
Methyl *cis,cis,cis*-9,12,15-octadecatrienoate	1.78	1.52	1.809	1.804	—	—	1.67	1.675	1.668	0.769	0.785	1.757	1.757	1.757
Methyl *cis,cis,cis,cis*-6,9,12,15-octadecatetraenoate	—	—	—	2.123	—	—	1.87	1.874	—	—	0.701	—	—	—

[b] Methyl octadecanoate

Table 32 (continued)
METHYL ESTERS OF FATTY ACIDS: REFERENCE AND NATURAL MIXTURES EMPHASIZING DIVERSE ISOMERIC MONOETHYLENIC FATTY ACIDS, ON POLAR OR NONPOLAR LIQUID PHASES, REPORTED AS RELATIVE RETENTION TIMES

rMethyl octadecanoate

Compound											
Methyl nonadecanoate	1.80	1.27	—	—	—	—	1.40	1.430	—	—	—
Methyl icosanoate	1.92	1.92	1.505	—	—	1.930	1.96	2.053	1.370	1.301	1.261
Methyl cis-5-icosenoate	—	—	—	—	—	2.046	—	—	—	—	—
Methyl cis-9-icosenoate	2.11	2.05	—	1.717	2.194	2.152	2.14	2.026	—	—	—
Methyl cis-11-icosenoate	—	—	1.745	1.744	2.232	—	2.20	2.278	—	—	—
Methyl cis-13-icosenoate	—	—	1.787	1.783	2.308	—	2.25	—	—	—	—
Methyl cis-15-icosenoate	—	—	—	—	2.706	—	—	—	—	—	—
Methyl cis,cis-11,14-icosadienoate	2.54	2.35	2.175	2.175	—	—	2.63	2.662	—	—	—
Methyl cis,cis,cis-5,8,11-icosatrienoate	2.66	2.38	—	—	—	—	—	—	—	—	—
Methyl cis,cis,cis-8,11,14-icosatrienoate	2.88	2.55	—	—	—	—	2.92	3.02	—	—	—
Methyl cis,cis,cis-11,14,17-icosatrienoate	3.23	2.88	—	2.889	—	—	3.27	3.42	—	—	—
Methyl cis,cis,cis,cis-5,8,11,14-icosatetraenoate	3.23	2.73	—	3.323	—	—	3.11	3.21	—	—	—
Methyl cis,cis,cis,cis-8,11,14,17-icosatetraenoate	4.17	3.30	—	3.789	—	—	3.63	3.77	—	—	—
Methyl cis,cis,cis,cis,cis-5,8,11,14,17-icosapentaenoate	2.66	2.63	—	—	—	—	3.89	4.00	—	—	—
Methyl henicosanoate	3.71	3.66	2.365	—	—	3.729	4.11	—	1.927	1.744	1.625
Methyl docosanoate	—	—	—	—	—	—	—	—	—	—	—
Methyl cis-11-docosenoate	—	—	2.783	2.698	—	—	4.11	—	2.174	2.017	1.852
Methyl cis-13-docosenoate	3.99	3.83	2.836	2.744	—	4.135	4.27	—	—	—	—
Methyl cis-15-docosenoate	—	—	—	2.818	—	—	4.40	—	2.092	1.890	1.769
Methyl trans-13-docosenoate	—	—	3.458	—	—	—	—	—	—	—	—
Methyl cis,cis-13,16-docosadienoate	—	—	—	—	—	—	6.21	—	—	—	—
Methyl cis,cis,cis,cis-7,10,13,16-docosatetraenoate	—	—	—	—	—	—	6.56	—	—	—	—
Methyl cis,cis,cis,cis,cis-4,7,10,13,16,19-docosapentaenoate	—	—	—	6.246	—	—	7.71	—	—	—	—
Methyl cis,cis,cis,cis,cis-7,10,13,16,19-docosapentaenoate	—	—	—	7.027	—	—	8.19	—	—	—	—
Methyl -cis,cis,cis,cis,cis,cis-4,7,10,13,16,19-docosahexaenoate	—	7.00	—	—	—	—	7.55	—	—	—	—
Methyl tetracosanoate	—	7.01	3.839	—	—	—	8.33	—	2.829	2.428	2.201
Methyl cis-15-tetracosenoate	—	7.27	4.438	4.436	—	7.895	—	—	—	—	—

Column packing

P1	=	EGSS-X (ethylene glycol succinate combined with methyl silicone) in form of wall coating in stainless steel tubing
P2	=	FFAP (Carbowax-20M terminated with 2-nitroterephthalic acid) in form of wall coating in stainless steel tubing
P3	=	SP2340 (Supelco 100% cyanopropyl silicone) in form of wall coating on glass tubing
P4	=	SP2340 (Supelco 100% cyanopropyl silicone) in form of wall coating on glass tubing
P5	=	Poly (1,4-butanediol succinate) in form of wall coating in stainless steel tubing
P6	=	SILAR-5CP (50% cyanopropyl, 50% phenyl, siloxane) in form of wall coating in stainless steel tubing
P7	=	SILAR-5CP (50% cyanopropyl, 50% phenyl, siloxane) in form of wall coating in stainless steel tubing
P8	=	SILAR-5CP (50% cyanopropyl, 50% phenyl, siloxane) in form of wall coating in stainless steel tubing
P9	=	SILAR-5CP (50% cyanopropyl, 50% phenyl, siloxane) in form of wall coating in stainless steel tubing
P10	=	Apiezon-L in form of wall coating in stainless steel tubing
P11	=	Apiezon-L in form of wall coating in stainless steel tubin
P12	=	15% OV-275 (100% cyanoethyl siloxane) on Gas-Chrom RZ (100-120 mesh)
P13	=	15% OV-275 (100% cyanoethyl siloxane) on Gas-Chrom RZ (100-120 mesh)
P14	=	15% OV-275 (10% cyanoethyl siloxane) on Gas-Chrom RZ (100-120 mesh)

REFERENCES

1. **Kingsbury, K. J., Morgan, D. M. L., and Brett, C. G.,** *Biochem. Med.*, 18, 37, 1977.
2. **Ackman, R. G. and Eaton, C. A.,** *Fette Seifen Anstrichm.*, 80, 21, 1978.
3. **Ackman, R. G.,** unpublished results.
4. **Joseph, J. D.,** unpublished results.
5. **Ackman, R. G.,** unpublished results.
6. **Sebedio, J.-L. and Ackman, R. G.,** *J. Am. Oil Chem. Soc.*, 56, 15, 1979.
7. **Itabashi, Y. and Takagi, T.,** *Yukagaku*, 27, 219, 1978.

Table 33
METHYL ESTERS OF FATTY ACIDS OF SELECTED UNCOMMON MONOETHYLENIC AND DIETHYLENIC STRUCTURES, ON POLAR AND NONPOLAR COLUMNS, REPORTED AS EQUIVALENT CHAIN LENGTHS

	P1	P2	P3	P4	P5	P6	P7	P8	P9	P10	P11	P12	P13	P14	P15	P16
Column packing																
Temperature (°C)	170	180	170	170	180	185	180	180	160	160	220	na	200	220	230	240
Gas	He	He	He	He	He	He	He	N$_2$	N$_2$	N$_2$	He	na	He	N$_2$	N$_2$	N$_2$
Flow rate (ml/min)	5	5	5	5	5	5	5	na	na	na	30	na	30	8	8	8
Column Length (cm)	4700	4700	4700	4700	4700	4700	4700	4500	4500	10,000	478	na	478	600	600	600
Diameter (cm I.D.)	0.025	0.025	0.025	0.025	0.025	0.025	0.025	0.025	0.025	0.025	na	na	na	0.3	0.3	0.3
(cm O.D.)	na	na	na	na	na	na	na	na	na	na	0.318	na	0.318	na	na	na
Form	WCOT	WCOT	WCOT	WCOT	WCOT	WCOT	WCOT	WCOT	WCOT	WCOT	Helix	na	Helix	na	na	na
Material	SS	SS	SS	SS	SS	SS	SS	SS	SS	SS	SS	na	SS	SS	SS	SS
Detector	FID	FID	FID	FID	FID	FID	FID	FID	FID	FID	FID	FID	FID	FID	FID	FID
References	1	2	3	3	3	4	3	5	5	5	6	7	8	9	9	9
Compound								**ECL**								
Methyl tetradecanoate	14.00	14.00	14.00	14.00	14.00	14.00	14.00	14.00	14.00	14.00	14.00	14.00	14.00	14.00	14.00	14.00
Methyl cis-5-tetradecenoate	—	14.22	—	—	—	13.67	—	—	—	—	—	—	—	—	—	—
Methyl cis-7-tetradecenoate	—	14.41	14.35	—	—	13.69	—	—	—	—	—	—	—	—	—	—
Methyl cis-9-tetradecenoate	—	—	14.50	—	13.81	13.79	—	—	—	—	—	14.50	—	—	—	—
Methyl trans-3-tetradecenoate	—	—	14.47	—	—	—	13.90	—	—	—	—	—	—	—	—	—
Methyl pentadecanoate	15.00	15.00	15.00	15.00	15.00	15.00	15.00	15.00	15.00	15.00	15.00	15.00	15.00	15.00	15.00	15.00
Methyl cis-5-pentadecenoate	—	15.21	—	—	—	14.64	—	—	—	—	—	—	—	—	—	—
Methyl cis-8-pentadecenoate	—	15.35	—	—	—	14.65	—	—	—	—	—	—	—	—	—	—
Methyl cis-9-pentadecenoate	—	—	—	—	—	—	—	14.67	15.45	15.25	—	—	—	—	—	—
Methyl trans-5-pentadecenoate	—	15.21	—	—	—	14.78	—	—	—	—	—	—	—	—	—	—
Methyl hexadecanoate	16.00	16.00	16.00	16.00	16.00	16.00	16.00	16.00	16.00	16.00	16.00	16.00	16.00	16.00	16.00	16.00
Methyl cis-6-hexadecenoate	16.25[a]	16.31	16.32	16.31	15.63	15.65	—	—	—	—	—	—	—	—	—	—
Methyl cis-7-hexadecenoate	—	16.40	16.39	16.36	15.71	15.64	—	—	—	—	17.00	16.40	—	—	—	—
Methyl cis-9-hexadecenoate	—	16.51	16.50	16.48	15.71	15.70	15.73	15.71	16.49	16.35	—	—	—	—	—	—
Methyl cis-11-hexadecenoate	—	—	16.44	16.54	15.90	15.78	15.89	—	—	—	—	—	—	—	—	—
Methyl trans-3-hexadecenoate	16.25[a]	—	—	—	—	15.74	—	—	—	—	—	—	—	—	—	—
Methyl trans-6-hexadecenoate	—	—	—	—	—	15.76	—	—	—	—	—	—	—	—	—	—
Methyl trans-9-hexadecenoate	—	—	—	—	—	—	—	—	16.41	16.31	16.55	16.55	—	—	—	—
Methyl heptadecanoate	17.00	17.00	17.00	17.00	17.00	17.00	17.00	17.00	17.00	17.00	17.00	17.00	17.00	17.00	17.00	17.00
Methyl cis-9-heptadecenoate	—	—	—	—	—	—	—	16.65	17.53	17.30	—	—	—	—	—	—
Methyl trans-9-heptadecenoate	—	—	—	—	—	—	—	17.00	—	17.26	—	—	—	—	—	—
Methyl octadecanoate	18.00	18.00	18.00	18.00	18.00	18.00	18.00	18.00	18.00	18.00	18.00	18.00	18.00	18.00	18.00	18.00
Methyl cis-5-octadecenoate	18.16	—	—	18.19	—	—	—	—	—	—	—	—	—	—	—	—
Methyl cis-6-octadecenoate	18.22[a]	—	—	—	—	—	—	—	—	—	—	—	18.75	—	—	—
Methyl cis-7-octadecenoate	—	—	—	—	—	17.62	—	—	—	—	—	—	—	—	—	—

Compound																
Methyl *cis*-9-octadecenoate	18.27,(20)[a]	18.28	18.31	18.25	17.62	17.62	17.63	17.61	18.35	18.27	18.90	18.37	18.80	18.92	19.02	19.08
Methyl *cis*-11-octadecenoate	18.34,(28)[a]	18.34	18.37	18.33	17.68	17.68	17.68	17.70	—	—	—	—	—	—	—	—
Methyl *cis*-12-octadecenoate	—	—	—	18.46	—	—	—	—	—	—	—	—	18.90	—	—	—
Methyl *cis*-13-octadecenoate	18.46	—	—	—	—	—	—	—	—	—	—	—	—	—	—	—
Methyl *cis*-15-octadecenoate	—	—	—	—	—	—	—	—	—	—	—	—	19.15	—	—	—
Methyl *trans*-6-octadecenoate	18.22[a]	—	—	—	—	17.71	—	—	—	—	—	—	18.50	—	—	—
Methyl *trans*-9-octadecenoate	—	18.26	—	—	—	—	—	17.71	18.33	18.24	18.65	—	18.55	18.57	18.63	18.66
Methyl *trans*-12-octadecenoate	—	—	—	—	—	—	—	—	—	—	—	—	18.55	—	—	—
Methyl *trans*-15-octadecenoate	—	—	—	—	—	—	—	—	—	—	—	—	18.70	—	—	—
Methyl *cis*-9,12-octadecadienoate	18.71[a]	18.78	18.83	17.50	17.50	17.42	—	—	—	—	20.00	19.00	19.90	20.13	20.33	20.47
Methyl *cis,trans*-9,12-octadecadienoate	—	—	—	—	—	—	—	—	—	—	19.85	—	19.70	—	—	—
Methyl *trans,cis*-9,12-octadecadienoate	—	—	—	17.50	—	—	—	—	—	—	19.85	—	19.70	—	—	—
Methyl *trans,trans*-9,12-octadecadienoate	—	—	—	—	—	—	—	—	—	—	19.40	—	19.30	—	—	—
Methyl *cis,cis,cis*-9,12,15-octadecatrienoate	19.38[a]	19.38	19.46	17.45	17.50	17.45	—	—	—	—	21.40	19.80	—	21.46	21.72	22.00
Methyl icosanoate	20.00	20.00	20.00	20.00	20.00	20.00	20.00	20.00	20.00	20.00	20.00	20.00	20.00	20.00	20.00	20.00
Methyl *cis*-5-icosenoate	20.11	—	20.14	—	—	—	—	—	—	—	—	—	—	—	—	—
Methyl *cis*-9-icosenoate	20.15	—	20.20	—	—	—	—	—	—	—	—	—	—	—	—	—
Methyl *cis*-11-icosenoate	20.22	20.30	20.29	—	—	—	—	—	—	—	20.90	—	—	—	—	—
Methyl *cis*-13-icosenoate	20.31	—	—	—	—	—	—	—	—	—	—	20.26	—	—	—	—
Methyl *cis*-15-icosenoate	20.44	20.25	20.41	—	—	—	—	—	—	—	—	—	—	—	—	—
Methyl *trans*-11-icosenoate	—	—	—	—	—	—	—	—	—	—	—	—	—	—	—	—
Methyl *cis,cis*-5,11-icosadienoate	20.36	—	—	—	—	—	—	—	—	—	20.60	—	—	—	—	—
Methyl *cis,cis*-5,13-icosadienoate	20.42	—	—	—	—	—	—	—	—	—	—	—	—	—	—	—
Methyl *cis,cis*-7,10-icosadienoate	20.50	—	—	—	—	—	—	—	—	—	—	—	—	—	—	—
Methyl docosanoate	22.00	22.00	22.00	22.00	22.00	22.00	22.00	22.00	22.00	22.00	22.00	22.00	22.00	22.00	22.00	22.00
Methyl *cis*-5-docosenoate	22.05	—	—	—	—	—	—	—	—	—	—	—	—	—	—	—
Methyl *cis*-11-docosenoate	22.09	—	—	—	—	—	—	—	—	—	—	—	—	22.67	22.74	22.88
Methyl *cis*-13-docosenoate	22.16	22.28	—	—	—	—	—	—	—	—	22.85	—	—	—	—	—
Methyl *cis*-15-docosenoate	22.25	22.24	—	—	—	—	—	—	—	—	22.55	—	—	22.43	22.49	22.56
Methyl *trans*-13-docosenoate	—	—	—	—	—	—	—	—	—	—	—	—	—	—	—	—
Methyl *cis,cis*-7,13-docosadienoate	22.31	—	—	—	—	—	—	—	—	—	—	—	—	—	—	—
Methyl *cis,cis*-7,15-docosadienoate	22.39	—	—	—	—	—	—	—	—	—	—	—	—	—	—	—

[a]   Values from column of lower polarity (Ackman, Lit. 1).

Table 33 (continued)
METHYL ESTERS OF FATTY ACIDS OF SELECTED UNCOMMON MONOETHYLENIC AND DIETHYLENIC STRUCTURES, ON POLAR AND NONPOLAR COLUMNS, REPORTED AS EQUIVALENT CHAIN LENGTHS

Column packing		
P1	=	Poly (1,4-butanediol succinate) in form of wall coating in stainless steel tubing
P2	=	SILAR-5CP (50% cyanopropyl, 50% phenyl siloxane) in form of wall coating in stainless steel tubing
P3	=	SILAR-5CP (50% cyanopropyl, 50% phenyl, siloxane) in form of wall coating in stainless steel tubing
P4	=	Poly (1,4-butanediol succinate) in form of wall coating in stainless steel tubing
P5	=	Apiezon-L in form of wall coating in stainless steel tubing
P6	=	Apiezon-L in form of wall coating in stainless steel tubing
P7	=	Apiezon-L in form of wall coating in stainless steel tubing
P8	=	Apiezon-L in form of wall coating in stainless steel tubing
P9	=	Poly (diethylene glycol succinate) in form of wall coating in stainless steel tubing
P10	=	Poly (1,4-butanediol succinate) in form of wall coating in stainless steel tubing
P11	=	10% SILAR-10C (100% cyanopropyl siloxane) on Gas-Chrom Q (100-120 mesh)
P12	=	25% Poly(ethylene glycol adipate) on Celite (100-210 mesh)
P13	=	10% SILAR 10C (100% cyanopropyl siloxane) on Gas-Chrom Q (100-120 mesh)
P14	=	15% OV-275 (100% cyanoethyl siloxane) on Gas-Chrom RZ (100-120 mesh)
P15	=	15% OV-275 (100% cyanoethyl siloxane) on Gas-Chrom RZ (100-120 mesh)
P16	=	15% OV-275 (100% cyanoethyl siloxane) on Gas-Chrom RZ (100-120 mesh)

REFERENCES

1. **Ackman, R. G. and Hooper, S. N.,** *Comp. Biochem. Physiol.,* 46 B, 153, 1973; **Ackman, R. G.,** *Progr. Chem. Fats Other Lipids,* 12, 165, 1972.
2. **Sebedio, J.-L. and Ackman, R. G.,** *J. Am. Oil Chem. Soc.,* 56, 15, 1979.
3. **Ackman, R. G.,** unpublished.
4. **Sebedio, J.-L. and Ackman, R. G.,** *J. Am. Oil Chem. Soc.,* 56, 15, 1979; **Ackman, R. G.,** *Progr. Chem. Fats Other Lipids,* 12, 165, 1972.
5. **Hrivnak, J., Sojak, L., Krupcik, J., and Duchesne, Y. P.,** *J. Am. Oil Chem. Soc.,* 50, 68, 1973.
6. **Conacher, H. B. S., Iyengar, J. R., and Beare-Rogers, J. L.,** *J. Assoc. Off. Anal. Chem.,* 60, 899, 1977.
7. **Haigh, W. G., Safford, R., and James, A. T.,** *Biochim. Biophys. Acta,* 176, 647, 1969.
8. **Conacher, H. B. S. and Iyengar, J. R.,** *J. Assoc. Off. Anal. Chem.,* 61, 702, 1978.
9. **Itabashi, Y. and Takagi, T.,** *Yukagaku,* 27, 219, 1978.

Note added in proof: For additional RRT data for fatty acids of these types analysed on the cyanosilicone liquid phases: (A) SILAR-10C; (B) SP-2340; (C) SILAR-9CP; (D) OV-275; on packed columns of different loadings on Chromosorb W/HP (80-100 mesh) see: **Dittmar, K. E. J., Heckers, H., and Melcher, F. W.,** *Fette. Seifen. Anstrichm.,* 80, 279, 1978. For ECL data for methyl esters of a variety of mono- and polymethylbranched fatty acids on several liquid phases: (A) BDS; (B) SE-54; (C) SIL-47-CNP; (D) Apiezon-L; wall-coated in open-tubular columns, see: (A) **Massart-Leen, A. M., De Pooter, H., Decloedt, M., and Schamp, N.,** *Lipids,* 16, 286, 1981; (A) **Ackman, R. G.,** in *Progress in the Chemistry of Fats and Other Lipids,* Vol. 12, Holman, R T., Ed., Pergamon Press, New York, 1972, 165; (B) **Ayanoglu, E., Walkup, R. D., Seca, D., and Djerassi, C.,** *Lipids,* 17, 617, 1982. (C) **Gillan, F. T., Johns, R. B., Verheyen, T. V., Volkmen, J. K., and Bavor, H. J.,** *Appl. Env. Microbiol.,* 41, 849, 1981; and (D) **Ackman, R. G.,** in *Progress in the Chemistry of Fats and Other Lipids,* Vol. 12, Holman, R. T., Ed., Pergamon Press, New York, 1972, 165.

Table 34
METHYL ESTERS OF FATTY ACIDS: *cis*-UNDECENOATES, REPORTED AS EQUIVALENT CHAIN LENGTHS

Column packing	P1	P2	P3	P4	P5	P6	P7	P8
Temperature (°C)	175	130	150	190	160	155	190	125
Gas	N_2	N_2	N_2	N_2	N_2	N_2	N_2	N_2
Flow rate (mℓ/min)	40	60	45	80	45	60	50	45
Column								
Length (cm)	200	200	200	200	200	200	200	200
Diameter								
(cm I.D.)	na	0.6	0.3	0.6	0.3	0.3	na	na
(cm O.D.)	na	na	na	na	na	na	na	na
Form	na	na	na	na	na	na	na	na
Material	na	na	na	na	na	na	na	na
Detector	FID	FID	FID	FID	FID	FID	FID	FID
References	1	2	2	2	2	2	1	1

Compound				ECL				
Methyl *cis*-2-undecenoate	10.98	10.86	10.88	10.92	11.16	11.17	11.28	11.03
Methyl *cis*-3-undecenoate	10.87	10.89	10.94	11.24	11.48	11.45	11.88	11.70
Methyl *cis*-4-undecenoate	10.76	10.82	10.85	11.06	11.25	11.27	11.62	11.44
Methyl *cis*-5-undecenoate	10.75	10.77	10.78	11.08	11.21	11.20	11.58	11.48
Methyl *cis*-6-undecenoate	10.79	10.86	10.88	11.27	11.38	11.41	11.73	11.73
Methyl *cis*-7-undecenoate	10.83	10.88	10.87	11.30	11.41	11.44	11.81	11.77
Methyl *cis*-8-undecenoate	10.89	10.96	10.94	11.35	11.49	11.54	11.95	11.86
Methyl *cis*-9-undecenoate	11.10	11.13	11.18	11.61	11.82	11.84	12.29	12.19
Methyl-10-undecenoate	10.89	10.88	10.92	11.33	11.52	11.48	11.93	11.88

Column packing P1 = 5% Apiezon-L on Chromosorb W (80-100 mesh)

P2 = 3% SE-30 (methyl silicone) on Chromosorb W (80-100 mesh)

P3 = 1.5% OV-101 (methyl silicone) on Chromosorb W (80-100 mesh)

P4 = 20% XE-60 (25% cyanoethyl, with methyl, silicone) on Chromosorb W (80-100 mesh)

P5 = 10% FFAP (Carbowax 20M modified with 2-nitroterephthalic acid) on Chromosorb W (80-100 mesh)

P6 = 10% Carbowax 20M on Chromosorb W (80-100 mesh)

P7 = 20% Poly (diethylene glycol succinate) on Chromosorb W (80-100 mesh)

P8 = 10% SILAR-10C (100% cyanopropyl siloxane) on Chromosorb W (80-100 mesh)

REFERENCES

1. **Lie Ken Jie, M. S. F. and Lam, C. H.,** *J. Chromatogr.,* 97, 165, 1974.
2. **Lie Ken Jie, M. S. F.,** *J. Chromatogr.,* 111, 189, 1975.

Table 35A

METHYL ESTERS OF FATTY ACIDS: *cis*- OCTADECENOATES, ON POLAR COLUMNS, REPORTED IN EQUIVALENT CHAIN LENGTHS

	P1	P2	P3	P4	P5	P6	P7	P8	P9
Column packing									
Temperature °C	190	190	190	165	200	190	190	170	180
Gas	N_2	N_2	N_2	Ar	Ar	Ar	N_2	He	He
Flow rate (ml/min)	na	na	na	2	0.5	2.7	na	na	na
Column									
Length (cm)	5000	200	5000	6100	6100	4700	5000	4700	4700
Diameter									
(cm I.D.)	0.05	na	0.05	0.025	0.025	0.025	0.05	0.025	0.025
(cm O.D.)	na	0.64	na	na	na	na	na	na	na
Form	WCOT	Helix	WCOT	WCOT	WCOT	WCOT	WCOT	WCOT	WCOT
Material	SS	SS	SS	SS	SS	SS	SS	SS	SS
Detector	FID	FID	FID	Argon	Argon	Argon	FID	FID	FID
References	1	1	2	3	3	3	1	4	4
Compound					ECL				
Methyl *cis*-2-octadecenoate	18.07	18.32	18.23	—	—	—	17.93	—	—
Methyl *cis*-3-octadecenoate	18.42	18.74	18.72	—	—	—	18.23	18.50	18.46
Methyl *cis*-4-octadecenoate	18.14	18.42	18.58	—	—	—	18.02	—	—
Methyl *cis*-5-octadecenoate	18.13	18.34	18.41	18.28	18.30	17.93	18.11	18.26	18.30
Methyl *cis*-6-octadecenoate	18.14	18.45	18.43	18.38	18.40	17.99	18.04	18.26	18.23
Methyl *cis*-7-octadecenoate	18.14	18.45	18.43	—	—	—	18.01	18.24	—
Methyl *cis*-8-octadecenoate	18.12	18.45	18.41	18.38	18.44	17.98	18.02	—	—
Methyl *cis*-9-octadecenoate	18.14	18.49	18.45	18.39	18.44	18.01	18.07	18.26	18.32
Methyl *cis*-10-octadecenoate	18.19	18.53	18.49	18.44	18.44	18.02	18.11	—	—
Methyl *cis*-11-octadecenoate	18.23	18.58	18.53	18.49	18.51	18.08	18.15	18.37	18.39
Methyl *cis*-12-octadecenoate	18.27	18.62	18.56	18.54	18.60	18.13	18.18	18.41	18.46
Methyl *cis*-13-octadecenoate	18.32	18.76	18.71	—	—	—	18.24	18.49	18.47
Methyl *cis*-14-octadecenoate	18.40	18.86	18.76	—	—	—	18.32	—	—
Methyl *cis*-15-octadecenoate	18.41	19.01	18.83	18.80	18.76	18.30	18.37	18.64	18.64
Methyl *cis*-16-octadecenoate	18.75	19.38	19.22	—	—	—	18.63	—	—
Methyl-17-octadecenoate	18.49	18.91	18.84	18.83	—	18.32	18.34	—	—

Column packing
P1 = Poly (neopentyl glycol succinate) in form of wall coating in stainless steel tubing
P2 = 20% Poly (diethylene glycol succinate) on Chromosorb W, HMDS (80-100 mesh)
P3 = Poly (diethylene glycol succinate) in form of wall coating in stainless steel tubing
P4 = Poly (diethylene glycol succinate plus Arquad 2HT-75) in form of wall coating in stainless steel tubing
P5 = GE 238-149-99 (50% cyanoethyl, with methyl, silicone) in form of wall coating in stainless steel tubing
P6 = Six-ring polyphenyl ether (plus Arquad 2HT-75) in form of wall coating in stainless steel tubing
P7 = XE-60 (25% cyanoethyl, with methyl, silicone) in form of wall coating in stainless steel tubing
P8 = Poly (1,4-butanediol succinate) in form of wall coating in stainless steel tubing
P9 = SILAR-5CP (50% cyanopropyl, 50% phenyl, siloxane) in form of wall coating in stainless steel tubing

REFERENCES

1. **Gunstone, F. D., Ismail, I. A., and Lie Ken Jie, M.,** *Chem. Phys. Lipids,* 1, 376, 1967.
2. **Barve, J. A., Gunstone, F. D., Jacobsberg, F. R., and Winlow, P.,** *Chem. Phys. Lipids,* 8, 117, 1972.
3. **Scholfield, C. R. and Dutton, H. J.,** *J. Am. Oil Chem. Soc.,* 47, 1, 1970; see also **Scholfield, C. R.,** *J. Am. Oil Chem. Soc.,* 58, 662, 1981.
4. **Ackman, R. G., Manzer, A., and Joseph, J.,** *Chromatographia,* 7, 107, 1974.

Table 35B
METHYL ESTERS OF FATTY ACIDS: *cis*-OCTADECENOATES, ON NONPOLAR OPEN-TUBULAR COLUMNS, REPORTED IN EQUIVALENT CHAIN LENGTHS

Column packing	P1	P2	P3	P4
Temperature (°C)	200	180	200	180
Gas	N_2	He	Ar	—
Flow rate (mℓ/min)	na	na	na	na
Column				
Length (cm)	5000	4700	6100	8000
Diameter				
(cm I.D.)	0.05	0.025	0.05	na
(cm O.D.)	na	na	na	na
Form	WCOT	WCOT	WCOT	WCOT
Material	SS	SS	SS	na
Detector	FID	FID	Argon	FID
References	1	2	3	4

Compound	ECL			
Methyl *cis*-2-octadecenoate	17.98	—	—	—
Methyl *cis*-3-octadecenoate	17.87	—	—	—
Methyl *cis*-4-octadecenoate	17.72	—	—	—
Methyl *cis*-5-octadecenoate	17.66	—	—	—
Methyl *cis*-6-octadecenoate	17.65	17.65	—	—
Methyl *cis*-7-octadecenoate	17.64	—	—	—
Methyl *cis*-8-octadecenoate	17.64	—	—	—
Methyl *cis*-9-octadecenoate	17.63	17.65	17.66	17.63
Methyl *cis*-10-octadecenoate	17.66	—	—	17.67
Methyl *cis*-11-octadecenoate	17.68	—	—	17.71
Methyl *cis*-12-octadecenoate	17.73	17.73	17.76	17.76
Methyl *cis*-13-octadecenoate	17.80	—	—	17.83
Methyl *cis*-14-octadecenoate	17.86	—	—	17.92
Methyl *cis*-15-octadecenoate	17.89	17.93	17.96	17.99
Methyl *cis*-16-octedecenoate	18.14	—	—	—
Methyl-17-octadecenoate	17.89	—	—	—

Column packing
P1 = Apiezon-L in form of wall coating in stainless steel tubing
P2 = Apiezon-L in form of wall coating in stainless steel tubing
P3 = Apiezon-L in form of wall coating in stainless steel tubing
P4 = OV-101 (methyl silicone) in form of wall coating in stainless steel tubing

REFERENCES

1. **Gunstone, F. D., Ismail, I. A., and Lie Ken Jie, M.,** *Chem. Phys. Lipids,* 1, 376, 1967.
2. **Ackman, R. G. and Hooper, S. N.,** *J. Chromatogr.,* 86, 73, 1973.
3. **Scholfield, C. R. and Dutton, H. J.,** *J. Am. Oil Chem Soc.,* 48, 228, 1971.
4. **Alexander, G., Lukacs, P., and Jeranek, M.,** *Olaj Szappan Kozmetica,* 24, 72, 1977.

Table 36
METHYL ESTERS OF FATTY ACIDS: *trans*-OCTADECENOATES, REPORTED AS EQUIVALENT CHAIN LENGTHS

	P1	P2	P3	P4	P5	P6	P7	P8	P9	P10
Column packing										
Temperature (°C)	200	210	180	200	180	190	180	165	200	190
Gas	N$_2$	N$_2$	He	Ar	—	N$_2$	N$_2$	Ar	Ar	Ar
Flow rate (ml/min)	na	na	na	na	na	na	na	2	0.5	2.7
Column Length (cm)	5000	5000	4700	6100	8000	5000	5000	6100	6100	4700
Diameter (cm I.D.)	0.05	0.05	0.025	0.025	na	0.05	0.05	0.025	0.025	0.025
(cm O.D.)	na	na	na	na	na	na	na	na	na	na
Form	WCOT	WCOT	WCOT	WCOT	WCOT	WCOT	WCOT	WCOT	WCOT	WCOT
Material	SS	SS	SS	SS	SS	SS	SS	SS	SS	SS
Detector	FID	FID	FID	Argon	FID	FID	FID	Argon	Argon	Argon
References	1	2	3	4	5	1	2	5	5	5
Compound						ECL				
Methyl *trans*-2-octadecenoate	18.80	18.61	—	—	—	19.34	19.57	19.50	19.84	19.68
Methyl *trans*-3-octadecenoate	17.87	17.86	—	—	—	18.42	18.67	18.69	—	18.30
Methyl *trans*-4-octadecenoate	17.72	17.75	—	—	—	18.10	18.32	—	—	—
Methyl *trans*-5-octadeconate	17.79	17.76	—	—	—	18.13	18.34	18.25	18.11	18.08
Methyl *trans*-6-octadecenoate	17.75	17.72	—	—	—	18.14	18.35	18.33	18.13	18.10
Methyl *trans*-7-octadecenoate	17.73	17.71	—	—	—	18.14	18.33	—	—	—
Methyl *trans*-8-octadecenoate	17.74	17.71	—	—	—	18.12	18.36	18.30	18.19	18.10
Methyl *trans*-9-octadecenoate	17.74	17.72	17.74	17.73	17.72	18.14	18.35	18.33	18.20	18.10
Methyl *trans*-10-octadecenoate	17.75	17.74	—	—	17.75	18.14	18.35	18.33	18.20	18.10
Methyl *trans*-11-octadecenoate	17.78	17.75	—	—	17.77	18.19	18.38	18.34	18.22	18.14
Methyl *trans*-12-octadecenoate	17.73	17.78	17.80	17.79	17.80	18.23	18.40	18.36	18.22	18.16
Methyl *trans*-13-octadecenoate	17.80	17.81	—	—	17.85	18.21	18.45	18.41	18.27	18.19
Methyl *trans*-14-octadecenoate	17.83	17.81	—	—	17.89	18.26	18.50	—	—	—
Methyl *trans*-15-octadecenoate	17.84	17.86	17.88	17.88	17.93	18.26	18.52	—	—	—
Methyl *trans*-16-octadecenoate	18.03	18.00	—	—	—	18.32	18.60	18.60	18.37	18.26
Methyl 17-octadecenoate	17.89	17.89	—	—	—	18.53	18.89	—	—	—
						18.49	18.84	18.83	—	18.32

Table 36 (continued)

METHYL ESTERS OF FATTY ACIDS: *trans*-OCTADECENOATES, REPORTED AS EQUIVALENT CHAIN LENGTHS

Column packing
P1 = Apiezon-L in form of wall coating in stainless steel tubing
P2 = Apiezon-L in form of wall coating in stainless steel tubing
P3 = Apiezon-L in form of wall coating in stainless steel tubing
P4 = Apiezon-L in form of wall coating in stainless steel tubing
P5 = OV-101 (dimethyl silicone) in form of wall coating in stainless steel tubing
P6 = Poly (neopentyl glycol succinate) in form of wall coating in stainless steel tubing
P7 = Poly (diethylene glycol succinate) in form of wall coating in stainless steel tubing
P8 = Poly (diethylene glycol succinate; plus Arquad 2HT-75) in form of wall coating in stainless steel tubing
P9 = General Electric 238-149-99 (50% cyanoethyl, with methyl silicone) in form of wall coating in stainless steel tubing
P10 = Six-ring polyphenyl ether (plus Arquad 2HT-75) in form of wall coating in stainless steel tubing

REFERENCES

1. **Gunstone, F. D., Ismail, I. A., and Lie Ken Jie, M.**, *Chem. Phys. Lipids*, 1, 376, 1967.
2. **Barve, J. A., Gunstone, F. D., Jacobsberg, F. R., and Winlow, P.**, *Chem. Phys. Lipids*, 8, 117, 1972.
3. **Ackman, R. G. and Hooper, S. N.**, *J. Chromatogr.*, 86, 73, 1973.
4. **Scholfield, C. R. and Dutton, H. J.**, *J. Am. Oil Chem. Soc.*, 48, 228, 1971.
5. **Alexander, G., Lukacs, P., and Jeranek, M.**, *Olaj. Szappon Kozmetika*, 24, 72, 1975.
6. **Scholfield, C. R. and Dutton, H. J.**, *J. Am. Oil Chem. Soc.*, 47, 1, 1970, see also **Scholfield, C. R.**, *J. Am. Oil Chem. Soc.*, 58, 662, 1981.

Table 37A
METHYL ESTERS OF FATTY ACIDS: ISOMERIC METHYLENE-INTERRUPTED *cis,cis*-OCTADECADIENOATES, ON POLAR COLUMNS, REPORTED AS EQUIVALENT CHAIN LENGTHS

	P1	P2	P3	P4	P5	P6	P7	P8	P9	P10	P11	P12
Column packing												
Temperature (°C)	200	190	190	180	234	180	240	180	180	170	170	190
Gas	N₂	N₂	N₂	N₂	He	Ar	He	N₂	N₂	He	He	Ar
Flow rate (mℓ/min)	50	2	50	50	120	60	70	50	50	5	5	na
Column												
Length (cm)	153	5000	153	153	360	200	250	213	213	4700	4700	4700
Diameter												
(cm I.D.)	na	0.025	na	na	0.3	0.4	0.3	na	na	0.025	0.025	0.025
(cm O.D.)	0.64	na	0.64	0.64	na	na	na	0.64	0.64	na	na	na
Form	na	WCOT	na	na	na	U-tube	Helix	na	na	WCOT	WCOT	WCOT
Material	na	SS	na	na	na	G	SS	na	na	SS	SS	SS
Detector	FID	FID	FID	FID	TC	Argon	TC	FID	FID	FID	FID	Argon
References	1	1	1	1	3	3	3	1	1	2	2	4
Compound					ECL							
Methyl *cis,cis*-2,5-octadecadienoate	18.25	18.14	18.37	18.64	—	—	—	18.65	18.72	—	—	—
Methyl *cis,cis*-3,6-octadecadienoate	18.67	18.65	18.94	19.36	—	—	—	19.54	19.68	—	—	—
Methyl *cis,cis*-4,7-octadecadienoate	18.45	18.36	18.69	19.05	—	—	—	19.28	19.42	—	—	—
Methyl *cis,cis*-5,8-octadecadienoate	18.38	18.38	18.71	19.06	—	—	—	19.27	19.40	—	—	—
Methyl *cis,cis*-6,9-octadecadienoate	18.46	18.44	18.81	19.19	—	—	—	19.37	19.57	18.68	18.66	—
Methyl *cis,cis*-7,10-octadecadienoate	18.46	18.46	18.80	19.23	—	—	—	19.42	19.63	—	—	—
Methyl *cis,cis*-8,11-octadecadienoate	18.51	18.53	18.87	19.28	—	19.43	19.46	19.47	19.65	—	—	—
Methyl *cis,cis*-9,12-octadecadienoate	18.57	18.60	18.95	19.38	—	—	—	19.55	19.75	18.82	18.83	18.19
Methyl *cis,cis*-10,13-octadecadienoate	18.60	18.70	19.03	19.46	19.37	19.55	19.55	19.69	19.81	—	—	—
Methol *cis,cis*-11,14-octadecadienoate	18.75	18.82	19.15	19.62	19.47	19.73	19.60	19.83	20.00	—	18.66	—
Methyl *cis,cis*-12,15-octadecadienoate	18.88	18.90	19.28	19.75	19.50	19.63	19.63	19.97	20.16	19.16	19.10	18.46
Methyl *cis,cis*-13,16-octadecadienoate	19.20	19.27	19.68	20.25	—	—	—	20.37	20.60	—	—	—
Methyl *cis*-14,17-octadecadienoate	18.95	19.04	19.33	19.78	—	—	—	19.96	20.18	—	—	—

Table 37A (continued)

METHYL ESTERS OF FATTY ACIDS: ISOMERIC METHYLENE-INTERRUPTED *cis,cis*-OCTADECADIENOATES, ON POLAR COLUMNS, REPORTED AS EQUIVALENT CHAIN LENGTHS

Column packing
- P1 = 5% FFAP (Carbowax 20 M-modified with 2-nitroterephthalic acid) on Chromosorb G (80-100 mesh)
- P2 = Poly (neopentyl slycol succinate) in form of wall coating in stainless steel tubing
- P3 = 15% Poly (ethylene glycol adipate) on Gas-Chrom Z (70-80 mesh)
- P4 = 20% Poly (diethylene glycol succinate) on Gas-Chrom Z (70-80 mesh)
- P5 = 20% β-cyclodextrin acetate on Gas-Chrom P (30-60 mesh)
- P6 = 20% Poly (ethylene glycol succinate) on Gas-Chrom P (80-100 mesh)
- P7 = 20% Poly(diethylene glycol succinate) on Gas Chrom P (80-100 mesh)
- P8 = 20% Poly (ethylene glycol succinate) on Chromosorb W (100-120 mesh)
- P9 = 20% Poly (ethylene glycol succinate; modified with 2% H₃PO₄) on Gas-Chrom P (80-100 mesh)
- P10 = Poly (1,4-butanediol succinate) in form of wall coating in stainless steel tubing
- P11 = SILAR-5CP (50% cyanopropyl, 50% phenyl, siloxane) in form of wall coating in stainless steel tubing
- P12 = Six-ring polyphenyl ether (plus Arquad 2HT-75) in form of wall coating in stainless steel tubing

REFERENCES

1. **Christie, W. W.,** *J. Chromatogr.,* 37, 27, 1968.
2. **Ackman, R. G., Manzer, A., and Joseph, J.,** *Chromatographia,* 7, 107, 1974.
3. **Hofstetter, H. H., Sen, N., and Holman, R. T.,** *J. Am. Oil Chem. Soc.,* 42, 537, 1965.
4. **Scholfield, C. R. and Dutton, H. J.,** *J. Am. Oil Chem. Soc.,* 48, 228, 1971; see also **Scholfield, C. R.,** *J. Am. Oil Chem. Soc.,* 58, 662, 1981.

Table 37B
METHYL ESTERS OF FATTY ACIDS: ISOMERIC METHYLENE-INTERRUPTED *cis,cis*-OCTADECADIENOATES, ON NONPOLAR COLUMNS, REPORTED AS EQUIVALENT CHAIN LENGTHS

	P1	P2	P3	P4	P5	P6
Column packing						
Temperature (°C)	220	180	200	240	200	180
Gas	N_2	He	N_2	He	Ar	N_2
Flow rate (mℓ/min)	2	5	50	70	na	1
Column Length (cm)	5000	4700	153	250	6100	6000
Diameter (cm I.D.)	0.025	0.025	na	0.3	0.025	na
(cm O.D.)	na	na	0.64	na	na	na
Form	WCOT	WCOT	na	Helix	WCOT	WCOT
Material	SS	SS	na	SS	SS	G
Detector	FID	FID	FID	TC	Argon	FID
References	1	2	1	3	4	5

Compound	P1	P2	P3	P4	P5	P6
			ECL			
Methyl *cis,cis*-2,5-octadecadienoate	17.68	—	17.64	—	—	—
Methyl *cis,cis*-3,6-octadecadienoate	17.62	—	17.57	—	—	—
Methyl *cis,cis*-4,7-octadecadienoate	17.47	—	17.42	—	—	—
Methyl *cis,cis*-5,8-octadecadienoate	17.43	—	17.38	—	—	—
Methyl *cis,cis*-6,9-octadecadienoate	17.46	17.38	17.40	—	—	—
Methyl *cis,cis*-7,10-octadecadienoate	17.44	—	17.38	—	—	—
Methyl *cis,cis*-8,11-octadecadienoate	17.48	—	17.42	17.48	—	—
Methyl *cis,cis*-9,12-octadecadienoate	17.50	17.47	17.47	17.53	17.50	17.57
Methyl *cis,cis*-10,13-octadecadienoate	17.60	—	17.56	17.57	—	—
Methyl *cis,cis*-11,14-octadecadienoate	17.68	—	17.63	17.62	—	—
Methyl *cis,cis*-12,15-octadecadienoate	17.78	17.72	17.72	17.75	17.83	17.87
Methyl *cis,cis*-13,16-octadecadienoate	18.00	—	17.95	—	—	—
Methyl *cis*-14,17-octadecadienoate	17.80	—	17.75	—	—	—

Table 37B (continued)

METHYL ESTERS OF FATTY ACIDS: ISOMERIC METHYLENE-INTERRUPTED *cis,cis*-OCTADECADIENOATES, ON NONPOLAR COLUMNS, REPORTED AS EQUIVALENT CHAIN LENGTHS

Column packing		
P1	=	Apiezon-L in form of wall coating in stainless steel tubing
P2	=	Apiezon-L in form of wall coating in stainless steel tubing
P3	=	5% Apiezon-L on Gas-Chrom Z (70-80 mesh)
P4	=	20% Apiezon-L on Gas-Chrom P (100-140 mesh)
P5	=	Apiezon-L in form of wall coating in stainless steel tubing
P6	=	Apiezon-L in form of wall coating in glass capillary tubing

REFERENCES

1. Christie, W. W., *J. Chromatogr.*, 37, 27, 1968.
2. Ackman, R. G., Manzer, A., and Joseph, J., *Chromatographia*, 7, 107, 1974.
3. Hofstetter, H. H., Sen, N., and Holman, R. T., *J. Am. Oil Chem. Soc.*, 42, 537, 1965.
4. Scholfield, C. R. and Dutton, H. J., *J. Am. Oil Chem. Soc.*, 48, 228, 1971.
5. Strocchi, A., Piretti, M., and Capella, P., *Riv. Ital. Sost. Grasse*, 46, 80, 1969.

Table 38A
METHYL ESTERS OF FATTY ACIDS: ISOMERIC DIMETHYLENE-INTERRUPTED *cis,cis*-OCTADECADIENOATES REPORTED AS EQUIVALENT CHAIN LENGTHS

	P1	P2	P3	P4	P5	P6	P7	P8	P9
Column packing									
Temperature (°C)	215	185	190	195	195	200	195	170	170
Gas	N_2	N_2	N_2	N_2	N_2	N_2	N_2	N_2	N_2
Flow rate (mℓ/min)	90	50	50	90	50	50	75	50	50
Column									
Length (cm)	na	na	na	na	na	na	na	na	na
Diameter									
(cm I.D.)	6.2	3.1	3.1	6.2	3.1	3.1	6.2	3.1	6.2
(cm O.D.)	na	na	na	na	na	na	na	na	na
Form	na	na	na	na	na	na	na	na	na
Material	na	na	na	na	na	na	na	na	na
Detector	FID	FID	FID	FID	FID	FID	FID	FID	FID
References	1	1	1	1	1	1	1	1	1
Compound				ECL					
Methyl *cis,cis*-2,6-octadecadienoate	17.69	17.62	17.62	17.87	18.32	18.43	18.56	18.55	18.49
Methyl *cis,cis*-3,7-octadecadienoate	17.54	17.64	17.68	18.17	18.63	18.62	18.93	19.10	19.15
Methyl *cis,cis*-4,8-octadecadienoate	17.42	17.55	17.56	17.96	18.37	18.33	18.68	18.82	18.87
Methyl *cis,cis*-5,9-octadecadienoate	17.39	17.47	17.52	18.05	18.32	18.38	18.66	18.78	18.90
Methyl *cis,cis*-6,10-octadecadienoate	17.41	17.47	17.51	18.14	18.46	18.44	18.81	18.95	19.06
Methyl *cis,cis*-7,11-octadecadienoate	17.41	17.51	17.52	18.16	18.46	18.50	18.85	18.97	19.09
Methyl *cis,cis*-8,12-octadecadienoate	17.47	17.55	17.60	18.23	18.52	18.60	18.90	19.00	19.21
Methyl *cis,cis*-9,13-octadecadienoate	17.49	17.63	17.63	18.32	18.59	18.67	19.01	19.14	19.30
Methyl *cis,cis*-10,14-octadecadienoate	17.59	17.71	17.72	18.42	18.72	18.71	19.09	19.20	19.36
Methyl *cis,cis*-11,15-octadecadienoate	17.68	17.78	17.76	18.50	18.82	18.85	19.18	19.28	19.50
Methyl *cis,cis*-12,16-octadecadienoate	17.88	17.95	17.99	18.79	19.16	19.18	19.57	19.84	19.89
Methyl *cis*-13,17-octadecadienoate	17.66	17.72	17.82	18.50	18.82	18.90	19.25	19.49	19.57

Table 38A (continued)

METHYL ESTERS OF FATTY ACIDS: ISOMERIC DIMETHYLENE-INTERRUPTED *cis,cis*-OCTADECADIENOATES REPORTED AS EQUIVALENT CHAIN LENGTHS

Column packing
P1 = 5% Apiezon-L on Chromosorb W (80-100 mesh)
P2 = 1.5% OV-101 (methyl silicone) on Chromosorb W (80-100 mesh)
P3 = 3% SE-30 (methyl silicone) on Chromosorb W (80-100 mesh)
P4 = 20% XE-60 (25% cyanothyl, with methyl, silicone) on Chromosorb W (80-100 mesh)
P5 = 10% Carbowax 20M on Chromosorb W (80-100 mesh)
P6 = 10% FFAP (Carbowax 20M terminated with 2-nitroterephthalic acid) on Chromosorb W (80-100 mesh)
P7 = 10% Poly (diethylene glycol adipate) on Chromosorb W (80-100 mesh)
P8 = 10% Poly (diethylene glycol succinate) on Chromosorb W (80-100 mesh)
P9 = 10% SILAR-10C (100% cyanopropyl siloxane) on Chromosorb W (80-100 mesh)

REFERENCE

1. **Lam, C. H. and Lie Ken Jie, M. S. F.**, *J. Chromatogr.*, 117, 365, 1976; see also **Scholfield, C. R.**, *J. Am. Oil Chem. Soc.*, 58, 662, 1981.

Table 38B
METHYL ESTERS OF FATTY ACIDS: ISOMERIC DIMETHYLENE-INTERRUPTED *trans,trans*-OCTADECADIENOATES AND *trans,cis*-2,6-OCTADECADIENOATE, REPORTED AS EQUIVALENT CHAIN LENGTHS

	P1	P2	P3	P4	P5	P6	P7	P8	P9	P10
Column packing										
Temperature (°C)	200	190	190	190	200	195	195	165	140	170
Gas	N_2	N_2	N_2	N_2	N_2	N_2	N_2	N_2	N_2	N_2
Flow rate (mℓ/min)	40	40	40	40	40	40	80	70	40	50
Column										
Length (cm)	200	200	200	200	200	200	200	200	200	200
Diameter (cm I.D.)	3.1	3.1	3.1	6.2	3.1	3.1	6.2	6.2	3.1	6.2
(cm O.D.)	na	na	na	na	na	na	na	na	na	na
Form	na	na	na	na	na	na	na	na	na	na
Material	na	na	na	na	na	na	na	na	na	na
Detector	FID	FID	FID	FID	FID	FID	FID	FID	FID	FID
References	1	1	1	1	1	1	1	1	1	1
Compound					ECL					
Methyl *trans,trans*-2,6-octadecadienoate	18.19	18.17	18.14	18.84	19.28	19.32	19.63	19.64	19.59	19.81
Methyl *trans,trans*-3,7-octadecadienoate	17.47	17.66	17.74	17.98	18.64	18.63	18.82	18.75	18.99	18.92
Methyl *trans,trans*-4,8-octadecadienoate	17.42	17.59	17.55	17.84	18.36	18.42	18.50	18.44	18.59	18.58
Methyl *trans,trans*-5,9-octadecadienoate	17.43	17.61	17.59	17.90	18.44	18.41	18.55	18.54	18.62	18.65
Methyl *trans,trans*-6,10-octadecadienoate	17.40	17.56	17.59	17.93	18.45	18.46	18.59	18.60	18.68	18.69
Methyl *trans,trans*-7,11-octadecadienoate	17.41	17.59	17.60	17.96	18.43	18.45	18.61	18.64	18.68	18.78
Methyl *trans,trans*-8,12-octadecadienoate	17.43	17.65	17.66	18.02	18.52	18.47	18.67	18.71	18.74	18.78
Methyl *trans,trans*-9,13-octadecadienoate	17.49	17.68	18.68	18.07	18.55	18.51	18.73	18.77	18.80	18.87
Methyl *trans,trans*-10,14-octadecadienoate	17.59	17.70	17.71	18.10	18.56	18.55	18.78	18.80	18.88	18.93
Methyl *trans,trans*-11,15-octadecadienoate	17.61	17.70	17.73	18.16	18.64	18.68	18.86	18.92	18.99	18.97
Methyl *trans,trans*-12,16-octadecadienoate	17.72	17.86	17.84	18.26	18.88	18.88	19.16	19.22	19.25	19.10
Methyl *trans*-13,17-octadecadienoate	17.65	17.76	17.79	18.28	18.78	18.86	19.13	19.21	19.24	19.22
Methyl *trans,cis*-2,6-octadecadienoate	18.14	18.13	18.12	18.99	19.26	19.31	19.74	19.84	19.69	19.93

Table 38B (continued)

METHYL ESTERS OF FATTY ACIDS:ISOMERIC DIMETHYLENE-INTERRUPTED *trans,trans*-OCTADECADIENOATES AND *trans,cis*-2,6-OCTADECADIENOATE, REPORTED AS EQUIVALENT CHAIN LENGTHS

Column packing   P1   =   5% Apiezon-L on Chromosorb W (80-100 mesh)

            P2   =   1.5% OV-101 (methyl silicone) on Chromosorb W (80-100 mesh)

            P3   =   3% SE-30 (methyl silicone) on Chromosorb W (80-100)

            P4   =   20% XE-60 (25% cyanoethyl, with methyl, silicone) on Chromosorb W (80-100 mesh)

            P5   =   10% Carbowax 20M on Chromosorb W (80-100 mesh)

            P6   =   10% FFAP (Carbowax 20M terminated with 2-nitroterephthalic acid) on Chromosorb W (80-100 mesh)

            P7   =   10% Poly (diethylene glycol adipate) on Chromosorb W (80-100 mesh)

            P8   =   10% Poly (diethylene glycol succinate) on Chromosorb W (80-100 mesh)

            P9   =   10% SILAR-10 C (100% cyanopropyl siloxane) on Chromosorb W (80-100 mesh)

          P10   =   10% SP-2330 (90% cyanopropyl, 10% phenyl, siloxane) on Chromosorb W (80-100 mesh)

REFERENCE

1. **Lam, C. H. and Lie Ken Jie, M. S. F.**, *J. Chromatog.*, 121, 303, 1976.

Table 39A
METHYL ESTERS OF FATTY ACIDS: ISOMERIC -*cis-cis*-OCTADECADIENOATES, ON POLAR COLUMNS, REPORTED AS EQUIVALENT CHAIN LENGTHS

	P1	P2	P3	P4	P5	P6	P7	P8	P9	P10	P11	P12
Column packing												
Temperature (°C)	190	190	234	180	240	190	190	190	170	170	190	190
Gas	N_2	N_2	He	Ar	He	N_2	N_2	N_2	He	He	N_2	N_2
Flow rate (mℓ/min)	na	na	120	60	70	na	na	na	5	5	na	na
Column												
Length (cm)	5000	5000	360	200	250	153	200	200	4700	4700	200	200
Diameter												
(cm I.D.)	0.05	0.05	0.3	0.4	0.3	na	na	na	0.025	0.025	na	na
(cm O.D.)	na	na	na	na	na	0.63	na	na	na	na	na	na
Form	WCOT	WCOT	na	U-tube	Helix	Helix	na	na	WCOT	WCOT	na	na
Material	SS	SS	na	G	SS	na	na	na	SS	SS	na	na
Detector	FID	FID	TC	Argon	TC	FID	FID	FID	FID	FID	FID	FID
References	1	1	2	2	2	1	3	3	4	4	3	3
Compound					ECL							
Methyl *cis,cis*-5,12-octadecadienoate	17.94	18.92	—	—	—	19.10	—	—	—	—	—	—
Methyl *cis,cis*-6,12-octadecadienoate	17.96	18.98	—	—	—	19.21	—	—	18.68	18.75	—	—
Methyl *cis,cis*-7,12-octadecadienoate	17.89	19.04	—	—	—	19.13	—	—	—	—	—	—
Methyl *cis,cis*-8,12-octadecadienoate	18.03	19.04	—	—	—	19.27	—	—	—	—	—	—
Methyl *cis,cis*-9,12-octadecadienoate	18.11	19.16	19.23	19.22	19.30	19.34	—	—	18.82	18.83	—	—
Methyl *cis,cis*-10,12-octadecadienoate	18.76	—	—	—	—	21.01	—	—	—	—	—	—
Methyl *cis,cis*-6,8-octadecadienoate	18.58	20.25	—	—	—	20.69	—	—	—	—	—	—
Methyl *cis,cis*-6,9-octadecadienoate	17.93	19.01	—	—	—	19.22	—	—	18.68	18.66	—	—
Methyl *cis,cis*-6,10-octadecadienoate	17.94	18.97	—	—	—	19.14	—	—	—	—	—	—
Methyl *cis,cis*-6,11-octadecadienoate	17.86	18.92	—	—	—	19.10	19.12	19.34	—	—	—	—
Methyl *cis,cis*-7,15-octadecadienoate	18.14	19.21	—	—	—	19.40	—	—	—	—	—	—
Methyl *cis,cis*-8,15-octadecadienoate	18.15	19.23	—	—	—	19.49	—	—	—	—	—	—
Methyl *cis,cis*-9,15-octadecadienoate	18.17	19.31	19.33	19.43	19.46	19.48	—	—	18.95	18.95	—	—
Methyl *cis,cis*-5,11-octadecadienoate	—	—	18.93	18.90	19.03	—	—	—	—	—	—	—
Methyl*cis,cis*-8,11-octadecadienoate	—	—	19.40	19.43	19.46	—	—	—	—	—	—	—
Methyl *cis,cis*-10,13-octadecadienoate	—	—	19.37	19.55	19.55	—	—	—	—	—	—	—
Methyl *cis,cis*-11,14-octadecadienoate	—	—	19.47	19.73	19.60	—	—	—	—	—	—	—

Table 39A (continued)
METHYL ESTERS OF FATTY ACIDS: ISOMERIC -cis-cis-OCTADECADIENOATES, ON POLAR COLUMNS, REPORTED AS EQUIVALENT CHAIN LENGTHS

Compound	ECL						
Methyl cis,cis-12,15-octadecadienoate	19.50	19.63	19.63	19.16	19.10	—	—
Methyl cis,cis-2,7-octadecadienoate	—	—	—	—	—	18.74	18.73
Methyl cis,cis-3,8-octadecadienoate	—	—	—	—	—	19.25	19.40
Methyl cis,cis-4,9-octadecadienoate	—	—	—	—	—	18.98	19.04
Methyl cis,cis-5,10-octadecadienoate	—	—	—	—	—	19.11	19.34

Column packing
P1 = XE-60 (25% cyanoethyl, with methyl, silicone) in form of wall coating in stainless steel tubing
P2 = Poly (diethylene glycol succinate) in form of wall coating in stainless steel tubing
P3 = 20% Poly (β-cyclodextrin acetate) on Gas-Chrom P (30-60 mesh)
P4 = 20% Poly (ethylene glycol succinate) on Gas-Chrom P (80-100 mesh)
P5 = 20% Poly (diethylene glycol succinate) on Gas-Chrom P (80-100 mesh)
P6 = 20% Poly (diethylene glycol succinate) on Chromosorb W (80-100 mesh)
P7 = 20% Poly (diethylene glycol succinate) on Chromosorb W (80-100 mesh)
P8 = 10% SILAR-10C (100% cyanopropyl siloxane)
P9 = Poly (1,4-butanediol succinate); in form of wall coating in stainless steel tubing
P10 = SILAR-5CP (50% cyanopropyl, 50% phenyl, siloxane) in form of wall coating in stainless steel tubing
P11 = 20% Poly (diethylene glycol succinate) on Chromosorb W (80-100 mesh)
P12 = 10% SILAR-10C (100% cyanopropyl siloxane) on Chromosorb W (80-100 mesh)

REFERENCES

1. **Gunstone, F. D. and Lie Ken Jie, M.**, *J. Chromatogr.*, 4, 131, 1970.
2. **Hofstetter, H. H., Sen, N., and Holman, R. T.**, *J. Am. Oil Chem. Soc.*, 42, 537, 1965.
3. **Lie Ken Jie, M. S. F.**, *J. Chromatogr.*, 109, 81, 1975.
4. **Ackman, R. G., Manzer, A., and Joseph, J.**, *Chromatographia*, 7, 107, 1974; see also **Scholfield, C. R.**, *J. Am. Oil Chem. Soc.*, 58, 662, 1981.

Table 39B
METHYL ESTERS OF FATTY ACIDS: ISOMERIC *cis,cis*-OCTADECADIENOATES, ON NONPOLAR COLUMNS, REPORTED AS EQUIVALENT CHAIN LENGTHS

	P1	P2	P3	P4	P5	P6	P7	P8
Column packing								
Temperature (°C)	190	210	240	190	180	200	180	180
Gas	N$_2$	N$_2$	He	N$_2$	He	Ar	He	N$_2$
Flow rate (mℓ/min)	na	na	70	95	5	na	5	1
Column Length (cm)	153	5000	250	200	4700	6100	4700	6000
Diameter (cm I.D.)	na	0.05	0.3	na	0.025	0.025	0.025	na
(cm O.D.)	0.63	na	na	na	na	na	na	na
Form	Helix	WCOT	Helix	na	WCOT	WCOT	WCOT	WCOT
Material	na	SS	SS	na	SS	SS	SS	G
Detector	FID	FID	TC	FID	FID	Argon	FID	FID
References	1	1	2	3	4	5	6	7
Compound				ECL				
Methyl *cis,cis*-5,12-octadecadienoate	17.38	17.42	—	—	—	—	—	—
Methyl *cis,cis*-6,12-octadecadienoate	17.39	17.41	—	—	17.43	—	17.38	—
Methyl *cis,cis*-7,12-octadecadienoate	17.33	17.31	—	—	—	—	—	—
Methyl *cis,cis*-8,12-octadecadienoate	17.41	17.44	—	—	—	—	—	—
Methyl *cis,cis*-9,12-octadecadienoate	17.48	17.48	17.53	—	17.50	17.50	17.47	17.57
Methyl *cis,cis*-10,12-octadecadienoate	18.67[a]	18.68[a]	—	—	—	—	—	—
Methyl *cis,cis*-6,8-octadecadienoate	18.54[a]	18.55[a]	—	—	—	—	—	—
Methyl *cis,cis*-6,9-octadecadienoate	17.38	17.42	—	—	17.43	—	17.38	—
Methyl *cis,cis*-6,10-octadecadienoate	17.36	17.39	—	—	—	—	—	—
Methyl *cis,cis*-6,11-octadecadienoate	17.33	17.34	—	17.29	—	—	—	—
Methyl *cis,cis*-7,15-octadecadienoate	17.52	17.54	—	—	—	—	—	—
Methyl *cis,cis*-8,15-octadecadienoate	17.53	17.55	—	—	—	—	—	—
Methyl *cis,cis*-9,15-octadecadienoate	17.55	17.58	17.60	—	17.61	17.66	17.57	17.66
Methyl *cis,cis*-5,11-octadecadienoate	—	—	17.40	—	—	—	—	—
Methyl *cis,cis*-12,15-octadecadienoate	—	—	17.75	—	17.73	17.83	—	17.87
Methyl *cis,cis*-2,7-octadecadienoate	—	—	—	17.57	—	—	—	—

Table 39B (continued)

METHYL ESTERS OF FATTY ACIDS: ISOMERIC *cis,cis*-OCTADECADIENOATES, ON NONPOLAR COLUMNS, REPORTED AS EQUIVALENT CHAIN LENGTHS

Compound					ECL				
Methyl *cis,cis*-3,8-octadecadienoate	—	—	—	—	17.43	—	—	—	—
Methyl *cis,cis*-4,9-octadecadienoate	—	—	—	—	17.32	—	—	—	—
Methyl *cis,cis*-5,10-octadecadienoate	—	—	—	—	17.30	—	—	—	—

ᵃ   Arbitrarily increased by R.G. Ackman, from value published, by 1 ECL unit.

Column packing
P1 =   5% Apiezon-L on Chromosorb W, HNDS (80-100 mesh)
P2 =   Apiezon-L in form of wall coating in stainless steel tubing
P3 =   20% Apiezon-L on Gas-Chrom P (100-140 mesh)
P4 =   5% Apiezon-L on Chromosorb W (80-100 mesh)
P5 =   Apiezon-L in form of wall coating in stainless steel tubing
P6 =   Apiezon-L in form of wall coating in stainless steel tubing
P7 =   Apiezon-L in form of wall coating in stainless steel tubing
P8 =   Apiezon-L in form of wall coating in stainless steel tubing

REFERENCES

1. **Gunstone, F. D. and Lie Ken Jie, M.,** *J. Chromatogr.,* 4, 131, 1970.
2. **Hofstetter, H. H., Sen, N., and Holman, R. T.,** *J. Am. Oil Chem. Soc.,* 42, 537, 1965.
3. **Lie Ken Jie, M. S. F.,** *J. Chromatogr.,* 109, 81, 1975.
4. **Ackman, R. G. and Hooper, S. N.,** *J. Chromatogr.,* 86, 73, 1973.
5. **Scholfield, C. R. and Dutton, H. J.,** *J. Am. Oil Chem. Soc.,* 48, 228, 1971.
6. **Ackman, R. G., Manzer, A., and Joseph, J.,** *Chromatographia,* 7, 101, 1974.
7. **Strocchi, A., Piretti, M., and Capella, P.,** *Riv. Ital. Sost. Grasse,* 46, 80, 1969.

Table 40A
METHYL ESTERS OF FATTY ACIDS: *trans, trans*-OCTADECADIENOATES, ON POLAR COLUMNS, REPORTED AS EQUIVALENT CHAIN LENGTHS

	P1	P2	P3	P4	P5	P6	P7	P8	P9	P10
Column packing										
Temperature (°C)	170	180	190	190	190	190	165	170	200	160
Gas	He	He	N_2	N_2	N^2	Ar	Ar	Ar	Ar	Ar
Flow rate (mℓ/min)	5	5	na	na	na	na	na	na	na	na
Column										
Length (cm)	4700	4700	5000	153	5000	4600	6100	6100	6100	6100
Diameter										
(cm I.D.)	0.025	0.025	0.05	na	0.05	0.025	0.025	0.025	0.025	0.025
(cm O.D.)	na	na	na	0.63	na	na	na	na	na	na
Form	WCOT	WCOT	WCOT	Helix	WCOT	WCOT	WCOT	WCOT	WCOT	WCOT
Material	SS	SS	SS	na	SS	SS	SS	SS	SS	SS
Detector	FID	FID	FID	FID	FID	Argon	Argon	Argon	Argon	Argon
Reference	1	1	2	2	2	3	3	3	3	3
Compound					ECL					
Methyl *trans,trans*-5,12-octadecadienoate	—	—	17.92	19.00	18.83	—	—	—	—	—
Methyl *trans,trans*-6,12-octadecadienoate	—	—	17.87	18.98	18.91	—	—	—	—	—
Methyl *trans,trans*-7,12-octadecadienoate	—	—	17.96	19.12	18.96	—	—	—	—	—
Methyl *trans,trans*-8,12-octadecadienoate	—	—	17.87	19.02	18.85	—	—	—	—	—
Methyl *trans,trans*-9,12-octadecadienoate	18.82	18.77	17.98	19.20	19.04	18.41	19.00	18.67	18.69	19.14
Methyl *trans,trans*-10,12-octadecadienoate	—	—	19.16	20.99	20.61	—	—	—	—	—
Methyl *trans,trans*-6,8-octadecadienoate	—	—	19.25	21.05	20.54	—	—	—	—	—
Methyl *trans,trans*-6,9-octadecadienoate	—	—	17.89	19.13	18.98	—	—	—	—	—
Methyl *trans,trans*-6,10-octadecadienoate	—	—	17.79	18.92	18.78	—	—	—	—	—
Methyl *trans,trans*-6,11-octadecadienoate	—	—	17.98	19.10	18.91	—	—	—	—	—
Methyl *trans,trans*-9,15-octadecadienoate	18.81	18.72	—	—	—	18.40	18.98	18.68	—	19.13
Methyl *trans,trans*-12,15-octadecadienoate	19.05	18.91	—	—	—	18.55	19.23	—	—	19.34

Table 40A (continued)
METHYL ESTERS OF FATTY ACIDS: *trans, trans*-OCTADECADIENOATES, ON POLAR COLUMNS, REPORTED AS EQUIVALENT CHAIN LENGTHS

Column packing

P1 = Poly (1,4-butanediol succinate) in form of wall coating in stainless steel tubing

P2 = SILAR-5CP (50% cyanopropyl, 50% phenyl, siloxane) in form of wall coating in stainless steel tubing

P3 = XE-60 (25% cyanoethyl, with methyl, silicone) in form of wall coating in stainless steel tubing

P4 = 20% Poly (diethylene glycol succinate)

P5 = Poly (diethylene glycol succinate) in form of wall coating in stainless steel tubing

P6 = Six-ring polyphenyl ether (plus Arquad 2HT-75) in form of wall coating in stainless steel tubing

P7 = Poly (diethylene glycol succinate) in form of wall coating in stainless steel tubing

P8 = General Electric XF-1150 (50% cyanoethyl, with methyl, silicone) in form of wall coating in stainless steel tubing

P9 = General Electric 238-149-00 (50% cyanoethyl, with methyl, silicone) in form of wall coating in stainless steel tubing

P10 = TCPE (tetracyanoethylated pentaerythritol) in form of wall coating in stainless steel tubing

REFERENCES

1. **Ackman, R. G. and Hooper, S. N.,** *J. Chromatogr.,* 86, 83, 1973.
2. **Gunstone, F. D. and Lie Ken Jie, M.,** *Chem. Phys. Lipids,* 4, 131, 1970.
3. **Scholfield, C. R. and Dutton, H. J.,** *J. Am. Oil Chem. Soc.,* 48, 228, 1971; see also **Scholfield, C. R.,** *J. Am. Oil Chem. Soc.,* 58, 662, 1981.

Table 40B
METHYL ESTERS OF FATTY ACIDS: *trans, trans-*OCTADECADIENOATES, ON NONPOLAR COLUMNS, REPORTED AS EQUIVALENT CHAIN LENGTHS

	P1	P2	P3	P4
Column packing				
Temperature (°C)	190	210	180	200
Gas	N_2	N_2	He	Ar
Flow rate (mℓ/min)	na	na	5	na
Column				
Length (cm)	153	5000	4700	6100
Diameter				
(cm I.D.)	na	0.05	0.025	0.025
(cm O.D.)	0.63	na	na	na
Form	Helix	WCOT	WCOT	WCOT
Material	na	SS	SS	SS
Detector	FID	FID	FID	Argon
Reference	1	1	2	3

Compound		ECL		
Methyl *trans,trans*-5,12-octadecadienoate	17.53	17.46	—	—
Methyl *trans,trans*-6,12-octadecadienoate	17.51	17.53	—	—
Methyl *trans,trans*-7,12-octadecadienoate	17.58	17.59	—	—
Methyl *trans,trans*-8,12-octadecadienoate	17.48	17.50	—	—
Methyl *trans,trans*-9,12-octadecadienoate	17.60	17.60	17.63	17.65
Methyl *trans,trans*-10,12-octadecadienoate	18.59	18.62	—	—
Methyl *trans,trans*-6,8-octadecadienoate	—	18.60	—	—
Methyl *trans,trans*-6,9-octadecadienoate	17.57	17.59	—	—
Methyl *trans,trans*-6,10-octadecadienoate	17.43	17.46	—	—
Methyl *trans,trans*-6,11-octadecadienoate	17.57	17.57	—	—
Methyl *trans,trans*-9,15-octadecadienoate	—	—	17.65	17.68
Methyl *trans,trans*-12,15-octadecadienoate	—	—	17.73	17.85

Column packing P1 = 5% Apiezon-L on Chromosorb W, HMDS (80-100 mesh)
P2 = Apiezon-L in form of wall coating in stainless steel tubing
P3 = Apiezon-L in form of wall coating in stainless steel tubing
P4 = Apiezon-L in form of wall coating in stainless steel tubing

REFERENCES

1. **Gunstone, F. D. and Lie Ken Jie, M.,** *Chem. Phys. Lipids,* 4, 131, 1970.
2. **Ackman, R. G. and Hooper, S. N.,** *J. Chromatogr.,* 86, 73, 1973.
3. **Scholfield, C.R. and Dutton, H. J.,** *J. Am. Oil Chem. Soc.,* 48, 228, 1971.

Table 41A
METHYL ESTERS OF FATTY ACIDS: GEOMETRICALLY ISOMERIC 9.12-, 9.15-, AND-12.15-OCTADECADIENOATES, ON POLAR COLUMNS, REPORTED IN EQUIVALENT CHAIN LENGTHS

	P1	P2	P3	P4	P5	P6	P7	P8	P9	P10	P11	P12
Column packing												
Temperature (°C)	160	170	180	165	170	200	190	160	175	170	170	200
Gas	He	He	He	Ar	Ar	Ar	Ar	Ar	Ar	N_2	N_2	He
Flow rate (mℓ/min)	15	5	5	na	na	na	na	na	0.7	10	20	30
Column Length (cm)	4700	4700	4700	6100	6100	6100	4700	6100	4700	400	365	478
Diameter (cm I.D.)	0.025	0.025	0.025	0.025	0.025	0.025	0.025	0.025	0.025	0.2	0.2	na
(cm O.D.)	na	na	na	na	na	na	na	na	na	na	na	0.318
Form	WCOT	WCOT	WCOT	WCOT	WCOT	WCOT	WCOT	WCOT	WCOT	na	Helix	na
Material	SS	SS	SS	SS	SS	SS	SS	SS	SS	G	SS	SS
Detector	FID	FID	FID	FID	FID	FID	FID	FID	FID	FID	FID	FID
Reference	1	1	1	2	2	2	2	2	3	4	5	6
Compound						ECL						
Methyl *cis,cis*-9,12-octadecadienoate	18.50	18.84	18.81	19.06	19.08	19.16	18.19	19.49	19.46	19.71	19.79	19.90
Methyl *cis,trans*-9,12-octadecadienoate	18.50	18.84	18.80	—	18.95	18.96	18.35	19.36	19.46	19.58	19.60	19.70
Methyl *trans,cis*-9,12-octadecadienoate	18.61	18.95	18.90	—	—	—	18.43	19.42	19.55	19.65	19.66	19.70
Methyl *trans,trans*-9,12-octadecadienoate	18.49	18.82	18.77	19.00	18.67	18.69	18.41	19.14	19.35	19.31	19.32	19.30
Methyl *cis,cis*-9,15-octadecadienoate	18.61	18.99	18.95	19.24	19.22	19.29	18.32	19.59	—	—	—	20.00
Methyl *cis,trans*-9,15-octadecadienoate	18.50	18.83	18.79	19.04	18.89	—	18.30	19.32	—	—	—	—
Methyl *trans,cis*-9,15-octadecadienoate	18.63	19.01	18.92	19.18	19.05	—	18.45	19.46	—	—	—	—
Methyl *trans,trans*-9,15-octadecadienoate	18.47	18.81	18.72	18.98	18.68	19.48	18.40	19.13	—	—	—	20.25
Methyl *cis,cis*-12,15-octadecadienoate	18.61	18.99	18.95	19.45	19.41	—	18.46	19.82	—	—	—	—
Methyl *cis,trans*-12,15-octadecadienoate	18.71	19.10	18.99	19.30	—	—	18.51	19.56	—	—	—	—
Methyl *trans,cis*-12,15-octadecadienoate	18.81	19.22	19.06	19.38	—	—	18.61	19.68	—	—	—	—
Methyl *trans,trans*-12,15-octadecadienoate	18.67	19.05	18.91	19.23	—	—	18.55	19.34	—	—	—	—

Column packing P1 = Poly (1,4-butanediol succinate) in form of wall coating in stainless steel tubing

P2 = Poly (1, 4-butanediol succinate) in form of wall coating in stainless steel tubing

P3 = SILAR-5CP (50% cyanopropyl, 50% phenyl, siloxane) in form of wall coating in stainless steel tubing

P4 = Poly (diethylene glycol succinate) in form of wall coating in stainless steel tubing

P5 = XF-1150 (50% cyanoethyl, with methyl, silicone) in form of wall coating in stainless steel tubing

P6 = General Electric 238-149-99 (50% cyanoethyl, with methyl, silicone) in form of wall coating in stainless steel tubing

P7 = Six-ring polyphenyl ether (plus Arquad 2HT-75) in form of wall coating in stainless steel tubing

P8 = TCPE (tetracyanoethylated pentaerythritol) in form of wall coating in stainless steel tubing

P9 = Poly (diethylene glycol succinate) in form of wall coating in stainless steel tubing

P10 = 12% SILAR-10C (100% cyanopropyl siloxane) on Chromosorb W, HP (80-100 mesh)

P11 = 10% SILAR-10C (100% cyanopropyl siloxane) on Gas-Chrom Q (100-120 mesh)

P12 = 10% SILAR-10C (100% cyanopropyl siloxane) on Gas-Chrom Q (100-120 mesh)

REFERENCES

1. **Ackman, R. G. and Hooper, S. N.**, *J. Chromatogr. Sci.*, 12, 131, 1974; see also **Scholfield, C. R.**, *J. Am. Oil Chem. Soc.*, 58, 662, 1981.
2. **Scholfield, C. R. and Dutton, H. J.**, *J. Am. Oil Chem. Soc.*, 48, 228, 1971.
3. **Litchfield, C., Isbell, A. F., and Reiser, R.**, *J. Am. Oil Chem. Soc.*, 39, 330, 1962.
4. **Heckers, H., Dittmar, K., Melcher, F. W., and Kalinowski, H. O.**, *J. Chromatogr.*, 135, 93, 1977.
5. **Parodi, P. W.**, *J. Am. Oil Chem. Soc.*, 53, 530, 1976.
6. **Conacher, H. B. S. and Iyengar, J. R.**, *J. Ass. Off. Anal. Chem.*, 61, 702, 1978.

Note added in proof: For additional elution order information on the four geometrically different 9,12-octadecadienoates (9t,12t; 9c,12t; 9t,12c; 9c,12c) on the popular cyanosilicone liquid phases: (A) OV-275; (B) SS-4; (C) SP-2340; wall-coated in open-tubular columns, see: (A) **Devinat, G., Scamaroni, L., and Naudet, M.**, *Rev. Franc. Corps Gras*, 27, 283, 1980; (B) **Kobayashi, T.**, *J. Chromatogr.*, 194, 404, 1980; and (C) **Lanza, E., Zyren, J., and Slover, H. T.**, *J. Agric. Food Chem.*, 28, 1182, 1980.

Table 41B
METHYL ESTERS OF FATTY ACIDS: GEOMETRICALLY ISOMERIC 9,12-, 9,15-, AND 12,15-OCTADECADIENOATES, ON NONPOLAR COLUMNS, REPORTED AS EQUIVALENT CHAIN LENGTHS

Column packing	P1	P2	P3	P4	P5	P6	P7	P8
Temperature (°C)	180	2	200	200	180	220	200	180
Gas	He	N_2	Ar	Ar	N_2	N_2	N_2	—
Flow rate (mℓ/min)	5	na	na	0.9	1	2	2	na
Column								
Length/cm)	4700	5000	6100	6100	6000	5000	153	3000
Diameter								
(cm I.D.)	0.025	0.05	0.025	0.025	na	0.025	na	0.03
(cm O.D.)	na	na	na	na	na	na	0.64	na
Form	WCOT	WCOT	WCOT	WCOT	WCOT	WCOT	Helix	Helix
Material	SS	SS	SS	SS	G	SS	na	G
Detector	FID	FID	Argon	Argon	FID	FID	FID	FID
Reference	1	2	3	4	5	6	6	7

Compound	ECL							
Methyl *cis,cis*-9,12-octadecadienoate	17.50	17.50	17.50	17.48	17.57	17.50	17.47	17.48
Methyl *cis,trans*-9,12-octadecadienoate	—	—	17.59	17.59	—	—	—	17.59
Methyl *trans,cis*-9,12-octadecadienoate	—	—	—	17.64	—	—	—	17.69
Methyl *trans,trans*-9,12-octadecadienoate	17.63	17.63	17.65	17.64	17.70	—	—	17.69
Methyl *cis,cis*-9,15-octadecadienoate	17.61	—	17.66	—	17.66	—	—	—
Methyl *cis,trans*-9,15-octadecadienoate	—	—	17.60	—	—	—	—	—
Methyl *trans,cis*-9,15-octadecadienoate	—	—	17.75	—	—	—	—	—
Methyl *trans,trans*-9,15-octadecadienoate	17.65	—	17.68	—	—	—	—	—
Methyl *cis,cis*-12,15-octadecadienoate	17.73	—	17.83	—	17.87	17.78	17.72	—
Methyl *trans,trans*-12,15-octadecadienoate	—	—	—	—	—	—	—	—
Methyl *trans,cis*-12,15-octadecadienoate	—	—	—	—	—	—	—	—
Methyl *trans,trans*-12,15-octadecadienoate	17.76	—	17.85	—	—	—	—	—

Column packing	P1 =	Apiezon-L in form of wall coating in stainless steel tubing
	P2 =	Apiezon-L in form of wall coating in stainless steel tubing
	P3 =	Apiezon-L in form of wall coating in stainless steel tubing
	P4 =	Apiezon-L in form of wall coating in stainless steel tubing
	P5 =	Apiezon-L in form of wall coating in stainless steel tubing
	P6 =	Apiezon-L in form of wall coating in stainless steel tubing
	P7 =	5% Apiezon-L on GAS Chrom Z (70-80 mesh)
	P8 =	Apiezon-L in form of wall coating in glass capillary tubing

REFERENCES

1. **Ackman, R.G. and Hooper, S.N.,** *J. Chromatogr.,* 86, 73, 1973.
2. **Barve, J.A., Gunstone, F.D., Jacobsberg, F.R., and Winlow, P.,** *Chem. Phys. Lipids,* 8, 117, 1972.
3. **Scholfield, C.R. and Dutton, H.J.,** *J. Am. Oil Chem. Soc.,* 48, 228, 1971.
4. **Litchfield, C., Isbell, A.F., and Reiser, R.,** *J. Am. Oil Chem. Soc.,* 39, 330, 1962.
5. **Strocchi, A., Piretti, M., and Capella, P.,** *Riv. Ital. Sost. Grasse,* 46, 80, 1969.
6. **Christie, W.W.,** *J. Chromatogr.,* 37, 27, 1968.
7. **Cartoni, G., Liberti, A., and Ruggieri, G.,** *Riv. Ital. Sost. Grasse,* 40, 482, 1963.

Table 42A

METHYL ESTERS OF FATTY ACIDS: GEOMETRICALLY ISOMERIC 9,12,15-OCTADECATRIENOATES, ON POLAR LIQUID PHASES, REPORTED AS EQUIVALENT CHAIN LENGTHS

	P1	P2	P3	P4	P5	P6	P7	P8	P9	P10
Column packing										
Temperature (°C)	160	170	180	165	170	200	190	160	175	170
Gas	He	He	He	Ar	Ar	Ar	Ar	Ar	Ar	N_2
Flow rate (ml/min)	5	5	5	na	na	na	na	na	0.7	10
Column Length(cm)	4700	4700	4700	6100	6100	6100	4700	6100	4700	400
Diameter (cm I.D.)	0.025	0.025	0.025	0.025	0.025	0.025	0.025	0.025	0.025	0.2
(cm O.D.)	na	na	na	na	na	na	na	na	na	na
Form	WCOT	WCOT	WCOT	WCOT	WCOT	WCOT	WCOT	WCOT	WCOT	Helix
Material	SS	SS	SS	SS	SS	SS	SS	SS	SS	G
Detector	FID	FID	FID	Argon	Argon	Argon	Argon	Argon	Argon	FID
Reference	1	1	1	2	2	2	2	2	3	4
Compound					ECL					
Methyl *cis,cis,cis*-9,12,15-octadecatrienoate	19.10	19.61	19.44	19.91	19.90	20.00	18.52	20.49	20.44	20.91
Methyl *cis,cis,trans*-9,12,15-octadecatrienoate	19.01	19.52	19.35	—	—	—	—	—	—	—
Methyl *cis,trans,trans*-9,12,15-octadecatrienoate	19.12	19.59	19.41	—	—	—	—	—	—	—
Methyl *trans,trans,trans*-9,12,15-octadecatrienoate	19.04	19.52	19.35	19.75	19.14	19.12	18.83	19.91	20.23	20.17
Methyl *cis,trans,cis*-9,12,15-octadecatrienoate	19.18	19.68	19.51	—	19.91	19.80	18.55	—	—	—
Methyl *trans,trans,cis*-9,12,15-octadecatrienoate	19.15	19.68	19.50	—	—	—	—	—	—	—
Methyl *trans,cis,cis*-9,12,15-octadecatrienoate	19.20	19.70	19.51	—	—	—	—	—	—	—
Methyl *trans,cis,trans*-9,12,15-octadecatrienoate	19.13	19.57	19.41	—	19.50	19.52	18.79	20.15	—	—

Table 42A (continued)
METHYL ESTERS OF FATTY ACIDS: GEOMETRICALLY ISOMERIC 9,12,15-OCTADECATRIENOATES, ON POLAR LIQUID PHASES, REPORTED AS EQUIVALENT CHAIN LENGTHS

Column packing
P1 = Poly (1,4-butanediol succinate); in form of wall coating in stainless steel tubing
P2 = Poly (1,4-butanediol succinate); in form of wall coating in stainless steel tubing
P3 = SILAR-5CP (50% cyanopropyl, 50% phenyl, siloxane); in form of wall coating in stainless steel tubing
P4 = Poly (diethylene glycol succinate); in form of wall coating in stainless steel tubing
P5 = XF-1150 (50% cyanoethyl, with methyl, silicone); in form of wall coating in stainless steel tubing
P6 = General Electric 238-149-99 (50% cyanoethyl, with methyl, silicone) in form of wall coating in stainless steel tubing
P7 = Six-ring polyphenyl ether (plus Arquad 2HT-75) in form of wall coating in stainless steel tubing
P8 = TCPE (tetracyanoethylated pentaerythritol) in form of wall coating in stainless steel tubing
P9 = Poly (diethylene glycol succinate) in form of wall coating in stainless steel tubing
P10 = 12% SILAR-10C (100% cyanopropyl siloxane) on Chromosorb W, HP (80-100 mesh)

REFERENCES

1. **Ackman, R. G. and Hooper, S. N.,** *J. Chromatogr. Sci.,* 12, 131, 1974; see also **Scholfield, C. R.,** *J. Am. Oil Chem. Soc.,* 58, 662, 1981;
2. **Scholfield, C. R. and Dutton, H. J.,** *J. Am. Oil Chem. Soc.,* 48, 228, 1971.
3. **Litchfield, C., Reiser, R., and Isbell, A. F.,** *J. Am. Oil Chem. Soc.,* 40, 302, 1963.
4. **Heckers, H., Dittmar, K., Melcher, F. W., and Kalinowski, H. O.,** *J. Chromatogr.,* 135, 93, 1977; see also **Rakoff, H. and Emden, E. A.,** *Chem. Phys. Lipids,* 31, 215, 1982.

Table 42B
**METHYL ESTERS OF FATTY ACIDS: GEOMETRICALLY ISOMERIC
9,12,15-OCTADECATRIENOATES, ON NONPOLAR COLUMNS,
REPORTED AS EQUIVALENT CHAIN LENGTHS**

Column packing	P1	P2	P3	P4
Temperature (°C)	180	210	180	200
Gas	He	N_2	N_2	Ar
Flow rate (mℓ/min)	5	na	1	0.9
Column				
Length (cm)	4700	5000	6000	6100
Diameter				
(cm I.D.)	0.025	0.05	na	0.025
(cm O.D.)	na	na	na	na
Form	WCOT	WCOT	WCOT	WCOT
Material	SS	SS	G	SS
Detector	FID	FID	FID	Argon
Reference	1	2	3	4

Compound		ECL		
Methyl *cis,cis,cis*-9,12,15-octadecatrienoate	17.53	17.49	17.57	17.48
Methyl *cis,cis,trans*-9,12,15-octadecatrienoate	17.53	17.47	17.57	—
Methyl *cis,trans,trans*-9,12,15-octadecatrienoate	17.61	—	17.66	—
Methyl *trans,trans,trans*-9,12,15-octadecatrienoate	17.63	—	17.67	17.56
Methyl *cis,trans,cis*-9,12,15-octadecatrienoate	17.68	17.60	17.72	—
Methyl *trans,trans,cis*-9,12,15-octadecatrienoate	17.69	—	17.72	—
Methyl *trans,cis,cis*-9,12,15-octadecatrienoate	17.71	17.64	17.73	—
Methyl *trans,cis,trans*-9,12,15-octadecatrienoate	17.73	—	17.75	—

Column packing
P1 = Apiezon-L in form of wall coating in stainless steel tubing
P2 = Apiezon-L in form of wall coating in stainless steel tubing
P3 = Apiezon-L in form of wall coating in stainless steel tubing
P4 = Apiezon-L in form of wall coating in stainless steel tubing

REFERENCES

1. **Ackman, R. G. and Hooper, S. N.,** *J. Chromatogr.*, 86, 73, 1973.
2. **Barve, J. A., Gunstone, F. D., Jacobsberg, F. R., and Winlow, P.,** *Chem. Phys. Lipids*, 8, 117, 1972.
3. **Strocchi, A., Piretti, M., and Capella, P.,** *Riv. Ital. Sost. Grasse*, 46, 80, 1969.
4. **Litchfield, C., Reiser, R., and Isbell, A. F.,** *J. Am. Oil Chem. Soc.*, 39, 302, 1963.

Table 43A
METHYL ESTERS OF FATTY ACIDS: CONJUGATED OCTADECADIENOATE ISOMERS ON POLAR COLUMNS, REPORTED AS EQUIVALENT CHAIN LENGTHS

	P1	P2	P3	P4	P5	P6	P7	P8	P9	P10	P11	P12	P13
Column packing													
Temperature (°C)	190	175	190	195	170	165	190	190	160	170	180	185	170
Gas	N$_2$	—	Ar	H$_2$	Ar	Ar	N$_2$	N$_2$	Ar	He	N$_2$	He	He
Flow rate (mℓ/min)	na	na	na	1.8	na	na	na	na	na	5	30	5	5
Column													
Length (cm)	5000	3000	4600	3900	6100	6100	5000	153	6100	4700	183	4700	4700
Diameter													
(cm I.D.)	0.05	0.03	0.025	0.03	0.025	0.025	0.05	na	0.025	0.025	na	0.025	0.025
(cm O.D.)	na	na	na	na	na	na	na	0.64	na	na	0.318	na	na
Form	WCOT	Helix	WCOT	WCOT	WCOT	WCOT	WCOT	Helix	WCOT	WCOT	Helix	WCOT	WCOT
Material	SS	G	SS	G	SS	SS	SS	na	SS	SS	SS	SS	SS
Detector	FID	FID	Argon	FID	Argon	Argon	FID	FID	Argon	FID	FID	FID	FID
Reference	1	2	3	4	3	3	1	1	3	5	6	5	5
Compound						ECL							
Methyl *cis,cis*-9,11-octadecadienoate	—	18.69	19.31	—	—	20.32	—	—	20.87	—	—	—	—
Methyl *cis,trans*-9,11-octadecadienoate	—	18.99	19.04	19.85	20.10	20.10	—	—	20.60	19.69	19.78	19.80	20.11
Methyl *trans,cis*-9,11-octadecadienoate	—	18.95	—	—	20.15	—	—	—	20.65	—	—	—	—
Methyl *trans,trans*-9,11-octadecadienoate	—	19.49	19.60	20.48	20.59	20.70	—	—	21.19	20.29	20.39	20.22	20.62
Methyl *cis,cis*-10,12-octadecadienoate	18.76	—	19.32	—	—	—	—	21.01	—	—	—	—	—
Methyl *cis,trans*-10,12-octadecadienoate	—	18.95	19.07	—	—	—	—	—	—	—	—	—	—
Methyl *trans,cis*-10,12-octadecadienoate	—	19.20	19.16	19.97	20.24	20.30	—	—	20.77	19.82	19.85	19.91	20.32
Methyl *trans,trans*-10,12-octadecadienoate	19.16	19.49	19.61	20.48	20.57	20.72	20.61	20.99	21.16	20.56	(20.56)	20.68	21.04
Methyl *cis,cis*-6,8-octadecadienoate	18.58	—	—	—	—	—	20.25	20.69	—	—	—	—	—
Methyl *trans,trans*-6,8-octadecadienoate	19.25	—	—	—	—	—	20.59	21.05	—	—	—	—	—

Column packing P1 = XE-60 (25% cyanoethyl, with methyl, silicone) in form of wall coating in stainless steel tubing
P2 = Poly (neopentyl glycol adipate) in form of wall coating in glass capillary tubing
P3 = Six-ring polyphenyl ether (plus Arquad 2HT-75) in form of wall coating in stainless steel tubing
P4 = Poly (1,4-butanediol succinate) in form of wall coating in glass capillary tubing
P5 = GE XF-1150 (50% cyanoethyl, with methyl, silicone) in form of wall coating in stainless steel tubing
P6 = Poly (diethylene glycol succinate) in form of wall coating in stainless steel tubing
P7 = Poly (diethylene glycol succinate) in form of wall coating in stainless steel tubing
P8 = 20% Poly (diethylene glycol succinate) on Chromosorb W, HNDS (80-100 mesh)
P9 = Tetracyanoethylated pentaerythritol in form of wall coating in stainless steel tubing
P10 = Poly (1,4-butanediol succinate) in form of wall coating in stainless steel tubing
P11 = 8% Poly (1,4-butanediol succinate) on Chromosorb W, HMDS
P12 = SILAR-5CP (50% cyanopropyl, 50% phenyl, silicone) in form of wall coating in stainless steel tubing
P13 = SILAR-7CP (70% cyanopropyl, 30% phenyl, silicone) in form of wall coating in stainless steel tubing

REFERENCES

1. **Gunstone, F. D. and Lie Ken Jie, M.**, *Chem. Phys. Lipids*, 4, 131, 1970.
2. **Cartoni, G., Lieberti, A., and Ruggieri, G.**, *Riv. It al. Sost. Grasse*, 40, 482, 1963.
3. **Scholfield, C. R. and Dutton, H. J.**, *J. Am. Oil Chem. Soc.*, 48, 228, 1971.
4. **Holmbom, B.**, *J. Am. Oil Chem. Soc.*, 54, 289, 1977.
5. **Ackman, R. G., Eaton, C. A., Sipos, J. C., and Crewe, N.**, *Can. Inst. Food Sci. Technol. J.*, 14, 103, 1981; see also **Scholfield, C. R.**, *J. Am. Oil Chem. Soc.*, 58, 662, 1981.
6. **Jamieson, G. R. and Reid, E. H.**, *J. Chromatogr.*, 20, 232, 1965.

Table 43B
METHYL ESTERS OF FATTY ACIDS: CONJUGATED OCTADECADIENOATE ISOMERS ON NONPOLAR COLUMNS, WITH THE *cis,cis*-9,12 ISOMER, REPORTED AS EQUIVALENT CHAIN LENGTHS

	P1	P2	P3	P4	P5	P6	P7	P8	P9	P10
Column packing										
Temperature (°C)	180	200	200	240	200	210	180	197	180	190
Gas	N_2	Ar	N_2	He	N_2	N_2	—	N_2	He	He
Flow rate (mℓ/min)	1	1	1.6	70	na	na	na	10	5	5
Column Length (cm)	6000	6100	5000	250	153	5000	3000	116	4700	4700
Diameter (cm I.D.)	na	0.025	na	0.3	na	0.05	0.03	0.4	0.025	0.025
(cm O.D.)	na	na	na	na	0.64	na	na	na	na	na
Form	WCOT	WCOT	WCOT	Helix	Helix	WCOT	Helix	na	WCOT	WCOT
Material	G	SS	G	na	SS	SS	G	na	SS	SS
Detector	FID	Argon	FID	TC	FID	FID	FID	na	FID	FID
Reference	1	2	3	4	5	5	6	7	8	8
Compound					ECL					
Methyl *cis,cis*-9,11-octadecadienoate	18.53	18.3	—	—	—	—	—	—	—	—
Methyl *cis,trans*-9,11-octadecadienoate	18.28	18.2	18.05	—	—	—	18.12	18.18	18.10	18.18
Methyl *trans,cis*-9,11-octadecadienoate	18.31	—	—	—	—	—	18.20	—	—	—
Methyl *trans,trans*-9,11-octadecadienoate	18.80	18.6	18.49	18.68	—	—	18.61	18.62	18.61	18.68
Methyl *cis,cis*-10,12-octadecadienoate	18.53	18.3	—	—	18.67[a]	18.68[a]	—	—	—	—
Methyl *cis,trans*-10,12-octadecadienoate	18.31	—	—	—	—	—	18.20	—	—	—
Methyl *trans,cis*-10,12-octadecadienoate	18.35	18.1	18.14	—	—	—	18.25	18.18	18.20	18.28
Methyl *trans,trans*-10,12-octadecadienoate	18.80	18.6	18.49	—	18.59	18.62	18.61	18.62	18.61	18.78
Methyl *cis,cis*-6,8-octadecadienoate	—	—	—	—	18.54[a]	18.55[a]	—	—	—	—
Methyl *trans,trans*-6,8-octadecadienoate	—	—	—	—	—	18.60	—	—	—	—
Methyl *cis,cis*-9,12-octadecadienoate	17.57	17.4	17.64	17.53	17.48	17.48	—	17.49	17.48	17.51

[a] Arbitrary increase of 1 ECL unit by R. G. Ackman.

Column packing P1 = Apiezon L in form of wall coating in glass capillary tubing
P2 = Apiezon-L in form of wall coating in stainless steel tubing
P3 = SE-30 (dimethyl silicone) in form of wall coating in glass capillary tubing
P4 = 20% Apiezon-L on Gas-Chrom P (100-140 mesh)
P5 = 3% Apiezon-L on Chromosorb W, HMDS (80-100 mesh)
P6 = Apiezon-L in form of wall coating in stainless steel column
P7 = Apiezon-L in form of wall coating in class capillary tubing
P8 = 10% Apiezon-L on Celite 545
P9 = Apiezon-L in form of wall coating in stainless steel tubing
P10 = Apiezon-L in form of wall coating in stainless steel tubing

REFERENCES

1. **Strocchi, A., Piretti, M., and Capella, P.,** *Riv. Ital. Sost. Grasse,* 46, 80, 1969.
2. **Litchfield, C., Reiser, R., and Isbell, A. F.,** *J. Am. Oil Chem. Soc.,* 40, 302, 1963.
3. **Holmbom, B.,** *J. Am. Oil Chem. Soc.,* 54, 289, 1977.
4. **Hofstetter, H. H., Sen, N., and Holman, R. T.,** *J. Am. Oil Chem. Soc.,* 42, 537, 1965.
5. **Gunstone, F. D. and Lie Ken Jie, M.,** *Chem. Phys. Lipids,* 4, 131, 1970.
6. **Cartoni, G., Liberti, A., and Ruggieri, G.,** *Riv. Ital. Sost. Grasse,* 40, 482, 1963.
7. **Beerthuis, R. K., Dijkstra, G. D., Keppler, J. G., and Recourt, J. H.,** *Ann. N.Y. Acad. Sci.,* 72, 616, 1959.
8. **Ackman, R. G., Eaton, C. A., Sipes, J. C., and Crewe, N.,** *Can. Inst. Food Sci. Technol. J.,* 14, 103, 1981.

Table 44

**METHYL ESTERS OF FATTY ACIDS: GEOMETRIC AND POSITIONAL
ISOMERS OF OCTADECATRIENOATES AND OCTADECATETRAENOATES
INCLUDING CONJUGATION, REPORTED AS EQUIVALENT CHAIN LENGTHS**

	P1	P2	P3	P4
Column packing				
Temperature (°C)	180	197	210	190
Gas	N_2	N_2	N_2	—
Flow rate (mℓ/min)	1	10	na	na
Column				
Length (cm)	6000	116	5000	na
Diameter				
(cm I.D.)	na	0.4	0.05	na
(cm O.D.)	na	na	na	na
Form	WCOT	na	WCOT	na
Material	FID	na	FID	FID
Detector	1	2	3	4
Reference				

Compound	ECL			
Methyl *cis,trans,cis*- 9,11,15-octadecatrienoate	18.22	—	—	—
Methyl *cis,trans,cis*-9,13,15-octadecatrienoate	18.22	—	—	—
Methyl *cis,trans,trans*-9,11,13-octadecatrienoate	—	19.12	19.1	21.97
Methyl *trans,trans,trans*-9,11,13-octadecatrienoate	—	19.47	19.5	22.40
Methyl *cis,trans,trans,cis*-9,11,13,15-octadecatetraenoate	—	—	19.9	—
Methyl *trans,trans,trans,trans*-9,11,13,15-octadecatetraenoate	—	—	20.3	—

Column packing P1 = Apiezon-L In form of wall coating in glass capillary tubing
                P2 = 10% Apiezon-L on Celite 545
                P3 = Apiezon-L in form of wall coating in stainless steel tubing
                P4 = Poly (1,4-butanediol succinate) on Chromosorb W (80-100 mesh)

REFERENCES

1. **Strocchi, A., Piretti, M., and Capella, P.,** *Riv. Ital. Sost. Grasse,* 46, 80, 1969.
2. **Beerthuis, R. K., Dijkstra, G. D., Keppler, J. G., and Recourt, J. H.,** *Ann. N.Y. Acad. Sci.,* 72, 616, 1959.
3. **Gunstone, F. D. and Subbarao, R.,** *Chem. Phys. Lipids,* 1, 349, 1967.
4. **Jamieson, G. R.,** *Topics in Lipid Chemistry,* Vol. I, Gunstone, F. D., Ed., Logos Press, London, 1970, 107; see also **Scholfield, C. R.,** *J. Am. Oil Chem. Soc.,* 58, 662, 1981.

Note added in proof: For additional ECL data recently reported for the dimethyl silicone liquid phase OV-1 wall-coated in an open-tubular column see **Takogi, T. and Itabashi, Y.,** *Lipids,* 16, 546, 1981.

Table 45
METHYL ESTERS OF FATTY ACIDS: ISOMERIC NONYNOATES, REPORTED AS RETENTION TIMES RELATIVE TO METHYL NONANOATE

	P1	P2
Column packing	P1	P2
Temperature (°C)	126	136
Gas	Ar	Ar
Flow rate (ml/min)	na	na
Column		
Length (cm)	3050	2600
Diameter		
(cm I.D.)	0.025	0.025
(cm O.D.)	na	na
Form	WCOT	WCOT
Material	SS	SS
Detector	Argon	Argon
Reference	1	1

Compound	ECL	
Methyl-2-nonynoate	1.62	3.51
Methyl-3-nonynoate	1.22	3.31
Methyl-4-nonynoate	1.11	2.52
Methyl-5-nonynoate	1.11	2.59
Methyl-6-nonynoate	1.20	2.99
Methyl-7-nonynoate	1.50	4.13
Methyl-8-nonynoate	1.12	3.74

Column packing P1 = Apiezon-L in form of wall coating in stainless steel tubing
P2 = Poly (diethylene glycol succinate) in form of wall coating in stainless steel tubing

REFERENCE

1. **Anderson, R. E. and Rakoff, H.,** *J. Am. Oil Chem. Soc.,* 42, 1102, 1965.

Table 46
METHYL ESTERS OF FATTY ACIDS: UNDECYNOATES, REPORTED AS EQUIVALENT CHAIN LENGTHS

Column packing	P1	P2	P3	P4	P5	P6	P7	P8
Temperature (°C)	175	130	150	190	160	155	190	125
Gas	N_2	N_2	N_2	N_2	N_2	N_2	N_2	N_2
Flow rate (mℓ/min)	40	60	45	80	45	60	50	45
Column								
Length (cm)	200	200	200	200	200	200	200	200
Diameter								
(cm I.D.)	na	0.6	0.3	0.6	0.3	0.3	na	na
(cm O.D.)	na	na	na	na	na	na	na	na
Form	na	na	na	na	na	na	na	na
Material	na	na	na	na	na	na	na	na
Detector	FID	FID	FID	FID	FID	FID	FID	FID
Reference	1	2	2	2	2	2	1	1

Compound	ECL							
Methyl 2-undecynoate	11.61	11.63	11.62	13.10	13.61	13.47	14.49	14.39
Methyl 3-undecynoate	11.17	11.31	11.34	12.10	13.31	13.03	14.12	13.65
Methyl 4-undecynoate	11.02	11.18	11.16	11.78	12.67	12.64	13.38	12.99
Methyl 5-undecynoate	11.02	11.18	11.17	11.84	12.67	12.62	13.43	13.12
Methyl 6-undecynoate	11.05	11.18	11.18	11.92	12.80	12.78	13.64	13.38
Methyl 7-undecynoate	11.07	11.24	11.18	12.05	12.86	12.84	13.72	13.45
Methyl 8-undecynoate	11.14	11.33	11.34	12.15	13.10	13.01	13.99	13.61
Methyl 9-undecynoate	11.52	11.59	11.63	12.65	13.77	13.65	14.87	14.38
Methyl 10-undecynoate	11.00	11.15	11.16	12.51	13.42	13.34	14.60	14.18

Column packing P1 =   5% Apiezon-L on Chromosorb W (80-100 mesh)

P2 =   3% SE-30 (methyl silicone) on Chromosorb W (80-100 mesh)

P3 =   1.5% OV-101 (methyl silicone) on Chromosorb W (80-100 mesh)

P4 =   20% XE-60 (25% cyanoethyl, with methyl, silicone) on Chromosorb W (80-100 mesh)

P5 =   10% FFAP (Carbowax 20M modified with 2-nitroterephthalic acid) on Chromosorb W (80-100 mesh)

P6 =   10% Carbowax 20M on Chromosorb W (80-100 mesh)

P7 =   20% Poly (diethylene glycol succinate) on Chromosorb W (80-100 mesh)

P8 =   10% SILAR-10C (100% cyanopropyl siloxane) on Chromosorb W (80-100 mesh)

REFERENCES

1. **Lie Ken Jie, M. S. F. and Lam, C. H.,** *J. Chromatogr.,* 97, 165, 1974.
2. **Lie Ken Jie, M. S. F.,** *J. Chromatogr.,* 111, 189, 1975.

Table 47
METHYL ESTERS OF FATTY ACIDS: DODECYNOATES, REPORTED IN EQUIVALENT CHAIN LENGTHS

Column packing	P1	P2	P3
Temperature (°C)	150	150	190
Gas	N_2	N_2	N_2
Flow rate (mℓ/min)	na	na	na
Column			
Length (cm)	5000	5000	200
Diameter			
(cm I.D.)	0.05	0.05	na
(cm O.D.)	na	na	0.318
Form	WCOT	WCOT	Helix
Material	SS	SS	SS
Detector	FID	FID	FID
Reference	1	1	1

Compound	ECL		
Methyl 5-dodecynoate	12.05	13.23	14.47
Methyl 6-dodecynoate	12.00	13.34	14.68
Methyl 7-dodecynoate	12.03	13.42	14.78
Methyl 8-dodecynoate	12.06	13.48	14.88
Methyl 9-dodecynoate	12.15	13.62	15.13
Methyl 10-dodecynoate	12.52	14.26	16.00
Methyl 11-dodecynoate	12.01	14.00	15.74

Column packing P1 = Apiezon-L in form of wall coating in stainless steel tubing
P2 = Poly (neopentylglycol succinate) in form of wall coating in stainless steel tubing
P3 = Poly (diethyleneglycol succinate) in form of wall coating in stainless steel tubing

REFERENCE

1. **Gunstone, F. D., Ismail, I. A., and Lie Ken Jie, M.,** *Chem. Phys. Lipids,* 1, 376, 1967.

Table 48
METHYL ESTERS OF FATTY ACIDS: OCTADECYNOATES, REPORTED AS EQUIVALENT CHAIN LENGTHS

	P1	P2	P3	P4	P5	P6	P7	P8	P9	P10	P11
Column packing											
Temperature (°C)	200	210	240	190	180	165	200	190	234	226	180
Gas	N₂	N₂	He	N₂	N₂	Ar	Ar	Ar	He	He	Ar
Flow rate (mℓ/min)	na	na	70	na	na	2	0.5	2.7	120	60	60
Column											
Length (cm)	5000	5000	250	5000	5000	6100	6100	4700	360	200	200
Diameter											
(cm I.D.)	0.05	0.05	0.3	0.05	0.05	0.025	0.025	0.025	0.3	0.3	0.4
(cm O.D.)	na	na	na	na	na	na	na	na	na	na	na
Form	WCOT	WCOT	Helix	WCOT	WCOT	WCOT	WCOT	WCOT	na	na	U-tube
Material	SS	SS	SS	SS	SS	SS	SS	SS	na	na	G
Detector	FID	FID	TC	FID	FID	Argon	Argon	Argon	TC	TC	Argon
Reference	1	2	3	1	2	4	4	4	3	3	3
Compound						ECL					
Methyl 2-octadecynoate	18.60	18.57	—	20.47	21.13	—	—	—	—	—	—
Methyl 3-octadecynoate	—	18.05	—	—	20.65	—	—	—	—	—	—
Methyl 4-octadecynoate	17.91	17.88	—	19.02	19.97	—	—	—	—	—	—
Methyl 5-octadecynoate	17.87	17.86	—	19.07	19.90	19.85	19.58	18.88	—	—	—
Methyl 6-octadecynoate	17.85	17.81	17.84	19.16	19.98	19.94	19.60	18.96	20.03	20.33	20.40
Methyl 7-octadecynoate	17.84	17.80	17.82	19.20	20.03	—	—	—	20.04	20.35	20.45
Methyl 8-octadecynoate	17.83	17.80	17.80	19.22	20.02	19.93	19.74	18.99	20.04	20.40	20.45
Methyl 9-octadecynoate	17.83	17.78	17.70	19.23	20.02	19.98	19.78	19.03	20.12	20.44	20.48
Methyl 10-octadecynoate	17.87	17.82	—	19.25	20.09	20.03	19.86	19.09	—	—	—
Methyl 11-octadecynoate	—	17.86	17.87	19.28	20.17	20.10	19.91	19.14	20.17	20.53	20.60
Methyl 12-octadecynoate	—	17.93	—	19.39	20.25	20.20	20.02	19.19	—	—	—
Methyl 13-octadecynoate	—	17.99	—	—	20.33	—	—	—	—	—	—
Methyl 14-octadecynoate	—	18.01	—	—	20.42	—	—	—	—	—	—
Methyl 15-octadecynoate	—	18.12	—	—	20.64	20.65	—	19.42	—	—	—
Methyl 16-octadecynoate	—	18.50	—	—	21.46	—	—	—	—	—	—
Methyl 17-octadecynoate	—	17.98	—	—	21.16	—	—	—	—	—	—

Column packing P1 = Apiezon-L in form of wall coating in stainless steel tubing
P2 = Apiezon-L in form of wall coating in stainless steel tubing
P3 = 20% Apiezon-L on Gas-Chrom P (100-140 mesh)
P4 = Poly (neopenthyl glycol succinate) in form of wall coating in stainless steel tubing
P5 = Poly (diethylene glycol succinate) in form of wall coating in stainless steel tubing
P6 = Poly (diethylene glycol succinate; plus Arquad 2HT-75) in form of wall coating in stainless steel tubing
P7 = GE 238-149-99 (50% cyanoethyl, with methyl, silicone) in form of wall coating in stainless steel tubing
P8 = Six-ring polyphenyl ether (plus Arquad 2 HT-75) in form of wall coating in stainless steel tubing
P9 = 20% β-cyclodextrin acetate on Gas-Chrom P (30-60 mesh)
P10 = 20% Poly (diethylene glycol succinate)n on Gas-Chrom P (80-100 mesh)
P11 = 20% Poly (ethylene glycol succinate) on Gas-Chrom P (80-100 mesh)

REFERENCES

1. **Gunstone, F. D., Ismail, I. A., and Lie Ken Jie, M.**, *Chem. Phys. Lipids*, 1, 376, 1967.
2. **Barve, J. A., Gunstone, F. D., Jacobsberg, F. R., and Winlow, P.**, *Chem. Phys. Lipids*, 8, 117, 1972.
3. **Hofstetter, H. H., Sen, N., and Holman, R. T.**, *J. Am. Oil Chem. Soc.*, 42, 537, 1965.
4. **Scholfield, C. R. and Dutton, H. J.**, *J. Am. Oil Chem. Soc.*, 47, 1, 1970.

Table 49

METHYL ESTERS OF FATTY ACIDS: ISOMERIC METHYLENE-INTERRUPTED OCTADECADIYNOATES, REPORTED AS EQUIVALENT CHAIN LENGTHS

	P1	P2	P3	P4	P5	P6	P7	P8	P9	P10
Column packing										
Temperature (°C)	220	200	205	190	205	205	190	190	190	190
Gas	N_2	N_2	N_2	N_2	N_2	N_2	N_2	N_2	N_2	N_2
Flow rate (mℓ/min)	150	50	50	125	50	50	150	50	60	60
Column										
Length (cm)	200	200	200	200	200	200	200	200	200	200
Diameter										
(cm I.D.)	3.1	3.1	3.1	6.2	3.1	3.1	6.2	6.2	3.1	na
(cm O.D.)	na	na	na	na	na	na	na	na	na	na
Form	na	na	na	na	na	na	na	na	na	na
Material	na	na	na	na	na	na	na	na	na	na
Detector	FID	FID	FID	FID	FID	FID	FID	FID	FID	FID
Reference	1	1	1	1	1	1	1	1	1	1
Compound					ECL					
Methyl 2,6-octadecadiynoate	18.40	18.61	18.63	20.55	—	22.09	22.62	23.18	23.38	23.92
Methyl 3,7-octadecadiynoate	18.16	18.35	18.38	19.90	—	—	22.14	—	22.61	—
Methyl 4,8-octadecadiynoate	17.90	18.30	18.27	19.52	21.13	21.22	21.59	22.37	21.93	22.44
Methyl 5,9-octadecadiynoate	17.87	18.27	18.25	19.60	21.10	21.19	21.64	22.40	22.06	22.62
Methyl 6,10-octadecadiynoate	17.90	18.26	18.25	19.67	21.19	21.28	21.80	22.59	22.29	22.79
Methyl 7,11-octadecadiynoate	17.93	18.27	18.29	19.80	21.21	21.31	21.87	22.63	22.45	22.95
Methyl 8,12-octadecadiynoate	17.96	18.32	18.32	19.83	21.31	21.36	21.99	22.72	22.53	23.05
Methyl 9,13-octadecadiynoate	18.03	18.33	18.37	19.90	21.39	21.47	22.03	22.80	22.63	23.14
Methyl 10,14-octadecadiynoate	18.05	18.38	18.40	20.01	21.44	21.54	22.13	22.86	22.72	23.24
Methyl 11,15-octadecadiynoate	18.14	18.44	18.48	20.09	21.64	21.79	22.37	23.05	22.96	23.43
Methyl 12,16-octadecadiynoate	18.51	18.82	18.82	20.61	22.28	21.47	23.19	23.98	23.70	24.29
Methyl 13,17-octadecadiynoate	17.96	18.20	18.21	20.51	21.86	22.01	22.81	23.59	23.31	23.95

Column packing
P1 = 5% Apiezon-L on Chromosorb W (80-100 mesh)
P2 = 1.5% OV-101 (methyl silicone) on Chromosorb W (80-100 mesh)
P3 = 3% SE-30 (methyl silicone) on Chromosorb W (80-100 mesh)
P4 = 20% XE-60 (25% cyanoethyl, with methyl, silicone) on Chromosorb W (80-100 mesh)
P5 = 10% Carbowax 20M on Chromosorb W (80-100 mesh)
P6 = 10% FFAP (Carbowax 20M modified with 2-nitroterephthalic acid) on Chromosorb W (80-100 mesh)
P7 = 10% Poly (diethylene glycol adipate) on Chromosorb W (80-100 mesh)
P8 = 20% Poly (diethylene glycol succinate) on Chromosorb W (80-100 mesh)
P9 = 10% SILAR-10C (100% cyanopropyl siloxane, aged) on Chromosorb W (80-100 mesh)
P10 = 10% SILAR-10C (100% cyanopropyl siloxane, new) on Chromosorb W (80-100 mesh)

REFERENCE

1. **Lam, C. H. and Lie Ken Jie, M. S. F.,** *J. Chromatogr.,* 115, 559, 1975.

Table 50
METHYL ESTERS OF FATTY ACIDS: ISOMERIC OCTADECADIYNOATES, REPORTED AS EQUIVALENT CHAIN LENGTHS

Column packing	P1	P2	P3	P4	P5	P6	P7	P8
Temperature (°C)	190	190	210	190	190	190	190	190
Gas	N_2	N_2	N_2	N_2	N_2	N_2	N_2	N_2
Flow rate (mℓ/min)	na	95	na	na	na	na	115	50
Column								
Length (cm)	153	200	5000	5000	153	5000	200	200
Diameter								
(cm I.D.)	na	na	0.05	0.05	na	0.05	na	na
(cm O.D.)	0.63	0.64	na	na	0.63	na	0.64	0.64
Form	Helix	Helix	WCOT	WCOT	Helix	WCOT	Helix	Helix
Material	na	G	SS	SS	na	SS	G	G
Detector	FID	FID	FID	FID	FID	FID	FID	FID
Reference	1	2	1	1	1	1	2	2

Compound				ECL				
Methyl 5,12-octadecadiynoate	17.90	—	17.85	19.28	22.76	22.29	—	—
Methyl 6,12-octadecadiynoate	17.89	—	17.83	19.27	22.84	22.35	—	—
Methyl 7,12-octadecadiynoate	18.00	—	17.89	19.33	22.88	22.40	—	—
Methyl 8,12-octadecadiynoate	17.93	—	17.90	19.39	22.93	22.43	—	—
Methyl 9,12-octadecadiynoate	18.34	—	18.23	19.76	23.91	23.39	—	—
Methyl 10,12-octadecadiynoate	19.63	—	19.60	21.58	25.56	24.94	—	—
Methyl 6,8-octadecadiynoate	19.72	—	19.61	21.57	25.53	25.02	—	—
Methyl 6,9-octadecadiynoate	18.19	—	18.18	19.72	23.78	23.21	—	—
Methyl 6,10-octadecadiynoate	17.95	—	17.85	19.24	22.75	22.27	—	—
Methyl 6,11-octadecadiynoate	17.95	17.93	17.88	19.29	22.84	22.32	22.90	22.86
Methyl 7,15-octadecadiynoate	18.02	—	17.94	19.51	23.25	22.72	—	—
Methyl 8,15-octadecadiynoate	17.97	—	17.95	19.57	23.34	22.82	—	—
Methyl 9,15-octadecadiynoate	18.03	—	17.95	19.49	23.27	22.77	—	—
Methyl 2,7-octadecadiynoate	—	18.44	—	—	—	—	23.48	24.05
Methyl 3,8-octadecadiynoate	—	18.07	—	—	—	—	23.40	23.12
Methyl 4,9-octadecadiynoate	—	17.89	—	—	—	—	22.68	22.69
Methyl 5,10-octadecadiynoate	—	17.42	—	—	—	—	22.89	22.87

Column packing P1 = 5% Apiezon-L on Chromosorb W, HMDS (80-100 mesh)
P2 = 5% Apiezon-L on Chromosorb W (80-100 mesh)
P3 = Apiezon-L in form of wall coating in stainless steel tubing
P4 = XE-60 (25% cyanoethyl, with methyl, siloxane) in form of wall coating in stainless steel tubing
P5 = 20% Poly (diethylene glycol succinate) on Chromosorb W HMDS (80-100 mesh)
P6 = Poly (diethylene glycol succinate) in form of wall coating in stainless steel tubing
P7 = 20% Poly (diethylene glycol succinate) on Chromosorb W (80-100 mesh)
P8 = 10% SILAR-10C (100% cyanopropyl siloxane) on Chromosorb W (80-100 mesh)

REFERENCES

1. **Gunstone, F. D. and Lie Ken Jie, M.,** *J. Chromatogr.,* 4, 131, 1970.
2. **Lie Ken Jie, M. S. F.,** *J. Chromatogr.,* 109, 81, 1975.

Table 51
METHYL ESTERS OF FATTY ACIDS: ISOMERIC DIMETHYLENE-INTERRUPTED OCTADEC-*cis*-ENYNOATES, REPORTED AS EQUIVALENT CHAIN LENGTHS

Column packing	P1	P2	P3	P4	P5
Temperature (°C)	190	190	190	190	190
Gas	N_2	N_2	N_2	N_2	N_2
Flow rate (mℓ/min)	50	50	70	50	60
Column					
Length (cm)	200	200	200	200	200
Diameter					
(cm I.D.)	0.31	0.31	0.62	0.31	0.62
(cm O.D.)	na	na	na	na	na
Form	na	na	na	na	na
Material	na	na	na	na	na
Detector	FID	FID	FID	FID	FID
Reference	1	1	1	1	1

Compound			ECL		
Methyl octadec-*cis*-6-ene-2-ynoate	—	20.68	20.29	21.47	22.31
Methyl octadec-*cis*-7-ene-3-ynoate	—	—	—	—	—
Methyl octadec-*cis*-8-ene-4-ynoate	19.78	19.85	20.23	20.46	20.68
Methyl octadec-*cis*-9-ene-5-ynoate	19.76	19.93	20.26	20.51	20.78
Methyl octadec-*cis*-10-ene-6-ynoate	19.75	19.95	20.34	20.51	21.00
Methyl octadec-*cis*-11-ene-7-ynoate	19.78	19.98	20.40	20.68	21.06
Methyl octadec-*cis*-12-ene-8-ynoate	19.92	20.02	20.39	20.73	21.08
Methyl octadec-*cis*-13-ene-9-ynoate	20.00	20.05	20.55	20.82	21.16
Methyl octadec-*cis*-14-ene-10-ynoate	20.17	20.24	20.55	20.92	21.34
Methyl octadec-*cis*-15-ene-11-ynoate	20.28	20.26	20.51	20.98	21.51
Methyl octadec-*cis*-16-ene-12-ynoate	20.56	20.59	21.01	21.58	21.70
Methyl octadec-*cis*-17-ene-13-ynoate	20.28	20.20	20.68	21.01	21.27
Methyl octadec-*cis*-2-ene-6-ynoate	19.72	19.82	19.97	20.18	20.34
Methyl octadec-*cis*-3-ene-7-ynoate	—	—	—	—	—
Methyl octadec-*cis*-4-ene-8-ynoate	19.78	19.85	20.23	20.46	20.68
Methyl octadec-*cis*-5-ene-9-ynoate	19.76	19.93	20.26	20.51	20.78
Methyl octadec-*cis*-6-ene-10-ynoate	19.75	19.95	20.34	20.51	21.00
Methyl octadec-*cis*-7-ene-11-ynoate	19.78	19.98	20.40	20.68	21.06
Methyl octadec-*cis*-8-ene-12-ynoate	19.92	20.02	20.39	20.73	21.08
Methyl octadec-*cis*-9-ene-13-ynoate	20.00	20.05	20.55	20.82	21.16
Methyl octadec-*cis*-10-ene-14-ynoate	20.17	20.24	20.68	20.92	21.34
Methyl octadec-*cis*-11-ene-15-ynoate	20.28	20.26	20.67	20.98	21.51
Methyl octadec-*cis*-12-ene-16-ynoate	21.12	21.55	21.64	22.19	22.28

Column packing P1 = 10% Carbowax 20M on Chromosorb W (80-100 mesh)
P2 = 10% FFAP (Carbowax 20M modified with 2-nitroterephthalic acid) on Chromosorb W (80-100 mesh)
P3 = 10% Poly (diethylene glycol adipate) on Chromosorb W (80-100 mesh)
P4 = 20% Poly (diethylene glycol succinate) on Chromosorb W (80-0 mesh)
P5 = 10% SILAR-10C (100% cyanopropyl siloxane) on Chromosorb W (80-100 mesh)

REFERENCE

1. **Lie Ken Jie, M. S. F.,** *J. Chromatogr.,* 131, 239, 1977.

Table 52
METHYL ESTERS OF FATTY ACIDS: VARIOUS COMBINATIONS OF *cis* AND *trans* ETHYLENIC UNSATURATION WITH ACETYLENIC UNSATURATION IN C_{18} AND C_{20} CHAINS, INCLUDING NATURAL PLANT FATTY ACIDS, REPORTED AS EQUIVALENT CHAIN LENGTHS

Column packing	P1	P2	P3	P4	P5	P6	P7	P8	P9	P10	P11	P12	P13	P14
Temperature (°C)	210	240	190	234	226	180	232	220	~180	~220	~180	180	180	170
Gas	N_2	He	N_2	He	He	Ar	He	He	He	He	He	N_2	N_2	N_2
Flow rate (mℓ/min)	na	70	na	120	60	60	na	na	na	na	na	5	5	5
Column Length (cm)	5000	250	5000	360	200	200	184	184	184	184	184	1520	1520	5000
Diameter (cm I.D.)	0.05	0.03	0.05	0.3	0.3	0.4	0.8	na	na	na	na	0.075	0.05	0.05
(cm O.D.)	na	na	na	na	na	na	na	0.318	0.318	0.318	0.318	na	na	na
Form	WCOT	Helix	WCOT	na	na	U-tube	Helix	Helix	Helix	Helix	Helix	SCOT	SCOT	WCOT
Material	SS	na	SS	na	na	G	AL	AL	AL	AL	AL	SS	SS	SS
Detector	FID	TC	FID	TC	TC	Argon	FID	FID	FID	FID	FID	FID	FID	FID
Reference	1	2	1	2	2	2	3	3	3	4	4	5	5	5
Compound									ECL					
Methyl octadec-*cis*-9-ene-6-ynoate	17.63	—	20.52	—	—	—	—	—	—	—	—	19.80	20.42	20.73
Methyl octadec-*trans*-12-ene-6-ynoate	17.66	—	20.54	—	—	—	—	—	—	—	—	—	—	—
Methyl octadec-*trans*-12-ene-7-ynoate	17.97	—	20.91	—	—	—	—	—	—	—	—	—	—	—
Methyl octadec-*cis*-12-ene-9-ynoate	18.00	—	20.97	—	—	—	—	—	—	—	—	—	—	—
Methyl octadec-*trans*-12-ene-9-ynoate	17.91	—	20.88	—	—	—	—	—	—	—	—	—	—	—
Methyl octadec-*cis*-9-ene-12-ynoate	18.03	17.90	21.19	20.97	21.48	21.23	—	—	—	—	—	—	—	—
Methyl octadec-*trans*-9-ene-12-ynoate	18.47	—	23.02	—	—	—	—	—	—	—	—	—	—	—
Methyl octadec-9,12-diynoate	—	18.30	—	23.75	24.10	23.74	—	—	—	—	—	—	—	—
Methyl octadec-6,9,12-triynoate	—	18.70	—	27.33	27.80	27.90	—	—	—	—	—	—	—	—
Methyl octadec-*cis,cis*-9,12-diene-6-ynoate	—	—	—	—	—	—	—	—	—	21.6	22.8	—	—	—
Methyl octadec-*cis,cis,cis*-9,12,15-triene-6-ynoate	—	—	—	—	—	—	19.0	20.5	23.5	—	—	—	—	—
Methyl icosa-7-ene-13-ynoate	—	19.50	—	22.63	23.10	22.75	—	—	—	—	—	20.09	20.87	21.31
Methyl icosa-13-ene-7-ynoate	—	19.50	—	22.63	23.10	22.75	—	—	—	—	—	20.65	21.68	22.17
Methyl icosa-7,13-diynoate	—	19.77	—	24.20	24.92	24.47	—	—	—	—	—	—	—	—

Compound	P1	P2	P3	P4	P5	P6	P7	P8	P9	P10	P11	P12	P13
Methyl icosa-*cis,cis*-11,14-diene-8-ynoate	—	—	—	—	—	—	—	—	23.7	24.8	22.02	22.81	23.21
For reference:-													
Methyl *cis,cis,cis*-9,12,15-octadecatrienoate	17.49	17.51	20.06	20.10	20.40	20.13	—	19.0	20.6	—	19.00	19.50	19.79

Column packing
P1 = Apiezon-L in form of wall coating in stainless steel tubing
P2 = 20% Apiezon-L on Gas-Chrom P (100-140 mesh)
P3 = Poly (diethylene glycol succinate) in form of wall coating in stainless steel tubing
P4 = 20% β-cyclodextrin acetate on Gas-Chrom P (30-60 mesh)
P5 = 20% Poly (diethylene glycol succinate) on Gas-Chrom P (80-100 mesh)
P6 = 20% Poly (ethylene glycol succinate) on Gas-Chrom P (80-100 mesh)
P7 = 15% Cycloheptaamylose valerate on Chromosorb W, HMDS (60-80 mesh)
P8 = 20% Cycloheptaamylose propionate on Chromosorb W (60-100 mesh)
P9 = 15% Poly (ethylene glycol succinate) on Chromosorb W (100-120 mesh)
P10 = 20% Poly (diethylene glycol succinate) on Chromosorb W (100-120 mesh)
P11 = 15% Poly (ethylene glycol succinate) on Chromosorb W (100-120 mesh)
P12 = EGSS-X (ethylene glycol succinate combined with methyl silicone, old) in form of wall coating in stainless steel tubing
P13 = EGSS-X (ethylene glycol succinate combined with methyl silicone, new) in form of wall coating in stainless steel tubing

REFERENCES

1. **Barve, J. A., Gunstone, F. D., Jacobsberg, F. R., and Winlow, P.,** *Chem. Phys. Lipids,* 8, 117, 1972.
2. **Hofstetter, H. H., Sen, N., and Holman, R. T.,** *J. Am. Oil Chem. Soc.,* 42, 537, 1965.
3. **Andersson, B., Anderson, W. H., Chipault, J. R., Ellison, E. C., Fenton, S. W., Gellermann, J. L., Hawkins, J. M., and Schlenk, H.,** *Lipids,* 9, 506, 1974.
4. **Anderson, W. H., Gellermann, J. L., and Schlenk, H.,** *Lipids,* 10, 501, 1975.
5. **Jamieson, G. R. and Reid, E. H.,** *J. Chromatogr.,* 128, 193, 1976.

Table 53
ETHYL ESTERS OF FATTY ACIDS: REPORTED AS EQUIVALENT CHAIN LENGTHS BASED ON METHYL ESTERS OF SATURATED STRAIGHT-CHAIN FATTY ACIDS

	P1	P2	P3	P4
Column packing				
Temperature (°C)	234	226	180	240
Gas	He	He	Ar	He
Flow rate (mℓ/min)	120	60	60	70
Column				
Length (cm)	360	200	200	250
Diameter				
(cm I.D.)	0.3	0.3	0.4	0.3
(cm O.D.)	na	na	na	na
Form	na	na	U-tube	Helix
Material	na	na	G	SS
Detector	TC	TC	Argon	TC
Reference	1	1	1	1

Compound	ECL			
Ethyl octadecanoate	18.30	18.23	18.36	18.62
Ethyl *cis*-9-octadecenoate	18.75	18.62	18.75	18.30
Ethyl *cis,cis*-9,12-octadecadienoate	19.60	19.65	19.70	18.13
Ethyl *cis,cis,cis*-9,12,15-octadecatrienoate	20.46	20.65	20.75	18.17
Ethyl *cis,cis,cis,cis*-5,8,11,14-icosatetraenoate	22.43	22.63	22.60	19.60
Ethyl *cis,cis,cis,cis,cis*-4,7,10,13,16-docosapentaenoate	24.97	25.04	25.18	21.24

Column packing P1 = 20% Poly β-cyclodextrin acetate on Gas-Chrom P (30-60 mesh)
    P2 = 20% Poly (diethylene glycol succinate) on Gas-Chrom P (80-100 mesh)
    P3 = 20% Poly (ethylene glycol succinate) on Gas-Chrom P (80-100 mesh)
    P4 = 20% Apiezon-L on Gas-Chrom P (100-140 mesh)

REFERENCE

1. **Hofstetter, H. H., Sen, N., and Holman, R. T.,** *J. Am. Oil Chem. Soc.,* 42, 537, 1965.

Table 54A
'FREE' FATTY ACIDS: ON COLUMNS WITH SPECIAL SUPPORTS OR WITH POLYGLYCOL-TYPE LIQUID PHASES, REPORTED AS RETENTION TIMES RELATIVE TO OCTADECANOIC ACID

Column packing	P1	P2	P3	P4	P5	P6	P7	P8	P9
Temperature (°C)	205	200	177	200	280	200	225	200	225
Gas	He	He	N_2	N_2	N_2	N_2	N_2	N_2	N_2
Flow rate (mℓ/min)	45	60	67	na	25	25	25	25	25
Column Length (cm)	153	173	38	160	180	92	92	92	92
Diameter (cm I.D.)	na	0.217	na	0.25	0.3	0.2	0.2	0.2	0.2
(cm O.D.)	na	0.318	0.64	na	na	na	na	na	na
Form	na	na	na	na	U-tube	U-tube	U-tube	U-tube	U-tube
Material	na	SS	G	G	G	G	G	G	G
Detector	TC	FID	FID	FID	FID	FID	FID	FID	FID
Reference	1	2	3	4	5	6	6	6	6

Compound	$r_{Octadecanoic\ acid}$								
Hexanoic acid	0.12	0.02	—	—	—	—	—	—	—
Octanoic acid	0.15	0.03	—	—	—	—	—	—	—
Decanoic acid	0.20	0.07	0.04	—	—	—	—	—	—
Dodecanoic acid	0.26	0.13	0.09	—	—	—	—	—	—
Tetradecanoic acid	0.38	0.25	0.20	—	0.39	0.27	0.32	0.26	0.30
Hexadecanoic acid	0.60	0.50	0.45	—	0.62	0.52	0.56	0.51	0.55
cis-9-Hexadecenoic acid	—	—	—	—	0.68	0.58	0.63	0.51	0.55
Octadecanoic acid	1.00	1.00	1.00	1.00	1.00	1.00	1.00	1.00	1.00
cis-9-Octadecenoic acid	1.10	1.00	—	1.09	1.07	1.09	1.10	1.07	1.08
cis,cis-9,12-Octadecadienoic acid	—	1.00	—	1.24	1.20	1.27	1.26	1.25	1.23
cis,cis,cis-9,12,15-Octadecatrienoic acid	—	—	—	1.54	1.42	1.58	1.53	1.54	1.49
Icosanoic acid	1.76	—	—	—	1.59	1.94	1.78	1.92	1.79

Column packing
P1 = 0.1% Poly (ethylene glycol succinate plus 0.05% H_3PO_4) on glass heads (80-100 mesh)
P2 = 3.0% Poly (1,4-butanediol succinate) on Teflon 6 powder
P3 = 5% Carbowax 20,000 plus 5% isophthalic acid on Embacel, HMDS (60-80 mesh)
P4 = 5-7% Carbowax 20M plus 1% H_3PO_4 on Chromosorb W (60-80 mesh)
P5 = 10% SP-1000 (Carbowax plus substituted terephthalic acid) on Chromosorb W, AW (100-120 mesh)
P6 = 10% FFAP (Carbowax 20M terminated with 2-nitroterephthalic acid) on Supelcoport (80-100 mesh)
P7 = 10% FFAP (Carbowax 20M terminated with 2-nitroterephthalic acid) on Supelcoport (80-100 mesh)
P8 = 10% 20M-TPA (Carbowax 20M terminated with terephthalic acid) on Supelcoport (80-100 mesh)
P9 = 10% 20M-TPA (Carbowax 20M terminated with terephthalic acid) on Supelcoport (80-100 mesh)

REFERENCES

1. **Metcalfe, L. D.**, *J. Gas Chromatogr.*, 1, 7, 1963.
2. **Kabot, F. J. and Ettre, L. S.**, *J. Gas Chromatogr.*, 1(9), 17, 1963.
3. **Clark, J. R. P. and Fredericks, K. M.**, *J. Gas Chromatogr.*, 5, 99, 1967.
4. **Hrivnak, J. and Palo, V.**, *J. Gas Chromatogr.*, 5, 325, 1967.
5. **Sampson, D. and Hensley, W. J.**, *Clin. Chim. Acta*, 61, 1, 1975.
6. **Ottenstein, D. M. and Supina, W. R.**, *J. Chromatogr.*, 91, 119, 1971.

Table 54B
'FREE' FATTY ACIDS: ON COLUMNS WITH POLYESTER-TYPE LIQUID PHASES, REPORTED AS RETENTION TIMES RELATIVE TO OCTADECANOIC ACID

	P1	P2	P3	P4	P5	P6	P7	P8	P9	P10	P11	P12	P13	P14	P15
Column packing															
Temperature (°C)	225	212	200	200	200	200	200	200	200	200	200	200	200	200	225
Gas	N$_2$	He	N$_2$	N$_2$	N$_2$	N$_2$	N$_2$	N$_2$	N$_2$	N$_2$	N$_2$	N$_2$	N$_2$	N$_2$	N$_2$
Flow rate (mℓ/min)	50	na	na	na	na	na	na	na	25	20	20	20	20	20	20
Column															
Length (cm)	122	183	160	160	160	160	160	160	100	92	92	92	92	92	92
Diameter															
(cm I.D.)	na	0.318	0.25	0.25	0.25	0.25	0.25	0.25	0.2	0.2	0.2	0.2	0.2	0.2	0.2
(cm O.D.)	na	na	na	na	na	na	na	na	na	na	na	na	na	na	na
Form	na	Helix	na	na	na	na	na	na	U-tube	U-tube	U-tube	U-tube	U-tube	U-tube	U-tube
Material	na	SS	G	G	G	G	G	G	G	G	G	G	G	G	G
Detector	TC	FID	FID	FID	FID	FID	FID	FID	FID	FID	FID	FID	FID	FID	FID
Reference	1	2	3	3	3	3	3	3	4	5	5	5	5	5	5
Compound									$r_{\text{Octadecanoic acid}}$						
Decanoic acid	0.12	—	—	—	—	—	—	—	—	—	—	—	—	—	—
Dodecanoic acid	0.20	—	—	—	—	—	—	—	—	—	—	—	—	—	—
Tetradecanoic acid	0.34	—	—	—	—	—	—	—	0.40	0.41	0.41	0.34	0.34	0.29	0.34
Hexadecanoic acid	0.58	0.60	—	—	—	—	—	—	0.62	0.64	0.63	0.59	0.59	0.54	0.58
cis-9-Hexadecenoic acid	—	—	—	—	—	—	—	—	0.78	0.80	0.77	0.71	0.70	0.62	0.66
Octadecanoic acid	1.00	1.00	1.00	1.00	1.00	1.00	1.00	1.00	1.00	1.00	1.00	1.00	1.00	1.00	1.00
cis-9-Octadecenoic acid	—	1.19	1.11	1.09	1.05	1.03	1.09	1.09	1.21	1.23	1.18	1.20	1.19	1.12	1.14
cis,cis-9,12-Octadecadienoic acid	—	1.51	1.38	1.33	1.21	1.16	1.30	1.30	1.60	1.60	1.52	1.49	1.49	1.35	1.35
cis,cis,cis-9,12,15-Octadecatrienoic acid	—	2.01	1.81	1.66	1.47	1.39	1.63	1.66	2.20	2.22	2.04	1.96	1.96	1.74	1.68
Icosanoic acid	—	—	—	—	—	—	—	—	1.73	1.60	1.52	1.71	1.70	1.87	1.71

Column packing
P1 = 20% Poly (diethyleneglycol succinate plus 3% H_3PO_4) on Chromosorb W (60-80 mesh)
P2 = 20% Poly (ethylene glycol succinate with formic acid in the carrier gas) on Chromosorb W (80-100 mesh)
P3 = 5-7% Poly (ethylene glycol succinate) on Chromosorb W (60-80 mesh)
P4 = 5-7% Poly (1,4-butanediol succinate) on Chromosorb W (60-80 mesh)
P5 = 5-7% Poly (neopentyl glycol succinate) on Chromosorb W (60-80 mesh)
P6 = 5-7% Poly (cyclohexane dimethanol succinate) on Chromosorb W (60-80 mesh)
P7 = 5-7% Poly (ethylene glycol adipate) on Chromosorb W (60-80 mesh)
P8 = 5-7% Reoplex 400 (commercial poly propylene glycol adipate on Chromosorb W (60-80 mesh)
P9 = 10% SP-216-PS (proprietary polyester with H_3PO_4) on Supelcoport (100-120 mesh)
P10 = 10% SP-216-PS (proprietary polyester with H_3PO_4) on Supelcoport (80-100 mesh)
P11 = 5% SP-216-PS (proprietary polyester with H_3PO_4) on Supelcoport (80-100 mesh)
P12 = 10% Poly (diethylene glycol succinate plus H_3PO_4) on Supelcoport (80-100 mesh)
P13 = 5% Poly (diethylene glycol succinate plus H_3PO_4) on Supelcoport (80-100 mesh)
P14 = 10% Poly (diethylene glycol succinate plus H_3PO_4) on Supelcoport (80-100 mesh)
P15 = 10% Poly (diethylene glycol succinate plus H_3PO_4) on Supelcoport (80-100 mesh)

REFERENCES

1. **Metcalfe, L. D.,** *J. Gas Chromatogr.,* 1, 7, 1963.
2. **Ackman, R. G., Burgher, R. D., and Sipos, J. C.,** *Nature (London),* 200, 777, 1963.
3. **Hrivnak, J. and Palo, V.,** *J. Gas Chromatogr.,* 5, 325, 1967.
4. **Sampson, D. and Hensley, W. J.,** *Clin. Chim. Acta,* 61, 1, 1975.
5. **Ottenstein, D. M. and Supina, W. R.,** *J. Chromatogr.,* 91, 119, 1974.

REFERENCES

1. **Ackman, R. G. and Castell, J. D.**, *J. Gas Chromatogr.*, 5, 489, 1967.
2. **Ackman, R. G. and Eaton, C. A.**, *Fette Seifen Anstrichm.*, 80, 21, 1978.
3. **Jamieson, G. R.**, in *Topics in Lipid Chemistry*, Vol. 1, Gunstone, F. D., Ed., Logos Press, London, 1970, 107.
4. **Ackman, R. G.**, *J. Gas Chromatogr.*, 3, 15, 1965.
5. **Ackman, R. G. and Hooper, S. N.**, *J. Chromatogr. Sci.*, 7, 549, 1969.
6. **Ackman, R. G.**, *Methods Enzymol.*, 14, 329, 1969.
7. **Ackman, R. G.**, in *Progress in the Chemistry of Fats and Other Lipids*, Vol. 12, Holman, R. T., Ed., Pergamon Press, New York, 1972, 165.
8. **Woodford, F. P. and Van Gent, C. M.**, *J. Lipid Res.*, 1, 188, 1960.
9. **Miwa, T. K., Mikolajczyk, K. L., Earle, F. R., and Wolff, I. A.**, *Anal. Chem.*, 32, 1739, 1960.
10. **Miwa, T. K.**, *J. Am. Oil Chem. Soc.*, 40, 309, 1963.
11. **Ettre, L. S.**, *Chromatographia*, 6, 489, 1973.
12. **Jamieson, G. R.**, *J. Chromatogr. Sci.*, 13, 491, 1975.
13. **Nelson, G. J.**, *Lipids*, 9, 254, 1974.
14. **Ackman, R. G.**, *J. Am. Oil Chem. Soc.*, 40, 558, 1963.
15. **Ackman, R. G.**, *J. Gas Chromatogr.*, 1, 11, 1963.
16. **Ackman, R. G.**, *Lipids*, 2, 502, 1967.
17. **Jamieson, G. R. and Reid, E. H.**, *J. Chromatogr.*, 42, 304, 1969.
18. **Orr, C. H. and Callen, J. E.**, *J. Am. Chem. Soc.*, 80, 249, 1958.
19. **Lipsky, S. R. and Landowne, R. A.**, *Biochim. Biophys. Acta*, 27, 666, 1958.
20. **Lipsky, S. R., Landowne, R. A., and Godet, M. R.**, *Biochim. Biophys. Acta*, 31, 336, 1959.
21. **Craig, B. M.**, in *Gas Chromatography, Proc., 3rd Int. Symp. Michigan State*, Brenner, N., Callen, J. E., and Weiss, M. G., Eds., Academic Press, New York, 1962, 37.
22. **Seher, A. and Josephs, P.**, *Fette Seifen Anstrichm.*, 71, 749, 1969.
23. **Haken, J. K.**, *J. Chromatogr. Sci.*, 13, 430, 1975.
24. **Haken, J. K.**, *J. Chromatogr.*, 73, 419, 1972.
25. **Haken, J. K.**, *J. Chromatogr.*, 141, 247, 1977.
26. **Ackman, R. G. and Burgher, R. D.**, *J. Am. Oil Chem. Soc.*, 42, 38, 1965.
27. **Ackman, R. G., Manzer, A., and Joseph, J.**, *Chromatographia*, 7, 107, 1974.
28. **Beroza, M. and Hornstein, I.**, *J. Agric. Food Chem.*, 21, 7A, 1973.

Chapter 7

BRANCHED-CHAIN FATTY ACIDS

D. R. Body

I. INTRODUCTION

One of the sources that provides a wide range of branched-chain fatty acids is bacterial lipids.[1] In the *Bacillus* species,[2] over 90% of the fatty acid composition is made up of branched-chain fatty acids, so little treatment is necessary to prepare a concentrate of branched-chain fatty acids for chromatographic analysis. These are predominantly of the *iso* and *anteiso*-monomethyl branched-chain structure.

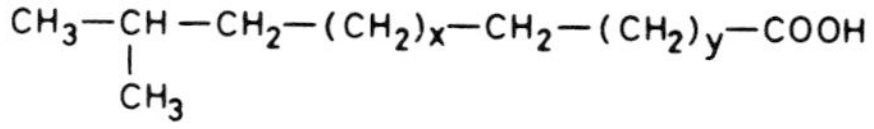

$$CH_3-CH-CH_2-(CH_2)_x-CH_2-(CH_2)_y-COOH$$
$$|$$
$$CH_3$$

iso Fatty acids

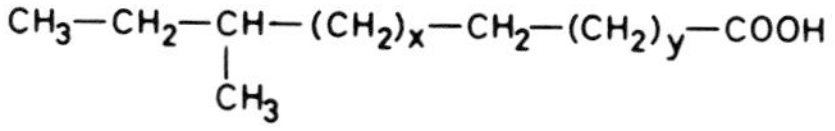

$$CH_3-CH_2-CH-(CH_2)_x-CH_2-(CH_2)_y-COOH$$
$$|$$
$$CH_3$$

anteiso Fatty acids

$$CH_3-CH_2-CH_2-(CH_2)_x-CH-(CH_2)_y-COOH$$
$$|$$
$$CH_3$$

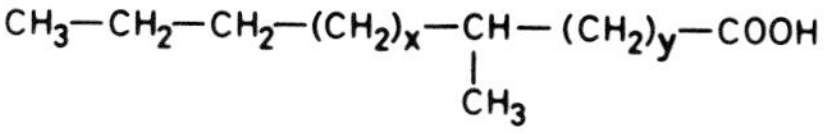

other monomethyl-series

Branched-chain fatty acids are also major constituents of the wax esters isolated by silicic acid liquid chromatography (LC) of the uropygial gland secretion matter from many birds.[3-8] These are not of the *iso-* or *anteiso*-branched structure, but consist of both other mono- and multibranched-chain fatty acids. The mono-branched substitute may be either methyl- or ethyl-, according to the species under investigation.[4-8] Furthermore, the principal mono-branched-chain fatty acid series commences from the 2-position of the molecule.[4-8]

Other groups of branched-chain isomers include dimethyl- or ethyl-, and methyl-di-branched fatty acids.[4-8] These substituents are generally situated in even positions, and these are often spaced by more than one methylene group.[9] Polybranched-chain fatty acids (either tri- or tetrabranched-chain) are also present.[4-8] Series of these branched-chain fatty acids may provide over 90% of the fatty acid composition of the free-flowing preen gland waxes, and these may be directly analyzed chromatographically.

Branched-chain fatty acid isomers of mono- (including saturated and unsaturated *iso-* and *anteiso-*),[10] di-, and tribranched fatty acids[11] may make up as much as 16% of the total fatty acid composition of human skin surface waxes.[12] This group of *iso-* and *anteiso-* derivatives may be extended up to the range of C_{26} and C_{31},[10] and these ranges also form 21 and 11% of the total fatty acids detected in the fecal lipid extracts from rats and sheep, respectively.[13,14]

Total lipid extracts from sheep adipose tissues can offer fatty acid compositions that contain only between 1 to 5% *iso-* and *anteiso*-branched-chain isomers.[15-18] Other mono-methyl-derivatives are also present, but these are naturally at low levels. This overall concentration of branched-chain fatty acids may be increased to 20% after the sheep are experimentally fed with rich barley-grain diets.[19] Nevertheless, when only low levels of branched-chain fatty acids are provided, there are a number of procedures that may be applied to obtain sufficient enriched branched-chain fatty acid fractions.

Initially, the total lipid extracts can be readily subdivided into their main specific lipid classes by means of well-established silicic acid LC[20] or TLC[21,22] techniques, i.e., sterol esters, triacylglycerols, fatty acids, and phospholipids. The esterified fatty acids from any particular group under investigation can be released and converted into their methyl esters as described in Chapter 2. These fatty acid methyl esters can be worked up to provide fractions suitably enriched with branched-chain components for further analysis.

In some cases, these branched-chain fatty acids are mainly associated with a single lipid constituent. A good example of this was demonstrated by Baron and Blough[23] who showed that 68% of the constituent fatty acid of the major cholesteryl esters isolated from bovine meibomian secretions were *iso-* and *anteiso-*branched-chain isomers (C_{15} to C_{27}).

Many operations are available to concentrate branched-chain fatty acid methyl esters, but, individually, these have been modified to cope with groups of branched-chain isomers according to their carbon chain lengths as described below.

Throughout, the information presented does not necessarily represent the original work in this field. However, the authors cited supply suitable chromatographic data that enabled the preparation of the tabulated chromatographic properties of branched-chain fatty acids in this chapter.

II. VOLATILE SHORT-LENGTH BRANCHED-CHAIN FATTY ACIDS (C_2 to C_5)

This range of branched-chain fatty acids can be directly prepared from biological materials by steam distillation associated with a micro-Kjeldahl apparatus.[24,25] The distillate residue, made slightly alkaline (soaps of volatile fatty acid), can be concentrated by evaporation under reduced pressure. These soaps can be directly injected on to the gas chromatography (GC) column when formic acid is incorporated with the carrier gas.[26,27] However, if formic acid is not included, the volatile fatty acid soaps must be acidified with metaphosphoric acid,[28] H_3PO_4,[29] dilute HCl,[30,31] or formic acid[30,32] before being applied to the GC column.

A. Gas Chromatography (GC)

One of the difficulties that has arisen from the GC analysis of volatile fatty acids is the ghosting phenomenon. In this case, a portion of the previously loaded sample has remained on the surface of the packed column and this is released by the application of the next sample. This problem has been examined by Geddes and Gilmour[33] and van Eenaeme et al.[34,35] However, it could be satisfactorily overcome if the sample is injected directly into the column packing, rather than using the alternative flash evaporation in the preheated vapor phase region ahead of the column.[36] This is essential when analyses of radioactive (^{14}C) volatile fatty acids mixtures are required.[27] Finally, it is particularly important to calibrate the response of the GC detector for the proportions of these individual standard volatile fatty acids, because these can vary in the C_2 to C_5 range. However, when concerned only with the common branched-chain isomers, namely 2-methylpropanoic (*iso*butyric) and 3-methylbutanoic (*iso*valeric) acids, their responses are similar.

To fully resolve a mixture of branched-chain fatty acids there is a choice of liquid phases and conditions available. Hrivnak et al.[37] (Table 1*) used three different liquid phases with

---

*   All tables appear at the end of text (p. 247).

capillary column GC and found that Ucon LB-500-X was the most suitable agent to separate a full range (C_2 to C_5) of branched-chain isomers.

When using packed GC columns, the common branched-chain isomers (*iso*butyric and *iso*valeric) can be separated from their *n*-counterparts by various liquid phases. The degree of such separations achieved with various available liquid phases has been compared by Nikelly,[38] Doelle,[39] and Ottenstein and Bartley.[40] Individually, many of these liquid phases have been successfully applied elsewhere, e.g., FFAP (Table 1),[27] PPGS (polypropylene glycol sebacate),[29] NPGA,[31,32] and Carbowax 20M.[41]

Individually, columns packed with porous polymer beads have been employed, and these are of some merit because of the short time involved in the analysis of branched-chain and *n*-fatty acid mixtures, e.g., Porapak N (Table 1),[26,30] Chromosorb 101 (Table 1),[26,42] Porapak QS,[28] Porapak Q,[43] and Phaspak Q.[44] Any undesirable adsorption of these volatile fatty acids with the column packings or poor detector response problems, such as peak tailing, can be reduced if the solid supports (or polymer beads) are treated with H_3PO_4 prior to the coating of the liquid phase, when applicable.[41,45]

When 2-methylbutanoic and 3-methylbutanoic acids are present, they cannot be resolved using GC columns packed with either liquid phases on solid supports of diatomaceous earth or porous polymer beads. However, if graphitized carbon blacks are included as the solid support with liquid phases such as Carbowax 20M after H_3PO_4 treatment, both isomers can be resolved.[46]

Esterified forms of these volatile branched-chain fatty acids can be analyzed by GC. By condensing them with longer chain-length alcohols, any evaporation loss of the short chain-length acid materials is reduced, and any former precautions are not so necessary. Benzyl esters of branched-chain fatty acids are analyzed using packed GC columns with different liquid phases, e.g., EGSS-X,[47] SE-30,[47] or E-301.[48] Butylated volatile fatty acid esters can also be used. In this case, different liquid phases with packed columns are suitable, i.e., Dexsil 300 GC,[49,50] or OV-1.[51]

Under the latter conditions, the GC instrument is coupled with a mass spectrometer (GC/MS) and the mass fragments produced can provide sufficient information to identify the branched-chain fatty acid derivatives under investigation.

B. Thin-Layer Chromatography (TLC)

Partition TLC has been used preparatively to analyze mixtures of short chain-length (C_1 to C_9) *n*-fatty acids. For example, silica gel coated plates are impregnated with 10% ethylene glycol, and these are developed twice with petroleum hydrocarbon — acetone (2:1 v/v) saturated with ethylene glycol.[52] The fractions are detected with a methyl red reagent. Alternatively, silanized silica gel HF 254 can be used to separate short chain length fatty acids by development with either dioxane-water-formic acid (60:35:5 v/v) or methanol-water (40:60 v/v).[53] The spots are detected by spraying the plates with a saturated methyl red solution in 0.2% methanolic NaOH. If branched-chain isomers of fatty acids within the carbon chain length (C_2 to C_5) are chromatographed under identical TLC conditions, their chromatographic properties will be similar to those of the equivalent *n*-fatty acids as reported above. In both cases, the distance of migration corresponds directly to the basic carbon chain length of the substances applied.

III. VOLATILE MEDIUM-LENGTH BRANCHED-CHAIN FATTY ACIDS (C_6 to C_{10})

The medium-length branched-chain fatty acids are often associated with flavor chemistry. Sheep meat lipid constituents have been the subject of recent study.[54] By the application of a continuous steam distillation diethyl ether extraction process on a 2-g aliquot of saponified lipid extracts, a sufficient concentration of volatile medium branched-chain fatty acids is provided.[55] These fatty acids can be converted into their methyl esters by standard methylation methods prior to GC analysis. However, they can also be directly esterified by silver oxide-methyl iodide reactions as described by Johnson and Wong.[56] Submicro levels of fatty acid mixtures are injected into stainless steel tubes (15.0 × 0.32 cm O.D.) filled with silver oxide powder which have been designed as suitable accessories to the GC equipment. Esterification of these trapped fatty acids is accomplished by injecting 1 $\mu\ell$ of methyl iodide and carrying out the reaction at 100°C for 2 min. After cooling, the reaction tube is connected to the injection port of the chromatograph, and the carrier gas elutes methyl esters into the GC column for analysis.

To reduce the time involved for the GC analysis of the C_6 to C_{10} range of methyl esters included with the steam distillates, a back-flushing device is introduced to the GC instrument.[55] This is done to speedily remove from the GC column a large quantity of nonvolatile fatty acids (up to C_{18}) that is carried over during the steam distillation.

For the analysis of monomethyl branched-chain fatty acids, packed GC columns with either polar (stabilized EGA)[54] or nonpolar (OV-101)[8,55] liquid phases operated under programmed temperature conditions (Table 2) are found to be suitable. With the instrument coupled with a mass spectrometer (GC/MS), the identification of the components is accomplished by examining their mass spectral fragments. However, only groups of dimethyl branched-isomers of the same carbon chain length can be resolved.[54]

Compared with any other chromatographic technique, i.e., LC, TLC, etc. GC/MS[8,54-56] is the most suitable tool to analyze the low levels (5 ppm) of medium-length branched-chain fatty acids present in normal animal lipid extracts.

IV. NONVOLATILE BRANCHED CHAIN-FATTY ACIDS (C_{11} AND ABOVE)

Branched-chain fatty acids (C_{11} and above) are extracted together with the nonvolatile *n*-fatty acids that constitute a major proportion of general lipid extracts. However, the quantity of branched-chain fatty acids is often limited and must be concentrated in order to enable their identification by chromatographic techniques.

Originally, fractional distillation and low temperature crystallizations of the methyl esters recovered from acids of the constituent fatty ruminant lipid extracts yielded *iso-* and *anteiso-* odd-numbered and *iso*-even-numbered fatty acids between C_{13} and C_{18} isomers.[57] Some later fractions, containing other monomethyl branched-chain fatty acids, were thoroughly examined by GC, and it was found that, a part from even–numbered terminal *iso*—derivatives, natural monomethyl branched-chain fatty acids have only their methyl branches substituted on even-numbered carbon sites from the carboxyl group.[58]

More recently, other forms of fractionation have become available to yield concentrates of branched-chain fatty acids and methyl esters. Saturated branched-chain fatty acids are associated with the *n*-saturated derivatives that may be enriched by either silver nitrate impregnated TLC[59-61] or LC.[10,62-65] Similarly, mercuric acetate complexes can be applied and the resulting products resolved by TLC.[66-69] In both approaches, the branched-chain derivatives are concentrated with the *n*-saturated fatty acids (which may be readily crystallized out of solution by complexing with urea) from branched-chain derivatives that are retained in the filtrate.[10,69-72] Urea has also been used in TLC or LC for the purpose of enriching branched-chain fatty acid fractions.[73]

A. Gas Chromatography

The GC properties of the methyl esters of many monobranched-chain fatty acids on either polar or nonpolar liquid phases are recorded in Tables 3 and 4, respectively. Whenever possible these findings are expressed as equivalent carbon-length (ECL) values,[74,75] and these may be compared with the properties of the methyl esters of the *n*-saturated fatty acids chromatographed under identical conditions.

Table 5 illustrates a range of GC properties of various homologous series of multibranched-chain methyl esters. These series are not complete but, from these recorded ECL values the approximate ECL values of the missing members can be predicted:

4,8	-	Dimethyltridecanoate	13.90
4,8	-	Dimethyltetradecanoate	14.85
4,8	-	Dimethylpentadecanoate	15.82
4,8	-	Dimethyloctadecanoate	18.76

Therefore, it is quite likely that the ECL values for 4,8-dimethylhexadecanoate would be 16.8 and for 4,8-dimethylheptadecanoate 17.8.

The methyl esters of monoenoic branched-chain fatty acids have also been analyzed by GC. According to the literature, some of their published ECL values have been reported on either polar or nonpolar liquid phases. These findings, together with appropriate references, are shown in Table 6.

Methyl esters of isoprenoid multibranched-chain fatty acids have constant spacings between their methyl substituents, and their GC properties are listed in Tables 7 and 8. In special circumstances, capillary GC can even specifically resolve isoprenoid multibranched-chain components of different diastereoisomers (Table 8).[76,77] These findings have been dealt with separately from other forms of branched-chain isomers because their origins are quite different.

Isoprenoid fatty acids are derived from the phytol moiety of chlorophyll,[78] whereas *iso-* and *anteiso*-branched-chain isomers are largely obtained from bacterial sources.[79] Similarly, the other mono- and multi-branched series may be considered as products from animal synthetic processes on exogenous materials.[4,10,71]

There is an extensive range of choice of GC conditions available to analyze mixtures of branched-chain fatty acids. For simple branched-chain fatty acid extracts, typical of bacterial origin, i.e., *iso-* and *anteiso*-derivatives, packed GC columns with polar liquid phases (Table 3) are adequate. On the other hand, although not completely suitable for defining *iso-* and *anteiso*-components, nonpolar packed columns can easily indicate the presence of other monomethyl branched-chain isomers (Table 4).

When complete series of branched-chain fatty acids have to be analyzed capillary GC columns coated with polar or nonpolar liquid phases are satisfactory (Tables 3 and 4).

The ECL values of the methyl esters of *iso-* and *anteiso*-branched-chain fatty acids are found to be similar for both packed and capillary GC conditions (see Tables 3 and 4). However, the other individual monomethyl branched-chain isomers are resolved better on capillary columns coated with either polar or nonpolar liquid phases.

Unfortunately, capillary GC columns do have some limitations. When operated at maximum efficiency, these columns can work only with a minimal load of injected sample. This reduces the amount of any minor constituents observed and these are not sufficient to be directly analyzed by a mass spectrometer (MS) connected to the GC equipment. This can be compensated for by using a packed glass column, i.e., 1000 cm × 0.28 cm I.D. with 3 to 5% OV-101 on Gas-Chrom Q (Table 4), for general purposes.[7]

Throughout, the operating temperatures can vary according to the composition of the branched-chain fatty acids under investigation. If the range represents an extensive homologous series, e.g., C_{12} to C_{26}, programmed temperatures are recommended. However, when interest is focused only on groups of branched-chain isomers within small ranges, e.g., C_{11} to C_{16}, C_{16} to C_{20}, or C_{20} to C_{26}, different isothermal operating conditions may be applied for each group.

Determination of mass spectra of individual branched-chain fatty acids observed is an important aspect of their identification. This may be readily accomplished by coupling the GC equipment directly with the MS. Lists of specific mass spectral fragments have been well tabulated elsewhere.[3-7,70,80-89]

B. Thin-Layer Chromatography

Reverse phase partition TLC has been used to separate methyl esters of fatty acids of different chain lengths and degrees of unsaturation.[66] The method has been modified to develop current high pressure column liquid chromatographic (HPLC) systems, and similar separations were achieved.[90-93] No single application of any of the above methods was found in the literature for measuring the chromatographic properties of branched-chain esters alone. However, these would probably follow the established procedures and, no doubt, branched-chain isomers could also be resolved according to their chain length and unsaturated structure. All details have been described in Chapter 6.

Standard TLC plates coated with silica gel G may be used to separate some of the fatty acids. Methyl esters of chain lengths greater than C_{19} are not as firmly bound to the silica gel as those of C_{18} and below.[78,94] This effect also occurs with lipid constituents on TLC plates, and their migration patterns are governed to some extent by the chain lengths of their esterified fatty acid moieties, e.g., in sphingomyelins,[95] triacylglycerols,[96,97] and wax esters.[23,98]

Specific evidence of this behavior with branched-chain fatty acids has been shown by Baron and Blough.[23] These authors found that the cholesteryl esters containing branched-chain fatty acids (16-methyloctadecanoic to 24-methylhexacosanoic acids) migrated further on the silica gel G plates after being developed with diethyl ether-benzene-ethanol-acetic acid (40:50:2:0.2) and diethyl ether-hexane (6:94,v/v)[99] than those with other acyl-groups between 12-methyltetradecanoic and 16-methylheptadecanoic acids.

Using argentation TLC, saturated and unsaturated *n*-fatty acids can be readily resolved.[59] However, with branched-chain fatty acids, it is clear (Table 10) that separations of saturated and unsaturated branched-chain fatty acids with the double bond adjacent to the methyl substitute can be achieved only if the plates are repeatedly developed with the appropriate solvent system.[61] This also applies to the isoprenoid fatty acid derivatives and to unsaturated isoprenoid hydrocarbons with either *cis* or *trans* configurations discussed elsewhere.[100] In all cases, this is caused by some steric hindrance applied to the double bond by the methyl branch that tends to inhibit the formation of $AgNO_3$-complexes. Therefore, resolutions between these saturated and unsaturated branched-chain fatty acids may be limited.

Table 1
METHYL ESTERS OF SATURATED BRANCHED-CHAIN CARBOXYLIC ACIDS (C$_2$-C$_5$) (GC)

	P1	P1	P2	P2	P3	P3	P4	P5	P6	P7
Column packing										
Temperature (°C)	120	140	110	130	125	145	130	140	160	200
Gas	N$_2$	N$_2$	N$_2$	N$_2$	N$_2$	N$_2$	N$_2$	N$_2$	He	He
Flow rate (ml/min)	a	a	a	a	a	a	16[b]	16[b]	97[b]	77[b]
Column										
Length (cm)	450	450	450	450	450	450	225	150	165	165
Diameter (cm I.D.)	0.025	0.025	0.025	0.025	0.025	0.025	0.25	0.4	0.317	0.317
(cm O.D.)										
Form	Capillary	Capillary	Capillary	Capillary	Capillary	Capillary	Coiled	Coiled	Coiled	Coiled
Material	SS	SS	SS	SS	SS	SS	Glass	Glass	SS	SS
Detector	FID	FID	FID	FID	FID	FID	FID	FID	FID	FID
Reference	1	1	1	1	1	1	2	2	3	3
Compound (Fatty Acids)	Retention time[c]						Retention times (min)			
Mono-branched										
2 - Methylpropanoic (*iso*butyric)	0.405	0.449	0.374	0.416	0.428	0.461	0.45	0.45	3.6	3.5
2 - Methylbutanoic	0.759	0.795	0.717	0.746	0.776	0.796	—	—	}7.2	}6.8
3 - Methylbutanoic (*iso*valeric)	0.697	0.736	0.694	0.725	0.742	0.763	0.72	0.72		
2 - Ethylbutanoic	1.286	1.274	1.267	1.241	1.308	1.280				
2 - Methylpentanoic	1.381	1.360	1.333	1.296	1.340	1.306				
3 - Methylpentanoic	1.431	1.426	1.449	1.407	1.427	1.386				
4 - Methylpentanoic	1.524	1.490	1.540	1.480	1.485	1.436				
Di-branched										
2,2 - Dimethylbutanoic	1.080	1.079	0.933	0.943	1.040	1.039				
3,3 - Dimethylbutanoic	1.215	1.212	1.605	1.539	1.542	1.489				
2,3 - Dimethylbutanoic	1.236	1.229	1.154	1.148	1.215	1.200				

Table 1 (continued)
METHYL ESTERS OF SATURATED BRANCHED-CHAIN CARBOXYLIC ACIDS (C₂-C₅) (GC)

Compound (Fatty Acids)	Retention time[c]					Retention times (min)
Tri-branched						
Trimethylacetic	0.519	0.555	0.441	0.479	0.521	0.552

a   Flow-rate not reported.
b   Formic acid vapor included with the carrier gas.
c   Relative to methyl *n*-pentanoate.

Column packing	P1 =	Trimer Acid (coating)
	P2 =	Tricresyl phosphate (coating)
	P3 =	Ucon LB-550-X (coating)
	P4 =	20% FFAP on Chromosorb W (60/70 mesh)
	P5 =	20% FFAP + 2% (v/w) H_3PO_4 on Chromosorb W (60/70 mesh)
	P6 =	Chromosorb 101
	P7 =	Porapak - N

REFERENCES

1. **Hrivnak, J., Sojak, L., Beska, E., and Janák, J.,** *J. Chromatogr.,* 68, 55, 1972.
2. **Kirk, R. D., Body, D. R., and Hawke, J. C.,** *J. Sci. Food. Agric.,* 22, 620, 1971.
3. **Ackman, R. G.,** *J. Chromatogr. Sci.,* 10, 560, 1972.

Table 2
METHYL ESTERS OF SATURATED BRANCHED-CHAIN CARBOXYLIC ACIDS (C_6-C_{10}) (GC)

Column packing	P1	P2	P3	P3
Temperature (°C)	75	50 - 200 at 2/min	50 - 200 at 2/min	160 - 240 at 1/min
Gas	He	N_2	N_2	N_2
Flow-rate (mℓ/min)	35	50	50	a
Column				
Length (cm)	609.6	250	250	900
Diameter				
(cm I.D.)				
(cm O.D.)	0.317	0.32	0.32	
Form	Coiled	Coiled	Coiled	Coiled
Material	SS	SS	SS	Glass
Detector	TC	FID	FID	FID
Reference	1	2	2	3

Compound (Methyl Esters of Fatty Acids)	Retention Time[b]		ECL[c]	
Mono-branched				
2 - Ethylbutanoate	1.14	—	—	—
2 - Methylpentanoate	1.20	—	—	—
3 - Methylpentanoate	1.45	—	—	—
4 - Methylpentanoate	1.54	—	—	—
2 - Methylhexanoate	—	—	6.42	—
4 - Methylhexanoate (*anteiso*)	—	6.70	6.71	—
2 - Methylheptanoate	—	—	7.40	—
6 - Methylheptanoate (*iso*)	—	7.55	7.63	—
2 - Methyloctanoate	—	—	8.42	—
4 - Methyloctanoate (Hircinoate)	—	8.52	8.58	—
6 - Methyloctanoate (*anteiso*)	—	8.66	8.69	—
4 - Methylnonanoate	—	9.51	9.61	—
8 - Methylnonanoate (*iso*)	—	9.57	9.65	—
2 - Methyldecanoate	—	—	10.39	—
8 - Methyldecanoate (*anteiso*)	—	10.65	10.70	—
9 - Methyldecanoate	—	10.52	10.66	—
10 - Methylundecanoate (*iso*)	—	11.56	11.65	—
Di-branched				
2,2 - Dimethylbutanoate	0.84	—	—	—
3,3 - Dimethylbutanoate	0.84	—	—	—
2,3 - Dimethylbutanoate	1.07	—	—	—
2 - Ethyl - 4 - methyloctanoate	—	—	—	9.70
2 - Ethyl - 6 - methyloctanoate	—	—	—	9.90
4,6 - Dimethyloctanoate	—	9.36	9.45	—
4,6 - Dimethylnonanoate	—	10.41	10.55	—
2 - Ethyl - 6 - methyldecanoate	—	—	—	11.60
4,8 - Dimethyldecanoate	—	}11.26	}11.40	—
6,8 - Dimethyldecanoate	—			—
4,6 - Dimethyldecanoate	—	11.44	11.54	—

a   Carrier gas flow-rate not reported.
b   Relative to methyl *n*-pentanoate.
c   Expressed in relationship with *n*-saturated-chain fatty acid methyl esters.

Column packing   P1 =   3% XE - 60 on Chromosorb W (80-100 mesh)
                  P2 =   10% Stabilized EGA on Gas-Chrom Q (100-120 mesh)
                  P3 =   10% OV - 101 on Gas-Chrom Q (100-120 mesh)

Table 2 (continued)
METHYL ESTERS OF SATURATED BRANCHED-CHAIN CARBOXYLIC ACIDS (C$_6$-C$_{10}$) (GC)

REFERENCES

1. **Sniegoski, P. J.,** *J. Chromatogr. Sci.,* 10, 644, 1972.
2. **Wong, E., Nixon, L. N., and Johnson, C. B.,** *J. Agric. Food Chem.,* 23, 495, 1975.
3. **Jacob, J.,** *J. Lipid Res.,* 19, 148, 1978.

Table 3
METHYL ESTERS OF SATURATED MONO-BRANCHED-CHAIN CARBOXYLIC ACIDS (C₁₁ AND ABOVE) (GC, USING POLAR LIQUID PHASES)

	P1	P2	P3	P4	P5	P6	P7	P8	P8	P9
Column packing										
Temperature (°C)	216	170	170	170	185	185	160—190 at 1/min	150	170	190
Gas		N_2	N_2	He	He	He	He	He	He	N_2
Flow rate (mℓ/min)		15	10	2—4	a	60	15—20	40 lbs/in²	40 lbs/in²	2
Column										
Length (cm)		210	400	600	225	304.8	1523	5000	10000	10000
Diameter (cm I.D.)		0.25	0.2	0.3	0.15	0.635	0.051	0.025	0.025	0.050
(cm O.D.)										
Form		Coiled		Coiled	Coiled	Coiled	Capillary	Capillary	Capillary	Capillary
Material		Glass	Glass	Copper	Aluminum	SS		SS		SS
Detector		FID	FID	FID	FID	TC	FID	FID	TC	FID
Reference	1—2	3—4	5	6—7	8—9	10	11	2,12	13	14
Compound **(Methyl Esters of Fatty Acids)**						ECL[b]				
Mono-branched										
4 - Methylundecanoate	—	—	—	—	—	—	—	—	—	—
9 - Methylundecanoate	—	—	—	—	—	—	—	11.71	11.41	—
10 - Methylundecanoate (*iso*)	—	—	—	11.62	—	—	—	11.58	—	—
3 - Methyldodecanoate	—	—	—	—	—	—	—	12.12	—	—
4 - Methyldodecanoate	—	—	—	—	—	—	—	12.42	12.40	—
10 - Methyldodecanoate (*anteiso*)	—	12.85	12.56	12.72	—	—	—	—	—	—
11 - Methyldodecanoate (*iso*)	—	12.70	—	12.58	—	—	—	—	—	—
4 - Methyltridecanoate	—	—	—	—	—	—	—	—	—	—
12 - Methyltridecanoate (*iso*)	—	13.66	13.43	13.60	—	—	—	—	13.39	—
11 - Ethyltridecanoate	—	—	—	14.68	—	—	14.68	—	—	13.53
2 - Methyltetradecanoate	—	—	—	—	—	—	—	—	13.80	—
4 - Methyltetradecanoate	—	—	—	—	—	—	—	14.40	14.37	—
5 - Methyltetradecanoate	—	—	—	—	—	—	—	14.32	—	—
6 - Methyltetradecanoate	—	14.35	—	—	—	—	—	14.34	14.30	—

Table 3 (continued)
METHYL ESTERS OF SATURATED MONO-BRANCHED-CHAIN CARBOXYLIC ACIDS (C$_{11}$ AND ABOVE) (GC, USING POLAR LIQUID PHASES)

Compound (Methyl Esters of Fatty Acids)	ECL[b]							
8 - Methyltetradecanoate	—	—	—	—	—	—	14.34	—
9 - Methyltetradecanoate	—	—	—	—	—	14.38	—	—
10 - Methyltetradecanoate	—	—	—	—	—	14.45	14.42	—
11 - Methyltetradecanoate	—	—	—	—	14.65	—	—	—
12 - Methyltetradecanoate (*anteiso*)	14.72	14.56	14.74	—	14.75	14.69	14.70	2
13 - Methyltetradecanoate (*iso*)	14.52	—	14.60	—	14.62	16.59	—	—
4 - Methylpentadecanoate	—	—	—	—	—	—	—	—
6 - Methylpentadecanoate	—	—	—	—	—	—	15.36	—
8 - Methylpentadecanoate	15.30	—	—	—	—	15.39	—	—
10 - Methylpentadecanoate	—	—	—	—	—	—	—	—
12 - Methylpentadecanoate	—	—	—	—	—	—	—	—
14 - Methylpentadecanoate (*iso*)	15.50	15.43	15.57	—	—	15.55	15.80	15.53
2 - Methylhexadecanoate	—	—	—	15.72	—	—	—	—
4 - Methylhexadecanoate	—	—	—	—	—	16.31	16.36	—
6 - Methylhexadecanoate	—	—	—	—	—	16.33	16.28	—
7 - Methylhexadecanoate	—	—	—	—	—	16.33	—	—
8 - Methylhexadecanoate	16.32	—	—	—	—	16.37	16.32	—
9 - Methylhexadecanoate	—	—	—	—	—	—	—	—
10 - Methylhexadecanoate	—	—	—	—	—	—	—	—
12 - Methylhexadecanoate	—	—	—	—	—	—	—	—
14 - Methylhexadecanoate (*anteiso*)	16.73	16.56	16.58	—	—	16.71	16.42	16.67
15 - Methylhexadecanoate (*iso*)	16.48	16.48	16.73	—	—	16.56	16.70	16.53
4 - Methylheptadecanoate	—	—	—	—	—	—	17.35	—
6 - Methylheptadecanoate	—	—	—	—	—	—	17.28	—
8 - Methylheptadecanoate	17.26 }	—	—	—	—	—	17.28	—
9 - Methylheptadecanoate	17.26 }	—	—	—	—	—	—	—
10 - Methylheptadecanoate	17.26 }	—	—	—	—	17.32	17.28	—
12 - Methylheptadecanoate	17.58 }	—	—	—	—	17.34	—	—
14 - Methylheptadecanoate	17.58 }	—	—	—	—	17.55	—	—

Compound	(1)	(2)	(3)	(4)	(5)	(6)	(7)
15 - Methylheptadecanoate (*anteiso*)	17.53	—	—	—	—	—	17.70
16 - Methylheptadecanoate (*iso*)	—	—	—	—	17.43	17.46	—
2 - Methyloctadecanoate	—	17.59	—	17.78	—	—	17.91
3 - Methyloctadecanoate	—	17.89	—	—	—	—	18.13
4 - Methyloctadecanoate	—	18.13	—	—	—	—	18.41
5 - Methyloctadecanoate	—	18.39	—	—	—	—	18.30
6 - Methyloctadecanoate	—	—	—	—	—	—	18.29
7 - Methyloctadecanoate	—	—	—	—	—	—	18.30
8 - Methyloctadecanoate	—	—	—	—	—	—	18.28
9 - Methyloctadecanoate	—	—	—	—	—	18.40	18.30
10 - Methyloctadecanoate	—	18.30	—	—	—	—	18.30
11 - Methyloctadecanoate	—	18.31	—	—	—	—	18.33
12 - Methyloctadecanoate	—	18.32	—	—	—	—	18.34
13 - Methyloctadecanoate	—	—	—	—	—	—	18.33
14 - Methyloctadecanoate	—	18.35	—	—	—	—	18.44
15 - Methyloctadecanoate	—	—	—	—	—	—	—
16 - Methyloctadecanoate (*anteiso*)	—	18.53	—	—	18.56	18.70	18.64
17 - Methyloctadecanoate (*iso*)	—	18.70	—	—	—	—	18.58
18 - Methylnonadecanoate (*iso*)	—	—	—	—	—	—	19.54
2 - Methyleicosanoate	—	—	—	19.85	—	—	—
5 - Methyleicosanoate	—	—	20.30[c]	—	—	—	—
10 - Methyleicosanoate	—	—	20.20	—	—	—	—
13 - Methyleicosanoate	—	—	20.30	—	—	—	—
17 - Methyleicosanoate	—	—	20.57	—	—	—	—
18 - Methyleicosanoate (*anteiso*)	—	—	20.77	—	—	—	20.73
19 - Methyleicosanoate (*iso*)	—	—	20.58	—	—	—	—
20 - Methylheneicosanoate (*iso*)	—	—	—	—	—	—	21.40
20 - Methyldocosanoate (*anteiso*)	—	—	—	—	—	—	22.75
22 - Methyltricosanoate (*iso*)	—	—	—	—	—	—	23.53
22 - Methyltetracosanoate (*anteiso*)	—	—	—	—	—	—	24.67
24 - Methylpentacosanoate (*iso*)	—	—	—	—	—	—	25.53
24 - Methylhexacosanoate (*anteiso*)	—	—	—	—	—	—	26.70
26 - Methylheptacosanoate (*iso*)	—	—	—	—	—	—	27.60
26 - Methyloctacosanoate (*anteiso*)	—	—	—	—	—	—	28.65
28 - Methylnonacosanoate (*iso*)	—	—	—	—	—	—	29.50
28 - Methyltricontanoate (*anteiso*)	—	—	—	—	—	—	30.70
30 - Methylhentriacontanoate (*iso*)	—	—	—	—	—	—	31.55

Table 3 (continued)
METHYL ESTERS OF SATURATED MONO-BRANCHED-CHAIN CARBOXYLIC ACIDS (C_{11} AND ABOVE) (GC, USING POLAR LIQUID PHASES)

a   Carrier gas flow-rate not reported.

b   Expressed in relationship with n - saturated-chain fatty acid methyl esters.

c   Calculated from the published retention time data relative to methyl n - eicosanoate.

Column packing   P1 =   Reoplex 400, no other details available

P2 =   10% EGSS-X on Gas-Chrom Z (60-70 mesh)

P3 =   12% Silar 10C on Chromosorb W HP (80-100 mesh)

P4 =   7% EGA on Anakrom ABS (60-80 mesh)

P5 =   20% EGA on Chromosorb W (100-120 mesh)

P6 =   15% EGS on Gas-Chrom P (80-100 mesh)

P7 =   EGA (coating)

P8 =   BDS (coating)

P9 =   FFAP (coating)

REFERENCES

1. **Abrahamsson, S., Ställberg-Stenhagen, S., and Stenhagen, E.,** *Progress in the Chemistry of Fats and Other Lipids,* Vol. 7, Part 1, Holman, R. T. and Malkin, T., Eds., Pergamon Press, Oxford, 1963, 1.
2. **Ackman, R. G.,** *Progress in the Chemistry of Fats and Other Lipids,* Vol. 12, Holman, R. T., Ed., Pergamon Press, Oxford, 1972, 167.
3. **Body, D. R. and Hansen, R. P.,** *J. Sci. Food Agric.,* 29, 107, 1978.
4. **Body, D.R.,** unpublished.
5. **Heckers, H., Dittmar, K., Melcher, F. W., and Kalinowski, H. O.,** *J. Chromatogr.,* 135, 93, 1977.
6. **Kaneda, T.,** *J. Bacteriol.,* 95, 2210, 1968.
7. **Kaneda, T.,** *Bacteriol. Rev.,* 41, 391, 1977.
8. **Gerson, T. and Schlenk, H.,** *Chem. Phys. Lipids,* 2, 213, 1968.
9. **Gerson, T.,** personal communication.
10. **Do, U.H. and Sprecher, H.,** *Chem. Phys. Lipids,* 16, 255, 1976.
11. **Kaneda, T.,** *Biochemistry,* 10, 340, 1971.
12. **Ackman, R. G.,** *J. Chromatogr. Sci.,* 10, 243, 1972.
13. **Duncan, W. R. H., Lough, A. K., Garton, G. A., and Brooks, P.,** *Lipids,* 9, 669, 1974.
14. **Flanzy, J., Boudon, M., Leger, C., and Pihet, J.,** *J. Chromatogr. Sci.,* 14, 17, 1976.

Table 4
METHYL ESTERS OF SATURATED MONO-BRANCHED-CHAIN CARBOXYLIC ACIDS (C_{11} AND ABOVE) (GC, USING NONPOLAR LIQUID PHASES)

Column packing	P1	P2	P3	P4	P4	P4	P5	P6	P7	P8
Temperature (°C)	244	170	240	150-270 at 2/min	200	160-240 at 1/min	130-260 at 5/min	120-240 at 2/min	190	170-210 at 0.5/min
Gas		N_2	N_2	He	He	N_2	He	He	He	He
Flow-rate (mℓ/min)		30	30	40	40	a	20	13	80 lbs/in²	8
Column Length (cm)		183	183	274	183	1000	150	183	5000	30480
Diameter (cm I.D.)		0.15	0.4	0.32	0.64	0.28	0.32	0.32	0.025	0.076
(cm O.D.)										
Form		Coiled	Coiled	Coiled	Coiled	Coiled	Coiled	Coiled	Capillary	Capillary
Material		SS	Glass	SS	SS	Glass	SS	SS	SS	SS
Detector		FID	FID	FID	FID	FID	FID	FID	FID	FID
References	1-4	5,6	5,6	7	7	8-10	11,12	13	2-4	14,15

Compound (Methyl Esters of Fatty Acids) — ECL[b]

Mono-branched

Compound	P1	P2	P3	P4	P4	P4	P5	P6	P7	P8
4 - Methylundecanoate	—	—	—	11.49	—	—	—	—	—	11.51
6 - Methylundecanoate	—	—	—	—	—	—	—	—	—	11.47
8 - Methylundecanoate	—	—	—	—	—	—	—	—	—	11.55
9 - Methylundecanoate	—	—	—	—	—	—	—	—	11.70	—
10 - Methylundecanoate (*iso*)	—	—	—	11.64	—	—	11.6	—	11.62	11.62
3 - Methyldodecanoate	—	—	—	—	—	—	—	—	12.37	—
4 - Methyldodecanoate	—	—	—	12.50	—	—	—	—	12.47	12.51
6 - Methyldodecanoate	—	—	—	—	—	—	—	—	—	12.45
8 - Methyldodecanoate	—	—	—	—	—	—	—	—	—	12.51
10 - Methyldodecanoate (*anteiso*)	—	—	—	12.69	—	—	12.7	—	—	12.72
11 - Methyldodecanoate (*iso*)	—	—	—	—	—	—	12.6	—	—	12.62
4 - Methyltridecanoate	—	13.40	—	13.45	—	—	—	—	—	13.50
6 - Methyltridecanoate	—	{	—	—	—	—	—	—	—	13.43
8 - Methyltridecanoate	—		—	—	—	—	—	—	—	13.47

Table 4 (continued)
METHYL ESTERS OF SATURATED MONO-BRANCHED-CHAIN CARBOXYLIC ACIDS (C_{11} AND ABOVE) (GC, USING NONPOLAR LIQUID PHASES)

Compound (Methyl Esters of Fatty Acids)	ECL[b]						
10 - Methyltridecanoate	13.59	—	—	—	—	—	13.58
12 - Methyltridecanoate (*iso*)	—	13.59	—	13.6	13.52	—	13.62
11 - Ethyltridecanoate	—	—	—	14.68	—	—	—
2 - Methyltetradecanoate	—	—	—	—	—	—	14.28
4 - Methyltetradecanoate	—	14.50	—	—	—	14.40	14.50
5 - Methyltetradecanoate	—	—	—	—	—	14.38	—
6 - Methyltetradecanoate	14.31	—	—	—	—	14.38	14.40
8 - Methyltetradecanoate	—	—	—	—	—	—	14.41
9 - Methyltetradecanoate	—	—	—	—	—	14.43	—
10 - Methyltetradecanoate	—	—	—	—	—	14.48	14.50
12 - Methyltetradecanoate (*anteiso*)	—	14.73	—	14.7	14.70	14.71	14.72
13 - Methyltetradecanoate (*iso*)	14.65	14.62	—	14.6	—	14.62	14.62
2 - Methylpentadecanoate	—	—	—	—	—	—	15.28
4 - Methylpentadecanoate	—	15.43	—	—	—	—	15.48
6 - Methylpentadecanoate	—	—	—	—	—	—	15.40
8 - Methylpentadecanoate	—	—	—	—	—	15.37	15.40
10 - Methylpentadecanoate	—	—	—	—	—	—	15.45
12 - Methylpentadecanoate (*iso*)	15.61	15.58	—	15.6	15.60	15.63	15.57
14 - Methylpentadecanoate (*iso*)	—	—	—	—	—	—	15.62
2 - Methylhexadecanoate	—	—	—	—	—	—	16.28
4 - Methylhexadecanoate	—	16.46	—	—	—	—	16.48
6 - Methylhexadecanoate	—	—	16.44	—	—	16.34	16.40
8 - Methylhexadecanoate	16.33	—	16.44	—	—	16.37	16.40
9 - Methylhexadecanoate	—	—	—	—	—	16.39	16.42
10 - Methylhexadecanoate	—	—	—	—	—	—	16.48
12 - Methylhexadecanoate	—	—	—	—	—	—	—
14 - Methylhexadecanoate (*anteiso*)	—	16.70	16.73	16.7	16.62	16.71	16.72
15 - Methylhexadecanoate (*iso*)	16.62	—	—	16.6	—	16.62	16.62

Compound	1	2	3	4	5	6	7	8	9
2 - Methylheptadecanoate	—	—	—	—	—	—	—	—	17.28
4 - Methylheptadecanoate	—	—	17.45	—	—	—	—	—	17.50
6 - Methylheptadecanoate	—	—	—	—	—	—	—	—	17.40
8 - Methylheptadecanoate	—	—	—	—	—	—	—	—	17.40
10 - Methylheptadecanoate	—	—	—	—	—	—	—	17.35	17.40
12 - Methylheptadecanoate	—	—	—	—	—	—	—	—	17.40
14 - Methylheptadecanoate	17.54	—	—	—	—	—	—	17.52	17.45
15 - Methylheptadecanoate (*anteiso*)	—	—	—	—	—	—	—	—	—
16 - Methylheptadecanoate (*iso*)	17.63	—	17.63	—	—	—	—	17.60	17.62
2 - Methyloctadecanoate	18.22	—	—	18.33	18.35	—	—	—	—
3 - Methyloctadecanoate	18.28	—	—	18.39	18.40	—	—	18.12	—
4 - Methyloctadecanoate	18.41	—	—	18.51	18.50	—	—	18.27	—
5 - Methyloctadecanoate	18.29	—	—	—	—	—	—	18.42	18.49
6 - Methyloctadecanoate	18.30	—	—	18.42	18.41	—	—	—	—
7 - Methyloctadecanoate	18.31	—	—	18.40	—	—	—	—	18.39
8 - Methyloctadecanoate	18.26	—	—	18.40	18.40	—	—	—	—
9 - Methyloctadecanoate	18.26	18.35	—	18.41	18.41	—	—	18.31	—
10 - Methyloctadecanoate	18.29	—	—	—	—	—	—	18.31	—
11 - Methyloctadecanoate	18.31	—	—	18.41	—	—	—	18.31	—
12 - Methyloctadecanoate	18.35	—	—	18.44	18.44	—	—	—	—
13 - Methyloctadecanoate	18.35	—	—	—	—	—	—	18.36	—
14 - Methyloctadecanoate	18.45	—	—	18.52	18.55	—	—	—	—
15 - Methyloctadecanoate	—	—	—	18.58	—	—	—	—	—
16 - Methyloctadecanoate (*anteiso*)	18.69	18.66	—	18.73	18.73	—	—	18.51	—
17 - Methyloctadecanoate (*iso*)	18.56	—	—	18.63	18.63	—	—	—	18.72
2 - Ethyloctadecanoate	—	—	—	—	—	—	—	—	18.62
2 - Propyloctadecanoate	—	—	—	19.00	—	—	—	—	—
2 - Butyloctadecanoate	—	—	—	19.75	—	—	—	—	—
18 - Methylnonadecanoate (*iso*)	19.56	19.6	—	20.68	—	—	—	—	19.62
10 - Methylicosanoate	—	—	—	20.38	—	—	—	—	—
12 - Methylicosanoate	—	—	—	20.40	—	—	—	—	—
14 - Methylicosanoate	—	—	—	20.44	—	—	—	—	—
16 - Methylicosanoate	—	—	—	20.50	—	—	—	—	—
18 - Methylicosanoate (*anteiso*)	20.7	—	—	20.75	—	—	—	—	—
19 - Methylicosanoate (*iso*)	—	—	—	20.65	—	—	—	—	—
20 - Methylhenicosanoate (*iso*)	21.6	—	—	—	—	—	—	—	—
20 - Methyldocosanoate (*anteiso*)	22.7	—	—	—	—	—	—	—	—

Table 4 (continued)
METHYL ESTERS OF SATURATED MONO-BRANCHED-CHAIN CARBOXYLIC ACIDS (C_{11} AND ABOVE) (GC, USING NONPOLAR LIQUID PHASES)

Compound (Methyl Esters of Fatty Acids)						ECL						
22 - Methyltricosanoate (*iso*)	—	—	—	—	—	23.6	—	—	—	—	—	—
22 - Methyltetracosanoate (*anteiso*)	—	—	—	—	—	24.7	—	—	—	—	—	—
24 - Methylpentacosanoate (*iso*)	—	—	—	—	—	25.6	—	—	—	—	—	—
24 - Methylhexacosanoate (*anteiso*)	—	—	—	—	—	26.7	—	—	—	—	—	—
26 - Methylheptacosanoate (*iso*)	—	—	—	—	—	27.6	—	—	—	—	—	—
26 - Methyloctacosanoate (*anteiso*)	—	—	—	—	—	28.7	—	—	—	—	—	—
28 - Methylnonacosanoate (*iso*)	—	—	—	—	—	29.6	—	—	—	—	—	—
28 - Methyltricontanoate (*anteiso*)	—	—	—	—	—	30.7	—	—	—	—	—	—
30 - Methylhentriacontanoate (*iso*)	—	—	—	—	—	31.6	—	—	—	—	—	—

ᵃ Carrier gas flow-rate not reported.

ᵇ Expressed in relationship with *n*-saturated-chain fatty acid methyl esters.

Column packing P1 = Silicone, no other details available
P2 = 3% ApL on Gas-Chrom Z (60-70 mesh)
P3 = 3% JXR on Gas-Chrom Q (100-120 mesh)
P4 = 3% OV-101 on Gas-Chrom Q (100-120 mesh)
P5 = 2.5% SE-30 on Chromosorb G - AW - DMCS (80-100 mesh)
P6 = 2.5% SE-30 on Gas-Chrom G - AW - DMCS
P7 = ApL (coating)
P8 = Pentasil + 5% Igepal - CO - 880 (coating)

REFERENCES

1. **Abrahamsson, S., Ställberg-Stenhagen, S., and Stenhagen, E.,** *Progress in the Chemistry of Fats and Other Lipids*, Vol. 7, Part 1, Holman, R. T. and Malkin, T., Eds., Pergamon Press, Oxford, 1963, 1.
2. **Ackman, R. G.,** *Progress in the Chemistry of Fats and Other Lipids*, Vol. 12, Holman, R. T., Ed., Pergamon Press, Oxford, 1972, 167.
3. **Ackman, R. G. and Hooper, S. N.,** *J. Am. Oil Chem. Soc.,* 47, 525, 1970.
4. **Ackman, R. G.,** *J. Chromatogr. Sci.,* 10, 243, 1972.

5. **Body, D. R. and Hansen, R. P.,** *J. Sci. Food Agric.,* 29, 107, 1978.
6. **Body, D. R.,** unpublished.
7. **Nicolaides, N.,** *Lipids,* 6, 901, 1971.
8. **Poltz, J. and Jacob, J.,** *Biochim. Biophys. Acta,* 360, 348, 1974.
9. **Jacob, J.,** *J. Chromatogr. Sci.,* 13, 415, 1975.
10. **Jacob, J.,** *J. Lipid Res.,* 19, 148, 1978.
11. **Kaneda, T.,** *J. Bacteriol.,* 95, 2210, 1968.
12. **Kaneda, T.,** *Bacteriol. Rev.,* 41, 391, 1977.
13. **Kaneda, T.,** *J. Chromatogr.,* 136, 323, 1977.
14. **Apon, J. M. B. and Nicolaides, N.,** *J. Chromatogr. Sci.,* 13, 467, 1975.
15. **Nicolaides, N., Apon, J. M. B., and Wong, D. H.,** *Lipids,* 11, 781, 1976.

Table 5

METHYL ESTERS OF SATURATED MULTI-BRANCHED-CHAIN CARBOXYLIC ACIDS (C_{11} AND ABOVE) (GC)

	P1	P1	P2	P3	P4	P5	P6	P7
Column packing Temperature (°C)	210	160-250 at 1/min	170	170	130-260 at 5/min	170-210 at 0.5/min	180-250 at 5 min after 25 min	180
Gas	N_2	N_2	He	He	He	He	N_2	N_2
Flow-rate (ml/min)	a	a	2-4	40 lbs/in²	20	8	10	20
Column Length (cm)	1000	900	600	10,000	150	30,480	120	370
Diameter (cm I.D.)	0.28	0.2	0.3	0.025	0.3	0.076	0.3	0.3
(cm O.D.)								
Form	Coiled	Coiled	Coiled	Capillary	Coiled	Capillary	Coiled	Coiled
Material	Glass	Glass	Copper		SS	SS	Glass	SS
Detector	FID	FID	FID	TC	FID	FID	FID	FID
References	1,2	3,4	5	6	5	7	8	8
Compound (Methyl Esters of Fatty Acids)				ECL[b]				
Di-Branched								
4,8 - Dimethylundecanoate	—	—	—	11.84	—	12.03	—	—
4,10 - Dimethylundecanoate	—	—	—	—	—	12.13	—	—
6,10 - Dimethylundecanoate	—	—	—	—	—	12.05	—	—
4,8 - Dimethyldodecanoate	—	—	—	12.64	—	12.96	—	—
4,10 - Dimethyldodecanoate	—	—	—	—	—	13.20	—	—
4,11 - Dimethyldodecanoate	—	—	—	—	—	13.12	—	—
6,10 - Dimethyldodecanoate	—	—	—	—	—	13.12	—	—
6,11 - Dimethyldodecanoate	—	—	—	—	—	13.05	13.63[c]	12.86[c]
2,4 - Dimethyltridecanoate	—	—	—	—	—	—	13.68[c]	13.06[c]
2,6 - Dimethyltridecanoate	—	—	—	—	—	—	13.74	13.11

Compound	1	2	3	4	5	6	7
2,8 - Dimethyltridecanoate	13.24	13.85	—	—	—	—	—
2,12 - Dimethyltridecanoate	13.87[c]	14.63[c]	13.92	—	—	—	—
4,6 - Dimethyltridecanoate	—	—	13.92	—	—	—	—
4,8 - Dimethyltridecanoate	—	—	13.90	—	—	13.58	—
4,10 - Dimethyltridecanoate	—	—	14.01	—	—	—	—
4,12 - Dimethyltridecanoate	—	—	14.12	—	—	—	—
6,12 - Dimethyltridecanoate	—	—	14.03	—	—	—	—
8,12 - Dimethyltridecanoate	—	—	14.03	—	—	—	—
11,11 - Dimethyltridecanoate	—	—	—	14.38	—	—	14.40
12,12 - Dimethyltridecanoate	—	—	—	⇅	—	—	14.10
2,4 - Dimethyltetradecanoate	14.03[c]	14.69[c]	—	—	—	—	—
2,6 - Dimethyltetradecanoate	—	—	14.70	—	—	—	—
2,8 - Dimethyltetradecanoate	—	—	14.70	—	—	—	—
2,12 - Dimethyltetradecanoate	—	—	14.99	—	—	—	—
2,13 - Dimethyltetradecanoate	—	—	14.89	—	—	—	—
4,8 - Dimethyltetradecanoate	—	—	14.85	—	—	14.64	—
4,10 - Dimethyltetradecanoate	—	—	14.99	—	—	—	—
4,12 - Dimethyltetradecanoate	—	—	15.19	—	—	—	—
4,13 - Dimethyltetradecanoate	—	—	15.10	—	—	—	—
6,12 - Dimethyltetradecanoate	—	—	15.10	—	—	—	—
6,13 - Dimethyltetradecanoate	—	—	15.02	—	—	—	—
2,6 - Dimethylpentadecanoate	—	—	15.70	—	—	—	—
2,8 - Dimethylpentadecanoate	—	—	15.70	—	—	—	—
2,12 - Dimethylpentadecanoate	—	—	15.77	—	—	—	—
2,14 - Dimethylpentadecanoate	—	—	15.89	—	—	—	—
4,6 - Dimethylpentadecanoate	—	—	15.82	—	—	—	—
4,8 - Dimethylpentadecanoate	—	—	17.82	—	15.59	—	—
4,10 - Dimethylpentadecanoate	—	—	15.93	—	—	—	—
4,12 - Dimethylpentadecanoate	—	—	15.95	—	—	—	—
4,14 - Dimethylpentadecanoate	—	—	16.10	—	—	—	—
8,14 - Dimethylpentadecanoate	—	—	15.93	—	—	—	—
10,14 - Dimethylpentadecanoate	—	—	16.05	—	—	—	—
2,6 - Dimethylhexadecanoate	—	—	16.70	—	—	—	—
2,8 - Dimethylhexadecanoate	—	—	16.70	—	—	—	—
2,10 - Dimethylhexadecanoate	—	—	16.70	—	—	—	—
2,12 - Dimethylhexadecanoate	—	—	16.73	—	—	—	—

Table 5 (continued)
METHYL ESTERS OF SATURATED MULTI-BRANCHED-CHAIN CARBOXYLIC ACIDS (C_{11} AND ABOVE) (GC)

Compound (Methyl Esters of Fatty Acids)	ECL[b]		
2,14 - Dimethylhexadecanoate	—	—	17.00
2,15 - Dimethylhexadecanoate	—	—	16.90
4,12 - Dimethylhexadecanoate	—	—	—
4,14 - Dimethylhexadecanoate	—	—	17.20
4,15 - Dimethylhexadecanoate	—	—	17.10
6,14 - Dimethylhexadecanoate	—	—	17.10
6,15 - Dimethylhexadecanoate	—	—	17.00
8,14 - Dimethylhexadecanoate	—	—	17.10
8,15 - Dimethylhexadecanoate	—	—	16.95
2 - Ethyl - 6 - Methylhexadecanoate	—	17.53	—
2 - Ethyl - 8 - Methylhexadecanoate	—	17.62	—
2 - Ethyl - 14 - Methylhexadecanoate	—	17.87	—
4 - Ethyl - 14 - Methylhexadecanoate	—	18.25	—
2,6 - Dimethyloctadecanoate	18.71	—	—
2,8 - Dimethyloctadecanoate	18.72	—	—
2,10 - Dimethyloctadecanoate	18.73	—	—
2,12 - Dimethyloctadecanoate	18.77	—	—
2,14 - Dimethyloctadecanoate	18.84	—	—
2,16 - Dimethyloctadecanoate	19.07	—	—
4,6 - Dimethyloctadecanoate	18.70	—	—
4,8 - Dimethyloctadecanoate	18.76	—	—
4,10 - Dimethyloctadecanoate	18.80	—	—
4,12 - Dimethyloctadecanoate	18.88	—	—
4,14 - Dimethyloctadecanoate	18.96	—	—
4,16 - Dimethyloctadecanoate	19.20	—	—
10,14 - Dimethyloctadecanoate	18.89	—	—
10,16 - Dimethyloctadecanoate	19.12	—	—
2 - Ethyl - 6 - methyloctadecanoate	19.45	19.45	—
2 - Ethyl - 8 - methyloctadecanoate	—	19.45	—
2 - Ethyl - 10 - methyloctadecanoate	19.60	19.60	—
2 - Ethyl - 16 - methyloctadecanoate	19.80	19.80	—

Tri-Branched

2 - Ethyl - 4,8 - dimethylundecanoate	—	13.05	—	—	—	—	—
2,6,12 - Trimethyltridecanoate	14.28	—	—	—	—	—	—
4,8,12 - Trimethyltridecanoate	14.48	—	—	—	—	—	—
2,4,6 - Trimethyltetradecanoate	15.10	—	—	—	—	—	—
2 - Ethyl - 6,10 - dimethyltetradecanoate	—	16.00	—	—	—	—	—
4,8,14 - Trimethylpentadecanoate	16.42	—	—	—	—	—	—
2,6,14 - Trimethylhexadecanoate	—	17.47	—	—	—	—	—
2,6,10 - Trimethyloctadecanoate	—	19.08	—	—	—	—	—
2,6,12 - Trimethyloctadecanoate	—	19.10	—	—	—	—	—
2,6,14 - Trimethyloctadecanoate	—	19.22	19.20	—	—	—	—
2,6,16 - Trimethyloctadecanoate	—	19.45	19.45	—	—	—	—
2 - Ethyl - 6,14 - dimethyloctadecanoate	—	19.93	—	—	—	—	—
2 - Ethyl - 6,16 - dimethyloctadecanoate	—	20.20	—	—	—	—	—

Tetra-Branched

2,4,6,8 - Tetramethyldecanoate	—	11.75	—	—	—	—	—
2,4,6,8 - Tetramethylundecanoate	—	12.50	—	—	—	—	—
2,4,6,8 - Tetramethyldodecanoate	—	13.87	—	—	—	—	—
2,4,6,8 - Tetramethyltridecanoate	—	14.60	—	—	—	—	—
2 - Ethyl - 4,8,12 - trimethylpentadecanoate	—	17.35	—	—	—	—	—
2,6,10,14 - Tetramethyloctadecanoate	—	19.77	—	—	—	—	—

a   Carrier gas flow-rate not reported.

b   Expressed in relationship with *n*-saturated-chain fatty acid methyl esters.

c   Mixtures of diastereoisomers.

Column packing P1 = 3% OV-101 on Gas-Chrom Q (100-120 mesh)

P2 = 7% EGA on Anakrom ABS (60-80 mesh)

P3 = BDS (coating)

P4 = 2.5% SE-30 on Chromosorb G-AW-DMCS (80-100 mesh)

P5 = Pentasil + 5% Igepal-Co-880 (coating)

P6 = 3% OV-101 on Gas-Chrom Q (100-120 mesh)

P7 = 10% EGA on Chromosorb W AW (80-100 mesh)

Table 5 (continued)
METHYL ESTERS OF SATURATED MULTI-BRANCHED-CHAIN CARBOXYLIC ACIDS (C$_{11}$ AND ABOVE) (GC)

REFERENCES

1. **Jacob, J.,** *J. Chromatogr. Sci.,* 13, 415, 1975.
2. **Jacob, J.,** *Hoppe - Seyler's Z. Physiol. Chem.,* 357, 609, 1976.
3. **Poltz, J. and Jacob, J.,** *Biochim. Biophys. Acta,* 360, 348, 1974.
4. **Jacob, J.,** *J. Lipid Res.,* 19, 148, 1978.
5. **Kaneda, T.,** *Bacteriol. Rev.,* 41, 391, 1977.
6. **Duncan, W. R. H., Lough, A. K., Garton, G. A., and Brooks, P.,** *Lipids,* 9, 669, 1974.
7. **Nicolaides, N., Apon, J. M. B., and Wong, D. H.,** *Lipids,* 11, 781, 1976.
8. **Julák, J., Tureček, F., and Mikova, Z.,** *J. Chromatogr.,* 190, 183, 1980.

Table 6
METHYL ESTERS OF MONOENOIC BRANCHED-CHAIN CARBOXYLIC ACIDS (GC)

	P1	P2	P3	P4	P5	P6	P7	P8	P6	P7	P9	P10
Column packing Temperature (°C)	185	200	170	130—260 at 5/min	150	170	160	120	170	175	190	140—200 at 2/min
Gas	N₂	N₂	He	He	He	He	He	He	He	He	He	He
Flow rate (mℓ/mn)	12	25	2—4	20	24	20	30	60	50	50	80	6
Column Length (cm)	200	200	610	150	200	4600	4600	4600	4600	4600	4600	2500
Diameter (cm I.D.)	0.23	0.4	0.3	0.3	0.32	0.025	0.025	0.025	0.025	0.025	0.025	0.048
(cm O.D.)												
Form	Coiled	Coiled	Coiled	Coiled	Coiled	Capillary	Capillary	Capillary	Capillary	Capillary	Capillary	Capillary
Material	SS	Aluminum	Copper	SS	SS	SS	SS	SS	SS	SS	SS	Glass
Detector	FID	FID	FID	FID	FID	FID	FID	FID	FID	FID	FID	FID
Reference	1	1	2,3	2,3	4	4	4	4	5	5	4.5	6,7
Compound (Methyl Esters of Fatty Acids)					ECL[a]							
cis - Methyldodec - 2 - enoate	—	—	—	—	12.53	12.83	12.79	12.72	—	—	12.40	—
cis - Methyldodec - 3 - enoate	—	—	—	—	12.31	12.46	12.46	12.49	—	—	12.15	—
trans - Methyldodec - 2 - enoate	—	—	—	—	13.12	13.80	13.66	13.64	—	—	13.28	—
trans - Methyldodec - 3 - enoate	—	—	—	—	12.57	12.95	12.87	12.91	—	—	12.70	—
3 - Methylenedodecanoate	—	—	—	—	12.41	—	—	12.85	—	—	12.31	—
5 - Methyltetradec - 4 - enoate	—	—	—	—	—	—	—	—	14.88	14.90	14.50	—
12 - Methyltetradec - 7 - enoate	—	—	—	—	—	—	—	—	—	—	—	14.5
13 - Methyltetradec - 7 - enoate	—	—	—	—	—	—	—	—	—	—	—	14.4

Table 6 (continued)
METHYL ESTERS OF MONOENOIC BRANCHED-CHAIN CARBOXYLIC ACIDS (GC)

Compound (Methyl Esters of Fatty Acids)	ECL[a]										
14 - Methylpentadec - 9 - enoate	—	—	—	—	—	—	—	—	—	—	15.4
14 - Methylpentadec - 10 - enoate	—	15.90	15.4	—	—	—	—	—	—	—	—
7 - Methylhexadec - 6 - enoate	—	—	—	—	—	—	—	16.59	16.62	16.20	—
7 - Methylhexadec - 7 - enoate	—	—	—	—	—	—	—	16.90	16.92	16.62	—
14 - Methylhexadec - 9 - enoate	—	—	—	—	—	—	—	—	—	—	16.4
14 - Methylhexadec - 10 - enoate	—	16.97	16.5	—	—	—	—	—	—	—	—
15 - Methylhexadec - 9 - enoate	—	—	—	—	—	—	—	—	—	—	16.3
15 - Methylhexadec - 10 - enoate	—	16.88	16.4	—	—	—	—	—	—	—	—
11 - Methyloctadec - 11 - enoate	18.22	18.20	—	—	—	—	—	—	—	—	—

[a] Expressed in relationship with *n*-saturated-chain fatty acid methyl esters.

Column packing

P1 = 5% EGSS-X on Chromosorb W (100-120 mesh)
P2 = 40% EGA on Chromosorb W (100-120 mesh)
P3 = 7% EGA on Anakrom ABS (60-80 mesh)
P4 = 2.5% SE-30 on Chromosorb G-AW-DMCS (80-100 mesh)
P5 = 5% SE-30 on Chromosorb W (80-100 mesh)
P6 = BDS (Coating)
P7 = Silar - 5CP (Coating)
P8 = Silar - 7CP (Coating)
P9 = ApL (Coating)
P10 = OV-101 (Coating)

REFERENCES

1. **Gerson, T., Patel, J. J., and Nixon, L. N.,** *Lipids,* 10, 134, 1975
2. **Kaneda, T.,** *J. Bacteriol.,* 95, 2210, 1968.
3. **Kaneda, T.,** *Bacteriol. Rev.,* 41, 391, 1977.
4. **Ackman, R. G. and Eaton, C. A.,** *J. Chromatogr. Sci.,* 13, 509, 1975.
5. **Pascal, J. C. and Ackman, R. G.,** *Lipids,* 10, 478, 1975.
6. **Boon, J. J., de Leeuw, J. W., van de Hoek, G. J., and Vosjan, J. H.,** *J. Bacteriol.,* 129, 1183, 1977.
7. **Nieberg-van Velzen, E. and Boon, J. J.,** personal communication.

Table 7
METHYL ESTERS OF SYNTHETIC POLYUNSATURATED BRANCHED-CHAIN CARBOXYLIC ACIDS (GC)

	P1	P2
Column packing		
Temperature (°C)	185	185
Gas	He	He
Flow-rate (mℓ/min)	60	[a]
Column		
Length (cm)	304.8	225
Diameter		
(cm I.D.)		0.15
(cm O.D.)	0.635	
Form	Coiled	Coiled
Material	SS	Aluminum
Detector	TC	FID
Reference	1	2,3

Compound (Methyl Esters of Fatty Acids)	ECL[b]	
5 - Methylicosa - 8,11,14 - trienoate	22.27[c]	—
10 - Methylicosa - 8,11,14 - trienoate	20.90	—
13 - Methylicosa - 8,11,14 - trienoate	21.04	—
17 - Methylicosa - 8,11,14 - trienoate	22.54	—
18 - Methylicosa - 8,11,14 - trienoate	22.72	—
19 - Methylicosa - 8,11,14 - trienoate	22.52	—
2 - Methylicosa - 5,8,11,14 - tetraenoate	—	21.40

[a] Carrier gas flow-rate not reported.

[b] Expressed in relationship with *n*-saturated-chain fatty acid methyl esters.

[c] Calculated from the published retention time data relative to methyl *n*-icosanoate.

Column packing	P1 =	15% EGS on Gas-Chrom P (80-200 mesh)
	P2 =	20% EGA on Chromosorb W-AW (100-120 mesh)

REFERENCES

1. **Do, U. H. and Sprecher, H.,** *Chem. Phys. Lipids,* 16, 255, 1976.
2. **Gerson, T. and Schlenk, H.,** *Chem. Phys. Lipids,* 2, 213, 1968.
3. **Gerson, T.,** personal communication.

Table 8
METHYL ESTERS OF SATURATED MULTI-BRANCHED ISOPRENOID CARBOXYLIC ACID DIASTEREOISOMERS (GC)

	P1	P2	P3	P3	P4
Column packing					
Temperature (°C)	172	170	150	150	180
Gas	N_2	He	He	He	He
Flow rate (ml/min)	14	40	60	40	60
Column					
Length (cm) Diameter	200	5000	5000	4600	5000
(cm I.D.)	0.4	0.025	0.025	0.025	0.025
(cm O.D.)					
Form					
Material	Coiled Glass	Capillary SS	Capillary SS	Capillary SS	Capillary SS
Detector	FID			FID	
Reference	1	1	1	2	1
Compound (Saturated) (Methyl Esters of Fatty Acids)			ECL[a]		
Multi-branched					
4D, 8D, 12 - Trimethyltridecanoate				14.12	
2L, 6D, 10D, 14 - Tetramethylpentadecanoate (Pristanate)	—	—	—	15.76	—
2D, 6D, 10D, 14 - Tetramethylpentadecanoate (Pristanate)	15.70	15.68	15.85	15.80	—
3L, 7D, 11D, 15 - Tetramethylhexadecanoate (Phytanate)	—	—	—	16.98	16.40
3D, 7D, 11D, 15 - Tetramethylhexadecanoate (Phytanate)	16.97	17.01	17.11	17.02	17.47
4D, 8D, 12D, 16 - Tetramethylheptadecanoate	18.24	18.34	18.35	—	18.57
5D, 9D, 13D, 17 - Tetramethyloctadecanoate	19.11	19.30	19.24	—	19.48

[a] Expressed in relationship with *n*-saturated-chain fatty acid methyl esters.

Table 8 (continued)

METHYL ESTERS OF SATURATED MULTI-BRANCHED ISOPRENOID CARBOXYLIC ACID DIASTEREOISOMERS (GC)

Column packing
- P1 = 10% BDS on Gas-Chrom A
- P2 = DEGS (coating)
- P3 = BDS (coating)
- P4 = ApL (coating)

REFERENCES

1. **Kates, M., Hancock, A. J., Ackman, R. G., and Hooper, S. N.,** *Chem. Phys. Lipids,* 8, 32, 1972.
2. **Ackman, R. G. and Hansen, R. P.,** *Lipids,* 2, 357, 1967.

Table 9
METHYL ESTERS OF MONOENOIC MULTI-BRANCHED ISOPRENOID CARBOXYLIC ACIDS (GC)

Column packing	P1	P2	P3	P4	P5
Temperature (°C)	165	180	180		
Gas		180	180		186
Gas		He	He	He	
Flow rate (mℓ/min)		a	a	a	
Column					
Length (cm)	365.8	180	240	3000	365.8
Diameter					
(cm I.D.)		0.3	0.3	0.025	
(cm O.D.)					
Form					
Material		Coiled	Coiled	Capillary	
Detector		FID	FID	FID	
Reference	1-2	3	3	4	1-2

Compound (Unsaturated) (Methyl Esters of Fatty Acids)			ECL[b]		
Multi-branched					
cis - 3,7,11,15 - Tetramethylhexadec - 2 - enoate	17.92	—	—	—	17.87
cis - 3,7,11,15 - Tetramethylhexadec - 3 - enoate	17.62	—	—	—	17.51
trans - 3,7,11,15 - Tetramethylhexadec - 2 - enoate (Phytenate)	19.00	18.9	18.8	17.22	18.95
trans - 3,7,11,15 - Tetramethylhexadec - 3 - enoate	18.21	—	—	—	17.89
3 - Methylene - 7,11,15 - trimethylhexadecanoate	18.07	—	—	—	17.88

[a] No carrier gas flow-rate recorded.

[b] Expressed in relationship with *n*-saturated-chain fatty acid methyl esters.

Column packing
P1 = 17% EGS (solid support not reported)
P2 = 15% EGS on Gas-Chrom P (100–120 mesh)
P3 = 15% DEGS on Chromosorb W
P4 = SP2250 (coating)
P5 = 2% SE-30 (solid support not reported)

REFERENCES

1. **Baxter, J. H. and Milne, G. W. A.,** *Biochim. Biophys. Acta,* 176, 265, 1969.
2. **Ackman, R. G. and Eaton, C. A.,** *J. Chromatogr. Sci.,* 13, 509, 1975.
3. **Gellerman, J. L., Anderson, W. H., and Schlenk, H.,** *Lipids,* 10, 656, 1975.
4. **Boon, J. J., Rijpstra, W.I.C., de Leeuw, J.W., and Schenck, P.A.,** *Nature (London),* 258, 414, 1975.

Table 10
METHYL ESTERS OF BRANCHED-CHAIN CARBOXYLIC ACIDS (TLC)

	L1	L2	L3	L4	L3	L3	L3	L5
Layer	L1	L2	L3	L4	L3	L3	L3	L5
Solvent	S1	S2	S3	S4	S5	S6	S7	S8
Technique	T1	T2	T3	T2	T2	T2	T2	T4
Reference	1	2	3	3	4	4	5	5

Compound
(Methyl Esters of Fatty Acids) R_f

5 - Methyl - 4 - tetradecenoate	0.53	—	—	—	—	—	—	—
7 - Methyl - 6 - hexadecenoate	0.48	—	—	—	—	—	—	—
7 - Methyl - 7 - hexadecenoate	0.42	—	—	—	—	—	—	—
cis - 3,7,11,14 - Tetramethylhexadec - 2 - enoate	—	0.62	—	—	—	—	—	—
trans - 3,7,11,14 - Tetramethylhexadec - 2 - enoate	—	0.53	0.70	0.27	—	—	—	—
cis - 3,7,11,14 - Tetramethylhexadec - 3 - enoate	—	0.53	—	—	—	—	—	—
trans - 3,7,11,14 - Tetramethylhexadec - 3 - enoate	—	0.48	—	—	—	—	—	—
3 - Methylene - 7,11,14 - trimethylhexadecanoate	—	0.42	—	—	—	—	—	—
2,6,10,13 - Tetramethylpentadecanoate	—	—	—	—	0.70	0.55	0.58	—
3,7,11,14 - Tetramethylhexadecanoate	—	0.53	0.70	0.29	0.70	0.55	0.56	0.57
4,8,12,15 - Tetramethylheptadecanoate	—	—	—	—	0.70	0.55	0.53	—
5,9,13,16 - Tetramethyloctadecanoate	—	—	—	—	0.70	0.55	0.54	—
(Dodecanoate)	0.70	—	—	—	—	—	—	—
(Octadecanoate)	—	0.47	0.63	0.22	—	—	—	0.54
(*cis*-Octadec-9-enoate)	—	—	0.63	0.09	—	—	—	0.46

Layer		
	L1 =	Supelcosil - 12D impregnated with 15% $AgNO_3$
	L2 =	Silica gel G impregnated with 9.3% $AgNO_3$
	L3 =	Silica gel H
	L4 =	Silica gel impregnated with 10% $AgNO_3$
	L5 =	Silica gel G
Solvent (v/v)	S1 =	Benzene - hexane (3:7)
	S2 =	Benzene - hexane (2:1)
	S3 =	(i) isopropyl ether - acetic acid (96:4); (ii) petroleum hydrocarbon - diethyl ether- acetic acid (90:10:1)
	S4 =	Hexane - diethylether (97:5:2.5)
	S5 =	Petroleum hydrocarbon - diethyl ether - acetic acid (90:10:1)
	S6 =	Benzene
	S7 =	Hexane - diethyl ether (95:5)
	S8 =	Petroleum hydrocarbon - diethyl ether (195:5)
Detection		See Section II.I, Volume Two
Technique	T1 =	Develop chromatogram three times
	T2 =	20 cm distance
	T3 =	(i) 13-14 cm distance; (ii) 19-19.5 cm distance
	T4 =	14 cm distance

REFERENCES

1. **Pascal, J. C. and Ackman, R. G.,** *Lipids,* 10, 478, 1975.
2. **Baxter, J. H. and Milne, G. W. A.,** *Biochim. Biophys. Acta,* 176, 265, 1969.
3. **Boon, J. J., Rijpstra, W. I. C., de Leeuw, J. W., and Schenck, P. A.,** *Nature (London),* 258, 414, 1975.
4. **Kates, M., Hancock, A. J., Ackman, R. G., and Hooper, S. N.,** *Chem. Phys. Lipids,* 8, 32, 1972.
5. **Patton, S. and Benson, A. A.,** *Biochim. Biophys. Acta,* 125, 22, 1966.

Table 11
METHYL ESTERS OF SATURATED MONO-BRANCHED-CHAIN HYDROXYCARBOXYLIC ACIDS (GC)

Column packing	P1
Temperature (°C)	180
Gas	N_2
Flow rate (mℓ/min)	30
Column	
Length (cm)	180
Diameter (cm I.D.)	0.2
Form	Coiled
Material	Glass
Detector	FID
Reference	1

Compound (Methyl Esters of Fatty Acids)	ECL[a]
Mono-branched	
13-Methyl-2-hydroxytetradecanoate	19.22
13-Methyl-3-hydroxytetradecanoate	20.42
15-Methyl-2-hydroxyhexadecanoate	21.28
15-Methyl-3-hydroxyhexadecanoate	22.48

[a] Expressed in relationship with *n*-saturated-chain fatty acid methyl esters calculated from the published data.

Column packing	P1 =	10% EGSS-X on Gas-Chrom P (100-120 mesh)

REFERENCE

1. **Fautz, E., Rosenfelder, G., and Grotjahn, L.,** *J. Bacteriol.*, 140, 852, 1979.

Table 12
METHYL ESTERS OF BRANCHED-CHAIN HYDROXY-CARBOXYLIC ACIDS (TLC)

Layer	L1
Solvent	S1
Technique	T1
Reference	1

Compound (Methyl Esters of Fatty Acids)	R_f
(*n*-Octadecanoate)	0.76
13-Methyl-2-hydroxytetradecanoate	0.22
13-Methyl-3-hydroxytetradecanoate	0.16
15-Methyl-2-hydroxyhexadecanoate	0.22
15-Methyl-3-hydroxyhexadecanoate	0.16

Layer	L1 =	Silica gel 60 F_{254}
Solvent	S1 =	Petroleum hydrocarbon - diethyl ether (85:15 v/v)
Detection		Chapter 28, Reagent No. 38
Technique	T1 =	17 cm distance

REFERENCE

1. **Fautz, E., Rosenfelder, G., and Grotjahn, L.,** *J. Bacteriol.*, 140, 852, 1979.

REFERENCES

1. **Asselineau, J.,** *The Bacterial Lipids,* Hermann, Paris, 1966, 82.
2. **Kaneda, T.,** *Bacteriol. Rev.,* 41, 391, 1977.
3. **Odham, G.,** *Arkiv Kemi,* 21, 379, 1963.
4. **Jacob, J. and Poltz, J.,** *J. Lipid Res.,* 15, 243, 1974.
5. **Poltz, J. and Jacob, J.,** *Biochim. Biophys. Acta,* 360, 348, 1974.
6. **Jacob, J. and Poltz, J.,** *Lipids,* 10, 1, 1975.
7. **Jacob, J.,** *J. Chromatogr. Sci.,* 13, 415, 1975.
8. **Jacob, J.,** *J. Lipid Res.,* 19, 148, 1978.
9. **Odham, G.,** *Arkiv Kemi,* 27, 295, 1967.
10. **Nicolaides, N. and Ray, T.,** *J. Am. Oil Chem. Soc.,* 42, 702, 1965.
11. **Nicolaides, N.,** *Science,* 186, 19, 1974.
12. **Nicolaides, N.,** *J. Am. Oil Chem. Soc.,* 42, 708, 1965.
13. **Grigor, M. R., Dunckley, G. G., and Purves, H. D.,** *Biochim. Biophys. Acta,* 218, 400, 1970.
14. **Body, D. R. and Hansen, R. P.,** *J. Sci. Food Agric.,* 29, 107, 1978.
15. **Downing, D. T.,** *J. Lipid Res.,* 5, 210, 1964.
16. **Shorland, F. B., Body, D. R., and Gass, J. P.,** *Biochim. Biophys. Acta,* 125, 217, 1966.
17. **Duncan, W. R. H. and Garton, G. A.,** *J. Sci. Food Agric.,* 18, 99, 1967.
18. **Hansen, R. P. and Czochanska, Z.,** *N. Z. J. Sci.,* 19, 413, 1976.
19. **Garton, G. A.,** *Biochem. Soc. Trans.,* 2, 1200, 1974.
20. **Carroll, K. K. and Serdarevich, B.,** in *Lipid Chromatographic Analysis,* Vol. 1, Marinetti, G. V., Ed., Marcel Dekker, New York, 1967, 205.
21. **Mangold, H. K.,** *J. Am. Oil Chem. Soc.,* 38, 708, 1961.
22. **Skipski, V. P., Good, J. J., Barclay, M., and Reggio, R. B.,** *Biochim. Biophys. Acta,* 152, 10, 1968.
23. **Baron, C. and Blough, H. A.,** *J. Lipid Res.,* 17, 373, 1976.
24. **Annison, E. F.,** *Biochem. J.,* 57, 400, 1954.
25. **Edwards, G. B., McManus, W. R., and Bigham, M. L.,** *J. Chromatogr.,* 63, 397, 1971.
26. **Ackman, R. G.,** *J. Chromatogr. Sci.,* 10, 560, 1972.
27. **Kirk, R. D., Body, D. R., and Hawke, J. C.,** *J. Sci. Food Agric.,* 22, 620, 1971.
28. **Kellogg, D. W.,** *J. Dairy Sci.,* 52, 1690, 1969.
29. **Ziolecki, A. and Kwiatkowska, E.,** *J. Chromatogr.,* 80, 250, 1973.
30. **Du Preez, J. C. and Lategan, P. M.,** *J. Chromatogr.,* 124, 63, 1976.
31. **Gibbs, B. F., Itiaba, K., Crawhall, J. C., Cooper, B. A., and Mamer, O. A.,** *J. Chromatogr.,* 81, 65, 1973.
32. **Gardner, J. W. and Thompson, G. E.,** *Analyst,* 99, 326, 1974.
33. **Geddes, D. A. M. and Gilmour, M. N.,** *J. Chromatogr. Sci.,* 8, 394, 1970.
34. **van Eenaeme, C., Bienfait, J. M., Lambot, O., and Pondant, A.,** *J. Chromatogr. Sci.,* 12, 398, 1974.
35. **van Eenaeme, C., Bienfait, J. M., Lambot, O., and Pondant, A.,** *J. Chromatogr. Sci.,* 12, 404, 1974.
36. **Dressman, R. C.,** *J. Chromatogr. Sci.,* 8, 265, 1970.
37. **Hrivnak, J., Sojak, L., Beska, E., and Janák, J.,** *J. Chromatogr.,* 68, 55, 1972.
38. **Nikelly, J. G.,** *Anal. Chem.,* 36, 2244, 1964.
39. **Doelle, H. W.,** *J. Chromatogr.,* 39, 398, 1969.
40. **Ottenstein, D. M. and Bartley, D. A.,** *Anal. Chem.,* 43, 952, 1971.
41. **Remesy, C. and Demigne, C.,** *Biochem. J.,* 141, 85, 1974.
42. **Robinson, P. G.,** *J. Chromatogr.,* 74, 13, 1972.
43. **Mahadevan, V. and Stenroos, L.,** *Anal. Chem.,* 39, 1652, 1967.
44. **Tyler, J. E. and Dibdin, G. H.,** *J. Chromatogr.,* 105, 71, 1975.
45. **Hrivnak, J.,** *J. Chromatogr. Sci.,* 8, 602, 1970.
46. **Di Corcia, A. and Samperi, R.,** *Anal. Chem.,* 46, 140, 1974.
47. **Klemm, H.-P., Hintze, U., and Gercken, G.,** *J. Chromatogr.,* 75, 19, 1973.
48. **Corina, D. L. and Dunstan, P. M.,** *Anal. Biochem.,* 53, 571, 1973.
49. **Lambert, M. A. and Moss, C. W.,** *J. Chromatogr.,* 74, 335, 1972.
50. **Salanitro, J. P. and Muirhead, P. A.,** *Appl. Microbiol.,* 29, 374, 1975.
51. **Choudhary, G. and Moss, C. W.,** *J. Chromatogr.,* 128, 261, 1976.
52. **Lupton, C. J.,** *J. Chromatogr.,* 104, 223, 1975.
53. **Rodrigues de Miranda, J. F. and Eikelboom, T. D.,** *J. Chromatogr.,* 114, 274, 1975.
54. **Wong, E., Nixon, L. N., and Johnson, C. B.,** *J. Agric. Food Chem.,* 23, 495, 1975.
55. **Johnson, C. B., Wong, E., Birch, E. J., and Purchas, R. W.,** *Lipids,* 12, 340, 1977.
56. **Johnson, C. B. and Wong, E.,** *J. Chromatogr.,* 109, 403, 1975.
57. **Shorland, F. B. and Hansen, R. P.,** *Dairy Sci. Abstr.,* 19, 168, 1957.
58. **Ackman, R. G., Hooper, S. N., and Hansen, R. P.,** *Lipids,* 7, 683, 1972.

59. **Morris, L. J.,** *Chem. Ind. (London).* p.1238, 1962.
60. **Gunstone, F. D., Ismail, I. R., and Lie Ken Jie, M.,** *Chem. Phys. Lipids,* 1, 376, 1967.
61. **Pascal, J. C. and Ackman, R. G.,** *Lipids,* 10, 478, 1975.
62. **de Vries, B.,** *J. Am. Oil Chem. Soc.,* 40, 184, 1963.
63. **Ansari, M. N. A., Nicolaides, N., and Fu, H. C.,** *Lipids,* 5, 838, 1970.
64. **Emken, E. A., Scholfield, C. R., and Dutton, H. J.,** *J. Am. Oil Chem. Soc.,* 41, 388, 1964.
65. **Scholfield, C. R. and Mounts, T. L.,** *J. Am. Oil Chem. Soc.,* 54, 319, 1977.
66. **Mangold, H. K.,** in *Thin-Layer Chromatography. A Laboratory Handbook,* Stahl, E., Ed., Springer-Verlag, Berlin, 1969, 363.
67. **Hofheinz, W. and Grisebach, H.,** *Z. Naturforsch.,* 20 b, 45, 1965.
68. **Kishimoto, Y., Williams, M., Moser, H. W., Hignite, C., and Biemann, K.,** *J. Lipid Res.,* 14, 69, 1973.
69. **Duncan, W. R. H., Lough, A. K., Garton, G. A., and Brooks, P.,** *Lipids,* 9, 669, 1974.
70. **Egge, H., Murawski, U., Ryhage, R., Gyorgy, P., Chatranon, W., and Zilliken, F.,** *Chem. Phys. Lipids,* 8, 42, 1972.
71. **Garton, G. A., Hovell, F. D. DeB., and Duncan, W. R. H.,** *Br. J. Nutr.,* 28, 409, 1972.
72. **Smith, A. and Lough, A. K.,** *J. Chromatogr. Sci.,* 13, 486, 1975.
73. **Hradec, J. and Mensik, P.,** *J. Chromatogr.,* 32, 502, 1968.
74. **Woodford, F. P. and van Gent, C. M.,** *J. Lipid Res.,* 1, 188, 1960.
75. **Miwa, T. K., Mikolajczak, K. L., Earle, F. R., and Wolff, I. A.,** *Anal. Chem.,* 32, 1739, 1960.
76. **Ackman, R. G. and Hansen, R. P.,** *Lipids,* 2, 357, 1967.
77. **Kates, M., Hancock, A. J., Ackman, R. G., and Hooper, S. N.,** *Chem. Phys. Lipids,* 8, 32, 1972.
78. **Patton, S. and Benson, A. A.,** *Biochim. Biophys. Acta,* 125, 22, 1966.
79. **El-Shazly, K.,** *Biochem. J.,* 51, 647, 1952.
80. **Bertelsen, O.,** *Arkiv Kemi,* 32, 17, 1970.
81. **Nicolaides, N.,** *Lipids,* 6, 901, 1971.
82. **Jacob, J.,** *Fette Seifen Anstrichm.,* 76, 241, 1974.
83. **Nicolaides, N., Apon, J. M. B., and Wong, D. H.,** *Lipids,* 11, 781, 1976.
84. **Apon, J. M. B. and Nicolaides, N.,** *J. Chromatogr. Sci.,* 13, 467, 1975.
85. **Gerson, T., Patel, J. J., and Nixon, L. N.,** *Lipids,* 10, 134, 1975.
86. **Jacob, J.,** *Hoppe-Seyler's Z. Physiol. Chem.,* 357, 609, 1976.
87. **Boon, J. J., de Leeuw, J. W., van de Hoek, G. J., and Vosjan, J. H.,** *J. Bacteriol.,* 129, 1183, 1977.
88. **Boon, J. J., van de Graaf, B., Schuyl, P. J. W., de Lange, F., and de Leeuw, J. W.,** *Lipids,* 12, 717, 1977.
89. **Nicolaides, N. and Apon, J. M. B.,** *Biomed. Mass Spectrom.,* 4, 337, 1977.
90. **Scholfield, C. R.,** *J. Am. Oil Chem. Soc.,* 52, 36, 1975.
91. **Pei, T.-S. P., Henly, R. S., and Ramachandran, S.,** *Lipids,* 10, 152, 1975.
92. **Borch, R. F.,** *Anal. Chem.,* 47, 2437, 1975.
93. **Hoffman, N. E. and Liao, J. C.,** *Anal. Chem.,* 48, 1104, 1976.
94. **Body, D. R.,** *Phytochemistry,* 13, 1527, 1974.
95. **Wood, P. D. S. and Holton, S.,** *Proc. Soc. Exp. Biol. Med.,* 115, 990, 1964.
96. **Karlsson, K.-A., Nilsson, K., and Pascher, I.,** *Lipids,* 3, 389, 1968.
97. **Laurell, S., Nilsen, R., and Norden, A.,** *Clin. Chim. Acta,* 36, 169, 1972.
98. **Body, D. R.,** *Biochim. Biophys. Acta,* 380, 45, 1975.
99. **Freeman, C. P. and West, D.,** *J. Lipid Res.,* 7, 324, 1966.
100. **Body, D. R.,** *Lipids,* 12, 204, 1977.

Chapter 8

CARBOCYCLIC AND FURANOID FATTY ACIDS

M. S. F. Lie Ken Jie

I. INTRODUCTION

Carbocyclic and furan fatty acids are less common in nature than the acyclic saturated and unsaturated fatty acids. But where such cyclic derivatives of fatty acids have been discovered in the lipid extracts of seed or animal fats, these compounds have shown extraordinary chemical, physical, physiological, and therapeutic properties. The precise role of many of these cyclic fatty acids remains a matter of speculation with the exception of the prostaglandins. This chapter refers briefly to the natural and synthetic sources of three-, five-, (excluding the prostaglandins) and six-membered carbocyclic fatty acids and also the furan fatty acids. Methods of isolation and detection are described. Their chromatographic properties are presented.

II. PREPARATION OF SAMPLES

A. Cyclopropane Fatty Acids

Some bacterial and insect lipid extracts and a number of the seed oils of the plant family Sapindaceae are sources of natural cyclopropane fatty acids.[1-7] Lactobaccilic (Structure 1), dihydrosterculic (Structure 2), and meromycolic acid (Structure 3) are a few of the more commonly encountered cyclopropane fatty acids.

$$CH_3-(CH_2)_n-\overset{\displaystyle \overset{CH_2}{\diagup \ \diagdown}}{CH-CH}-(CH_2)_m-COOH$$

$$n = 5, m = 9 \ (1)$$
$$n = 7, m = 7 \ (2)$$

$$CH_3-(CH_2)_{17}-\overset{\displaystyle \overset{CH_2}{\diagup \ \diagdown}}{CH-CH}-(CH_2)_{14}-\overset{\displaystyle \overset{CH_2}{\diagup \ \diagdown}}{CH-CH}-(CH_2)_{17}-COOH$$

$$(3)$$

The synthesis of cyclopropane fatty acids from ethylenic acids has been accomplished by the Simmons-Smith reaction.[8] *cis-* and *trans-* Cyclopropane fatty acids are readily prepared from the corresponding *cis-* and *trans-* ethylenic fatty acids, respectively.[9,10] Acid catalyzed cyclization reactions involving methyl epoxyoleate and some derivatives of methyl linoleate and ricinoleate have also been reported to give cyclopropane derivatives.[11-14]

Procedure for the Isolation of the Methyl Esters of Cyclopropane Fatty Acids from a Seed Oil[2]
(Details of the interesterification of a seed oil to the corresponding methyl ester derivatives are described in Chapter 2.)
Dissolve 150 mg of the methyl ester mixture in 3 mℓ of n-hexane and apply this solution as a narrow streak on a silver ion impregnated thin-layer chromatography (TLC) plate (10% AgNO₃ on silica gel (w/w), 20 × 20 cm, 1.0 mm wet thickness and activated for 2 hr at 110°C before use). Develop the plate with a mixture of diethyl ether- n-hexane (15:85 v/v). Spray the developed plate with a 0.1% solution of 2',7'-dichlorofluorescein in ethanol and visualize the plate under a uv light source (260 nm). The saturated

esters (including the cyclopropane esters) appear as a prominent yellow band (top most) at R_f = 0.6 to 0.7. Scrape off this band of silica into a 50 mℓ conical flask and extract the silica scraping with 3 × 10 mℓ of diethyl ether. Filter the ethereal extract and evaporate the solvent under reduced pressure.

Add 2 g of urea and 10 mℓ of methanol to the isolated saturated fatty ester fraction and warm on a water bath until solution of the urea crystals is complete. Allow the solution to cool to room temperature and leave at −5°C for 6 hr. Filter off the urea adduct under vacuum using a sintered glass funnel. Do not rinse the urea adduct. Collect the filtrate in a separating funnel and add 20 mℓ of 5% brine to the filtrate. Extract the aqueous methanolic mixture with 3 × 20 mℓ of n-hexane. Dry the hexane extract over anhydrous sodium sulfate, filter the solution, and concentrate for GC and other analyses. This extract contains the methyl ester derivatives of the cyclopropane fatty acids and also any trace of methyl-branched fatty esters that may be present in the oil.

1. Detection of Cyclopropane System in Fatty Acid by Proton Magnetic Resonance (PMR) Spectroscopy[9]

Run a proton magnetic resonance (PMR) spectrum of the isolated methyl esters, suspected to contain cyclopropane system, in CCl_4 with tetramethylsilane as standard (δ = 0). The presence of a cyclopropane system is confirmed by a relatively weak signal (multiplet) at −0.26 (one proton, H_b) and another at +0.64 (three protons, H_a, H_c, and H_d).

Procedure for the Determination of the Position of a Cyclopentane Ring in an Acyl Chain[15]

Prepare a solution of BF_3 in anhydrous methanol (1:1 w/w) and mix 1 mℓ of this solution with 35 mg of the methyl esters of cyclopropane fatty acids dissolved in 3 mℓ of dichloromethane in a stoppered flask. Allow the mixture to remain at room temperature for 3 hr. Add 5 mℓ of water, shake the reaction mixture vigorously, and isolate the dichloromethane layer. Wash the extract with 3 mℓ of water and dry over anhydrous sodium sulfate. Subject the mixture of products to mass spectral analysis or use GC/MS to analyze each component. The products of this reaction for $CH_3(CH_2)_nCHCH_2CH(CH_2)_mCOOCH_3$ *are*

---

* Characteristic mass spectral fragmentation patterns are shown by dotted lines.

B. Cyclopropene Fatty Acids

Cyclopropene fatty acids have been reported as constituents in the seed oils of many species of the families Malvaceae, Sterculiaceae, and Tilliaceae.[3,16] Sterculic (Structure 4) and malvalic acid (Structure 5) are the more commonly encountered cyclopropene fatty acids.

$$CH_3-(CH_2)_n-\overset{\displaystyle CH_2}{\overset{\diagup\diagdown}{C=C}}-(CH_2)_m-COOH$$

$$n = 7, m = 7 \quad (4)$$
$$n = 7, m = 6 \quad (5)$$

The synthesis of cyclopropene fatty acids from acetylenic fatty acids has recently been reviewed.[17] The high reactivity of the cyclopropene system towards silver nitrate makes argentation TLC unacceptable for the initial isolation and purification of this class of fatty acids.[18] However, deliberate treatment with silver nitrate converts the cyclopropene fatty acids to a mixture of ether and keto derivatives suitable for chromatography.[19] Direct GC analysis of cyclopropene fatty esters tends to cause isomerization and decomposition of fatty esters as they pass through the GC column.[20]

Procedure for the Detection of Cyclopropene Fatty Acids in Seed Oils (Halphen Test)[21]
Prepare a solution of 1% sulfur in carbon disulfide and then add an equal volume of 1-pentanol. Mix 10 mℓ of the seed oil with an equal volume of the sulfur solution in a test tube. Shake and heat gently in hot water (70 to 80°C) for a few minutes until the carbon disulfide is boiled off and the sample stops foaming. Place the tube in a bath maintained at 110 to 115°C and hold for 1 to 2 hr. A red color at the end of this period indicates the presence of cyclopropene fatty acids in the oil.

Procedure for the Quantitative Determination of Cyclopropene Fatty Acids in Seed Oils[22,23]
Prepare a standard solution of 0.1 M hydrogen bromide by adding the appropriate amount of glacial acetic acid to a commercially available high concentration solution of anhydrous hydrogen bromide in glacial acetic acid. This reagent is standardized daily against anhydrous sodium carbonate (0.1 g is dissolved in 5 mℓ glacial acetic acid and titrated to the blue-green endpoint of crystal violet indicator).
Dissolve 25 g of the crude seed oil in 25 mℓ of petroleum hydrocarbon (b.p. 40 to 60°C) and percolate this solution through an alumina column (100 g of activated alumina in a column of internal diameter of 16 mm containing petroleum hydrocarbon) and elute with an additional amount of 225 mℓ of the same solvent.* Evaporate the solvent completely under reduced pressure at <60°C on a rotary evaporator.
Weigh accurately about 7.0 g of the solvent-free alumina-treated oil sample into a 50 mℓ conical flask, and dissolve the sample in 5 mℓ of benzene and 15 mℓ of glacial acetic acid. Add a few drops of a 0.1% solution of crystal violet in acetic acid and place a small Teflon coated magnet into the flask. Attach the flask to a titration buret graduated in 0.05 mℓ divisions and cool the flask externally in an ice bath on top of a magnetic stirrer for 15 min until its content reaches 3°C. Titrate the solution until a blue-green endpoint of at least 30 sec duration is observed. Note the volume (V_1) of the HBr solution consumed.
Replace the ice bath with a water bath maintained at 55°C. Allow the sample to reach this temperature and continue the titration to the same endpoint. Record the volume (V_2) of HBr solution used. Correct the two volumes for a solvent blank titration.
Calculate the cyclopropene fatty acid concentration in terms of sterculic acid or malvalic acid as follows:

---

* For crude cotton seed oil a second or third alumina treatment may be necessary.

1. % Sterculic acid = 29.45 × MV$_2$/W
2. % Malvalic acid = 28.04 × MV$_2$/W
3. M = molarity of HBr solution
4. V$_2$ = volume in mℓ of HBr solution used
5. W = weight of oil sample used

Procedure for the Indirect Quantitative Determination of Cyclopropene Fatty Acids by GC Analysis[19]
Interesterify 100 mg of the seed oil and isolate the methyl ester derivatives. Treat the methyl esters with 15 mℓ of anhydrous methanol saturated with silver nitrate at 30°C for 2 hr or for 20 hr at room temperature. Dilute the methanolic mixture with 30 mℓ of water and extract with 2 × 10 mℓ petroleum hydrocarbon. Dry the combined petroleum hydrocarbon fractions over anydrous sodium sulfate and filter the solution. For oils containing substantial amounts of cyclopropene fatty acids, the reaction product after evaporation of the solvent is suitable for direct injection into the GC instrument.
For oils containing low levels of cyclopropenes (<5%), the petroleum hydrocarbon extracts of the silver nitrate treated methyl esters is loaded on a chromatographic column (12 mm × 450 mm column size, containing 8 g of freshly prepared Woelm neutral alumina, activity grade II). Elute the column with 200 mℓ of 1% diethyl ether in petroleum hydrocarbon. Continue the elution with 150 mℓ of chloroform. Add a known amount of methyl henicosanoate and tricosanoate as internal standards to both eluates and concentrate the solutions for GC analysis. The ether and keto derivatives are contained in the chloroform fractions, while most of the normal fatty esters are in the petroleum hydrocarbon fraction. Inject suitable quantities of each into a GC column (15% DEGS on Anakrom A, 3 m × 0.3 mm O.D., 190°C, 40 mℓ N$_2$/min).
The peaks at equivalent chain length (ECL) values of 21.7 and 22.7 from the chloroform extract are due to the ether derivatives of methyl malvalate and sterculate, respectively. The peaks with ECL values of 24.5 and 25.5 represent the corresponding keto derivatives of malvalate and sterculate. Measure the area of each peak and determine the percentage of methyl malvalate and sterculate with reference to the internal standards.

C. ω-Cyclopentenyl and ω-Cyclohexyl Fatty Acids

ω-Cyclopentenyl fatty acids are found in the seed oils of the family Flacourtiaceae.[3] Hydnocarpic (Structure 6), chaulmoogric (Structure 7), and gorlic acid (Structure 8) are the more common isomers isolated. A recent study reveals the presence of a higher homolog of chaulmoogric acid and also several C$_{16}$, C$_{18}$, and C$_{20}$ homologs of gorlic acid with the unsaturated center located at different positions of the fatty acid chain.[24]

$\bigcirc\!\!\!-$ (CH$_2$)$_n$–COOH	n = 10 (6)
	n = 12 (7)
$\bigcirc\!\!\!-$ (CH$_2$)$_6$–CH=CH–(CH$_2$)$_4$–COOH	(8)

Various strains of acidothermophilic bacteria also contain high percentages (up to 65%) of ω-cyclohexyl fatty acids (9).[25,26] Trace quantities of similar fatty acids have been found in milk fat and in the perinephric fat of sheep.[27-29]

$\bigcirc\!\!\!-$ (CH$_2$)$_n$–COOH	n = 10 or 12 (9)

Isolation of cyclopentenyl fatty acids can be accomplished by following the procedure described for cyclopropane fatty acids.

Procedure for the Isolation of ω-Cyclohexyl Fatty Acids from Bacteria[26]
*Separate the bacterial cells from the culture medium by centrifugation and wash with
0.1 M sodium chloride solution. Extract the lyophilized cells with a mixture of chloroform-methanol (1:1 v/v) using a Soxhlet extractor. Saponify the crude lipid extract
with a 10% solution of potassium hydroxide in methanol for 6 hr and esterify the fatty
acids by a standard method. Examine the methyl esters by GC.*

D. Furanoid Fatty Acids

Morris et al.[30] have claimed an isolation of a C_{18} furan fatty acid (Structure 10) from the
seed oil of *Exocarpus cupressiformis* Labill. Crundwell and Cripps[31] believe this acid to
have arisen as an artifact of the isolation procedure from a C_{18} acid containing a hydroxy-
cis-enzyme (−CHOH−CH=CH−C≡C−) system, which is apt to furan formation with alkali
or on treatment with silver nitrate.

$$CH_3(CH_2)_5 \overset{\text{[furan]}}{\diagup\diagdown}{}_{O} (CH_2)_7COOH \tag{10}$$

The entire series of C_{18} furan acids has been synthesized from furan, furfural, or furoic
acid,[32,33] and some of these positional isomers have also been obtained through cyclization
reactions involving linoleic, ricinoleic, and diacetylenic acids.[34-38]
However, a series of closely related methyl-substituted furan fatty acids (Structure 11)
have been identified from the liver and testes of fresh water[39-41] and marine fish oils.[42,43]

$$CH_3-(CH_2)_n \overset{R\quad\quad R}{\diagup\diagdown}{}_{O} (CH_2)_m-COOH \tag{11}$$

R = H or CH₃
m + n = 10, 12,
14 or 16

Another source of furan fatty acids has been reported by Fawcett et al.[44] from shoots of
the broad bean that furnish an unsaturated keto derivative of a C_{14} furan acid known as
wyerone acid (Structure 12).

$$CH_3-CH_2-\overset{cis}{CH=CH}-C≡C-\underset{\overset{||}{O}}{C}\overset{\text{[furan]}}{\diagup\diagdown}{}_{O}\overset{trans}{CH=CH}-COOH \tag{12}$$

This acid has been noted for its fungitoxic properties[45] and in vitro with *Botrytis cinerea*
and *B. fabae* gives the corresponding hydroxy derivative.[46] An epoxy derivative of wyerone
acid has also been isolated from a similar source and shown to be metabolized by fungi to
several hydroxy derivatives.[47]

Procedure for the Detection of Furanoid Fatty Acids by TLC[41]
*Spot about 1 to 2 mg of the methyl ester mixture (containing saturated, unsaturated,
and furan fatty esters) on a silver nitrate impregnated layer (10% $AgNO_3$ in silica gel
G, 0.3 mm wet thickness, activated at 110°C for 1 hr before use), and develop the
plate with diethyl ether-n-hexane (1:9 v/v).
Spray the developed plate with a 2% solution of tetracyanoethylene in acetone. The
furan esters appear as blue spots on a pale yellow background. The reagent is nondestructive and is sensitive to above 5% of furan esters in the mixture of fatty esters.*

Procedure for the Separation of Furanoid Fatty Acids from Polyunsaturated Fatty Acids[41]
Fill a 1 mℓ tuberculin syringe with silica gel G containing 5% silver nitrate (w/w) to the 0.6 mℓ mark. Clamp the syringe in a vertical position and dispense 50 µℓ of a 5 to 10% solution of methyl esters dissolved in n-hexane or methanol-benzene (1:1 v/v) on to the top of the dry column. As soon as the liquid has gone into the column, fill the space over it with n-hexane-diethyl ether (3:2 v/v). Collect the first 0.2 mℓ of eluent, which contains all saturated fatty esters including the furan esters.

Procedure for the Quantitative Analysis of Furanoid Fatty Acids in Fish Oils[43]
Weigh 150 to 180 mg of oil into a quickfit test tube and add 1 mℓ of standard squalane solution in methanol (0.25 mg/mℓ) and 6 mℓ of sodium methoxide (0.5 M in methanol) solution. Stopper the tube and heat it at 50°C for 20 min, cool the contents, and transfer this to a 50 mℓ separating funnel. Add 10 mℓ of 5% NaCl solution and 0.5 mℓ of glacial acetic acid, and then extract twice with 10 mℓ of n-hexane. Combine the n-hexane extracts, wash once with 5 mℓ of water, and dry over anhydrous Na₂SO₄. Filter the solution and concentrate by evaporation to about 0.5 mℓ.
Apply this solution as a narrow streak to a silver nitrate-impregnated layer (10% silver nitrate in silica gel G (w/w), 20 × 20 cm, 1 mm wet thickness, activated at 110°C for 2 hr) and develop the plate with diethyl ether-n-hexane (1:9 v/v). Spray the developed plate with a 1% ethanolic solution of 2',7'-dichlorofluorescein and visualize under uv light source (260 nm). The monoene band is prominent at R_f = 0.4 to 0.5. Mark the plate just below the front of this band and scrape off all the silica above it. Extract the silica with 4 × 15 mℓ of diethyl ether and evaporate the extract to dryness.
Transfer the extract to a small conical flask with 3 mℓ of methanol, add 0.5 g of urea, dissolve it by warming and leave the solution to crystallize at −15°C for 3 hr. Filter the solution through a sintered glass funnel into a small quickfit test tube (without washing the precipitate) and add 6 mℓ 5% NaCl solution to the filtrate. Shake this twice with 0.5 mℓ of n-hexane, drawing the hexane extract off with a Pasteur pipet each time. Dry the extract over anhydrous sodium sulfate and concentrate as required before GC analysis.

Procedure for the Isolation of Wyerone from Broad Beans[44]
Soak 4 kg broad beans in water for 24 hr and allow to germinate for 8 days in the dark between steam-sterilized wet sacks. Collect 10 kg of seedlings and cool to 2°C during 24 hr period. Mince the seedling and steep in 10 ℓ benzene for 24 hr at 15°C. Filter the benzene extract, dry over anhydrous magnesium sulfate and evaporate the solvent under reduced pressure.
Load the extract on to two silica gel (acid washed) columns (60 × 6 cm) wrapped with black cloth or paper. Elute the columns with diethyl ether-benzene mixtures and collect fractions of 50 mℓ each. (Fraction 1 to 15 with 0 to 5% diethyl ether and continue thereafter with 5% diethyl ether in benzene.) Pool fractions 24 to 31 and evaporate the solvent. Recrystallize the extract from n-hexane at −20°C several times until the pale yellow needles of wyerone attain a melting point of 63.5 to 64.5°C.

Recently, a new group of furanoid acids has been found in human blood and urine. These "urofuran" acids are dicarboxylic acids with a total of ten or twelve carbon atoms, including those of the furane ring.[48-50] One of the carboxyl groups is attached to the furan ring, whereas the other is present as ethylcarboxyl moiety.

III. CHROMATOGRAPHY

Gas chromatography is the method of choice for the analysis of methyl esters of the various cyclic fatty acids.

The equivalent chain length (ECL) values of methyl esters of cyclopropane fatty acids are presented in Tables 1 to 3; the ECL values of various methyl esters of ω-cyclopentenyl and ω-cyclohexyl fatty acids in Table 4; and the ECL and R_f values of methyl esters of furan fatty acids are presented in Tables 5 to 7. Reference is made to a recent publication on the analysis of furanoid fatty acids including urofuran acids.[51]

Table 1
METHYL ESTERS OF *cis* MONOCYCLOPROPANE FATTY ACIDS (GC)

	P1	P2	P3	P4	P5	P6	P7	P8	P9	P10	P11	P12
Column packing												
Temperature (°C)	190	180	190	220	190	190	150	190	180	220	na	na
Gas	N_2	N_2	N_2	N_2	N_2	N_2	na	na	Ar	He	na	na
Flow rate (mℓ/min)	20 psi	50 psi	50 psi	25 psi	60	50	na	na	70	50	na	na
Column												
Length (m)	50	2	2	50	2	2	na	na	2.4	1.8	na	na
Diameter (cm I.D.)	0.025	0.6	0.6	0.025	0.6	0.6	na	na	0.6	0.95	na	na
Form	Capillary	Coiled	Coiled	Capillary	Coiled	Coiled	Capillary	Capillary	Coiled	na	na	na
Material	Stainless steel	na	na	Stainless steel	Glass	Glass	na	na	Glass	Aluminum	na	na
Detector	FID	FID	FID	FID	FID	FID	na	na	β-ionization detector	FID	na	na
Reference	1	1	1	1	2	2	3	3	4	4	5	5
Compound					ECL							
Methyl *cis*-9,10-methyleneundecanoate	—	—	—	—	—	—	12.88	12.22	—	—	—	—
Methyl *cis*-10,11-methyleneundecanoate	—	—	—	—	—	—	—	—	13.70	11.20	—	—
Methyl *cis*-3,4-methylenedodecanoate	—	—	—	—	—	—	13.35	12.86	—	—	—	—
Methyl *cis*-5,6-methylenetetradecanoate	—	—	—	—	—	—	15.28	14.82	—	—	—	—
Methyl *cis*-7,8-methylenetetradecanoate	—	—	—	—	—	—	15.29	—	—	—	—	—
Methyl *cis*-9,10-methylenetetradecanoate	—	—	—	—	—	—	15.38	—	—	—	—	—
Methyl *cis*-11,12-methylenetetradecanoate	—	—	—	—	—	—	15.55	—	—	—	—	—
Methyl *cis*-8,9-methyleneheptadecanoate	—	—	—	—	—	—	—	—	—	—	17.70	18.40
Methyl *cis*-7,5-methylenehexadecanoate	—	—	—	—	—	—	17.24	16.75	—	—	—	—
Methyl *cis*-9,10-methylenehexadecanoate	—	—	—	—	17.50	16.90	17.25	—	17.57	16.85	—	—
Methyl *cis*-11,12-methylenehexadecanoate	—	—	—	—	—	—	17.33	—	—	—	—	—
Methyl *cis*-2,3-methyleneoctadecanoate	19.38	19.64	19.55	18.98	—	—	—	—	—	—	—	—
Methyl *cis*-3,4-methyleneoctadecanoate	19.32	19.62	19.50	18.85	—	—	—	—	—	—	—	—
Methyl *cis*-4,5-methyleneoctadecanoate	19.30	19.58	19.46	18.84	—	—	—	—	—	—	—	—
Methyl *cis*-5,6-methyleneoctadecanoate	19.20	19.51	19.36	18.76	—	—	—	—	—	—	—	—
Methyl *cis*-6,7-methyleneoctadecanoate	19.20	19.55	19.37	18.76	—	—	—	—	19.56	18.93	—	—

Methyl ester	P1	P2	P3	P4	P5	P6	P7	P8	P9	P10	P11	
Methyl cis-7,8-methyleneoctadecanoate	19.18	19.52	19.35	18.75	—	—	—	—	—	—	—	—
Methyl cis-8,9-methyleneoctadecanoate	19.15	19.51	19.37	18.73	—	—	—	—	—	—	—	—
Methyl cis-9,10-methyleneoctadecanoate	19.24	19.55	19.37	18.73	19.50	18.90	19.18	18.72	19.46	18.77	18.70	19.40
Methyl cis-10,11-methyleneoctadecanoate	19.25	19.55	19.40	18.78	—	—	—	—	19.57	18.88	—	—
Methyl cis-11,12-methyleneoctadecanoate	19.25	19.58	19.44	18.78	—	—	—	—	—	—	—	—
Methyl cis-12,13-methyleneoctadecanoate	19.29	19.65	19.48	18.80	—	—	—	—	—	—	—	—
Methyl cis-13,14-methyleneoctadecanoate	19.41	19.71	19.56	18.85	—	—	—	—	—	—	—	—
Methyl cis-14,15-methyleneoctadecanoate	19.45	19.82	19.65	18.96	—	—	—	—	—	—	—	—
Methyl cis-15,16-methyleneoctadecanoate	19.50	19.94	19.80	19.02	—	—	—	—	—	—	—	—
Methyl cis-16,17-methyleneoctadecanoate	19.80	20.30	20.19	19.22	—	—	—	—	—	—	—	—
Methyl cis-17,18-methyleneoctadecanoate	20.04	20.54	20.38	19.38	—	—	—	—	—	—	—	—
Methyl cis-11,12-methyleneeicosanoate	—	—	—	—	—	—	—	—	21.31	20.85	—	—
Methyl cis-9,10-methylenedocosanoate	—	—	—	—	—	—	—	—	23.38	22.80	—	—

Column packing P1 = Neopentyl glycol succinate (coating)
P2 = 20% Diethylene glycol succinate on Gas Chrom Z (70/80 mesh)
P3 = 15% Polyethylene glycol adipate on Gas Chrom Z (70/80 mesh)
P4 = Apiezon L (coating)
P5 = 10% Silar 10C on Chromosorb W (80/100 mesh)
P6 = 1.5% OV-101 on Chromosorb W (80/100 mesh)
P7 = Butanediol succinate (coating)
P8 = Apiezon L (coating)
P9 = 20% Ethylene glycol succinate and 2% phosphoric acid on Gas Chrom P (80/100 mesh)
P10 = Apiezon L
P11 = Resoflex - 446

REFERENCES

1. **Christie, W. W., Gunstone, F. D., Ismail, I. A., and Wade, L.,** *Chem. Phys. Lipids,* 2, 196, 1968.
2. **Lie Ken Jie, M. S. F. and Chan, M. F.,** *J. Chem. Soc., Chem. Commun.,* 78, 1977.
3. **Ackman, R.G.,** *Progr. Chem. Fats Other Lipids,* 12, 165, 1972.
4. **Christie, W.W. and Holman, R.T.,** *Lipids,* 1, 176, 1966.
5. **Bohannon, M.B. and Kleiman, R.,** *Lipids,* 13, 270, 1978.

Table 2
METHYL ESTERS OF *trans*-MONOCYCLOPROPANE FATTY ACIDS (GC)

Column packing	P1	P2	P3	P4
Temperature (°C)	na	na	180	220
Gas	—	—	Ar	He
Flow rate (mℓ/min)	na	na	70	50
Column				
Length (m)	2	2	2.4	1.8
Diameter (cm I.D.)	0.6	0.6	0.6	0.6
Form	Coiled	Coiled	na	na
Material	na	na	Glass	Aluminum
Detector	FID	FID	β-ionization detector	FID
Reference	1	1	2	2

Compound	ECL			
Methyl *trans*-2,3-methyleneoctadecanoate	19.67	18.98	—	—
Methyl *trans*-3,4-methyleneoctadecanoate	18.93	18.45	—	—
Methyl *trans*-4,5-methyleneoctadecanoate	18.91	18.47	—	—
Methyl *trans*-5,6-methyleneoctadecanoate	18.92	18.45	—	—
Methyl *trans*-6,7-methyleneoctadecanoate	18.90	18.40	18.92	18.57
Methyl *trans*-7,8-methyleneoctadecanoate	18.88	18.40	—	—
Methyl *trans*-8,9-methyleneoctadecanoate	18.90	18.42	—	—
Methyl *trans*-9,10-methyleneoctadecanoate	18.93	18.40	18.80	18.52
Methyl *trans*-10,11-methyleneoctadecanoate	18.91	18.41	—	—
Methyl *trans*-11,12-methyleneoctadecanoate	18.96	18.44	—	—
Methyl *trans*-12,13-methyleneoctadecanoate	19.03	18.46	—	—
Methyl *trans*-13,14-methyleneoctadecanoate	19.07	18.52	—	—
Methyl *trans*-14,15-methyleneoctadecanoate	19.17	18.59	—	—
Methyl *trans*-15,16-methyleneoctadecanoate	19.24	18.66	—	—
Methyl *trans*-16,17-methyleneoctadecanoate	19.41	18.72	—	—
Methyl *trans*-17,18-methyleneoctadecanoate	20.56	19.35	—	—

Column packing   P1 =   20% Diethylene glycol succinate on HMDS Chromosorb W
                                P2 =   3% Apiezon L on AW DMCS Chromosorb G
                                P3 =   20% Ethylene glycol succinate and 2% phosphoric acid on Gas-Chrom P (80—100 mesh)
                                P4 =   20% Apiezon L on Gas-Chrom P (80—100 mesh)

REFERENCES

1. **Gunstone, F. D. and Perera, B. S.,** *Chem. Phys. Lipids,* 10, 303, 1973.
2. **Christie, W. W. and Holman, R. T.,** *Lipids,* 1, 176, 1966.

Table 3
METHYL ESTERS OF POLYCYCLOPROPANE FATTY ACIDS (GC)

Column packing	P1	P2	P3	P4
Temperature (°C)	210	180	180	220
Gas	N_2	N_2	Ar	He
Flow rate (mℓ/min)	8 p.s.i.	6 p.s.i.	70	50
Column				
Length (m)	50	50	2.4	1.8
Diameter (mm I.D.)	0.05	0.05	0.6	0.9
Form	Coiled	Coiled	na	na
Material	SS	SS	Glass	Aluminum
Detector	FID	FID	β-ionization detector	FID
Reference	1	1	2	2

Compound	ECL			
Methyl all *cis*-4,5;7,8-dimethyleneoctadecanoate	—	—	20.09	19.68
Methyl all *cis*-5,6;8,9-dimethyleneoctadecanoate	—	—	20.83	19.56
Methyl all *cis*-5,6;12,13-dimethyleneoctadecanoate	19.54	21.00	—	—
Methyl all *cis*-6,7;9,10-dimethyleneoctadecanoate	19.47[a]	20.94[a]	—	—
	19.39	20.82	—	—
Methyl all *cis*-6,7;10,11-dimethyleneoctadecanoate	19.44	20.89	—	—
Methyl all *cis*-6,7;11,12-dimethyleneoctadecanoate	19.48	20.95	—	—
Methyl all *cis*-6,7;12,13-dimethyleneoctadecanoate	19.50	20.97	—	—
Methyl all *cis*-7,8;12,13-dimethyleneoctadecanoate	19.51	20.99	—	—
Methyl all *cis*-8,9;12,13-dimethyleneoctadecanoate	19.50	20.96	—	—
Methyl all *cis*-9,10;12,13-dimethyleneoctadecanoate	19.53[a]	21.06[a]	21.07	19.60
	19.46	20.97		
Methyl all *cis*-10,11;13,14-dimethyleneoctadecanoate	—	—	21.05	19.63
Methyl all *cis*-9,10;12,13;15,16-trimethyleneoctadecanoate	—	—	22.95	20.58
Methyl all *cis*-5,6;8,9;11,12;14,15-tetramethyleneeicosanoate	—	—	25.68	23.01
Methyl all *trans*-9,10;11,12-dimethyleneoctadecanoate	—	—	20.02	18.12
Methyl all *trans*-9,10;12,13-dimethyleneoctadecanoate	—	—	19.90	18.95

[a]   Stereoisomers.

Column packing P1 = Apiezon L (coating)
P2 = Diethylene glycol succinate (coating)
P3 = 20% Ethylene glycol succinate and 2% phosphoric acid on Gas-Chrom P (80–100 mesh)
P4 = 20% Apiezon L on Gas-Chrom P (80–100 mesh)

REFERENCES

1. **Gunstone, F. D., Lie Ken Jie, M. S. F., and Wall, R. T.**, *Chem. Phys. Lipids,* 6, 147, 1971.
2. **Christie, W. W. and Holman, R. T.**, *Lipids,* 1, 176, 1966.

Table 4
METHYL ESTERS OF ω-CYCLIC FATTY ACIDS (GC)

	P1	P2	P3	P4	P5	P6	P7	P8	P9
Column packing									
Temperature (°C)	179	179	na	na	200	170	235	240	190
Gas	—	—	—	—	—	—	—	N_2	N_2
Flow rate (mℓ/min)	na	na	na	na	na	na	na	35	35
Column									
Length (m)	12	12	50	50	2	2	2	0.8	1.6
Diameter (cm I.D.)	na	na	na	na	0.5	0.5	0.5	na	na
Form	na	na	Capillary	Capillary	na	na	na	na	na
Material	na	na	na	na	na	na	na	na	na
Detector	na	na	na	na	na	na	na	FID	FID
Reference	1	1	2	2	3	3	3	4	4

Compound				ECL					
cyclobutyl—$(CH_2)_{10}COOCH_3$	16.10	15.60	—	—	16.22	15.40	15.64	—	—
cyclopentyl—$(CH_2)_9OOCH_3$	16.50	15.80	16.50	15.80	16.65	15.60	16.04	—	—
cyclopentyl—$(CH_2)_{10}COOCH_3$	—	—	18.65	16.85	—	—	—	16.85	18.70
cyclopentyl—$(CH_2)_{12}COOCH_3$	—	—	20.50	18.90	—	—	—	18.90	20.64
cyclopentyl—$(CH_2)_6CH=CH(CH_2)_4COOCH_3$	—	—	—	—	—	—	—	18.68	21.20
cyclohexyl—$(CH_2)_8COOCH_3$	16.70	15.90	16.70	15.90	16.90	16.80	16.32	—	—
cyclohexyl—$(CH_2)_{10}COOCH_3$	—	—	18.40	17.92	17.90	17.80	17.32	—	—
cycloheptyl—$(CH_2)_{10}COOCH_3$	—	—	—	—	20.60	18.92	19.68	—	—

Column packing P1 = 20% Reoplex 400 on Celite 545
P2 = 3% Apiezon L on Aeropak 30
P3 = Butanediol succinate or Neopentyl glycol succinate (coating)
P4 = Apiezon L (coating)
P5 = 10% Diethylene giycol succinate on silanized Chromosorb P (100–120 mesh)
P6 = 5% SE-30 on silanized Chromosorb P (100–120 mesh)
P7 = 10% GAL on silanized Chromosorb P (100–120 mesh)
P8 = 20% Apiezon M on Celite (0.2/0.3 mm)
P9 = 20% Polyethylene glycol succinate on Celite (0.2/0.3 mm)

REFERENCES

1. **Sen Gupta, A. K. and Peters, H.,** *Chem. Phys. Lipids,* 3, 371, 1969.
2. **Ackman, R. G.,** *Progr. Chem. Fats Other Lipids,* 207, 1972.
3. **De Rosa, M. and Gambacorta, A.,** *Phytochemistry,* 14, 209, 1975.
4. **Zeman, I. and Pokorny, J.,** *J. Chromatogr.,* 10, 15, 1963.

Table 5
METHYL ESTERS OF FURAN FATTY ACIDS (GC)

Column packing	P1	P2	P3	P4	P5	P6	P7	P8
Temperature (°C)	220	225	235	190	195	na	195	190
Gas Flow rate (mℓ/min)	120	55	100	80	30	na	40	40
Column Length (m)	2	2	2	2	2	na	2	2
Diameter (cm I.D.)	0.31	0.62	0.62	0.62	0.62	na	0.2	0.31
Form	Coiled	Coiled	Coiled	Coiled	Coiled	na	Coiled	Coiled
Material	Glass	Glass	Glass	Glass	Glass	na	SS	SS
Detector	FID	FID	FID	FID	FID	na	FID	FID
Reference	1	1	1	1	1	2	3	4

Compound — **ECL**

$CH_3(CH_2)_n$—furan—$(CH_2)_m COOCH_3$

n	m								
12	0	19.39	19.26	22.00	23.17	23.87	—	—	—
11	1	18.12	18.22	20.46	21.49	21.72	—	—	—
10	2	18.09	18.20	20.11	21.06	21.23	—	—	—
9	3	18.00	18.10	20.06	21.02	21.35	—	—	—
8	4	17.89	18.08	20.12	21.16	21.59	—	—	—
7	5	17.87	18.12	20.12	21.19	21.63	—	—	—
6	6	17.99	18.13	20.16	21.21	21.72	—	—	—
5	7	18.02	18.14	20.23	21.23	21.82	18.15	—	—
4	8	18.06	18.18	20.28	21.35	21.87	—	—	—
3	9	18.14	18.24	20.39	21.45	21.93	—	—	—
2	10	18.21	18.31	20.51	21.57	22.03	—	—	—
1	11	18.43	18.52	20.82	21.97	22.45	—	—	—
0	12	18.66	18.73	21.38	22.50	23.05	—	—	—
—	13	18.94	18.97	21.83	23.08	23.84	—	—	—

$CH_3(CH_2)_n$—(R, CH_3 furan)—$(CH_2)_m COOCH_3$

R	n	m								
H	4	8	—	—	—	—	—	—	21.53	—
H	4	10	—	—	—	—	—	—	23.46	—
H	5	7	—	—	—	—	—	—	—	18.63
CH₃	2	8	—	—	—	—	—	—	20.56	—
CH₃	4	8	—	—	—	—	—	—	22.07	—
CH₃	2	10	—	—	—	—	—	—	22.57	—
CH₃	4	10	—	—	—	—	—	—	24.04	—
CH₃	5	7	—	—	—	—	—	—	—	19.26

Column packing P1 = 5% Apiezon L on Chromosorb W (80—100 mesh)
P2 = 10% SE-30 on Chromosorb W (80—100 mesh)
P3 = 10% FFAP on Chromosorb W (80—100 mesh)
P4 = 20% Diethylene glycol succinate on Chromosorb W (80—100 mesh)
P5 = 10% Silar 10C on Chromosorb W (80—100 mesh)
P6 = SE-30
P7 = 15% Diethylene glycol succinate on Chromosorb W AW (80—100 mesh)
P8 = 1.5% OV-101 on Chromosorb W (80—100 mesh)

Table 5 (continued)
METHYL ESTERS OF FURAN FATTY ACIDS (GC)

REFERENCES

1. **Lie Ken Jie, M. S. F. and Lam, C. H.,** *J. Chromatogr.,* 138, 373, 1977.
2. **Morris, L. J., Marshall, M. O., and Kelley, W.,** *Tetrahedron Lett.,* 36, 4249, 1966.
3. **Scrimgeour, C. M.,** *J. Am. Oil Chem. Soc.,* 54, 210, 1977.
4. **Lie Ken Jie, M. S. F. and Lam, C. H.,** *Chem. Phys. Lipids,* 20, 1, 1977.

See also **Sand, D. M., Schlenk, H., Thoma, H., and Spiteller, G.,** *Biochim. Biophys. Acta,* 751, 455, 1983.

Table 6
METHYL ESTERS OF UNSATURATED FURAN FATTY ACIDS (GC)

Column packing	P1	P2	P3	P4	P5
Temperature (°C)	220	225	235	190	195
Gas					
Flow rate (mℓ/min)	120	55	100	80	30
Column					
Length (cm)	2	2	2	2	2
Diameter (cm I.D.)	0.31	0.62	0.62	0.62	0.62
Form	Coiled	Coiled	Coiled	Coiled	Coiled
Material	Glass	Glass	Glass	Glass	Glass
Detector	FID	FID	FID	FID	FID
Reference	1	1	1	1	1

Compound		ECL			
(furan)–CH=CH–(CH$_2$)$_9$COOCH$_3$ (c)	17.43	17.36	21.04	22.77	23.86
(furan)–CH=CH–(CH$_2$)$_9$COOCH$_3$ (t)	17.81	17.68	21.61	23.46	24.67
(furan)–CH=CH–(CH$_2$)$_{11}$COOCH$_3$ (c)	19.43	19.35	23.11	24.68	25.79
(furan)–CH=CH–(CH$_2$)$_{11}$COOCH$_3$ (t)	19.80	19.65	23.65	25.32	26.54
CH$_3$(CH$_2$)$_4$–(furan)–CH=CH(CH$_2$)$_6$COOCH$_3$ (c)	18.43	—	21.39	22.73	23.72
CH$_3$(CH$_2$)$_4$–(furan)–CH=CH(CH$_2$)$_6$COOCH$_3$ (t)	18.43	—	22.45	22.73	25.03
CH$_3$(CH$_2$)$_{10}$–(furan)–CH=CH–COOCH$_3$ (t)	19.96	19.77	22.80	24.00	24.78
H$_2$C=HC–(furan)–(CH$_2$)$_{11}$COOCH$_3$	18.98	18.94	22.24	23.67	24.73

Column packing P1 = 5% Apiezon L on Chromosorb W (80—100 mesh)
                  P2 = 10% SE-30 on Chromosorb W (80—100 mesh)
                  P3 = 10% FFAP on Chromosorb W (80—100 mesh)
                  P4 = 20% Diethylene glycol succinate on Chromosorb W (80—100 mesh)
                  P5 = 10% Silar 10C on Chromosorb W (80—100 mesh)

REFERENCE

1. **Lie Ken Jie, M. S. F. and Lam, C. H.,** *J. Chromatogr.,* 138, 446, 1977.

Table 7

METHYL ESTERS OF FURAN FATTY ACIDS (TLC)

	L1	L2	L3	L1	L4	L1	L5
Layer							
Solvent	S1	S2	S3	S4	S5	S6	S7
Technique	T1	T1	T2	T1	T1	na	na
Reference	1	1	2	3	4	5	6

Compound $R_F \times 100$

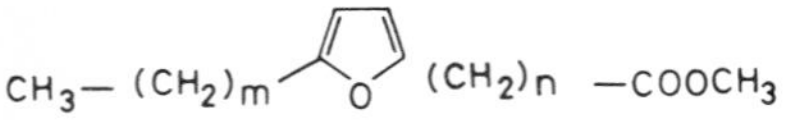

$$CH_3-(CH_2)_m \overset{\text{furan}}{} (CH_2)_n -COOCH_3$$

m	n							
12	0	46	66	—	—	—	—	—
11	1	43	60	—	—	—	—	—
10	2	46	65	—	—	—	—	—
9	3	46	65	—	—	—	—	—
8	4	45	65	—	—	—	—	—
7	5	47	63	—	—	—	—	—
6	6	47	63	—	—	—	—	—
5	7	48	63	85	—	—	—	—
4	8	49	63	—	—	—	—	—
3	9	50	64	—	—	—	—	—
2	10	50	64	—	—	—	—	—
1	11	50	64	—	—	—	—	—
0	12	48	65	—	—	—	—	—
—	13	48	64	—	—	—	—	—

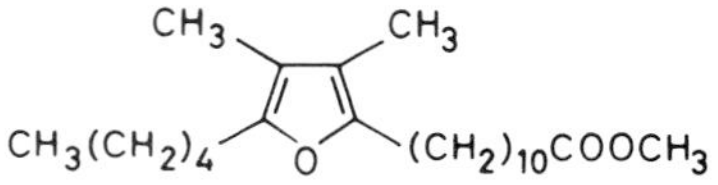

$$CH_3(CH_2)_4 \overset{CH_3 \quad CH_3}{\text{furan}} (CH_2)_{10}COOCH_3 \qquad — \quad — \quad — \quad 70 \quad — \quad — \quad —$$

$$CH_3CH_2CH\overset{c}{=}CHC\equiv C-\underset{O}{\overset{\parallel}{C}}-\overset{t}{\text{furan}}-CH=CHCOOCH_3 \qquad — \quad — \quad — \quad — \quad 45 \quad 44 \quad 86$$

$$CH_3CH_2CH\overset{c}{=}CHC\equiv C-\underset{OH}{CH}-\overset{t}{\text{furan}}-CH=CHCOOCH_3 \qquad — \quad — \quad — \quad — \quad — \quad 35 \quad —$$

$$CH_3CH_2\overset{O}{CH}-CHC\equiv C-\underset{O}{\overset{\parallel}{C}}-\overset{t}{\text{furan}}-CH=CHCOOCH_3 \qquad — \quad — \quad — \quad — \quad — \quad — \quad 78$$

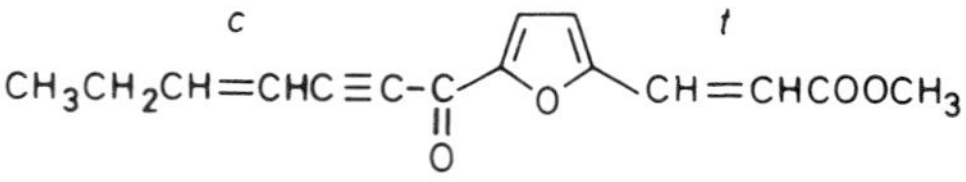

$$CH_3CH_2\overset{O}{CH}-CHC\equiv C-\underset{OH}{CH}-\overset{t}{\text{furan}}-CH=CHCOOCH_3 \qquad — \quad — \quad — \quad — \quad — \quad 49 \quad —$$

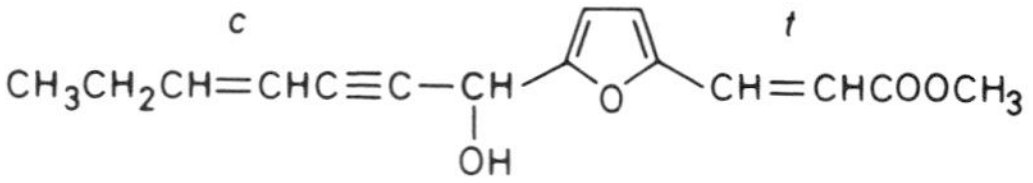

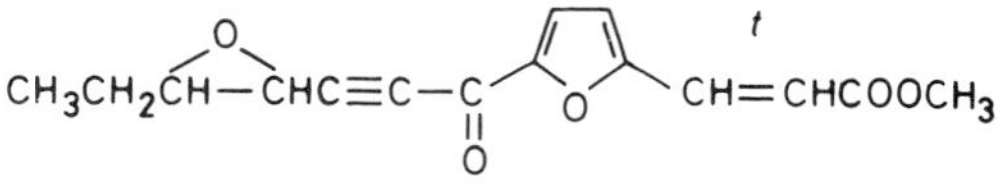

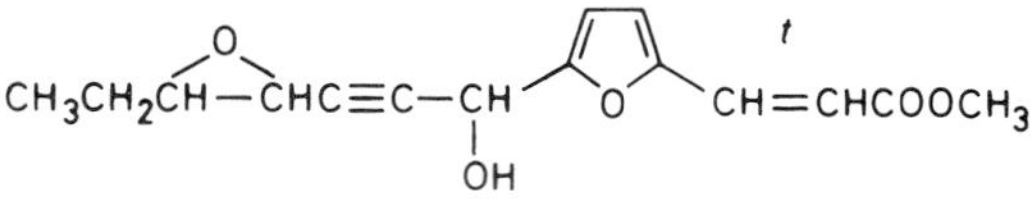

$$CH_3CH_2\underset{OH}{CH}-\underset{OH}{CH}-C\equiv C-\underset{OH}{CH}-\overset{t}{\text{furan}}-CH=CHCOOCH_3 \qquad — \quad — \quad — \quad — \quad — \quad 27 \quad —$$

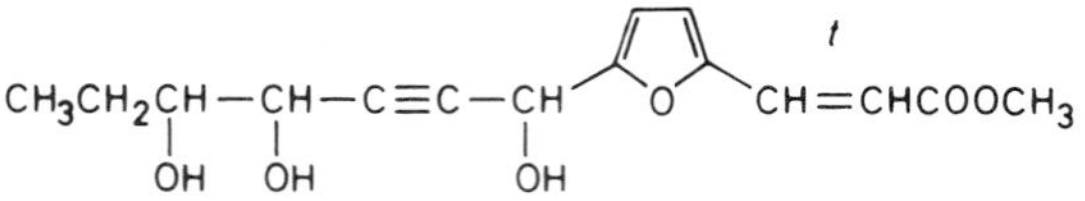

Layer		
	L1 =	Silica gel G
	L2 =	20% AgNO$_3$ on silica gel G
	L3 =	10% AgNO$_3$ on silica gel G
	L4 =	Silica gel HF$_{254}$
	L5 =	Silica gel 60 F$_{254}$

Table 7 (continued)
METHYL ESTERS OF FURAN FATTY ACIDS (TLC)

Solvent (v/v)	S1 = Petroleum hydrocarbon-diethyl ether (22:3)
	S2 = Petroleum hydrocarbon-diethyl ether (13:2)
	S3 = Toluene at −15°C
	S4 = Acetic acid -diethyl ether − hexane (1:15:85)
	S5 = Diethyl ether - benzene (1:19)
	S6 = Hexane - acetone (2:1)
	S7 = Hexane - acetone (2:1), second development with chloroform - hexane (2:1)
Technique	T1 = 18 cm distance
	T2 = Develop chromatogram twice
Detection	Chapter 28, Reagent No. 8

REFERENCES

1. **Lie Ken Jie, M. S. F. and Lam, C. H.,** *J. Chromatogr.*, 138, 373, 1977.
2. **Morris, L. J., Marshall, M. O., and Kelley, W.,** *Tetrahedron Lett.*, 36, 4249, 1966.
3. **Glass, R. L., Krick, T. P., Sand, D. M., Rahn, C. H., and Schlenk, H.,** *Lipids*, 10, 695, 1975.
4. **Fawcett, C. H., Spencer, D. M., Wain, R. L., Fallis, A. G., Jones, E. R. H., LeQuan, M., Page, C. B., Thaller, V., Shubrook, D. C., and Whitham, P. M.,** *J. Chem. Soc. (C)*, 2455, 1968.
5. **Hargreaves, J. A., Mansfield, J. W., and Coxon, D. T.,** *Phytochemistry*, 15, 651, 1976.
6. **Hargreaves, J. A., Mansfield, J. W., Coxon, D. T., and Price, K. R.,** *Phytochemistry*, 15, 1119, 1970.

REFERENCES

1. **Hofmann, K. and Lucas, R. A.,** *J. Am. Chem. Soc.,* 72, 4328, 1950.
2. **Lie Ken Jie, M. S. F. and Chan, M. F.,** *J. Chem. Soc. Chem. Commun.,* 1977, 78.
3. **Smith, C. R., Jr.,** *Prog. Chem. Fats Other Lipids,* 11, 137, 1971.
4. **Asselineau, J.,** *The Bacterial Lipids,* Hermann, Holden-Day, San Francisco, 1962.
5. **Gensler, W. J., Marshall, J. R., Langone, J. J., and Chen, J. C.,** *J. Org. Chem.,* 42, 118, 1977.
6. **Oudejans, R. C. H. M., van der Horst, D. J., and van Dongen, J. P. C. M.,** *Biochemistry,* 10, 4938, 1971.
7. **van der Horst, D. J. and Oudejans, R. C. H. M.,** *Comp. Biochem. Physiol.,* 46B, 277, 1973.
8. **Simmons, H. E., Cairns, T. L., Vladuchick, S. A., and Hoiness, C. M.,** *Organic Reactions,* 20, 1, 1973.
9. **Christie, W. W., Gunstone, F. D., Ismail, I. A., and Wade, L.,** *Chem. Phys. Lipids,* 2, 196, 1968.
10. **Gunstone, F. D. and Perera, B. S.,** *Chem. Phys. Lipids,* 10, 303, 1973.
11. **Conacher, H. B. S. and Gunstone, F. D.,** *Chem. Phys. Lipids,* 3, 203, 1969.
12. **Ucciani, E., Vantillard, A., and Naudet, M.,** *Chem. Phys. Lipids,* 4, 225, 1970.
13. **Gunstone, F. D. and Said, A. I.,** *Chem. Phys. Lipids,* 7, 121, 1971.
14. **Olson, E. S.,** *J. Am. Oil Chem. Soc.,* 54, 51, 1977.
15. **Minneken, D. E.,** *Lipids,* 7, 398, 1972.
16. **Bohannon, M. B. and Kleiman, R.,** *Lipids,* 13, 270, 1978.
17. **Lie Ken Jie, M. S. F.,** *Chem. Phys. Lipids,* 24, 407, 1979.
18. **Kirchner, H. W.,** *J. Am. Oil Chem. Soc.,* 42, 899, 1965.
19. **Schneider, E. L., Loke, S. P., and Hopkins, D. T.,** *J. Am. Oil Chem. Soc.,* 45, 585, 1968.
20. **Recourt, J. H., Jurriens, G., and Schmitz, M.,** *J. Chromatogr.,* 30, 35, 1967.
21. American Oil Chemists' Society Official Method, CB 1-25, Halphen Test, American Oil Chemist's Society, Champaign, Ill., approved 1973.
22. **Harris, J. A., Magne, F. C., and Skau, E. L.,** *J. Am. Oil Chem. Soc.,* 41, 309, 1964.
23. **Harris, J. A., Magne, F. C., and Skau, E. L.,** *J. Am. Oil Chem. Soc.,* 40, 718, 1963.
24. **Spener, F. and Mangold, H. K.,** *Biochemistry,* 13, 2241, 1974.
25. **De Rosa, M., Gambacorta, A., Minale, L., and Bu'Lock, J. D.,** *J. Chem. Soc. Chem. Commun.,* 619, 1971.
26. **De Rosa, M., Gambacorta, A., and Bu'Lock, J. D.,** *J. Bacteriol.,* 117, 212, 1974.
27. **Schogt, J. C. M. and Begemann, P. H.,** *J. Lipid Res.,* 6, 466, 1965.
28. **Egge, H., Murawski, U., Gyorgy, P., and Zilliken, F.,** *FEBS Lett.,* 2, 255, 1969.
29. **Hansen, R. P. and Gerson, T.,** *J. Sci. Food Agric.,* 18, 225, 1967.
30. **Morris, L. J., Marshall, M. O., and Kelley, W.,** *Tetrahedron Lett.,* 36, 4249, 1966.
31. **Crundwell, E. and Cripps, A. L.,** *Chem. Phys. Lipids,* 11, 39, 1973.
32. **Lie Ken Jie, M. S. F. and Lam, C. H.,** *Chem. Phys. Lipids,* 21, 275, 1978.
33. **Elix, J. A. and Sargent, M. V.,** *J. Chem. Soc.,* C, 595, 1968.
34. **Abbot, G. G. and Gunstone, F. D.,** *Chem. Phys. Lipids,* 7, 290, 1971.
35. **Abbot, G. G., Gunstone, F. D., and Hoyes, S. D.,** *Chem. Phys. Lipids,* 4, 351, 1970.
36. **Lie Ken Jie, M. S. F. and Lam, C. H.,** *Chem. Phys. Lipids,* 19, 275, 1977.
37. **Lie Ken Jie, M. S. F. and Lam, C. H.,** *Chem. Phys. Lipids,* 20, 1, 1977.
38. **Ranganathan, S., Ranganathan, D., and Mehrotra, M. M.,** *Synthesis,* 838, 1977.
39. **Glass, R. L., Krick, T. P., and Eckhardt, A. E.,** *Lipids,* 9, 1004, 1974.
40. **Glass, R. L., Krick, T. P., Sand, D. M., Rahn, C. H., and Schlenk, H.,** *Lipids,* 10, 695, 1975.
41. **Glass, R. L., Krick, T. P., Olson, D. L., and Thorson, R. L.,** *Lipids,* 12, 828, 1977.
42. **Gunstone, F. D., Wijesundera, R. C., Love, R. M., and Ross, D.,** *J. Chem. Soc. Chem. Commun.,* 630, 1976.
43. **Scrimgeour, C. M.,** *J. Am. Oil Chem. Soc.,* 54, 210, 1977.
44. **Fawcett, C. H., Spencer, D. M., Wain, R. L., Fallis, A. G., Jones, E. R. H., Le Quan, M., Page, C. B., Thaller, V., Shubrook, D. C., and Whitman, P. M.,** *J. Chem. Soc.,* C, 1968, 2455.
45. **Letcher, R. M., Widdowson, D. A., Deverall, B. J., and Mansfield, J. W.,** *Phytochemistry,* 9, 249, 1970.
46. **Hargreaves, J. A., Mansfield, J. W., and Coxon, D. T.,** *Phytochemistry,* 15, 651, 1976.
47. **Hargreaves, J. A., Mansfield, J. W., Coxon, D. T., and Price, K. R.,** *Phytochemistry,* 15, 1119, 1976.
48. **Spiteller, M. and Spiteller, G.,** *J. Chromatogr.,* 164, 253, 1979.
49. **Spiteller, M., Spiteller, G., and Hoyer, G.-A.,** *Chem. Ber.,* 113, 699, 1980.
50. **Pfordt, J., Thoma, H., and Spiteller, G.,** *Justus Liebigs Ann. Chem.,* 2298, 1981.
51. **Sand, D. M., Schlenk, H., Thoma, H., and Spiteller, G.,** *Biochem. Biophys. Acta,* 751, 455, 1983.

Chapter 9

HYDROXY-, EPOXY-, AND KETO-ACIDS

E. Vioque

I. INTRODUCTION

Oxygenated (hydroxy-, epoxy-, and keto-) fatty acids are mostly present in the vegetable kingdom as constituents of triacylglycerols, waxes, cerebrosides, and other lipids. Until recently, the number of such acids known as naturally occurring was rather scarce. With the help of new analytical methods it has become possible to discover many new oxygenated fatty acids, even in cases where they are present in very small proportions. The following three oxygenated fatty acids are constituents of triacylglycerols occurring in seed oils.

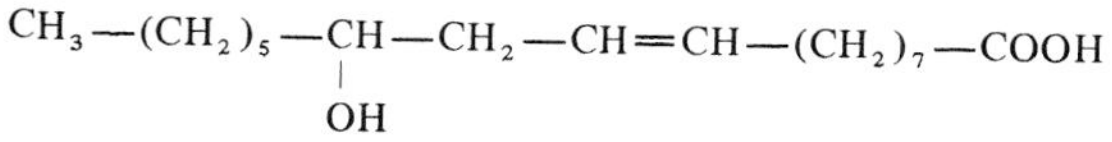

$$CH_3-(CH_2)_5-\underset{\underset{OH}{|}}{CH}-CH_2-CH=CH-(CH_2)_7-COOH$$

Ricinoleic acid

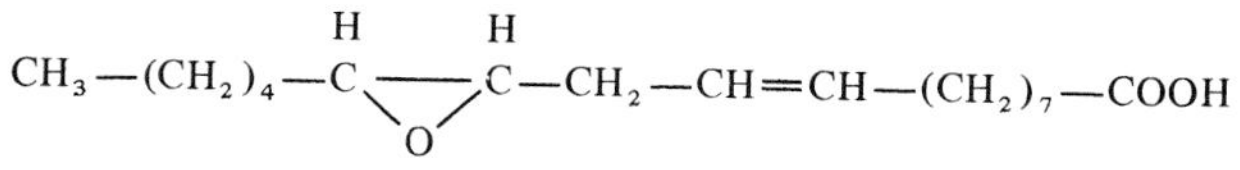

$$CH_3-(CH_2)_4-\overset{H}{C}\underset{O}{\diagdown}\overset{H}{C}-CH_2-CH=CH-(CH_2)_7-COOH$$

Vernolic acid

$$CH_3-(CH_2)_3-CH=CH-CH=CH-CH=CH-(CH_2)_4-\underset{\underset{O}{\|}}{C}-(CH_2)_2-COOH$$

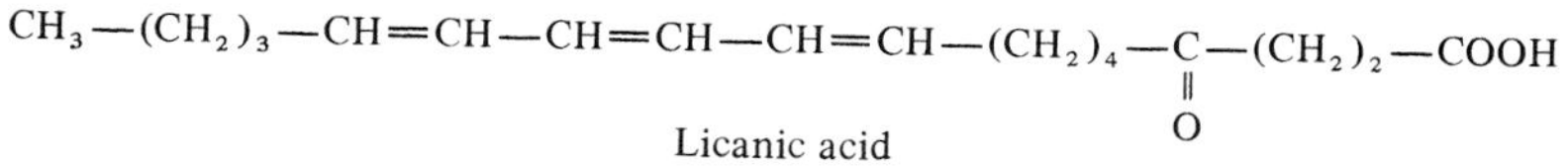

Licanic acid

The literature concerning oxygenated acids found in animals, plants, and microorganisms, as well as the methods employed for their detection, isolation, and structure determination have been reviewed by numerous authors.[1-14] The present chapter deals with methods based on paper, column, thin-layer, and gas chromatography.

The oxygenated fatty acids must first be isolated from the natural product by hydrolysis according to standard methods. They may be studied as free acids, or by transforming them into derivatives that facilitate their separation. Some oxygenated fatty acids contain more than one oxygenated function. For the total elucidation of the structure of such compounds, several chromatographic methods must be used successively. This is illustrated at the end of this chapter in a discussion of the various chromatographic techniques.

II. PREPARATION OF SAMPLES

A. Extraction

In the extraction of seeds containing triacylglycerols of oxygenated fatty acids, special precautions must be considered to avoid alteration of epoxy groups and moieties having a system of conjugated double bonds.

Procedure for the Extraction of Oxygenated Seed Oils
The ground seeds are slurried with hexane (1:3 w/v) for 90 min at room temperature. After decanting, the seeds are ground again and once more extracted under the same conditions. A third extraction may be carried out with diethyl ether to recover very

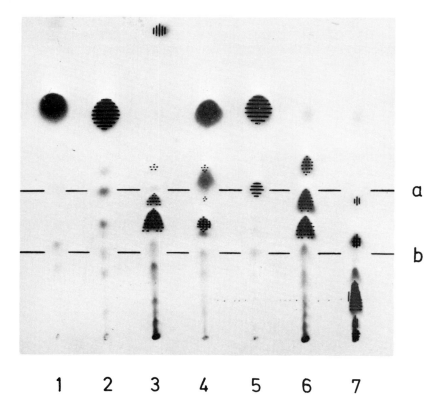

FIGURE 1.   Thin-layer chromatogram of seed oils 1, *Olea europaea*; 2, *Malope trifida*; 3, *Vernonia anthelmintica*; 4, *Artemisia absinthium*; 5, *Ceiba pentandra*; 6, *Cephalocroton cordofanus*; 7, *Dimorphotheca aurantiaca* (200 μg, each). Adsorbent: Silica gel G; Solvent: Hexane - diethyl ether - acetic acid (70:30:2); Detection: Charring after spraying with chromic-sulfuric acid solution; The fractions marked by horizontal lines respond to heating with a solution of sulfur in carbon disulfide (Halphen reagent) those marked by dotted lines react with picric acid solution (Reagent No. 28), and those marked by vertical lines quench fluorescence in UV light. a and b indicate the levels to which fatty acids (a) and sterols (b) migrate. (Courtesy of Drs. L. J. Morris and H. K. Mangold.)

polar constituents. The extracts are combined and, after evaporation of the solvents in vacuo, the oil is analyzed by thin-layer chromatography (TLC) under the conditions specified in Figure 1. The triacylglycerols of oxygenated fatty acids are detected by their rate of migration and by specific color reactions.

The saponification of seed oils containing epoxy acids and the recovery of these acids must be done with great care. A pH between five and six should be maintained in extracting the acids to avoid their destruction.

B. Derivatives
Procedures for the esterification of fatty acids are given in Chapter 2; additional methods used in work on oxygenated fatty acids are described here.

Procedure for the Preparation of Acetyl Derivatives
Methyl 4- and 5-acetoxystearates are prepared by treating the hydroxy esters with an excess of acetic anhydride and pyridine (1:1 v/v) at 0°C for 18 hr and then removing the agents at 0°C under reduced pressure. The acetates of all other hydroxy esters are

prepared by refluxing with acetic anhydride for 1 hr and evaporating the excess at 70°C under reduced pressure.

Procedure for the Preparation of Trifluoroacetyl Derivatives
To 5 to 10 mg of hydroxy esters in a vial are added 0.1 mℓ of chloroform and 0.05 mℓ of trifluoroacetic anhydride (TFAA). The solution is shaken for 30 sec and allowed to stand for 5 min at room temperature. Excess TFAA and solvent are removed by evaporating the solution under a stream of nitrogen. The residue is diluted to a 10% solution with chloroform, prior to gas chromatography (GC) analysis.

Procedure for the Preparation of Heptafluorobutyryl Derivatives
A solution of the methyl ester of a hydroxy acid (about 1 mg) is heated for 2 hr at 60°C with 0.2 mℓ of N-heptafluorobutyryl imidazole. The derivatives are extracted three times with hexane, and the hexane solution is used directly for GC analysis.

Procedure for the Preparation of Trimethylsilyl Ethers
5 to 10 mg of hydroxy ester, 0.5 mℓ of pyridine, 0.1 mℓ hexamethyldisilazane and 0.04 mℓ trimethylchlorosilane are placed in a 1 mℓ vial. The contents are shaken for 60 sec, and the mixture is analyzed directly or further purified as follows: the mixture is filtered through a fine, sintered glass funnel, and volatiles are removed from the filtrate by evaporation under reduced pressure. After diluting the residue with chloroform, the solution is analyzed.

Procedure for the Preparation of Isopropylidene Derivatives
One to 10 mg of di- or polyhydroxy esters are placed in a screwcap vial with 1 mℓ of acetone and 1 μℓ of 60% perchloric acid. After 15 min the perchloric acid is neutralized with an excess of ammonium hydroxide. The total mixture is used directly for GC analysis, with the exception of the isopropylidene derivatives of trihydroxyesters; these samples are evaporated to dryness, and the remaining hydroxyl group is trifluoroacetylated before GC analysis.

Procedure for the Preparation of R(−) Menthyloxycarbonyl Derivatives
The methyl ester of an unsaturated hydroxy acid, 0.1 to 0.2 mg, is dissolved in 50 μℓ of benzene and treated successively with 20 μℓ of a 1 mM solution of R(−) menthylchloroformate in toluene and 5 μℓ of dry pyridine. After mixing, the solution is kept for 90 min at 20°C and then applied as a band on a thin-layer plate that is developed with hexane-diethyl ether (9:1 v/v). The band of the derivative of the conjugated ester, easily visible under UV light at 254 nm, is scraped off and eluted with 2 mℓ of diethyl ether.

Procedure for the Preparation of N-(1-Phenylethyl)urethanes
Reagent solution: R-α-phenylethylamine (17 nmol) in dry toluene (13 mℓ) is slowly added to a solution of 71 nmol of phosgene in 62 mℓ of toluene. The mixture is kept at 80°C for 16 hr and then under reflux for 2 hr. After cooling, the excess of phosgene is removed by bubbling dry N_2 through the solution for 24 hr. The solution of R-1-phenylethylisocyanate thus obtained (0.3 nmol/mℓ) is stored at +4°C over $CaCO_3$. — (R)-1-Phenylethylisocyanate (30 nmol; 0.1 mℓ of the reagent solution) is added to the hydroxy compound (3 to 7 nmol) and the mixture is kept at 120°C for 3 hr under argon. The solvent is removed under reduced pressure, and the residue is dissolved in ethyl acetate and analyzed by GC.

Procedures of the Preparation of 2,4-Dinitrophenylhydrazones (DNPH)
(a) A solution of 2,4-dinitrophenylhydrazine (0.2% w/v) in 2 N HCl is added to an

aqueous solution of the keto acid or its sodium salt. The derivative is extracted into ethyl acetate, then into aqueous NaHCO₃, and finally, after acidification, into ethyl acetate; the solution is stored at 4°C.

(b) To 50 mℓ of deproteinized sample, 2.0 mℓ of 0.5% 2,4-DNPH in 6 N HCl is added and the mixture is allowed to stand for 30 min at 25°C. It is then transferred to a separatory funnel, and the hydrazones are extracted from the aqueous solution with three 15 mℓ portions of chloroform-ethanol (4:1 v/v), or with ethyl acetate. The combined organic layers are then extracted with 1 N Na₂CO₃ (15 mℓ), the organic phase is discarded, and the sodium carbonate extract, containing the hydrazones, is washed with 10 mℓ of chloroform-ethanol or with ethyl acetate and acidified in the cold (0 to 4°C) with 6 N HCl (6 mℓ). The hydrazones are then extracted from the aqueous layer with three successive portions of chloroform-ethanol (or ethyl acetate), 10, 5, and 5 mℓ, respectively. The extracts containing the keto acid derivatives are combined and evaporated by a gentle nitrogen stream.

Procedure for the Preparation of Rhodanine Derivatives
A solution, 2.0 mℓ, of the keto acid (5% in ethanol) is mixed with 4.0 mℓ of cool reagent solution (250 mg of Rhodanine, 4-oxothionthiazolidine, five drops of 25% ammonia and 50 mg of ammonium chloride in 100 mℓ of ethanol). The mixture is heated for 10 min under reflux in a water bath.

Procedures for the Preparation of O-Trimethylquinoxalinols
(a) The quinoxalinols are prepared by reaction of the keto acids with O-phenylenedi-amine in methanol-acetic acid.
(b) The silylation of 1 to 2 mg samples is carried out in 0.1 mℓ of pyridine with either 0.1 mℓ of bis(trimethylsilyl)trifluoroacetamide or with 0.1 mℓ of d₉-bis(trimethylsilyl)acetamide, plus 0.05 mℓ of d₉-trimethylchlorosilane at 70°C for 30 min.

III. CHROMATOGRAPHY

A. Paper Chromatography (PC)

Oxygenated fatty acids with hydroxy, epoxy, or keto groups can be separated using paper chromatography. Table 1* illustrates the kind of separations achieved by reversed phase partition chromatography. The hydroxy fatty acids may be resolved either as free acids or as methyl esters.

Short-chain hydroxy fatty acids, such as 2-hydroxyvaleric, 2-hydroxycaproic and 3-hydroxycaproic acids have been resolved on nontreated Whatman No. 1 paper.[15] Kaufmann and Nitsch[16] have separated di- and trihydroxystearic acids as well as stearic, 2-hydroxy-, and 12-hydroxystearic acids by reversed phase paper chromatography.

Epoxy acids and their methyl esters have been separated by reversed phase chromatography (Table 1), but this technique has found little application, mainly because other chromatographic techniques yield better resolutions.

Keto fatty acids, as free acids or as 2,4-DNPHs, have been resolved using different solvent systems, as illustrated in Table 2. Magasanik and Umbarger[17] have achieved good separations of short-chain keto acids by chromatography on plain paper. After spraying the chromatogram with 2,4-DNPH·HCl, the hydrazones formed can be estimated colorimetrically.

B. Column Chromatography (CC)

Separations of normal from hydroxylated fatty acids have been worked out in trying to analyze the fatty acids of castor oil. The triacylglycerols of castor oil contain high proportions

---

* The tables will appear following the text.

of ricinoleic acid, a hydroxy fatty acid, together with smaller amounts of dihydroxy fatty acids.

The fractionation by CC of the three types of fatty acids has been described by several authors. Thus Chobanov,[18] using simple CC, has obtained a clear separation of normal and hydroxy fatty acids. Table 3 includes three examples of the application of CC.

Many workers have made use of adsorption, partition, and reversed phase partition chromatography in studying the hydroxy fatty acids occurring in cerebrosides,[19] bacterial lipids,[20] vegetable oils,[21-23] waxes,[24,25] castor oil,[26,27] and also the 9- and 13-hydroxystearates,[28] and hydroxy linoleates that are obtained by reduction of the hydroperoxides formed during the autoxidation of methyl linoleate.[29]

Morris et al.,[30] in a thorough investigation of the different chromatographic methods for the isolation of epoxy acids and their esters, have employed both adsorption and partition chromatography in columns. These authors used silicic acid as the adsorbent and mixtures of hexane with diethyl ether as solvents. In partition chromatography, they have used acetonitrile as the stationary phase and hexane or trimethyl pentane as eluting solvents. In a study of monoepoxydation of methyl linoleate to methyl vernolate and methyl coronate, Ferrari et al.[31] have used CC on silica gel-celite and ligroin (petroleum hydrocarbon) with increasing proportions of diethyl ether for elution.

The rearrangement of epoxy esters to keto esters has been studied by Walens et al.,[32] who have separated the products of the reaction by adsorption chromatography on silica gel columns. Elution was started with benzene, continued with benzene-diethyl ether mixtures, and finished with diethyl ether. The compounds eluted in different fractions were identified by their infrared spectra.

In an investigation of the biosynthesis of unsaturated fatty acids of cow's milk, Keeney et al.[33] used CC on magnesia to separate the 2,4-DNPHs of keto esters, as a whole, followed by chromatography on grade III alumina to separate the hydrazones of the different keto esters from each other. In this way the authors were able to identify a series of 8-keto- to 13-keto stearic acids, after regeneration of the keto stearic esters from the hydrazones followed by GC analysis, infrared spectrophotometry, and other methods.

C. Thin-Layer Chromatography (TLC)

A great number of publications dealing with the application of TLC in the lipid field are devoted to hydroxy fatty acids. Complexing agents such as boric acid and sodium arsenite have been introduced to facilitate the separation of polyhydroxy fatty acids or their esters. Tables 4 through 7 include examples of chromatography on plain adsorbents, as well as chromatography on adsorbents containing complexing agents; reversed phase partition systems are also included.

In many instances, TLC has been found to be very useful. Besides some papers describing the general behavior of isomeric hydroxy fatty acids,[34-42] several selected examples of applications to particular problems of lipid research can be mentioned. Thus Morris et al.,[43,44] Mikolajczak and Smith,[45] and Phillips et al.[46] have applied TLC in order to detect, separate, and determine the structure of several new hydroxy fatty acids in seed oils. The occurrence of the oxygenated fatty acids as constituents of cutins[47-49] (see Chapter 10), bacteria,[50] etc. has also been investigated using such simple techniques. The 2-hydroxy fatty acids present in the white matter of the brain of infants and adults have been investigated[51] (see Chapter 17).

The epoxy stearates, from the 2,3- to the 17,18-compounds, have been synthesized by Gunstone and Jacobsberg,[52] and the chromatographic behavior of these isomers has been investigated. There is a clear separation between the *cis* and *trans* isomers except for the 2,3-epoxides.

Table 8 includes several examples of TLC separation of mono- and diepoxy fatty acids and their esters using adsorption or reversed phase partition chromatography. Vioque et al.[53,54] have made use of chromatography on silica gel for the study of epoxy fatty acids in seed oils. The determination of the geometric configuration of epoxides can be carried out using ion exchange resins, to open the oxirane ring, followed by TLC of the reaction products on boric acid impregnated silica gel.[55]

Keto fatty acids, as their rhodanine derivatives, can be resolved on Cellulose MN 300 Ac using different solvent systems (Table 9). Ronkainen[56] has separated some short-chain keto acids in the form of their 2,4-DNPH methyl esters using silica gel G as adsorbent.

D. Gas Chromatography (GC)

Hydroxy fatty acids can be chromatographed as methyl esters, but the use of derivatives such as acetates, trifluoroacetates, trimethylsilyl ethers, phenylurethanes, etc. facilitates their separation. Tables 10 through 13 list examples of GC of derivatives of mono-, di-, and trihydroxy fatty acids. Wood et al.[57] have described a quantitative method for the gas chromatographic analysis of mono- and polyhydroxystearates as their trimethylsilyl ether derivatives.

The study of 2-hydroxy fatty acids is of special interest due to their presence in many biological tissues. The qualitative and quantitative composition of such acids in the lipids of sheep brain has been studied by Downing[58] who reduced the hydroxy acids to hydrocarbons prior to GC analysis. The same author has investigated the aliphatic constituents of wool wax.[59] Kishimoto and Radin[60] have isolated 2-hydroxy fatty acids as copper chelates, which they transformed to the acetates of methyl esters using dimethoxypropane and isopropylacetate.

O'Brien and Rouser[61] have studied the separation of the methyl esters of all the isomeric hydroxy palmitic acids, as acetyl derivatives. The same authors have compared the separations obtained with those of the nonderivated methyl esters of hydroxy palmitic acids using different stationary phases.

The methyl esters of epoxy- and keto fatty acids have been separated by GC as shown in Tables 14 through 16. The keto esters can be analyzed, e.g., in the form of their 2,4-DNPHs or *O*-trimethylsilylquinoxalinol derivatives.

E. Complementary Use of Different Chromatographic Techniques

As a rule, no single chromatographic technique is capable of elucidating the total composition of a complex mixture of isomeric oxygenated fatty acids present in a sample. For this reason, it is usually compulsory to apply different principles of chromatography that complement each other. The final structure determination of the isolated compounds is achieved by means of some physical methods such as mass spectrometry, nuclear magnetic resonance, etc.

As a corollary to this chapter on the chromatographic analysis of hydroxy-, epoxy-, and keto-acids, several examples related to the complementary function of various chromatographic methods should be mentioned. The fatty acids composition of *Vernonia anthelmintica*, *Artemisia absinthium* and castor oils,[62] *Chysobalanus icaco* and *Parinarum lauricum*,[63] *Stenacherium*,[64] cutins,[65,66] and liverwort species[67] has been obtained using successively TLC and GC. Also, counter-current distribution, CC, TLC, and GC have been used in succession.[68-73] New oxygenated fatty acids have been discovered in some cases.

ACKNOWLEDGMENT

The author gratefully acknowledges the help of his colleague Dr. M. P. Maza in the preparation of this chapter.

Table 1
HYDROXY AND EPOXY ACIDS AND METHYL ESTERS (PC)

Paper	P1	P2	P3	P4	P4	P4	P5	P5	P5	P5
Solvent	S1	S2	S3	S4	S5	S6	S7	S8	S9	S10
Technique	T1	T2	T3	T4	T4	T4	T5	T5	T5	T5
Reference	1	2	3	4	4	4	5	5	5	5

Compound	$R_f \times 100^b$						$R_f \times 100^a$		$R_f \times 100^b$	
2-Hydroxymyristic acid	89	—	—	—	—	—	—	—	—	—
2-Hydroxypalmitic acid	82	—	—	—	—	—	—	—	—	—
2-Hydroxystearic acid	64	—	—	—	—	—	—	—	—	—
12-Hydroxystearic acid	—	—	—	51	60	69	—	—	—	—
2-Hydroxylignoceric acid	0	—	—	—	—	—	—	—	—	—
12-Hydroxy-*cis*-9-oleic acid	—	—	—	60	66	74	50	65	72	88
Methyl 2-hydroxymyristate	—	—	78	—	—	—	—	—	—	—
Methyl 2-hydroxypalmitate	—	—	65	—	—	—	—	—	—	—
Methyl 2-hydroxyheptadecanoate	—	—	55	—	—	—	—	—	—	—
Methyl 2-hydroxystearate	—	—	44	—	—	—	—	—	—	—
Methyl 2-hydroxynonadecanoate	—	—	33	—	—	—	—	—	—	—
Methyl 2-hydroxyarachidate	—	—	24	—	—	—	—	—	—	—
9,10-Dihydroxystearic acid	—	65	—	90	91	93	—	—	—	—
9,10,12,13-Tetrahydroxystearic acid	—	52	—	—	—	—	—	—	—	—
9,10,12,13,15,16-Hexahydroxystearic acid	—	24	—	—	—	—	—	—	—	—
9,10-Epoxystearic acid	—	—	—	—	—	—	28	51	48	74
12,13-Epoxyoleic acid	—	—	—	—	—	—	33	55	59	78
12,13-Dihydroxyoleic acid	—	—	—	—	—	—	80	92	91	99
9,10,12,13-Diepoxystearic acid	—	—	—	—	—	—	61	67	84	95

[a]   Esters.
[b]   Fatty acids.

Paper		
	P1 =	Whatman 1 or Whatman 3MM impregnated (12-20%) with paraffin oil
	P2 =	Schleicher and Schüll 2043b
	P3 =	Schleicher and Schüll 2043 bM
	P4 =	Schleicher and Schüll 2043 bMgl
	P5 =	Whatman 1, siliconized by immersion in a solution of 4% of Dow Silicone 200 in diethyl ether

Solvent (v/v)		
	S1 =	Acetic acid-water, 65:35
	S2 =	*n*-Butanol-abs. ethanol-water-conc. ammonia (75:15:8:2)
	S3 =	Acetic acid-water (70:30) saturated with paraffin (190-220°C)
	S4 =	Acetic acid-water (60:40)
	S5 =	Acetic acid-water (65:35)
	S6 =	Acetic acid-water (70:30)
	S7 =	Acetonitrile-water (50:50)
	S8 =	Tetrahydrofuran-water (75:25)
	S9 =	Acetonitrile-water-acetic (45:53:2)
	S10 =	Tetrahydrofuran-water (80:20)

Technique		
	T1 =	Reversed phase, ascending; 5 hr at room temperature; 1-1.5 hr at 35-40°C
	T2 =	Descending
	T3 =	Descending, 5 hr
	T4 =	Ascending, 13 hr; 20 ± 1°C
	T5 =	Reversed phase

Detection   See Chapter 28, Reagents Nos. 8 and 38

REFERENCES

1. **Skipski, V.P., Arfin, S.M., and Rapport, M.M.,** *Arch. Biochem. Biophys.,* 87, 259, 1960.
2. **Winsauer, K.,** *Mikrochim. Acta,* 480, 1957.
3. **Baykut, F. and Bolayir, S.,** *Rev. Fac. Sci. Univ. Istanbul, Ser. C.,* 23, 77, 1958.
4. **Kaufmann, H.P. and Ko, Y.S.,** *Fette Seifen Anstrichm.,* 64, 434, 1962.
5. **Morris, L.J., Holman, R.T., and Fontell, K.,** *J. Lipid Res.,* 1, 68, 1961.

Table 2
KETO ACIDS AND DERIVATIVES (PC)

Paper	P1	P1	P2	P2	P2	P2	P3	P3	P4
Solvent	S1	S2	S3	S4	S5	S6	S7	S8	S9
Technique	T1	T1	T2	T2	T2	T2	—	—	—
Detection	—	—	—	—	—	—	—	—	D1
Reference	1	1	1	1	1	1	2	2	2
Compound					$R_f \times 100$				
2,4-DNPH[a] of									
2-Ketobutyric acid	64	22,30,60	52,57	44	85	79	63	80	—
2-Ketovaleric acid	73	34,58	56,60	48	87	81	—	—	—
2-Ketoisovaleric acid	74	45	62	50	87	82	80	86	—
2-Ketocaproic acid	76	30,55	61,65	48	88	81	—	—	—
2-Ketoisocaproic acid	77	63	62	48	88	82	—	—	—
2-Keto-3-methylvaleric acid	77	47	67	51	89	84	77	86	—
2-Keto-4-methylthiolbutyric acid	68	32	51	39	83	74	—	—	—
As free acids									
2-Ketobutyric acid	—	—	—	—	—	—	—	—	76
3-Ketoisovaleric acid	—	—	—	—	—	—	—	—	83

[a]   See Chapter 2.

Paper	P1 =	Whatman No. 20
	P2 =	Whatman SG-81
	P3 =	Whatman No. 1,2,3 MM or 4
	P4 =	Eaton and Dickeman No. 613
Solvent	S1 =	sec - Butanol-water (4:1)
(v/v)	S2 =	5% aqueous sodium bicarbonate solution
	S3 =	Benzene-acetic acid (19:1)
	S4 =	Benzene-formic acid (19:1)
	S5 =	Chloroform-acetic acid (39:1)
	S6 =	2-Nitropropane-formic acid (99:1)
	S7 =	*n*-Butanol-water-ethanol (50:40:10)
	S8 =	Methanol-benzene-*n*-butanol-water (40:20:20:20)
	S9 =	Water saturated *n*-butanol-formic acid (95:5)
Technique	T1 =	Ascending; room temperature (20-25°C); 20-25 cm distance, overnight
	T2 =	Ascending; room temperature (20-25°C); 20-25 cm, 2 hr
Detection	D1 =	See Chapter 28, Reagents Nos. 27, 34

REFERENCES

1. **Smith, P.,** *J. Chromatogr.*, 30, 273, 1966.
2. **Sherma, J. and Zweig, G.,** *Paper Chromatography and Electrophoresis,* Vol. II, *Paper Chromatography,* Academic Press, New York, 1971, 186.

Table 3
HYDROXY ACIDS (CC)

Packing	P1	P2	P3	P4	P5	P5	P5	P5
Column								
Length (cm)	—	—	—	50	40	40	40	40
Diameters (cm)	1.35	1.35	1.65	1.8	0.12	0.12	0.12	0.12
Solvent	S1	S2	S3	S4	S5	S6	S7	S3
Temperature	18	18	23	—	—	—	—	—
Detection	D1	D1	D1	—	D2	D2	D2	D2
Technique	T1	T1	T1	T2	T1	T1	T1	T1
Reference	1	1	1	2	3	3	3	3

Compound				Elution volume (mℓ)			
9,10,12,13-Tetrahydroxystearic acid	35	28	40	—	—	—	—
9,10-Dihydroxy stearic acid	70	45	60	—	—	—	—
Normal saturated fatty acids[a]	—	—	—	50—130	—	—	—
Normal unsaturated fatty acids[a]	—	—	—	140—170	—	—	—
Hydroxy saturated fatty acids[a]	—	—	—	180—200	—	—	—
Hydroxy unsaturated fatty acids[a]	—	—	—	240—300	—	—	—
2-Acetoxy stearic	—	—	—	—	50—90	—	—
2-Acetoxy arachidic	—	—	—	—	—	100—140	—
2-Acetoxy behenic	—	—	—	—	—	—	150—190
2-Acetoxy lignoceric	—	—	—	—	—	—	—	200—230

[a]   = As methyl esters.

Packing	P1 =	7 g Castor oil on 14 g Kieselguhr
	P2 =	8 g Castor oil on 14 g Kieselguhr
	P3 =	8 g Castor oil on 16 g Kieselguhr
	P4 =	Silica gel, Merck, 0.05—0.5 mm, impregnated with 10% silver nitrate
	P5 =	Mineral oil-siliconized Celite
Solvent	S1 =	Acetone-water (66:34)
(v/v)	S2 =	Acetone-water (74:26)
	S3 =	Acetone-water (70:30)
	S4 =	100 mℓ chloroform, then chloroform-diethyl ether (9:1)
	S5 =	Acetone-water (55:45)
	S6 =	Acetone-water (60:40)
	S7 =	Acetone-water (65:35)
Detection	D1 =	Titration with 0.005 N KOH
	D2 =	Titration with 0.02 N KOH
Technique	T1 =	Reversed phase
	T2 =	Complexing agent added to layer

REFERENCES

1. **Savary, P. and Desnuelle, P.,** *Bull. Soc. Chim. France,* 939, 1953.
2. **Wagner, H., Goetschel, J. D., and Nesch, P.,** *Helv. Chim. Acta,* 46, 2986, 1963.
3. **Fulco, A. J. and Mead, J. F.,** *J. Biol. Chem.,* 236, 2416, 1961.

Table 4
MONO-HYDROXY FATTY ACIDS (TLC)

	L1	L1	L1	L1	L2	L3	L3	L3
Layer	L1	L1	L1	L1	L2	L3	L3	L3
Solvent	S1	S2	S3	S4	S5	S6	S7	S8
Technique	T1	T1	T1	T1	—	T2	T2	T2
Detection	D1	D2	D2	D3	D4	D5	D5	D5
Reference	1	2	2	2	3	4	4	4
Compound				$R_f \times 100$				
3-Hydroxy propionic acid	27	—	—	—	—	—	—	—
2-Hydroxy butyric acid	42	86	93	37	—	—	—	—
3-Hydroxy butyric acid	38	82	90	31	—	—	—	—
4-Hydroxy butyric acid	28	—	—	—	—	—	—	—
2-Hydroxy isobutyric acid	—	85	85	35	—	—	—	—
2-Hydroxy valeric acid	55	—	—	—	—	—	—	—
2-Hydroxyisovaleric acid	—	90	100	48	—	—	—	—
2-Hydroxy caproic acid	66	—	—	—	—	—	—	—
2-Hydroxyisocaproic acid	—	92	100	62	—	—	—	—
2-Hydroxyheptanoic acid	77	—	—	—	—	—	—	—
6-Hydroxystearic acid	—	—	—	—	36	—	—	—
7-Hydroxystearic acid	—	—	—	—	39	—	—	—
8-Hydroxystearic acid	—	—	—	—	43	—	—	—
9-Hydroxystearic acid	—	—	—	—	45	14	55	27
10-Hydroxystearic acid	—	—	—	—	50	—	—	—
12-Hydroxystearic acid	—	—	—	—	56	16	62	27
18-Hydroxystearic acid	—	—	—	—	40	—	—	—
12-Hydroxy oleic acid	—	—	—	—	—	22	58	27
12-Hydroxyelaidic acid	—	—	—	—	—	22	64	28
9-Hydroxy-10,12-octadecadienoic	—	—	—	—	—	16	46	27

Layer	L1 =	Cellulose powder HR 300 MN
	L2 =	Silica gel G
	L3 =	5% Starch-bonded silicic acid, 1% zinc-cadmium sulfide and 1% zinc silicate
Solvent	S1 =	n-Butanol-diethylamine-water (85:1:14)
(v/v)	S2 =	Amyl alcohol-acetic acid-water (40:40:2)
	S3 =	n-Butanol-acetic acid-water (60:10:20)
	S4 =	sec-Butanol-2 N NH$_4$OH solution (80:20)
	S5 =	Petroleum-hydrocarbon-diethyl ether-acetic or formic acid (60:40:2)
	S6 =	Skellysolve F-diethyl ether (70:30)
	S7 =	Skellysolve F-diethyl ether (50:50)
	S8 =	Benzene-methanol (85:15)
Technique	T1 =	10 cm Distance
	T2 =	1.2 × 14 cm glass strips
Detection	D1 =	After spraying with ninhydrin, the plates were warmed for 30 min at 60°C
	D2 =	See Chapter 28, Reagent No. 6
	D3 =	See Chapter 28, Reagent No. 38
	D4 =	See Chapter 28, Reagent No. 38
	D5 =	See Chapter 28, Reagent No. 8

REFERENCES

1. **Gütlbauer, F.,** J. Chromatogr., 45, 104, 1969.
2. **Dittmann, J.,** J. Chromatogr., 34, 407, 1962.
3. **Subbarao, R. and Achaya, K. T.,** J. Chromatogr., 16, 235, 1964.
4. **Applewhite, T. H., Diamond, M. J., and Goldblatt, L. A.,** J. Am. Oil Chem. Soc., 38, 609, 1961.

Table 5
METHYL ESTERS OF MONO- AND DIHYDROXY FATTY ACIDS (TLC)

	L1	L1	L2	L2	L2	L2	L1	L3	L4
Layer	L1	L1	L2	L2	L2	L2	L1	L3	L4
Solvent	S1	S2	S3	S4	S5	S6	S7	S8	S9
Technique	—	—	T1	T1	T1	T1	T2	T3	T4
Detector	—	D1	D2	D2	D2	D2	—	—	—
Reference	1	2	3	3	3	3	4	4	4
Compound					$R_f \times 100$				
3-Hydroxycaprate	40	—	—	—	—	—	—	—	—
2-Hydroxylaurate	48	—	—	—	—	—	—	—	—
3-Hydroxylaurate	40	—	—	—	—	—	—	—	—
3-Hydroxymyristate	40	—	—	—	—	—	—	—	—
6-Hydroxystearate	—	32	—	—	—	—	—	—	—
7-Hydroxystearate	—	37	—	—	—	—	—	—	—
8-Hydroxystearate	—	40	—	—	—	—	—	—	—
9-Hydroxystearate	—	45	43	—	—	—	—	—	—
10-Hydroxystearate	—	50	—	—	—	—	—	—	—
12-Hydroxystearate	—	55	50	—	87	45	—	—	—
18-Hydroxystearate	—	37	—	—	—	—	—	—	—
12-Hydroxyoleate	—	—	49	72	—	45	—	—	—
9-Hydroxy-10,12-octadecadienoic acid	—	—	43	67	84	41	—	—	—
erythro-6,7-Dihydroxystearate	—	—	—	—	—	—	20	29	78
threo-6,7-Dihydroxystearate	—	—	—	—	—	—	24	55	78
erythro-9,10-Dihydroxystearate	—	—	—	—	—	—	29	33	78
threo-9,10-Dihydroxystearate	—	—	—	—	—	—	33	60	78
erythro-2,3-Dihydroxybehenate	—	—	—	—	—	—	55	56	—
threo-2,3-Dihydroxybehenate	—	—	—	—	—	—	60	64	—
erythro-13,14-Dihydroxybehenate	—	—	—	—	—	—	39	43	64
threo-13,14-Dihydroxybehenate	—	—	—	—	—	—	46	84	64

Layer
L1 = Silica gel G
L2 = 5% Starch-bonded silicic acid, 1% zinc-cadmium sulfide and 1% zinc silicate
L3 = Silica gel G with 12.3% boric acid
L4 = Silica gel G. The dried, coated plate was uniformly impregnated with silicone oil by allowing a 5% solution in ether to ascend the length of the plate in a developing chamber

Solvent (v/v)
S1 = Petroleum hydrocarbon (40-60°)-diethyl ether (60:40)
S2 = Petroleum hydrocarbon-diethyl ether-acetic or formic acid (75:25:2)
S3 = Skellysolve F-diethyl ether (70:30)
S4 = Skellysolve F-diethyl ether (50:50)
S5 = Benzene-diethyl ether (50:50)
S6 = Benzene-methanol (85:15)
S7 = Petroleum hydrocarbon-diethyl ether (50:50)
S8 = Petroleum hydrocarbon-diethyl ether (60:40)
S9 = Acetonitrile-acetic acid-water (70:10:20)

Technique
T1 = 1.2 × 14 cm chromatostrips
T2 = Adsorption TLC
T3 = Complexing agent added to layer
T4 = Reversed phase

Detection
D1 = Charring, see Chapter 28, Reagent No. 38
D2 = 2′,7′-Dichlorofluorescein, see Chapter 28, Reagent No. 8

REFERENCES

1. **Hancock, I. C., Humphreys, S. G. D., and Meadow, P. M.,** *Biochim. Biophys. Acta,* 202, 389, 1970.
2. **Subbarao, R. and Achaya, K. T.,** *J. Chromatogr.,* 16, 235, 1964.
3. **Applewhite, T. H., Diamond, M. J., and Goldblatt, L. A.,** *J. Am. Oil Chem. Soc.,* 38, 609, 1961.
4. **Roomi, M. W., Subbaram, M. R., and Achaya, K. T.,** *J. Chromatogr.,* 24, 93, 1966.

Table 6
DI- AND TETRAHYDROXY FATTY ACIDS (TLC)

	L1	L2	L3
Layer	L1	L2	L3
Solvent	S1	S2	S3
Technique	T1	T2	T3
Detection	D1	D2	—
Reference	1	1	1

Hydroxy Fatty Acid	$R_f \times 100$		
erythro-6,7-Dihydroxystearic acid	5	30	84
threo-6,7-Dihydroxystearic acid	8	42	84
erythro-9,10-Dihydroxystearic acid	10	35	85
threo-9,10-Dihydroxystearic acid	14	48	85
erythro-2,3-Dihydroxybehenic acid	0	0	—
threo-2,3-Dihydroxybehenic acid	0	0	—
erythro-13,14-Dihydroxybehenic acid	22	48	72
threo-13,14-Dihydroxybehenic acid	25	61	72
erythro,erythro-9,10,12,13-Tetrahydroxystearic acid	0	5	94
threo,threo-9,10,12,13-Tetrahydroxystearic acid	0	10	94

Layer	L1 =	Silica gel G
	L2 =	Silica gel G with 12.3% boric acid
	L3 =	Silica gel G. The dried, coated plate was impregnated with silicone oil by allowing a 5% solution in ether to ascend the length of the plate in a developing chamber
Solvent	S1 =	Petroleum hydrocarbon-diethyl ether (50:50)
(v/v)	S2 =	Petroleum hydrocarbon-diethyl ether (60:40)
	S3 =	Acetonitrile-acetic acid-water (70:10:20)
Technique	T1 =	Adsorption TLC
	T2 =	Complexing agent added to layer
	T3 =	Reversed phase
Detection	D1 =	Charring, see Chapter 28, Reagent No. 38

REFERENCE

1. **Roomi, M. W., Subbaram, M. R., and Achaya, K. T.,** *J. Chromatogr.,* 24, 93, 1966.

Table 7
METHYL ESTERS OF TRI- AND TETRA- HYDROXY FATTY ACIDS (TLC)

Layer	L1	L2	L2	L2	L1	L3
Solvent	S1	S2	S3	S4	S5	S6
Technique	T1	—	T2	T3	T1	T4
Detection	D1	—	D2	—	—	—
Reference	1	2	3	4	4	4

Methyl Ester			$R_f \times 100$			
erythro-9,10,18-Trihydroxystearate	8	—	—	—	—	—
threo-9,10,18-Trihydroxystearate	22	—	—	—	—	—
erythro-9,10-*erythro*-10,12-Trihydroxystearate[a,b]	—	43	—	—	—	—
erythro-9,10-*threo*-10,12-Trihydroxystearate[a]	—	27	—	—	—	—
threo-9,10-*erythro*-10,12-Trihydroxystearate[a]	—	50	—	—	—	—
threo-9,10-*threo*-10,12-Trihydroxystearate[a]	—	40	—	—	—	—
erythro,erythro,erythro-9,10,12,13-Tetrahydroxystearate	—	—	35	—	—	—
erythro,threo,erythro-9,10,12,13-Tetrahydroxystearate	—	—	40	—	—	—
threo,erythro,threo-9,10,12,13-Tetrahydroxystearate	—	—	44	—	—	—
threo,threo,threo-9,10,12,13-Tetrahydroxystearate	—	—	48	—	—	—
erythro,erythro-9,10,12,13-Tetrahydroxystearate	—	—	—	0	10	87
threo,threo-9,10,12,13-Tetrahydroxystearate	—	—	—	0	17	87

[a] As isopropylidene derivatives.
[b] Showed a double spot.

Layer L1 = Silica gel G with 12.3% boric acid
L2 = Silica gel G
L3 = Silica gel G; the dried, coated plate was impregnated with silicone oil by allowing a 5% solution in ether to ascend the length of the plate in a developing chamber
Solvent (v/v) S1 = Chloroform-ethyl acetate (70:30)
S2 = Chloroform
S3 = Chloroform-methanol-acetic acid (45:5:1)
S4 = Petroleum hydrocarbon-diethyl ether (50:50)
S5 = Petroleum hydrocarbon-diethyl ether (60:40)
Technique T1 = Complexing agent
T2 = 10 cm distance
T3 = Adsorption TLC
T4 = Reversed phase
Detector D1 = See Chapter 28, Reagent No. 21
D2 = See Chapter 28, Reagent No. 8

REFERENCES

1. **Holloway, P. J.**, *Chem. Phys. Lipids*, 9, 158, 1972.
2. **Wood, R.**, *Lipids*, 2, 199, 1967.
3. **Sgoutas, D. and Kummerow, F. A.**, *J. Am. Oil Chem. Soc.*, 40, 138, 1963.
4. **Roomi, M. W., Subbaram, M. R., and Achaya, K.T.**, *J. Chromatogr.*, 24, 93, 1966.

Table 8
EPOXY FATTY ACIDS AND METHYL ESTERS (TLC)

	L1	L2	L1	L2	L1	L1	L3	L4
Layer	L1	L2	L1	L2	L1	L1	L3	L4
Solvent	S1	S2	S1	S2	S3	S4	S5	S6
Technique	T1	T2	T1	T2	T3	T4	—	—
Detection	—	—	—	—	D1	D1	D2	D2
Reference	1	1	1	1	2	3	4	4

Compound	$R_f \times 100$							
cis-6,7-Epoxystearic acid	41	76	—	—	—	—	—	—
trans-6,7-Epoxystearic acid	44	76	—	—	—	—	—	—
cis-9,10-Epoxystearic acid	45	77	—	—	—	—	41	—
trans-9,10-Epoxystearic acid	48	77	—	—	61	—	39	—
cis-13,14-Epoxybehenic acid	48	64	—	—	69	—	—	—
trans-13,14-Epoxybehenic acid	58	64	—	—	79	—	—	—
Methyl cis,cis-9,10,12,13-diepoxystearate	32	96	—	—	—	—	—	—
Methyl cis-6,7-epoxystearate	—	—	48	50	—	—	—	—
Methyl trans-6,7-epoxystearate	—	—	54	50	—	—	—	—
Methyl cis-9,10-epoxystearate	—	—	54	51	—	52	88	79
Methyl trans-9,10-epoxystearate	—	—	60	51	—	—	87	86
Methyl cis-13,14-epoxybehenate	—	—	72	33	—	—	—	—
Methyl trans-13,14-epoxybehenate	—	—	80	33	—	—	—	—
Methyl cis,cis-9,10,12,13-diepoxystearate	—	—	3	81	—	—	—	—
Methyl cis-12,13-epoxyoleate	—	—	—	—	—	55	—	—

Layer	L1 =	Silica gel G
	L2 =	Silica gel G. The dried, coated plate was impregnated with silicone oil by allowing a 5% solution in ether to ascend the plate in a developing chamber
	L3 =	Silica gel H
	L4 =	Silicic acid
Solvent	S1 =	Petroleum hydrocarbon-diethyl ether (70:30)
(v/v)	S2 =	Acetonitrile-acetic acid-water (70:10:20)
	S3 =	Petroleum hydrocarbon (40:60)-diethyl hydrocarbon-acetic acid (70:30:1)
	S4 =	Pentane-diethyl ether (80:20)
	S5 =	Petroleum hydrocarbon-dioxane-formic acid (76.5:22:1.5)
	S6 =	Petroleum hydrocarbon-diethyl ether-formic acid (78.5:19.5:2)
Technique	T1 =	Adsorption TLC
	T2 =	Reversed phase
	T3 =	15 cm distance
	T4 =	1.2 × 14 cm chromatostrips, development in stoppered 1.8 × 15 cm test tube
Detection	D1 =	Charring, See Chapter 28, Reagent No. 38
	D2 =	Spray with a 5% ammonium molybdate solution in 20% sulfuric acid

REFERENCES

1. **Roomi, M. W., Subbaram, M. R., and Achaya, K. T.**, *J. Chromatogr.*, 24, 93, 1966.
2. **Subbarao, R., Roomi, M. W., Subbaram, M. R., and Achaya, K. T.**, *J. Chromatogr.*, 9, 295, 1962.
3. **Binder, R. G., Applewhite, T. H., Diamond, M. J., and Goldblatt, L. A.**, *J. Am. Oil. Chem. Soc.*, 41, 108, 1964.
4. **Sliwiok, J., Kowalska, T., Rzepa, J., and Biernat, A.**, *Microchem. J.*, 18, 207, 1973.

Table 9
KETO FATTY ACIDS (TLC)

Layer	L1	L1	L1	L1
Solvent	S1	S2	S3	S4
Detection	D1	D1	D1	D1
Reference	1	1	1	1

Compound	$R_f \times 100$			
Rhodanine derivatives[a] of				
2-Keto propionic acid	55	22	32	51
2-Keto butyric acid	67	34	44	59
2-Keto valeric acid	73	51	57	70
2-Keto isovaleric acid	71	47	53	65
4-Keto valeric acid	59	26	41	54
2-Keto caproic acid	78	65	68	78
2-Keto isocaproic acid	78	70	67	77
2-Keto heptanoic acid	83	79	77	85
2-Keto caprylic acid	87	83	85	91
2-Keto pelargonic acid	74	62	50	75

[a] See Chapter No. 2

Layer	L1 =	Cellulose MN 300 Ac
Solvent (v/v)	S1 =	n-Propanol-n-butanol-ammonium carbonate solution (this solution includes two parts of a 10% solution ammonium carbonate in water and one part of 5 N ammonia) (40:20:30)
	S2 =	n-Propanol-butanol-ammonium carbonate solution (30:30:30)
	S3 =	n-Propanol-butanol-ammonium carbonate solution (35:25:30)
	S4 =	n-Propanol-ammonium carbonate solution (2:1)
Detection	D1 =	Ultraviolet, 365 nm

REFERENCE

1. **Rink, M. and Herrmann, S.,** *J. Chromatogr.,* 14, 523, 1964.

Table 10
METHYL ESTERS OF MONOHYDROXY FATTY ACIDS (GC)

	P1	P1	P2	P2	P3	P4
Column packing	P1	P1	P2	P2	P3	P4
Temperature (°C)	230	230	T1	T1	250	190
Gas	He	He	N_2	N_2	He	He
Flow rate (mℓ/min)	40	40	40	40	30	100
Column						
Length (cm)	150	150	366	366	275	200
Diameter (cm O.D.)	0.3	0.3	0.4	0.4	0.6	0.6
Form	—	—	W	W	U	U
Material	SS	SS	Glass	Glass	Glass	Glass
Detector	FID	FID	D1	D1	TC	TC
Reference	1	1	2	2	3	3

Methyl Ester	Retention time (min)		Methylene units		ECL	
	a	b	c	d		
5-Hydroxycaprylate[c]	16.50	15.44	—	—	—	—
2-Hydroxycaprate[c]	18.04	18.13	—	—	—	—
5-Hydroxycaprate[c]	18.76	18.17	—	—	—	—
4-Hydroxyundecanoate[c]	—	20.16	—	—	—	—
5-Hydroxyundecanoate[c]	—	20.08	—	—	—	—
5-Hydroxylaurate[c]	21.78	21.82	—	—	—	—
5-Hydroxymyristate[c]	26.50	26.42	—	—	—	—
5-Hydroxypentadecanoate[c]	27.88	29.26	—	—	—	—
5-Hydroxypalmitate[c]	30.65	32.11	—	—	—	—
5-Hydroxystearate[c]	38.37	38.21	—	—	—	—
2-Hydroxystearate	—	—	23.21	21.85	—	—
2-Hydroxyarachidate	—	—	25.17	23.74	—	—
2-Hydroxybehenate	—	—	27.15	25.64	—	—
2-Hydroxytricosanoate	—	—	28.15	26.58	—	—
2-Hydroxylignocerate	—	—	29.14	27.53	—	—
2-Hydroxypentacosanoate	—	—	30.14	28.49	—	—
2-Hydroxycerotate	—	—	31.12	29.46	—	—
2-Hydroxy-*cis*-15-tetracosenoate	—	—	28.87	27.25	—	—
2-Hydroxypentacosenoate	—	—	29.88	28.24	—	—
2-Hydroxyhexacosenoate	—	—	30.88	29.24	—	—
2-Hydroxyoleate	—	—	—	—	18.7	22.3
2-Hydroxylinoleate	—	—	—	—	18.7	23.8
2-Hydroxylinolenate	—	—	—	—	18.7	24.3

[a]   Retention time, minutes.
[b]   Authentic retention time, minutes.
[c]   TMSi-ethers.
[d]   Heptafluorobutyryl esters.

Packing	P1 = 20% Apiezon L on 100-120 mesh Celite 545
	P2 = 5% Se-30 on 80-100 mesh acid washed-hexamethyldisilanized Celite
	P3 = 20% Apiezon L on 100-150 mesh Celite 545
	P4 = 20% Resoflex 446 on 100-150 mesh Celite 545
Detector	D1 = FID
Temperature	T1 = Temperature programming; starting temperature 200°C; 2°C/min

REFERENCES

1. **Wyatt, C.J., Pereira, R.L., and Day, E.A.,** *Lipids*, 2, 208, 1967.
2. **Gasparrini, G., Horning, E.C., and Horning, M.G.,** *Lipids*, 3, 1, 1969.
3. **Bohannon, M.B. and Kleiman, R.,** *Lipids*, 10, 703, 1975.

Table 11
MONO- AND DI-HYDROXY FATTY ACIDS (GC)

	P1	P2	P3	P1	P2	P3	P4	P5	P6	P7	P8	P9
Column packing												
Temperature (°C)	220	202	224	220	202	224	—	200	190	—	—	250
Gas	He	He	He	He	He	He	N₂	He	He	—	—	He
Flow rate (mℓ/min)	40	40	60	40	40	60	—	60	100	—	—	30
Column												
Length (m)	90	180	366	90	180	366	—	180	200	120	360	275
Diameter (cm I.D.)	0.6	0.5	0.5	0.6	0.5	0.5	—	0.6	0.6	0.6	0.6	0.6
Material	Cu	Cu	Cu	Cu	Cu	Cu	—	SS	Glass	Glass	Glass	Glass
Detector	TC	TC	TC	TC	TC	TC	A	—	TC	—	—	TC
Reference	1	1	1	1	1	1	2	3	4,6	5	5	6

Compound	ECL^a			ECL^b			ECL^a					
2-Hydroxy stearic acid	19.25	20.10	23.55	20.30	21.95	23.50	—	—	—	—	—	—
3-Hydroxy stearic acid	19.50	20.80	24.45	20.35	22.35	23.95	—	—	—	—	—	—
4-Hydroxy stearic acid	19.85^c	24.55^d	27.15^c	20.40	22.45	24.40	—	—	—	—	—	—
5-Hydroxy stearic acid	20.10	25.10	28.00	20.45	22.60	24.45	—	—	—	—	—	—
6-Hydroxy stearic acid	19.85	21.70	26.15	20.50	22.85	24.50	—	—	—	—	—	—
7-Hydroxy stearic acid	19.90	21.75	26.20	20.50	22.95	24.60	—	—	—	—	—	—
8-Hydroxy stearic acid	19.95	21.75	26.20	20.50	23.00	24.60	—	—	—	—	38	—
9-Hydroxy stearic acid	20.00	21.80	26.20	20.50	23.00	24.65	—	25.3	24.2	—	—	—
10-Hydroxystearic acid	20.00	21.80	26.25	20.50	23.05	24.65	—	—	—	—	—	—
11-Hydroxystearic acid	20.00	21.80	26.25	20.55	23.05	24.70	—	—	—	—	—	—
12-Hydroxystearic acid	20.00	21.80	26.25	20.55	23.10	24.70	—	—	—	—	—	—
13-Hydroxystearic acid	20.05	21.80	26.30	20.60	23.15	24.80	—	—	24.4	—	—	—
14-Hydroxystearic acid	20.05	21.80	26.30	20.65	23.20	24.90	—	—	—	—	—	—
15-Hydroxystearic acid	20.05	21.80	26.35	20.75	23.30	25.10	—	—	—	—	—	—
16-Hydroxystearic acid	20.10	21.90	26.50	20.95	23.60	25.50	—	—	—	—	—	—
17-Hydroxystearic acid	20.10	22.10	26.95	21.15	23.90	25.95	—	—	—	—	—	—
18-Hydroxystearic acid	21.00	23.05	—	21.90	24.90	27.40	—	—	—	—	—	—
12-Hydroxyoleic acid	—	—	—	—	—	—	25.8	—	24.7	—	—	—
12-Hydroxyelaidic acid	—	—	—	—	—	—	25.8	—	—	—	—	19.4
9-Hydroxy-*cis*-12-octadecenoic acid	—	—	—	—	—	—	26.8	—	—	—	—	—

Table 11 (continued)
MONO- AND DI-HYDROXY FATTY ACIDS (GC)

Compound	ECL	ECL	ECL	ECL	ECL
9-Hydroxy-*trans*-10,*trans*-12-octadecadienoic acid	—	—	—	28.8 27.2	— 21.2	— 20.3
14-Hydroxy-*cis*-11-*cis*-17-icosadienoic	—	—	—	— —	21.2 27.2	— 20.3
12-Hydroxy-*cis*-9-*cis*-15-octadecadienoic acid	—	—	—	— —	19.2 25.2	— —
9,10-Dihydroxystearic acid	—	—	—	— 29.8	— —	— 21.3

a   Methyl esters.

b   Methyl acetyl esters.

c,d,e   The ECL of 4-stearolactone were 19.80, 24.55, and 27.15, respectively.

Column packing   P1 = 14.3% SE-30 on 60-80 mesh acid-washed Celite
                     P2 = 14.3% QF-1 on 60-70 mesh Anachrom AB
                     P3 = 20.0% Ethylene glycol succinate on 60-80 mesh acid-washed Chromosorb W
                     P4 = 20% Diethylene glycol succinate
                     P5 = 15% Diethylene glycol succinate on 60-80 mesh Gas Chrom P
                     P6 = 20% Resoflex 446 on 100-150 mesh Celite 545
                     P7 = 5% Apiezon L on Chromosorb W-dimethylchlorosilanized
                     P8 = 5% Resoflex 446 on Chromosorb W-dimethylchlorosilanized
                     P9 = 20% Apiezon L on 100-150 mesh Celite 545

REFERENCES

1.   **Tulloch, A. P.**, *J. Am. Oil Chem. Soc.*, 41, 833, 1964.
2.   **Abbot, G. G., Gunstone, F. D., and Hoyes, S. D.**, *Chem. Phys. Lipids*, 4, 351, 1970.
3.   **Binder, R. G., Applewhite, T. H., Diamond, M. J., and Goldblatt, L. A.**, *J. Am. Oil Chem. Soc.*, 41, 108, 1964.
4.   **Tallent, W. H., Harris, J., Spencer, G. F., and Wolff, I. A.**, *Lipids*, 3, 425, 1968.
5.   **Kleiman, R., Spencer, G. F., Earle, F. R., Nieschlag, H. J., and Barclay, A. S.**, *Lipids*, 7, 660, 1972.
6.   **Miwa, T. K., Mikolajczak, K. L., Earle, I. R., and Wolff, I. A.**, *Anal. Chem.*, 32, 1739, 1960.

Table 12
MONO-, DI-, AND TRIHYDROXY FATTY ACIDS (GC)

Column packing	P1	P2	P3	P4	P5	P6	P7	P8	P9
Temperature (°C)	250	245	220	210	220	180	250	T1	T2
Gas	He	He	He	He	He	A	A	N_2	N_2
Flow rate (mℓ/min)	30	30	50	50	50	50	50	35	35
Column									
Length (cm)	300	300	240	90	15	200	300	180	150
Diameter (cm O.D.)	0.3	0.3	0.3	0.3	0.3	0.3	0.3	0.6	0.6
Material	SS,Al	SS,Al	SS,Al	SS,Al	SS,Al	—	—	SS	SS
Detector	TC	TC	TC	TC	TC	FID	FID	FID	FID
Reference	1	1	1	1	1	2	2	3	3

Compound	Retention time					Relative retention time			
						a	a	b	c
As methyl esters:									
12-Hydroxystearic acid	27.8	—	6.6	5.0	15.8	—	—	—	—
12-Hydroxyoleic acid	25.5	—	6.4	5.0	15.8	—	—	—	—
erythro-9,10-Dihydroxystearic acid	—	—	—	—	—	—	9.34	—	—
threo-9,10-Dihydroxystearic acid	—	—	—	—	—	—	9.12	—	—
As TMSi- ethers:									
12-Hydroxystearic acid	22.1	14.6	2.2	1.8	4.3	—	—	—	—
12-Hydroxyoleic acid	20.3	13.5	2.2	1.8	4.3	—	—	—	—
erythro-9,10-Dihydroxystearic acid	—	—	—	—	—	10.40	6.21	—	—
threo-9,10-Dihydroxystearic acid	—	—	—	—	—	9.65	6.05	—	—
As TFA derivatives:									
12-Hydroxystearic acid	11.6	8.3	2.1	1.4	3.7	—	—	—	—
12-Hydroxyoleic acid	10.5	7.1	2.1	1.4	3.7	—	—	—	—
16-Hydroxypalmitic acid	—	—	—	—	—	—	—	0.64	0.42
20-Hydroxyarachidic acid	—	—	—	—	—	—	—	1.51	0.87
22-Hydroxybehenic acid	—	—	—	—	—	—	—	2.00	1.22
10,16-Dihydroxypalmitic acid	—	—	—	—	—	—	—	0.79	0.81
10,18-Dihydroxystearic acid	—	—	—	—	—	—	—	1.21	1.03
erythro-9,10,18-Trihydroxystearic acid	—	—	—	—	—	—	—	1.15	1.21
threo-9,10,18-Trihydroxystearic acid	—	—	—	—	—	—	—	1.16	1.23

Note: a = Relative to methyl stearate; b = Relative to tricosane; c = Relative to dotricosane.

Column packing		
P1 =	20% Apiezon L on 60-80 mesh Gas-Chrom P
P2 =	2% Apiezon L on 70-80 mesh acid-washed-dimethyldichlorosilanized Chromosorb G
P3 =	15% Diethylene glycol succinate on 70-80 mesh Gas Chrom P
P4 =	5% Carbowax 20M on 60-80 mesh hexamethyldisilanized Chromosorb W
P5 =	10% FFAP on 70-80 mesh AW-dimethyldichlorosilanized Chromosorb W
P6 =	10% EGSS-X on 100-120 mesh Gas-Chrom Q
P7 =	2% SE-30 on 60-80 mesh acid-washed-dimethyldichlorosilanized Chromosorb P
P8 =	5% E-301 on 80-100 mesh acid-washed-dimethylchlorosilanized Chromosorb W
P9 =	5% QF-1 on 80-100 mesh acid-washed-dimethylchlorosilanized Chromosorb W

Temperature		
T1 =	Temperature programming from 190 to 250°C, 2°C/min
T2 =	Temperature programming from 190 to 230°C, 2°C/min

REFERENCES

1. **Freedman, B.,** *J. Am. Oil Chem. Soc.,* 44, 113, 1967.
2. **Sliwiok, J., Kowalska, T., Rzepa, J., and Biernat, A.,** *Microchem. J.,* 18, 207, 1973.
3. **Baker, E. A. and Holloway, P. J.,** *Phytochemistry,* 9, 1557, 1970.

Table 13
HYDROXY FATTY ACIDS (GC)

Column packing	P1	P1	P2	P3	P3
Temperature (°C)	210	210	180	188	188
Gas	He	He	He	N_2	N_2
Flow rate (mℓ/min)	—	—	60	—	—
Column					
Length (cm)	180	180	150	200	200
Diameter (cm O.D.)	—	—	0.3	0.3	0.3
Reference	1	1	2	3	3

Compound	Retention time (min)		rRt[a]	ECL	
Urethanes of					
17-Hydroxystearic acid	8.4ˢ	9.1ʳ	—	—	—
19-Hydroxyarachidic acid	13.6	14.7	—	—	—
21-Hydroxybehenic acid	21.8	23.6	—	—	—
Isopropylidene-TFA					
Isopropylidene-TFA derivatives of					
threo-9,10-*erythro*-10,12-Trihydroxystearic acid	—	—	1.00	—	—
threo-9,10-*threo*-10,12-Trihydroxystearic acid	—	—	1.12	—	—
erythro-9,10-*threo*-10,12-Trihydroxystearic acid	—	—	1.25	—	—
erythro-9,10-*erythro*-10,12-Trihydroxystearic acid	—	—	1.46	—	—
R(-)-Menthyloxycarbonyl derivatives of					
2-Hydroxyenanthic acid	—	—	—	19.4ᴸ	19.7ᴰ
2-Hydroxycapric acid	—	—	—	22.3ᴸ	22.6ᴰ
2-Hydroxydecadienoic acid	—	—	—	22.4ᴸ	22.7ᴰ
2-Hydroxydodecendioic acid	—	—	—	—	24.7ᴰ

Note: S = Diasteroisomer S; R = Diasteroisomer R; L = L-form; and D = D-form.

[a]   Relative retention time to *threo*-9,10-*erythro*-10,12-Trihydroxystearic acid. Retention time of this compound = 15 minutes.

Column packing   P1 =   1% QF-1 on Gas Chrom Q
                   P2 =   15% Ethylene glycol succinate siloxane-X on 100-120 mesh Gas Chrom P
                   P3 =   5% QF-1 on Diatoport S

REFERENCES

1. **Hamberg, M.,** *Chem. Phys. Lipids,* 6, 152, 1971.
2. **Wood, R.,** *Lipids,* 2, 199, 1967.
3. **Nugteren, D. H.,** *Biochim. Biophys. Acta,* 380, 299, 1975.

Table 14
METHYL ESTERS OF EPOXY ACIDS (GC)

Column packing	P1	P2	P3	P4	P5	P6	P7	P8	P9
Temperature (°C)	180	250	150	190	190	210	—	—	T1
Gas	A	A	He	He	—	—	—	—	N_2
Flow rate (mℓ/min)	50	50	30	100					30
Column									
Length	200	300	275	200	150	150	365	120	200
Diameter									
(cm O.D.[a])	0.3[a]	0.3[a]	0.6[b]	0.6[b]	0.6[a]	0.6[a]	0.6[a]	0.6[a]	0.4[b]
(cm I.D.[b])									
Material	—	—	Glass	Glass	—	—	Glass	Glass	Glass
Detector	FID	FID	TC	TC	A	A	—	—	—
Reference	1	1	2	2	3	3	4	4	5

Methyl Ester	Retention time[c]		ECL						
cis-9,10-Epoxystearate	5.00	2.94	—	—	—	—	—	—	—
trans-9,10-Epoxystearate	4.65	2.72	—	—	—	—	—	—	—
cis-9L,10L-Epoxystearate	—	—	19.6	22.7	—	—	—	—	—
cis-9L,10L-Epoxy-*cis*-12-octadecenoate	—	—	19.0	23.1	—	—	—	—	—
cis-12D,13D-Epoxyoleate	—	—	19.0	23.1	24.6	19.1	—	—	—
cis-12,13-Epoxystearate	—	—	—	—	24.0	19.3	—	—	—
cis-9,10-Epoxy-octadec-12-ynoate	—	—	—	—	26.0	19.1	—	—	—
cis-12,13-Epoxy-*trans*-6-*cis*-9-octadecadienoate	—	—	—	—	—	—	23.4	19.3	—
cis-12,13-Epoxy-*cis*-6-*cis*-9-octadecadienoate	—	—	—	—	—	—	23.5	19.3	—
cis,cis-9,10,12,13-Diepoxystearate	—	—	—	—	—	—	—	—	20.70
									21.07[e]
cis,cis,cis-9,10,12,13,15,16-Triepoxystearate	—	—	—	—	—	—	—	—	22.46

Note: [c] Relative to methyl C18:0:; [d] *anti*-isomer; [e] *syn*-isomer.

Column packing		
P1 =	10% Ethylene glycol succinate 5-X on 100-120 mesh Gas Chrom Q
P2 =	2% SE-30, on 60-80 mesh acid-washed dimethylchlorosilanized Chromosorb P
P3 =	20% Apiezon L on 100-150 mesh Celite 545
P4 =	20% Resoflex 446 on 100-150 mesh Celite 545
P5 =	20% Diethylene glycol succinate on 70-80 mesh Gas Chrom Z
P6 =	5% Apiezon L on 70-80 Gas Chrom Z
P7 =	5% Resoflex 446 on 60-80 mesh acid-washed dimethylchlorosilanized Chromosorb W
P8 =	5% Apiezon L on 60-80 mesh acid-washed dimethylchlorosilanized Chromosorb W
P9 =	3% JXR on 80-100 mesh Gas Chrom P

Temperature   T1 =   Temperature programming from 180°C to 280°C at 5°C/min

REFERENCES

1. **Sliwiok, J., Kowalska, T., Rzepa, J., and Biernat, T.,** *Microchem. J.,* 18, 207, 1973.
2. **Powell, R. G., Smith, C. R. Jr., and Wolff, I. A.,** *Lipids,* 2, 172, 1967.
3. **Conacher, H. B. S. and Gunstone, F. D.,** *Chem. Phys. Lipids,* 3, 203, 1969.
4. **Spencer, G. F.,** *Phytochemistry,* 16, 282, 1977.
5. **Strocchi, A., Bonaga, G., and Capitani, P. N.,** *Riv. Ital. Sost. Grasse,* 53, 40, 1976.

Table 15
KETO ACIDS (SHORT CHAIN) (GC)

Column packing	P1	P2	P2	P3	P4	P5
Temperature (°C)	175	220	T1	T2	T2	T2
Gas	A	—	—	N_2	N_2	N_2
Flow rate (mℓ/min)	30	30	30	60	60	60
Column						
Length (cm)	120	—	—	183	183	183
Diameter (cm O.D.)	0.4	—	—	0.3	0.3	0.3
Form	—	—	—	U	U	U
Material	—	—	—	Glass	Glass	Glass
Detector	A	FID	FID	FID	FID	FID
Reference	1	2	2	3	3	3

Compound	Relative retention times			Methylene units[c]		
5-Decanolide	0.40[a]	—	—	—	—	—
Methyl 5-ketocaprate	0.30[a]	—	—	—	—	—
5-Decanolide	1.00[a]	—	—	—	—	—
Methyl 5-ketolaurate	0.74[a]	—	—	—	—	—
2,4-DNPH[d] of						
2-Ketobutyric acid	—	1.00[b]	1.00[b]	—	—	—
2-Keto-3-methylbutyric acid	—	1.06[b]	1.06[b]	—	—	—
2-Keto-3-methylvaleric acid	—	1.31[b]	1.31[b]	—	—	—
2-Keto-4-methylvaleric acid	—	1.32[b]	1.32[b]	—	—	—
O-trimethylsilylquinoxalinol derivatives[b] of						
2-Ketobutyric acid	—	—	—	15.78	17.17	16.06
2-Ketoisovaleric acid	—	—	—	16.00	17.24	16.20
2-Ketovaleric acid	—	—	—	16.46	17.84	16.73
2-Keto-3-methylvaleric acid	—	—	—	16.65	17.90	16.81
2-Ketoisocaproic acid	—	—	—	16.77	18.07	16.95
2-Ketocaprylic acid	—	—	—	19.23	20.68	19.50
2-Keto-4-methylthiobutyric acid	—	—	—	19.48	21.83	20.11

[a]   Relative to 5-dodecanolide.
[b]   Relative to the derivative of 2-ketobutyric acid.
[c]   The methylene units were determined with linear temperature programs as described by **Dalgliesh, C. E.** et al., *Biochem J.*, 101, 792, 1966.
[d]   See Chapter 2.

Column packing	P1 = 10% Apiezon L on Celite (150-178 μ)
	P2 = 4% SE-30 on acid-washed trimethyl silylated Chromosorb W (80-100 mesh)
	P3 = 3% OV-1 on 100-120 mesh Gas-Chrom Q
	P4 = 3% OV-17 on 100-120 mesh Gas-Chrom Q
	P5 = 3% Dexsil 300 GC on 100-120 Supelcoport
Temperature	T1 = 220°C for 17 minutes, then linear programming to 260°C at the rate of 4°C/min
	T2 = Linear temperature programming from 50°C to 180°C at 2°C/min

REFERENCES

1. **Smits, P.**, *J. Chromotogr.*, 20, 397, 1965.
2. **Kallio, H. and Linko, R. R.**, *J. Chromatogr.*, 76, 229, 1973.
3. **Langenbeck, U., Möhring, H. U., and Dieckmann, K. P.**, *J. Chromatogr.*, 115, 65, 1975.

Table 16
METHYL ESTERS OF KETO ACIDS (GC)

	P1	P2	P3	P4	P5	P6
Column packing						
Temperature (°C)	220	202	224	190	210	242
Gas	He	He	He	—	—	He
Flow rate (mℓ/min)	40	40	60	—	—	—
Column						
Length (cm)	90	180	366	150	150	305
Diameter (cm O.D.)	0.6	0.5	0.5	0.6	0.6	0.6
Material	Cu	Cu	Cu	—	—	—
Detector	TC	TC	TC	Argon	Argon	TC
Reference	1	1	1	2	2	3

Methyl Ester	ECL					Retention time (min)
	P1	P2	P3	P4	P5	P6
2-Ketostearate	18.95	21.00	22.55	—	—	—
4-Ketostearate	19.40	21.95	24.70	—	—	—
5-Ketostearate	19.50	22.40	24.85	—	—	—
6-Ketostearate	19.65	22.80	25.20	—	—	—
7-Ketostearate	19.65	22.90	25.25	—	—	—
8-Ketostearate	19.65	23.00	25.30	—	—	—
9-Ketostearate	19.70	23.05	25.30	—	—	6.7
10-Ketostearate	19.70	23.05	25.35	—	—	6.7
11-Ketostearate	19.70	23.10	25.35	—	—	—
12-Ketostearate	19.75	23.15	25.40	24.9	19.4	7.0
13-Ketostearate	19.80	23.15	25.40	—	—	—
14-Ketostearate	19.85	23.15	25.50	—	—	—
15-Ketostearate	19.90	23.15	25.60	—	—	—
16-Ketostearate	19.95	23.36	25.90	—	—	—
17-Ketostearate	20.00	23.75	26.40	—	—	—
4-Ketononanoate	—	—	—	15.2	10.3	—
12-Keto-*trans*-10-stearate	—	—	—	26.8	19.9	—
cis-9,10-Methylene-12-ketoheptadecanoate	—	—	—	25.6	19.2	—

Table 16 (continued)
METHYL ESTERS OF KETO ACIDS (GC)

Methyl Ester	ECL						Retention time (min)
trans-9,10-Methylene-12-ketoheptadecanoate	—	—	24.8	18.8	—	—	—
10,11-Methylene-12-ketoheptadecanoate	—	—	25.2	19.1	—	—	—
cis-9,10-Methylene-12-ketostearate	—	—	26.6	20.2	—	—	—
trans-9,10-Methylene-12-ketostearate	—	—	25.8	19.8	—	—	—
12-Ketooleate	—	—	25.3	19.1	—	—	7.4
12-Keto-*trans*-10-octadecenoate	—	—	—	—	—	—	7.8

Column packing   P1 = 14.3% SE-30 on 60-80 mesh acid-washed Celite
                      P2 = 14.3% QF-1 on 60-70 mesh Anachrom AB
                      P3 = 20% Ethylene glycol succinate on 60-80 mesh acid-washed Chromosorb W
                      P4 = 20% Diethylene glycol succinate on 70-80 mesh Gas Chrom Z
                      P5 = 5% Apiezon L on 70-80 mesh Gas Chrom Z
                      P6 = 25% Silicone rubber on 42-60 mesh, acid-washed fire brick

REFERENCES

1. **Tulloch, A. P.**, *J. Am. Oil Chem. Soc.*, 41, 833, 1964.
2. **Conacher, H. B. S. and Gunstone, F. D.**, *Chem. Phys. Lipids*, 3, 203, 1969.
3. **Kitagawa, I., Sucai, M., and Kummerow, F. A.**, *J. Am. Oil Chem. Soc.*, 34, 217, 1962.

REFERENCES

1. **Downing, D. T.**, *Rev. Pure Appl. Chem.*, 11, 196, 1961.
2. **O'Leary, W. M.**, *Bacteriol. Revs.*, 26, 421, 1962.
3. **Applewhite, T. H.**, *J. Am. Oil Chem. Soc.*, 42, 321, 1965.
4. **Radin, N. S.**, *J. Am. Oil Chem. Soc.*, 42, 569, 1965.
5. **Tallent, W. H., Cope, D. G., Hagemann, J. W., Earle, F. R., and Wolff, I. A.**, *Lipids*, 1, 335, 1966.
6. **Wolff, I. A.**, *Science*, 154, 1140, 1966.
7. **Stodola, F. H., Deinema, M. H., and Spencer, J. F. T.**, *Bacteriol. Revs.*, 31, 194, 1967.
8. **Hopkins, C. Y. and Chisholm, M. J.**, *J. Am. Oil Chem. Soc.*, 45, 176, 1968.
9. **Krewson, C. F.**, *J. Am. Oil Chem. Soc.*, 45, 250, 1968.
10. **Earle, F. R.**, *J. Am. Oil Chem. Soc.*, 47, 510, 1970.
11. **Nicolaides, N., Fu, H. C., and Ansari, M. N. A.**, *Lipids*, 5, 299, 1970.
12. **Tulloch, A. P.**, *Lipids*, 5, 247, 1970.
13. **Pohl, P. and Wagner, H.**, *Fette Seifen Anstrichm.*, 74, 541, 1972.
14. **Kolattukudy, P. E. and Walton, T. J.**, *Prog. Chem. Fats Other Lipids*, 13, 119, 1972.
15. **Libermann, L. A., Zaffaroni, A., and Stotz, E.**, *J. Am. Oil Chem. Soc.*, 73, 1387, 1951.
16. **Kaufmann, H. P. and Nitsch, W. H.**, *Fette Seifen Anstrichm.*, 58, 234, 1956.
17. **Magasanik, B. and Umbarger, H. E.**, *J. Am. Chem. Soc.*, 72, 2308, 1950.
18. **Chobanov, D., Popov, A., and Mitzev, I.**, *C. R. Acad. Sci. Bug.*, 14, 463, 1961.
19. **Kishimoto, Y. and Radin, N. S.**, *J. Lipid Res.*, 1, 72, 1959.
20. **O'Leary, W. M.**, *J. Bacteriol.*, 77, 367, 1959.
21. **Gunstone, F. D. and Sykes, P. I.**, *J. Chem. Soc.*, p. 5050, 1960.
22. **Badami, R. C. and Gunstone, F. D.**, *J. Sci. Food Agric.*, 14, 863, 1963.
23. **Gunstone, F. D. and Sykes, P. J.**, *J. Sci. Food Agric.*, 12, 115, 1961.
24. **Downing, D. T., Kranz, Z. H., Lamberton, J. A., Murray, K. E., and Redcliffe, A. H.**, *Aust. J. Chem.*, 14, 253, 1961.
25. **Mazliak, P.**, *Phytochemistry*, 2, 253, 1963.
26. **Binder, R. G., Applewhite, T. H., Kohler, G. O., and Goldblatt, L. A.**, *J. Am. Oil Chem. Soc.*, 39, 513, 1962.
27. **Frankel, E. N., Mc Conell, D. G., and Evans, C. D.**, *J. Am. Oil Chem. Soc.*, 39, 297, 1962.
28. **Dolev, A., Rohwedder, W. K., and Dutton, H. J.**, *Lipids*, 1, 231, 1966.
29. **Chan, H. W. S. and Levett, G.**, *Lipids*, 12, 99, 1977.
30. **Morris, L. J., Hayes, H., and Holman, R. T.**, *J. Am. Oil Chem. Soc.*, 38, 316, 1961.
31. **Ferrari, M., Ghisalberti, E. L., Pagnoni, U. M., and Pelizzoni, F.**, *J. Am. Oil Chem. Soc.*, 45, 649, 1968.
32. **Walens, H. A., Koob, R. P., Ault, W. C., and Maerker, G.**, *J. Am. Oil Chem. Soc.*, 42, 126, 1965.
33. **Keeney, M., Katz, I., and Schwartz, A. P.**, *Biochim. Biophys. Acta*, 62, 615, 1962.
34. **Kaufmann, H. P. and Makus, Z.**, *Fette Seifen Anstrichm.*, 62, 1014, 1960.
35. **Mangold, H. K., Kammereck, R., and Malins, D. C.**, *Microchem. J. Symp.*, II, 697, 1962.
36. **Morris, L. J.**, *Chem. Ind. (London)*, p. 1238, 1962.
37. **Morris, L. J.**, *J. Chromatogr.*, 12, 321, 1963.
38. **Morris, L. J.**, *Lab. Pract.*, 13, 284, 1964.
39. **Wood, R., Bever, E. L., and Snyder, F.**, *Lipids*, 1, 399, 1966.
40. **Morris, L. J. and Wharry, D. W.**, *J. Chromatogr.*, 20, 27, 1965.
41. **Osman, F., Subbaram, M. R., and Achaya, K. T.**, *J. Chromatogr.*, 26, 286, 1967.
42. **Morris, L. J., Wharry, D. M., and Hammond, E. W.**, *J. Chromatogr.*, 33, 471, 1968.
43. **Morris, L. J., Holman, R. T., and Fontell, K.**, *J. Am. Oil Chem. Soc.*, 37, 323, 1960.
44. **Morris, L. J. and Hall, S. W.**, *Chem. Ind. (London)*, p. 32, 1967.
45. **Mikolajczak, K. L. and Smith, C. R., Jr.**, *Lipids*, 2, 261, 1967.
46. **Phillips, B. E., Smith, C. R., Jr., and Tjarks, L. W.**, *Biochim. Biophys. Acta*, 210, 353, 1970.
47. **Baker, E. A. and Holloway, P. J.**, *Phytochemistry*, 9, 1557, 1970.
48. **Croteau, R. and Fagerson, I. S.**, *Phytochemistry*, 11, 353, 1972.
49. **Eglinton, G. and Hunneman, D. H.**, *Phytochemistry*, 7, 318, 1968.
50. **Knoche, H. W. and Shively, J. M.**, *J. Biol. Chem.*, 244, 4773, 1969.
51. **Eng, L. F., Gerstl, B., Hayman, R. B., Lee, Y. L., Tietsort, R. W., and Smith, J. K.**, *J. Lipid Res.*, 6, 135, 1965.
52. **Gunstone, F. D. and Jacobsberg, F. R.**, *Chem. Phys. Lipids*, 9, 26, 1972.
53. **Vioque, E., Morris, L. J., and Holman, R. T.**, *J. Am. Oil Chem. Soc.*, 38, 489, 1961.
54. **Vioque, E. and Maza, M. P.**, *Grasas Aceites*, 22, 25, 1971.
55. **Vioque, E.**, *J. Chromatogr.*, 39, 235, 1969.

56. **Ronkainen, P.,** *J. Chromatogr.,* 20, 404, 1965.
57. **Wood, R. D., Raju, P. K., and Reiser, R.,** *J. Am. Oil Chem. Soc.,* 42, 81, 1965.
58. **Downing, D. T.,** *Austr. J. Chem.,* 14, 150, 1961.
59. **Downing, D. T., Kranz, Z. H., and Murray, K. E.,** *Aust. J. Chem.,* 13, 80, 1960.
60. **Kishimoto, Y. and Radin, N. S.,** *J. Lipid Res.,* 4, 130, 1963.
61. **O'Brien, J. S. and Rouser, G.,** *Anal. Biochem.,* 7, 288, 1964.
62. **Vioque, E. and Holman, R. T.,** *J. Am. Oil Chem. Soc.,* 39, 63, 1962.
63. **Gunstone, F. D. and Subbarao, R.,** *Chem. Phys. Lipids,* 1, 349, 1967.
64. **Kleiman, R., Spencer, G. F., Tjarks, L. W., and Earle, F. R.,** *Lipids,* 6, 617, 1971.
65. **Kolattukudy, P. E.,** *Lipids,* 8, 90, 1973.
66. **Doas, A. H. B., Baker, E. A., and Holloway, P. J.,** *Phytochemistry,* 13, 1901, 1974.
67. **Caldicott, A. B. and Eglinton, G.,** *Phytochemistry,* 15, 1139, 1976.
68. **Powell, R. G., Smith, C. R., Jr., and Wolff, I. A.,** *J. Am. Oil Chem. Soc.,* 42, 165, 1965.
69. **Smith, C. R., Jr.,** *Lipids,* 1, 268, 1966.
70. **Capella, P., Galli, G., and Fumagalli, R.,** *Lipids,* 3, 431, 1968.
71. **Weihrauch, J. L., Bremington, C. R., and Schwartz, D. P.,** *Lipids,* 9, 883, 1974.
72. **Holloway, P. J.,** *Phytochemistry,* 13, 2201, 1974.
73. **Gardner, H. W. and Kleiman, R.,** *Lipids,* 12, 941, 1977.

Chapter 10

CUTINS AND SUBERINS, THE POLYMERIC PLANT LIPIDS

P. J. Holloway

I. INTRODUCTION

Cutins and suberins are the lipid biopolymers that provide the framework of the protective coverings on the outer surfaces of all higher plants. They are differentiated chiefly on the basis of their location in the plant and the type of covering in which they occur. Cutin is the characteristic component of the cuticular membrane or "cuticle" that occurs as a thin continuous extracellular layer on the surface of the epidermal cells of leaves, fruits, and nonwoody stems. Suberin is found mainly as an intracellular component in the walls of cork cells which comprise the periderm layer of woody plants. Both polymers, in association with waxes and probably phenolics, are the insulating materials of the surface layers giving them their tough, durable, and waterproof properties.

The physical and chemical properties of plant cutins and suberins are similar, all being insoluble high molecular weight polyesters composed of various long-chain fatty and hydroxy fatty acid monomeric units (summarized in Table 1). Most of these acids, however, are not found in any other class of plant lipid and since, at the present time, the polymers cannot be analyzed without prior depolymerization, they are the only means of identification and structural determination. It is impossible to be more precise in defining the chemistry of either polymer because the monomeric composition of each varies considerably according to species, with many of the monomers occurring as important constituents in both (Table 1). Cutins generally lack any monomers greater than C_{18} in chain length and most contain varying proportions of both C_{16} and C_{18} compounds. Some are composed almost exclusively of C_{16} monomers, e.g., tomato fruit (*Lycopersicon esculentum*), broad bean leaf (*Vicia faba*), or C_{18} monomers, e.g., spinach leaf (*Spinacia oleracea*), while others sometimes contain a high proportion of unsaturated C_{18} monomers, e.g., apple fruit (*Malus pumila*), *Gasteria planifolia* leaf. The most abundant C_{16} acids are 9,16- and 10,16-dihydroxyhexadecanoic and the C_{18} acids 9,10-epoxy-18-hydroxy- and 9,10,18-trihydroxyoctadecanoic.

Suberins show a more diverse variation in composition than cutins and can be classified into several types according to the chain lengths and structures of the predominant monomers. The first type comprises mainly C_{16} and C_{18} compounds, e.g., stems of Ribes species, parsnip root (*Pastinaca sativa*); the second almost entirely C_{18} compounds, e.g., skins of potato tubers (*Solanum tuberosum*), outer bark of conifers; and the third predominantly C_{18} and C_{22} compounds, e.g., *Quercus suber* bark, apple tree bark. The most abundant acids in type 1 are 1,16-hexadecanedioic, 16-hydroxyhexadecanoic, 9-octadecene-1,18-dioic, 1,18-octadecanedioic and 18-hydroxyoctadec-9-enoic; in type 2, 9-octadecene-1,18-dioic and 18-hydroxyoctadec-9-enoic; and in type 3, 18-hydroxyoctadec-9-enoic, 9,10-epoxy-18-hydroxy- and 9,10,18-trihydroxyoctadecanoic (phloionolic), 9,10-epoxy- and 9,10-dihydroxy-octadecane-1,18-dioic (phloionic), and 22-hydroxydocosanoic (phellonic). Unlike cutins, most suberins also yield a small neutral fraction after depolymerization, the most common components of which are 1-alkanols.[30-37,43]

It should be apparent from this brief summary that cutin and suberin are closely related chemically and that there may not always be a clear distinction between them. Although some qualitative differences do occur, others are quantitative so that monomers that are minor components in cutins become major ones in suberins and vice versa. Much also depends on the particular plant species involved. The problems of analysis, however, are common to both and concern (1) the isolation and purification of the polymers from plant

Table 1
CLASSES OF MONOMERIC FATTY ACIDS FOUND IN PLANT CUTINS AND SUBERINS

Class		Cutins	References[a]	Suberins	References[a]
Monobasics	$CH_3-(CH_2)_n-CO_2H$	Common $n = 14 \rightarrow 34$ Even	1—15	Common $n = 14 \rightarrow 28$ Even	30—37
α,ω-Dibasics	$HO_2C-(CH_2)_n-CO_2H$	Uncommon $n = 14, 16\Delta^9, 16\Delta^{9,12}$	1,3,4,8—10,14	Common $n = 14^c, 16\Delta^{9c}, 18, 20^c$	30—40
Monohydroxymonobasics	$CH_3-(CH_2)_4-\overset{\overset{\displaystyle OH^b}{\displaystyle\vert}}{CH}-(CH_2)_8-CO_2H$	Rare	7,9,16,17	Not reported	—
ω-Monohydroxymonobasics	$HOCH_2-(CH_2)_n-CO_2H$	Common $n = 14, 16\Delta^9, 16\Delta^{9,12}$	1—20	Common $n = 14^c, 16, 16\Delta^{9c}, 18, 20^c, 22^c$	30—41
Monohydroxydibasics	$HO_2C-(CH_2)_5-\overset{\overset{\displaystyle OH^b}{\displaystyle\vert}}{CH}-(CH_2)_8-CO_2H$	Common	7—11,15,16	Not reported	—
	$HO_2C-(CH_2)_8-\overset{\overset{\displaystyle OH}{\displaystyle\vert}}{CH}-(CH_2)_7-CO_2H$	Not reported	—	Rare	38

Monohydroxyoxomonobasics

Structure				
$OHC-(CH_2)_6-\overset{\overset{OH^b}{\mid}}{CH}-(CH_2)_7-CO_2H$	Rare	21—23	Not reported	—
$HOCH_2-(CH_2)_5-\overset{\overset{O^b}{\parallel}}{C}-(CH_2)_8-CO_2H$	Uncommon	24	Not reported	—

Monohydroxyepoxymonobasics

Structure				
$HOCH_2-(CH_2)_7-HC\overset{O}{\underset{\diagup\diagdown}{}}CH-(CH_2)_7-CO_2H$	Common^c	4,5,8—10,20,25—27	Common^c	27,31,42
$HOCH_2-(CH_2)_4-HC=CH-CH_2HC\overset{O}{\underset{\diagup\diagdown}{}}CH-(CH_2)_7-CO_2H$	Uncommon	4,5,8,10,27	Not reported	—

Epoxydibasics

Structure				
$HO_2C-(CH_2)_7-HC\overset{O}{\underset{\diagup\diagdown}{}}CH-(CH_2)_7-CO_2H$	Rare	10	Common	27,42

Dihydroxymonobasics

Structure				
$HOCH_2-(CH_2)_5-\overset{\overset{OH^b}{\mid}}{CH}-(CH_2)_8-CO_2H$	Common^c	1—13,14—20,25,28,29	Rare	29,31,32
$HOCH_2-(CH_2)_7-\overset{\overset{OH^b}{\mid}}{CH}-(CH_2)_8-CO_2H$	Uncommon	4,8,9,18,20	Rare	31

Table 1 (continued)
CLASSES OF MONOMERIC FATTY ACIDS FOUND IN PLANT CUTINS AND SUBERINS

Class	Cutins	References[a]	Suberins	References[a]
Dihydroxydibasics				
$HOCH_2—(CH_2)_7—\overset{HO}{HC}—\overset{OH}{CH}—(CH_2)_7—CO_2H$	Rare	1,3,4,9,10,20	Common	27,31,36,38
Trihydroxymonobasics				
$HOCH_2—(CH_2)_7—\overset{HO}{HC}—\overset{OH}{CH}—(CH_2)_7—CO_2H$	Common[c]	1—5,8—11, 16—20,25,27	Common[c]	27,31,35,36
$HOCH_2—(CH_2)_4—HC≡CH—CH_2—\overset{HO}{HC}—\overset{OH}{CH}—(CH_2)_7—CO_2H$	Uncommon	1,4,5,8—11,27	Not reported	—
Pentahydroxymonobasics				
$HOCH_2—(CH_2)_4—\overset{HO}{HC}—\overset{OH}{CH}—CH_2—\overset{HO}{HC}—\overset{OH}{CH}—(CH_2)_7—CO_2H$	Rare	17	Not reported	—

[a] Only modern references are included that describe chromatographic methods; structures of many acids known before advent of chromatography.

[b] Other positional isomers of this substituent also occur.

[c] Often major components of cutins and suberins.

materials, (2) depolymerization of the purified polymer, and (3) separation and identification of the individual components of the complex mixtures of substituted fatty acids that result from depolymerization. The main purpose of the present chapter is to discuss the application of chromatographic methods to stage (3) of the analysis. During the past 10 years, the use of chromatography in conjunction with other techniques such as mass spectrometry has enabled considerable advances to be made in our knowledge of these unique biological materials. A detailed account of the first two stages of the analytical process is included for the benefit of those not familiar with working with this type of lipid.

II. PREPARATION OF SAMPLES

It is difficult to obtain cutins and suberins in a pure state because unlike other plant lipids they cannot be extracted by any solvent. Although the polymers can be analyzed from the depolymerization of whole plant material, the yield of monomers is invariably very small. Much greater yields of monomers (20 to 80% of dry weight) are obtained if isolated cuticular membrane (for cutin) or cork layer (for suberin) preparations are used that are essentially free from other tissues and contain the polymer in a reasonably pure form. Both preparation techniques are similar and involve detachment by chemical or mechanical means, removal of adhering cell debris, and removal of all soluble material by exhaustive extraction with organic solvents.

A. Cutins
1. Isolated Membrane Preparations from Fresh Material
For most leaves and fruits it is convenient to remove 1 to 2 cm^2 discs of tissue with a cork borer or suitable punch. These are easy to manipulate as well as providing a known area of surface for quantitative assessment of cutin content. If the tissue is fleshy, cut the discs in half and dissect away most of the flesh with a scalpel or sharp blade. If removal of discs is not practicable, cut suitably sized strips of tissue; for leaf tissue, ensure that the margins are cut away. Small fruits can be cut in half and the flesh scraped from inside.

It is not always easy to estimate the amount of material that will be required for analysis because the cutin content of different plants is so variable, ranging from as little as 10 μg/cm^2 in leaves to more than 1 mg/cm^2 in some fruits. Provision should also be made for any losses that might occur during the subsequent isolation procedure. As a general rule, remove an area of tissue equivalent to 500 to 1000 cm^2 for delicate leaves and 100 to 200 cm^2 for thicker leaves and fruits.

To ensure adequate wetting by the reagents used for membrane isolation, dewax all samples by immersion in chloroform for 1 to 2 min. Allow the solvent to evaporate from the material by spreading it out on filter paper (fume hood). When dry, place in distilled water, allow to soak for at least 30 min, and then transfer to solution for membrane isolation.

Three macerating reagents are commonly used to detach cuticular membranes from plant tissues, and these are listed below together with some recommendations for use. The main effect of the reagents on the prepared tissue is the disruption of cellular organization, the ammonium oxalate-oxalic acid and enzyme solutions acting on pectinaceous components, zinc chloride-hydrochloric acid dissolving cellulose. The choice of reagents is largely determined by trial and error, and their relative efficiencies may vary according to plant species.

Procedure for the Isolation of Membranes by the Oxalate-Oxalic Acid Method[44]
Reagent: Ammonium oxalate (COONH$_4$)$_2$·H$_2$O (1.6% w/v) and oxalic acid (COOH)$_2$·2H$_2$O (0.4% w/v) in distilled water.
Method: Use 2 to 5 mℓ reagent per 2 cm^2 disc of tissue and incubate at 35 to 40°C in conical flasks fitted with a cotton wool plug. The separation process can be assisted

by gentle rotation of the contents carried out continuously or intermittently. Continue incubation until the membranes (transparent) are seen to separate away from the remainder of the tissue (green), which will by then have usually begun to disintegrate. The reagent may yield whole membranes that float off by themselves or which can readily be detached by pulling with fine forceps, or membrane fragments resulting from disintegration of whole membranes. The time taken for separation may vary from as little as 3 hr up to 2 to 3 weeks or longer, and depends upon the plant species; age of tissue (young tissues decompose more rapidly than older tissues); and position on the plant (the membrane on the lower surfaces of leaves is usually released faster than that from the corresponding upper surface).

After isolation, membranes are washed free of reagent with distilled water. Thick membranes can be conveniently washed in bulk in a Buchner funnel without any filter paper, more delicate ones by transferring through several changes of water by decantation. As they are easily damaged, membranes are best handled individually with fine forceps or picked up on the tip of a clean dissecting needle.

Variations of method: When dealing with tissues possessing a substantial membrane, e.g., fruits, the separation can be accelerated by gentle refluxing at 100°C, which yields membranes normally within 30 min to 1 hr. The heating time should be kept to a minimum, otherwise membranes which have separated may be broken up. If separations are slow at lower temperatures, they can be assisted by occasional immersion of the flasks containing the preparations in an ultrasonic cleaning bath for 5 to 10 min; longer periods of sonication cause membrane disintegration.

An alternative reagent that gives similar separations under the same conditions as ammonium oxalate-oxalic acid is a 0.05 M solution of ethylenediaminetetraacetic acid disodium salt [CH₂·N (CH₂·COOH)·CH₂·COONa]₂·2H₂O), in distilled water.

Procedure for the Isolation of Membranes by an Enzymatic Method[45]
Reagent: Commercial pectinase (pectin glucosidase, poly-α-1,4-galacturonide:glycanohydrolase E.C. 3.2.1.15) (2 to 5% w/v) dissolved in McIlvaine's citric acid-phosphate or Walpole's acetate buffer at pH 3.5 to 4. Add chloroform (0.25% v/v) or phenylmercuric nitrate (0.001% w/v) to prevent growth of microorganisms.
Method: Proceed as described for Method 1 and incubate at 35 to 40°C.
Variations of method: Separation may be assisted by the addition of commercial cellulase (β-2,4-glucan: glucanhydrolase E.C. 3.2.1.4) (0.1 to 0.5% w/v) to the solution[46] or by sonication.

Procedure for the Isolation of Membranes by the Zinc Chloride-Hydrochloric Acid Method[47]
Reagent: Zinc chloride, ZnCl₂ (granular), dissolved in concentrated hydrochloric acid at 1 g/1.7 mℓ of acid.
(Caution! corrosive reagent, work in fume hood.)
Method: Use 4 to 5 mℓ of reagent per 2 cm² disc of tissue at room temperature. The reaction is considered complete if membranes are released on placing a test sample into a large excess of distilled water. This test should be carried out at regular intervals. Separation is normally rapid, occurring within 1 to 12 hr; tissues that fail to release membranes using Methods 1 and 2 can be transferred to this reagent. Tissues should be exposed to the reagent for the minimum period of time because fragmentation of membranes and some depolymerization of cutin can occur.
Isolated membranes are washed well with distilled water until they are free from acid as described for Method 1.

Apart from zinc chloride-hydrochloric acid, the macerating agents yield isolated cuticular membranes containing variable amounts of attached cellular material. In many cases the complete epidermal layer is recovered; since it contains no chloroplasts, it has a membranous appearance. Several methods can be used to remove this cellular contamination. For substantial membranes, mechanical methods such as brushing under water or brief treatments in water using an ultrasonic cleaning bath are often successful in eliminating gross contamination. The chemical methods of cleanup use reagents that dissolve cellulose, and membranes are transferred to either zinc chloride-hydrochloric acid (see above) for 4 to 5 min, 72% w/w sulfuric acid for 1 to 2 min or incubated at 35 to 40°C with cellulase (see previous Procedure) for 12 to 24 hr. (Do not use Schweitzer's reagent, cuoxam, for this purpose because it is alkaline and will result in some loss of cutin from the membranes.) After such treatments and washing, membranes ideally should be checked for freedom from contamination by light or scanning electron microscopy.[48] Some preparations, however, will still contain cellulose because, if the epidermal cell wall itself is also cutinized, the cutin will protect the cellulose in the wall and prevent its removal by the reagents. Such cellulose, therefore, is an integral part of the cuticular membrane.

The final stage of the preparation process is the extraction of soluble components from the cleaned membranes. The most important one to remove is cuticular wax embedded within the membranes that is not completely removed in the initial dewaxing step. If not removed, this wax will be carried over into the cutin depolymerization products and give erroneous results. An exhaustive extraction with hot chloroform followed by hot methanol or hot chloroform-methanol (1:1 v/v) mixture is adequate for most membranes. It is important to check gravimetrically that the extraction is complete as this may take up to 24 hr continuous extraction. Substantial membranes can be air dried on filter papers and then placed in the solvents and refluxed gently. Delicate membranes are best not dried at this stage because they may form intractable aggregates that are difficult to extract completely. Such membranes can be transferred from water through several changes of methanol to remove most of the water and then treated as described above. Any contact with weak alkaline solutions of sodium carbonate, sodium sulfite or potassium hydroxide, which have been used in the past to extract phenolics from membranes, should be avoided because hydrolysis of cutin inevitably takes place.

Cleaned, extractive-free membrane preparations can be conveniently stored in methanol prior to depolymerization or dried and powdered (Wiley Mill) if so required.

2. Preparations of Fresh or Dried Material without Membrane Isolation

Fresh material (known weight or surface area) is cut up or homogenized and exhaustively extracted with hot chloroform and methanol (Procedure above). The extractive-free residue is used for analysis. Dried material (known weight) should be finely powdered and treated as described above for fresh material.

B. Suberins

1. Isolated Cork Preparations from Fresh Material

It is not always possible to obtain a good preparation of cork cells, much depending on the age of tissue and the growth characteristics of the particular plant. Cork layers on plants range from thin layers only five to ten cells thick to massive formations such as those found on the cork oak (*Quercus suber*). With some species, e.g., silver birch (*Betula pendula*), *Acer griseum*, collection is easy because the cork layers exfoliate naturally from the stems and can be picked off by hand. On others, e.g., apple trees, the layers can sometimes be detached fairly easily by peeling, especially from the younger stems. For older stems and trunks, pieces of outer bark are removed with a machete or axe, cut into small strips, and boiled with ammonium oxalate-oxalic acid solution (Section A.1) until the cork layers (usually dark brown) float away or can be lifted off with the aid of forceps. Any adhering tissue can be scraped or brushed off or removed by ultrasonic treatment, or if difficult, by further treatment with ammonium oxalate-oxalic acid. The cleaned isolated layers are washed well with distilled water, dried, and reduced to a fine powder (Wiley Mill). The washing stage is omitted for layers obtained by exfoliation or peeling. The absence of cells from tissues other than cork should be verified by examination of the prepared samples under the light microscope. The suberin content of cork preparations is variable and may range from 10 to 70% of the dry weight.

Most cork cells contain considerable quantities of wax, sometimes up to 30 to 40% of the dry weight, which must be removed before suberin analysis. Exhaustive extraction of the cleaned, powdered preparations is carried out as described for cutins (Section A.1). As suberin is an intracellular component, the presence of variable amounts of cellulose and lignin in the cleaned, extractive-free preparations is inevitable. However, it is not necessary to remove these components because they do not interfere with the suberin determination. In addition, cork usually contains substantial amounts of polyphenolic materials deposited in the walls and lumen of the cells, which are difficult to extract with organic solvents. However, no attempt should be made to extract them from preparations by treatment with aqueous solutions of sodium sulfite or sodium carbonate; otherwise considerable losses of suberin will also occur by hydrolysis. The bulk of the polyphenolic material is liberated from cork cells only after the removal of suberin, and then only in an aqueous medium, so that it does not usually interfere with the suberin analysis. Simple phenols, however, such as ferulic and cinnamic acids are soluble in organic solvents and are commonly recovered together with the aliphatic compounds in the depolymerization products of suberins.

2. Preparations of Fresh or Dried Material without Cork Isolation

Proceed as described for cutins (Section A.1) using the materials in a finely divided state. The extractive-free residue is used for analysis.

C. Depolymerization of Samples

As cutins and suberins are polyester in nature they can be depolymerized by any of the common reagents used to cleave ester bonds; a detailed account of the preparation and use of such reagents is given in Chapter 2 of this volume. The reactions proceed at different rates and generally are much slower than with other types of ester lipids, such as triacylglycerol. Different products are obtained according to the reagent used (summarized in Table 2), hydrolysis (aqueous and alcoholic KOH) yielding acids, transesterification (NaOMe-MeOH, HCl-MeOH, BF_3-MeOH), the corresponding methyl esters and reduction ($LiAlH_4$) leads to alcohols. Depolymerization occurs most efficiently in alcoholic solutions of bases. Apart from the different compounds obtained from ester fission, the qualitative and quantitative composition of the final depolymerization products is further influenced by additional reactions with other functional groups that may take place during the depolymerization itself or during the subsequent work-up. The most important and most reactive group is epoxide, which occurs commonly in both cutins and suberins. A summary of reactions relevant to cutin and suberin analyses is given in Table 3; these should be noted carefully before carrying out depolymerizations and in interpreting results therefrom.

It is not possible to recommend the use of any one reagent for cutin and suberin depolymerization. All that can be offered is some general practical advice and comments on the merits and limitations of the various methods. Ideally, a sample should be analyzed by more than one method and the results compared. A known weight of dry powdered sample or material isolated from a known surface area is used and depolymerization is continued with an excess of reagent until no further yield of monomers is obtained. The cutin/suberin content of the sample is assessed from the weight of monomers.

Diethyl ether is recommended as the solvent for recovery of monomers because it is easily purified, can be rapidly dried with anhydrous sodium sulfate, and gives few problems with

Table 2
PRODUCTS DERIVED FROM THE DEPOLYMERIZATION OF 10,16-DIHYDROXYHEXADECANOIC ACID FULLY ESTERIFIED IN A CUTIN POLYMER

$$\begin{array}{c} O-CUTIN \\ | \\ CUTIN-O-CH_2(CH_2)_5CH(CH_2)_8CO_2-CUTIN \end{array}$$

Method of depolymerization	Rate	Monomeric product
Aqueous KOH	Slow	$\begin{array}{c} OH \\ | \\ HOH_2C(CH_2)_5CH(CH_2)_8CO_2H \end{array}$
Alcoholic KOH	Rapid	
NaOMe-MeOH	Rapid	$\begin{array}{c} OH \\ | \\ HOH_2C(CH_2)_5CH(CH_2)_8CO_2Me \end{array}$
HCl-MeOH	Slow	
BF$_3$-MeOH	Slow	
LiAlH$_4$ in THF	Slow	$\begin{array}{c} OH \\ | \\ HOH_2C(CH_2)_5CH(CH_2)_8CH_2OH \end{array}$
LiAlD$_4$ in THF	Slow	$\begin{array}{c} OD \\ | \\ DOH_2C(CH_2)_5CH(CH_2)_8CD_2OD \end{array}$

Note: MeOH = methanol; THF = tetrahydrofuran.

emulsions in the separating funnel. Monomers can be extracted directly from alcoholic solutions with this solvent after the addition of five volumes of saturated sodium chloride solution. For hydrolysis methods, neutralization with strong mineral acids should be done carefully (pH 7 to 8) to avoid excess of acid. Hydrochloric acid in particular is very reactive towards epoxides, giving the corresponding chlorohydrins (Table 3). Since unsaturated compounds occur quite commonly in cutins and suberins, precautions should be taken to exclude oxygen from the depolymerization reaction and also from the resultant products. Small scale reactions can often be carried out in sealed Pyrex tubes, and on a larger scale, nitrogen can be passed continuously through the reaction mixture. Depolymerization products after removal of solvent may be stored for extended periods of time in ampoules sealed under nitrogen. To prevent possible reesterification, it is advisable to convert any free acid monomers obtained to the corresponding methyl esters (excess diazomethane in diethyl ether: sample dissolved in diethyl ether-methanol 9:1 v/v[49]).

Specific comments on individual depolymerization reagents are listed below. Alcoholic solutions of potassium hydroxide and sodium methoxide have been used most extensively; the reductive method introduced in 1972 by Kolattukudy and Walton[5] has not been adopted by other workers. Each reaction yields different products from epoxide-containing polymers and not all provide a reliable estimate of original epoxide content.

Procedure for Depolymerization with Aqueous Potassium Hydroxide Solution
(1) Use weak solutions (circa 1%) and heat at 100°C; depolymerization is slow and may take up to 18 hr. (2) Cannot be used for the quantitative determination of epoxide-containing cutins and suberins. Epoxides are slowly hydrolyzed to the corresponding vic-diols (Table 3), which themselves occur in the polymers; complete conversion is achieved in 12 to 18 hr. (3) Liberates much phenolic material from cork preparations, which may cause some interface problems during subsequent recovery of monomers. (4) Neutral components from suberin depolymerizations can be conveniently isolated

Table 3
REACTIVITY OF EPOXIDE GROUP TOWARDS REAGENTS COMMONLY USED IN THE ANALYSIS OF CUTINS AND SUBERINS

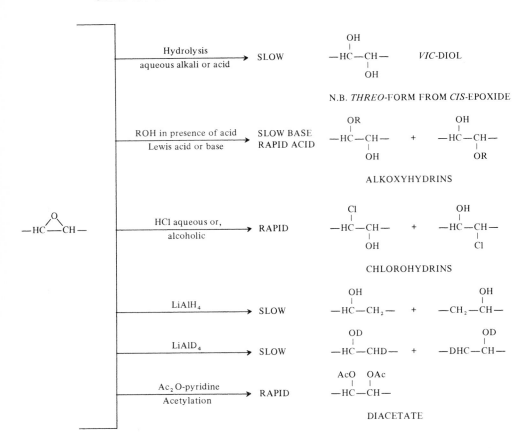

using this method. Shake the ether solution of total monomers obtained after neutralization with a 1% aqueous solution of KOH previously saturated with ether. The neutrals remain in the ether layer, and the acids are recovered from the aqueous phase by acidification and ether extraction. The back extraction of acids gives a better separation of neutrals than that obtained by direct extraction of the original KOH solution used for depolymerization.

Procedure for Depolymerization with Alcoholic Potassium Hydroxide Solution
(1) Use 1 to 5% solutions in methanol or ethanol and heat at reflux temperatures; depolymerization time is 30 min to 3 hr. (2) Reaction proceeds initially by transesterification, since most of the dissolved hydroxide is converted to alkoxide in alcoholic solutions, followed by hydrolysis of the esters formed by hydroxide.[50] (3) Cannot be used for the quantitative determination of epoxides because of partial conversion to alkoxyhydrins (Table 3) and some hydrolysis to vic-diols. (4) Phenolic material not liberated from cork preparations. (5) Neutral suberin fractions can be recovered as described for aqueous KOH.

Procedure for Depolymerization with Sodium Methylate
(1) Main advantage is formation of monomers as methyl esters directly. (2) Use weak solutions 0.1 to 0.5 M (0.2 to 1.15 g Na metal to 100 mℓ dry methanol) (see Chapter

2), *and heat at reflux temperature taking precautions to exclude moisture; depolymerization time is 30 min to 3 hr. (3) For making the reagent, use specially dried grades of methanol or dry it by passage through a column of Molecular Sieve 4A immediately before use. (4) Several good commercial preparations of this reagent are available in sealed ampules ready for use. (5) In practice the reaction is very difficult to carry out under anhydrous conditions; hydrolysis of the methyl esters formed usually occurs after 10 to 30 min from hydroxide generated by the reaction of methoxide with water. If the reaction is terminated with dry methanol-sulfuric acid (9:1 v/v),[27] reesterification of the acids prior to recovery can be achieved. Alternatively, hydrolysis can be kept to a minimum by renewing the reagent every 5 to 10 min until depolymerization is complete; each aliquot removed is immediately neutralized with acetic acid. The depolymerization products are recovered in the usual way from the combined aliquots. (6) The reaction can be carried out under essentially anhydrous conditions in sealed tubes if dry methyl acetate is added to the methoxide solution as a hydroxide scavenger.[51] Add a volume equivalent to 50% neutralization, assuming that all the methoxide is present as hydroxide. (7) Epoxides are partially converted to corresponding methoxyhydrins (Table 3), but no hydrolysis to vic-diols occurs even under conditions which result in hydrolysis of methyl esters. If dry methanol-sulfuric acid (9:1 v/v) is used for termination, any unchanged epoxides are rapidly converted to methoxyhydrins, so that the latter may be used for determination of original epoxide content of cutins and suberins. (8) Neutral suberin fractions cannot be removed separately. (9) Phenolic material is not liberated from cork.*

Procedure for Depolymerization with Boron Trifluoride in Methanol
(1) Gives monomers as methyl esters. (2) Use commercially available 14% reagent (or a dilution thereof with dry methanol), and heat at reflux temperature; depolymerization is slow, taking up to 48 hr. (3) Epoxides are rapidly converted to corresponding methoxyhydrins (Table 3), so that they can be used for determination of original epoxide content. (4) There is the possibility of side reactions with unsaturated compounds to give methoxy artifacts.

Procedure for Depolymerization with Methanolic Hydrochloric Acid Solution
(1) Gives monomers as methyl esters. (2) Prepare reagent from dry methanol and acetyl chloride (see Chapter 2) at concentrations of 0.5 to 3 N (1.7 to 10% w/v HCl) and heat at reflux temperatures taking precautions to exclude moisture; depolymerization is slow, requiring up to 48 hr. (3) Epoxides are rapidly converted to corresponding chlorohydrins (Table 3), which can be used for determination of original epoxide content.

Procedure for Depolymerization with Lithium Aluminum Hydride
(1) Use a two to threefold excess by weight of hydride in dry, freshly distilled tetrahydrofuran[5] (there is little reaction if dry ether is used) and heat at reflux temperatures taking precautions to exclude moisture. (2) Epoxides are converted to corresponding alcohols (Table 3), but the results from $LiAlH_4$ alone are ambiguous because several of the original polymeric acids yield the same reduced monomer (Table 8). Although these monomers are distinguishable after $LiAlD_4$ reduction, mass spectrometry is essential for both identification and quantitative assessment.

III. CHROMATOGRAPHY

Before the advent of the modern chromatographic approach, fractionation of the complex mixtures of monomers obtained from cutins and suberins relied on traditional methods of recrystallization from different solvents, the differential solubilities of the corresponding sodium or calcium salts in water, and counter-current distribution, all requiring considerable quantities of starting material. Nevertheless, many of the important monomers were isolated by such methods, and their structures determined with remarkable accuracy considering the identification procedures then available. As with other types of lipids, the application of chromatographic techniques has improved the efficiency of separation, reduced the time taken for analysis and, with the greatly increased sensitivity available from TLC and GLC, requires only a few milligrams of monomers for complete analysis. Details of these techniques are described below under the general headings of column (CC), thin-layer (TLC), and gas-liquid (GC) chromatography. Older methods such as reversed phase paper which are now only of historical interest, are not included. At the time of writing, there were no reports of analyses of cutin and suberin monomers by high pressure liquid chromatography.

A. Column Chromatography (CC)

Conventional adsorption CC on alumina, Florisil, or silica gel has been used to separate methyl esters of cutin and suberin monomers, but few workers have attempted to use it for the analysis of total depolymerization products. The technique has been most widely used for the large scale purification of crude fractions obtained by other nonchromatographic methods. Published elution schemes, therefore, are suitable mainly for the separation of only a limited number of compounds; some useful systems are summarized below:

1. Lipid mixture 10 g, silica gel 200 g, elution with hexane-diethyl ether (1:1 v/v), 25 mℓ fractions collected; fractions 7 to 12 (α,ω-dibasic methyl esters), fractions 15 to 22 (ω-hydroxymonobasic methyl esters).[34]
2. Lipid mixture 1 g, Florisil 100 g, elution with benzene (monobasic methyl esters), benzene-diethyl ether (19:1 v/v) (α,ω-dibasic methyl esters), and diethyl ether (ω-hydroxymonobasic methyl esters).[30]
3. No sample details, silica gel, elution with benzene-diethyl ether (5:1 v/v) (dimethyl *cis*-9,10-epoxyoctadecane-1,18-dioate) and benzene-diethyl ether (2:1 v/v) (methyl *cis*-9,10-epoxy-18-hydroxyoctadecanoate).[42]
4. Mixture 1 g, silica gel 200 g, 5 mℓ fractions collected, elution with chloroform fractions 21 to 32 (methyl 9,10-epoxy-18-hydroxyoctadecanoate); elution with chloroform-ethyl acetate (3:1) fractions 55 to 69 (methyl 10,16-dihydroxyhexadecanoate) and fractions 75 to 86 (methyl 9,10,18-trihydroxyoctadecanoate).[51]

For the CC separation of methyl esters of total depolymerization products on silica gel or Florisil, the following systems can be recommended (arrows indicate increasing polarity of eluting solvent):

1. Hexane $\rightarrow$ Hexane containing increasing proportions of diethyl ether $\rightarrow$ Diethyl ether
2. Benzene $\rightarrow$ Benzene containing increasing proportions of diethyl ether $\rightarrow$ Diethyl ether
3. Benzene $\rightarrow$ Benzene containing increasing proportions of chloroform* $\rightarrow$ Chloroform $\rightarrow$ Chloroform containing increasing proportions of ethyl acetate

The selection of the correct polarity of eluting solvents is essential for the analysis, otherwise complete separation of constituents will not be achieved. The composition of the sample revealed by analytical TLC can be used to advantage at this stage because the pattern of separation in the column will usually follow that on TLC. Fractionation proceeds strictly according to the polarity of the constituents, and the R_f values (Table 4) can be used to predict the order of elution from a column. The higher R_f components will be eluted from

---

* Throughout this chapter, chloroform refers to the commercial product that contains circa 2% ethanol added as stabilizer.

Table 4
CUTIN AND SUBERIN MONOMERS OBTAINED AFTER DEPOLYMERIZATION BY HYDROLYSIS AND TRANSESTERIFICATION METHODS (TLC)

	L1	L1	L1	L1	L1	L1
Layer						
Solvent	S1	S2	S3	S4	S5	S6
Technique	T1	T1	na	na	T1	na
Derivative	ME	ME	ME	FA	FA	FA
Reference	1	1	2,3	4	5,1	6
Compound (s)	$R_f \times 100$					
ACIDS						
Monobasics[b]						
C_{16}	68[a]	70	94	—	80[a]	88
C_{20}	72	72	94	—	85	—
C_{26}	75	75	—	—	90	—
α,ω-Dibasics[b]						
C'_6	45[a]	70	85	87	75[a]	58
C_{18}	47[b]	71	85	—	77	—
C_{22}	53	72	—	—	80	—
cis-9,10-Epoxyoctadecane-1,18-dioic	20	66	—	—	—	—
ω-Hydroxymonobasics[b]						
C_{16}	9[a]	45[a]	60	76	60[a]	58
C_{18}	10[b]	47	60	—	62[b]	64
C_{22}	13	50	—	—	65	—
7-Hydroxyhexadecane-1,16-dioic	4	50	—	—	45	—
8-Hydroxyhexadecane-1,16-dioic	4	50	—	—	45	—
9-Hydroxy-10-methoxyoctadecane-1,18-dioic[c]	3	50	—	—	—	—
cis-9,10,-Epoxy-18-hydroxyoctadecanoic	2	40	—	56	52	—
cis-9,10,-Epoxy-18-hydroxyoctadec-12-enoic	2	40	—	—	52	—
16-Hydroxy-10-oxohexadecanoic	0	34	—	—	—	—
16-Hydroxy-9-oxohexadecanoic	0	34	—	—	—	—
erythro-9,10-Dihydroxyoctadecane-1,18-dioic	0	26	34	—	20	27
threo-9,10-Dihydroxyoctadecane-1,18-dioic	0	26	34	—	20	27
9,18-Dihydroxyoctadecanoic	0	26	—	—	35	—
10,18-Dihydroxyoctadecanoic	0	23	—	—	32	—
9,18-Dihydroxy-10-methoxyoctadecanoic[c]	0	19	—	—	—	—
10,18-Dihydroxy-9-methoxyoctadecanoic[c]	0	19	—	—	—	—
9,18-Dihydroxy-methoxyoctadec-12-enoic[c]	0	19	—	—	—	—
10,18-Dihydroxy-9-methoxyoctadec-12-enoic[c]	0	19	—	—	—	—
9,16-Dihydroxyhexadecanoic	0	17	30	—	26	—
10,16-Dihydroxyhexadecanoic	0	14	30	34	23	32
erythro-9,10,18-Trihydroxyoctadecanoic	0	7	16	16	12	16
threo-9,10,18-Trihydroxyoctadecanoic	0	7	16	16	12	16
9,10,18-Trihydroxyoctadec-12-enoic	0	7	16	—	12	—
NEUTRALS						
1-Alkanols						
C_{20}	23	53	—	—	—	—
C_{26}	26	55	—	—	—	—
α,ω-Diols						
C_{16}	0	22	—	—	—	—
C_{24}	0	26	—	—	—	—

[a] Separate spots may be obtained if fraction contains homologs differing in chain length by 4 or more carbon atoms.

[b] Saturated and unsaturated homologs not resolved.

[c] Methoxyhydrin derivative of corresponding epoxide formed when methanol used in depolymerization reaction.

Layer   L1 = Silica gel G

Table 4 (continued)
CUTIN AND SUBERIN MONOMERS OBTAINED AFTER DEPOLYMERIZATION BY HYDROLYSIS AND TRANSESTERIFICATION METHODS (TLC)

Solvent	S1	= Chloroform-benzene (1:1)
(v/v)	S2	= Chloroform-ethyl acetate (7:3)
	S3	= Diethyl ether-*n*-hexane-methanol (40:10:1)
	S4	= Diethyl ether-petroleum hydrocarbon (b.p. 50-70°C)-acetone-acetic acid (21:7:3:1)
	S5	= Chloroform-acetic acid (9:1)
	S6	= Chloroform-methanol-formic acid (20:1:3)
Derivative	ME = methyl ester, FA = fatty acid	
Detection	Charring with H_2SO_4	
Technique	T1 = 12 cm distance, na = not quoted	

REFERENCES

1. Typical values obtained in author's laboratory from runs of all the compounds listed on the same 20 × 20 cm plate.
2. **Eglinton, G. and Hunneman, D. H.,** *Phytochemistry,* 7, 313, 1968.
3. **Croteau, R. and Fagerson, I. S.,** *Phytochemistry,* 11, 353, 1972.
4. **Brieskorn, C. H. and Böss, J.,** *Fette Seifen Anstrich.,* 66, 925, 1964.
5. **Baker, E. A. and Martin, J. T.,** *Ann. Appl. Biol.,* 60, 313, 1967.
6. **Kabelitz, L.,** Thesis, Julius-Maximilians-Universität, Würzburg, West Germany, 1970.

adsorbents such as silica gel using the less polar solvent mixtures, and those with lower R_f values by increasing the polarity of the mixture. Fractions of small volume should always be collected, their composition monitored regularly by TLC and checked against that of the unfractionated sample. The polarity of eluting solvents must also be increased gradually (10% at a time) when necessary.

The CC method is tedious and time consuming, and on its own provides little information about the identity of any of the components. It is useful only for the large scale isolation of cutin/suberin components and, since the fractions that are obtained can be accurately weighed, provides a reliable means of quantitation. However, with modern methods of structural determination, it is rarely necessary to use large quantities of material, and for isolations on a milligram scale CC has largely been replaced by preparative TLC or (preparative-layer chromatography) which offer the advantage of speed and greater efficiency of separation.

B. Thin-Layer Chromatography (TLC)
1. Analytical Thin-Layer Chromatography

The total products derived from any depolymerization method (Section C) can be analyzed without further derivatization by ascending TLC on 0.25 mm thick layers of commercial grades of silica gel G or other types of silica gel; other adsorbents such as aluminum oxide do not appear to have been widely used in this type of work. Plates prepared by standard methods are activated at 110°C for 1 hr, allowed to cool, and stored in a desiccator. Cutin and suberin samples are applied to the layers with fine glass capillary pipettes from solutions in chloroform (5 to 10 mg/mℓ); for free acid monomers it may be necessary to add methanol or ethanol to improve solubility. Analyses are carried out in glass tanks with their walls lined with filter paper and the running solvent (mobile phase) is placed inside for at least 1 hr beforehand to ensure that the tank atmosphere is saturated with solvent vapor. After development and subsequent removal of running solvent (important), plates are sprayed with either sulfuric acid-water (1:1 v/v) (Caution! corrosive reagent, use fume hood) or a 10% w/v solution of phosphomolybdic acid (approximately $H_3PO_4·12 MoO_3·24 H_2O$) in ethanol water (19:1 v/v). Both are general reagents capable of detecting a minimum of 1 to 2 µg of monomers; for sulfuric acid spots are visualized after charring at 180°C (black/white background) and for phosphomolybdate by heating at 100 to 120°C (dark blue/yellow background).

A specific reagent is available for epoxy monomers, if they are present, and is based on the formation of an adduct with picric acid (2,4,6-trinitrophenol);[52] it is sensitive to 5 to 10 μg of epoxide. Plates are sprayed with 0.05 M picric acid in ethanol-water (19:1 v/v) and placed immediately in a tank saturated with ether-ethanol-water-acetic acid (80:19:1:1 v/v). The plates are left for 30 min and then exposed to pyridine vapor; epoxides appear as orange spots on a yellow background.

Several different running solvents have been used for the TLC analysis of cutin/suberin monomers but unfortunately few workers have quoted any R_f data. Adequately documented systems for the separation of free acid and methyl ester monomers are given in Table 4 and those for reduced monomers in Table 5 (it was necessary to calculate some of the R_f values from illustrations of chromatograms). A more comprehensive list of compounds determined in the author's laboratory is also included. The R_f values recorded in the Tables should not be regarded as absolute because they are not exactly reproducible. However, they do establish the order of elution of the compounds in a given system and the degree of separation that can be achieved.

Because the separation of cutin/suberin monomers on layers of silica gel takes place chiefly by adsorption, it is ultimately governed by the polarity of the compounds relative to one another. The major factor influencing mobility, therefore, is the nature and number of functional groups present giving a general separation into major classes of compounds, with further smaller movements within classes arising from differences in chain length and position of the functional groups. On unmodified layers there is no resolution between saturated and unsaturated analogs or between the diastereoisomeric forms of *vic*-disubstituted compounds. Monomers that contain one or two functional groups are fractionated mainly according to class, and only minor variations in R_f value occur according to chain length. The chain length variations, however, are sufficiently great to give separate spots if the class fraction contains substantial quantities of homologs differing by more than four carbon atoms. The longer chain homolog has the higher R_f value. This situation occurs commonly in the α,ω-dibasic and ω-hydroxymonobasic acid fractions of suberins (C_{18} and C_{22} type), and any such differences are accentuated by reducing the polarity of the running solvent (S1, Table 4).

For monomers that have three or more functional groups there is a clear separation within each class between homologs differing by only two carbon atoms, the longer chain homolog again having the greater R_f value. Small but measurable differences in R_f value may also occur within such classes between positional isomers, e.g., 10,16- and 9,16-dihydroxyhexadenoic acids and the corresponding methyl esters.

The order of elution of monomers as acids or as methyl esters is similar except for those that have an α,ω-dibasic acid function; as acids their R_f values relative to other monomers are different from those of the corresponding methyl esters (compare S2 and S5, Table 4).

In some of the systems a number of monomers and monomer classes are not well resolved from one another and may co-elute or elute very close together. In the general solvent S2 (Table 4), the following have similar or identical R_f values: (1) short chain (circa C_{16}) monobasic and longer chain (circa C_{22}) methyl esters of α,ω-dibasic acids; (2) methyl esters of monohydroxydibasic and ω-hydroxymonobasic acids (and the methoxyhydrins of dimethyl 9,10-epoxyoctadecane-1,18-dioate, if present), and (3) dimethyl 9,10-dihydroxyoctadecane-1,18-dioate and positional isomers of methyl dihydroxyoctadecanoate. If the neutral fraction is not removed from suberin samples, any 1-alkanols will elute close to the compounds described in (2) and α,ω-diols, if present, co-elute with those in (3). Some of these problems can be overcome by using the alternative less polar solvent S1 (Table 4). The comparatively

Table 5
CUTIN AND SUBERIN MONOMERS OBTAINED AFTER DEPOLYMERIZATION BY REDUCTION METHODS (TLC)

	L1	L1
Layer	L1	L1
Solvent	S1	S2
Technique	T1	T2
Derivative	A	A
Reference	1	2

Compounds	$R_f \times 100$	
1-Alkanols		
C_{16}[a]	72	62
C_{18}	72	64
C_{22}	—	66
α,ω-Diols		
C_{16}[a]	56	42
C_{18}	56	44
C_{22}	—	48
1,9,18-Octadecanetriol	42	26
12-Octadecene-1,9,18-triol	42	26
1,7,16-Hexadecanetriol	33	20
1,8,16-Hexadecanetriol	33	20
1,9,10,18-Octadecanetetraol	26	10
12-Octadecene-1,9,10,18-tetraol	26	10

[a] Separate spots may be obtained if fraction contains homologs differing in chain length by 4 or more carbon atoms.

Layer	L1	=	Silica gel G
Solvent	S1	=	Diethyl ether-*n*-hexane-methanol (8:2:1)
	S2	=	Chloroform-ethyl acetate (3:7)
Derivative	A	=	Free alcohols
Detection			Charring with H_2SO_4
Technique	T1	=	20 cm distance
	T2	=	12 cm distance

REFERENCES

1. **Walton, T. J. and Kolattukudy, P. E.,** *Biochemistry*, 11, 1885, 1972.
2. Typical values obtained in author's laboratory from runs of all the compounds listed on the same 20 × 20 cm plate.

few classes of monomers obtained by reductive depolymerization are well resolved from one another by TLC (Table 5).

In cutin or suberin work, R_f values alone can never be taken as definite proof of identity even if authentic compounds are used. However, because the monomers are comprised of a limited number of specific compounds, R_f values do give a reliable indication of either their class or, in the case of the more highly substituted monomers, their probable identity. GC is needed to confirm the chain length of homologs and GC and MS to determine the exact positions of functional groups especially where nonterminal.

As with other types of fatty acids containing the same groups, unsaturated cutin/suberin monomers can be separated from one another, and from their saturated analogs, on layers of silica gel impregnated with silver nitrate, and the diastereoisomeric forms of any *vic*-dihydroxy monomers can be resolved on the same adsorbent impregnated with boric acid. The techniques, however, have not been widely employed, and are poorly documented in the case of silver nitrate where no R_f values are available. The separation of the *cis*- and *trans*- forms of dimethyl 9-octadecene-1,18-dioate is reported[53] with an unspecified concentration of silver nitrate using hexane-ether (1:1 v/v), and the following systems are recommended[5] for the analysis of unsaturated components in classes of reduced monomers on layers prepared from 4 g silver nitrate/ 30 g silica gel G(1) monools, ether-hexane (4:1 v/v); (2) diols, diethyl-hexane-methanol (20:5:1 v/v), (3) triols, diethyl ether-hexane-methanol (8:2:1 v/v); and (4) tetraols, diethyl ether-hexane-methanol (4:1:1 v/v).

On boric acid layers [2.8 g boric acid/25 g silica gel G,[54] using S2 (Table 4)], a good separation of methyl ester of diastereoisomers is obtained.[31] For a 12-cm run, $R_f \times 100$ values are dimethyl 9,10-dihydroxyoctadecane-1,18-dioate *erythro* 46, *threo* 30; methyl 9,10,18-trihydroxyoctadecanoate *erythro* 22; and *threo* 8.

It is advisable to use both techniques only on isolated fractions because if total depolymerization products are examined, any separations that occur may be obscured by other components present in the sample.

2. Preparative Layer Chromatography

The separation of mixtures of cutin/suberin monomers achieved by adsorption TLC using the solvents described in Tables 4 and 5 can be simply adapted for preparative purposes to provide discrete fractions that are invaluable for further chromatographic analysis and complete identification. Methyl ester and reduced monomers are most commonly used for such purposes, free acid monomers being difficult to recover from the adsorbent. On a 20 × 20 cm plate the usual 0.25 mm thick silica gel layers can be used for separations of up to 1 mg of starting material — preparative TLC (prep-TLC) — but up to 10 mg of material can be fractionated on the same size plate spread at 0.75 mm thickness — preparative layer chromatography (PLC). The optimum amount of sample that can be applied without overloading the layer will vary according to its composition and is largely determined by trial and error. Overloading causes streaking resulting in poor resolution between components. The results of analytical TLC should be used for guidance, and samples that have one major component or that have predominantly high R_f constituents will require a lower loading than those having an even distribution of components.

Prior to analysis, plates should be developed the full 20 cm distance in the chosen mobile phase in order to elute any impurities in the silica gel to the top of the plate; before loading, all traces of solvent must be allowed to evaporate from the layers. Samples dissolved in a minimum volume of solvent (warmed if necessary) are applied to the layers as a band on the starting line. The efficiency of the separation is strongly influenced by the width of the starting band because the bands that separate as fractionation proceeds inevitably spread to become wider than the original application, with those having higher R_f showing the greatest enlargement. If the starting bands are wide, the separated bands will be too wide and consequently poorly resolved, especially those that have only small differences in R_f value. Narrow starting bands are obtained if samples are loaded onto layers which have been previously warmed to ensure that there is rapid evaporation of solvent from the sample solution; plates must be allowed to cool to room temperature before development if this method is employed.

After development and removal of all traces of running solvent (important!), the bands are visualized using a nondestructive reagent. Suitable methods are exposure to iodine vapor in a tank or spraying the layers with a 0.1% aqueous solution of fluorescein[5] or a 0.05%

aqueous solution of Rhodamine 6G;[4] the positions of the zones are marked. Alternatively, a narrow lane down each side of the plate is sprayed with phosphomolybdate (Chapter 28, Reagent No. 21) and the color developed using localized heat from a hot air blower. The unsprayed portion of the layer is protected by a glass plate placed on top throughout this procedure. The silica gel corresponding with the position of each marked zone is removed from the plate and the compounds recovered from the adsorbent with diethyl ether. Elution can be performed using the adsorbent packed into a small column (25 to 50 mℓ solvent), or the silica gel can be refluxed with the solvent for 30 min followed by filtration. It is important at this stage to use high purity solvent, clean apparatus, and to extract filter papers, glass wool, etc. with eluting solvent to eliminate any contamination. After evaporation, check the purity of each isolated fraction by analytical TLC against the original sample and repeat the prep-TLC or PLC if necessary.

Although the R_f values of individual fractions already give some information about their probable identity, confirmation can now be obtained using a variety of techniques, e.g., GC, MS, IR, NMR, or by carrying out further reactions to verify the nature of functional groups.

C. Gas Chromatography (GC)

The GC analysis of cutin/suberin monomers has been carried out mainly on short packed columns (1 to 3 m length, 1.5 to 3 mm O.D.) with low loadings (1 to 5%) of silicone-type stationary phases. The choice of phase is limited by the high temperatures necessary for elution of all the various compounds and only those materials capable of operation up to 275 to 300°C can be used, e.g., SE-30, OV-1, Dexsil 300, OV-17, OV-210, OV-225, or their commercial equivalents. Temperature programming is also normally carried out, particularly where analysis of total depolymerization products is required. Flame ionization has been most frequently used as the means of detection.

Before analysis it is necessary to derivatize any free carboxyl and hydroxyl groups of monomers; other functional groups such as epoxy and oxo can be analyzed directly. Monomers containing carboxyl groups are obtained only from depolymerization by hydrolysis methods (Section C), and are converted into the corresponding methyl esters (Chapter 2); these are obtained directly by the transesterification methods. For the hydroxyl groups a number of derivatives are potentially available, but in modern work preference is given to those that combine volatility with a readily interpretable mass spectrum for subsequent identification by combined GC/MS. Acetates or trifluoroacetates were used in earlier work,[2,28,30] but although good derivatives for GC, they do not always provide a satisfactory mass spectrum for identification purposes. Acetylation also cannot be used if epoxide monomers are present in the sample because they are converted into the corresponding diacetyl derivative (Table 3); this is the same derivative as that formed from the naturally occurring *vic*-dihydroxy compound. Trimethylsilyl ethers (TMSi), however, satisfy all the analytical requirements with the added advantage of easy preparation in one step without any necessity for removal of excess reagents. Although these derivatives can be prepared from cutin/suberin monomers using any of the usual silylating reagents, the most reliable method in the author's laboratory has been reaction with BSA (*N,O-bis* trimethylsilylacetamide)-anhydrous pyridine (2:1 or 3:1 w/v) at 50°C for 1 hr. Monomers are freely soluble in this reagent, and a suitable concentration for most GC work is 1 mg sample to 100 to 200 µℓ reagent (injection volume 1 to 1.5 µℓ). It is better to work with a concentrated solution at first and dilute with BSA, if necessary. For transesterification products it is important to ensure that no acids are present in the sample, otherwise they will be converted into TMSi esters during silylation and give extra GC peaks.

GC retention data for cutin/suberin monomers on two silicone phases of different polarity, SE-30 and OV-17, are given in Tables 6 and 7 for methyl ester TMSi ether derivatives and

Table 6
MORE COMMON CUTIN MONOMERS OBTAINED AFTER DEPOLYMERIZATION BY HYDROLYSIS AND TRANSESTERIFICATION METHODS (GC)

Column packing	P1	P2
Temperature (°C)		
Initial	130	130
Final	300	285
Program rate (°C/min)	6	6
Injection port (°C)	250	250
Detector	300	300
Gas	N_2	N_2
Flow rate (mℓ/min)	30	35
Column		
Length (m)	2	2
Diameter (mm O.D.)	2	2
Form	Coil	Coil
Material	SS	SS
Detector	FI	FI
Reference	1	1

Compound	Retention Index[a]	
Methyl hexadecanoate	1915	2030
Methyl 16-hydroxyhexadecanoate TMSi	2280	2385
Dimethyl 7-hydroxyhexadecane-1,16-dioate TMSi (and similar hydroxy positional isomers)	2425	2625
Methyl 16-hydroxy-10-oxohexadecanoate TMSi (and similar oxo positional isomers)	2425	2645
Methyl 18-hydroxyoctadec-9-enoate TMSi	2430	2580
Methyl 10,16-dihydroxyhexadecanoate *bis* TMSi (and similar hydroxy positional isomers)	2475	2555
Methyl *cis*-9,10,-epoxy-18-octadec-12-enoate TMSi	2570	2815
Methyl *cis*-9,10-epoxy-18-hydroxyoctadecanoate TMSi	2620	2820
Methyl 10,18-dihydroxyoctadecanoate *bis* TMSi (and similar hydroxy positional isomers)	2675	2755
Methyl 9,18-dihydroxy-10-methoxyoctadec-12-enoate *bis* TMSi[b]	2700	2835
Methyl 10,18-dihydroxy-9-methoxyoctadec-12-enoate *bis* TMSi[b]	2700	2835
Methyl 9,18-dihydroxy-10-methoxyoctadecanoate *bis* TMSi[b]	2750	2840
Methyl 10,18-dihydroxy-9-methoxyoctadecanoate *bis* TMSi[b]	2750	2840
Methyl *erythro*-9,10,18-trihydroxyoctadec-12-enoate *tris* TMSi	2745	2835
Methyl *-threo*-9,10,18-trihydroxyoctadec-12-enoate *tris* TMSi	2750	2835
Methyl *erythro*-9,10,18-trihydroxyoctadecanoate *tris* TMSi	2795	2840
Methyl *threo*-9,10,18-trihydroxyoctadecanoate *tris* TMSi	2800	2840

[a] $$\text{Retention Index} = 100 \frac{R_t\text{ Compound} - R_t\text{ Alkane eluting before compound}}{R_t\text{ Alkane eluting after compound} - R_t\text{ Alkane eluting before compound}} + 100 \times \text{Carbon number of alkane eluting before compound}$$

[b] Methoxyhydrin derivatives of corresponding epoxide formed when methanol was used in the depolymerization reaction.

Column packing	P1 = 1% SE30 on Gas Chrom Q (80–100 mesh)
	P2 = 1% OV17 on Gas Chrom Q (100–20 mesh)

REFERENCE

1. Typical values (expressed to the nearest 5 index units) obtained in the author's laboratory from runs of all the compounds listed using co-injection with a mixture of *n*-alkanes from "Parafilm". (See **Gaskin, P., MacMillan, J., Firn, R. D., and Pryce, R. J.**, *Phytochemistry,* 10, 1155, 1971.)

in Table 8 for TMSi ether derivatives of reduced monomers. For ease of reference, Table 6 lists those compounds most likely to be encountered in cutin analyses and Table 7 those

Table 7
MORE COMMON SUBERIN MONOMERS OBTAINED AFTER DEPOLYMERIZATION BY HYDROLYSIS AND TRANSESTERIFICATION METHODS (GC)

Column packing	P1	P2
Conditions and literature as Table 6		

Compound	Retention Index	
	P1	P2
Dimethyl 1,16-hexadecanedioate	2225	2455
Methyl 16-hydroxyhexadecanoate TMSi	2280	2385
Methyl icosanoate	2315	2430
1-Icosanol TMSi	2360	2360
Dimethyl 9-octadecene-1, 18-dioate	2385	2650
Dimethyl 1,18-octadecanedioate	2425	2655
Methyl 18-hydroxyoctadec-9-enoate TMSi	2430	2580
Methyl 18-hydroxyoctadecanoate TMSi	2480	2585
Methyl docosanoate	2515	2630
1-Docosanol TMSi	2560	2560
Dimethyl *cis*-9,10-epoxyoctadecane-1,18-dioate	2575	2900
Methyl *cis*-9,10-epoxy-18-hydroxyoctadecanoate TMSi	2620	2820
Dimethyl 1,20-icosanedioate	2625	2855
Methyl 20-hydroxyicosanoate TMSi	2680	2785
Dimethyl 9-hydroxy-10-methoxyoctadecane-1,18-dioate TMSi[a]	2710	2920
Methyl tetracosanoate	2715	2830
Methyl 9,18-dihydroxy-10-methoxyoctadecanoate TMSi[a]	2750	2840
Methyl 10,18-dihydroxy-9-methoxyoctadecanoate TMSi[a]	2750	2840
Dimethyl *erythro*-9,10-dihydroxyoctadecane-1,18-dioate *bis* TMSi	2755	2910
Dimethyl *threo*-9,10-dihydroxyoctadecane-1,18-dioate *bis* TMSi	2760	2910
1-Tetracosanol TMSi	2760	2760
Methyl *erythro*-9,10,18-trihydroxyoctadecanoate *tris* TMSi	2795	2840
Methyl *threo*-9,10,18-trihydroxyoctadecanoate *tris* TMSi	2800	2840
Dimethyl 1,22-docosanedioate	2825	3055
Methyl 22-hydroxydocosanoate TMSi	2880	2985
Methyl hexacosanoate	2915	3030
1-Hexacosanol TMSi	2960	2960
Methyl 24-hydroxytetracosanoate TMSi	3080	3185

[a]   Methoxyhydrin derivatives of corresponding epoxide formed when MeOH used in depolymerization reaction.

Column packing	P1 =	1% SE30 on Gas Chrom Q (80–100 mesh)
	P2 =	1% OV17 on Gas Chrom Q (100–120 mesh)

in suberin analyses; both tables should be used in conjunction with the survey of constituents given in Table 1. The data are expressed as retention indices (RI), which, because they are not as reproducible as those determined by isothermal analysis, are given to the nearest five units. These should not be regarded as absolute values because they may vary slightly with the chromatographic conditions used, but they do establish the order of elution of compounds on a particular phase and the degree of separation that can be expected between them. Ideally, each GC column should always be calibrated using the relevant reference compounds. Although some retention information on cutin/suberin compounds has already been published, it is not included in the GC tables because the papers either do not list a sufficient number of compounds or provide enough details about the chromatographic conditions. The few RI values available for methyl ester TMSi ethers on SE-30 type phases,[9,13,14,16] however, are in good agreement with those given in Tables 6 and 7. On columns impregnated with 1 to 3% of OV-210 or OV-225, the overall separation of monomers is very similar to that

Table 8
CUTIN AND SUBERIN MONOMERS OBTAINED AFTER DEPOLYMERIZATION BY REDUCTION METHODS (GC)

Column packing
Conditions and literature as Table 6

Compound	Retention Index		Corresponding monomeric compound in original cutin or suberin
	P1	P2	
1-Hexadecanol TMSi	1960	1960	Hexadecanoic acid, neutral component suberin
1-Octadecanol TMSi	2160	2160	Octadecanoic acid, neutral component suberin
1,16-Hexadecanediol bis-TMSi	2320	2310	1,16-Hexadecanedioic, 16-hydroxyhexadecanoic acids
1-Icosanol TMSi	2360	2360	Icosanoic acid, neutral component suberin
9-Octadecene-1,18-diol bis TMSi	2480	2500	9-Octadecene-1,18-dioic, 18-hydroxyoctadec-9-enoic acids
1,18-Octadecanediol bis TMSi	2520	2510	1,18-Octadecanedioic, 18-hydroxyoctadecanoic acids, neutral component suberin
1,7,16-Hexadecanetriol tris TMSi	2520	2485	10,16-Dihydroxyhexadecanoic, 7-hydroxyhexadecane-1,16-dioic, 16-hydroxy-10-oxohexadecanoic acids
1,8,16-Hexadecanetriol tris TMSi	2520	2485	9,16-Dihydroxyhexadecanoic,8-hydroxyhexadecane-1,16-dioic, 16-hydroxy-9-oxohexadecanoic acids
1-Docosanol TMSi	2560	2560	Docosanoic acid, neutral component suberin
12-Octadecene-1,9,18-triol tris TMSi	2675	2660	9,10-Epoxy-18-hydroxyoctadec-12-enoic acid
1,9,18-Octadecanetriol tris TMSi	2715	2675	9,10-Epoxy-18-hydroxyoctadecanoic, 9,10-epoxyoctadecane-1,18-dioic, 10,18-dihydroxyoctadecanoic acids
1,20-Icosanediol bis TMSi	2720	2710	1,20-Icosanedioic, 20-hydroxyicosanoic acids, neutral component suberin
1-Tetracosanol TMSi	2760	2760	Tetracosanoic acid, neutral component suberin
12-Octadecene-1,9,10,18-tetraol *tetrakis* TMSi	2790	2755	9,10,18-Trihydroxyoctadec-12-enoic acid
1,9,10,18-Octadecanetetraol *tetrakis* TMSi	2830	2770	9,10,18-Trihydroxyoctadecanoic, 9,10-dihydroxyoctadecane-1,18-dioic acids
1,22-Docosanediol bis TMSi	2920	2910	1,22-Docosanedioic, 22-hydroxydocosanoic acids, neutral component suberin
1-Hexacosanol TMSi	2960	2960	Hexacosanoic acid, neutral component suberin
1,24-Tetracosanediol bis TMSi	3120	3110	1,24-Tetracosanedioic, 24-hydroxytetracosanoic acids, neutral component suberin

Column packing   P1 = 1% SE-30 on Gas Chrom Q (80–100 mesh)
                    P2 = 1% OV-17 on Gas Chrom Q (100–120 mesh)

on OV-17, but the RI values are greater by circa 200 units than the values for the same compounds on OV-17. However, both of these phases are not as stable as OV-17 at high temperatures (>250°C), and dual column operation is usually necessary to compensate for bleed.

The fractionation of cutin/suberin monomers by GC is primarily according to chain length, the nature, number, and position of any functional groups then producing additional changes in retention related to their polarity. The effects produced by functional groups in turn are influenced by the polarity of the stationary phase, and any differences can be used to advantage for the identification of particular groups in compounds, e.g., monomers containing a dibasic acid function on OV-17. A good separation between the majority of monomers is generally obtained by GC, but at the present time no single phase is capable of resolving all of them. A difference of 5 RI units between compounds (Tables 6, 7, and 8) signifies that they are poorly separated and likely to produce overlapping peaks or occur as shoulders on major peaks. On SE-30 the most important cutin/suberin compounds having a similar or identical RI values are (1) dimethyl 7-hydroxyhexadecane-1,16-dioate TMSi, methyl 16-hydroxy-10-oxohexadecanote TMSi, methyl 18-hydroxyoctadec-9-enoate TMSi and dimethyl 1,18-octadecanedioate; (2) methyl 10,16-dihydroxyhexadecanoate TMSi and methyl 18-hydroxyoctadecanoate TMSi; (3) methoxyhydrins of 9,10-epoxy-18-hydroxy-octadecanoate TMSi and dimethyl 9,10-dihydroxyoctadecane-1,18-dioate TMSi; and (4) 1,18-octadecanediol TMSi and 1,7,16-hexadecanetriol TMSi. These groups of compounds, however, can be separated on OV-17, although on this phase unsaturated compounds are generally poorly resolved from their saturated analogs, and the methoxyhydrin TMSi ethers of epoxy compounds have retention values similar to those of the corresponding *vic*-dihydroxy TMSi compounds. Closely related positional isomers, e.g., epoxide methoxyhydrins TMSi, methyl dihydroxyhexadecanoates TMSi, and hexadecanetriols TMSi, are not separated on either phase and elute as single mixed peaks. *Erythro* and *threo* isomers are also poorly resolved on both phases using temperature programming at 6°C/min, but on SE-30 they can be separated, if required, by isothermal analysis or by using a lower program rate; the *erythro* isomer is eluted before the corresponding *threo* compound. Analysis by capillary GC[55] (20 m × 0.5 mm glass column coated with OV-101, N_2 flow rate 2 mℓ/min) offers little overall improvement, and the same co-elutions as those on packed SE-30 columns are still evident.

If GC retention data are determined using both nonpolar and polar phases, identification of monobasic, α,ω-dibasic and ω-hydroxymonobasic acids, 1-alkanols and α,ω-diols can be made, i.e., those monomers containing only terminal functional groups. If prep-TLC or PLC fractions have been analyzed, further confirmation of the class of compound is also given by the R_f value. In other classes of monomer, GC retention does not provide unequivocal identification because it does not establish the exact position of any nonterminal functional groups, the various positional isomers having very similar or identical RI values. Although MS is needed for the full structural determination of such compounds (for details consult marked papers in reference list), the GC data are reliable indicators of the type of compound, because cutin/suberin analysis is concerned mainly with a limited number of specific compounds. Additional verification can also be obtained from R_f values if isolated fractions are analyzed. For unsaturated monomers, no information is obtained from GC about the position or configuration of the double bonds, since separation occurs only according to the degree of unsaturation. On SE-30 the resolution between saturated, monoene, and diene analogs is good with a 40 to 50 RI unit difference between them. The detailed structural determination of these compounds, however, can only be achieved by techniques such as osmylation followed by mass spectrometric analysis of the resultant *vic*-diol TMSi derivative.[1,4,5,27,36]

IV. CONCLUSIONS

Chromatography is a powerful and sensitive tool in the analysis of the depolymerization

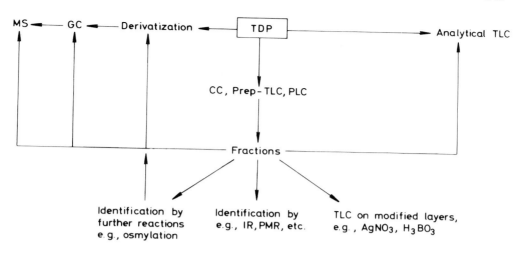

Scheme 1. Summary of the use of chromatographic techniques in the analysis of plant cutins and suberins (TDP = total depolymerization products).

products from plant cutins and suberins providing a rapid and generally an efficient means of separating the constituents. Although the data provided (TLC, R_f and GC, RI) cannot always be used for positive identification, confirmation in most cases can be simply obtained by combination with MS. The correctness of any structural assignments, however, should always be compared with the full chromatographic data because mistakes can be made.[55] Also in interpreting analytical results, due consideration must always be given to the method of depolymerization used because it can have a considerable influence on both the qualitative and quantitative composition of the final products.

A summary of the various chromatographic approaches to cutin/suberin analysis is given in Scheme 1.

V. SOURCES OF REFERENCE COMPOUNDS

Unfortunately, not many of the compounds cited in this article are available commercially, apart from monobasic acids, 1-alkanols, and a few of the α,ω-dibasic and ω-hydroxymonobasic acids. However, as an alternative, a number of plant cuticles or corks can be recommended as reliable sources of particular monomers or groups of monomers. These materials are usually available throughout the year, are easy to isolate, and the compounds are obtained in good yield after depolymerization:

1. Tomato fruit cuticles (*Lycopersicon esculentum*), green or red fruits — mainly 10,16-dihydroxyhexadecanoic acid[19]
2. Gooseberry fruit cuticles (*Ribes grossularia*), fresh or canned — mainly 10,16-dihydroxyhexadecanoic acid[28]
3. Apple fruit cuticles (*Malus pumila*), any variety — major constituents 10,16-dihydroxyhexadecanoic, 18-hydroxyoctadec-9,12-dienoic, 9,10-epoxy-18-hydroxyoctadec-12-enoic, 9,10-epoxyhydroxyoctadecanoic and 9,10,18-trihydroxyoctadecanoic acids[8]
4. *Agave americana* and *Sanseviera trifasciata* leaf cuticles, common ornamental pot plants in Europe — major constituents 9,10-epoxy-18-hydroxyoctadecanoic and 9,10,18-trihydroxyoctadecanoic acids[27]
5. Lemon fruit cuticles (*Citrus limon*) — mainly 16-hydroxy-10-oxohexadecanoic and 10,16-dihydroxyhexadecanoic acids[24]

6. Bottle cork (*Quercus suber*), commercially available — major constituents 9-octade-cene-1,18-dioic, 18-hydroxyoctadec-9-enoic, 9,10-dihydroxyoctadecane-1,18-dioic (and corresponding 9,10-epoxide), 9,10,18-trihydroxyoctadecanoic (and corresponding 9,10-epoxide), 1,22-docosanedioic and 22-hydroxydocosanoic acids[27,31]

7. Silver birch cork (*Betula pendula*), common tree, cork layers easily collected from trunk — major constituents 9,10,18-trihydroxyoctadecanoic (and corresponding 9,10-epoxide), 22-hydroxydocosanoic acid 1,22-docosanedioic acids[27,31]

8. Potato cork (*Solanum tuberosum*), skin of tubers — mainly 9-octadecene-1,18-dioic and 18-hydroxyoctadec-9-enoic acids[34-36]

ACKNOWLEDGMENTS

The author is extremely grateful to Professor I. Ribas-Marqúes (Santiago de Compostela, Spain), Professor E. Seoane (University of Valencia, Spain), and Dr. E. A. Baker (University of Bristol) who have generously provided authentic samples of many of the cutin/suberin monomers used in this work.

REFERENCES

1. **Eglinton, G. and Hunneman, D. H.,** *Phytochemistry*, 7, 313, 1968.
2. **Baker, E. A. and Holloway, P. J.,** *Phytochemistry*, 9, 1557, 1970.
3. **DeVries, H.A.M.A.,** *Acta Bot. Neerl.*, 19, 36, 1970.
4. **Croteau, R. and Fagerson, I. S.,** *Phytochemistry*, 11, 353, 1972.
5. **Walton, T. J. and Kolattukudy, P. E.,** *Biochemistry*, 11, 1885, 1972.
6. **Kolattukudy, P. E. and Walton, T. J.,** *Biochemistry*, 11, 1897, 1972.
7. **Holloway, P. J., Deas, A. H. B., and Kabaara, A. M.,** *Phytochemistry*, 11, 1443, 1972.
8. **Holloway, P. J.,** *Phytochemistry*, 12, 2913, 1973.
9. **Caldicott, A. B.,** Thesis, University of Bristol, 1973.
10. **Holloway, P. J.,** *Phytochemistry*, 13, 2201, 1974.
11. **Brieskorn, C. H. and Binnemann, P. H.,** *Phytochemistry*, 14, 1363, 1975.
12. **Baker, E. A. and Procopiou, J.,** *J. Sci. Food Agric.*, 26, 1347, 1975.
13. **Caldicott, A. B. and Eglinton, G.,** *Phytochemistry*, 14, 1799, 1975.
14. **Caldicott, A. B., Simoneit, B. R. T., and Eglinton, G.,** *Phytochemistry*, 14, 2223, 1975.
15. **Caldicott, A. B. and Eglinton, G.,** *Phytochemistry*, 15, 1139, 1976.
16. **Hunneman, D. H. and Eglinton, G.,** *Phytochemistry*, 11, 1989, 1972.
17. **Brieskorn, C. H. and Kabelitz, L.,** *Phytochemistry*, 10, 3195, 1971.
18. **Crisp, C. E.,** Thesis, University of California, Davis, 1965.
19. **Brieskorn, C. H. and Reinartz, H.,** *Z. Lebensm. Unters. Forsch.*, 135, 55, 1967.
20. **Haas, K.,** Thesis, Universität Hohenheim, West Germany, 1974.
21. **Kolattukudy, P. E.,** *Biochem. Biophys. Res. Commun.*, 49, 1040, 1972.
22. **Kolattukudy, P. E.,** *Biochemistry*, 13, 1354, 1974.
23. **Kolattukudy, P. E.,** *Lipids*, 8, 90, 1973.
24. **Deas, A. H. B., Baker, E. A., and Holloway, P. J.,** *Phytochemistry*, 13, 1901, 1974.
25. **Brieskorn, C. H. and Böss, A. J.,** *Fette Seifen Anstrichm.*, 66, 925, 1964.
26. **Kolattukudy, P. E., Walton, T. J., and Kushwaha, R. P. S.,** *Biochem. Biophys. Res. Commun.*, 42, 739, 1971.
27. **Holloway, P. J. and Deas, A. H. B.,** *Phytochemistry*, 12, 1721, 1973.
28. **Baker, E. A. and Martin, J. T.,** *Ann. Appl. Biol.*, 60, 313, 1967.
29. **Holloway, P. J. and Deas, A. H. B.,** *Phytochemistry*, 10, 2781, 1971.
30. **Duhamel, L.,** *Ann. Chim.*, 8, 315, 1963.
31. **Holloway, P. J.,** *Chem. Phys. Lipids*, 9, 158, 1972.
32. **Holloway, P. J.,** *Chem. Phys. Lipids*, 9, 171, 1972.
33. **Brieskorn, C. H. and Binnemann, P. H.,** *Tetrahedron Lett.*, 1127, 1972.
34. **Rodrígues-Miguéns, B. and Ribas-Marqués, I.,** *An. Real Soc. Espan. Fis. Quim.*, 68, 303, 1972.

35. **Kolattukudy, P. E. and Agrawal, V. P.,** *Lipids,* 9, 682, 1974.
36. **Brieskorn, C. H. and Binneman, P. H.,** *Z. Lebensm. Unters. Forsch.,* 154, 213, 1974.
37. **Kolattukudy, P. E., Kronman, K., and Poulose, A. J.,** *Plant Physiol.,* 55, 567, 1975.
38. **Gonzales-Gonzales, A., Sola, J., Iglesias-Martin, F., Rivas-Paris, V., and Ribas-Marqués, I.,** *Quimica Industria,* 15, 139, 1968.
39. **Swan, E. P.,** *Tech. Assoc. Pulp Paper Ind.,* 51, 301, 1968.
40. **Swan, E. P. and Naylor, A. F. S.,** *Bi-Monthly Res. Notes Dept. Fish Forest Can.,* 25, 32, 1969.
41. **Kirrmann, M. A. and Duhamel, L.,** *C. Rendu.,* 254, 1303, 1962.
42. **Seoane, E., Serra, M. C., and Agulló, C.,** *Chem. Ind.,* 662, 1977.
43. **Bescansa-Lopéz, J. L., Gil-Curbera, G., and Ribas-Marqués, I.,** *An. Real Soc. Espan. Fis. Quim. Ser. B.,* 62, 865, 1966.
44. **Huelin, F. E. and Gallop, R. A.,** *Aust. J. Sci. Res.,* B4, 526, 1951.
45. **Orgell, W. H.,** *Plant Physiol.,* 30, 78, 1955.
46. **Yamada, Y., Wittwer, S. H., and Bukovac, M. J.,** *Plant Physiol.,* 39, 28, 1964.
47. **Holloway, P. J. and Baker, E. A.,** *Plant Physiol.,* 43, 1878, 1968.
48. **Holloway, P. J. and Baker, E. A.,** in *Principles and Techniques of Scanning Electron Microscopy, Biological Applications,* Vol. 1, Hayat, M. A., Ed., Von Nostrand Reinhold, New York, 1974, 181.
49. **Schlenk, H. and Gellerman, J. L.,** *Anal. Chem.,* 32, 1412, 1960.
50. **Glass, R. L.,** *Lipids,* 6, 919, 1971.
51. **Holloway, P. J.,** unpublished work.
52. **Fioriti, J. A. and Sims, R. J.,** *J. Chromatogr.,* 32, 761, 1968.
53. **Rodríguez-Miguéns, B. and Ribas-Marqués, I.,** *An. Real Soc. Espan. Fis. Quim.,* 68, 1301, 1972.
54. **Morris, L. J.,** *J. Chromatogr.,* 12, 321, 1963.
55. **Cardoso, J. N., Eglinton, G., and Holloway, P. J.,** in *Advances in Organic Geochemistry,* Campos, R. and Goni, J., Eds., ENADIMFA, Madrid, 1977, 273.
56. **Holloway, P. J. and Deas, A. H. B.,** *Chem. Ind.,* p. 1971, 1140.

The following publications include details of the mass spectrometric identification of compounds:
1, 4—11, 13—17, 20—24, 26, 27, 29, 31—33, 35—37, 42, and 56.

Chapter 11

SURFACE LIPIDS OF PLANTS AND ANIMALS

P. J. Holloway

I. INTRODUCTION

The outer surface of most living organisms is covered with a protective coating of lipid material that has been synthesized by the cells, or by specialized groups of cells, in the outermost layers of the organism. This superficial material, commonly referred to as wax,* is generally markedly different in chemical composition from any lipids found elsewhere in the internal tissues of the same organism. Thus, unlike the major somatic lipids that are derived mainly from glycerol, those on the surface form a much more complex and heterogeneous group comprising a great variety of different classes of compounds, predominantly long chain and aliphatic in nature (summarized in Tables 1 and 2). The group as a whole is also extremely variable in composition with considerable differences occurring between species of both plants and animals in the number, nature, and relative amount of the classes present. Within each class, even greater variations are found in the chain length of individual components and in the positions of functional groups. A large number of the classes are specifically associated with the surface lipid, and some of them are confined either to particular groups or individual species of organism. The same compounds, however, may be present in the surface lipids obtained from widely different sources.

Our knowledge of natural surface lipids is far from complete, and most information is available about these lipids from four main groups of organisms:

- Mammals — skin of fur and hair-bearing species, products of the epidermal cells and sebaceous glands (review[1])
- Birds — feathers, products of the preen (uropygial) glands (review[2])
- Insects — epicuticle of exoskeleton, products of cells near surface probably oenocytes (reviews[3-5])
- Plants — cuticular membrane of aerial parts, products of epidermal cells (reviews[6-8]) — surface of fungal spores, origin unknown

Only in the case of birds and higher plants have a large number of individual species been examined. Because the chemistry is so diverse it is difficult, and indeed dangerous, to generalize about the composition of surface lipids although a comprehensive survey of species is outside the scope of the present article. However, a number of observations on classes and functional groups must be made for each group of organism, mainly for the benefit of workers unfamiliar with this type of lipid, but also to establish the compounds most likely to be encountered during any chromatographic work.

In the mammalian skin lipids that have been examined so far, monoesters (aliphatic and sterol), and diesters (1,2-alkanediol and/or 2-hydroxyalkanoic types) are generally the predominant classes, and both have a high content of branched chain and unsaturated components. Man is exceptional in containing larger quantities of triacylglycerols and fatty acids.

---

* In this article, the term "wax" is deliberately avoided because it has been so misused that it is impossible to define precisely or concisely in either physical or chemical terms. Because the classical chemical definition — an ester of a long-chain primary alcohol and a long-chain fatty acid — is obviously no longer adequate, it has become a convenient collective term for an assortment of lipid compounds not based on glycerol. The physical properties of such compounds may range from liquids to hard crystalline solids.

Table 1
SUMMARY OF THE MAIN CLASSES OF COMPOUND FOUND IN SURFACE LIPIDS OF ANIMALS AND PLANTS[a]

Class[a]	Homolog range or predominant compounds	Occurrence			
		Mammals	Birds	Insects	Plants
Hydrocarbons					
n-Alkanes	Odd C_{20}–C_{50}	Uncommon	Uncommon	Common[b]	Common[b]
br-Alkanes	Odd & even C_{20}–C_{40}	Rare	Not reported	Common	Uncommon
n-Alkenes	Odd C_{20}–C_{40}	Not reported	Not reported	Uncommon	Rare
Monoesters (see Table 2)	Odd & even C_{16}–C_{70}	Common	Common[b]	Common[b]	Common
Ketones	Odd C_{25}–C_{33}	Not reported	Not reported	Rare	Uncommon[b]
Diesters (see Table 2)	Odd & even C_{24}–C_{70}	Common[b]	Uncommon[b]	Rare	Uncommon[b]
Aldehydes	Even C_{20}–C_{40}	Not reported	Not reported	Rare	Common[b]
β-Diketones	Odd C_{27}–C_{35}	Not reported	Not reported	Not reported	Uncommon[b]
Secondary alcohols	Odd C_{13}–C_{35}	Not reported	Not reported	Rare[c]	Uncommon[b]
Ketols	Odd C_{29}–C_{31}	Not reported	Not reported	Rare[c]	Rare
Triesters (see Table 2)	Odd & even C_{20}–C_{80}	Common	Common	Uncommon	Uncommon[b]
Primary alcohols					
n-1-Alkanols	Even C_{10}–C_{40}	Common[c]	Common[c]	Uncommon[c]	Common[b]
br-1-Alkanols	Odd & even C_{10}–C_{40}	Uncommon[c]	Common[c]	Not reported	Rare
n-Alkene-1-ols	Even C_{10}–C_{40}	Uncommon[c]	Not reported	Rare[c]	Not reported
Pentacyclic triterpenols	β-Amyrin, lupeol	Not reported	Not reported	Rare[c]	Uncommon
Sterols	Cholesterol	Common[c]	Uncommon	Uncommon	Not reported
	Stigmasterol, β-sitosterol	Not reported	Not reported	Not reported	Uncommon
Hydroxy β-diketones	Odd C_{27}–C_{35}	Not reported	Not reported	Not reported	Uncommon
Diols	Odd & even C_{12}–C_{34}	Common[c]	Uncommon[c]	Uncommon[c]	Uncommon[c]
Fatty acids					
n-Alkanoic acids	Even C_{10}–C_{40}	Common[c]	Common[c]	Common	Common[b]
br-Alkanoic acids	Odd & even C_{10}–C_{40}	Common[c]	Common[c]	Not reported	Rare[c]
n-Alkenoic acids	Even C_{10}–C_{30}	Uncommon[c]	Not reported	Rare[c]	Rare
Hydroxy fatty acids	Odd & even C_{8}–C_{30}	Common[c]	Uncommon[c]	Rare[c]	Uncommon[c]

| Pentacyclic triterpenoid acids | Ursolic, oleanolic | Not reported | Not reported | Not reported | Uncommon[b] |
| Polyesters (see Table 2) | Even C_{80} upwards | Not reported | Not reported | Rare[c] | Uncommon[b] |

[a] The classes are arranged in approximate order of polarity.
[b] Major components of the surface lipids of certain species or groups of species.
[c] Found chiefly in an esterified form (see Table 2).

Table 2
TYPES OF ESTER FOUND IN THE SURFACE LIPIDS OF ANIMALS AND PLANTS

Class	Homolog range	Occurrence			
		Mammals	Birds	Insects	Plants
MONOESTERS					
n-Alkanoic acids + n-1-alkanols	Even C_{18}-C_{70}	Common	Common	Common	Common
br-Alkanoic acids + n-1-alkanols	Odd & even C_{18}-C_{50}	Common	Common	Not reported	Rare
n-Alkanoic acids + br-1-alkanols	Odd & even C_{18}-C_{50}	Common	Common	Not reported	Rare
br-Alkanoic acids + br-1-alkanols	Odd & even C_{18}-C_{50}	Common	Common	Not reported	Rare
n-Alkanoic acids + n-1-alkanols	Even C_{24}-C_{52}	Uncommon[10]	Not reported	Rare[11]	Uncommon[12]
n-Alkanoic acids + n-alkene-1-ols	Even C_{24}-C_{52}	Uncommon[10]	Not reported	Rare	Not reported
n-Alkenoic acids + n-alkene-1-ols	Even C_{24}-C_{46}	Uncommon[10]	Not reported	Not reported	Not reported
br-Alkenoic acids + n-alkene-1-ols	Odd & even C_{26}-C_{42}	Uncommon[10]	Not reported	Not reported	Not reported
n-Alkanoic acids + br-alkene-1-ols	Odd & even C_{26}-C_{42}	Uncommon[10]	Not reported	Not reported	Not reported
n-Alkenoic acids + n-alkene-1-ols	Odd & even C_{26}-C_{42}	Uncommon[10]	Not reported	Not reported	Not reported
br-Alkenoic acids + br-alkene-1-ols	Odd & even C_{26}-C_{42}	Uncommon[10]	Not reported	Not reported	Not reported
n-Alkanoic acids + n-2-alkanols	Odd C_{29}-C_{37}	Not reported	Not reported	Not reported	Uncommon[13]
n-Alkanoic acids + n-secondary alcohols	Odd C_{37}-C_{45}	Not reported	Not reported	Uncommon[14]	Uncommon
n-Alkanoic acids + sterols	Acids even C_{2}-C_{26}	Common	Uncommon	Uncommon	Uncommon
br-Alkanoic acids + sterols	Acids odd & even C_{14}-C_{24}	Uncommon	Not reported	Not reported	Not reported
n-Alkenoic acids + sterols	Acids even C_{14}-C_{24}	Uncommon	Not reported	Not reported	Not reported
n-Alkanoic acids + pentacyclic triterpenols	Acids even C_{2}-C_{26}	Not reported	Not reported	Not reported	Uncommon
n-Oxoalkanoic acid + n-oxoalkan-1-ol	C_{66}	Not reported	Not reported	Uncommon[15]	Not reported
n-(ω-1)-Hydroxyalkanoic acids + n-1-alkanols (1)	Even C_{40}-C_{48}	Not reported	Not reported	Rare[16]	Not reported
n-Alkanoic acids (1) + n-α,(ω-1)-alkanediols	Even C_{40}-C_{48}	Not reported	Not reported	Rare[16]	Not reported
n-Alkanoic acids (1) + n-α,ω-alkanediols	Even C_{38}-C_{50}	Not reported	Not reported	Not reported	Rare[17]
n-Alkanoic acids (1) + n-(ω-1)-hydroxyalkanoic acids	Even C_{32}-C_{44}	Not reported	Not reported	Rare[16]	Not reported
DIESTERS					
n-Alkanoic acids + n-2-hydroxyalkanoic acids + n-1-alkanols and various combinations with unsaturated and br-analogs	Odd & even C_{48}-C_{68}	Common	Not reported	Not reported	Not reported

Compound	Carbon range				
n-Alkanoic acids + n-3-hydroxyalkanoic acids + n-1-alkanols	Odd & even C_{24}–C_{42}	Not reported	Rare[18]	Not reported	Not reported
n-Alkanoic acids + n-(ω-1)-hydroxyalkanoic acids + n-1-alkanols	Even C_{56}–C_{70}	Not reported	Not reported	Rare[16]	Not reported
n-Alkanoic acids + n-ω-hydroxyalkanoic acids + n-1-alkanols	Even C_{36}–C_{66}	Rare	Not reported	Not reported	Uncommon[19]
n-Alkanoic acids (2) + n-1,2-alkanediols and various combinations with βρ-analogs	Odd & even C_{48}–C_{68}	Common	Uncommon	Not reported	Not reported
n-Alkanoic acids (2) + n-2,3-alkanediols	Odd & even C_{24}–C_{68}	Not reported	Uncommon	Not reported	Not reported
n-Alkanoic acids (2) + n-α,(ω-1)-alkanediols	Even C_{56}–C_{66}	Not reported	Not reported	Rare[16]	Not reported
n-Alkanoic acids (2) + n-α,ω-alkanediols	Odd & even C_{50}–C_{60}	Rare	Not reported	Rare	Rare[20,21]
n-Δ²-Alkenoic acids (2) + n-α,ω-alkanediols	Odd & even C_{50}–C_{60}	Not reported	Not reported	Not reported	Rare[20,21]

TRIESTERS

Compound	Carbon range				
Triacylglycerols	Acids even C_{6}–C_{22}	Common	Common	Uncommon	Rare
n-Alkanoic acids + n-alkylhydroxypropanedioic acids + n-1-alkanols (2)	Odd & even C_{35}–C_{78}	Not reported	Uncommon[22]	Not reported	Not reported
n-Alkanoic acids + n-(ω-1)-hydroxyalkanoic acids (2) + n-1-alkanols	Even C_{72}–C_{82}	Not reported	Not reported	Rare[16]	Not reported
n-Alkanoic acids (2) + n-(ω-1)-hydroxyalkanoic acids + n-α,(ω-1)-alkanediols	Even C_{72}–C_{80}	Not reported	Not reported	Rare[16]	Not reported
n-Alkanoic acids (2) + n-ω-hydroxyalkanoic acids + n-α,ω-alkanediols	Even C_{48}–C_{68}	Not reported	Not reported	Not reported	Uncommon[19]
n-Alkanoic acids + n-ω-hydroxyalkanoic acids (2) + n-1-alkanols	Even C_{48}–C_{82}	Not reported	Not reported	Not reported	Uncommon[19]

POLYESTERS

Compound	Carbon range				
n-Alkanoic acids (2) + n-ω-hydroxyalkanoic acids (2,3 or 4) + n-α,ω-alkanediols	Even C_{60}–C_{104}	Not reported	Not reported	Not reported	Uncommon[19]
n-Alkanoic acids + n-(ω-1)-hydroxyalkanoic acids (4) + n-α,(ω-1)-alkanediols	Even C_{104}–C_{114}	Not reported	Not reported	Rare[16]	Not reported
n-(ω-1)-Hydroxyalkanoic acids (4) + n-1-alkanols	Even C_{88}–C_{98}	Not reported	Not reported	Rare[16]	Not reported

Note: References to the more unusual types only included.

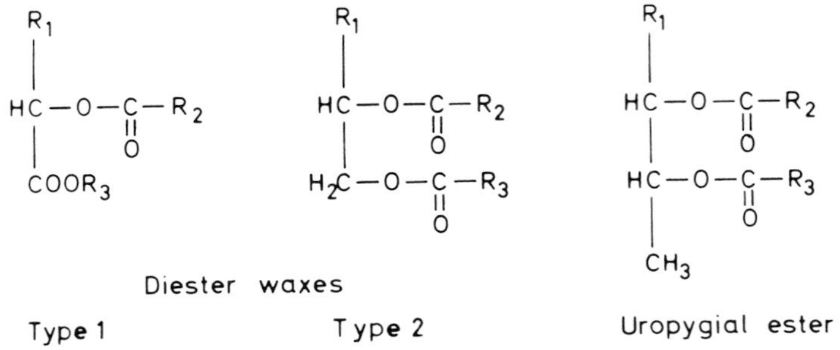

$$
\begin{array}{ccc}
R_1 & R_1 & R_1 \\
| & | & | \\
HC-O-\underset{\underset{O}{\|}}{C}-R_2 & HC-O-\underset{\underset{O}{\|}}{C}-R_2 & HC-O-\underset{\underset{O}{\|}}{C}-R_2 \\
| & | & | \\
COOR_3 & H_2C-O-\underset{\underset{O}{\|}}{C}-R_3 & HC-O-\underset{\underset{O}{\|}}{C}-R_3 \\
& & | \\
& & CH_3
\end{array}
$$

Diester waxes

Type 1   Type 2   Uropygial ester

FIGURE 1. Structures of some diester constituents of surface lipids. Type 1: 2-hydroxyalkanoic acid with the hydroxyl group esterified with an alkanoic acid and the carboxyl group esterified with a 1-alkanol. Type 2: 1,2-alkanediol with both hydroxy groups esterified with an alkanoic acid. Uropygial: 2,3-alkanediol with both hydroxy groups esterified with an alkanoic acid. R_1 is usually straight chain; R_2 and R_3 may be straight chain, branched chain, or unsaturated.

Aliphatic monoesters are commonly the major class in the secretions of bird uropygial glands, but some orders or species contain diesters of 1,2-alkanediols (rails), 2,3-alkanediols (hens, turkeys, pheasants), or 3-hydroxyalkanoic acids (ring dove), or triesters of alkylhydroxypropanedioic acids (geese, ducks). The presence of a wide variety of saturated single and multiple branched chain components is a characteristic feature of only the monoesters.

In contrast to the two previous groups of animals, the surface lipids of insects and plants show much more variation both in qualitative and quantitative composition, and the overall chain length of individual compounds is also generally longer. In insects several different types of epicuticular lipids are found that differ from one another mainly in the nature of the predominant class present. Some contain a large proportion of normal and/or branched-chain hydrocarbons (cockroaches, grasshoppers, crickets), others aliphatic ester classes (bees, scale insects), 1-alkanols (eri silkworm), diols (mealworm), or fatty acids (pea aphid). Of all the classes known in insects, hydrocarbons and fatty acids probably occur most frequently.

The epicuticular lipids of plants contain the largest variety of classes and many of them are found as major components of certain groups or species, e.g., hydrocarbons (peas, brassicas), aliphatic monoesters (*Portulaca oleracea*), ketones (leeks), aldehydes (sugar cane), β-diketones (*Eucalyptus globulus*, wheat), secondary alcohols (*Aquilegia* sp., Papaveraceae), primary alcohols (many grasses), and fatty acids (sorghum). However, unlike the surface lipids from other sources, compounds of a particular chain length or substitution tend to predominate in several of the classes irrespective of species; branching and unsaturation is rare. Thus C_{29} and C_{31} homologs are frequently the major components of hydrocarbon, ketone, and secondary alcohol fractions with any substituent groups being located either in the middle of the chain or at the 10-position. The monoesters comprise chiefly C_{42}, C_{44}, C_{46}, and C_{48} homologs, the aldehydes, primary alcohols and fatty acid C_{26}, C_{28}, and C_{30} homologs; and the β-diketones, if present, C_{29}, C_{31}, and C_{33} compounds with substitution mainly at the 12,14-, 14,16-, and 16,18-positions, respectively. Hydrocarbons, aliphatic monoesters, aldehydes, primary alcohols, and fatty acids are present in most plant surface lipids.

In addition to aliphatic components, cyclic constituents may be present in natural surface lipids. Sterols (chiefly cholesterol) and sterol monoesters occur commonly in mammals, but less frequently in birds and insects. Sterols (most commonly β-sitosterol and stigmasterol), pentacyclic triterpenoid alcohols (β-amyrin, lupeol), and acids (ursolic, oleanolic), together with their corresponding monoesters, are found in plants, and sometimes they are the major components of the surface lipid, e.g., in fruits of apple and grape.

Chromatographic methods are widely used in the routine analysis of natural surface lipids to (1) fractionate the total extract into its constituent classes and to provisionally identify them, (2) isolate individual classes from the total extract, and (3) separate and identify individual compounds in each class isolated. In the present chapter, however, the main emphasis will be on (1) and (2) because detailed accounts of chromatographic techniques suitable for the analysis of most of the surface lipid classes when isolated are found elsewhere in this volume. It will discuss how thin-layer (TLC), column (CC), and gas-liquid (GC) chromatography can be utilized in the qualitative and quantitative analysis of total unhydrolyzed surface lipid extracts. These methods, in general, have tended to evolve independently for each main group of organism that has been studied in detail despite the fact that many of the constituent classes are common to all. Many of the systems, therefore, differ only in detail but achieve the same objective.

As with other types of lipid analysis, high pressure-liquid chromatography has not yet been developed to its full potential for surface lipid work, mainly because of the lack of a detection system of adequate sensitivity for all classes of compounds. At the time of writing there is only one report[9] of its use, in the analysis of human sebum lipid mixtures on a microparticulate silica column in conjunction with a moving wire detector.

II. PREPARATION OF SAMPLES

The isolation of surface lipids normally presents few problems since the materials are soluble at room temperature in organic solvents of low to moderate polarity, e.g., *n*-hexane, benzene, methylene chloride, diethyl ether, chloroform, or acetone. The lipid is usually obtained by washing or immersing a known area or weight of the organism for short periods (10 to 30 sec) in an excess of solvent. The extracts are filtered, dried if necessary over anhydrous sodium sulfate, and the solvent removed under reduced pressure with a rotary evaporator. The residue obtained is then dried to constant weight. If the material is not to be analyzed immediately, it is advisable to store it without solvent either under vacuum in a dessicator or under nitrogen in a sealed tube to avoid any decomposition. This procedure is most important for samples that contain classes susceptible to atmospheric oxidation, particularly aldehydes (mainly plant surface lipids), and those that have any unsaturated components.

Occasionally the surface lipid on plants and animals is produced in sufficiently large quantities so that a solvent extraction is unnecessary, and it can be removed by scraping or brushing the surface, e.g., carnauba palm, silkworm larvae, or geese preen glands.

Problems of contamination sometimes arise during surface lipid isolation from (1) extraneous compounds introduced accidentally from solvents or apparatus, and (2) co-extraction of variable quantities of lipids that originate from the internal tissues of the organism. The first source of contamination can be eliminated by:

- Avoiding all contact of extracting solvents with rubber or plastic materials
- Always using glass apparatus that has been carefully cleaned in chromic acid
- Exhaustively extracting any filtering aids (paper, glass wool, sinters) with the solvent chosen for lipid isolation
- Using high quality redistilled solvents checked regularly by evaporating 100 to 200 mℓ to dryness, and analyzing the residue dissolved in a small volume (0.1 to 0.2 mℓ) of pure solvent by TLC and GC
- Removing any organic impurities from sodium sulfate, if it is used, by heating overnight at 500°C

It is not always easy either to prevent the contamination of a surface lipid extract with some of the internal lipids or to show conclusively that all the lipids extracted are derived exclusively from the surface of the organism. In all work with mammalian lipids,[1] it must be anticipated that some of the lipid isolated will be from within the epidermis (especially the stratum corneum). There is also the likelihood of contamination by extraneous lipids from the environment and also from the microbial degradation of the native surface lipids. Such problems can be largely overcome by making two consecutive extractions on different days; the second extract should then contain only those lipids secreted during the intervening interval. With insects and higher plants, an additional complication is that substantial amounts of lipid are usually embedded within the subsurface layers of the exoskeleton and cuticular membrane, respectively. These materials are not, strictly speaking, internal lipids because they are still located outside the cells of the organism. In the case of higher plants, however, it is known that the intracuticular lipid fraction is poorly extracted, if at all, by the mild procedures used for surface lipid isolation because it can only be obtained after extensive extraction of the isolated cuticular membrane with hot organic solvents (see Chapter 10). The chemical composition of this fraction is also usually markedly different from that obtained from the surface of the same plant. A similar situation has been demonstrated in insects from experiments with isolated exoskeletons obtained from the cast larval skins of certain insects.[23]

It is generally agreed that the possibility of extracting internal lipids is minimized by using a nonpolar solvent and avoiding prolonged contact of the surface of the organism with the solvent. The polarity of the solvent, however, must be sufficiently great to ensure that the surface lipids are rapidly dissolved; a hydrocarbon solvent, for instance, may not be the best one for a surface lipid containing large amounts of more polar classes such as primary alcohols or fatty acids. Ideally, the efficiency of a range of solvents should be evaluated for the particular organism being studied by monitoring the yield and composition of extracts obtained by successive short duration immersions or washings. This procedure enables a solvent to be selected that will give the maximum yield of surface lipid in the shortest possible contact time. If TLC is used to analyze each consecutive extract, the extent of any contamination can often be assessed by direct comparison of the composition, either with that of an extract known to be free of internal lipids or with one prepared by deliberately extracting them by prolonged contact of the surface with the same solvent, the hot solvent, or by using a more polar solvent such as chloroform-methanol (1:1 v/v). In many cases, small quantities of uncontaminated lipids may be obtained by gentle wiping of the surface of the organism with swabs of preextracted cotton wool or paper tissue soaked in solvent; the lipid is recovered from the swabs by soaking them in further quantities of solvent, filtering, and evaporating down to dryness. For higher plants or insects, where the surface lipid deposits often possess a crystalline or semicrystalline structure, the efficiency of the isolation method can also be verified by examination of the extracted surfaces in the scanning electron microscope.[24,25]

Some typical isolation procedures suitable for different types of plants and animals are summarized below.

A. Mammals

Procedures for the Isolation of Surface Lipids of Large Mammals
Horse, cow, sheep, goat, etc.;[26] animals are restrained, anesthetized, or freshly killed. Clip hair with scissors and press an open bell jar or cylinder (7 cm diameter) firmly onto the exposed skin. Quickly pour redistilled hexane into the jar or cylinder onto the skin surface, and irrigate solvent over the skin with the aid of a large pipet equipped with a rubber bulb (taking care no solvent enters the bulb!). Transfer extract to another container with the pipet, filter through a medium porosity sintered glass funnel, and remove solvent. In man this method yields 50 to 350 μg of lipid/cm² skin depending on subject and anatomical site.[1]

Human scalp:[27] *shampoo all subjects thoroughly with a 1% solution of sodium lauryl sulfate. On the following day, pour 500 mℓ of chloroform over the scalp and allow to flow into a basin. Remove all debris by filtration and evaporate the solvent. On the average circa 200 mg/scalp/day is obtained from a young adult male.*

Procedures for the Isolation of Surface Lipids of Small Mammals
Mice:[28] *lightly anesthetize each animal with diethyl ether. Hold animal by head and tail over a large beaker and irrigate skin with a fine jet of acetone for circa 10 sec (do not wash head, tail, and anal regions). Use circa 10 mℓ solvent for each animal and collect washings in a beaker. Combine extracts and remove solvent. Using this method, the toxic effects of the solvent are minimal and the animals survive. Washing of three mice on each of three successive days produced 31 mg of lipid (131 mg/kg body weight per washing).*
Rats, guinea pigs, hamsters, etc.:[26] *anesthetize animals with diethyl ether. Wipe tail, feet, and anus with cotton pledgets soaked in hexane. Dip each animal up to its neck in a beaker of redistilled hexane or roll it on its back in a shallow dish containing the same solvent. Yield of surface lipid obtained from the rat using this method, but with acetone as solvent was 130 to 170 mg/day/kg body weight.*[29]

B. Birds

Most of the isolation work has been carried out directly on the secretion from the uropygial gland itself and not on the lipid that is eventually deposited on the feathers from the gland. If extraction of feathers is required use any suitable washing or immersion method.

Procedure for the Isolation of Surface Lipids of Birds with Large Preen Glands
Geese: the gland is a paired organ situated at the base of the tail. A row of hair-like downy feathers is arranged in an oval around the two excretory openings, one for each lobe of the gland. In living birds these feathers are soaked with secretion that can be collected easily by scraping with a spatula. Dissolve collected secretion in distilled diethyl ether, filter, and wash three times with a small volume of distilled water to remove water-soluble material. Dry ether solution over anhydrous sodium sulfate and evaporate solvent. About 40 mg of crude secretion is obtained every day from each adult bird.
Turkey and chickens:[31] *remove preen glands from freshly slaughtered birds and freeze. Open before tissue has completely thawed so that the secretion can be removed as a solid without contamination by tissue fragments. Each chicken gland yields circa 75 mg lipid and those from turkey circa 50 mg lipid.*

Procedure for the Isolation of Surface Lipids of Birds with Small Preen Glands[32]
Carefully excise glands from freshly killed birds taking care to avoid contamination with surrounding tissue (rich in triacylglycerols). Homogenize excised glands in chloroform-methanol (2:1 v/v) and filter the extract. Add one volume of water and collect the lower layer that contains the extracted lipids. The amount of secretion from preen glands varies from 0.5 mg (Passeriformes) to circa 600 mg (gulls).[2]

C. Insects

Procedure for the Isolation of Surface Lipids of Whole Insects
Cockroaches:[33] *kill insects by freezing at −29°C for 20 min. Place in a Buchner*

funnel fitted with a medium porosity fritted disc and cover with hexane. Allow the solvent to pass through the disc without applying suction. Repeat twice with fresh portions of solvent. Combine extracts and evaporate. Yield is 1.3 to 2 mg lipid/insect. Chloroform can also be used for this method.[34]

Procedure for the Isolation of Surface Lipids of Cast Larval Skins[35]
Wash integuments (circa 0.5 to 1 g dry weight) three or four times, by decantation, with separate portions of solvent. Each wash is of 10 sec duration. Combine extracts, filter, and evaporate. Yields (dry weight) 1 to 2% Eristalis tenax, Calliphora vomitoria, Cydia pomonella; 2% Rhodnius prolixus; up to 3% Pieris brassicae (pupal case); 2 to 4% Schistocerca gregaria.

D. Higher Plants

Procedure for the Isolation of Surface Lipids of Leaves and Fruits[36-38]
Determine the area of leaf samples by tracing outlines on paper of known weight and area. Cut out areas of paper corresponding with the outlines of individual leaves and weigh. Place redistilled chloroform in three beakers or shallow dishes; the dimensions of the vessels and volumes of solvent should be sufficient to allow total immersion of individual leaves. (Diethyl ether is an equally efficient solvent but is inflammable and, therefore, more hazardous to use.) Immerse each leaf, held by the petiole or cut end with a pair of forceps, for 5 to 10 sec in each portion of solvent in turn. Allow solvent to drain from the leaf before immersing in the next portion of solvent, and avoid immersing any cut areas. Combine extractives, filter, and evaporate solvent. Dry the residue to constant weight at 40°C. Typical approximate yields from some leaves using this method ($\mu g/cm^2$) are 10 tomato, grape, spinach, 20 to 30 citrus species, peas, 30 to 70 apple cultivars, 70 to 80 brassica cultivars, 250 rhododendron, and 460 laurel.
If necessary, the surface lipid from either upper or lower surfaces of leaves can be isolated separately by washing the required surface with circa 5 mℓ of chloroform delivered from a fine-orifice buret. Hold individual leaves in position with forceps under the buret and allow the solvent to run down the surface and drain from the leaf tip into a beaker. Combine washings, filter and evaporate solvent.
For fruits, calculate surface areas from measurement of mean diameters (if spherical $4\pi r^2$) or from volumes determined by displacement of water (if spherical $^4/_3\pi r^3$). For extraction, proceed as described for leaves, immersing each fruit for up to 30 sec in successive portions of solvent. Typical yields from some fruits using this method ($\mu g/cm^2$) are 20 to 50 citrus species, 100 to 150 grapes, blackcurrants, and 600 to 1200 apple cultivars.

E. Fungal Spores

Procedure for the Isolation of Surface Lipids of Fungi[39]
Wash spores from 7-day-old cultures grown in Petri dishes with a fine jet of water (washbottle) into a beaker. Filter washings through fine cheese cloth and freeze-dry the spore suspension obtained. Suspend freeze-dried spores in hexane at room temperature for 1 min, centrifuge, and remove the supernatant. Filter through extracted cotton wool and remove solvent. Yields (% dry weight) are 0.15, Alternaria tenuis; 0.19, Botrytis fabae; 0.17, Neurospora crassa; 0.2, Rhizopus stolonifer.

Procedure for the Isolation of Surface Lipids of Pathogenic Fungi[21]
Sphaerotheca fuliginea:[21] *collect spores by brushing them off the surface of infected marrow leaves and remove any plant debris by passing through a 100 mesh sieve. Extract dried spores (100 mg) with hexane (6 mℓ) in a 10 mℓ glass stoppered centrifuge tube by gentle shaking for 1 min. Centrifuge for 1 min, filter the supernatant through extractive-free glass wool, and evaporate solvent. The yield is 5.6% of dry weight.*

III. CHROMATOGRAPHY

A. Thin-Layer Chromatography (TLC)

This technique is recommended as the first stage in the analysis of any natural surface lipid extract because it provides a simple, rapid, and sensitive means of determining the qualitative composition of the lipid mixture. Extracts can be examined directly by ascending TLC on 0.25 mm thick layers of aluminum oxide and silica gel prepared and activated by any of the conventional methods. Layers can also be obtained commercially precoated. Most modern work, however, is carried out on silica gel because it is now known that aluminum oxide is not suitable for the analysis of all classes of surface lipid, particularly aldehydes, that are decomposed by the adsorbent and acidic compounds that are strongly adsorbed and streak.

For analysis, lipid samples are dissolved in any suitable organic solvent (usually the same one used for isolation, e.g., chloroform) to give a concentration of circa to 5 to 10 mg/mℓ solvent, and 50 to 100 μg is applied to the layer using a narrow bore glass pipette or microsyringe. It is advantageous to apply a range of different amounts of extracts of unknown composition to the same plate so that the relative amounts of both major and minor components can be assessed in the mixture. Any authentic compounds used for reference markers should be treated in a similar fashion to the extracts, but a lower loading of 10 to 20 μg need only be applied. To ensure that results are reproducible, the following precautions are recommended:

- Use high quality solvents, redistilled before use, for development
- Line walls of the TLC tanks with filter paper
- Place the chosen running solvent (mobile phase) inside the tank at least 1 hr before using for development of plates to ensure saturation of tank atmosphere with solvent vapor
- Check that all solvent has evaporated from the layer after application of samples before placing the plate in the tank for development.

A number of different reagents can be used for the detection of surface lipids after TLC separation and these include several general types,[40-42] together with a few specific ones for certain classes or functional groups. Both destructive and nondestructive reagents are available, the latter being especially useful for subsequent preparative-TLC. A selection of reagents that have proved to be most reliable in the author's laboratory over the past 10 years or so for plant surface lipids is summarized below; they are all capable of detecting at least 1 to 2 μg of lipid. The same reagents, however, have also been widely used for surface lipids from other sources. Before attempting to detect compounds, always allow sufficient time for all traces of developing solvent to evaporate from the layer, otherwise sensitivity may be impaired or contrast lost between the spots and the background. Also remember to chromatograph reference compounds on the same plate as the samples, especially when using specific tests, to verify that the reagent is giving a satisfactory response.

Heating with sulfuric acid (Chapter 28, Reagent No. 38) or phosphomolybdic acid solution

(Reagent No. 21) is used for the detection of all lipids; iodine (Reagent No. 18) is particularly useful for visualizing β-diketones and all unsaturated compounds.

In contrast to these destructive reagents, solutions of Rhodamine 6G[41,43] (Reagent No. 32) and 2′,7′-dichlorofluorescein[44,45] (Reagent No. 9), which are also suitable as general spray reagents, do not alter the lipids. Thus the various fractions can be eluted from the adsorbent for further analysis.

A specific reagent for aldehydes is 4-amino-5-hydrazino-1,2,4-triazole-3-thiol,[46] whereas 2,4-dinitrophenylhydrazine[41] (Reagent No. 12) is suitable for the detection of both aldehydes and ketones. β-Diketones can be detected with Fast Blue B (tetraazotized di-*O*-anisidine) in cold sodium hydroxide solution.[47,48]

Isoprenoid lipids are best visualized with aqueous sulfuric acid (Reagent No. 38) as they develop different colors in the course of heating. Sterols and their esters turn red, whereas pentacyclic triterpenoid acids turn mauve. Various colors are also obtained using sprays of chlorosulfonic acid-glacial acetic acid,[49] and sulfuric acid-acetic anhydride[41] followed by short periods of heating at temperatures below 100°C. With the latter reagent a distinction can be made between stigmasterol and β-sitosterol on the basis of color response if the temperature is carefully controlled at 76°C.[50]

A very wide choice of developing solvents is available in the literature for the TLC fractionation of surface lipids, most of which have been formulated specifically for the separation of the lipids found within the major groups of organisms. Thus, for the analysis of mammalian lipids, workers have tended to adopt the multiple development systems designed by Nicolaides et al. and Downing et al., for birds the solvents recommended by Jacob,[32] and for plants those proposed by either Holloway and Challen or Tulloch. It should be emphasized, however, that the majority of these systems will work equally well for any surface lipid irrespective of source because several of the constituent classes are common to all. The choice of solvent is accordingly mainly one of personal preference, but in the author's opinion the use of multiple developments in different solvents offers little advantage for analytical TLC because one solvent can usually be found that will give the same separation in a single development.

A selection of useful solvent systems for plant and animal surface lipids is presented in Tables 3 and 4, respectively; in an article of this nature preference has to be given to those publications which quote R_f data or describe the separation of a substantial number of different compounds. Of all the solvents mentioned, the chloroform-ethanol system of Tulloch (S5 Table 3, S2 Table 4) is probably the best documented since it has been employed for the analysis of both plant and insect lipids and many synthetic compounds. Unless otherwise specified, chloroform refers to the commercial product stabilized with circa 2% of ethanol. An indication of the relative polarity, or eluting power, of the various systems is given by the R_f value obtained for triacylglycerols — the greater its value the greater the polarity of the developing solvent. The R_f values given in the tables will not always be exactly reproducible, but they do serve to establish for a given system the order of elution for different compounds and the degree of separation (resolution) that can be achieved between them. In practice, the performance and efficiency of any system must always be evaluated in the workers' own laboratory using the relevant reference compounds.

When dealing with a lipid sample of unknown composition it is recommended that the first analysis be made in a solvent of intermediate polarity, such as S2, S5 (Table 3); S2, S7, S8 (Table 4), which will give a general separation of constituents ranging from hydrocarbons to fatty acids (see Table 1). If after development the constituents appear predominantly at high R_f values, the analysis should be repeated using one of the less polar systems (e.g., S1 Table 3; S1, S3, S5 Table 4). Conversely, if the components are of low R_f and poorly resolved from the start line, another analysis is carried out in a polar system (e.g., S3, S4 Table 3; S4, S6 Table 4). As a general rule the relative merits of a number of different solvents should be evaluated for the particular lipid extract under investigation.

It is well established that the fractionation of surface lipids on thin layers of silica gel or aluminum oxide takes place predominantly by a process of selective adsorption that gives a separation of the constituents of a mixture according to their chemical class but not into individual compounds. The major factor governing the movement of classes in TLC, therefore, is their polarity relative to one another, which in turn is directly related to the nature, number, and position of the functional groups present. Hydrocarbons having no functional groups are the least strongly adsorbed and consequently always exhibit the highest R_f value. On substitution of the hydrocarbon skeleton, the R_f value is lowered, and this takes place progressively in the following approximate order of increasing polarity of functional groups: monoester < ketone < β-diketone < aldehyde < diester < secondary alcohol < triester < ketol < primary alcohol < hydroxy-β-diketone < diol < fatty acid. The R_f value, however, is unaffected by the presence of unsaturation or branching and remains identical to that of the corresponding saturated compound. The solubility of individual lipid classes in the developing solvent also plays some part in the overall fractionation process, and this is reflected in significant changes in the relative order of elution of certain classes in particular systems. The effect is most pronounced in those solvents containing a small proportion of ethanol or acetic acid, and it can be used to advantage for the potential identification of some classes, e.g., β-diketones, hydroxy-β-diketones, ketols, fatty acids.

Within classes, any variation in chain length usually produces insignificant changes in R_f values unless the difference is quite large, i.e., in the order of 15 to 20 carbon atoms; the longer chain homolog will then have a higher R_f value than that of the shorter chain homolog. A good example of this effect is shown by the R_f difference between a sterol acetate and a sterol hexadecanoate (Table 3), but such differences are much smaller between the homologs of other aliphatic classes. Single homologs belonging to a particular class normally appear on developed chromatograms as round spots, whereas mixtures of homologs of the same class tend to give an elongated spot.

The R_f values of classes that possess one or two functional groups in nonterminal positions, e.g., ketones and secondary alcohols, does vary significantly according to the position of the group. Chromatographic mobility decreases in a smooth curve from the centrally substituted isomers to the 2-substituted isomer.[51,52] Only small differences in R_f value are found between the most centrally substituted isomers, but these become larger when substitution approaches the terminal position, resulting in a definite separation between the 2,3, and 4 isomers and the remainder. For long-chain secondary alcohols chromatographed in benzene (S2 Table 3), the difference in $R_f \times 100$ value between the 2-isomers and the corresponding centrally substituted isomers is circa 25, the former having values close to those observed for primary alcohols. Further, small R_f differences may also occur within other classes, such as the separation between the 2-hydroxy alkanoic acid and 1,2-alkanediol types of aliphatic diesters (Table 4).

Although the separation between the majority of surface lipid classes is generally good, there is no universal system capable of separating all of them completely in one development. This is not surprising in view not only of the wide range of different polarities that occur between the classes but also the small differences that occur between some of them. Each solvent system, therefore, has some disadvantage for the analysis of particular classes: some may not migrate at all (remain on start line), others co-migrate with other classes or are poorly resolved from one another. Where lack of separation of classes for a solvent given in Tables 3 and 4 is apparent, an alternative that provides the separation required can usually be found elsewhere in the same Tables. The alternative, however, may only be suitable for the resolution of a limited number of classes. The use of specific spray reagents in conjunction with changes in developing solvent is also especially useful in overcoming some of these difficulties, e.g., separation of ketone/β-diketone/aldehyde; primary alcohol/pentacyclic triterpenol; hydroxy β-diketone/sterol mixtures.

Table 3
CLASSES OF CONSTITUENTS IN UNHYDROLYZED SURFACE LIPIDS OF PLANTS (TLC)

		Layer	L1	L1	L1	L1	L1	L1	L1
		Solvent	S1	S1	S2	S3	S4	S5	S6
		Technique	T1	T2	T2	T1	T1	na	na
Class	Homolog range or compound(s)	Reference	1	1,2	1,3	1	1	4,5	6,7
			$R_f \times 100$						
Hydrocarbons	C_{20}-C_{36}		73	90	80	78	78	83	86
Monoesters									
Fatty acids + primary alcohols	C_{40}-C_{60}		31	53	70	78	78	65	81
Δ^{2a} Fatty acids + primary alcohols	C_{40}-C_{50}		—	42	70	—	—	—	—
Fatty acids + 2-alkanols	C_{31}-C_{37}		29	50	68	78	78	—	—
Fatty acid methyl esters	C_{16}-C_{30}		16	30	49	70	72	47	—
Fatty acids + sterols	β-Sitosteryl hexadecanoate		35	57	72	78	78	—	—
Sterol acetates	β-Sitosteryl acetate		12	24	43	70	70	—	—
Fatty acids + triterpenols	β-Amyrenyl hexadecanoate		37	60	75	78	78	—	—
Triterpenol acetates	β-Amyrenyl acetate		14	27	48	72	72	—	—
Triterpenoid acid methyl esters	Methyl ursolate, methyl oleanolate		0	0	4	42	55	—	—
Diesters									
α,ω-Alkanediol type	C_{50}-C_{60}		—	5	49	—	—	47	—
Ketones	10- and 15-Nonacosanone, 16-hentriacontanone		22	40	63	78	78	53	—
β-Diketones	14,16-Hentriacontanedione		23	42	63	78	73	39	—
Aldehydes	C_{20}-C_{30}		19	35	52	72	78	42	69
Secondary alcohols	10- and 15-Nonacosanol, 16-hentriacontanol		5	11	35	65	72	36	—
Triacylglycerols	Tristearin		0	2	28	45	64	15	—
α-Ketols	14-Hydroxynonacosan-15-one		—	—	28	65	—	—	—
β-Ketols	13-Hydroxynonacosan-15-one		—	—	10	54	—	—	—
Primary alcohols	C_{20}-C_{30}		2	4	10	40	55	15	28
Triterpenols	β-Amyrin, lupeol		2	3	10	45	55	—	—
Sterols	β-Sitosterol, stigmasterol		0	2	6	33	48	—	—
Hydroxy β-diketones	25-Hydroxyhentriacontane-14,16-dione		0	0	6	56	66	5	—
Diols	5,10-Nonacosanediol		0	0	0	12	35	—	—

| Triterpenoid acids | Ursolic, oleanolic | 0 | 0 | 0 | 12[a] | 35[a] | — | 15 |
| Fatty acids | C_{16}-C_{30} | 0 | 0 | 1 | 8[a] | 20[a] | 1 | 46 |

[a] In the absence of an organic acid in the developing solvent, acidic classes of lipid give elongated spots when resolved from the start line, the R_f values of which are dependent on the amount originally applied to the layer. The values quoted are for circa 10 µg of the particular compounds.

Layer	L1 =	Silica gel G
Solvent	S1 =	Carbon tetrachloride
(v/v)	S2 =	Benzene
	S3 =	Chloroform
	S4 =	Chloroform - ethyl acetate (7:3)
	S5 =	Chloroform (ethanol free) - ethanol (99:1)
	S6 =	Petroleum hydrocarbon diethyl ether - acetic acid (70:30:1.5)
Detection		Charring with H_2SO_4
Technique	T1 =	12 cm distance
	T2 =	12 cm distance, two developments
	na =	not quoted

REFERENCES

1. Typical values obtained in author's laboratory from runs of all the compounds listed on the same 20 × 20 cm plate.
2. **Clark, T. and Watkins, D. A. M.,** *Phytochemistry,* 17, 943, 1978.
3. **Purdy, S. J. and Truter, E. V.,** *Proc. Roy. Soc. B,* 158, 536, 1963.
4. **Tulloch, A. P.,** *J. Chromatogr. Sci.,* 13, 403, 1975.
5. **Tulloch, A. P.,** in *Chemistry and Biochemistry of Natural Waxes,* Kolattukudy, P. E., Ed., Elsevier, Amsterdam, 255, 1976.
6. **Radler, F. and Horn, D. H. S.,** *Aust. J. Chem.,* 18, 1059, 1965.
7. **Radler, F.,** *Aust. J. Biol. Sci.,* 18, 1045, 1965.

Table 4
CLASSES OF CONSTITUENTS IN UNHYDROLYZED SURFACE LIPIDS OF ANIMALS (TLC)

$R_f \times 100$

		Layer: L1	L1	L1	L1	L1	L1	L1	L1
Class	**Homolog range of compound(s)**	Solvent: S1	S2	S3	S4	S5	S6	S7	S8
		Technique: na	na	na	na	na	na	T1	T2
		References: 1	1	2	3	4	4	5,6	7,8
Hydrocarbons									
Aliphatic	C20-C24	83	83	—	—	—	—	84	98
Cyclic	Squalene	—	—	—	—	—	—	79	82
Monoesters									
Fatty acids + primary alcohols	C32-C48	70	73	90	—	56	92	66	65
Fatty acids + sterols	Cholesteryl palmitate/oleate	—	—	—	—	65	—	—	72
Diesters									
2-Hydroxyalkanoic acid type	C48-C68	—	—	—	—	34	83	57—60	—
1,2-Alkanediol type	C48-C68	—	—	—	—	25	73	43—54	—
2,3-Alkanediol type	C30-C60	—	—	—	52	—	—	43—54	—
α,(ω-1)-Alkanediol type	C56-C64	49	57	—	—	—	—	—	—
(ω-1)-Hydroxyalkanoic acid type	C56-C64	49	57	—	—	—	—	—	—
Triesters									
Alkylhydroxypropanedioic acid type	C40-C80	—	—	65	—	—	—	—	—
(ω-1)-hydroxyalkanoic acid type	C70-C80	27	39	—	—	—	—	—	—
(ω-1)-hydroxyalkanoic acid + α (ω-1)-alkanediol type	C70-C80	27	39	—	—	—	—	—	—
Triacylglycerols	Tristearin, tripalmitin	—	22	50	30	—	19—21	34	37
Primary alcohols	C24	12	15	—	—	—	—	—	—
Sterols	Cholesterol	5	7	—	—	—	—	8	10
Hydroxymonoesters of (ω-1)-hydroxyalkanoic acids	C40-C48	2	3	—	—	—	—	—	—
Hydroxydiesters of (ω-1)-hydroxyalkanoic acids	C58	0	1	—	—	—	—	—	—
Fatty acids	C16-C24	—	—	—	—	—	—	18	16

Layer   L1 = Silica gel G

Solvent   S1 = Chloroform (ethanol free)

(v/v)   S2 = Chloroform (ethanol free) - ethanol (99:1)

         S3 = Chloroform - carbon tetrachloride (1:1)

         S4 = Chloroform

         S5 = *n*-Hexane - benzene (1:1)

         S6 = *n*-Hexane - benzene (3:7)

         S7 = A *n*-Hexane - diethyl ether - acetic acid (80:20:1)

               B *n*-Hexane - diethyl ether (19:1)

               C *n*-Hexane

         S8 = A *n*-Hexane

               B Benzene

               C *n*-Hexane - diethyl ether - acetic acid (70:30:1)

Detection   Charring with H_2SO_4

Technique   T1 = Triple development; i) solvent A 10 cm distance; ii) solvent B 20 cm distance iii) solvent C 20 cm distance

           T2 = Triple development; i) solvent A 19 cm distance; ii) solvent B 19 cm distance; iii) solvent C 10 cm distance

           na = not quoted

REFERENCES

1. **Tulloch, A. P.,** *Chem. Phys. Lipids,* 6, 235, 1971.
2. **Jacob, J. and Polz, J.,** *Z. Naturforsch.,* 29c, 236, 1974.
3. **Jacob, J. and Grimmer, G.,** *Z. Naturforsch.,* 25b, 689, 1970.
4. **Nikkari, T. and Haahti, E.,** *Biochim. Biophys. Acta,* 164, 294, 1968.
5. **Nicolaides, N., Fu, H. C., and Rice, G. R.,** *J. Invest. Dermatol.,* 51, 83, 1968 .
6. **Nicolaides, N., Fu, H. C., and Ansari, M. N. A.,** *Lipids,* 5, 299, 1970.
7. **Downing, D. T.,** *J. Chromatogr.,* 38, 91 1968.
8. **Downing, D. T., Strauss, J. S., and Poch, P. E.,** *J. Invest. Dermatol.,* 53, 322, 1969.

On layers of silica gel, a number of classes still remain difficult to separate or cause problems during chromatography. In surface lipid work the most important of these are (1) the partial separation at best of sterol esters from aliphatic monoesters (primary alcohol + fatty acid type), and (2) the poor resolution of classes containing a carboxyl group unless the solvent system contains a small proportion of acetic acid. For (1) the separation can be improved by TLC on aluminum oxide layers developed in hexane-benzene (99:1 v/v):[43] R_f × 100 for sterol ester = 26, for aliphatic monoester = 10. However, an even better resolution between the two classes is obtained on magnesium oxide layers (20 g adsorbent per 40 mℓ water to spread five 20 × 20 cm plates at 0.25 mm thickness, air dried 1 hr, and activated at 180°C for 2 hr) developed in hexane-acetone (99:1):[53] R_f × 100 for sterol ester = 46, aliphatic monoester = 21. The presence of sterol esters in any sample is, of course, readily detected by heating with sulfuric acid.

In solvent systems that lack an acetic acid component, classes of acidic surface lipid frequently give a poorly shaped tailing spot after development, the R_f value of which cannot be determined reliably because it varies with the amount of material originally applied to the layer, becoming higher with increasing concentrations of compound. If free acids are suspected in the sample after inspection of chromatograms obtained in such solvents, the analysis can be repeated in an acid-containing solvent, e.g., S6 Table 3; S7A, S8 Table 4, using the relevant acid reference compounds, but the resolution of the less polar constituents in the sample will inevitably be lost. Alternatively, a portion of the lipid sample can be methylated with diazomethane-diethyl ether (Chapter 2), and a chromatographic comparison made with the nonmethylated sample. [Do *not* use boron trifluoride-methanol for this purpose, otherwise additional reactions will occur if any aldehydes (formation of dimethyl acetals) and esters (transesterification) classes are present in the sample.] The appearance of new constituents in the methylated sample signifies that free acids are present in the original lipid extract, and their R_f values can then be compared with those obtained for methyl ester reference compounds.

In TLC, R_f values alone can never be taken as definite proof of identity of any individual surface lipid compound, even if it co-chromatographs with the authentic material. Although discrete spots corresponding with particular constituents may be observed for a lipid sample, each one could be a mixture of different homologs or positional isomers or contain branched-chain and/or unsaturated analogs. If the constituent class is an ester, several different types may also be present (see Table 2). The TLC technique, therefore, should be used only for the provisional identification of classes, although its reliability is greatly improved if confirmation can also be obtained from a specific spray reagent. R_f values should always be determined in more than one development system with preference being given to those that contain different organic solvent components. For positive identification, GC is needed to establish the chain lengths of all constituents, and other analytical procedures, especially GC/MS, are necessary in order to determine the nature and exact positions of the functional groups present, particularly where these are located in nonterminal positions. In the structural determination of ester classes, hydrolysis or transesterification is an additional requirement for the identification of the acid and alcohol components.

TLC on silica gel is commonly employed for the second stage in the analysis of surface lipid extracts to isolate discrete fractions from the mixture for further detailed analysis and final structural determination. The method is a simple adaptation of the basic TLC technique and makes use of the rapid and efficient fractionation that it provides (Tables 3 and 4). Samples dissolved in a minimum volume of a suitable volatile organic solvent are applied to the start lines of the preparative layers as a continuous band across the plate rather than as individual spots. Chromatography is carried out with the normal 0.25 mm thickness layers of adsorbent and is suitable for separations of a maximum of about 1 to 2 mg of starting material per 20 × 20 cm plate. If layer thickness of between 0.5 to 1 mm is used, the load of starting material can be increased up to about 10 mg per 20 × 20 cm plate. Problems can arise with cracked layers, particularly if plates are activated too soon after spreading. The layers are allowed to air dry for at least 1 hr before activation. Precoated preparative plates are also available from several commercial sources.

Because commercial sources of silica gel frequently contain impurities, it is important in any preparative work to remove them before hand because they might contaminate the fractions eventually recovered from the adsorbent. It is rarely necessary to treat the adsorbent itself, only to develop the prepared plates at least once to its full distance with the solvent system selected for analysis. This will elute most of the impurities in the layer to the top of the plate. (It is often convenient to carry out this procedure overnight.) The solvent is allowed to evaporate completely before loading the samples. After loading, the subsequent development must be stopped 2 to 3 cm below the prewash solvent front.

To ensure that the resolution of the preparative separation is comparable to that obtained by the analytical method note the following precautions:

1. Carry out the predevelopment procedures described above. If the atmosphere of the tank is not completely saturated with solvent vapor an edge effect will occur, and the separated bands will be carved rather than straight. The effect is more pronounced for components of high R_f value; the R_f value of the bands is lower at the center of the plate than that at the edges.
2. When loading samples, make the starting bands as narrow as possible (1 to 2 mm wide) because the constituent bands that separate during development always spread out to become wider than the initial band of starting material. Consequently, if the initial band is too wide, the resolution between separated bands will be reduced because they also will be wide, and have a tendency to overlap with each other. One of the simplest ways of achieving satisfactory starting bands is to warm the plates prior to loading so that when the sample solution is applied the solvent rapidly evaporates to leave a compact zone of material on the layer. Allow plates to cool to room temperature before development.
3. Do not overload the layers, otherwise separated bands will streak and become elongated to produce a similar loss of resolution. In practice the optimum load for each sample is determined largely by trial and error. It will vary according to the number of components, their R_f values, and their concentration in the mixture. Use the analytical TLC result for guidance.
4. Make sure that the level of solvent in the TLC tank is not higher than the position of the starting band on the plate. This would result in a dilute solution of the lipid sample, also chromatographing up the plate! The solvent level in the tank should be 1 to 2 cm below the start line.

In preparative analysis it may also be necessary to modify the composition of the developing solvent, or to use a multiple development procedure, in order to obtain the best separation between a particular group of constituents in a lipid sample. If the mixture is complex, it may even be advisable to carry out the fractionation in several stages using solvents of different polarity to ensure a complete separation of all components. The use of multiple development can often improve the overall resolution because it has a concentrating effect on the bands, making them more compact and, thus, better separated. Such developments are especially useful in the separation of mixtures where the major components are predominantly of high R_f value, e.g., plant or insect surface lipids having a high hydrocarbon content. With a single development of such samples it is very easy to overload the layers, but if one or two developments in *n*-hexane are carried out first, the hydrocarbons can be completely separated from all other components to a convenient position at the top of the plate. The location of the hydrocarbon band can be detected temporarily with iodine vapor and its position noted before continuing the development in another solvent to a position below that of the hydrocarbon band. This solvent will now separate the other components. Such a technique effectively increases the loading capacity of a given plate.

Any suitable nondestructive reagent (e.g., Reagents No. 6, 8, and 40) can be utilized for the detection of bands after the preparative separation, but it must not interfere with the final recovery of the compounds from the adsorbent. Although the fluorescent dyes have been widely used for this purpose, there is always a danger of small quantities of the detecting agents being subsequently eluted together with the lipid fractions. This imposes restrictions on the solvents that can be employed for recovery, and precludes the use of any in which the dyes are soluble, e.g., ethanol. For most purposes, therefore, iodine vapor is recommended because it can be rapidly removed from the layers after being used for detection. The position of the inner and outer limits of each detected band are marked accurately by lightly scoring the layer immediately after removal of the plates from the tank of vapor. The marked plate is then set aside or warmed in an oven until the iodine is seen to have evaporated from every band.

The marked areas of layer are carefully removed from the plates by scraping (spatula or razor blade) or by using a combined vacuum suction and extraction apparatus. The use of the latter minimizes any losses of adsorbent during the manipulative procedures. The lipid fractions are recovered from the adsorbent by elution or extraction with an organic solvent, usually diethyl ether or chloroform. The same precautions are observed as described earlier to prevent accidental contamination. Elution can be carried out with either hot or cold solvent after first packing the scraped adsorbent into a short column. The adsorbent is retained in the column by a plug of glass wool, and the solvent allowed to pass through at about 1 to 2 mℓ/min to give a total elution volume of 20 to 50 mℓ. Traces of adsorbent that may pass through can be removed by filtration. For extraction, the adsorbent is placed directly into an excess of solvent for 30 to 60 min with agitation (cold solvent) or gentle refluxing (hot solvent). If small quantities of adsorbent are involved, these may be removed by centrifugation at the end of the extraction period; filtration through Whatman No. 50 or equivalent paper is used for larger quantities. The fractions recovered are subsequently evaporated and stored as described for lipid isolation.

Before proceeding any further with the analysis, the purity of all the fractions is checked by analytical TLC, and the pre-TLC procedure repeated on any that are not homogeneous.

B. Column Chromatography (CC)

Before the introduction of TLC, conventional adsorption CC on alumina or silicic acid was the only chromatographic method available for the fractionation of complex surface lipid mixtures. Most of the early work, however, was carried out on samples that had been previously hydrolyzed, and it was not until the late 1950s that the method was successfully adapted for the analysis of whole unhydrolyzed extracts. CC is still widely employed today, but in the author's opinion it is essential only for the large scale isolation or purification of material such as for further biochemical or pharmacological investigation, special analytical techniques, e.g., ^{13}C NMR, optical rotation, or as a cleanup procedure during synthetic work. Preparative TLC is recommended for the routine isolation of surface lipid components, particularly when less than about 50 mg of starting material is available for investigation. Although the TLC method may ultimately provide less than a milligram of purified material from the lipid mixture, such amounts are generally more than adequate for a comprehensive analysis to be undertaken by a number of modern instrumental methods of structural determination, e.g., mass spectrometry (MS) or GC/MS.

It is not intended to give an extensive survey of elution schemes for the CC fractionation of intact surface lipids because such a large number exist in the literature. In any case, their basic principle is the same: elution is commenced with a hydrocarbon solvent, and the polarity of the eluent is gradually increased by the addition or progressively more polar solvents such as benzene, diethyl ether, or chloroform, usually terminating with a mixture containing a substantial proportion of methanol or acetic acid. Some examples of schemes that have been used for the complete separation of some lipid extracts from different animal and plant sources are summarized below, together with recommendations for carrying out the particular chromatographic analysis.

Procedure for the Fractionation of Human Skin Lipids[10,54]
Adsorbent silicic acid (100 to 120 mesh, acid washed, water content 14.5 to 15%). For samples of 200 to 1000 mg, adsorbent 20 to 30 g, column 20 to 30 × 2 cm I.D., collect 20 mℓ fractions; for samples of 5 to 75 mg, adsorbent 3 to 5 g, column 15 × 1 cm I.D., collect 10 mℓ fractions. Fraction 1 is eluted with hexane → ALIPHATIC HYDROCARBONS (tube numbers 2 to 6); Fraction 2 hexane → SQUALENE (tube numbers 8 to 14); Fraction 3 hexane-benzene (41:9 v/v) → ALIPHATIC AND STEROL MONOESTERS (+ ALDEHYDES) (tubes numbers 21 to 37); Fraction 4 hexane-benzene (2:3) → TRIACYLGLYCEROLS (+ 1-ALKANOLS) (tube numbers 48 to 57); Fraction 5 benzene → STEROLS (tube numbers 62 to 72); Fraction 6 chloroform → DIACYLGLYCEROLS (tube numbers 88 to 97); Fraction 7 methanol → MONOACYLGLYCEROLS, PHOSPHOLIPIDS (tubes numbers 98 to 100).

N.B. FATTY ACIDS removed from samples by extraction into 0.06 N NaOH prior to CC.

Procedure for the Fractionation of Uropygial Secretions from Birds[32]
Sample 20 to 50 mg: adsorbent 5 g silicic acid (particle size 0.1 to 0.16 mm, water content 9.1%). Fraction 1 is eluted with 70 mℓ cyclohexane → HYDRO-CARBONS; Fraction 2, 75 mℓ cyclohexane-benzene (9:1) → MONOESTERS; Fraction 3, 75 mℓ cyclohexane-benzene (3:2) → DIESTERS + TRIESTERS; Fraction 4, 50 mℓ benzene-chloroform (9:1) → TRIACYLGLYCEROLS; Fraction 5, 50 mℓ chloroform-methanol (9:1) → PRIMARY ALCOHOLS + DIOLS + STEROLS + FATTY ACIDS.

Procedure for the Fractionation of Beeswax[16]
Sample 5.2 g: adsorbent 200 g silicic acid (activated at 100°C for 18 hr). Fraction 1 is eluted with 1 ℓ hexane → HYDROCARBONS; Fraction 2, 3 ℓ hexane-chloroform (9:1) → MONOESTERS; Fraction 3, 2 ℓ hexane-chloroform (4:1) and 1 ℓ hexane-chloroform (7:3) → DIESTERS; Fraction 4, 1 ℓ hexane-chloroform (7:3) → DIESTERS + TRIESTERS + FATTY ACIDS; Fractions 5, 1 ℓ hexane-chloroform (7:3) → TRIESTERS + HYDROXYESTERS + FATTY ACIDS; Fraction 6, 1 ℓ hexane-chloroform (7:3) → HYDROXYESTERS + FATTY ACIDS; Fraction 7, 4 ℓ hexane-chloroform (3:2), 1 ℓ hexane-chloroform (1:1), 1 ℓ chloroform and 1.5 ℓ chloroform-methanol (9:1) → FATTY ACIDS + ACID ESTERS + unidentified material.

Procedure for the Fractionation of Kale Leaf Lipids[55]
Sample 51.3 mg: adsorbent 60.4 g silica gel H (activated), column 140 × 1.3 cm I.D., collect 10 mℓ fractions at flow rate of 0.2 mℓ/min. Fraction 1 is eluted with carbon tetrachloride → HYDROCARBONS (tube numbers 8 to 12); Fraction 2 carbon tetrachloride → MONOESTERS (tube numbers 16 to 27); Fraction 3 carbon tetrachloride → KETONES (tube numbers 30 to 37); Fraction 4 carbon tetrachloride → ALDEHYDES (tube numbers 36 to 43); Fraction 5 benzene → SECONDARY ALCOHOLS (tube numbers 78 to 83); Fraction 6 ethyl acetate →

1-ALKANOLS (tube numbers 102 to 104). Volumes of eluents (mℓ), carbon tetrachloride 693, benzene 288, ethyl acetate 291 mℓ. N.B., FATTY ACIDS are removed from the sample by preliminary CC on Florisil; acids are retained all other components are eluted with ethyl acetate. Acids are recovered from Florisil by subsequent elution with acetic acid-diethyl ether (1:24).

Procedure for the Fractionation of Lipids of Portulaca oleracea Leaves[1]
Sample 4.13 g: adsorbent 200 g silicic acid. Fraction 1 is eluted with 1.5 ℓ hexane → HYDROCARBONS; Fraction 2, 10 ℓ hexane-chloroform (19:1) → ALIPHATIC MONOESTERS; Fraction 3, 5 ℓ hexane-chloroform (9:1) → ALIPHATIC MONOESTERS + TRITERPENE MONOESTERS; Fraction 4, 2 ℓ hexane-chloroform (4:1) and 1 ℓ chloroform → FATTY ACIDS + 1-ALKANOLS + TRITERPENOLS + HYDROXY ESTERS + DIOLS. Fraction 4 is treated with excess diazomethane-diethyl ether and 0.78 g rechromatographed on 100 g of silicic acid. Fraction 4a is eluted with 2 ℓ hexane-diethyl ether (99:1) → FATTY ACID METHYL ESTERS; Fraction 4b, 4 ℓ hexane-diethyl ether (49:1) → COMPLEX MIXTURE: Fraction 4c, 2 ℓ hexane-diethyl ether (19:1) → 1-ALKANOLS + TRITERPENOLS; Fraction 4d, 2 ℓ hexane-diethyl ether (93:7) → ALCOHOLS + STIGMA-4-EN-3-ONE; Fraction 4e, 2 ℓ hexane-diethyl ether (17:3) → HYDROXY ESTERS; Fraction 4f, 3 ℓ hexane-diethyl ether (1:1) → DIOLS + UNIDENTIFIED MATERIAL.

Procedure for the Fractionation of Lipids of Rye Leaves[56]
Sample 2.42 g: adsorbent 200 g silicic acid. Fraction 1 is eluted with 1 ℓ hexane → HYDROCARBONS; Fraction 2, 2 ℓ hexane-diethyl ether (99:1) → MONOESTERS + β-DIKETONES; Fraction 3, 5 ℓ hexane-diethyl ether (99:1) → UNIDENTIFIED MATERIAL; Fraction 4, 2 ℓ hexane-diethyl ether → UNIDENTIFIED MATERIAL + DIESTERS (α,ω-diol type); Fraction 5, 3 ℓ hexane-diethyl ether (97:3) → UNIDENTIFIED MATERIAL: Fraction 6, 3 ℓ hexane-diethyl ether (24:1) → FATTY ACIDS; Fraction 7, 7 ℓ hexane-diethyl ether (24:1) → ALCOHOLS; Fraction 8, 2 ℓ hexane-diethyl ether (93:7) → UNIDENTIFIED MATERIAL; Fraction 9, 3 ℓ hexane-diethyl ether (23:2) → HYDROXY-β-DIKETONE; Fraction 10, 3 ℓ hexane-diethyl ether-ethanol (14:5:1) → UNIDENTIFIED MATERIAL.

When carrying out CC of surface lipids, the following general practical points should be noted together with a number of problems that may arise in the course of the analysis.

1. Alumina is not recommended as an adsorbent for the CC of surface lipids because sensitive compounds like aldehydes are decomposed and more polar oxygenated classes are too strongly adsorbed and consequently are difficult to elute.
2. Prepare columns by first making a slurry of the adsorbent in the solvent selected to commence the elution. Pour the slurry into the column and allow the adsorbent to settle with occasional light vibration of the walls of the column.
3. Difficulties are frequently encountered when loading samples onto columns because they may not always be completely soluble in the first eluting solvent. It may be necessary to evaporate a solution of the sample in another solvent onto a small amount of adsorbent which can then be transferred to the top of the column of adsorbent.
4. Fractions of small volume should always be collected from the column and the composition of each one carefully monitored by analytical TLC, preferably during the course of the elution. CC on silicic acid or Florisil may not separate surface lipid components in the same order as TLC on silica gel.[57,58] However, if the same adsorbent

is used for CC and TLC, the data obtained from the latter can be used in the selection of eluting solvents and also to predict the probable course of CC separation.[55]

5. Keep a strict control on the polarity of the eluent and do not increase it too rapidly, otherwise the resolution between components will be impaired. Increase the polarity of the eluting solvent only when analysis of consecutive fractions indicates that a particular component or group of components has been completely eluted from the column.

6. The presence of acids in surface lipid extracts causes problems for CC, especially on silicic acid, because the acids are not always easy to elute and may be spread out over a number of fractions containing other components. If acids contaminate any fractions, they can be removed by another CC separation after conversion to methyl esters. Alternatively, any acids can be removed from the extract prior to the CC separation.

C. Gas Chromatography (GC)

Although GC is a well-established technique for the separation and identification of individual compounds even up to C_{70} in chain length in most class fractions isolated from surface lipid mixtures, it is surprising that only a few workers have attempted to use it for the analysis of whole unhydrolyzed extracts. The major requirements for this type of analysis are an instrument oven, detection system, and stationary phase capable of operating up to 400°C. Temperature programing is essential in order to maintain an efficient separation between constituents because the sample may comprise components of widely different chain lengths. Most modern flame ionization (FI) detector instruments meet the first two requirements, but only since about 1970, with the introduction of the first carborane siloxane, Dexsil 300, have exceptionally stable high temperature phases been commercially available. Some degree of success, however, was achieved before this by analyses on short (0.5 to 2 m) low loaded packed columns of other silicone phases such as SE-30, SE-52 and OV-1 programed to temperatures above 300°C. A partial analysis of human skin lipids is reported by Haahti[10] on SE-30 following injection of the total lipid mixture previously treated with diazomethane; positive identification of fatty acids, hydrocarbons, aliphatic monoesters, and sterols was achieved, but sterol monoesters and triacylglycerols could not be resolved. On SE-52, Ludwig[59] was able to characterize the 1-alkanol and aliphatic monoester components in three commercial waxes (carnauba, montan, and ouricury) by direct injection of solutions of the lipids without any derivatization. With Dexsil 300, however, there is a considerable reduction in the bleed rate from the column at high temperatures, and a much greater proportion of the total lipid components can be chromatographed; in many plants an essentially complete analysis of the total surface lipid can be achieved. The first separation of an intact surface lipid on this phase was reported in 1972 by Tulloch,[60] who was able to identify and determine the relative amounts of circa 65% of the total components of beeswax. Further work by the same author using the same technique followed on a range of commercial waxes[61,62] (ouricury, carnauba, Chinese insect, Lac, esparto, candelilla, Japan wax, sugar cane) and plant epicuticular lipids (oats,[63] wheat,[64,65] *Portulaca oleracea*,[17] *Festuca ovina*,[57] *Agropyron intermedium*[65]). Despite the demonstration of its enormous potential for the analysis of these materials, the "Dexsil profile" has not as yet been evaluated for the surface lipids from other sources.

Before GC on Dexsil 300, most whole surface lipid samples usually require some form of derivatization to ensure that a complete analysis of all constituent classes is obtained with the maximum efficiency and sensitivity. This is essential if the sample contains acids because such classes are too polar to be analyzed directly by GC, and must be converted beforehand into less polar and more volatile derivatives such as methyl esters. Similarly, many hydroxy constituents give a poor peak shape and FI response when chromatographed as the free compounds, but their GC performance is greatly improved after acetylation or silylation. If both types of compound are present in a sample, a double derivatization will be necessary.

Table 5

DERIVATIVES OF CLASSES OF SURFACE LIPID COMPONENTS SUITABLE FOR THEIR GC ANALYSIS IN WHOLE UNHYDROLYZED EXTRACTS

Class	Derivative(s)
Hydrocarbons, monoesters, ketones, aldehydes[b], diesters[d], triesters[d], polyesters[d]	None[a]
β-Diketones	None;[a,b]; enol acetate[b]; enol TMSi ether[b]
Hydroxy-β-diketones	None[a,b]; enol acetate acetate[b]; enol TMSi ether TMSi ether[b]
Secondary alcohols, primary alcohols, ketols, sterols, pentacyclic triterpenols	None[a,c]; acetate; TMSi ether
Diols	Diacetate; *bis* TMSi ether
Fatty acids	Methyl ester; TMSi ester
Hydroxyfatty acids, pentacyclic triterpenoid acids viz. ursolic, oleanolic	Methyl ester acetate; methyl ester TMSi ether; TMSi ester TMSi ether

Note: TMSi = trimethylsilyl.

[a] Free compounds analyzed directly.
[b] Some decomposition during GC analysis.
[c] Much improved FID response and GC peak shape if derivatized.
[d] If chain length $> C_{70}$ not sufficiently volatile to permit GC analysis.

A summary of classes and their individual requirements for derivatization is given in Table 5.

Procedure for the Methylation and Acetylation of Whole Plant or Insect Surface Lipids[61]
Weigh circa 30 mg of lipid. Add 5 mℓ of a chloroform solution containing circa 1 mg of p-dioctylbenzene + circa 1 mg of octadecyl octadecanoate (internal standards for determination of retention [R_T] data and quantitative assessment) and an excess of a solution of diazomethane in methylene chloride. Evaporate solvents and treat the residue with 1 mℓ of acetic anhydride-pyridine (1:1 v/v) at 100°C. Remove reagents and dissolve the derivatized sample in chloroform for GC analysis on Dexsil 300.

In the author's laboratory a methylation-silylation procedure is preferred for the routine analysis of plant epicuticular lipids because (1) silylation is generally more convenient than acetylation because the removal of excess reagents is unnecessary, (2) the regular passage of excess silylating reagent through the chromatographic system removes potentially active sites of adsorption for classes of sensitive compounds and keeps the GC column in good condition; and (3) the mass spectra of the trimethylsilyl (TMSi) derivatives of the various classes of hydroxy compounds are characteristic and are very useful for both identification and structural determination when GC/MS is available.

Procedure for the Methylation and Silylation of Plant Surface Lipids
Place circa 1 mg of sample in a small tapered Pyrex glass tube fitted with a ground glass joint and stopper. This amount can be conveniently prepared by taking an aliquot from a solution of a larger weight of the lipid (30 to 50 mg) in chloroform. Add circa 0.1 mg of tetracosane and circa 0.1 mg of dotriacontane from bulk stock solutions containing 1 mg/mℓ in hexane or chloroform (internal standards for R_T data and/or quantitative assessment). Evaporate to dryness in a stream of nitrogen and dissolve the residue in a minimum volume of diethyl ether-methanol (9:1 v/v). Add an excess of a solution of diazomethane in diethyl

Table 6
COMPARISON OF THE ELUTION OF DIFFERENT CLASSES OF SURFACE LIPID CONSTITUENTS HAVING THE SAME CHAIN LENGTH n WITH THAT OF THE CORRESPONDING FATTY ACID METHYL ESTER (GC)[8,57]

(Methyl alkanoate), β-diketone	n
Alkane	$n - 3.5$
Monoester of alkanoic acid + 1-alkanol	$n - 2$
Ketone, secondary alcohol, secondary alcohol acetate (functional group in mid chain position)	$n - 1$
1-Alkanol	$n - 0.15$
Methyl Δ²¹-alkenoate, 1-alkanol acetate	$n + 1$
Methyl ω-hydroxyalkanoate acetate	$n + 4.25$
α,ω-Alkanediol diacetate	$n + 5.25$

Note: 1.5% Dexsil 300 on acid washed, silanized Chromosorb W (80-100 mesh); column stainless steel 1 m × 2 mm I.D.; carrier gas He 60 mℓ/min; temperature programing 125-400°C at 3° C/min; detection FI.

ether until a persistent yellow color is observed. (Caution! very toxic reagent, use fume chamber and gloves.) Evaporate to dryness and silylate the residue at 50°C for 1 hr with 75 μℓ of BSA (N,O-bistrimethylsilylacetamide) + 25 μℓ of anhydrous pyridine. Inject 1 to 1.5 μℓ of reaction mixture onto a Dexsil 300 column; if necessary dilute with more BSA.

The derivatization procedure is much simpler if silylation is used to convert both hydroxy and acid constituents into their corresponding TMSi derivatives. BSA-pyridine can be used for this purpose; omit the diazomethane step from the methylation-silylation method described above. An alternative reagent and method is recommended by Tulloch.[65] *Treat 5 mg of sample dissolved in 300 μℓ toluene-pyridine (5:1) with 25 μℓ of hexamethyldisilazane + 12 μℓ of trimethylchlorosilane at room temperature for 18 hr. Inject the reaction mixture directly onto a Dexsil column.*

Relative R_T data obtained on Dexsil 300 for some classes and individual components of surface lipids using the two methylation derivatization methods are given in Tables 6 and 7. The values quoted, however, should not be regarded as absolute because, in the author's experience with short low-loaded Dexsil columns, they can vary significantly not only between different columns of identical length filled with the same packing material but also from day to day on the same column. Nevertheless, they do serve to illustrate the relative order of elution of the various classes of constituents and the degree of separation that can be expected between individual compounds. Regular calibration with authentic compounds is, therefore, essential throughout any analysis on Dexsil to check column performance, and suitable internal standards should be co-injected with all samples for the determination of any retention values. *n*-Alkane homologs are a convenient source of internal standards, but any compound may be used provided that it does not have the same R_T as any of the components in the sample. In some cases it is possible to use an identified compound of

Table 7
SOME CONSTITUENTS OF PLANT EPICUTICULAR LIPIDS (GC)

Constituent	Chain length	RR_T (tetracosane)[a]	RR_T (dotriacontane)[a]
Hydrocarbons			
n-Heptacosane	C_{27}	1.34	0.74
n-Nonacosane	C_{29}	1.53	0.85
n-Hentriacontane	C_{31}	1.72	0.96
Aliphatic monoesters			
	C_{40}	2.55	1.42
Long chain alkanoic acids	C_{42}	2.68	1.49
esterified with long chain 1-alkanols	C_{44}	2.81	1.56
	C_{46}	2.94	1.63
	C_{48}	3.07	1.71
Ketones			
15-Nonacosanone			
14-Nonacosanone	C_{29}	1.75	0.97
10-Nonacosanone			
Aldehydes			
1-Hexacosanal	C_{26}	1.52	0.84
1-Octacosanal	C_{28}	1.71	0.95
1-Triacontanal	C_{30}	1.89	1.05
1-Dotriacontanal	C_{32}	2.07	1.15
Secondary alcohols (as TMSi ethers)			
15-Nonacosanol			
14-Nonacosanol	C_{29}	1.68	0.93
10-Nonacosanol			
16-Hentriacontanol	C_{31}	1.85	1.03
15-Hentriacontanol			
Ketols (as TMSi ethers)			
14-Hydroxynonacosan-15-one	C_{29}	1.80	1.0
15-Hydroxynonacosan-14-one			
13-Hydroxynonacosan-15-one			
14-Hydroxynonacosan-16-one	C_{29}	1.80	1.0
Primary alcohols (as TMSi ethers)			
1-Tetracosanol	C_{24}	1.37	0.76
1-Hexacosanol	C_{26}	1.56	0.87
1-Octacosanol	C_{28}	1.75	0.97
1-Triacontanol	C_{30}	1.93	1.07
1-Dotriacontanol	C_{32}	2.11	1.17
Fatty acids (as methyl esters)			
Hexadecanoic	C_{16}	0.48	0.27
Octadecanoic	C_{18}	0.72	0.40
Hexacosanoic	C_{26}	1.56	0.87
Octacosanoic	C_{28}	1.75	0.97
Triacontanoic	C_{30}	1.93	1.07
Dotriacontanoic	C_{32}	2.11	1.17
Triterpenoids			
β-Amyrin TMSi ether	C_{29}	1.95	1.08
α-Amyrin TMSi ether	C_{29}	1.92	1.06
Methyl oleanolate TMSi ether	C_{30}	2.15	1.19
Methyl ursolate TMSi ether	C_{30}	2.11	1.17

Note: 1% Dexsil 300 on Supelcoport (100—120 mesh); column stainless steel coil 1 m × 2 mm I.D.; carrier gas N_2 40 mℓ/min, temperature programing 130—400°C at 6°C/min; injector temperature 250°C; detector temperature 350°C; detection FI.

[a] Typical values obtained in the author's laboratory from runs of all the compounds listed using co-injection with the alkanes.

the sample itself for this purpose, e.g., squalene for human skin lipids[10] or *n*-nonacosane for plant epicuticular lipids. If the sample contains components with a very wide range of different chain lengths, the use of two standards is recommended.

The fractionation of surface lipid components by GC is primarily according to chain length, and any subsequent separation is governed by the nature, number, and position of the functional groups present, i.e., class. On Dexsil 300, the following is the order of increasing polarity (retention time) for compounds of the same chain length: alkane → aliphatic monoester → secondary alcohol TMSi → ketone + secondary alcohol + secondary alcohol acetate → aldehyde → primary alcohol → methyl ester of fatty acids + primary alcohol TMSi + β-diketone → primary alcohol acetate; not all of the classes or the corresponding derivatives are resolved. Branched chain and unsaturated analogs of any class are also eluted together just before the corresponding saturated compound. Because GC separates surface lipid mixtures not only into classes but also into individual compounds, it is inevitable that some more classes are poorly separated or are not resolved from one another. Thus on Dexsil 300, aldehydes co-elute with alkanes having three carbon atoms more, e.g., R_T 1-octadecanal = *n*-hentriacontane, and ketones with both methyl esters of fatty acids and 1-alkanol TMSi containing one carbon atom less, e.g., R_T 15-nonacosanone = methyl octacosanoate = 1-octacosanol TMSi. Other co-elutions occur between other classes or derivatives and these are indicated by any whole number increments of the *n* values in Table 6. Some of these problems arise because the Dexsil analysis of intact surface lipids has to be a working compromise between resolution and column length. If column length or loading of stationary phase is increased to improve efficiency, the ability to analyze very long chain components is lost. The length must be short enough and the loading sufficiently low to allow the elution of such compounds, and column efficiency has to be sacrificed accordingly.

In practice, however, several of the resolution problems do not arise because not all of the interfering compounds are present in the particular sample being analyzed. In plant epicuticular lipids, for example, the differences in chain lengths between the components of certain classes are sufficiently great to permit a complete separation, although the two homologous series actually overlap, e.g., aliphatic monoesters and methyl esters of fatty acids (Table 7). Also, in this type of surface lipid there is usually a marked odd or even carbon number preference between classes (Table 1), so that although constituents like β-diketones co-elute with methyl esters of the same length, this is rarely troublesome since the former are predominantly odd carbon numbered and the latter even numbered. Other separation problems can be overcome by the use of a particular derivative, and the class data previously obtained from analytical TLC of the sample can be used to advantage in the selection of reagents. For example, if the TLC results indicate that the lipid mixture contains a significant proportion of fatty acids and primary alcohols, the methylation-acetylation or silylation[65] methods should be used (the two classes overlap after methylation-silylation). Similarly, if ketones and secondary alcohols are both present, a silylation procedure must be employed (the two classes overlap after acetylation).

The identification of surface lipid compounds based solely on R_T data determined by GC of the total mixture on Dexsil 300 should be regarded as provisional until substantiated by other evidence. Although in some cases the R_f values in TLC may provide sufficient confirmatory evidence for the class, e.g., alkanes, aldehydes, primary alcohols, fatty acids, a more reliable identification of compounds would be possible if additional R_T information could be obtained on a stationary phase more polar than Dexsil 300.* For several compounds,

---

* Two new polar Dexsils, 400 and 410, have recently been introduced. They are very expensive and have not as yet been fully tested for total surface lipid work. A partial analysis of some whole surface lipids is possible on short low loaded columns of the more polar phases OV-17 (maximum 280°C), OV-210 (maximum 260°C) and OV-225 (maximum 250°C), but confirmation from R_T values can be obtained only for those classes of compound where the chain length is less than circa C_{32}.

however, any R_T information on packed columns provides information only on chain length and class and does not establish the exact location of the functional groups, e.g., all esters, ketones, secondary alcohols, β-diketones, ketols, cyclic constituents, branched chain and unsaturated compounds. GC/MS provides the most rapid and sensitive means for the positive identification and structural determination of these compounds in intact surface lipids.[65-67] Additional information about the nature of surface lipid components can also be obtained from the changes of R_T that occur in the Dexsil profile at various stages of derivatization (methylation, methylation-acetylation, etc.) using the underivatized sample for comparison.[57]

The areas or heights of the various peaks in the Dexsil profile cannot be used directly for the quantitative estimation of components because each surface lipid class has a different FI response factor. For any quantitative assessment these factors must be determined carefully with internal standards and authentic compounds (Chapter 2); otherwise considerable errors can be made in the estimation of certain classes. Calibration is particularly important for the analysis of sensitive compounds such as β-diketones (and their enols) and aldehydes (Table 5) where a very low response factor can result from decomposition both on the column and during derivatization.[64,65] An improvement in the GC performance of these classes can sometimes be achieved if the column is first preconditioned by repetitive injections of pure samples of the same class until an adequate response factor is obtained. An alternative procedure for aldehydes and β-diketones is to reduce them with sodium borohydride[68] to the corresponding primary alcohols and diols, respectively, the response factor and GC performance of the silyl or acetyl derivatives of the hydroxy compounds being much better than that of the original compounds. This technique, however, is more suitable for isolated class fractions than whole lipid mixtures because the reduction products themselves may also be natural components of the sample, and other classes, such as ketones, if present, will also be reduced.

Despite its limitations the Dexsil profile is still a rapid and sensitive means of identifying both the class and chain length of many components in whole surface lipid extracts; such information cannot be obtained by any other direct method of chromatographic analysis. The quality and reliability of the final analysis, however, is ultimately dependent on the composition of the extract, and a satisfactory separation of all components may not be obtained for every sample either because they are poorly resolved on the phase itself or are not sufficiently volatile for GC (limit on Dexsil 300 circa C_{70} chain length). When the Dexsil column is combined with MS, an unequivocal identification of all constituents that separate can be made because each surface lipid class has a characteristic mass spectrum. If a data system is available, the GC/MS analysis can be taken on a stage further by carrying out selective ion monitoring on all of the spectra collected during the Dexsil separation. This in turn provides a computer reconstructed chromatogram of the total mixture showing only those peaks comprising the particular class of compound selected for monitoring; m/e 74 is a diagnostic ion for all methyl esters of fatty acids, m/e 82 for aldehydes, m/e 71 for ketones, m/e 73 or 75 for silyl derivatives, etc. The Dexsil technique is especially useful when comparison of a large number of samples is required and for evaluating changes of surface lipid composition in a given organism during various stages of development[64] or those caused by mutation.[69,70]

D. Quantitative Chromatographic Analysis of Whole Surface Lipids

Each one of the direct methods of chromatographic analysis described in the present chapter can be used to provide some quantitative information about the composition of a surface lipid sample. It is not intended to deal with those aspects in any detail because they are standard techniques described in general textbooks of chromatography. Some general information about each procedure is summarized below.

1. Column Chromatography (CC)

The class composition can be estimated from the amounts of material recovered in the fractions collected from the CC of a known weight of sample. The reliability of the method is obviously dependent on the efficiency of the separation, and it is often used to make a preliminary quantitative assessment before the identity of the components has been definitely established.

2. Thin-Layer Chromatography (TLC)

When the qualitative composition of the lipid mixture is known, a densitometric or fluorometric assay of the constituent classes may be performed on the TLC plate after detection of the spots with a suitable spray reagent, e.g., sulfuric acid.[71] Calibration is made with the response obtained from known amounts of the authentic compounds or classes chromatographed under identical conditions. A more precise estimation is reported if the detecting agents are incorporated into the adsorbent layers[72,73] because inconsistencies of response caused by nonuniform coverage of the layer with reagent spray are eliminated.

3. Preparative Thin-Layer Chromatography (Prep-TLC)

A gravimetric procedure similar to that used for CC is frequently used for a direct quantitative assessment of classes but is not to be recommended since the fractions isolated are frequently less than 1 to 2 mg; such quantities cannot be accurately weighed unless a very sensitive balance is available. A more reliable method of estimation of such fractions is GC (see below), but a careful evaluation using authentic compounds is required to assess the efficiency of recovery for each class of compound.

4. Gas Chromatography (GC)

The internal standard method in combination with the Dexsil profile has been used for the quantitative estimation of individual compounds in several intact surface lipids.[60-64,67] The qualitative composition of the sample must be definitely established before hand because for quantitative assessment a response factor relative to the internal standard is needed for each constituent or class of constituent present in the mixture. These factors are determined at the time of analysis by injection of solutions of known amounts of both internal standard and authentic compounds. Ideally, the compounds used as calibration standards should be identical to those being assayed in the mixture, but if not they should be of the same class and similar in chain length. The relative response factor (RRF) is calculated from peak measurements and is given by the expression:

$$RRF = \frac{\text{Area or height (internal standard)} \times \text{Weight (calibration compound)}}{\text{Area or height (calibration compound)} \times \text{Weight (internal standard)}}$$

In high temperature work with low-loaded Dexsil columns, such values vary from day to day and are usually valid only for a given column and the particular instrument employed. The linearity of the system should also be verified for each class of constituent from determinations of the RRF for a range of different weight ratios of internal standard: calibration compound that cover the concentrations that can be expected in the sample. For some classes of compound, e.g., aldehydes, long-chain monoesters, a significant decrease of RRF can be observed for weight ratios less than one according to the condition of the column (probably adsorption effects). In such cases, it is important for the calculation of quantitative composition to use the RRF obtained from the calibration mixture having a similar peak height or area ratio relative to the internal standard as that observed for the same compound or class of compounds in the chromatogram of the sample.

The main requirement for the internal standard itself is that it should elute from the column in an area of the chromatogram where it will not interfere with the peaks of the sample; the

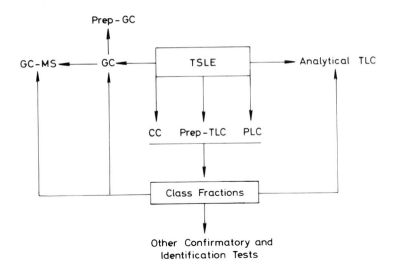

SCHEME 1. Summary of the use of chromatographic methods in the analysis of natural surface lipids (TSLE = total surface lipid extract).

choice of compound thus depends mainly on the composition of the sample. Preferably it should also have a similar structure to that of the compound(s) being assayed, and in surface lipid work this requirement can sometimes be met by using a lower or higher homolog belonging to the same class as that requiring quantitation but which is absent from the sample. Internal standards that are used in the quantitative analysis of various plant and insect surface lipid classes[60,61,63,64] are *p*-dioctylbenzene (hydrocarbons), icosane (hydrocarbons), methyl icosanoate (methyl esters of fatty acids), and octadecyl octadecanoate or octadecyl icosanoate (methyl esters, long-chain esters, alcohol acetates). In the author's laboratory,[67-70] tetracosane (all classes except long-chain esters) and dotriacontane (long-chain esters) are in routine use for the assay of both total mixtures and isolated fractions of plant epicuticular lipids. Some typical RRF values obtained for these compounds and calibration compounds using the Dexsil 300 column and conditions described in Table 7 are for tetracosane = 1.0, octacosane 0.75 to 0.86, 12-tricosanone 0.85 to 0.92, 12-tricosanol TMSi ether 0.92 to 0.97, 1-docosanol TMSi ether 0.97 to 1.0, 1-icosanal 0.36 to 0.45, methyl hexadecanoate 0.85 to 0.91, methyl hexacosanoate 0.71 to 0.80, and for dotriacontane = 1.0, octadecyl docosanoate 0.75 to 0.82.

For quantitative analysis, a known weight of internal standard is added to the sample, the mixture derivatized, and GC carried out (Tables 6 and 7). The weight of each identified component (W) is calculated by comparison of its peak height or area with that of the internal standard using the formula:

$$W = \frac{\text{Area or height (component)} \times \text{Weight (internal standard)} \times \text{RRF (component)}}{\text{Area or height (internal standard)}}$$

IV. CONCLUSIONS

Modern chromatographic methods offer a rapid and sensitive means for analyzing the complex mixtures of compounds that occur in natural surface lipids, and most of our present knowledge of these materials is derived from the application of such techniques. They provide a method for the identification (TLC, GC), preparative fractionation (CC, Prep-TLC), and quantitation of the majority of the constituents. The potential uses of chromatography for surface lipid work are summarized in Scheme 1. The most powerful analytical combination is GC/MS, and the following references are recommended for details of the diagnostic features of the mass spectra of the major surface lipid classes cited in Table 1: hydrocarbons,[74,75] aliphatic monoesters,[76,77] ketones,[78,79] aldehydes,[80] diesters and triesters,[19] β-diketones,[65,81,82] secondary alcohols,[14,67,83] ketols,[68,84] triacylglycerols,[85,86] primary alcohols,[87-89] pentacyclic triterpenoids,[90,91] sterols,[92-94] diols,[2,19,32,95] fatty acids,[2,32,96] and hydroxy fatty acids.[97]

Methods suitable for a more detailed chromatographic analysis of individual classes when they are isolated from a surface lipid extract are described in other chapters of this volume and are listed below for cross reference: Hydrocarbons (Chapter 4), Alcohols, Aldehydes and Ketones (Chapter 5), Fatty Acids (Chapters 6, 7, and 8), and Triacylglycerols (Chapter 12).

V. SOURCES OF REFERENCE COMPOUNDS

The key to any successful analysis by chromatography is a good supply of authentic reference compounds, and surface lipid work is no exception. A substantial number of the compounds or homologs of the classes of compound described in this article are commercially available, mainly from the specialist companies in chromatographic supplies. A range of hydrocarbons, monoesters of fatty acids and 1-alkanols, sterols and sterol monoesters, aldehydes, primary alcohols, triacylglycerols, fatty acids, hydroxy fatty acids and some secondary alcohols, ketones, and triterpenols are all obtainable.

An alternative, less expensive, source of reference compounds is found in a number of commercial waxes and natural surface lipids that are usually easy to obtain and whose composition is well documented. The following are recommended for general purposes or for particular compounds.

1. Beeswax.[16] Main components, alkanes C_{23} to C_{33}, aliphatic monoesters C_{40} to C_{52}, fatty acids C_{22} to C_{34}
2. Spermaceti.[98] Chiefly aliphatic monoesters C_{26} to C_{40}
3. Candelilla wax.[61,99] Rich source of hentriacontane (40 to 50% of wax); for details of other commercial waxes consult Tulloch[8] and Warth[100]
4. Epicuticular lipids of leaves of *Brassica* species and cultivars (cabbages, cauliflower, rape, etc.);[37,69,89] whole lipid mixtures strongly recommended for all TLC work viz. class composition alkanes, aliphatic monoesters, ketones, aldehydes, secondary alcohols, ketols, primary alcohols and β-amyrin, and fatty acids; major components nonacosane, 15-nonacosanone and 15-nonacosanol
5. Primary alcohols. Epicuticular lipids of oat (*Avena sativa*) leaves circa 50% of hexacosanol[17]
6. β-Diketones. Major constituents of leaf epicuticular lipids of carnation (*Dianthus caryophyllus*) (12,14-hentriacontane-dione + 12,14-nonacosane-dione),[101] *Eucalyptus globulus* (16,18-tritriacontane-dione)[101] and rye (*Secale cereale*) (14,16-hentriacontane-dione + 25-hydroxyhentriacontane-14,16-dione)[56]
7. Triterpenoid acids. Ursolic acid up to 50% of epicuticular lipids of cultivars of apple (*Malus pumila*) fruits,[38] oleanolic acid 40 to 70% of grape (*Vitis vinifera*) fruits[102]

Several of the surface lipid classes can also be prepared using standard methods of organic synthesis,[2,8] and such procedures may be necessary if a convenient commercial or natural source is not readily available, or when a large amount of a particular compound is required for GC calibration. The following references describe the preparation of some of the more unusual classes: branched chain alkanes,[75] diesters and triesters,[16,20,103] ketones,[52] and β-diketones.[81] Aliphatic methanesulfonates (mesylates) are also versatile intermediates for the synthesis of several important classes of surface lipid compounds.[104]

ACKNOWLEDGMENTS

The author is grateful to Dr. R. A. Franich, Forest Research Institute, Rotorua, New Zealand for a sample of 5,10-nonacosanediol; to Dr. C. E. Jeffree, University of Edinburgh for β-diketones and hydroxy-β-diketones; and to Dr. J. A. Lamberton, C. S. I. R. O., Melbourne, Australia for the triterpenoid samples used in this work.

REFERENCES

1. **Downing, D. T.**, in *Chemistry and Biochemistry of Natural Waxes,* Kolattukudy, P. E., Ed., Elsevier, Amsterdam, 1976, 17.
2. **Jacob, J.**, in *Chemistry and Biochemistry of Natural Waxes,* Kolattukudy, P. E., Ed., Elsevier, Amsterdam, 1976, 93.
3. **Jackson, L. L. and Baker, G. L.**, *Lipids,* 5, 239, 1970.
4. **Tulloch, A. P.**, *Lipids,* 5, 247, 1970.
5. **Jackson, L. L. and Blomquist, G. J.**, in *Chemistry and Biochemistry of Natural Waxes,* Kolattukudy, P. E., Ed., Elsevier, Amsterdam, 1976, 20.
6. **Martin, J. T. and Juniper, B. E.**, *The Cuticles of Plants,* Edward Arnold, Edinburgh, 1970, 121.
7. **Kolattukudy, P. E. and Walton, T. J.**, *Progress in the Chemistry of Fats and Other Lipids,* Vol. 13, Holman, R. T., Ed., Pergamon Press, Oxford, 1972, 119.
8. **Tulloch, A. P.**, in *Chemistry and Biochemistry of Natural Waxes,* Kolattukudy, P. E., Ed., Elsevier, Amsterdam, 1976, 235.
9. **Aitzetmüller, K. and Koch, J.**, *J. Chromatogr.,* 145, 195, 1978.
10. **Haahti, E.**, *Scand. J. Clin. Lab. Invest.,* 13 (Suppl. 59), 11, 1961.
11. **Stránský, K. and Streibl, M.**, *Coll. Czech. Chem. Commun.,* 36, 2267, 1971.
12. **Tulloch, A. P. and Hoffmann, L. L.**, *Phytochemistry,* 12, 2217, 1973.
13. **von Wettstein-Knowles, P. and Netting, A. G.**, *Lipids,* 11, 478, 1976.
14. **Blomquist, G. J., Soliday, C. L., Byers, B. A., Brakke, J. W., and Jackson, L. L.**, *Lipids,* 7, 356, 1972.
15. **Meinwald, J., Smolanoff, J., Chibnall, A. C., and Eisner, J.**, *J. Chem. Ecol.,* 1, 269, 1975.
16. **Tulloch, A. P.**, *Chem. Phys. Lipids,* 6, 235, 1971.
17. **Tulloch, A. P.**, *Lipids,* 9, 664, 1974.
18. **Jacob, J. and Zeman, A.**, *Hoppe-Seyler's Z. Physiol. Chem.,* 353, 492, 1972.
19. **Franich, R. A., Wells, L. G., and Holland, P. T.**, *Phytochemistry,* 17, 1617, 1978.
20. **Tulloch, A. P.**, *Lipids,* 6, 641, 1971.
21. **Clark, T. and Watkins, D. A. M.**, *Phytochemistry,* 17, 943, 1978.
22. **Jacob, J. and Grimmer, G.**, *Hoppe-Seyler's Z. Physiol. Chem.,* 354, 1648, 1973.
23. **Martin, J. T. and Batt, R. F.**, *Insect Biochem.,* 5, 379, 1975.
24. **Baker, E. A. and Holloway, P. J.**, *Micron,* 2, 364, 1971.
25. **Holloway, P. J. and Baker, E. A.**, in *Principles and Techniques of Scanning Electron Microscopy, Biological Applications,* Vol. 1, Hayat, M. A., Ed., Van Nostrand Reinhold, New York, 1974, 181.
26. **Nicolaides, N., Fu, H. C., and Rice, G. R.**, *J. Invest. Dermatol.,* 51, 83, 1968.
27. **Morello, A. M., Downing, D. T., and Strauss, J. S.**, *J. Invest. Dermatol.,* 66, 329, 1976.
28. **Wilkinson, D. I. and Karasek, M. A.**, *J. Invest. Dermatol.,* 47, 449, 1966.
29. **Nikkari, T. and Haahti, E.**, *Acta Chem. Scand.,* 18, 671, 1964.
30. **Odham, G.**, *Arkiv Kemi,* 21, 379, 1963.
31. **Hansen, I. A., Tang, B. K., and Edkins, E.**, *J. Lipid Res.,* 10, 267, 1969.
32. **Jacob, J.**, *J. Chromatogr. Sci.,* 13, 415, 1975.
33. **Jackson, L. L.**, *Lipids,* 5, 38, 1970.
34. **Tartivita, K. and Jackson, L. L.**, *Lipids,* 5, 35, 1970.
35. **Martin, J. T.**, *Insect Biochem.,* 5, 275, 1975.
36. **Martin, J. T.**, *J. Sci. Food Agric.,* 11, 635, 1960.
37. **Purdy, S. J. and Truter, E. V.**, *Proc. R. Soc. B,* 158, 536, 1963.
38. **Silva Fernandes, A. M. S., Batt, R. F., and Martin, J. T.**, *Ann. Appl. Biol.,* 53, 43, 1964.
39. **Fisher, D. J., Holloway, P. J., and Richmond, D. V.**, *J. Gen. Microbiol.,* 72, 71, 1972.
40. **Mangold, H. K.**, *J. Am. Oil Chem. Soc.,* 38, 708, 1961.

41. **Holloway, P. J. and Challen, S. B.,** *J. Chromatogr.,* 25, 336, 1966.
42. Dyeing Reagents for Thin-Layer and Paper Chromatography, E. Merck, Darmstadt, 1976.
43. **Haahti, E. and Nikkari, T.,** *Acta Chem. Scand.,* 17, 536, 1963.
44. **Mangold, H. K. and Malins, D. C.,** *J. Am. Oil Chem. Soc.,* 37, 383, 1960.
45. **Dunphy, P. J., Whittle, K. J., and Pennock, J. F.,** *Chem. Ind. (London),* 1965, p. 1217.
46. **Rahn, C. H. and Schlenk, H.,** *Lipids,* 8, 612, 1973.
47. **Dierickx, P. J. and Buffel, K.,** *Phytochemistry,* 11, 2654, 1972.
48. **Betts, T. J. and Holloway, P. J.,** *J. Pharm. Pharmacol.,* 19 (Suppl. 97S), 1967.
49. **Tschesche, R., Lampert, F., and Snatzke, G.,** *J. Chromatogr.,* 5, 217, 1961.
50. **Shepherd, I. S., Ross, L. F., and Morton, I. D.,** *Chem. Ind. (London),* 1966, 1706.
51. **Morris, L. J., Wharry, D. M., and Hammond, E. W.,** *J. Chromatogr.,* 33, 471, 1968.
52. **Streibl, M. and Stránský, K.,** *Fette Seifen Anstrichm.,* 74, 566, 1972.
53. **Nicolaides, N.,** *J. Chromatogr. Sci.,* 8, 717, 1970.
54. **Nicolaides, N. and Foster, R. C.,** *J. Am. Oil Chem. Soc.,* 33, 404, 1956.
55. **Netting, A. G.,** *J. Chromatogr. Sci.,* 53, 507, 1970.
56. **Tulloch, A. P. and Hoffmann, L. L.,** *Phytochemistry,* 13, 2535, 1974.
57. **Tulloch, A. P.,** *J. Chromatogr. Sci.,* 13, 403, 1975.
58. **Macey, M. J. K. and Barber, H. N.,** *Phytochemistry,* 9, 13, 1970.
59. **Ludwig, F. J.,** *Soap Chem. Specialities,* 43, 70, 1966.
60. **Tulloch, A. P.,** *J. Am. Oil Chem. Soc.,* 49, 609, 1972.
61. **Tulloch, A. P.,** *J. Am. Oil Chem. Soc.,* 50, 367, 1973.
62. **Tulloch, A. P.,** *Cosmet. Perfumery,* 89, 53, 1974.
63. **Tulloch, A. P. and Hoffmann, L. L.,** *Lipids,* 8, 617, 1973.
64. **Tulloch, A. P.,** *Phytochemistry,* 12, 2225, 1973.
65. **Tulloch, A. P. and Hogge, L. R.,** *J. Chromatogr.,* 157, 291, 1978.
66. **Evans, D., Knights, B. A., Math, V. B., and Ritchie, A. L.,** *Phytochemistry,* 14, 2447, 1975.
67. **Holloway, P. J., Jeffree, C. E., and Baker, E. A.,** *Phytochemistry,* 15, 1768, 1976.
68. **Holloway, P. J., and Brown, G. A.,** *Chem. Phys. Lipids,* 19, 1, 1977.
69. **Holloway, P. J., Brown, G. A., Baker, E. A., and Macey, M. J. K.,** *Chem. Phys. Lipids,* 19, 114, 1977.
70. **Holloway, P. J., Hunt, G. M., Baker, E. A., and Macey, M. J. K.,** *Chem. Phys. Lipids,* 20, 141, 1977.
71. **Downing, D. T.,** *J. Chromatogr.,* 38, 91, 1968.
72. **Korolczuk, J. and Kwásniewska, I.,** *J. Chromatogr.,* 88, 428, 1974.
73. **Segura, R. and Gotto, A. M.,** *J. Chromatogr.,* 99, 643, 1974.
74. **Kaneda, T.,** *Biochemistry,* 6, 2023, 1967.
75. **Pomonis, J. G., Fatland, C. F., Nelson, D. R., and Zaylskie, R. G.,** *J. Chem. Ecol.,* 4, 27, 1978.
76. **Aasen, A. J., Hofstetter, H. H., Iyengar, B. T. R., and Holman, R. T.,** *Lipids,* 6, 502, 1971.
77. **von Wettstein-Knowles, P. and Netting, A. G.,** *Lipids,* 11, 478, 1976.
78. **Wollrab, V.,** *Phytochemistry,* 8, 623, 1969.
79. **Netting, A. G. and Macey, M. J. K.,** *Phytochemistry,* 10, 1917, 1971.
80. **Christiansen, K., Mahadevan, V., Viswanathan, C. V., and Holman, R. T.,** *Lipids,* 4, 421, 1969.
81. **Trka, A. and Streibl, M.,** *Coll. Czech. Chem. Commun.,* 39, 468, 1974.
82. **Tulloch, A. P. and Hoffmann, L. L.,** *Phytochemistry,* 15, 1145, 1976.
83. **Ubik, K., Stránský, K., and Streibl, M.,** *Coll. Czech. Chem. Commun.,* 40, 1718, 1975.
84. **Schmid, H. H. O. and Bandi, P. C.,** *J. Lipid Res.,* 12, 198, 1971.
85. **Lauer, W. M., Aasen, A. J., Graff, G., and Holman, R. T.,** *Lipids,* 5, 861, 1970.
86. **Aasen, A. J., Lauer, W. M., and Holman, R. T.,** *Lipids,* 5, 869, 1970.
87. **Sharkey, A. G., Freidel, R. A., and Langer, S. H.,** *Anal. Chem.,* 29, 770, 1957.
88. **Karlsson, K. A., Samuelsson, B. E., and Steen, G. O.,** *Chem. Phys. Lipids,* 11, 17, 1973.
89. **Baker, E. A. and Holloway, P. J.,** *Phytochemistry,* 14, 2463, 1975.
90. **Budzikiewicz, H., Wilson, J. M., and Djerassi, C.,** *J. Am. Chem. Soc.,* 85, 3688, 1963.
91. **Wilkomirski, B.,** *Wiakomości Chemiczne,* 30, 831, 1976.
92. **Knights, B. A.,** *J. Gas Chromatogr.,* 5, 273, 1967.
93. **Brooks, C. J. W., Horning, E. C., and Young, J. S.,** *Lipids,* 3, 391, 1968.
94. **Brooks, C. J. W., Henderson, W., and Steel, G.,** *Biochim. Biophys. Acta,* 296, 431, 1973.
95. **Stránský, K., Streibl, M., and Kubelka, V.,** *Coll. Czech. Chem. Commun.,* 36, 2281, 1971.
96. **Abrahamsson, S., Ställberg-Stenhagen, Z., and Stenhagen, E.,** *Progress in the Chemistry of Fats and Other Lipids,* Vol. 7, Holman, R. T. and Malkin, T., Eds., Pergamon Press, Oxford, 1963, 1.
97. **Eglinton, G., Hunneman, D. H., and McCormick, A.,** *Org. Mass Spectrom.,* 1, 593, 1968.
98. **Holloway, P. J.,** *J. Pharm. Pharmacol.,* 20, 775, 1968.

99. **Stránský, K., Streibl, M., and Buděšinsky, Z.,** *Seifen-Öle-Fette-Wachse,* 102, 473, 1976.
100. **Warth, A. H.,** *The Chemistry and Technology of Waxes,* Reinhold, New York, 1956.
101. **Horn, D. H. S. and Lamberton, J. A.,** *Chem. Ind. (London),* 1962, 2036.
102. **Radler, F.,** *Am. J. Enology Vitic.,* 16, 159, 1965.
103. **Nikkari, T. and Haahti, E.,** *Biochim. Biophys. Acta,* 164, 294, 1968.
104. **Spener, F.,** *Chem. Phys. Lipids,* 11, 229, 1973.

Chapter 12

ACYLGLYCEROLS (GLYCERIDES)

A. Kuksis

I. INTRODUCTION

Acylglycerols constitute a major form of energy storage for plants and animals. These compounds provide an important source of food and industrial raw material for man. Acylglycerols also serve as intermediates in glycerolipid metabolism and are known to accumulate under abnormal physiological and disease conditions. Since the industrial and metabolic properties of the acylglycerols vary with the composition, numerous methods have been developed for the resolution and identification of acylglycerols. Due to the extremely complex nature and large number of molecular species of closely similar chemical and physical properties, only chromatographic methods have proved adequate for this purpose. Positional isomers and enantiomers of acylglycerols, however, remain yet to be resolved by chromatographic methods. The purpose of this chapter is to provide a comprehensive and critical review of the methods of chromatography of acylglycerols and to point out the best techniques for solving specific problems in acylglycerol resolution and identification.

$$
\begin{array}{lll}
\text{Triacylglycerol} & \text{Hydroxytriacylglycerol} & \text{Estolide acylglycerol}
\end{array}
$$

II. CHEMICAL AND PHYSICAL PROPERTIES OF ACYLGLYCEROLS

Acylglycerols may be defined as glycerol esters in which one or more hydroxyl groups have been combined with fatty acids.[1] One or two hydroxyl groups of a glyceryl ester may be combined through an ether linkage to an aliphatic alcohol or aldehyde. Such molecules are essentially neutral and may occur in nature in the free form or may be released from more complex structures by enzymatic or chemical means.[2-4]

The three hydroxyl groups of the glycerol molecule may be identified in an unambiguous manner using the stereospecific numbering (*sn*-) system (Chapter 1). The two ends of the glycerol molecule are not stereochemically identical in many enzymatic reactions. The *sn*-1- and *sn*-3-hydroxyls are easily distinguished when the molecule forms a three-point attachment to any surface.[5] This stereochemical nonidentity of the two primary hydroxyl groups is demonstrated by glycerol kinase, which esterifies phosphate only at the *sn*-3-position.[6] Likewise, during the biosynthesis, specific acyltransferases introduce fatty acids in specific positions, thus accounting for the nonrandom distribution of fatty acids in the three positions of natural triacyglycerols.[7]

Furthermore, triacylglycerol lipases release the fatty acids from the *sn*-1- and *sn*-3-positions in preference to the *sn*-2-position,[8] and some of the lipases may show specificity either for the *sn*-1- or the *sn*-3-position of the acylglycerol molecule.[9,10] The *sn*-1(3)- and *sn*-2-positions also differ in the chemical reactivity,[11] and the differences in the relative polarity of their esters are responsible for the chromatographic resolution of the sn-1,2(2,3)- and sn-1,3-diacylglycerols, and sn-1(3)- and sn-2-monoacylglycerols.[12]

When a mixture of *n* different fatty acids is esterified with glycerol, the number of possible triacylglycerols can be calculated to be n^3. If the optical isomers are not distinguished, the total number of different triacylglycerols formed is equal to $(n^3 + n^2)/2$. With no isomers distinguished, the total number of different triacylglycerols formed is equal to $(n^3 + 3n^2 + 2n)/6$. Most natural fats contain 10 to 40 different acids, which would form a possible 1,000 to 64,000 different triacylglycerols. Butterfat, which is one of the most complex natural fats, is known to contain at least 142 different fatty acids,[13] which could generate a total of 2,863,288 possible triacylglycerol species.[14]

As a result of the association of different fatty acids in the triacylglycerol molecules, there arise characteristic combinations of fatty acids, which for chromatographic purposes have been classified in various systematic schemes.[14] Thus, recognition is made of mono-, di-, and tri-acid derivatives, which represent triacylglycerols with only one, only two, or three types of different fatty acids in each molecule. Any one of these triacylglycerol types can be further segregated into triacylglycerols containing only saturated, monoenoic, dienoic, trienoic, tetraenoic, or more unsaturated species, or any combinations of acids giving a total of one, two, or more double bonds per triacylglycerol molecule. Both saturated and unsaturated fatty acids and the acylglycerol make-up of them can differ in the carbon number or molecular weight distribution.

Theoretically, the fatty acids may be distributed in the triacylglycerol molecules according to one or more of the statistical patterns described by a completely random (1,2,3-random), partially random (1,2-random-3-random) and (1,3-random-2-random), or restricted random (1-random-2-random-3-random) distributions, which may be calculated by appropriate algebraic equations.[14] These distributions may be compared to those determined experimentally and hypotheses advanced about their origin.

The positional distribution and molecular association of the fatty acids determine the physicochemical properties of the triacylglycerols and thus their chromatographic behavior. Up to the present time, differences only in the molecular weight (volatility), polarity (partition), and in the overall degree of unsaturation (π-bonding) have been specifically exploited in chromatographic separations. However, there is evidence that positional isomers, reverse isomers, and eventually stereoisomers will also be resolved by chromatographic methods when these features are present in partial acylglycerols. In specific instances these characteristics may also form sufficient basis for chromatographic separation of triacylglycerols.

In general the values for various physical constants of natural acylglycerols are unknown, but they appear to follow the smooth curves observed for standard synthetic acylglycerols when any one homologous series is considered.[15,16] Hence values for various natural triacylglycerols can be predicted by interpolation, and several authors[16] have developed empirical equations and nomographs for predicting various indices and constants of triacylglycerols consisting of straight chain C_{10}-C_{22} saturated acids and oleic/linoleic/linolenic unsaturated acids. Since these properties are not of direct interest to a compilation of empirical chromatographic separations, they are not considered here further.

III. SAMPLE PREPARATION

The preparation of acylglycerols for chromatographic analyses usually involves a prior extraction, isolation and derivatization. These procedures are carried out under conditions that prevent contamination from solvents and equipment, oxidation by air and elevated temperatures as well as minimize chemical alterations by avoiding prolonged contact with acids and alkalis. Detailed discussion of the preparation and purification of lipid extracts may be found in Chapter 2 and Reference 17. The problem of sample contamination and oxidation generally can be avoided by working with distilled solvents and glass containers with Teflon seals and performing all manipulations under nitrogen at low temperatures. The samples may be stored for short periods of time in the presence of inert solvents, antioxidants, and low temperatures.

A. Extraction

Provided the tissue has been effectively destroyed, complete extraction of triacylglycerols can be obtained using most neutral lipid solvents. For complete extraction of monoacylglycerols and diacylglycerols, chloroform-methanol (2:1 to 1:1 v/v) extraction is the preferred method.[18,19] Losses of monoacylglycerols during the removal of nonlipid material from the solvent extracts may be prevented if the solvent of the original extract is evaporated and the residue is passed through a column of Sephadex.[20-23]

B. Isolation

The acylglycerol extracts usually contain other neutral lipids along with variable amounts of polar lipids, which must be removed prior to effective chromatography. The isolation or purification of the acylglycerols is usually accomplished by preparative column[24] or thin-layer[25] adsorption chromatography. Preparative thin-layer chromatography is the most effective method, but it is restricted to small sample sizes (less than 50 mg). It must be employed when working with complex mixtures of neutral lipids. The triacylglycerols can be readily resolved from other neutral lipids and 'free' fatty acids on plain silica gel layers using a variety of solvent systems.

Larger amounts (over 50 mg) of acylglycerols are best isolated by column adsorption chromatography on Florisil.[24] Florisil (or magnesium silicate) is preferred because it permits faster flow rates than silicic acid and because fatty acids are more strongly adsorbed on Florisil, which avoids contamination of triacylglycerols. From Florisil columns the triacylglycerols are eluted with hexane-diethyl ether (85:15), the diacylglycerols with hexane-diethyl ether (50:50), and the monoacylglycerols with diethyl ether-methanol (90:10), or equivalent polarities of other solvent mixtures. Acylglycerols have also been isolated from neutral lipid mixtures using chromatography on aluminum oxide, but this procedure has been found to be less reliable due to residual alkalinity of the adsorbent.[24] In special instances the isolation and purification of acylglycerols may be accomplished in a simplified manner. Thus, the triacylglycerols of plasma may be isolated essentially free of monoacylglycerols, diacylglycerols, and fatty acids by batch adsorption on Zeolite from an isopropanol solution.[26,27] This method has given highly reproducible quantitative results and has been incorporated as part of an automated routine in the Auto Analyzer method of plasma triacylglycerol analysis.[28] Other methods of isolation of acylglycerols based on countercurrent distribution, reversed phase partition chromatography, and preparative gas-liquid chromatography are considered below along with methods of resolution of molecular species of acylglycerols.

Isomerization of the monoacylglycerols and diacylglycerols on the layers[29] or columns[30] is minimized by incorporation of boric acid in the adsorbent. Under these conditions, the elution of the diacyl- and monoacylglycerols requires solvent systems of slightly higher polarity.[31] Both chromatoplates and columns yield separate fractions for the *sn*-1,3- and *sn*-1,2(2,3)-diacyl-glycerols and for the *sn*-1(3)- and *sn*-2-monoacylglycerols.

Column and thin-layer adsorption chromatography yield slightly different elution volumes for acylglycerols differing in unsaturation and in the length of the acyl chains,[32] causing tailing and subfractionation of the acylglycerol mixtures. Furthermore, during the preliminary isolation of triacylglycerols, separations are also effected among triacylglycerols containing none, one, two, or three hydroxy, epoxy, keto, and estolide groups.[33] Also triacylglycerols containing acetic,[34] isovaleric,[35] and phytanic[36] acid residues may be resolved from other triacylglycerols containing only long chain acyl moieties.

Under appropriate conditions effective resolutions are likewise obtained among pure acyl, alkylacyl, and alkenylacyl glycerols.[37]

The isolated acylglycerol mixtures are characterized by a determination of their fatty acid composition, which is usually accomplished by gas-liquid chromatography; in special instances, GC/MS analyses may be desirable. The determination of fatty acids in natural oils and fats has been recently reviewed, and specific recommendations for improved results have been made.[38] The total amount of triacylglycerol in a purified sample may be obtained by weighing when dealing with sufficiently large amounts of material, but quantitative gas chromatographic analysis of the intact lipid[39] or of the component fatty acids[40] in the presence of appropriate internal standards may be required for smaller quantities of acylglycerols. Other methods based on colorimetry or titration of fatty acids released from the acylglycerols are less exact and subject to a variety of errors.[40]

C. Derivatization

Preparation of derivatives increases the chemical stability of the acylglycerol molecules and in many instances improves their chromatographic resolution. Triacylglycerols as a rule do not require derivatization for most chromatographic separations. However, hydrogenation of unsaturated triacylglycerols[41] or their ether analog[42] has been used to improve resolution during gas-liquid chromatography and to prevent the decomposition of polyunsaturated species. Catalytic hydrogenation of acylglycerols is usually accomplished with hydrogen and platinum oxide (Adams catalyst). For GC/MS work, the unsaturated acylglycerols may be subjected to a reduction with deuterium using Wilkinson's catalyst.[43] This procedure confines the entry of the deuterium to the carbons involved in the double bond formation,[44] thus retaining the identity of the fatty acids in the appropriate mass spectra. It is important that the reduction with either hydrogen or deuterium is accomplished under conditions that do not cause hydrolysis or isomerization of the acylglycerols.

The mercuric acetate adducts of acylglycerols have been prepared to permit separations based on degree of unsaturation by means of adsorption chromatography. The unsaturated acylglycerols react quantitatively at room temperature in a weakly acidic solution with mercuric acetate and methanol. The reaction is free of any serious isomerization and/or double bond migration.[14,45] In special instances an acylglycerol mixture may be subjected to bromination to remove any unsaturated species, the bromides of which possess markedly different solubilities and chromatographic properties than the saturated acylglycerols.[46]

In the free form, the 1,2(2,3)-diacylglycerols are subject to rapid isomerization to the 1,3-diacylglycerols, and the 2-monoacylglycerols to the 1(3)-monoacylglycerols. This transformation may be prevented by appropriate complexation with borate or arsenate. The borate and arsenate complexes, however, are not compatible with separations in all chromatographic systems, and other derivatives must be found. This can be accomplished by the preparation of the acetate and trimethyl and the *tert*-butyldimethylsilyl ether derivatives of the mono- and diacylglycerols. Acetylation is accomplished by heating the acylglycerols with acetic anhydride pyridine (10:1) at 80°C for one hour.[47] The trimethylsilyl ethers of acylglycerols may be prepared by reaction with hexamethyldisilazane and trimethylchlorosilane in the presence of pyridine at room temperature.[25,48] Certain silylation mixtures, however, are known to cause isomerization of acylglycerols and must be avoided. The silyl ethers possess excellent gas chromatographic properties, but their use in other chromatographic systems is limited due to high sensitivity to moisture.[49,50] Much more stable to moisture are the *tert*-butyldimethylsilyl ethers, which are prepared by reacting the acylglycerols with tertiary-butyldimethylsilyl chloride imidazole reagent at 80°C for about 20 minutes.[51] These derivatives also possess excellent chromatographic properties, are suitable for analyses by TLC, and have special advantages for GC/MS work.[51] The monoacylglycerols yield isopropylidenes by condensation with acetone, and they constitute the choice derivatives for many analyses of monoacylglycerols.[52] Other derivatives for chromatography of monoacylglycerols are the trifluoroacetates[53] and the boronates.[54]

All the derivatives, except the trimethylsilyl ethers, may be purified before chromatographic analysis. For this purification of the derivatives both TLC and column adsorption chromatography may be used, depending on the amount of material involved. The solvent systems employed are usually of somewhat lower polarity than those required in the initial isolation of the original acylglycerols.[25]

IV. CHROMATOGRAPHY

The separation of molecular species of acylglycerols is based on the general properties of the original molecules or their derivatives. These properties are the molecular weight, π-bonding (or degree of unsaturation), and polarity, which is a combined effect of molecular weight and π-bonding. The property of polarity is exploited in the separation of acylglycerols on columns or thin-layers of adsorbent or in liquid-liquid partition chromatography on a column or thin-layer support. The π-bonding is primarily exploited in argentation adsorption chromatography on thin layers of adsorbents, while differences in molecular weight of the acylglycerols are most effectively utilized in gas-liquid chromatography. Although each chromatographic system exploits primarily one or the other of these effects, none of the chromatographic systems is entirely free of the influence of any of the other effects. In the following, the various methods of acylglycerol separation have been classified on the basis of the general method of execution of the analysis because a clear distinction cannot be made among the various chromatographic systems.

A. Column Adsorption Chromatography (CC)

Adsorption chromatography on columns of silicic acid is the oldest method of acylglycerol resolution (isolation). At the present time it is largely confined to the separation of mono-, di-, and triacylglycerols. However, this technique is also capable of effective resolution of acylglycerols having major differences in polarity. It is of particular interest and importance in the segregation of acylglycerols containing both short and long chain fatty acids and the common oxygenated fatty acids.

1. Apparatus

A number of apparatus have been described for adsorption chromatography on columns of silicic acid, Florisil, and aluminum oxide.[24] The original methods involved the use of discrete solvent gradients under manual control. Modern methods of high performance adsorption chromatography involve the use of fully automated systems including continuous gradients maintained by appropriate mixing and pumping systems operated by computers.[55,56]

2. Solvents

The solvents for adsorption chromatography on columns are selected for their ability to desorb acylglycerols in an orderly fashion from the adsorbent surface, for the ease of evaporation during fraction recovery, and for their freedom from ultraviolet absorbing materials in the ester-region used for automated monitoring of the peaks in the automated chromatographic systems. The specific solvents or combinations of solvents chosen depend on the nature of the acylglycerols and the desired type of separation. A variety of commonly used solvents is noted in Tables 1 to 8*, which describe specific separations.

---

*   All tables appear following the text, see p. 410.

a. Separation by Polarity

Separations of the common triacylglycerols from the di- and monoacylglycerols are tabulated (Table 1). Comparable separations have been realized for the alkylacylglycerols.[57]

The higher members of a homologous series of natural triacylglycerols are eluted ahead of the lower members that are more polar and, therefore, are retained more strongly by the adsorbent (Table 2). This was first demonstrated by the elution of trimyristoylglycerol well ahead of tributyroylglycerol from an alumina column using benzene as the solvent.[58] Subsequently, mixtures of petroleum hydrocarbon and diethyl ether have been used to separate short and long chain triacylglycerols of butterfat on columns of silicic acid. Milk fat triacylglycerols have been resolved into as many as 17 major fractions using silicic acid columns in combination with the petroleum hydrocarbon-diethyl ether solvents.[59]

Silicic acid columns in combination with an appropriate series of eluting solvents of increasing polarity give discrete fractions for triacylglycerols containing none, one, two, and three epoxy (Table 3), keto (Table 4), and hydroxy (Table 5) groups.

Mixtures of estolide triacylglycerols can be separated on silicic acid according to the number of ester groups per molecule. The resulting tri- (normal), tetra-, penta-, and hexa-acyl glycerols give separate fractions under either gradient or step-wise elution (Table 6).

Separations comparable to those achieved for the triacylglycerols are possible on the silicic acid columns also with the alkylacylglycerols containing short and long and oxygenated chains (Tables 3 to 5).

b. Separation by Unsaturation

Of the triacylglycerols containing fatty acids of the same chain length, the more unsaturated molecules are adsorbed more strongly on silicic acid than the more saturated ones.[60,61] This effect is not great enough to separate molecular species according to the number of double bonds they contain, but the effect is sufficient to create nonhomogeneous peaks during column chromatography.

The separation of triacylglycerols differing in degree of unsaturation can be achieved more readily by preparing mercuric acetate adducts of the unsaturated esters (Table 7) or by complexing the esters via the π-electrons with silver nitrate (Table 8).

The addition of mercuric acetate to the double bonds of triacylglycerols enhances their polarity so that the molecules can be separated on silicic acid according to number of double bonds they originally contained. Triacylglycerols containing zero, one, two, and three double bonds are completely resolved, but molecules containing more than four double bonds are so strongly adsorbed by the silicic acid that they cannot be quantitatively eluted from the column.[45] The original triacylglycerols are regenerated from the eluted mercuric acetate adducts by reaction with dilute HCl. Similar separations of unsaturated triacylglycerols have also been obtained using columns of Florisil® or alumina. These separations resemble those obtained on silicic acid with incomplete resolution between the various peaks.

Separations of triacylglycerols differing in degree of unsaturation can be achieved most readily by taking advantage of the reversible π-bonding of the unsaturated acyl chains with silver ions[62,63] (AgNO$_3$-chromatography, argentation chromatography).

Saturated and unsaturated triacylglycerols can be separated by adsorption column chromatography following impregnation of the adsorbent with silver nitrate.[64] The Ag$^+$/olefin complex is of sufficiently low energy that it can be formed and broken during standard lipid chromatographic procedures. The separation of triacylglycerol mixtures containing zero to four double bonds by column chromatography on AgNO$_3$-impregnated silicic acid is accomplished by stepwise elution with petroleum hydrocarbon-benzene-diethyl ether mixtures (Table 8). The exact solvent ratios employed can vary slightly from one batch of adsorbent to another.

Most techniques for column chromatography of triacylglycerols on AgNO$_3$-silicic acid can resolve only six triacylglycerol fractions: 000, 001, 002, 011, 111 + 012, and more unsaturated molecules. Attempts to resolve more unsaturated triacylglycerol mixtures have been unsuccessful.[65,66]

3. Detection Methods

Since the common triacylglycerols are colorless and also do not possess any characteristic absorption bands in the UV or infrared regions, they cannot be detected in the effluent by spectrophotometric means. The most common and best method for the detection of triacylglycerols in the effluent, when working with sufficient mass, has been gravimetry.[24] The most popular and effective method for the detection of small amounts of acylglycerols in the effluent fractions has been thin-layer[67] and gas-liquid chromatographic[68] analysis, but these procedures cannot be applied on a continuous basis. However, moving-wire flame ionization detection[69] and ultraviolet absorption[70] at the shorter wavelengths may be effective although expensive alternatives. In special instances, the acylglycerols have been separated as derivatives exhibiting characteristic absorption bands (mercuric acetate adducts), when they have been detected spectrophotometrically.[45] After reaction with diphenylcarbazone hydroxytriacylglycerols can be detected by infrared[71] and acylglycerols containing fatty acids with conjugated double bonds can be monitored by ultraviolet[7] absorption spectroscopy.

B. Thin-Layer Adsorption Chromatography (TLC)

Adsorption chromatography on thin layers of silicic acid or alumina has now largely replaced adsorption chromatography on columns as the preferred analytical method for acylglycerol separation, except for certain preparative applications. Thin-layer chromatography offers several advantages over column chromatography for acylglycerol separation. These are speed, greatly increased efficiency of resolution, and ease of detection of separated components. The general methodology of thin-layer chromatography of neutral acylglycerols has been described elsewhere[73,74] and need not be recapitulated here. Specific separations are tabulated along with the appropriate solvent systems (see Tables 9 to 23).

1. Apparatus

A number of apparatus have been described for adsorption chromatography on thin layers of silicic acid or aluminum oxide.[14,73,74] Usually the open plate method has been used, with the sandwich type of development being employed only occasionally. Recently the high performance mode of thin-layer separation has been introduced for the separation of lipids, but its advantages for work with neutral acylglycerols have yet to be established. In most instances the plates have measured 20 × 20 cm, and the compounds have been applied either as bands across the entire width of the plate, as narrow bands or as spots, and development has proceeded in the ascending direction.[14,73,74] Another recent development, the Iatroscan, combines TLC with flame ionization detection. This combination extends the sensitivity of TLC into the range of GLC and HPLC.[75]

2. Solvents

The solvent types employed in the thin-layer adsorption chromatography on plain silica gel are similar to those used for work with adsorption columns. Improved separations can be achieved in thin-layer chromatography by means of a repeat development with the same or increasingly more polar solvent combinations.[73] Usually acetic or formic acid is added to the solvent formulation to reduce adsorption and tailing of the more polar solutes in the sample.[74] The general types of solvent mixtures used for separating acylglycerols on $AgNO_3$-impregnated layers have been restricted largely to 0 to 6% alcohol in chloroform and various proportions of benzene and diethyl ether.[14]

a. Separation by Polarity

A large variety of solvent systems are capable of effecting a complete resolution among the triacyl-, diacyl-, and monoacylglycerols (Table 9). The *sn*-1,2(2,3)- and *sn*-1,3-diacyl-glycerols are also readily resolved by thin-layer chromatography using petroleum hydrocarbon-diethyl ether (50:50 or 60:40 v/v) as solvent system,[76] but the *sn*-2- and *sn*-1(3)-monoacyl-glycerols may remain together at the origin. Both diacyl and monoacylglycerol isomers are resolved by employing more polar solvent systems in conjunction with adsorbents impregnated with 5 to 8% boric acid,[29,77] which also minimizes the acyl migration resulting from prolonged contact with the gel.

The higher molecular weight triacylglycerols migrate well ahead of the lower molecular weight triacylglycerols so that separations into short, medium, and long chain homologs are readily attained. A difference of 6 methylene units is required for a complete resolution of C_{24} to C_{66} triacylglycerols, but molecules differing by only one or two methylene units can be separated below C_{24}, when a solvent system consisting of petroleum hydrocarbon-diethyl ether (80:20), is employed.[14,78] Other workers have used comparable solvent systems for the separation of natural short and long chain triacylglycerols from bovine milk fat and from coconut oil (Table 10).

Triacylglycerols containing both long-chain and acetic acids (acetin fats) are easily separated according to acetate content by thin-layer chromatography (Table 11). Similarly, beluga whale head triacylglycerols containing isovaleric acid have been fractionated by thin-layer chromatography on the basis of the number of isovaleroyl moieties per molecule (Table 12). The presence of phytanic acid in a triacylglycerol molecule decreases its adsorption by silicic acid so that triacylglycerols containing one, two, or three phytanate groups can be resolved (Table 13).

Like silicic acid columns, the silica gel layers allow the separation of triacylglycerols containing none, one, two, and three epoxy (Table 14) and hydroxy (Table 15) triacylglycerols. These separations, however, are much more complete than those realized with the adsorption columns. Likewise, mixtures of estolide triacylglycerols are resolved by thin-layer chromatography according to the number of the ester groups (Table 16). In several instances these resolutions have included estolides with free hydroxyl or acetoxy groups or both.[79]

Thin-layer adsorption chromatography on plain silica gel is also effective in separating alkylacyl and alkenylacyl triradyl, diradyl and monoradylglycerols, including the *sn*-1,2- and *sn*-1,3, diradyl and *sn*-2- and *sn*-1(3)-monoradyl isomers (see Chapter 14). The solvent systems required for these separations are somewhat less polar than those needed for the separation of the corresponding acylglycerol derivatives. Effective separations are also obtained among the triradylglycerols containing long and short chains, and between those with and without hydroxyl or acetoxy groups. Similarly, thin-layer adsorption chromatography is effective in resolving monoalkylglycerols containing various oxygen functions, when analyzed in the form of the isopropylidenes (see Chapter 14). Differences have also been noted in the R_f values of monoalkylglycerols containing *O*- and *S*-ether bonds. In all instances the *S*-alkyl ethers moved ahead of the *O*-alkyl ethers. Reference is made to Chapter 14.

b. Separation by Unsaturation

Of the triacylglycerols containing fatty acids of the same chain length, the more unsaturated molecular species are adsorbed more strongly on silicic acid than the more saturated species.[60] The difference, however, is not great enough to separate triacylglycerols according to the number of double bonds they contain, but the effect is sufficient to result in nonhomogenous spots or bonds during thin-layer chromatography. Unsaturated triacylglycerols may be resolved according to number of double bonds following oxidation and complex formation with double bonds. Addition of ozone, mercuric acetate, or π-bonding with silver nitrate leads to a retardation of the migration of the unsaturated glycerol esters on the adsorbent surface. Thus, ozonization of triacylglycerols converts each double bond into an ozonide group. The products can then be separated on silicic acid either as the ozonides or after reduction of the ozonide to aldehyde. The triacylglycerol ozonides can be separated by adsorption thin-layer chromatography on silicic acid using petroleum hydrocarbon-diethyl ether (80:20 v/v) as the developing solvent (Table 17).

The reduced ozonides can be resolved using comparable solvent systems, and this method has been used in analytical work (Table 18). Because of the nonquantitative nature of the ozonization reaction, ozonide separations are rarely used in quantitative triacylglycerol analysis, but they are of theoretical interest.

Oxidized triacylglycerols can be directly resolved into SSS, SSA, SAA and AAA bands by thin-layer chromatography on silicic acid using hexane-diethyl ether (60:40) as the developing solvent.[80] Alternatively, the oxidized triacylglycerols can be reacted with diazomethane and fractionated into bands having three, four, five, and six ester groups per molecule.[81] A double development procedure was required for this separation: an initial development with diethyl ether until the solvent front reached 1 cm above the point of sample application, followed by full development with benzene-diethyl ether (90:10).

Superior separation of triacylglycerols based on degree of unsaturation is obtained by argentation-TLC.

A total of five to ten bands of triacylglycerols can be resolved on a single chromatoplate for a synthetic mixture of triacylglycerols containing monounsaturated to hexaunsaturated triacylglycerols using benzene-diisopropyl ether (85:15).[82] Comparable separations have been obtained on natural oil using benzene-diethyl ether (85:15) and chloroform-methanol (94:6) (Table 19).

For triacylglycerols containing more than six double bonds, more polar solvent systems must be employed,[83] which no longer resolve the less unsaturated species completely. Complete resolution of oligo- and polyunsaturated triacylglycerols requires several developments with two or more solvent systems. Triacylglycerols from 000 to 222 can be separated as single bands if no linolenic acid is present (Figure 1).[82] When a fat contains linolenic acid, triacylglycerols from 222 to 333 are only partially resolved, and preparative TLC usually leaves unresolved ternary and quaternary mixtures of triacylglycerols.[84] Natural fat triacylglycerols containing fatty acids with four, five, and six double bonds are relatively poorly resolved, and only a few distinct bands can be obtained.[83]

In addition to the resolution based on the total number of double bonds per acylglycerol molecule, argentation thin-layer chromatography also allows the separation of four types of isomeric triacylglycerols. Thus, separations may be realized between isomers containing double bonds in different chains of the triacylglycerol molecule. Excellent separations are seen for 111 and 012, 022 and 013, and 020 and 011 species using most of the solvent systems (Table 19). Molecular species of triacylglycerols differing in the positioning of the double bonds within a single acyl chain have also been resolved; isomeric oleoyl and petroselenoyl triacylglycerols are separated by a triple development at $-22°C$ using toluene-diethyl ether (75:25) as the solvent.[85] Subfractionation of $18:1\omega9$ and $20:1\omega11$ triacylglycerols has been observed during $AgNO_3$- thin layer chromatography of certain seed fats.[86] Similar distinction between $16:1\omega9$ and $18:1\omega11$ acylglycerols has also been made. In addition, the isomeric oleoyl and elaidoyl triacylglycerols also have been resolved by $AgNO_3$-thin layer chromatography.[87] This separation is related to the formation of much weaker π-complexes with the *trans* double bonds. Still other isomer separations observed by argentation thin-layer chromatography relate to the weaker π-bonding encountered with the unsaturated acyl moieties located at the *sn*-2-position of the triacylglycerol molecule. Thus separations of 001 and 010, 011 and 101, and 002 and 020 have been recorded (Table 19). These isomer separations are best observed when the triacylglycerol molecules have relatively few double bonds. Chain length isomers of triacylglycerols are not separated by argentation thin-layer chromatography unless the molecular weight differences approach those seen between normal and acetylacylglycerols.

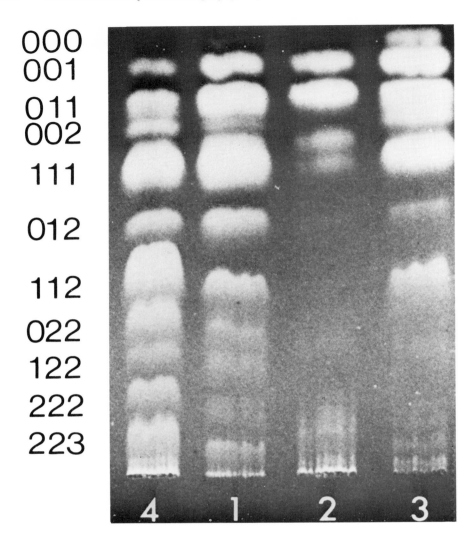

FIGURE 1.   AgNO$_3$-TLC of hepatic triacylglycerols of human infants receiving Intralipid (1,4) and controls (2,3). Triacylglycerol bands are identified by the component fatty acids with zero (0), one (1), two (2), or three (3) double bonds. Chromatography conditions: 20% AgNO$_3$ silica gel G (250 μ thick layer); chloroform-methanol 98:2; 2′,7′-dichlorofluorescein detection; (From Kuksis, A., unpublished results, 1980.)

Triacylglycerols containing cyclopropene fatty acids (malvalic and sterculic) cannot be separated by AgNO$_3$-thin-layer chromatography because the cyclopropene rings react with silver ions to produce a complex mixture of products.[88] Triacylglycerols of cyclopentenyl fatty acids are well resolved from those of the common fatty acids.[89]

Triacylglycerols containing oxygenated fatty acids are successfully separated on the basis of unsaturation using argentation thin-layer chromatography, but the polarity of the solvent systems must be increased.[84]

Alkyldiacylglycerols and alkenyldiacylglycerols are readily resolved according to the number of double bonds by AgNO$_3$-thin-layer chromatography using solvent systems comparable to those employed for triacylglycerol work.[90] Argentation thin-layer chromatography has been equally effective in the resolution of the diacylglycerols produced from triacylglycerols by partial degradation during structural studies or derived from diacylglycerophospholipids by dephosphorylation.[25] If handled rapidly, the diacylglycerols can be used directly for this purpose (Table 20), provided the polarity of the developing solvent is increased slightly. To avoid acyl migration it is preferable to acetylate the diacylglycerols. The resultant acetyldiacylglycerols can then be separated by the procedures described for the argentation thin-layer chromatography of triacylglycerols (Table 21). Since the *sn*-1,2(2,3)- and the *sn*-1,3-isomers are adsorbed to a different extent by the silicic acid, it is necessary to resolve them prior to argentation thin-layer chromatography. The acetyldiacylglycerols give essentially the same order of separation as the 'free' diacylglycerols. The degree of resolution observed with diacylglycerols is somewhat greater than that noted for triacylglycerols. Thus the reverse isomers of diacylglycerols are more completely resolved, when compared to triacylglycerols. Likewise, isomeric diacylglycerols differing in the positional placement and geometric configuration of double bonds of the component acids may be completely resolved. Furthermore, there is noticeable separation among diacylglycerols differing in chain length of the unsaturated component acids.

Even more complete are the $AgNO_3$-thin-layer chromatographic separations of the alkyl-acyl and alkenylacylglycerols. These separations approach those seen for methyl esters[63] when the long chains are completely uniform in their degree of unsaturation (Table 22). Since the alkylacyl and alkenylacyl glycerols are retained to different extents by silicic acid, it is necessary to resolve them on plain silica gel prior to $AgNO_3$-thin-layer chromatography. Alkylacyl and alkenylacylglycerols containing alkyl chains of varying degree of unsaturation require a separate assessment of the extent of unsaturation of the monoalkyl and monoalkenylglycerol moieties, which can be obtained by argentation-thin-layer chromatography with appropriate solvent systems (Table 23). Positional and chain length isomers of monoalkyl and monoalkenylglycerols usually are not resolved by argentation thin-layer chromatography, while oxygenated derivatives are well separated from the corresponding normal molecules. A combination of $AgNO_3$- and H_3BO_3-thin-layer chromatography[63] may allow a separation based on both positional placement and degree of unsaturation of the alkylglycerols.

1. Detection Methods

After development the chromatoplate is dried under a stream of nitrogen and the acylglycerols are located by any of a number of techniques. The most popular nondestructive method for locating acylglycerol spots or bands is spraying with 0.05% 2',7'-dichlorofluorescein solution in methanol-water (50:50) and viewing the plates under ultraviolet light, when the lipids appear as yellow spots against a purple background.[73] The highest sensitivity is obtained when the dichlorofluorescein solution contains 90% water or when the plate is humidified after spraying.[91] Rhodamine 6G, dibromofluorescein, and sodium fluorescein have been similarly used.[14]

Destructive methods for locating acylglycerol spots include[14] drawing the flame of a glass blower's torch across a plate, or spraying with 50% *o*-phosphoric acid or 50% sulfuric acid followed by charring in an oven at 200 to 400°C. Exposure to iodine vapor is a highly sensitive method of locating unsaturated acylglycerols on plain silica gel layers, but it is not suitable for locating acylglycerols in the presence of $AgNO_3$.[63] Several of the above methods can be used for quantitating of the acylglycerols directly on the TLC plate.[92] Alternately, the recovered acylglycerols can be determined by a variety of chemical methods specific for either the glycerol or acyl moiety of the acylglycerol molecule.[14,63,92]

Recently a hydrogen flame-TLC rod concept has been perfected in the Iatroscan system and it provides a major advancement in the detection and quantitation methods for TLC.[75]

The most effective methods for the identification and quantitation of acylglycerols on chromatoplates depend upon recovery of the acylglycerol bands from the adsorbent and are usually followed by a suitable quantitative analysis. The recovered acylglycerols can be quantitated by almost any of the micro methods of fatty acid and glycerol analysis. The method of choice is usually gas-liquid chromatography of the methyl esters with an internal standard[40,93] since it permits both the fatty acid composition and the amount of acylglycerol present to be determined simultaneously. An evaluation of the accuracy of the technique has found an error of about one percent absolute for major components when triplicate or quadruplicate analyses are averaged.[93] Single analyses have shown error as high as three to four percent absolute. Intact acylglycerols recovered from $AgNO_3$-impregnated layers may be quantitated by direct GC with an appropriate internal standard,[94] but the error of such analysis is about twice that noted for the fatty acids.

V. LIQUID-LIQUID CHROMATOGRAPHY

Liquid-liquid chromatography in various forms has long been used for the resolution of molecular species of acylglycerols.[14] Generally, the paper and thin-layer systems have been the most successful because they have permitted the easiest detection of the separated components. The methods of high performance liquid chromatography introduced more recently have greatly increased the efficiency of column separation and in specific instances have given separations approaching those realized in open tubular gas chromatography.[95]

A. Column Chromatography (CC)

In principle, two types of columns are presently employed in liquid-liquid chromatography. One type is provided by the conventional columns, which are of relatively large diameter and contain a liquid phase which is equilibrated between the support and the polar liquid phase.[96,97] The other type is the narrow bore column packed with a hydrophobic chemically bonded liquid phase over which the polar mobile phase is passed without significant displacement of the stationary liquid phase.[98,99] In addition, an intermediate type of column may be recognized that utilizes hydrophobic polymer particles as a support and liquid phase, with the polar phase having a limited solubility in the polymer.[100] A large variety of applications and solvent systems have been suggested or employed for acylglycerol separations, along with appropriate detection systems. The conventional columns are usually operated under gravity flow, while the narrow bore columns are operated under high pressure (see Tables 24 to 30).

1. Apparatus

The equipment employed to control the chromatographic columns for acylglycerol separation has ranged from simple gravity flow systems requiring manual change in solvents to fully automated systems with precisely regulated gradient flow. The continuous flow systems require specialized apparatus, which were originally built by the investigators themselves.[96,101,102] Recently, however, high quality instrumentation has become commercially available, and the methodology of liquid chromatography is now more generally accessible. The modern liquid chromatographic techniques have been described in recent reviews[95,103] and books[104] dealing with theory, instrumentation, and applications.

The improved control of the solvent composition and flow has permitted more efficient utilization of differential refractometry and short wavelength ultraviolet absorption as means of continuous monitoring of the effluent at high sensitivity.[95] These features are available as built-in modules in most of the commercial liquid chromatographs. Other methods of effluent analysis depend on accessory equipment and are discussed under Detection Methods.

A large variety of highly efficient liquid chromatographic columns are now commercially available for the separation of various lipid components. For the separation of acylglycerols, the various long-chain alkylsiloxane treated silica gel packings would appear to be of the greatest interest. In the past silanized celite had given excellent resolution of various triacylglycerols differing by a single methylene unit when the columns were operated under essentially gravity flow.[101] An effective early separation was achieved with alkylated Sephadex as the support.[105]

2. Solvents

Class separations of mono-, di-, and triacylglycerols can be accomplished usually with very simple solvent mixtures, while the resolution of positional isomers of mono- and diacylglycerols requires a more careful adjustment of solvent polarity and an optimum compatibility between the mobile and stationary phases. Even the simplest solvent systems employed in class separations usually provide also a considerable degree of resolution of molecular species within each acylglycerol class (Table 24).

The resolution of triacylglycerols by liquid-liquid chromatography depends on both the molecular weight and the number of double bonds in the molecule. Saturated triacylglycerols are separated on the basis of chain length or carbon number, with compounds of lower molecular weights eluting first when the mobile phase is polar or last if the mobile phase is nonpolar. The introduction of a double bond into a saturated triacylglycerol molecule changes its partition coefficient so that it elutes with the saturated triacylglycerols containing two carbon atoms less (Table 25). Both analytical[101] and preparative[96,97,105] type of columns gave comparable resolutions.

The order of elution of acylglycerols is reversed in gel permeation chromatography.[79,98,106-108] A critical examination[101] of the composition of the overlapping triacylglycerol species by means of high resolution techniques has shown that the "critical partners" actually have slightly different mobilities and that one double bond is really equivalent to 2.2 to 2.8 methylene units depending on the compounds compared and the operating conditions employed. A linear relationship exists between the log of the retention volume and the equivalent partition number for normal chain saturated triacylglycerols in liquid-liquid partition column chromatography. The equivalent partition number of an unsaturated triacylglycerol is estimated graphically from its elution volume.[14,100] Equivalent partition numbers of unsaturated triacylglycerols are known to vary slightly with chromatographic operating conditions, double bond configuration and location in the molecule.[14]

The hydrophobic gels separate triacylglycerols on the basis of molecular weight.[100] Since the log of the elution volume is proportional to the molecular weight, the gel permeation chromatography of saturated and unsaturated triacylglycerols would give approximately the same elution values.

Saturated and unsaturated triacylglycerols within a partition number can be resolved without prior chemical modification by HPLC using appropriately selected solvents systems (Table 26).

Saturated and unsaturated triacylglycerols have been resolved by liquid-liquid partition chromatography following oxidation.[109] A resolution of trisaturated, disaturated-azelaic and saturated-diazelaic, and triazelaic acid esters of glycerol was obtained by column partition chromatography using ethanol-water (90:10) as the polar stationary phase with petroleum hydrocarbon and diethyl ether as the eluting solvents. Triacylglycerols containing *trans* double bonds have slightly higher equivalent partition numbers than the corresponding *cis* isomers. Thus, trielaidoylglycerol has been found to elute after trilinoleoylglycerol in column partition chromatography.[101] In general, however, geometrical isomers are poorly separated by liquid-liquid partition chromatography, and isomer pairs such as trioleoylglycerol-dioleoylelaidoylglycerol cannot be resolved (Figure 2). For preparative separations, the use of HPLC columns containing silver ions has been explored (Table 27).

HPLC systems are also well suited for the resolution of the estolide acylglycerols, although the optimum solvent mixtures worked out for triacylglycerol resolution do not appear to have been tested with these compounds (Table 28).

Triacylglycerols containing epoxy groups are very easily separated according to the *number*

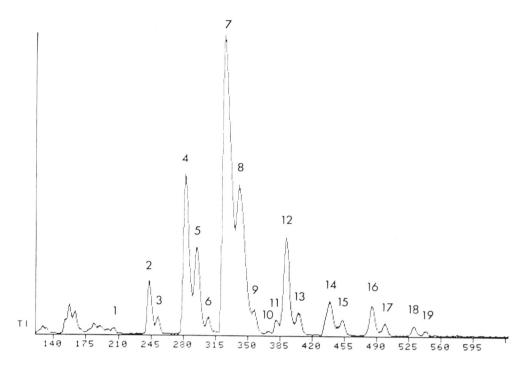

FIGURE 2.   LC/MS analysis of peanut oil triacylglycerols. Peak identity: 1, 18:2 18:2 18:2; 2, 18:1 18:2 18:2; 3, 16:0 18:2 18:2; 4, 20:1 18:2 18:2 + 18:1 18:1 18:2; 5, 18:2 18:2 18:0 + 16:0 18:1 18:2; 6, 16:0 16:0 18:2; 7, 18:1 18:2 20:1 + 18:1 18:1 18:1 + 18:2 18:2 18:0; 8, 18:1 18:2 18:0 + 16:0 18:1 18:2; 9, 16:0 18:0 18:2 + 16:0 16:0 18:1; 11, 18:1 18:1 20:1; 12, 18:2 18:2 22:0 + 18:1 18:2 20:0 + 18:0 18:1 18:1; 13, 16:0 18:1 18:0; 14, 18:2 18:2 24:0 + 18:1 18:2 22:0 + 18:1 18:1 20:0; 15, 16:0 18:1 20:0 + 18:0 18:1 18:0; 16, 18:1 18:2 24:0 + 18:1 18:1 22:0, 17, 16:0 18:1 22:0 + 18:0 18:1 20:0; 18, 18:1 18:1 24:0; 19, 16:0 18:1 24:0 + 18:0 18:1 22:0. Column: Supelcosil LC-18 reversed phase (25 cm × 0.46 cm); Mobile phase: 30-90% linear gradient of propionitrile in acetonitrile. Temperature: 30°C. Flow rate: 1.5 mℓ/min. Direct liquid inlet chemical ionization mass spectrometry using Hewlett-Packard quadrupole Model 5985B automatic mass spectrometer. (From Kuksis, A., Marai, L. and Myher, J. J., unpublished results, 1982.)

of epoxy groups per molecule using partition chromatography (Table 29). There is also a segregation of molecules within each triacylglycerol type based on the partition number.

Diacylglycerols can be effectively fractionated by liquid-liquid partition chromatography in the same manner as triacylglycerols. With an acetone-water (87.5:12.5) mobile phase and factice as the support, the sn-1(2,3)-diacylglycerols are eluted before the corresponding sn-1,3-diacylglycerol isomers, but diacylglycerol isomers are only poorly separated with an acetone-water (95:5) mobile phase.[96] The extent of separation of isomeric diacylglycerols in partition chromatography depends on the polarity of the mobile phase.

Free 1,2-diacyl-sn-glycerols could not be analyzed on modified Sephadex column without extensive isomerization, but the TMS ethers of sn-1,2-diacylglycerols were separated[50] on hydroxyalkoxy Sephadex prepared from Sephadex LH-20 and C_{11}-C_{14} alkyl epoxide (Table 30).[50] It was necessary to add 1% pyridine to the solvent system of acetone-water-heptane (87:13:10) to prevent acid-catalyzed hydrolysis of the silyl ethers. Tests indicated that 94.8 ± 0.3% of the TMS ethers remained unchanged after passing through the column. Diacylglycerols having the same number of double bonds per molecule were eluted in order of increasing chain lengths. Each additional double bond shifted the retention volume the equivalent of two methylene units.

3. Detection Methods

The application of liquid chromatography to lipid analysis has progressed very slowly, principally because detectors sensitive to all types of lipids have not been available. Since most lipids do not absorb UV light at 254 nm, a simple single wavelength detector is ineffective. A variable wavelength detector is more useful because it will register in the range of 190 to 220 nm where many lipid esters show a weak absorbance.[104] Although many solvents also absorb at these low wavelengths, some of the most popular, such as water, acetonitrile, hexane, and isooctane, do not interfere with lipid detection.

Refractive index detection is frequently employed for lipid analyses, but it has the disadvantages of being only one twentieth as sensitive as the UV detector, and of being greatly affected by changes in solvent concentration, so that an isocratic (constant composition) mobile phase rather than solvent programming must be used.[100,110]

The development of a moving wire flame ionization detector for liquid chromatography has eliminated most of these limitations and has been successfully employed in the liquid-liquid chromatographic analysis of triacylglycerols.[101,105]

It is obvious that the ultimate detector for liquid-liquid chromatographic columns will have to be a mass spectrometer,[111,112] preferably of the chemical ionization type, as already demonstrated for triacylglycerols. Suitable prototypes of such detection systems have already been proposed, and the initial versions of them are already on the market. At the present time, however, the cost of such detection systems keeps them out of reach of most laboratories. The use of light scattering as a detector principle has been described recently.[172]

B. Paper and Thin-Layer Chromatography

The use of thin layers instead of columns to support the stationary phase has the advantage of increased efficiency of separation, simplified detection of solutes, and speed of operation. These separations are limited to essentially analytical applications, but attempts have been made to scale them up for preparative purposes.

1. Apparatus

In principle, either paper or thin-layer systems may be employed with equal effectiveness for partition chromatography. In most laboratories, however, the thin-layer system has completely replaced the paper system because the process is much more rapid, e.g., 1 to 2 hours vs. 6 to 24 hours. Furthermore, the inorganic adsorbent layers may be subjected to a greater variety of corrosive sprays than paper chromatograms, which is a distinct advantage for the characterization of lipids.

The layers for partition work are prepared[16] in the usual manner using kieselguhr, silicic acid, or silanized silicic acid. Paper chromatograms are prepared from Whatman No. 1, Schleicher and Schull 2040b or 2043b (484) or similar grades of chromatographic papers. Papers or layers are impregnated with the nonpolar stationary phase either by immersing them into or by spraying them with a 0.5 to 15% solution of stationary phase in a volatile solvent. Alternatively, the original plate may be spread by slurrying the solid support in a solution of the stationary phase in a volatile solvent. The amount of stationary phase deposited, which is usually between 6 and 15% of the weight of the solid support, is controlled by varying the concentration of the impregnating solution. Silanized layers must be developed in methanol before they can be used with $AgNO_3$-containing solvents.[113] Paraffins are the stationary phases used most often, but silicone fluids are also satisfactory.[114] The amount of stationary phase impregnated into the chromatogram has a marked influence on the R_f values of specific triacylglycerols.[114]

The samples are applied from a micropipet near the bottom edge of the chromatoplate as a spot or a narrow band (2 to 10 μg per spot). The chromatogram is then developed in the appropriate mobile phase by the ascending technique in a suitable chromatography tank or sandwich apparatus.[73] Multiple development (two to five times) with the same mobile phase gives improved resolution.

2. Solvents

The selection of a suitable solvent system for separation of acylglycerols by paper or thin-layer partition chromatography is largely empirical. Many of the reversed phase partition systems have been worked out using countercurrent distribution[115] and these are directly transferable to separations on paper and thin layers. The highest resolutions have been obtained with the long-chain linear hydrocarbons.[116,117] Silicone fluids have occasionally been successfully substituted as the stationary phase.[118] Chemically bonded silicones are preferred since these stationary phases are not eluted with the developing solvent or extracted during sample recovery. The high resolution on long chain hydrocarbon or alkyl chains has been combined with the nonextractable nature of bonded silicones by using reagents such as hexadecyltrichlorosilane to produce a bonded long chain stationary phase.[119] The much more reactive octanyl derivative may give stationary phases that are as effective as those prepared with the longer chain derivative for the liquid-liquid chromatography of acylglycerols.[95] Polar phases such as acetone-acetonitrile (80:20 v/v), acetone-acetic acid (85:15 to 50:50) and methanol-water (80:20), have found application in most paper and thin-layer chromatographic systems. The liquid phases are equilibrated with the stationary phase.[120] In several instances,[113,121,122] the addition of $AgNO_3$ to the polar phase has been shown to cause Ag^+/olefin complexing to occur and to increase the solubility of the unsaturated triacylglycerols in the mobile phase.

Since the polar solvent is the mobile phase in reversed phase paper and thin-layer partition chromatography, triacylglycerols with the lower partition numbers move ahead of those with higher partition numbers. Thus, trilinoleoylglycerol migrates further than trioleoylglycerol, and trimyristoylglycerol moves ahead of tripalmitoylglycerol. Resolution becomes poorer as the partition number increases. Nevertheless, natural fat triacylglycerols containing only acids of even carbon numbers can be fully resolved into fractions of different partition numbers by paper and thin-layer reversed phase partition chromatography (Tables 31 and 32).

Using TLC with multiple development, it is sometimes possible to separate synthetic triacylglycerols having the same partition number, such as tripalmitoyl, dipalmitoylmono-oleoyl, dioleoylmonopalmitoyl, and trioleoylglycerols.[116] Similarly, the critical partners in natural fat triacylglycerols can sometimes be partially separated if the mixture is not too complex.[87,123] However, triacylglycerol isomers such as tripalmitoylglycerol and mono-lauroyldistearoylglycerol[114] or 1,3-dipalmitoyl-2-oleoyl and 1,2-dipalmitoyl-3-oleoylglycerols[124] cannot be separated by reversed-phase partition TLC.

Triacylglycerols containing *trans* double bonds have slightly higher equivalent partition numbers than the corresponding *cis* isomers, resulting in a partial resolution on reversed phase thin-layer chromatography.[116,125] As much as 40 mg of linseed oil triacylglycerols have been separated on a single paper partition chromatogram using circular development.[120]

The addition of $AgNO_3$ to the polar phase results in separation of acylglycerols on the basis of unsaturation rather than partition number,[121,122] as does the addition of bromine[126] or mercuric acetate[127] to the double bonds of the acylglycerol molecules. The more unsaturated molecules gave the higher R_f values with various critical partners now widely separated. The addition of bromine to the double bond in an unsaturated triacylglycerol decreases its partition number in a manner dependent on its fatty acid composition. Both partners of the critical pair stearoyloleoyllinoleoylglycerol and trioleoylglycerol convert to the hexa-bromides, but they can still be separated after bromination.

The mercuric acetate adducts of unsaturated triacylglycerols are resolved using methanol-acetic acid (83:17 v/v) as mobile phase on alkane- or tetralin-impregnated paper.[127] Results show a clearcut separation of triacylglycerols containing one, two, three, four, five, and six double bonds with some subfractionation of molecules such as palmitoyl and stearoyldi-oleoylglycerols by partition number. Long-chain saturated triacylglycerol molecules presumably remain at the origin with such a polar mobile phase.

Diacylglycerol acetates with partition numbers between 26 and 38 can be resolved[118] by developing silicone-impregnated paper with methanol-chloroform-water (71:24:5) or acetic acid-water (90:10). Similar separations have been obtained[128] using acetone-acetonitrile-water (65:35:5) with undecane-impregnated thin-layers.

3. Detection Methods

The location of the acylglycerol spots on the developed chromatoplate or paper chromatogram is complicated by the presence of the stationary phase, which is also lipid. The simplest method appears to be spraying with aqueous fluorescein[129] or Rhodamine B[130] solution. This method can be used only with kieselguhr as the solid support. The reason for this unique phenomenon with kieselguhr is not known. This reagent is nondestructive and extremely sensitive (less than 1 μg) and detects both saturated and unsaturated triacylglycerols. The unsaturated acylglycerols can be specifically detected[114,131] with iodine, iodine-starch, iodine-cyclodextrin, and phosphomolybdic acid reagents. These reagents destroy the sample and cannot be used for preparative work unless only part of the chromatogram is exposed to the reagent. In those instances where a relatively volatile stationary phase is used, the acylglycerols can be detected by conventional methods.

The ultimate identification of the acylglycerols resolved by reversed phase paper or thin-layer chromatography depends upon the determination of the fatty acid composition of the spots, and, if possible, a verification of the molecular weight distribution of the acylglycerols by means of high temperature gas chromatography. For this purpose[131] the triacylglycerols are extracted from the layer with diethyl ether. The excess stationary phase is removed by rechromatography of the lipid extract on a layer of ordinary silica gel. The methyl esters to be analyzed by gas-liquid chromatography are prepared either from the isolated fractions or from the total lipid extract. They usually require further purification by thin-layer chromatography. Gas chromatographic methods are suitable for quantitation of the various acylglycerols separated.

VI. GAS-LIQUID CHROMATOGRAPHY

Gas-liquid chromatography is capable of providing resolution of acylglycerols based on molecular weight, degree of unsaturation, and to a lesser extent, geometric configuration.[39,125] Enantiomers and reverse isomers of acylglycerols have not yet been separated by gas chromatographic methods. Both packed and open tubular columns can be used for acylglycerol resolution. In practice, the open tubular columns have given a higher level of performance than the packed columns.[133,134]

A. Packed Columns

Essentially all the separations of acylglycerols in the past have been obtained on packed columns.[39] The separation of natural triacylglycerols requires short columns filled with inert supports containing a thin film of liquid phase. Monoacyl and monoalkylglycerols, and, to a lesser extent, diacylglycerols can be separated on columns of conventional length and support coating, provided appropriate derivatives are employed.

1. Apparatus

Gas chromatography of acylglycerols is best performed on instruments that are capable of high temperature (up to 400°C) operation and which possess a linear temperature programmer along with a hydrogen flame ionization detector. The original methods of GLC analysis of neutral acylglycerols were based on the use of short columns of thin film packings prepared with thermostable liquid phases of low polarity.[132,135,136] More recently the column length has been significantly increased[137] and polar packings of moderate thermal stability have been employed for the separation of saturated and unsaturated species.[25,138] For routine analyses columns of 50 to 60 cm in length are best suited. Due to permanent leak-proof connections provided by metal columns, stainless steel, nickel, or glass-lined steel tubes have been preferred as column material to glass tubing in the past.

2. Liquid Phases

The liquid phases for GLC of acylglycerols must possess low vapor pressure over the entire range of operating temperatures (150 to 350°C). This requirement limits the choice of materials to the various silicone polymers, although certain other materials (e.g., Poly S-179) have also shown promise. The liquid phase is usually present as a 1 to 3% coating on an inert diatomaceous earth support, such as Gas Chrom Q, Supelcoport, or equivalent material in the 100 to 120 mesh range. The liquid phases employed for the separation of unsaturated acylglycerols by GLC are largely experimental at the present time, although significant results have been obtained. The best results have been achieved with SILAR 5 CP, SILAR 10C, and with the stabilized polyester phases.

a. Separation by Carbon Number

Gas-liquid chromatography is by far the best method for obtaining the resolution of acylglycerols based on carbon number. Both short and long chain triacylglycerols are completely resolved on several types of nonpolar liquid phases (Table 33) with linear temperature programming (100 to 400°C). Mixtures of triacylglycerols of a narrow range of molecular weights or carbon numbers can be effectively resolved using isothermal operating conditions (Table 34). Most of these separations are confined to the lower molecular weight species and in one instance involve a polyester liquid phase.[139] The latter phase also effects a resolution between acetyl and long-chain triacylglycerols of the same molecular weight. Good separations may also be obtained among normal-chain triacylglycerols and triacylglycerols containing normal and isovaleroyl chains (Table 35). On the longer columns the separations of natural triacylglycerols are further complicated by a partial resolution of saturated and unsaturated species. The latter separations arise from a difference in the polarity of these molecules of nearly identical molecular weight, and can be eliminated by hydrogenation of the sample.[14,41,140]

The presence of branched chain fatty acids in a triacylglycerol molecule also tends to decrease the elution temperature of the triacylglycerol, which emerges ahead of the corresponding normal chain triacylglycerol molecule.[137] Triacylglycerols containing fatty acids with terminal cyclopentane rings are eluted significantly later than the straight-chain analogs.[137]

The separations on the longer columns are in general characterized by low recoveries of the higher molecular weight components (as low as 50% of the total).[132] The longer columns also require higher relative elution temperatures for the triacylglycerols than do the shorter columns regardless of liquid phase and temperature program (Table 36).

The long chain trialkyl and mixed alkylacylglycerols are readily resolved under the conditions of triacylglycerol separation, with the alkylglycerols emerging ahead of the acylglycerols of corresponding chain length.

Triacylglycerols containing oxygen functions are retained longer than normal chain triacylglycerols. Thus, epoxytriacylglycerols have been separated[143] as their 1,3-dioxolane derivatives according to carbon number (Table 37). The conversion of the epoxy groups to the 1,3-dioxolane derivatives of cyclopentanone increases the molecular weight of the epoxytriacyglycerols so that they are eluted after carbon number 54, allowing nonepoxy and epoxytriacylglycerols to be clearly distinguished. Underivatized trivernoloylglycerol is eluted in the same peak as nonoxygenated triacylglycerols with carbon number 60. Triacylglycerols containing hydroxy and acetoxy fatty acids are also readily resolved by gas liquid chromatography;[142] even though they are eluted later than unsubstituted triacyclglycerols of corresponding carbon number, the gas chromatographic profile of a mixture of both types would be difficult to unscramble, and prior thin-layer separation is usually required.

The short columns allow an excellent resolution of the diacylglycerols in the form of a variety of derivatives, including acetates (Table 38).

The carbon number resolution of the diacylglycerol acetates has been extensively utilized in the characterization of the pairing of the fatty acid chains in the diacylglycerol moieties of diacylglycerophospholipids of natural origin. For this purpose the diacylglycerols are first resolved on the basis of unsaturation be argentation thin-layer chromatography (Table 39).

Long chain diacylglycerols as the acetate derivatives can be subjected to resolution on much longer columns and more complete separations obtained. Within each carbon number an effective resolution is obtained between the corresponding 1,2(2,3)- and the 1,3-diacyl-glycerol isomers, as well as between chain length isomers (Table 40).

Long-chain diacylglycerol derivatives other than acetates may also be readily resolved according to carbon number by gas liquid chromatography on the nonpolar liquid phases. Of particular interest and importance are the trimethylsilylether derivatives of diacylglycerols (Table 41). These derivatives also allow the separation of positional isomers. The resolution of saturated and unsaturated species is based largely on molecular weight when nonpolar packings are employed. Equally effective separations are realized with the *tert*-butyldi-methylsilyl ethers of the diacylglycerols.[144] The latter derivatives offer special advantages for gas chromatography with mass spectrometry because of the prominent M-57 ions produced.[51]

The alkylacyl and alkenylacyl ether analogs of diacylglycerols are readily separated with respect to carbon number either as the acetates or the trimethysilyl ethers. An effective resolution is achieved also within each carbon number, when run as trimethylsilyl ethers, but the dialkyl and alkylacyl acetates of the same carbon number tend to overlap. The trimethylsilyl ethers of the dialkylglycerols overlap with diacylglycerols having two meth-ylene units less, but the alkylacylglycerols are well resolved from the diacylglycerols.[25] Free diacylglycerols are subject to dehydration, isomerization, or transesterification on gas chro-matographic columns.[132]

The short gas chromatographic columns are not very well suited for the separation of monoacylglycerols unless they are converted into derivatives of relatively high molecular weight. Thus, while the diacetates and trimethylsilyl ethers of monoacylglycerols disappear in the solvent front, the diisovaleroylmonoacylglycerols are well resolved (Table 42). On account of the low molecular weight, acetyl and trimethylsilyl derivatives of the monoa-cylglycerols are readily recovered from gas chromatographic columns of 180 cm in length, and these are commonly employed for their separation.[145]

Unsaturated monoacylglycerols are eluted ahead of saturated ones from the nonpolar columns as are the 2-monoacylglycerols when compared to the 1-monoacylglycerols of the same carbon number.[146]

Monoacylglycerol, monoalkylglycerol, and monoalkenylglycerol derivatives are readily resolved on the basis of carbon number when run as the trimethylsilyl ethers. The 1-alk-1-enylglycerols are eluted ahead of the alkylglycerol of corresponding carbon number (Chapter 14). The separation factor between the alkenyl and alkyl ethers of the same carbon number is 1.12.[145] Good carbon number separations of monoalkylglycerols have been obtained on both nonpolar and polar columns with the acetate, trimethylsilyl ether, trifluoroacetate, and isopropylidene derivatives.

It is also possible to separate the 1-S-alkylglycerol ether from the corresponding 1-O-alkylglycerol ether when they are chromatographed as the isopropylidine derivatives. Relative to the saturated 1-O-alkylglycerols, the isopropylidine derivatives of the saturated 1-S-alkylglycerols are eluted as much as four methylene units earlier.[147,148]

b. Separation by Unsaturation

On nonpolar liquid phases, the separation of saturated and unsaturated acylglycerols is limited to the minor molecular weight discrepancies or any differences in the overall shape of the molecules. Thus partial resolution of trioleoylglycerol and tristearoylglycerol has been achieved on an 183 cm column of JXR or Apiezon L, with the unsaturated molecules being eluted slightly ahead of the saturated molecules.[137] Complete resolution of saturated and unsaturated triacylglycerols on nonpolar columns requires the presence of a minimum of six double bonds, which add up to a molecular weight difference of about 12 or of nearly one methylene unit. The minor differences in the relative retention times of the saturated and unsaturated triacylglycerols are apparent following gas-liquid chromatography of the subfractions of uniform degree of unsaturation recovered from $AgNO_3$-impregnated chromatoplates (Table 43). True separations of saturated and unsaturated triacylglycerols require the use of polar liquid phases, which possess low thermal stability. Certain cyanopropylphenyl and cyanopropylsiloxanes, however, have shown sufficient stability to demonstrate this separation in principle for high molecular weight triacylglycerols.[138] Thus standard saturated and unsaturated triacylglycerols have been resolved according to number of double bonds using SILAR 10C. These columns, however, were not sufficiently effective to resolve natural unsaturated triacylglycerols without prior argentation chromatography (Table 44).

Useful separations of natural mixtures of saturated and unsaturated triacylglycerols have been obtained, however, for the short and medium-chain length species. Thus the more volatile triacylglycerol fractions of butterfat derived by molecular distillation have been effectively segregated by gas chromatography on the basis of both molecular weight and degree of unsaturation using 3% EGSS-X as the liquid phase.[139] Although all of the peaks obtained were not fully characterized, it was obvious that the elution order followed a well-defined sequence (Table 45).

Effective separations of both synthetic and natural mixtures of saturated and unsaturated diacylglycerols can be obtained using currently available polar liquid phases. The original separations of these molecules were obtained on ethyleneglycol succinate polyester liquid phases.[149,150] The polyesters of succinic acid and higher alcohols provide liquid phases of higher temperature stability but give a poorer resolution between saturated and monounsaturated diacylglycerol species of the same carbon number. SILAR 5CP polymer has provided improved column stability and has given good resolution of all saturated and unsaturated diacylglycerols.[151] These separations are best accomplished with the trimethylsilyl ethers (Table 46). Only the oleoyllinoleoyl and dilinoleoyl species remain unresolved, whereas the TMS ethers of stearoyllinoleoyl and dilinoleoyl-glycerols are partially resolved. Likewise, 1-stearoyl-2-linoleoyl-*sn*-glycerol is only partially resolved from 1-palmitoyl-2-arachidonoyl-*sn*-glycerol. Reverse isomers, such as the TMS ethers of 1-oleoyl-2-linoleoyl-*rac*-glycerol and 1-linoleoyl-2-oleoyl-*rac*-glycerol and enantiomeric pairs are also unresolved. Temperature programming can be used for work with simple mixtures, but low gradients have proved necessary in order to allow time for solute equilibration. The acetates may also be employed for the resolution of diacylglycerols according to degree of unsaturation (Table 47).

On account of the lower polarity of the ether bond present in the alkylacyl and alk-1-enyl-acylglycerols, these compounds are especially well suited for gas chromatography (Tables 48 and 49).

The gas chromatographic separations of monoacylglycerols obtained on the polar liquid phases differ little from those realized with the corresponding methyl esters. Excellent separations of monoacetylglycerol diacetates have been obtained on EGSS-X and on SILAR 5CP (Table 50), while the trimethylsilyl ethers have been effectively chromatographed on SILAR 5CP and SILAR 10C (Table 51). Monoalkylglycerols have been chromatographed on EGSS-X as the trimethylsilyl ethers or as isopropylidenes and on SILAR 5CP and polyester liquid phases as the acetates. SILAR 5CP has a higher thermal stability than the other polar liquid phases such as EGSS-X and is, therefore, preferable for operation at elevated temperatures, especially for gas chromatography with mass spectrometry.[150]

3. Detection Methods

The most useful systems for detecting acylglycerols in the effluents of GLC columns are flame ionization,[151] mass spectrometric monitoring of total ion current,[144,152] and continuous counting of radioactivity[153,154] when dealing with radioactive acylglycerols. Most modern gas chromatographs are equipped with flame ionization detectors; the mass spectrometric and radioactivity measurements, however, require specialized equipment that must be specifically adapted for work at high temperatures.

The amount of sample required for GLC analysis with flame ionization detector depends upon the composition of the acylglycerol mixture. Usually a sample of 1 to 50 μg of total acylglycerol is sufficient. The usual sample size seen in triacylglycerol seperation is 0.5 to 2.0 μg for a single component, run at an attentuation of 100 to 300 times full sensitivity. The present limit of sensitivity for such work is about 0.1 μg. For most analyses it is convenient to dissolve 1 mg of the acylglycerol mixture in 1 ml of solvent and to inject 0.5 to 1 μℓ of the solution. These concentrations will give up to 50% full scale deflection at an attentuation of 100 to 200 times full sensitivity. The hydrogen flame ionization detector has a very wide linear range, but due to various other limitations of the chromatographic system, the practical range of linear operation is much more restricted (10 to 1000 μg). When working at mass concentrations below 10 μg, the on-column losses are proportionally higher, and a linear relationship may not be obtained. At concentrations exceeding 100 μg per peak, there is a severe loss in resolution due to column overloading, and normal peak shapes are no longer observed. GLC of the mono- and diacylglycerols as the TMS ethers on the polar siloxane columns or on polyester columns also requires somewhat higher amounts of material due to greater losses on the column.[145,146]

The above sample sizes are also adequate for the detection of natural acylglycerol peaks in the effluents of gas chromatographic columns by monitoring the total ion current in a mass spectrometer, as well as for recognizing the characteristic fragmentation patterns.[146,152] For this purpose, the column effluent is admitted directly to the mass spectrometer, and scans are made of any peak recognized in it at one or more times. The coupling of the GLC instrument with the mass spectrometer requires an effective carrier gas separator that will work satisfactorily with higher molecular weight solutes at high temperatures, without causing them to condense and produce memory effects.[144,152] Among the operating parameters that require careful attention are the temperature of the separator (350°C), the temperature of the transfer lines (350°C), and the temperature of the ionization chamber (⩾ 300°C).[155-157]

Each of the three detection systems may be combined with isothermal or temperature programmed operation, but all types of column packings are not equally suitable. Thus, combined gas chromatography-mass spectrometry requires columns of low bleed even for low temperature operation, and unchecked column bleed will rapidly exhaust the combustion train of a radio gas chromatographic monitoring system.

Preparative gas chromatographic systems are capable of isolating neutral acylglycerols in sufficient quantities for the determination of the component fatty acids as an aid in further identification and characterization of the acylglycerols.[141,156] Such instruments are usually equipped with an effluent splitter, so that column effluent can be monitored with a hydrogen flame ionization detector, while the fractions are collected manually or by means of automated fraction collectors. Triacylglycerol peaks from C_{24} to C_{54} have been collected in 95 to 97% purity.

Other detection systems suitable for the monitoring of acylglycerols in gas chromatographic effluents are based on argon ionization detection[52] and thermal conductivity,[53] but due to easy contamination these detectors are not recommended for work with high molecular weight materials at elevated temperatures.

B. Open Tubular Columns

Successful and highly efficient open tubular column separations were achieved nearly 20 years ago, but their versatility and usefulness for practical routine gas chromatographic analysis have been demonstrated only recently. Although the additional resolving power was clearly desirable, the poor precision and accuracy of the open tubular system excluded its consideration for all quantitative work with acylglycerols in the past. The great potential for resolution of complex mixtures in a single chromatographic step by open tubular chromatographic columns remains as a future development.

1. Apparatus

The unsatisfactory state of gas-liquid chromatography with open tubular column has stemmed largely from the unreliable column technology, difficulty of connection and installation of the columns in commercial instruments, and erratic sampling. In the last few years, considerable progress has been made in glass capillary column technology[133,157] including the development of the flexible quartz capillary.[158] The glass capillaries are etched with either dry HCl or a fluorinated ether, after which they are coated. The connection of the inlet end of the column to the injector block splitter unit and of the outlet end of the column to the detector must be made with a minimum of dead volume. The injector unit should be either heated to a sufficiently high temperature to obtain a representative introduction of sample, or the sample should be trapped on the cool column inlet and then displaced by high temperature programming.

Suitable splitters have been designed for both types of injection modes,[159] and several companies now offer special kits for adaptation of capillary columns to their instruments. However, the requirement for permanent glass to metal seals for high temperature gas chromatography has not been adequately solved and is a major obstacle to use of capillary columns even for qualitative work with acylglycerols.

It should be obvious that for precise control of capillary columns it is necessary to optimize the variable parameters of operation of conventional gas chromatographs (temperatures, gas flow rates and pressures). Likewise, adjustments may need to be made on the sampling devices, ovens, and detectors.

2. Liquid Phases

The general requirements of liquid phases for high temperature open tubular gas-liquid chromatography are the same as those for packed columns. OV-1, Dexil 300 GC, and Poly-S 179 with upper temperature limits of 300°, 350°, and 400°C, respectively, have been found suitable for work with triacylglycerols (Table 52). There is some variability observed in obtaining uniform films with the different nonpolar phases.[157] In some instances, improved coatings with nonpolar polymers have been obtained by prior application of a dilute solution of thermally stable strongly polar stationary liquid such as Carbowax 20M.[134]

An early separation of coconut oil on a 28 m glass capillary column coated with Dexil 300GC gave broad peaks and a number of shoulders[160] suggesting that the mixture was more complex than indicated by the simple chromatograms obtained for the oil on short packed columns.[136] However, the apparent low recovery of the higher molecular weight components from the capillary column distorted the characteristic coconut oil profile.

The separation of standard even-carbon-number triacylglycerols with 30 to 54 acyl carbons on a 20 m Poly-S 179 capillary column yielded sharp peaks with excellent baseline resolution.[134] The chromatograms obtained for the triacylglycerols of butterfat, palm oil, and a margarine under the same conditions indicated extensive further splitting of the carbon number peaks apparently resulting from the presence of odd carbon number, branched-chain and unsaturated acyl groups.[134] Again, the recovery of the higher molecular weight components was low and led to a distortion of the characteristic profile of butterfat triacylglycerols obtained on the short packed columns.[136] More recent work has shown better promise.[157,161]

A separation of *sn*-1,2-diacylglycerols as the trimethylsilyl ethers and of the *sn*-1(3)-

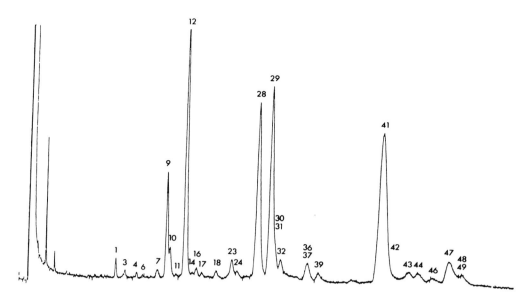

FIGURE 3. Resolution of t-BDMS ethers of *sn*-1,2-diacylglycerols of rat liver phosphatidylcholines by open tubular capillary GLC. Major peak identification: 1, 16:0 16:0; 3, 16:0 16:1ω7; 4, 14:0 18:2ω6; 9, 16:0 18:1ω9; 10, 16:0 18:1ω7; 16, 16:1ω9 18:2ω6; 23, 18:0 18:1ω9; 24, 18:0 18:1ω7; 28, 18:0 18:2ω6; 29, 16:0 20:4ω6; 36, 16:0 20:5ω3; 41, 18:0 20:4ω6; 43, 18:1ω9 20:4ω6; 44, 18:1ω7 20:4ω6; 46, 18:0 20:5ω3; 47, 16:0 22:6ω3; 49, 16:0 22:5ω3. Column: 10 m × 0.25 mm glass capillary coated with SP-2330. Split injection (75:1); Head pressure, 10 psl H$_2$; Temperature, 250°C, isothermal. (From Myher, J. J. and Kuksis, A., *Can. J. Biochem.*, 60, 638-650, 1982. With permission.)

monoacylglycerols as the methane boronates has been obtained on a 30 m OV-1 column.[54] In addition to an excellent resolution by carbon number, the oleoyllinoleoylglycerols were resolved from the stearoyllinoleoyl glycerols. The unsaturated monoacylglycerols were observed to move ahead of the saturated monoacylglycerols of the same carbon number.

On polar capillary columns, the saturated and unsaturated diacylglycerols are resolved with about the same efficiency as the saturated and unsaturated fatty acid methyl esters (Table 53). The SP-2330 liquid phase possesses just sufficient thermal stability (250°C) to permit the resolution and recovery of all natural diacylglycerols as the TMS, *t*-BDMS, and acetyl derivatives[162] (Figure 3).

3. Detection Methods

Only the flame ionization detector[134,159] and the mass spectrometer[155,157] have been thus far employed for the detection of acylglycerol solutes in the effluents of capillary columns. Because of the low rate of flow through the capillary column, a make-up gas is required to carry the components through the flame ionization detector. The hydrogen flow to the flame ionization detector can serve this purpose if appropriately diverted.[159] However, such arrangements may not be compatible with a convenient interchange of columns. To prevent adsorption and condensation of the solutes, the make-up gas should be preheated, as should be any transfer lines involved. A make-up gas is also required for the mass spectrometric detection of peaks in the effluents of capillary columns unless they are directly introduced into the source of the mass spectrometer. Thus, the flow to the two-stage jet of the LKB mass spectrometer has been made up with helium.[159] There have been reports[157] of a direct admission to the mass spectrometer of the acylglycerol peaks from a capillary gas chromatographic column.

VII. MULTIDIMENSIONAL CHROMATOGRAPHY

In general, complete analyses of complex mixtures of acylglycerols cannot be effected by any single chromatographic system. The segregation of the acylglycerol classes by means of preparatory chromatography either on adsorption columns or chromatoplates serves to simplify the task of analysis of homologous series of compounds by reversed phase or argentation thin-layer or by gas chromatography. Similar facilitation of analysis of homologous series of glycerol esters is accomplished by a preliminary segregation of the sn-1,2(2,3)- and sn-1,3-diacylglycerol isomers, and of the sn-2- and sn-1(3)-isomers of the monoacylglycerols. Likewise, a preliminary fractionation of alkylacyl, alkenylacyl, and acylglycerols by adsorption chromatography on columns or thin layers facilitates the separation and identification of individual components in any one of the chromatographic systems. For some homologous series of acylglycerols, it is necessary and profitable to effect a preliminary segregation of the components into long and short-chain subclasses. In most instances an effective subsequent segregation of the members of the various homologous series requires a complementary resolution in separate chromatographic systems based on molecular weight and degree of unsaturation. The combination of chromatographic techniques chosen for this purpose depends upon their inherent complementarity and ease of execution, and the choices may differ from one class of acylglycerols to another.[112]

A. Resolution of Triacylglycerols

The most useful combinations of chromatographic techniques for maximum resolution of triacylglycerols are argentation thin-layer and liquid-liquid partition chromatography[85,124] or argentation thin-layer and gas-liquid chromatography.[163-165] In both instances the argentation thin-layer chromatography serves as a means of initial segregation and preparation of greatly simplified triacylglycerol mixtures. Combinations of liquid-liquid chromatography and gas-liquid chromatography are also possible,[163] and this approach has been explored in practice.[162] Preparative gas-liquid chromatography of triacylglycerols up to 40 acyl carbons has also been employed for the initial isolation and segregation of simplified triacylglycerol mixtures,[141,156] to be followed by argentation thin-layer chromatography. Argentation thin-layer and liquid-liquid chromatography can be performed in the reverse order and on the same paper or thin-layer support by developing in one or two dimensions with two different solvent systems, one of which contains silver nitrate.[113,121,122,126] Recent progress made in liquid-liquid column chromatography[100,105] promises to make available large quantities of triacylglycerols of uniform molecular weight for further examination by the other chromatographic systems.

Executed to completion, these chromatographic methods will yield a complete separation of the molecular species of triacylglycerols on the basis of the total number of carbon atoms and double bonds per triacylglycerol molecule. Since the different chain lengths and different numbers of double bonds can be combined in a variety of ways to produce identical total carbon, double bond, and partition numbers, the resulting groups of molecular species of triacylglycerols may be still quite complex. The major combinations of fatty acids making up the component molecular species can be usually determined by a gas-liquid or mass spectrometric examination of the final fractions. These techniques, however, do not provide any information about the positional distribution of the fatty acids and they make no distinction between enantiomers. Further identification of the molecular species of triacylglycerols can be accomplished by a random deacylation of the triacylglycerols, a stereospecific separation of the diacylglycerol intermediates, and a complete determination of the structure of the sn-1,2- and sn-2,3-diacylglycerol enantiomers.[166] The above techniques are also suitable for the separation of triradylglycerols made up of alkylacyl and alkenylacyl moieties, but only a few such separations have been performed.[42,90]

B. Resolution of Diacylglycerols

A complete resolution of molecular species of diacylglycerols on the basis of carbon and double bond number is also readily obtained by the combination of argentation thin-layer chromatography with gas-liquid[167,168] and reversed phase liquid-liquid[128] chromatography techniques. For this purpose the diacylglycerols are usually converted into the acetates. Trimethylsilyl ethers, however, have been successfully employed in a combination of liquid-liquid and gas-liquid chromatography.[50] The more stable *tert*-butyldimethylsilyl ethers of diacylglycerols ought to be more suitable for this purpose as they are for argentation thin-layer and gas-liquid chromatography.[51] Furthermore, the *t*-BDMS ethers offer special advantages for the identification of the individual molecular species of diacylglycerols by mass spectrometry because of the prominent M-57 ions produced.[51]

Like triacylglycerols, the diacylglycerols cannot be resolved on the basis of positional distribution of the fatty acids, or a differentiation made between racemates and enantiomers. It has been demonstrated, however, that the reverse isomers of the fully saturated *rac*-1,2-diacylglycerols can be identified by mass spectrometry of their *t*-BDMS ethers.[51]

Many natural diacylglycerols can be resolved according to both carbon and double bond number by gas chromatography on polar liquid phases,[149,169,170] but positional isomers and enantiomers cannot be separated. The latter separation requires stereospecific analysis, using the preparation of suitable phosphorylated intermediates.[166]

The above combinations of chromatographic techniques are also suitable for the resolution by carbon and double bond number of the diradylglycerols made up of alkylacyl and alkenylacyl moieties,[171] and extensive work has been performed with both of these homologous series using combinations of argentation thin-layer and gas-liquid chromatography. Reference is made to Chapter 14.

C. Resolution of Monoacylglycerols

The resolution of monoacylglycerols is usually complete when performed by gas-liquid chromatography on polar liquid phases in the form of the isopropylidenes.[52] However, monoacylglycerols in the form of higher molecular weight or more polar derivatives present problems comparable to those encountered in work with the low molecular weight triacylglycerols. In such instances both argentation thin-layer and high temperature gas-liquid chromatography based on carbon number may be necessary. Most gas-liquid chromatography systems resolve the *sn*-1(3)- and the *sn*-2-isomers of monoacylglycerols, but argentation thin-layer chromatography does not unless boric acid is incorporated in the layer.[63]

A combination of capillary gas-liquid chromatography and mass spectrometry has shown some promise for the identification of the monoacylglycerols as the methaneboronates.[54] With this derivative, the *sn*-1(3)- and the *sn*-2-isomers yield distinctly different mass spectra.

Like the monoacylglycerols, the monoalkyl and the monoalkenyl-glycerols can be resolved by gas-liquid chromatography according to carbon and double bond numbers[145,150] and may not ordinarily require any other chromatographic routine, except for the initial isolation of the appropriate monoradylglycerol class.

ACKNOWLEDGMENT

The studies of the author and his collaborators referred to in this chapter were supported by funds from the Ontario Heart Foundation, Toronto, Ontario, the Hospital for Sick Children Foundation, Toronto, Ontario, the Medical Research Council of Canada, Ottawa, Ontario, and the Special Dairy Industry Board, Chicago, Illinois.

GENERAL REFERENCES

1. **Hilditch, T. P. and Williams, R. N.,** *The Chemical Constitution of Natural Fats,* 4th ed., Chapman and Hall, London, 1964.
2. **Snyder, F., Ed.,** *Ether Lipids, Chemistry and Biology,* Academic Press, New York, 1972.
3. **Mangold, H. K. and Paltauf, F., Eds.,** *Ether Lipids, Biochemical and Biomedical Aspects,* Academic Press, New York, 1983.
4. **Hawthorne, J. N. and Ansell, G. B., Eds.,** *Phospholipids,* Elsevier, New York, 1982.
5. **Hirschmann, H.,** *J. Biol. Chem.,* 235, 2762, 1960.
6. **Karnovsky, M. L., Hauser, G., and Elwyn, D.,** *J. Biol. Chem.,* 226, 881, 1957.
7. **Slakey, P. M. and Lands, W. E. M.,** *Lipids,* 3, 30, 1968.
8. **Brockerhoff, H. and Jensen, R. G.,** *Lipolytic Enzymes,* Academic Press, New York, 1974.
9. **Morley, N. and Kuksis, A.,** *J. Biol. Chem.,* 247, 6393, 1972.
10. **Paltauf, F., Esfandi, F., and Holasek, A.,** *FEBS Lett.,* 40, 119, 1974.
11. **Buchnea, D.,** in *Handbook of Lipid Research,* Vol. 1, Kuksis, A., Ed., Plenum Press, New York, 1978, 233.
12. **Privett, O. S. and Blank, M. L.,** *J. Lipid Res.,* 2, 37, 1961.
13. **Jensen, R. G., Quinn, J. G., Carpenter, D. L., and Sampugna, J.,** *J. Dairy Sci.,* 50, 119, 1967.
14. **Litchfield, C.,** *Analysis of Triglycerides,* Academic Press, New York, 1972.
15. **Holub, B. J. and Kuksis, A.,** *Adv. Lipid Res.,* 16, 1, 1978.
16. **Trimms, R. E.,** *Chem. Phys. Lipids,* 21, 113, 1978.
17. **Christie, W. W.,** *Lipid Analysis,* 2nd ed., Pergamon Press, Oxford, 1982.
18. **Folch, J., Lees, M., and Stanley, G. H. S.,** *J. Biol. Chem.,* 226, 497, 1957.
19. **Bligh, E. G. and Dyer, W. J.,** *Can. J. Biochem. Physiol.,* 37, 911, 1959.
20. **Wells, M. A. and Dittmer, J. C.,** *Biochemistry,* 2, 1259, 1963.
21. **Siakotos, A. N. and Rouser, G.,** *J. Am. Oil Chem. Soc.,* 42, 1259, 1965.
22. **Wuthier, R. E.,** *J. Lipid Res.,* 7, 558, 1966.
23. **Nelson, G. J.,** in *Blood Lipids and Lipoproteins,* Wiley-Interscience, New York, 1972, 3.
24. **Carroll, K. K.,** in *Lipid Chromatographic Analysis,* 2nd ed., Marinetti, G. V., Ed., Marcel Dekker, New York, 1976, 173.
25. **Myher, J. J.,** in *Handbook of Lipid Research,* Vol. 1, Kuksis, A., Ed., Plenum Press, New York, 1978, 123.
26. **Timms, A. R., Kelly, L. A., Spirito, J. A., and Engstrom, R. G.,** *J. Lipid Res.,* 9, 675, 1968.
27. **Foslien, E. and Musil, F.,** *J. Lipid Res.,* 11, 605, 1970.
28. **Anonymous,** *Manual of Laboratory Operations, Lipid Research Clinics Program,* Vol. 1, *Lipid Analysis,* National Heart and Lung Institute, National Institutes of Health, Bethesda, Md., 1974, 1.
29. **Thomas, A. E., III, Scharoun, J. E., and Ralston, H.,** *J. Am. Oil Chem. Soc.,* 42, 789, 1965.
30. **Serdarevich, B.,** *J. Am. Oil Chem. Soc.,* 44, 381, 1967.
31. **Serdarevich, B. and Carroll, K. K.,** *J. Lipid Res.,* 7, 277, 1966.
32. **Blank, M. L., and Privett, O. S.,** *J. Dairy Sci.,* 47, 481, 1964.
33. **Evans, D. C., McConnell, D. G., Hofmann, R. L., and Peters, H.,** *J. Am. Oil Chem. Soc.,* 44, 281, 1967.
34. **Gruger, E. H., Jr., Malins, D. C., and Gauglitz, E. J., Jr.,** *J. Am. Oil Chem. Soc.,* 37, 214, 1960.
35. **Litchfield, C., Ackman, R. G., Sipos, J. C., and Eaton, C. A.,** *Lipids,* 6, 674, 1971.
36. **Karlsson, K. A., Norrby, A., and Samuelsson, B.,** *Biochim. Biophys. Acta,* 144, 162, 1967.
37. **Snyder, F.,** *J. Chromatogr.,* 82, 7, 1973.
38. **Sheppard, A. J., Iverson, J. L., and Weihrauch, J. L.,** in *Handbook of Lipid Research,* Vol. 1, Kuksis, A., Ed., Plenum Press, New York, 1978, 341.
39. **Kuksis, A.,** in *Lipid Chromatographic Analysis,* 2nd ed., Volume 1, Marinetti, G. V., Ed., Marcel Dekker, 1976, 215.
40. **Christie, W. W., Noble, R. C., and Moore, J. H.,** *Analyst,* 95, 940, 1970.
41. **Litchfield, C., Harlow, R. D., and Reiser, R.,** *J. Am. Oil Chem. Soc.,* 42, 849, 1965.
42. **Wood, R. and Snyder, F.,** *Arch. Biochem. Biophys.,* 131, 478, 1969.
43. **Myher, J. J. and Kuksis, A.,** unpublished results, 1978.
44. **Emken, E. A.,** in *Handbook of Lipid Research,* Vol. 1, Kuksis, A., Ed., Plenum Press, New York, 1978, 77.
45. **Hirayama, O.,** *Nippon Nogei Kagaku Kaishi,* 35, 437, 1961.
46. **Kaufmann, H. P. and Khoe, T. H.,** *Fette Seifen Anstrichm.,* 64, 81, 1962.
47. **Privett, O. S. and Nutter, L. J.,** *Lipids,* 2, 149, 1967.
48. **Wood, R., Raju, P. K., and Reiser, R.,** *J. Am. Oil Chem. Soc.,* 42, 161, 1965.
49. **Curstedt, T. and Sjovall, J.,** *Biochim. Biophys. Acta,* 360, 24, 1974.

50. **Curstedt, T.,** *Biohim. Biophys. Acta,* 489, 79, 1977.
51. **Myher, J. J., Kuksis, A., Marai, L., and Yeung, S. K. F.,** *Anal. Chem.,* 50, 557, 1978.
52. **Hanahan, D. J., Ekholm, J., and Jackson, C. M.,** *Biochemistry,* 4, 630, 1963.
53. **Wood, R. and Snyder, F.,** *Lipids,* 1, 62, 1966.
54. **Gaskell, S. J. and Brooks, C. J. W.,** *J. Chromatogr.,* 142, 469, 1977.
55. **Saunders, D. L.,** *J. Chromatogr. Sci.,* 15, 372, 1977.
56. **Snyder, L. R. and Saunders, D. L.,** *J. Chromatogr. Sci.,* 7, 195, 1969.
57. **Shehata, A. A. Y., de Man, J. M., and Alexander, J. C.,** *Can. Inst. Food Technol. J.,* 5, 13, 1972.
58. **Kaufmann, H. P. and Wolf, W.,** *Fette Seifen,* 50, 519, 1943.
59. **Shehata, A. A. Y., de Man, J. M., and Alexander, J. C.,** *Can. Inst. Food Technol. J.,* 4, 61, 1971.
60. **Sahasrabudhe, M. R.,** *J. Am. Oil Chem. Soc.,* 38, 88, 1961.
61. **Marinetti, G. V.,** *J. Lipid Res.,* 7, 786, 1966.
62. **Padley, F. B.,** *Chromatogr. Rev.,* 8, 208, 1966.
63. **Morris, L. J.,** *J. Lipid Res.,* 7, 717, 1966.
64. **de Vries, B.,** *J. Am. Oil Chem. Soc.,* 41, 403, 1964.
65. **Dolev, A. and Olcott, H. S.,** *J. Am. Oil Chem. Soc.,* 42, 624, 1965.
66. **Dolev, A. and Olcott, H. S.,** *J. Am. Oil Chem. Soc.,* 42, 1046, 1965.
67. **Carroll, K. K.,** *J. Am. Oil Chem. Soc.,* 2, 135, 1961.
68. **Taylor, M. W. and Hawke, J. C.,** *N. Z. J. Dairy Sci. Technol.,* 10, 40, 1975.
69. **Stolyhwo, A. and Privett, O. S.,** *J. Chromatogr. Sci.,* 11, 20, 1973.
70. **Hettinger, J. and Majors, R. E.,** *Varian Instrument Applications,* 10, 6, 1976.
71. **Mikolajczak, K. L., Smith, C. R., Jr., and Wolff, I. A.,** *Lipids,* 3, 215, 1968.
72. **McNair, H. M. and Chandler, C. D.,** *J. Chromatogr. Sci.,* 14, 477 (1976).
73. **Mangold, H. K.,** in *Thin-layer Chromatography,* 2nd ed., Stahl, E., Ed., Springer-Verlag, New York, 1969, 363.
74. **Skipski, V. P.,** *Methods Enzymol.,* 14, 530, 1969.
75. **Ackman, R. G.,** *Methods Enzymol.,* 72, 205, 1981.
76. **Christie, W. W. and Moore, J. H.,** *Biochim. Biophys. Acta,* 176, 445, 1969.
77. **Yurkowski, M. and Brockerhoff, H.,** *Biochim. Biophys. Acta,* 125, 55, 1966.
78. **Privett, O. S., Nutter, L. J. and Gross, R. A.,** in *Dairy Lipids and Lipid Metabolism,* Kritchevsky, D. and Brink, M. F., Eds., AVI Publishing, Westport, Conn., 1967, 99.
79. **Kleiman, R., Spencer, G. F., Earle, F. R., Nieschlag, H. J., and Barclay, A. S.,** *Lipids,* 7, 660, 1972.
80. **Jurriens, G.,** *Anal. Charact. Oils, Fats, Fat Prod.,* 2, 237, 1968.
81. **Youngs, C. G. and Baker, C. D.,** as cited in Reference 14.
82. **den Boer, F. C.,** *Z. Physiol. Chem.,* 205, 308, 1964.
83. **Bottino, N. R.,** *J. Lipid Res.,* 12, 24, 1971.
84. **Gunstone, F. D. and Padley, F. B.,** *J. Am. Oil Chem. Soc.,* 42, 957, 1965.
85. **Wessels, H. and Rajagopal, N. S.,** *Fette Seifen Anstrichm.,* 71, 543, 1969.
86. **Gunstone, F. D. and Qureshi, M. I.,** *J. Am. Oil Chem. Soc.,* 42, 961, 1965.
87. **de Vries, B. and Jurriens, G.,** *Fette Seifen Anstrichm.,* 65, 725, 1963.
88. **Kircher, H. W.,** *J. Am. Oil Chem. Soc.,* 42, 899, 1965.
89. **Bandi, Z. L. and Mangold, H. K.,** *Sep. Sci.,* 4, 83, 1969.
90. **Snyder, F., Cress, E. A., and Stephens, N.,** *Lipids,* 1, 381, 1966.
91. **Van der Wal, R. J.,** *J. Am. Oil Chem. Soc.,* 42, 754, 1965.
92. **Privett, O. S., Dougherty, K. A., and Castell, J. D.,** *Am. J. Clin. Nutr.,* 24, 1265, 1971.
93. **Blank, M. L., Verdino, B., and Privett, O. S.,** *J. Am. Oil Chem. Soc.,* 42, 87, 1965.
94. **Kuksis, A.,** *Fette Seifen Anstrichm.,* 73, 332, 1971.
95. **Majors, R. E.,** *J. Chromatogr. Sci.,* 15, 334, 1977.
96. **Hirsch, J.,** *J. Lipid Res.,* 4, 1, 1963.
97. **Steiner, E. H. and Bonar, A. R.,** *Rev. Int. Choc.,* 20, 248, 1965.
98. **Aitzetmuller, K.,** *J. Chromatogr.,* 139, 61, 1977.
99. **Lawrence, J. G.,** *J. Chromatogr.,* 84, 299, 1973.
100. **Bombaugh, K. J., Dark, W. A., and Lavangie, R. F.,** *J. Chromatogr. Sci.,* 7, 42, 1969.
101. **Privett, O. S. and Erdahl, W. L.,** in *Analysis of Lipids and Lipoproteins,* Perkins, E. G., Ed., American Oil Chemists' Society, Champaign, Ill., 1975, 123.
102. **Nickel, E. C. and Privett, O. S.,** *Sep. Sci.,* 2, 307, 1967.
103. **Horvath, C. and Melander, W.,** *J. Chromatogr. Sci.,* 15, 393, 1977.
104. **Snyder, L. R. and Kirkland, J. J.,** *Introduction to Modern Liquid Chromatography,* 2nd ed., John Wiley & Sons, New York, 1979, 152.
105. **Lindqvist, B., Sjorgren, I., and Nordin, B.,** *J. Lipid Res.,* 15, 65, 1974.
106. **Ellingboe, J., Nystrom, E., and Sjövall, J.,** *Methods Enzymol.,* 14, 317, 1969.
107. **Fallon, W. E. and Schimizu, Y.,** *Lipids,* 12, 765, 1977.

108. **Kiuchi, K., Onta, T., and Ebine, H.,** *J. Chromatogr. Sci.,* 13, 461, 1975.
109. **Youngs, C. G.,** *J. Am. Oil Chem. Soc.,* 38, 62, 1961.
110. **Plattner, R. D., Wade, K., and Kleiman, R.,** *J. Am. Oil Chem. Soc.,* 55, 381, 1970.
111. **Arpino, P., Baldwin, M. A., and McLafferty, F. W.,** *Biomed. Mass Spectr.,* 1, 1, 1974.
112. **Kuksis, A., Marai, L., and Myher, J. J.,** *J. Chromatogr.,* 273, 43, 1983.
113. **Vereshchagin, A. G.,** *J. Chromatogr.,* 17, 382, 1965.
114. **Mangold, H. K., Lamp, B. G., and Schlenk, H.,** *J. Am. Chem. Soc.,* 77, 6070, 1955.
115. **Scholfield, C. R.,** in *Fatty Acids,* 2nd ed., Markley, K. S., Ed., Wiley-Interscience, New York, 1964, 2283.
116. **Kaufmann, H. P. and Das, B.,** *Fette Seifen Anstrichm.,* 64, 214, 1962.
117. **Kaufmann, H. P. and Wessels, H.,** *Fette Seifen Anstrichm.,* 66, 13, 1964.
118. **Mangold, H. K.,** *Fette Seifen Anstrichm.,* 61, 877, 1959.
119. **Kirkland, J. J. and De Stefano, J. J.,** *J. Chromatogr. Sci.,* 8, 309, 1970.
120. **Kaufmann, H. P., Wessels, H., and Viswanathan, C. V.,** *Fette Seifen Anstrichm.,* 64, 509, 1962.
121. **Ord, W. O. and Bamford, P. C.,** *Chem. Ind. (London),* 277, 1967.
122. **Novitskaya, G. V. and Maltseva, V. I.,** *Biokhimiya,* 31, 953, 1966.
123. **Anonymous,** *Gas-Chrom Newsletter,* 6(4), 2, 1965.
124. **Kaufmann, H. P. and Wessels, H.,** *Fette Seifen Anstrichm.,* 66, 81, 1964.
125. **Morris, L. J. and Nichols, B. W.,** in *Progress in Thin-Layer Chromatography and Related Methods,* Niederweiser, A. and Pataki, G., Eds., Ann Arbor-Humphrey Science, Ann Arbor, Mich., 1970, 75.
126. **Vereshchagin, A. G. and Novitskaya, G. V.,** *J. Am. Oil Chem. Soc.,* 42, 970, 1962.
127. **Noda, M. and Hirayama, O.,** *Yukugaku,* 10, 24, 1961.
128. **Åkesson, B.,** *Biochim. Biophys. Acta,* 218, 57, 1970.
129. **Whittle, K. J., Dauphy, P. J., and Pennock, J. F.,** *Chem. Ind. (London),* 1303, 1966.
130. **Chakrabarty, M. M., Bandyopandyay, C., Bhattacharyua, D., and Gayen, A. K.,** *J. Chromatogr.,* 36, 84, 1968.
131. **Litchfield, C.,** *Lipids,* 3, 170, 1968.
132. **Huebner, V. R.,** *J. Am. Oil Chem. Soc.,* 38, 628, 1961.
133. **Moseigny, A., Vigneron, P. V., Laracq, M., and Zwoboda, I.,** *Rev. Fr. Corps Gras,* 26, 107, 1979.
134. **Schomberg, G., Dielmann, Husmann, H., and Waeke, F.,** *J. Chromatogr.,* 122, 55, 1976.
135. **Huebner, C. R.,** *J. Am. Oil Chem. Soc.,* 38, 628, 1961.
136. **Kuksis, A. and McCarthy, M. J.,** *Can. J. Biochem. Physiol.,* 40, 679, 1962.
137. **Litchfield, C., Harlow, R. D., and Reiser, R.,** *Lipids,* 2, 363, 1967.
138. **Takagi, T. and Itabashi, Y.,** *Lipids,* 12, 1062, 1977.
139. **Kuksis, A., Marai, L., and Myher, J. J.,** *J. Am. Oil Chem. Soc.,* 50, 193, 1973.
140. **Farquhar, J. W., Insull, W., Rosen, P., Stoffel, W., and Ahrens, E. H., Jr.,** *Nutr. Abstr. Rev.,* (Suppl.) Part II, 8, 17, 1959.
141. **Kuksis, A. and Ludwig, J.,** *Lipids,* 1, 202, 1966.
142. **Powell, R. G., Kleiman, R., and Smith, C. R., Jr.,** *Lipids,* 4, 450, 1969.
143. **Fioriti, J. A., Kanuk, M. J., and Sims, R. J.,** *J. Chromatogr. Sci.,* 7, 448, 1969.
144. **Kuksis, A., Breckenridge, C. W., Myher, J. J., and Kakis, G.,** *Can. J. Biochem.,* 56, 630, 1978.
145. **Myher, J. J. and Kuksis, A.,** *Lipids,* 9, 382, 1974.
146. **Myher, J. J. Marai, L., and Kuksis, A.,** *J. Lipid Res.,* 15, 586, 1974.
147. **Ferrel, W. J.,** *Lipids,* 8, 234, 1973.
148. **Wood, R., Piantodosi, C., and Snyder, F.,** *J. Lipid Res.,* 10, 370, 1969.
149. **Kuksis, A.,** *J. Chromatogr. Sci.,* 10, 53, 1972.
150. **Myher, J. J., Marai, L., and Kuksis, A.,** *Anal. Biochem.,* 70, 302, 1976.
151. **Kuksis, A.,** in *Analysis of Lipids and Lipoproteins,* Perkins, E. G., Ed., American Oil Chemists' Society, Champaign, Ill., 1975, 36.
152. **Satouchi, K. and Saito, K.,** *Biomed. Mass Spectr.,* 3, 122, 1976.
153. **Breckenridge, W. C. and Kuksis, A.,** *Lipids,* 5, 342, 1970.
154. **Breckenridge, W. C. and Kuksis, A.,** *Can. J. Biochem.,* 53, 1184, 1975.
155. **Murata, T.,** *Anal. Chem.,* 49, 2209, 1977.
156. **Bugaut, M. and Bezard, J.,** *J. Chromatogr. Sci.,* 8, 380, 1970.
157. **Wakeham, S. G. and Frew, N. M.,** *Lipids,* 17, 831, 1982.
158. **Dandeneau, R. D. and Zerenner, E. H.,** *J. High Resol. Chromatogr. Chromatogr. Commun.,* 2, 351, 1979.
159. **Grob, K.,** *J. High Resol. Chromatogr. Chromatogr. Commun.,* 3, 493, 1980.
160. **Novotny, M., Segura, R., and Zlatkis, A.,** *Anal. Chem.,* 44, 9, 1972.
161. **Grob, K., Jr.,** *J. Chromatogr.,* 205, 289, 1981.
162. **Myher, J. J. and Kuksis, A.,** *Can. J. Biochem.,* 60, 638, 1982.
163. **McCarthy, M. J. and Kuksis, A.,** *J. Am. Oil Chem. Soc.,* 41, 527, 1964.

164. **Culp, T. W., Harlow, R. D., Litchfield, C., and Reiser, R.,** *J. Am. Oil Chem. Soc.,* 42, 974, 1965.
165. **Kuksis, A. and Breckenridge, W. C.,** *J. Am. Oil Chem. Soc.,* 42, 978, 1965.
166. **Myher, J. J. and Kuksis, A.,** *Can. J. Biochem.,* 57, 117, 1979.
167. **Kuksis, A., Breckenridge, W. C., Marai, L., and Stachnyk, O.,** *J. Am. Oil Chem. Soc.,* 45, 537, 1968.
168. **Kuksis, A. and Marai, L.,** *Lipids,* 2, 217, 1967.
169. **Kuksis, A.,** *Can. J. Biochem.,* 49, 1245, 1971.
170. **Myher, J. J. and Kuksis, A.,** *J. Chromatogr. Sci.,* 13, 138, 1974.
171. **Myher, J. J. and Kuksis, A.,** *J. Am. Oil Chemists' Soc.,* 59, 308A (Abstr. No. 294), 1982.
172. **Stolyhwo, A., Colin, H., and Guiochon, G.,** *J. Chromatogr.,* 265, 1, 1983.

TABLES OF CHROMATOGRAPHIC DATA

The tabulated material of this chapter has been collected in one section for ease of location and reference. The data are presented in an abbreviated and concise form, which includes some unusual conventions. Furthermore, much of the material in the tables has been derived by calculation from the chromatograms and drawings published by the original authors, and it requires qualification and explanation. For this reason a guide to the use of the tables and to the order of compilation of the data is presented below.

Arrangement of Tables

The tables have been arranged in the same order as the text, starting with adsorption chromatography in columns and ending with gas-liquid chromatography in open tubular columns. Within each chromatographic method the order of presentation is triacyl, diacyl, and monoacylglycerols. For ease of reference the general term "radyl" is employed to include acyl, alkyl, and alk-1-enylglycerols together (Chapter 1, Appendix). The titles of the tables indicate the type of chromatography employed by the following abbreviations: AC, adsorption chromatography in columns (Tables 1 to 7); $AgNO_3$-AC, adsorption chromatography in columns impregnated with silver nitrate (Table 8); TLC, adsorption thin-layer chromatography (Tables 9 to 18); $AgNO_3$-TLC, adsorption thin-layer chromatography on layers impregnated with silver nitrate (Tables 19 to 23); LLC, liquid-liquid chromatography on columns (Tables 24 to 30); LLC-PAPER, reversed phase liquid-liquid chromatography on paper (Table 31); LLC-TLC, reversed phase liquid-liquid chromatography on thin-layers (Table 32); GLC, gas-liquid chromatography (Tables 33 to 51) GLC, gas-liquid chromatography in open tubular columns (Tables 52 and 53) Under LLC, both direct and reversed phase separations have been included in the same tables.

Guide to Use of Tables

Inasmuch as possible, the tables have been composed to include the information necessary to reproduce the data. Thus the nature of the chromatographic adsorbent or support and the solvent system has been given along with the peculiarities of the technique, method of detection of the separated components, and a full reference to the source of the data. As a rule, at least one reference has been made to the original description of each technique, except where no chromatographic data were given either in the form of chromatograms or calculated values. In most instances only chromatograms were given, and the retention data included in the tables were calculated by the reviewer. Under these circumstances, allowance must be made for inaccuracies arising from measuring small distances on the reproduced chromatograms. In a few cases, elaborate descriptions of the experimental conditions were found not be be accompanied by any chromatographic data. In these cases, the reviewer has inserted his own data collected under conditions closely resembling those reported by the original authors (Tables 1 and 20).

Likewise, it was frequently necessary to calculate the relative retention times of the gas chromatographic peaks, as the original authors had provided only pictures of the chromatograms. Such calculations presented no special problems for isothermal runs, except for any inaccuracies in the measurement of the small distances in the published prints. However, the calculation of retention times from temperature programmed runs presented difficulties because of the uncertainty of the temperature program. Some of the runs had been programmed manually, and some of the temperature programs had been preceded and/or terminated by periods of isothermal operation of unknown duration. There was no alternative but to calculate the values while ignoring the deviation of the program from the linearity. Surprisingly, in many instances little difference was seen in the values calculated for truly linear and only approximately linear programs. The alterations in the temperature program are noted under "Technique". For the same reasons, it was also impractical to arrive at values for relative elution temperatures for the temperature programmed runs (Table 41). The calculation of these approximate retention time values appeared to provide the only manageable means for assembling the existing experimental data. In view of these difficulties, however, serious thought should be given to demanding that authors submit complete data on the chromatographic behavior of all known components analyzed for deposition and eventual publication in tabular form by the appropriate journals. Such a program appears to be presently carried out on a voluntary basis by the *Journal of Chromatography*.

The various compounds listed in the tables have been indicated using the *sn*-numbering system[3,4] and wherever appropriate listing the specific molecular species involved. This was not possible where only group separations were obtained, and abbreviations such as LCT, MCT, and SCT were employed to designate long, medium, and short-chain triacylglycerols, respectively (Table 2). In other instances, the number of specific functional groups in each acylglycerol molecule class has been indicated by none, one, two, three, or four as required (Tables 3 to 6, 8, 14 to 17). In both text and tables, use has been made of simple letter designations for specific fatty acids: A, azelaic; Ac, acetic; O, oleic; P, palmitic, S, stearic; L, linoleic; Ln, linolenic; Pe, petroselinic; Ph, phytanic; Vi, isovaleric; and Fa, any long chain fatty acid (Tables 7, 12, 13 and 37). In a few instances (Table 18), the abbreviations S and U have been used to designate any saturated and any unsaturated fatty acid, respectively. In most instances (Tables 10, 11, 20, 23, 31, 32, 34, 40, 41, 46, 47, 50, and 51), the component fatty acids have been identified by the number of carbon atoms followed by a colon and the number of double bonds.[140] The carbon numbers of the acylglycerols refer to the total number of acyl carbons in a given acylglycerol molecule including any alkyl or alk-1-enyl chains (Tables 33, 35, 36, 38, and 42). The partition number is defined as the carbon number minus two times the number of double bonds. In still other instances, the molecular species of acylglycerols or small groups thereof have been indicated by the total number of acyl carbons followed by a colon and the total number of double bonds per molecule (Tables 39, 43, 44, 48, and 49). In some tables, the acylglycerol species have been indicated by a series of numbers from 0 to 6, which refers to the number of double bonds per fatty acid residue, such as 000 for a fully saturated triacylglycerol and 222 for a triacylglycerol containing three fatty acids containing 2 double bonds each. A combination such as 001 indicates that a monounsaturated fatty acid is in a primary position, and 010 that a monounsaturated fatty acid is in the secondary position. No distinction is made between the two primary positions (Table 19).

In the diacylglycerol series, the number (0-6) on the left indicates the degree of unsaturation of the fatty acid in the primary position and that on the right (0-6) indicates the unsaturation of the acid in the secondary position of the glycerol molecule (Table 21). Finally, in a few instances the acylglycerols have been grouped together under the general terms saturates, monoenes, dienes, trienes, tetraenes, and polyenes to indicate the degree of unsaturation of the entire acylglycerol molecule rather than that of the component fatty acids (Table 22).

In describing the experimental conditions in the abbreviated form a special difficulty was encountered in identifying the eluting conditions for the various fractions resolved by column chromatography. In most instances, however, it was possible to describe the strength of the eluting solvents in terms of the percent content of the more polar solvent in the less polar one, assuming the order of the polarity observed for the common eluotropic series. Under such an abbreviated description, however, the volume of the solvent remains undefined and must be determined by consulting the original reference. In those instances where rigidly controlled solvent compositions and flow rates were used it was possible to define the elution order of the compounds in terms of relative retention times (Table 24 to 27), as was done for gas-liquid chromatography.

Table 1
ACYLGLYCEROLS (AC)

	A1	A2	A3	A4	A5	A6	A7	A8	A9	A9
Adsorbent	S1	S2	S1	S3	S4	S1	S5	S6	S1	S1
Solvent	T1	T2	T3	T4	T5	T5	T6	T7	T8	T8
Technique	D1	D1	D1	D1	D2	D2	D2	D2	D2	D2
Detection	1	2	3	4	5	5	6	7	8	8

Acylglycerol Class	% Polar Solvent in Nonpolar Solvent									
Triacylglycerols	5	15	15	15	—	—	20	50?	5?	5
1,3-Diacylglycerols	15	50	30	50	—	20	20	10?	15	18
1,2-Diacylglycerols	15	50	30	50	—	25	20	10?	15	18
2-Monoacylglycerols	100	10	100	3	2	—	100	50?	100	100
1-Monoacylglycerols	100	10	100	3	2	—	100	50?	100	100

Adsorbent
A1 = Silicic acid (fine mesh)
A2 = Florisil
A3 = Florisil (coarse mesh)
A4 = Deactivated (7% water) Florisil
A5 = Florisil with 10% boric acid
A6 = Acid treated Florisil
A7 = Silica gel G
A8 = Corasil treated with concentrated ammonia
A9 = Unisil

Solvent
(v/v)
S1 = Diethyl ether in hexane
S2 = Diethyl ether in petroleum hydrocarbon then 10% methanol in diethyl ether
S3 = Diethyl ether in petroleum hydrocarbon then 3% methanol in diethyl ether
S4 = Methanol in diethyl ether
S5 = Diethyl ether-ethanol, 19:1, in benzene
S6 = 25% Diethyl ether in petroleum hydrocarbon then chloroform in 25% diethyl ether

Technique
T1 = 1.5 × 20 cm column containing 30 g silicic acid
T2 = 2.0 x 17 cm column containing 30 g Florisil
T3 = 1.5 × 20 cm column containing 30 g Florisil
T4 = 18 g Florisil column
T5 = 1.5 × 43 cm column
T6 = 1.8 × 125 cm column
T7 = 0.3 × 100 cm column
T8 = 1.2 × 41 cm column

Detection
D1 = Gravimetric
D2 = Thin-layer chromatography

REFERENCES

1. **Hirsch, J. and Ahrens, E. H., Jr.,** *J. Biol. Chem.,* 233, 311, 1958.
2. **Carroll, K. K.,** *J. Lipid Res.,* 2, 135, 1961.
3. **Carroll, K. K.,** *J. Am. Oil Chem. Soc.,* 40, 413, 1963.
4. **Boudreau, A. and de Man, J. M.,** *Biochim. Biophys. Acta,* 98, 47, 1965.
5. **Serdarevich, B.,** *J. Am. Oil Chem. Soc.,* 44, 381, 1967.
6. **Hardy, R., Smith, J., and Mackie, P. R.,** *J. Chromatogr.,* 57, 142, 1971.
7. **Stolyhwo, A. and Privett, O. S.,** *J. Chromatogr. Sci.,* 11, 20, 1973.
8. **Carroll, K. K.,** in *Lipid Chromatographic Analysis,* Vol. 1, 2nd ed., Marinetti, G. V., Ed., Marcel Dekker, New York, 1976, 173.

Table 2
TRIACYLGLYCEROLS (AC)

Adsorbent	A1	A1	A1	A3	A1
Solvent	S1	S2	S3	S4	S4
Technique	T1	T2	T3	T4	T5
Detection	D1	D2	D3	D4	D2
Reference	1	2	3	4	5

Compound Class	% Polar Solvent in Nonpolar Solvent				
LCT	100	3	10	3	3
MCT	—	3	15	3	6
SCT	100	3	20	3	9

Adsorbent A1 = Aluminum oxide
              A2 = Silicic acid
              A3 = Florisil
Solvent S1 = Benzene
              S2 = Diethyl ether in petroleum hydrocarbon (b.p. 30-60°C)
              S3 = Diethyl ether in petroleum hydrocarbon
              S4 = Diethyl ether in hexane
Technique T1 = Column size
              T2 = 2.5 × 30 cm column; 10 mℓ fractions collected
              T3 = 1 × 20 cm column; 10 mℓ fractions collected
              T4 = 1 × 60 cm column; 10 mℓ fractions collected at the rate of 2 mℓ/min
Detection D1 = Gravimetric
              D2 = Thin-layer chromatography
              D3 = Gas-liquid chromatography of acylglycerols
              D4 = Gas-liquid chromatography of fatty acids

REFERENCES

1. **Kaufmann, H. P. and Wolf, W.,** *Fette Seifen Anstrichm.,* 50, 519, 1943.
2. **Blank, M. L. and Privett, O. S.,** *J. Dairy Sci.,* 47, 481, 1964.
3. **Kuksis, A.,** unpublished results, 1965.
4. **Shehata, A. A. Y., de Man, J. M., and Alexander, J. C.,** *Can. Inst. Food Technol. J.,* 4, 61, 1971.
5. **Taylor, M. W. and Hawke, J. C.,** *N. Z. J. Dairy Sci. Technol.,* 10, 40, 1975.

Table 3
EPOXYTRIACYLGLYCEROLS (AC)

Adsorbent	A1	A2
Solvent	S1	S2
Technique	T1	T2
Detection	D1	D1
Reference	1	2

Compound Class	% Polar Solvent in Nonpolar Solvent	
Zero-epoxy	20	10
Mono-epoxy	20	25
Di-epoxy	20	25
Tri-epoxy	20	50

Adsorbent	A1	=	Silicic acid
	A2	=	Silica gel, deactivated
Solvent	S1	=	Diethyl ether in hexane
	S2	=	Diethyl ether in petroleum hydrocarbon
Technique	T1	=	Gradient elution
	T2	=	Stepwise elution
Detection	D1	=	Thin-layer chromatography

REFERENCES

1. **Tallent, W. H., Cope, D. C., Hagemay, J. W., Earle, F. R., and Wolff, I. A.,** *Lipids,* 1, 335, 1966.
2. **Fioriti, J. F., Buide, N., and Sims, R. J.,** *J. Am. Oil Chem. Soc.,* 46, 108, 1969.

Table 4
KETOTRIACYLGLYCEROLS (AC)

Adsorbent	A1	A2
Solvent	S1	S2
Technique	T1	T2
Detection	D1	D2
Reference	1	2

Compound Class	% Polar Solvent in Nonpolar Solvent	
Zero-keto	2	0
Mono-keto	2	10
Di-keto	2	—
Tri-keto	2	—

Adsorbent A1 = Deactivated silicic acid
A2 = Deactivated silica gel
Solvent S1 = Methanol in benzene
S2 = Diethyl ether in benzene
Technique T1 = Gradient elution
T2 = Deactivated with 15% water
Detection D1 = Gas-liquid chromatography of fatty acids
D2 = Thin layer chromatography, gravimetry

REFERENCES

1. **Evans, C. D., McConnell, D. G., Hoffman, R. L., and Peters, H.,** *J. Am. Oil Chem. Soc.,* 44, 281, 1967.
2. **Franzke, N. K., Kretzschmann, F., Rustow, B., and Rugenstein, H.,** *Pharmazie,* 22, 487, 1967.

Table 5
HYDROXYTRIACYLGLYCEROLS (AC)

	A1	A2	A3	A2
Adsorbent	A1	A2	A3	A2
Solvent	S1	S2	S3	S4
Technique	T1	T2	T1	T2
Detection	D1	D2	D3	D2
Reference	1	2	3	4

Compound Class	% Polar Solvent in Nonpolar Solvent			
Zero-hydroxy	2	70	10	5
Monohydroxy	2	—	30—35	10
Monoacetoxy-monohydroxy	—	70	—	—
Dihydroxy	2	—	40-50	—
Trihydroxy	2	—	50-60	—

Adsorbent	A1	= Deactivated silicic acid
	A2	= Adsorbosil CAB
	A2	= Silica gel CH
Solvent	S1	= Methanol in benzene
	S2	= Diethyl ether in petroleum hydrocarbon
	S3	= Diethyl ether in hexane
	S4	= Benzene in hexane then diethyl ether in benzene
Technique	T1	= Gradient elution
	T2	= Stepwise elution
Detection	D1	= Gas-liquid chromatography of fatty acids
	D2	= IR spectra
	D3	= Thin layer chromatography

REFERENCES

1. **Evans, C. D., McConnell, D. G., Hoffman, R. L., and Peters, H.,** *J. Am. Oil Chem. Soc.,* 44, 281, 1967.
2. **Mikolajczak, K. L., Smith, C. R., Jr., and Wolff, I. A.,** *Lipids,* 3, 215, 1968.
3. **Pokorny, J., Hladik, J., and Zeman, I.,** *Pharmazie,* 23, 332, 1968.
4. **Kleiman, R., Spencer, G. F., Earle, F. R., Nieschlag, H. J., and Barclay, A. S.,** *Lipids,* 7, 660, 1972.

Table 6
ESTOLIDE ACYLGLYCEROLS (AC)

	A1	A2	A2
Adsorbent	A1	A2	A2
Solvent	S1	S1	S2
Technique	T1	T1	T2
Detection	D1	D1	D1
Reference	1	2	3

Compound Class	% Polar Solvent in Nonpolar Solvent		
Zero estolide	5	10	0 est
Mono-estolide	10	—	—
Monoestolide-monohydroxy	—	25	20 est
Monoestolide-monoacetoxy monohydroxy	—	30	—
Monoestolide-dihydroxy	—	30	50 est
Monoestolide-trihydroxy	—	40	—

Adsorbent A1 = Silicic acid
            A2 = Adsorbosil CA 14
Solvent S1 = Diethyl ether in petroleum hydrocarbon
            S2 = Diethyl ether in benzene
Technique T1 = Stepwise elution
            T2 = Gradient
Detection D1 = Thin-layer chromatography

REFERENCES

1. **Maier, R. and Holman, R. T.**, *Biochemistry*, 3, 270, 1964.
2. **Mikolajczak, K. L., and Smith, C. R., Jr.**, *Biochim. Biophys. Acta*, 152, 244, 1968.
3. **Kleiman, R., Spencer, G. F., Earle, F. R., Nieschlag, H. J., and Barclay, A. S.**, *Lipids*, 7, 660, 1972.

Table 7
MERCURIC ACETATE ADDUCTS OF
TRIACYLGLYCEROLS (AC)

	A1	A2	A3
Adsorbent	S1	S2	S3
Solvent	T1	T2	T3
Technique	D1	D2	D3
Detection	1	2	3
References			

Compound Class	% Polar Solvent in Nonpolar Solvent		
SSS	20	0	20
SSO	40	—	—
SOO	70	10	—
OOO	20	—	100
SOL	20	20	—

Adsorbent	A1	=	Silicic acid
	A2	=	Alumina
	A3	=	Florisil
Solvent (v/v)	S1	=	Diethyl ether in petroleum hydrocarbon and methanol in petroleum hydrocarbon
	S2	=	Acetic acid in diethyl ether
	S3	=	Diethyl ether in hexane then ethanol-chloroform-conc. HCl (9:8:1)
Technique	T1	=	2.1 × 30 cm column containing 15 g silicic acid; 508 mg fat applied and eluted at 0.5 mℓ/min; 100 drop fractions collected
	T2	=	2.0 × 30 cm column
	T3	=	2.0 × 40 cm column
Detection	D1	=	Spectrophotometrical
	D2	=	Chapter 28, Reagent No. 14
	D3	=	Gas-liquid chromatography of methyl esters

REFERENCES

1. **Hirayama, O.,** *Nippon Nogei Kagaku Kaishi,* 35, 437, 1961.
2. **Inkpen, J. A. and Quakenbush, F. W.,** *J. Am. Oil Chem. Soc.,* 39(Abstr. 23), 1962.
3. **Kerkhoven, E. and deMan, J. M.,** *J. Chromatogr.,* 24, 56, 1966.

Table 8
TRIACYLGLYCEROLS (AgNO₃-AC)

Adsorbent	A1	A1	A2	A3	A3
Solvent	S1	S2	S3	S3	S3
Technique	T1	T2	T3	T4	T5
Detection	D1	D2	D3	D4	D5
Reference	1	2	3	4	5

Compound Class	% Polar Solvent in Nonpolar Solvent				
Zero-double bonds	40	—	40	0.5	0.5
One double bond	55—60	—	60	10	10
Two double bonds	80	10	80	25	25
Three double bonds	100	20	100	50	50
Four or more double bonds	—	100	100	100	100

Adsorbent A1 = 30% AgNO₃ silicic acid
             A2 = 33% AgNO₃ silicic acid
             A3 = 45% AgNO₃ silica gel G
Solvent   S1 = Benzene in petroleum hydrocarbon
             S2 = Diethyl ether in petroleum hydrocarbon
             S3 = Benzene in petroleum hydrocarbon then ether in benzene
Technique T1 = 1 × 40 cm column, operated at 15°C
             T2 = 1 × 40 cm column, operated at room temperature
             T3 = 1.6 × 32 cm column, operated at room temperature
             T4 = 1 × 40 cm column operated at 10°C
             T5 = 2 × 31 cm column, operated at 12°C
Detection D1 = Gravimetrically
             D2 = Gas-liquid chromatography of oxidized triacylglycerols
             D3 = Thin-layer chromatography
             D4 = Gas-liquid chromatography of fatty acids
             D5 = Thin-layer chromatography

REFERENCES

1. **de Vries, B.,** *J. Am. Oil Chem. Soc.,* 41, 403, 1964.
2. **Subbaram, M. R. and Youngs, C. G.,** *J. Am. Oil Chem. Soc.,* 41, 445, 1964.
3. **Gunstone, F. D., Hamilton, R. J., and Qureshi, M. I.,** *J. Chem. Soc.,* 319, 1965.
4. **Dolev, A. and Olcott, H. S.,** *J. Am. Oil Chem. Soc.,* 42, 624, 1965.
5. **Dolev, A. and Olcott, H. S.,** *J. Am. Oil Chem. Soc.,* 42, 1046, 1965.

Table 9

ACYLGLYCEROLS (TLC)

Layer	L1	L2	L3	L3	L4	L3	L3	L5
Solvent	S1	S2	S3	S4	S5	S6	S7	S8
Technique	T1	T2	T1	T3	T2	T2	T1	T4
Detection	D1	D1	D2	D3	D1	D1	D4	D5
Reference	1	2	3	4	5	6	7	8

Compound Class				$R_f \times 100$				
Triacyl	70	97	67	71	54	47	60	64
1,3-Diacyl	48	55	56	64	24	—	24	52
1,2-Diacyl	42	42	50	54	20	28	—	42
2-Monoacyl	—	10	14	7	5	11	3	17
1-Monoacyl	12	6	14	7	5	6	3	3

Layer	L1	= Silica gel without $CaSO_4$
	L2	= Silica gel GF impregnated with 5% H_3BO_3
	L3	= Silica gel G
	L4	= Silica gel G with 0.01 M sodium carbonate
	L5	= Chromarod impregnated from 3% boric acid solution
Solvent (v/v)	S1	= Isopropyl ether-acetic acid (96:4)(13-14 cm), then petroleum hydrocarbon diethyl ether-acetic acid (90:10:1)
	S2	= Chloroform-acetone (96:4)
	S3	= Diethyl ether-benzene-ethanol-acetic acid (40:50:2:0.2) (25 cm) then diethyl ethers-hexane (6:94)
	S4	= Benzene-diethyl ether-ethyl acetate-acetic acid (80:10:10:0.2)
	S5	= Hexane-ethyl acetate-formic acid (125:25:5)
	S6	= Hexane-diethyl ether-acetic acid (85:15:1)
	S7	= Hexane-diethyl ether (98:2) then hexane-diethyl ether-acetic acid, (50:50:1)
	S8	= Chloroform-acetone (96:4)
Technique	T1	= Double development
	T2	= Single development
	T3	= Nonpresaturated chamber
	T4	= Iatroscan
Detection	D1	= Charring
	D2	= 1% Ultraphor reagent
	D3	= Dichromate reduction
	D4	= Rhodamine 6G
	D5	= Flame ionization

REFERENCES

1. **Skipski, V. P., Smolowe, A. F., Sullivan, R. C., and Barclay, M.,** *Biochim. Biophys. Acta,* 106, 386, 1965.
2. **Thomas, A. E., Scharoun, J. E., and Ralston, H.,** *J. Am. Oil Chem. Soc.,* 42, 789, 1965.
3. **Freeman, C. P. and West D.,** *J. Lipid Res.,* 1, 324, 1966.
4. **Storry, J. E. and Tuckley, P.,** *Lipids,* 2, 501, 1967.
5. **Duthie, A. H. and Atherton, H. V.,** *J. Chromatogr.,* 51, 319, 1970.
6. **Kramer, J. K. G., Hulan, H. W., Mahadevan, S., and Sayer, F.,** *Lipids,* 10, 505, 1975.
7. **Schlotzhauer, P. F., Ellington, J. J., and Schepartz, A. I.,** *Lipids,* 12, 239, 1977.
8. **Tanaka, M., Itch, T., and Kaneko, H.,** *Lipids,* 15, 872, 1981.

Table 10
TRIACYLGLYCEROLS (TLC)

Layer			L1	L2	L3	L4	L4	L5	L2
Solvent			S1	S2	S3	S4	S5	S1	S6
Technique			T1	T2	T1	T1	T1	T1	T3
Detection			D1	D2	D3	D1	D1	D4	D1
Reference			1	2	3	4	4	5	6
Molecular Species						$R_f \times 100$			
22:0	22:0	22:0	—	—	—	—	—	78	—
20:0	20:0	20:0	—	—	—	—	—	75	—
18:0	18:0	18:0	81	70	89	65	54	72	83
16:0	16:0	16:0	75	70	89	65	54	69	—
14:0	14:0	14:0	69	58	77	—	—	65	—
18:0	6:0	18:0	—	—	77	—	—	—	71
18:0	3:0	18:0	—	—	77	—	—	—	64
18:0	2:0	18:0	—	—	77	—	—	—	59
12:0	12:0	12:0	63	58	77	56	50	61	—
18:0	6:0	6:0	—	—	—	—	—	—	71
10:0	10:0	10:0	54	—	—	—	—	58	—
18:0	3:0	3:0	—	—	—	—	—	—	60
8:0	8:0	8:0	50	45	—	—	—	53	—
18:0	2:0	2:0	—	—	—	—	—	—	55
6:0	6:0	6:0	39	—	—	—	—	46	—
4:0	4:0	4:0	25	—	—	—	—	34	—
2:0	2:0	2:0	—	—	—	—	—	9	—

Adsorbent   A1 = Silicic acid
A2 = Adsorbosil-3
A3 = Silica gel G
A4 = Silica gel G layer prepared in 0.01 m sodium carbonate
A5 = Adsorbosil-1

Solvent (v/v)   S1 = Petroleum hydrocarbon diethyl ether (80:20)
S2 = Heptane-isopropyl ether-acetic acid (60:40:4)
S3 = Petroleum hydrocarbon diethyl ether-acetic acid (65:35:1)
S4 = Petroleum hydrocarbon diethyl ether-formic acid (90:10:1)
S5 = Hexane-ethyl acetate-formic acid (175:25:5)
S6 = Benzene and benzene-ethyl acetate (5:1)

Technique   T1 = Analytical
T2 = Preparative
T3 = Gradient development

Detection   D1 = Charring
D2 = 2′,7′-Dichlorofluorescein
D3 = Iodine vapor
D4 = Rhodamine 6G impregnated adsorbent

REFERENCES

1. **Privett, O. S., Nutter, L. J., and Gross, R. A.,** in *Dairy Lipids and Lipid Metabolism,* AVI Publishing, Westport, Conn., 1967, 99.
2. **Breckenridge, W. C. and Kuksis, A.,** *Lipids,* 3, 291, 1968.
3. **Glass, R. L., Jenness, R., and Lohse, L. W.,** *Comp. Biochem. Physiol.,* 28, 783, 1969.
4. **Duthie, A. H. and Atherton, H. V.,** *J. Chromatogr.,* 51, 319, 1970.
5. **Litchfield, C.,** *Analysis of Triglycerides,* Academic Press, New York, 1972, 151.
6. **Tarr, G. E. and Fairbairn, D.,** *Lipids,* 8, 303, 1973.

Table 11
ACETYLACYLGLYCEROLS (TLC)

Layer Solvent Technique Detection Reference	L1 S1 T1 D1 1	L2 S2 T2 D2 2	L3 S3 T3 D3 3	L4 S4 T4 D3 4	L5 S5 T4 D3 5	L2 S6 T1 D3 6
Compound Class			$R_f \times 100$			
18:0 18:0 18:0	63	90	80	93	71	—
16:0 16:0 16:0	63	90	80	93	—	—
14:0 14:0 14:0	—	—	—	—	65	—
18:0 2:0 18:0	36	—	72	56	56	—
18:0 18:0 2:0	—	52	65	50	50	—
16:0 2:0 16:0	36	—	—	—	—	—
16:0 16:0 2:0	—	52	—	—	—	—
14:0 2:0 14:0	—	—	—	—	52	—
14:0 14:0 2:0	—	—	—	—	45	—
10:0 2:0 10:0	—	—	—	—	—	—
10:0 10:0 2:0	—	—	—	—	—	29
12:0 2:0 8:0	—	—	—	—	—	29
12:0 8:0 2:0	—	—	—	—	—	—
14:0 2:0 6:0	—	—	—	—	—	—
14:0 6:0 2:0	—	—	—	—	—	28
16:0 2:0 4:0	—	—	—	—	—	—
16:0 4:0 2:0	—	—	—	—	—	27
18:0 2:0 2:0	—	—	—	—	—	24
2:0 18:0 2:0	19	—	—	—	—	—
16:0 2:0 2:0	—	—	—	—	—	—
2:0 16:0 2:0	19	—	—	—	—	—
12:0 2:0 2:0	—	—	—	—	28	—
2:0 12:0 2:0	—	—	—	—	—	—
2:0 2:0 2:0	—	—	—	—	2	—

Layer	L1	= Silicic acid
	L2	= Silica gel G
	L3	= Silica gel G with boric acid
	L4	= Silica gel F - 254
	L5	= Silica gel 60
Solvent	S1	= Petroleum hydrocarbon-diethyl ether (92:8)
(v/v)	S2	= Hexane-ethyl acetate (88:12)
	S3	= Hexane-diethyl ether (70:30)
	S4	= Petroleum hydrocarbon-diethyl ether (2:1)
	S5	= Petroleum hydrocarbon-diethyl ether-acetic acid (60:40:1)
	S6	= Heptane-isopropylether-acetic acid (60:40:4)
Technique	T1	= Analytical
	T2	= Preparative
	T3	= "Sandwich" chamber
	T4	= Precoated plates
Detection	D1	= Iodine vapor
	D2	= 2′,7′-Dichlorofluorescein and Rhodamine B
	D3	= 2′,7′-Dichlorofluorescein

Table 11 (continued)
ACETYLACYLGLYCEROLS (TLC)

REFERENCES

1. **Gruger, E. H., Jr., Malins, D. C., and Gauglitz, E. J., Jr.,** *J. Am. Oil Chem. Soc.,* 37, 214, 1960.
2. **Parodi, P. W.,** *J. Chromatogr.,* 111, 223, 1975.
3. **Kleiman, R., Miller, R. W., Earle, F. R., and Wolff, I. A.,** *Lipids,* 2, 473, 1967.
4. **Smith, C. R., Jr., Madrigal, R. V., Weisleder, D., and Plattner, R. D.,** *Lipids,* 12, 736, 1977.
5. **Neissner, R.,** *Fette Seifen Anstrichm.,* 79, 24, 1977.
6. **Myher, J. J. and Kuksis, A.,** unpublished results, 1978.

Table 12
ISOVALEROYLACYLGLYCEROLS (TLC)

Layer	L1	L2	L2	L1	L1	L3	L3
Solvent	S1	S2	S3	S3	S4	S5	S5
Technique	T1	T2	T1	T3	T1	T4	T4
Detection	D1	D1	D2	D3	D4	D3	D3
Reference	1	2	3	4	5	6	7

Compound Class				$R_G \times 100$			
Fa Fa Fa	51	62	81	—	35	53	60
Fa Vi Fa							
Fa Fa Vi	42	48	71	—	30	45	54
Fa Vi Vi							
Vi Fa Vi	32	—	60	—	19	30	48
Vi Vi Vi							

Layer	L1	= Adsorbosil 1
	L2	= Silica gel G
	L3	= Silica gel H
Solvent	S1	= Petroleum hydrocarbon-diethyl ether-acetic acid (87:12:1)
(v/v)	S2	= Benzene
	S3	= Hexane-diethyl ether (90:10)
	S4	= Petroleum hydrocarbon-diethyl ether (90:10)
	S5	= Petroleum hydrocarbon-diethyl ether-acetic acid (80:20:1)
Technique	T1	= Single development
	T2	= Double development
	T3	= Single development
	T4	= Plate not activated
Detection	D1	= Charring
	D2	= Rhodamine G
	D3	= Iodine
	D4	= Rhodamine B in adsorbent

REFERENCES

1. **Litchfield, C., Ackman, R. G., Sipos, J. C., and Eaton, C. A.,** *Lipids,* 6, 674, 1971.
2. **Blank, M. L., Kasama, K., and Snyder, F.,** *J. Lipid Res.,* 13, 390, 1972.
3. **Varanasi, U. and Malins, D. C.,** *Science,* 176, 926, 1972.
4. **Varanasi, U., Everitt, M., and Malins, D. C.,** *Int. J. Biochem.,* 4, 373, 1973.
5. **Litchfield, C. and Greenberg, A. J.,** *Comp. Biochem. Physiol.,* 47B, 401, 1974.
6. **Blomberg, J.,** *Lipids,* 9, 461, 1974.
7. **Blomberg, J. and Lindholm, L. E.,** *Lipids,* 11, 153, 1976.

Table 13
PHYTANOYLACYLGLYCEROLS (TLC)

Layer	L1	L1	L2
Solvent	S1	S2	S3
Technique	T1	T2	T2
Detection	D1	D2	D2
Reference	1	2	3

Compound Class	$R_f \times 100$		
PhPhPh	50	92	42
PhFaPh	44	86	—
FaPhPh	44	84	37
PhFaFa	—	79	30
FaPhFa	—	77	—
FaFaFa	37	71	24

Layer	L1	= Silica gel G
	L2	= Silicic acid
Solvent	S1	= Hexane-diethyl ether-acetic acid (90:10:1)
(v/v)	S2	= Heptane-diethyl ether (80:20)
	S3	= Petroleum hydrocarbon-diethyl ether-acetic acid (90:10:1)
Technique	T1	= 0.15 mm thick layer
	T2	= 0.25 mm thick layer
Detection	D1	= Iodine vapor
	D2	= Charring

REFERENCES

1. **Karlsson, K. A., Nilsson, K., and Pascher, I.,** *Lipids,* 3, 389, 1968.
2. **Laurell, S.,** *Biochim. Biophys. Acta,* 152, 75, 1968.
3. **Sezille, G., Jaillard, J., Scherpereel, P., and Biserte, G.,** *Clin. Chim. Acta,* 29, 335, 1970.

Table 14
EPOXY TRIACYLGLYCEROLS (TLC)

Layer	L1	L2	L2	L2
Solvent	S1	S2	S3	S4
Technique	T1	T2	T3	T4
Detection	D1	D2	D2	D3
Reference	1	2	3	4

Compound Class	**$R_f \times 100$**			
Zero-epoxy	73	81	92	81
Mono-epoxy	52	73	84	73
Di-epoxy	42	63	67	64
Tri-epoxy	35	47	50	55

Layer	L1	= Silicic acid
	L2	= Silica gel G

Solvent (v/v) S1 = Petroleum hydrocarbon-diethyl ether-acetic acid (80:20:1) then (90:10:1)

S2 = Hexane-diethyl ether, 70:30

S3 = Petroleum hydrocarbon-diethyl ether-acetic acid (60:40:1)

S4 = Petroleum hydrocarbon-diethyl ether (65:35)

Technique T1 = Double development

T2 = Single development

T3 = Nonactivated plates

T4 = Preparative

Detection D1 = Iodine vapor

D2 = Picric acid

D3 = 2',7'-Dichlorofluorescein

REFERENCES

1. **Kleiman, R., Smith, C. R., Jr., Yates, S. G., and Jones, Q.,** *J. Am. Oil Chem. Soc.,* 42, 169, 1965.
2. **Spencer, G. F., Kleiman, R., Earle, F. R., and Wolff, I. A.,** *Lipids,* 4, 99, 1969.
3. **Fioriti, J. A., Buide, N., and Sims, R. J.,** *J. Am. Oil Chem. Soc.,* 46, 108, 1969.
4. **Phillips, B. E., Smith, C. R., Jr., and Hageman, J. W.,** *Lipids,* 4, 473, 1969.

Table 15
HYDROXYTRIACYLGLYCEROLS (TLC)

Layer	L1	L2	L2	L2
Solvent	S1	S1	S2	S1
Technique	T1	T2	T2	T2
Detection	D1	D2	D1	D2
Reference	1	2	3	4

Compound Class	$R_f \times 100$			
Hydroxytriacylglycerols	75	92	82	75
Mono-hydroxytriacylglycerols	44	47	65	44
Di-hydroxytriacylglycerols	20	22	37	20
Tri-hydroxytriacylglycerols	7	6	15	7

Layer	L1	= Silica gel G impregnated with boric acid
	L2	= Silica gel G
Solvent	S1	= Hexane-diethyl ether (70:30)
(v/v)	S2	= Chloroform-methanol (99:1)
Technique	T1	= Sandwich development
	T2	= Open development
Detection	D1	= Iodine vapor
	D2	= Charring

REFERENCES

1. **Miller, R. W., Earle, F. R., Wolff, I. A., and Jones, Q.,** *J. Am. Oil Chem. Soc.,* 42, 817, 1965.
2. **Mikolajczak, K. L. and Smith, C. R., Jr.,** *Lipids,* 2, 261, 1967.
3. **Serck-Hanssen, K.,** *Acta Chem. Scand.,* 21, 301, 1967.
4. **Kleiman, R., Spencer, G. F., Earle, F. R., Nieschlag, H. J., and Barclay, A. S.,** *Lipids,* 7, 660, 1972.

Table 16
ESTOLIDE TRIACYLGLYCEROLS (TLC)

Layer	L1	L2	L2
Solvent	S1	S2	S3
Technique	T1	T2	T3
Detection	D1	D2	D3
Reference	1	2	3

Compound Class	\multicolumn{3}{c}{$R_f \times 100$}		
Zero-estolide	67	67	65
Mono-estolide	57	50	50
Di-estolide	50	—	—
Tri-estolide	40	—	—

Layer	L1	= Silicic acid
	L2	= Silica gel G
Solvent	S1	= Hexane-diethyl ether (92.5:7.5)
(v/v)	S2	= Petroleum hydrocarbon diethyl ether-acetic acid (80:20:2)
	S3	= Hexane-diethyl ether (80:20)
Technique	T1	= Developed three times
	T2	= Developed once
	T3	= Preparative
Detection	D1	= Charring with H_2SO_4
	D2	= Iodine vapor
	D3	= 2′,7′-Dichlorofluorescein

REFERENCES

1. **Morris, L. J. and Hall, S. W.,** *Lipids,* 1, 188, 1966.
2. **Maier, R. and Holman, R. T.,** *Biochemistry,* 3, 270, 1964.
3. **Christie, W. W.,** *Biochim. Biophys. Acta,* 187, 1, 1969.

Table 17
OZONOTRIACYLGLYCEROLS (TLC)

Layer	L1
Solvent	S1
Technique	T1
Detection	D1
Reference	1

Compound Class	$R_f \times 100$
Saturated triacylglycerols	83
Mono-ozonide	62
Di-ozonide	53
Tri-ozonide	50
Tetra-ozonide	42
Penta-ozonide	32
Hexa-ozonide	25

Layer	L1	= Silicic acid
Solvent	S1	= Petroleum hydrocarbon-diethyl ether (80:20 v/v)
Technique	T1	= 0.25 mm thick layer
Detection	D1	= Charring

REFERENCE

1. **Privett, O. S. and Blank, M. L.,** *J. Am. Oil Chem. Soc.*, 40, 70, 1963.

Table 18
OXIDIZED TRIACYLGLYCEROLS (TLC)

Layer	L1	L1	L1	L2	L2
Solvent	S1	S2	S2	S3	S4
Technique	T1	T1	T1	T2	T3
Detection	D1	D1	D1	D2	D3
Reference	1	1	2	3	4

Compound Class	$R_f \times 100$				
SSS	—	69	83		
SSU	67	12	49		
SUU	28	—	0		
UUU	0	—	0		

Layer	L1	= Silica gel G
	L2	= Silicic acid
Solvent	S1	= Petroleum hydrocarbon diethyl ether (65:35)
(v/v)	S2	= Petroleum hydrocarbon-diethyl ether (85:15)
	S3	= Hexane-diethyl ether (60:40)
	S4	= Diethyl ether to 1 cm then benzene-diethyl ether (90:10)
Technique	T1	= Aldehyde cores from reduced ozonides
	T2	= Fully oxidized triacylglycerols
	T3	= Methyl esters derived from oxidized triacylglycerols
Detection	D1	= Charring
	D2	= 2′,7′-Dichlorofluorescein
	D3	= Gas-liquid chromatography

REFERENCES

1. **Privett, O. S. and Blank, M. L.,** *J. Lipid Res.,* 2, 37, 1961.
2. **Privett, O. S., Blank, M. L., and Schmit, J. A.,** *J. Food Sci.,* 27, 463, 1962.
3. **Jurriens, G.,** *Anal. Character. Oils, Fats Prod.,* 2, 237, 1968.
4. **Youngs, C. G. and Baker, C. D.,** 1964, cited in **Litchfield, C.,** *Analysis of Triglycerides,* Academic Press, New York, 1972, 157.

Table 19
TRIACYLGLYCEROLS (AgNO₃ TLC)

	P1	P2	P3	P4	P5	P6	P7	P8	P9
Layer	P1	P2	P3	P4	P5	P6	P7	P8	P9
Solvent	S1	S2	S3	S4	S5	S6	S7	S8	S9
Detection	D1	D1	D1	D1	D1	D1	D1	D1	D1
Technique	T1	T2	T3	T4	T5	T6	T7	T8	T9
Reference	1	2	3	3	4	5	4	4	6

cis-Isomers of Fatty Acids — $R_f \times 100$

	P1	P2	P3	P4	P5	P6	P7	P8	P9
000	—	—	100	—	—	83	—	—	—
010	96	—	—	—	—	—	—	78	—
001	85	77	98	—	—	63	—	69	—
101	78	—	—	—	—	—	—	—	—
011	71	68	96	—	34	44	—	—	—
020	71	—	—	—	—	—	—	—	—
002	—	68	92	—	—	—	—	—	—
111	57	62	88	—	15	22	72	—	—
012	44	62	80	—	—	—	—	—	—
112	23	47	—	—	—	—	—	—	—
022	13	38	34	—	—	—	—	—	—
003	—	—	—	—	—	—	—	—	—
122	5	25	23	—	—	—	—	—	—
013	—	25	21	—	—	—	—	—	—
222	2	12	9	—	—	—	—	—	88
113	—	12	11	—	—	—	—	—	—
023	—	12	—	73	—	—	—	—	—
123	—	5	—	71	—	—	—	—	—
223	—	2.5	—	32	—	—	—	—	—
033	—	—	—	7	—	—	—	—	—
133	—	—	—	6	—	—	—	—	—
233	—	—	—	5	—	—	—	—	—
333	—	—	—	4	—	—	—	—	—
014	—	—	—	—	—	—	—	—	22
115	—	—	—	—	—	—	—	—	61
116	—	—	—	—	—	—	—	—	53
056	—	—	—	—	—	—	—	—	33
156	—	—	—	—	—	—	—	—	00
	—	—	—	—	—	—	—	—	00

trans-Isomers of Fatty Acids

	P1	P2	P3	P4	P5	P6	P7	P8	P9
001	—	—	—	—	—	75	—	—	—
011	—	—	—	—	58	69	—	—	—
111	—	—	—	—	48	56	—	—	—

Positional Isomers of Unsaturated Fatty Acids

	P1	P2	P3	P4	P5	P6	P7	P8	P9
Pe 11	—	—	—	—	—	—	58	—	—
PePe 1	—	—	—	—	—	—	42	—	—
PePePe	—	—	—	—	—	—	25	—	—

Layer
- L1 = 10% ? AgNO₃-silica gel G (1 mm layer)
- L2 = 30% AgNO₃-silicic acid (1 mm layer)
- L3 = 10—30% AgNO₃-silicic acid (0.5 mm layer)
- L4 = 10—30% AgNO₃-silicic acid (0.5 mm layer)
- L5 = 10% AgNO₃ silica gel G (0.4 mm layer)
- L6 = 30% AgNO₃ silica gel G (0.3 mm layer)
- L7 = 23% AgNO₃ silica gel G (0.5 mm layer)
- L8 = 23% AgNO₃ silica gel G (0.5 mm layer)
- L9 = 8% AgNO₃-Adsorbosil-1 (0.5 mm layer)

Table 19 (continued)
TRIACYLGLYCEROLS (AgNO₃ TLC)

Solvent   S1 = Benzene-diisopropyl ether (85:15)
              S2 = Benzene-diethyl ether (85:15)
              S3 = Chloroform-methanol (94:6)
              S4 = Chloroform-methanol (98.6:1.4)
              S5 = Benzene or chloroform-methanol (99.2:0.8)
              S6 = Benzene
              S7 = Toluene-diethyl ether (75:25)
              S8 = Toluene-diethyl ether (97:3)
              S9 = Chloroform-methanol (94:6)
Detection   D1 = 2',7'-Dichlorofluorescein
Technique   T1 = 40 cm distance
              T2 = 40 cm distance
              T3 = 17.5 cm distance
              T4 = 17.5 cm distance
              T5 = 17.5 cm distance, developed two times
              T6 = 17.5 cm distance
              T7 = 27 cm distance
              T8 = 17.5 cm distance, developed three times
              T9 = 17.5 cm distance

REFERENCES

1. **den Boer, F. C.,** *Z. Anal. Chem.,* 205, 308, 1964.
2. **Jurriens, G.,** *Chem. Weekbl.,* 61, 257, 1965.
3. **Roehm, J. N. and Privett, O. S.,** *Lipids,* 5, 353, 1970.
4. **Wessels, H. and Rajagopal, N. S.,** *Fette Seifen Anstrichm.,* 71, 543, 1969.
5. **de Vries, B. and Jurriens, G.,** *Fette Seifen Anstrichm.,* 65, 727, 1963.
6. **Bottino, N.,** *J. Lipid Res.,* 12, 24, 1971.

Table 20
DIACYLGLYCYCEROLS (AgNO₃-TLC)

Layer	L1	L1	L2	L3	L3	L3
Solvent	S1	S2	S3	S4	S5	S6
Technique	T1	T1	T1	T1	T1	T1
Detection	D1	D1	D1	D2	D2	D2
Reference	1	2	3	4	5	5
1,2(2′,3)-Isomers			$R_f \times 100$			
18:0 18:0	92	—	93	92	—	—
18:0 16:0	92	—	93	92	95	100
16:0 16:0	92	—	93	92	95	100
16:0 17:0	—	59	—	—	—	—
16:0 18:1	90	—	—	90	90	100
18:0 18:1	90	—	83	90	90	100
16:0 16:1	85	37	83	—	—	—
16:0 18:1	85	43	83	—	85	—
16:1 18:1	72	23	71	—	85	—
18:1 18:1	72	28	71	—	85	—
16:0 18:2	65	—	66	70	70	95
18:0 18:2	65	—	66	70	70	95
18:1 18:2	—	—	57	—	—	—
16:0 20:3	40	—	39	—	—	—
18:0 20:3	40	—	39	—	—	—
16:0 20:4	20	—	20	10	15	50
18:0 20:4	20	—	20	10	15	50
18:1 20:3	20	—	20	—	—	—
16:0 22:6	—	—	0	2	0	10
18:0 22:6	—	—	0	2	0	10
	(author)			(author)	(author)	(author)

Layer   L1 = 30% AgNO₃-silicic acid
        L2 = 25% AgNO₃-silica gel G
        L3 = 18% AgNO₃-silica gel H
Solvent S1 = Chloroform-ethanol (93:7)
(v/v)   S2 = Chloroform-ethanol (98:2)
         S3 = Chloroform-ethanol (94:6)
         S4 = Chloroform-methane (95:5)
         S5 = Benzene-chloroform-methanol (84:15:1)
         S6 = Chloroform-methanol-water (89:10:1)
Technique T1 = Preparative
Detection D1 = Rhodamine 6G
           D2 = 2′,7′-Dichlorofluorescein

REFERENCES

1. **van Golde, L. M. G. and Van Deenen, L. L. M.**, *Biochim. Biophys. Acta,* 125, 496, 1966.
2. **van Golde, L. M. G. and Van Deenen, L. L. M.**, *Chem. Phys. Lipids,* 1, 157, 1967.
3. **van Golde, L. M. G., Pieterson, W. A., and Van Deenen, L. L. M.**, *Biochim. Biophys. Acta,* 152, 84, 1968.
4. **Akesson, B.**, *Eur. J. Biochem.,* 9, 463, 1969.
5. **Akesson, B., Elovson, J., and Arvidson, G.**, *Biochim. Biophys. Acta,* 218, 44, 1970.

Table 21
ACETYLDIACYLGLYCEROLS (AgNO₃-TLC)

Layer	L1	L1	L2	L3	L4	L5
Solvent	S1	S2	S3	S4	S5	S6
Technique	T1	T2	T2	T2	T2	T1
Detection	D1	D1	D2	D2	D2	D1
Reference	1	2	3	4	5	6

1,2(2,3)-Isomers $R_f \times 100$

1,2(2,3)-Isomers						
00	83	89	100	48	90	88
01	75	80	100	24	68	76
11	63	62	100	13	46	—
02	52	62	100	—	31	62
12	33	44	100	—	—	45
22	—	—	—	—	—	39
03	—	28	—	—	—	35
04	17	18	—	—	—	29
14	—	5	—	—	—	24
05	0	5	70	—	—	18
06	0	5	70	—	—	4
24	—	5	70	—	—	—
15	—	5	70	—	—	—
55	—	—	20	—	—	—
56	—	—	20	—	—	—
66	—	—	20	—	—	—

Layer	L1	=	20% AgNO₃-silica gel G
	L2	=	10% AgNO₃-silica gel G
	L3	=	Silica gel G thinned in aqueous 10% AgNO₃
	L4	=	30% AgNO₃-silica gel G
	L5	=	25% AgNO₃
Solvent (v/v)	S1	=	Chloroform-methanol (99.3:0.7)
	S2	=	Chloroform-methanol (99.2:0.8)
	S3	=	Chloroform-methanol-water (65:25:4)
	S4	=	Benzene-chloroform-methanol (98:2:0.1)
	S5	=	Benzene-chloroform (9:1)
	S6	=	Petroleum hydrocarbon-diethyl ether-acetic acid (75:35:1)
Technique	T1	=	Developed two times
	T2	=	Developed once
Detection	D1	=	2',7'-Dichlorofluorescein
	D2	=	Rhodamine 6 G

REFERENCES

1. **Kuksis, A. and Marai, L.,** *Lipids,* 2, 217, 1967.
2. **Kuksis, A., Breckenridge, W. C., Marai, L., and Stachnyk, O.,** *J. Am. Oil Chem. Soc.,* 45, 537, 1968.
3. **Renkonen, O.,** *Lipids,* 3, 191, 1968.
4. **Hasegawa, K. and Suzuki, T.,** *Lipids,* 10, 667, 1975.
5. **Renkonen, O.,** *J. Am. Oil Chem. Soc.,* 42, 298, 1965.
6. **Das, A. K., Ghosh, R., and Data, J.,** *J. Chromatogr.,* 234, 472, 1982.

Table 22
ACETYLDIRADYLGLYCEROLS
(AgNO₃-TLC)

Layer	L1	L2	L2
Solvent	S1	S2	S3
Technique	T1	T1	T1
Detection	D1	D2	D2
Reference	1	2	3

1,2-Isomers	**$R_f \times 100$**		
Alkylacyl			
Saturates	72	75	80
Monoenes	38	70	77
Dienes	—	60	70
Trienes	—	40	50
Tetraenes	—	5	10
Hexaenes	—	0	0
Alkenylacyl			
Saturates	52	60	—
Monoenes	12	50	—
Dienes	—	20	—

Layer	L1	=	Silica gel G
	L2	=	Silica gel H
Solvent	S1	=	Benzene-chloroform (9:1)
(v/v)	S2	=	Chloroform-methanol (98:2)
	S3	=	Chloroform-methanol (97:3)
Technique	T1	=	Preparative
Detection	D1	=	Charring
	D2	=	2′,7′-Dichlorofluorescein

REFERENCES

1. **Renkonen, O.,** in *Progress in Thin-Layer Chromatography and Related Methods,* Vol. 2, Niederweiser, A. and Pataki, G., Eds., Ann Arbor Science, Ann Arbor, Mich., 1971, 143.
2. **Marai, L. and Kuksis, A.,** *Can. J. Biochem.,* 51, 1248, 1973.
3. **Yeung, S. K. F. and Kuksis, A.,** *Can. J. Biochem.,* 52, 830, 1974.

Table 23
MONOALKYLGLYCEROLS (AgNO₃-TLC)

Layer	L1	L2	L3
Solvent	S1	S2	S3
Technique	T1	T2	T2
Detection	D1	D2	D1
Reference	1	2	3

Compound Class	$R_f \times 100$		
1(3)-Isomers			
14:0	—	—	75
16:0	—	66	75
18:0	71	66	75
20:0	—	—	75
16:1	—	—	60
18:1	64	—	60
18:2	55	—	40
1-0-(2-Methoxy) 16:0	—	42	—
1-0-(2-Methoxy) 16:1	—	26	—
1-0-(2-Methoxy) 18:1	—	26	—
2-Isomers			
18:0	—	—	75
18:1	—	—	60
18:2	—	—	40

Layer	L1	=	20% AgNO₃-silica gel G
	L2	=	Silica gel G impregnated with AgNO₃
	L3	=	10% AgNO₃-silica gel G
Solvent	S1	=	Chloroform-ethanol (90:10)
(v/v)	S2	=	Trimethylpentane-ethyl acetate-methanol (50:40:7)
	S3	=	Chloroform-methanol (95:5)
Technique	T1	=	Double development
	T2	=	Single development
Detection	D1	=	2′,7′-Dichlorofluorescein
	D2	=	Charring

REFERENCES

1. **Wood, R. and Snyder, F.,** *Lipids,* 1, 62, 1966.
2. **Hallgren, B. and Stellberg, G.,** *Acta Chem. Scand.,* 21, 1519, 1967.
3. **Myher, J. J. and Kuksis, A.,** unpublished results, 1974.

Table 24
ACYLGLYCEROLS (LLC)

Column	C1	C1	C2	C3	C4	C5	C7	C7	C8	C9
Solvent	S1	S2	S3	S4	S5	S6	S7	S8	S9	S10
Technique	T1	T1	T1	T2	T3	T4	T5	T6	T7	T8
Detection	D1	D1	D2	D2	D2	D1	D3	D4	D5	D6
Reference	1	1	2	3	3	4	5	6	7	8
Compound Class					**RRT × 100**					
1-Monooctanoyl	—	—	—	—	28	—	—	—	—	—
1-Monodecanoyl	—	—	—	—	41	—	—	—	—	—
1-Monomyristoyl	—	—	—	—	60	—	—	—	—	—
1-Monopalmitoyl	7	—	70	65	77	163	10	—	250	314
1-Monostearoyl	10	—	—	94	100	—	—	—	—	—
1,3-dilauroyl	20	—	—	—	—	—	—	—	—	—
1,2-dipalmitoyl	—	76	—	—	—	—	—	—	116	372
1,3-dipalmitoyl	27	100	42	124	—	108	50	—	113	214
1,3-Distearoyl	43	—	—	162	—	—	—	—	—	—
Trioctanoyl	17	—	—	—	—	—	—	—	—	—
Tridecanoyl	27	—	—	—	—	—	—	—	—	—
Trilauroyl	43	—	—	—	—	—	—	—	—	—
Trimyristoyl	70	—	—	—	—	—	—	—	—	—
Tripalmitoyl	—	—	38	308	—	100	100	—	100	100
Tristearoyl	—	—	—	432	—	—	—	—	—	—
Trioleoyl	100	—	—	—	—	—	—	—	—	—
Sorboyldipalmitoyl	—	—	—	—	—	—	—	100	—	—
Sorboylmyristoyl-palmitoyl	—	—	—	—	—	—	—	106	—	—
Sorboyldimyristoyl	—	—	—	—	—	—	—	112	—	—

Column C1 = Factice
C2 = Methyl Sephadex G-25, 36% methoxyl
C3 = Hydroxyalkoxypropyl Sephadex G-50, 52% hydroxyalkyl (C_{11}-C_{14})
C4 = Sephadex LH-20, 170-325 mesh
C5 = Sephadex LH-20
C6 = Poragel 200A
C7 = Zorbax SIL
C8 = Lichrosorb S1-60
C9 = Partisil PX5 10/25 PAC

Solvent (v/v) S1 = Acetone-water (98:2)
S2 = Acetone-water (87.5:12.5)
S3 = Heptane-chloroform-methanol (2:1:1)
S4 = Heptane-acetone-water (4:15:1)
S5 = Heptane-chloroform-isopropanol-water (1:9:70:70)
S6 = Tetrahydrofuran
S7 = Acetone-water (95:5)
S8 = Methylene chloride-cyclohexane (2:1)
S9 = Carbon tetrachloride-isooctane (34:66)/chloroform-dioxane-hexane (40:11:49)/chloroform-methanol-diisopropyl ether (34:36:30)
S10 = Hexane-chloroform (60:65) and acetonitrile-hexane-chloroform (25:25:65) from 2 to 95% as linear gradient in 20 min at 2 mℓ/min

Technique T1 = 20°C
T2 = Gel permeation (lipophilic Sephadex)
T3 = Gel permeation (lipophilic Sephadex)
T4 = Sephadex SR 25/100 column
T5 = 0.4 × 100 cm column
T6 = 0.25 × 25 cm column; 2000 psig gel permeation
T7 = Mixed solvent gradient gel permeation
T8 = Linear gradient of mixed solvents

Table 24 (continued)
ACYLGLYCEROLS (LLC)

Detection   D1 = Differential refractometry
                 D2 = FI, moving chain
                 D3 = Thin-layer chromatography
                 D4 = Ultraviolet absorption
                 D5 = FI, moving chain (methane reduction)
                 D6 = IR, 5.72 um

REFERENCES

1. **Hirsch, J.,** *J. Lipid Res.,* 4, 1, 1963.
2. **Ellingboe, J., Nystrom, E., and Sjövall, J.,** *Methods Enzymol.,* 14, 317, 1969.
3. **Ellingboe, J., Nystrom, E., and Sjövall, J.,** *J. Lipid Res.,* 11, 266, 1970.
4. **Kleiman, R., Spencer, G. F., Earle, F. R., Nieschlag, H. J., and Barclay, A. S.,** *Lipids,* 7, 660, 1972.
5. **Lawrence, J. G.,** *J. Chromatogr.,* 84, 299, 1973.
6. **Fallon, W. E. and Schimizu, Y.,** *Lipids,* 12, 765, 1977.
7. **Aitzetmüller, K.,** *J. Chromatogr.,* 139, 61, 1977.
8. **Payne-Wahl, K., Spencer, G. F., Plattner, R. D., and Butterfield, R. O.,** *J. Chromatogr.,* 209, 61, 1981.

Table 25
TRIACYLGLYCEROLS (LLC)

Column	C1	C1	C2	C3	C4	C5	C6	C7	C7	C8
Solvent	S1	S1	S2	S3	S4	S5	S6	S7	S8	S9
Technique	T1	T1	T2	T3	T4	T5	T6	T7	T7	T8
Detection	D1	D1	D2	D3	D1	D4	D1	D1	D1	D1
Reference	1	2	3	4	5	6	7	8	8	9

Partition Number					**RRT × 100**					
12	3	—	—	—	—	—	—	—	—	—
18	6	—	—	—	—	30	—	—	—	—
24	12	—	—	—	—	—	7	50	—	—
28	—	—	—	14	—	—	—	80	—	—
30	21	—	—	24	100	61	27	100	—	—
32	—	36	—	36	—	71	—	130	—	—
34	—	73	—	60	—	85	—	160	—	—
36	41	100	—	100	92	100	100	210	100	100
38	51	—	—	168	—	116	—	260	125	—
40	62	—	—	284	—	134	—	—	155	89
42	100	—	—	460	86	152	—	—	190	—
44	—	—	—	—	—	171	—	—	240	—
46	—	—	—	—	—	189	—	—	300	81
48	—	—	47	—	80	205	—	—	366	—
50	—	—	74	—	—	218	—	—	466	—
52	—	—	100	—	—	231	—	—	566	—
54	—	—	—	—	76	245	—	—	—	74
56	—	—	—	—	—	255	—	—	—	—
60	—	—	—	—	70	—	—	—	—	68

Column
C1 = Factice
C2 = Cellulose powder-parrafin oil 50:9
C3 = Silanized Celite
C4 = Styragel 500A
C5 = Hydroxy (C_{15}-C_{18}) alkoxypropyl Sephadex (45% substituted)
C6 = VYDAC reversed phase support (Applied Science)
C7 = u Bondapak C_{18}
C8 = u Porasil

Solvent (v/v)
S1 = Acetone-water (95:5)
S2 = Acetone-methanol, (60:40)
S3 = Acetonitrile-methanol, (85:15)
S4 = Tetrahydrofuran
S5 = Isopropanol-chloroform-heptane-water (115:15:2:35) and heptane-acetone-water (4:15:1)
S6 = Methanol-water (90:10)
S7 = Acetonitrile
S8 = Acetonitrile-acetone (2:1)
S9 = Isooctane-ethyl ether-acetic acid (98:1:1)

Technique
T1 = 20°C
T2 = ''Dry column''
T3 = Very slow flow rates
T4 = Gel permeation
T5 = 40°C
T6 = 60°C, 1000 psig
T7 = Reversed phase HPLC
T8 = Normal phase HPLC

Detection
D1 = Differential refractometry
D2 = Iodine absorption
D3 = Gas-liquid chromatography
D4 = FI, moving wire

Table 25 (continued)
TRIACYLGLYCEROLS (LLC)

REFERENCES

1. **Hirsch, J.,** *J. Lipid Res.,* 4, 1, 1963.
2. **Knittle, J. L. and Hirsch J.,** *J. Lipid Res.,* 6, 565, 1965.
3. **Steiner, E. H. and Bonar, A. R.,** *Rev. Int. Choc.,* 20, 248, 1965.
4. **Nutter, L. J. and Privett, O. S.,** *J. Dairy Sci.,* 50, 1194, 1967.
5. **Bombaugh, K. J., Dar, W. A., and Levangie, R. F.,** *J. Chromatogr. Sci.,* 7, 42, 1969.
6. **Lindquist, B., Sjogren, I., and Nordin, R.,** *J. Lipid Res.,* 15, 65, 1974.
7. **Pei, P. T. S., Henly, R. S., and Ramachandran, S.,** *Lipids,* 10, 152 (1975).
8. **Plattner, R. D., Spencer, G. F., Kleiman, R.,** *J. Am. Oil Chem. Soc.,* 54, 511, 1977.
9. **Plattner, R. D.,** *Methods Enzymol.,* 72, 21, 1981.

Table 26
TRIACYLGLYCEROLS (LLC)

		C1	C2	C3	C3	C4	C5	C1	C3
Column		S1	S2	S3	S3	S3	S4	S5	S6
Solvent		T1	T2	T1	T1	T1	T1	T1	T3
Technique		D1	D1	D1	D1	D1	D1	D1	D2
Detection		1	2	3	4	4	5	6	7
Reference									
Molecular Species	**Partition Number**				RRT × 100				
18:3 18:3 18:3	36	—	174	28	29	26	—	—	—
18:3 18:3 18:2	38	100	—	33	—	—	—	—	—
13:0 13:0 13:0	39	—	—	—	41	37	—	—	—
18:3 18:3 18:1	40	125	—	41	—	—	—	—	—
18:3 18:3 16:0	40	—	—	41	—	—	—	—	—
14:0 14:0 14:0	42	—	—	—	58	45	—	—	—
18:2 18:2 18:2	42	159	131	49	—	—	—	60	62
18:2 18:2 18:1	44	209	—	60	—	—	—	67	73
18:2 18:2 16:0	44	—	—	64	—	—	—	—	76
15:0 15:0 15:0	45	—	—	—	82	83	—	—	—
18:2 18:1 18:1	46	261	—	77	—	—	—	81	85
16:0 18:2 18:1	46	—	—	81	—	—	—	—	89
16:0 16:0 18:2	46	—	—	87	—	—	—	—	—
16:0 18:1 18:1	48	—	—	—	—	—	108	—	103
18:1 18:1 18:1	48	340	100	100	100	100	100	100	100
18:1 18:1 18:1tr	48	—	—	—	110	116	—	—	—
18:0 16:0 18:2	48	—	—	107	—	—	—	—	—
16:0 16:0 18:1	48	—	—	114	112	116	116	—	100
18:0 18:1 18:2	48	—	—	—	—	—	—	123	—
16:0 16:0 16:0	48	—	—	136	124	127	—	—	—
18:0 18:1 18:1	50	—	—	—	138	152	—	—	119
16:0 18:1 18:0	50	—	—	—	147	190	—	—	123
18:0 16:0 18:0	52	—	—	—	165	—	—	—	—
18:0 18:1 18:0	52	545	—	—	188	—	—	—	—
18:0 18:0 18:0	54	—	95	—	—	—	—	—	—
18:0 18:2 22:0	54	—	—	—	—	—	—	152	—
18:1 18:1 22:0	54	—	—	—	—	—	—	188	—
18:1 18:1 24:0	56	—	—	—	—	—	—	323	—

Column
C1 = u Bondapak C-18
C2 = u Porasil
C3 = Supelcosil LC-18
C4 = Zorbax ODS
C5 = Lichrosorb RP-18

Solvent (v/v)
S1 = Acetonitrile-acetone (2:1)
S2 = Isooctane-ethyl ether-acetic acid (98:1:1)
S3 = Acetone-acetonitrile (63.6:36.4)
S4 = Acetonitrile-tetrahydrofuran-n-hexane (224:123.2:39.6)
S5 = Acetonitrile-acetone (42:58)
S6 = Linear gradient acetonitrile-propionitrile (70:30 to 10:90)

Detection
D1 = Refractive index
D2 = CI mass spectrometry

Technique
T1 = Reversed phase HPLC
T2 = Ordinary phase HPLC
T3 = Linear gradient reversed phase HPLC (60 min run)

Table 26 (continued)
TRIACYLGLYCEROLS (LLC)

REFERENCES

1. **Plattner, R. D., Spencer, G. F., and Kleiman, R.,** *J. Am. Oil Chem. Soc.,* 54, 511 (1977).
2. **Plattner, R. D.,** *Methods Enzymol.,* 72, 21, 1981.
3. **Perkins, E. G., Hendren, D. J., Pelick, N., and Bauer, J. E.,** *Lipids,* 17, 460, 1982.
4. **El-Hamdy, A. H. and Perkins, E. G.,** *J. Am. Oil Chem. Soc.,* 58, 867, 1981.
5. **Jensen, G. W.,** *J. Chromatogr.,* 204, 407, 1981.
6. **Bezard, J. A. and Ouedraogo, M. A.,** *J. Chromatogr.,* 196, 279, 1980; see also **Schulte, E.,** *Fette. Seifen, Anstrichm.,* 83, 289, 1981.
7. **Kuksis, A., Marai, L., and Myher, J. J.,** *J. Chromatogr.,* 273, 43, 1983.

Table 27
TRIACYLGLYCEROLS (AgNO₃-TLC)

Column	C1	C1	C2	C3
Solvent	S1	S1	S2	S3
Technique	T1	T2	T3	T4
Detection	D1	D1	D1	D1
Reference	1	1	2	3

Molecular Species	RRT × 100			
18:0 18:0 18:0	100	100	—	—
18:0 18:1 18:0	158	148	—	—
18:0 18:0 18:1	233	180	—	—
18:0 18:2 18:0	295	312	—	—
18:0 18:1 18:1	487	368	—	—
18:1 18:1 18:1	—	496	100	100
16:0 18:1 18:1	—	—	109	—
16:0 18:1 16:0	—	—	121	—
16:0 16:0 16:0	—	—	149	—
18:2 18:2 18:2	—	—	—	145
18:3 18:3 18:3	—	—	—	350

Column	C1 =	Partisil 5 with 10% AgNO₃ loading
	C2 =	Zorbax ODS
	C3 =	XN1010 resin (Na+ form) containing 32% silver
Solvent (v/v)	S1 =	Benzene
	S2 =	Acetonitrile-tetrahydrofuran-methylene chloride (60:20:20) with 0.2 N AgNO₃
	S3 =	Acetone (0.2 mℓ/min)
Technique	T1 =	Ambient temperature
	T2 =	6.8°C
	T3 =	0.2 N AgNO₃ added to mobile phase
	T4 =	0.6 × 72 cm glass column
Detection	D1 =	Refractive index

REFERENCES

1. **Smith, E. C., Jones, A. D., and Hammond, E. W.,** *J. Chromatogr.,* 188, 205, 1980.
2. **Plattner, R. D.,** *J. Am. Oil Chem. Soc.,* 58, 638, 1981.
3. **Adlof, R. O. and Emken, E. A.,** *J. Am. Oil Chem. Soc.,* 58, 99, 1981.

Table 28
ESTOLIDE TRIACYLGLYCEROLS (LLC)

Column	C1	C1	C2	C3
Solvent	S1	S2	S3	S4
Technique	T1	T1	T1	T2
Detection	D1	D1	D1	D1
Reference	1	1	1	1

Compound Class	Retention Volume (mℓ)			
Monohydroxy-triacylglycerols	—	15—20	—	—
Nonhydroxy-triacylglycerols	—	20—40	—	—
Multiacylglycerols: 70 carbons, 4 acyl groups	28—40	40—60	29—38	18—25
Multiacylglycerols: 82 carbons, 4 acyl groups	43—73	—	34—52	16—22
Multiacylglycerols: 96 carbons, 5 acyl groups	43—65	80—100	44—53	26—35
Multiacylglycerols: 104 carbons, 5 acyl groups	95—150	—	65—90	27—37
Multiacylglycerols: 112 carbons, 6 acyl groups	80—85	—	60—70	37—49
Multiacylglycerols: 126 carbons, 6 acyl groups	150—180	160—180	132—154	34—45

Column	C1 =	u Bondapak C-18
	C2 =	Waters triglyceride column
	C3 =	u Porasil
Solvent	S1 =	Acetone-acetonitrile (3:1)
(v/v)	S2 =	Acetonitrile-acetone (2:1)
	S3 =	Acetonitrile-tetrahydrofuran (3:1)
	S4 =	Isooctane-ethyl ether-acetic acid (98:2:1)
Technique	T1 =	Reversed phase HPLC
	T2 =	Normal phase HPLC
Detection	D1 =	Refractive index

REFERENCE

1. **Payne-Wahl, K., Plattner, R. D., Spencer, G. F., and Kleiman, R.,** *Lipids,* 14, 601, 1979.

Table 29
EXPOXYTRIACYLGLYCEROLS
(LLC)

Column	C1	C1
Solvent	S1	S2
Technique	T1	T2
Detection	D1	D2
Reference	1	2

Compound Class	**RRT × 100**	
Trivernoyl	36	36
Divernoyl-monolinoleoyl	39	39
Divernoyl-monooleoyl	50	50
Divernoyl-monostearoyl	60	60
Monovernoyl-dioleoyl	60	60
Trioleoyl	100	100

Column	C1 =	u Bondapak C_{18} column, 60 cm × 0.565 cm O.D.	
Solvent (v/v)	S1 =	Acetonitrile-acetone (2:1)	
	S2 =	Acetonitrile-acetone (2:1)	
Technique	T1 =	Chromatography	
	T2 =	Liquid chromatography	
Detection	D1 =	Differential refractometer	
	D2 =	Differential refractometer	

REFERENCES

1. **Plattner, R. D., Wade, K., and Kleiman, R.,** *J. Am. Oil Chem. Soc.,* 55, 381, 1978.
2. **Plattner, R. D., Spencer, G. F., and Kleiman, R.,** *J. Am. Oil Chem. Soc.,* 54, 511, 1977.

Table 30
TRIMETHYLSILYL AND TERTIARYBUTYLDIMETHYLSILYL DIACYLGLYCEROLS (LLC)

	C1	C2
Column	C1	C2
Solvent	S1	S2
Technique	T1	T2
Detection	D1	D2
Reference	1	2

Molecular Species	**RRT × 100**	
16:0 22:6	68	70
18:0 22:6	—	100
16:0 20:4	85	85
18:0 20:4	100	112
16:0 18:2	100	100
18:1 18:1	—	—
18:0 18:2	123	134
18:1 18:2	—	—
16:0 18:1	123	134
18:0 18:1	—	178

Column	C1 =	Hydroxyalkoxypropyl Sephadex
	C2 =	Supelcosil LC-18
Solvent	S1 =	Acetone-water-heptane (87:13:10) containing 1% pyridine
	S2 =	Linear gradient of acetonitrile-propionitrile (70:30 to 10:90)
Technique	T1 =	Reversed phase column chromatography
	T2 =	Reversed phase HPLC (60 min run)
Detection	D1 =	Gas-liquid chromatography after peak collection
	D2 =	CI mass spectrometry

REFERENCES

1. **Curstedt, T. and Sjovall, J.,** *Biochim. Biophys. Acta,* 360, 24, 1974.
2. **Kuksis, A., Marai, L., and Myher, J. J.,** unpublished results.

Table 31
TRIACYLGLYCEROLS (LLC-PAPER)

Paper	P1	P1	P1	P1	P1	P1	P2
Solvent	S1	S1	S2	S3	S4	S5	S6
Technique	T1	T2	T3	T3	T3	T3	
Detection	D1	D1	D1	D1	D1	D1	D2
Reference	1	1	2	2	2	2	3

Molecular Species				$RH^a \times 100$			
18:3 18:3 18:3	—	—	78	—	—	—	74
18:2 18:3 18:3	—	—	63	—	—	—	62
18:2 18:2 18:3	—	—	42	251	—	—	50
18:1 18:3 18:3	—	—	42	251	—	—	50
16:0 18:3 18:3	—	—	42	74	—	—	50
18:2 18:2 18:2	32	69	32	102	—	—	38
18:1 18:2 18:3	32	—	32	102	—	—	38
18:0 18:3 18:3	32	—	32	79	—	—	38
16:0 18:2 18:3	32	—	32	71	—	—	38
18:1 18:2 18:2	25	57	25	—	93	—	30
18:1 18:1 18:3	25	—	25	—	93	—	30
18:0 18:2 18:3	25	—	25	—	71	—	30
16:0 18:1 18:3	25	—	25	—	58	—	30
16:0 18:2 18:2	25	47	—	—	—	—	30
18:1 18:1 18:2	21	47	21	—	51	—	21
18:0 18:1 18:3	21	—	21	—	38	—	21
18:0 18:2 18:2	21	42	—	—	—	—	21
16:0 18:1 18:2	21	36	21	—	27	—	21
16:0 16:0 18:2	21	29	—	—	—	—	21
18:1 18:1 18:1	18	34	18	—	—	29	11
18:0 18:1 18:2	18	30	—	—	—	—	11
16:0 18:1 18:1	18	27	18	—	—	17	11
16:0 18:0 18:2	18	25	—	—	—	—	11
16:0 16:0 18:1	18	21	—	—	—	—	11
18:0 18:1 18:1	—	—	16	—	—	7	—

ª   RH = distance traveled by triacylglycerol/distance traveled by butylhexabromostearate.

Paper	P1 =	Paper impregnated with aliphatic hydrocarbons
	P2 =	Paper impregnated with tetradecane
Solvent	S1 =	Acetone-acetic acid (85:15)
(v/v)	S2 =	Acetone-acetic acid (50:50)
	S3 =	Methanol-water-AgNO₃ (80:20:saturated)
	S4 =	Methanol-water-AgNO₃ (95:5:saturated)
	S5 =	Methanol-water-AgNO₃ (98:2:saturated)
	S6 =	Acetone-acetonitrile (80:20)
Technique	T1 =	Reversed-phase chromatography
	T2 =	Brominated compounds
	T3 =	AgNO₃ in mobile phase
	T4 =	Developed 12 hours
Detection	D1 =	Sudan black reagent
	D2 =	Iodine vapor

Solvent S3 = Methanol-water-$AgNO_3$ (80:20:saturated)
S4 = Methanol-water-$AgNO_3$ (95:5:saturated)
S5 = Methanol-water-$AgNO_3$ (98:2:saturated)
Technique T3 = $AgNO_3$ in mobile phase

Table 31 (continued
TRIACYLGLYCEROLS (LLC-PAPER)

REFERENCES

1. **Vereshchagin, A. G.,** *Biokhimiya,* 27, 866, 1962.
2. **Novitskaya, G. V. and Maltseva, V. I.,** *Biokhimiya,* 3, 953, 1966.
3. **Kaufmann, H. P., Wessels, H. P., and Viswanathan, C. V.,** *Fette Seifen Anstrichm.,* 64, 509, 1962.

Table 32
TRIACYLGLYCEROLS AND DIMETHYLBORATE (DMB) DIACYLGLYCEROLS (LLC-TLC)

Layer	L1	L1	L1	L2	L2	L1	L3	L4
Solvent	S1	S2	S1	S3	S4	S5	S6	S7
Technique	T1	T1	T2	T3	T3	T4	T5	T6
Detection	D1	D1	D1	D2	D2	D3	D3	D2
Reference	1	1	2	3	3	4	5	6
Molecular Species				**$R_f \times 100$**				
18:2 18:2 18:3	—	—	—	—	—	89	65	—
18:1 18:2 18:3	—	92	—	—	—	—	—	—
18:2 18:2 18:2	86	—	—	—	61	80	60	—
18:1 18:2 18:2	—	84	—	—	48	67	47	—
18:1 18:1 18:3	—	84	—	—	—	—	—	—
18:0 18:2 18:2	74	72	—	—	31	—	—	—
16:0 18:2 18:2	70	80	—	—	—	67	47	—
18:1 18:1 18:2	65	74	—	—	—	—	—	—
18:0 18:2 18:2	61	—	—	—	—	54	36	—
16:0 18:1 18:2	63	69	—	—	—	54	36	—
16:0 16:0 18:2	61	—	—	—	—	—	—	—
18:1 18:1 18:1	55	63	64	62	27	44	27	—
18:0 18:1 18:2	51	61	—	—	—	44	27	—
16:0 18:0 18:2	49	—	—	—	—	—	—	—
16:0 18:1 18:1	55	59	60	49	23	44	27	—
16:0 16:0 18:1	47	55	55	36	—	—	—	—
16:0 16:0 16:0	—	—	50	—	—	—	—	—
18:0 18:1 18:1	41	51	—	44	20	33	—	—
18:0 18:0 18:2	—	—	—	—	17	—	—	—
16:0 18:0 18:1	—	47	—	—	—	33	—	—
18:0 18:0 18:1	29	41	—	28	—	—	—	—
16:0 18:0 18:1	37	—	—	—	—	—	—	—
18:0 18:0 DMB	—	—	—	—	—	—	—	20
18:1 18:1 DMB	—	—	—	—	—	—	—	28
18:0 18:2 DMB	—	—	—	—	—	—	—	36
18:1 18:1 DMB	—	—	—	—	—	—	—	36
18:1 18:2 DMB	—	—	—	—	—	—	—	44
18:2 18:2 DMB	—	—	—	—	—	—	—	44
18:3 18:1 DMB	—	—	—	—	—	—	—	52
18:3 18:2 DMB	—	—	—	—	—	—	—	60
18:3 18:3 DMB	—	—	—	—	—	—	—	68

Layer
L1 = Kieselguhr G, impregnated with liquid paraffin
L2 = Silanized kieselguhr
L3 = Silanized silicic acid impregnated with 8% paraffin
L4 = Silanized kieselguhr impregnated with 10% tetradecane

Solvent
(v/v)
S1 = Acetone-acetonitrile (8:2), 80% saturated with paraffin
S2 = Chloroform-methanol (99.2:0.8), 80% saturated with paraffin
S3 = Acetone-ethanol-water-acetonitrile-AgNO$_3$ (72:18:8:2:saturated)
S4 = Acetone-ethanol-water-acetonitrile-AgNO$_3$ (83:8:7:2:saturated)
S5 = Acetone-acetic acid (70:30), 80% saturated with paraffin
S6 = Acetone-acetonitrile (70:30)
S7 = Methanol-trimethylborate (92.7:7.3), saturated with n-tetradecane

Technique
T1 = Developed two times
T2 = Developed three times
T3 = AgNO$_3$ in polar phase
T4 = Developed three times, 30 minutes each time
T5 = Developed three times
T6 = Home-made plates with a permanent support layer

Table 32 (continued)
TRIACYLGLYCEROLS AND DIMETHYLBORATE (DMB)
DIACYLGLYCEROLS (LLC-TLC)

Detection   D1 = Cyclodextrin-iodine
               D2 = Phosphomolybdic acid
               D3 = Iodine vapor

REFERENCES

1. **Wessels, H. and Rajagopal, N. S.,** *Fette Seifen Anstrichm.,* 71, 543, 1969.
2. **Kaufmann, H. P. and Das, B.,** *Fette Seifen Anstrichm.,* 64, 214, 1962.
3. **Ord, W. O. and Bamford, P. C.,** *Chem. Ind. (London),* 277, 1967.
4. **Kwapniewski, Z. and Sliwiok, J.,** *Microchim. Acta,* 616, 1964.
5. **Anonymous,** *Gas Chrom. Newsletter,* 6, (4), 2, 1965.
6. **Pchelkin, V. P. and Vereshchagin, A. G.,** *J. Chromatogr.,* 209, 49, 1981.

Table 33
TRIACYLGLYCEROLS (GLC)

Column packing	P1	P2	P3	P4	P3	P3	P5	P6
Temperature program	T1	T2	T3	T4	T5	T6	T7	T8
Gas	N_2	N_2	H_2	N_2	H_2	N_2	N_2	Ar
Flow rate (mℓ/min)	100	100	100	25	100	150	100	20
Column								
Length(cm)	50	150	56	35	183	60	50	90
Diameter								
(cm I.D.)	0.2	0.2	0.24	0.4	0.24	0.2	0.2	0.2
(cm O.D.)	0.3	0.3	0.635	0.635	0.635	0.3	0.3	0.3
Form	Coil	Coil	U-tube		U-tube	Coil	U-tube	Coil
Material	SS	SS	Glass	Glass	Glass	SS	SS	SS
Detector	FI	FI	FI	FI	FI	FI	FI	FI
Reference	1	2	3	4	5	6	7	8

Carbon Number				RRT × 100				
12	—	—	—	—	—	11	—	—
14	—	—	—	—	—	21	—	—
16	—	—	—	—	—	30	—	—
18	—	—	—	—	—	39	—	—
20	—	—	—	—	—	48	5	—
22	—	68	—	—	—	56	10	—
24	2	76	33	—	—	62	19	11
25	—	80	—	—	—	—	—	—
26	8	84	—	—	—	69	30	19
27	—	88	—	—	—	—	—	—
28	14	92	—	15	—	75	43	29
29	—	96	—	—	—	—	—	—
30	24	100	67	25	—	82	57	43
31	—	104	—	—	—	—	—	—
32	34	108	—	33	—	88	71	61
33	—	112	—	—	—	—	—	—
34	44	116	—	41	—	95	86	81
35	—	120	—	—	—	—	—	—
36	57	124	100	51	—	100	100	100
37	—	—	—	—	—	—	—	—
38	63	—	—	59	—	105	112	119
39	—	—	—	—	—	—	—	—
40	71	—	—	68	—	111	122	140
41	—	—	—	—	—	—	—	—
42	78	—	128	76	76	116	133	156
43	—	—	—	—	80	—	—	—
44	85	—	—	84	84	120	144	177
45	—	—	—	—	88	—	—	—
46	92	—	—	91	92	125	156	193
47	—	—	—	—	96	—	—	—
48	100	—	154	100	100	129	166	206
49	—	—	—	—	104	—	—	—
50	107	—	—	107	108	134	175	224
51	—	—	—	—	112	—	—	—
52	115	—	—	116	116	139	184	239
53	—	—	—	—	120	—	—	—
54	121	—	178	125	123	145	193	254
55	—	—	—	—	126	—	—	—
56	127	—	—	—	130	—	204	—
57	—	—	—	—	134	—	—	—
58	132	—	—	—	138	—	—	—
59	—	—	—	—	143	—	—	—

Table 33 (continued)
TRIACYLGLYCEROLS (GLC)

Carbon Number					RRT × 100			
60	138	—	—	—	146	—	—	—
62	—	—	—	—	150	—	—	—
64	142	—	—	—	155	—	—	—
66	—	—	—	—	160	—	—	—

Column packing	P1 =	2.25% SE-30 on Chromosorb W (60-80 mesh)
	P2 =	5% SE-30 on silanized Chromosorb W (60-80 mesh)
	P3 =	3% JXR on Gas Chrom Q (100-120 mesh)
	P4 =	2% OV-17 on Shimalite W (80-100 mesh)
	P5 =	3% OV-1 on Gas Chrom Q (100-120 mesh)
	P6 =	2% Dexil 300 on Anachrom Q (60-70 mesh)
	P7 =	Open tubular Poly-S179
	P8 =	1% OV-1 on 100-120 mesh Gas Chrom Q
Temperature program	T1 =	200-320°C at 3°C/min
	T2 =	200-375°C at 4°C/min
	T3 =	170-305°C at 3°C/min
	T4 =	200-310°C at 4°C/min
	T5 =	210-375°C at 4°C/min
	T6 =	90-270°C at 4°C/min
	T7 =	200-320°C at 3°C/min
	T8 =	230-350°C at 1°C/min

REFERENCES

1. **Kuksis, A. and McCarthy, M. J.,** *Can. J. Biochem. Physiol.,* 40, 679, 1962.
2. **Kuksis, A. and Breckenridge, W. C.,** *J. Am. Oil Chem. Soc.,* 42, 978, 1965.
3. **Litchfield, C., Harlow, R. D. and Reiser, R.,** *J. Am. Oil Chem. Soc.,* 42, 849, 1965.
4. **Sato, K., Matsui, M., and Ikekawa, N.,** *Bunseki Kaguku,* 16, 1160, 1967.
5. **Litchfield, C., Harlow, R. D., and Reiser, R.,** *Lipids,* 2, 363, 1967.
6. **Breckenridge, W. C. and Kuksis, A.,** *J. Lipid Res.,* 9, 388, 1968.
7. **Kuksis, A., Marai, L., and Myher, J. J.,** *J. Am. Oil Chem. Soc.,* 50, 193, 1973.
8. **Lin, C. Y., Smith, S., and Abraham, S.,** *J. Lipid Res.,* 17, 647, 1976.

Table 34
TRIACYLGLYCEROLS (GLC)

Column packing	P1	P2	P2	P2	P3
Temperature program	275	285	305	330	280
Gas	N_2	N_2	N_2	N_2	He
Flow rate (mℓ/min)	100	35	35	35	40
Column					
Length(cm)	244	53	53	53	180
Diameter					
(cm I.D.)	0.2	0.3	0.3	0.3	0.2
(cm O.D.)	0.3	0.6	0.6	0.6	0.3
Form	Coil	Coil	Coil	Coil	U-tube
Material	SS	SS	SS	SS	SS
Detector	FI	FI	FI	FI	FI
Reference	1	2	2	2	3

Rac-Triacylglycerols

RRT × 100

			P1	P2	P2	P2	P3
4:0	16:0	4:0	60	—	—	—	—
16:0	4:0	4:0	64	—	—	—	—
16:0	10:0	4:0	—	—	—	—	32
6:0	16:0	6:0	80	—	—	—	—
16:0	6:0	6:0	82	—	—	—	—
14:0	14:0	4:0	—	—	—	—	45
14:0	14:0	6:0	—	—	—	—	—
12:0	14:0	6:0	—	—	—	—	55
16:0	14:0	4:0	—	—	—	—	66
12:0	12:0	12:0	—	71	—	—	—
16:0	4:0	16:0	—	82	—	—	—
16:0	16:0	4:0	—	72	—	—	—
16:0	14:0	6:0	—	—	—	—	82
16:0	16:0	4:0	—	—	—	—	100
16:0	16:1	4:0	—	—	—	—	106
18:0	16:0	4:0	—	—	—	—	138
14:0	18:1	6:0	—	—	—	—	138
14:0	18:2	6:0	—	—	—	—	147
16:0	18:1	4:0	—	—	—	—	147
16:0	18:2	4:0	—	—	—	—	155
18:0	18:0	4:0	—	—	—	—	207
18:0	4:0	18:0	—	—	81	—	—
18:1	4:0	18:1	—	—	74	—	—
18:1	18:1	4:0	—	—	71	—	—
16:0	18:1	6:0	—	—	—	—	207
16:0	18:2	6:0	—	—	—	—	227
18:0	18:1	4:0	—	—	—	—	227
18:0	18:2	4:0	—	—	—	—	246
18:1	18:1	18:1	—	—	—	201	—
18:0	18:0	18:0	—	—	—	213	—
18:2	18:2	18:2	—	—	—	201	—
18:3	18:3	18:3	—	—	—	201	—

Column packing	P1 = 5% SE-30 on Chromosorb W (60-80 mesh)
	P2 = 10% SE-30 on Gas Chrom Q (100-120 mesh)
	P3 = 3% EGSS-X on Gas Chrom Q (100-120 mesh)

Table 34 (continued)
TRIACYLGLYCEROLS (GLC)

REFERENCES

1. **Kuksis, A. and Breckenridge, W. C.,** *J. Am. Oil Chem. Soc.* 42, 978, 1965.
2. **Watts, R. and Dils, R.,** *J. Lipid Res.,* 9, 40, 1968.
3. **Kuksis, A. Marai, L., and Myher, J. J.,** *J. Am. Oil Chem. Soc.,* 50, 193, 1973.

Table 35
ISOVALEROYL DIACYLGLYCEROLS (GLC)

Column packing	P1	P2	P1
Temperature program	T1	T2	T3
Gas	He	He	He
Flow rate (mℓ/min)	100	20	100
Column			
Length(cm)	56	30	60
Diameter			
(cm I.D.)	0.25	0.2	0.2
(cm O.D.)	0.635	0.3	0.3
Form	U-tube	Coil	U-tube
Material	SS	SS	SS
Detector	FI	FI	FI
Reference	1	2	3

Mixed Positional Isomers		**RRT × 100**		
31		71	—	—
33		83	—	60
35		94	95	62
36	(12:0 12:0 12:0)	100	100	—
37		106	106	72
38		112	—	—
39		118	117	78
40		125	—	—
41		128	125	82
43		137	—	87
45		148	—	—
47		158	—	—
48	(16:0 16:0 16:0)	162	—	100

Column packing	P1 =	3% JXR on Gas Chrom Q (100-120 mesh)
	P2 =	2.5% SE-30 on Aeropak 30 (100-120 mesh)
Temperature program	T1 =	160-350°C at 4°C/min
	T2 =	250-300°C at 5°C/min
	T3 =	160-360°C at 4°C/min

REFERENCES

1. **Litchfield, C., Ackman, R. G., Sipos, J. C., and Eaton, C. A.,** *Lipids,* 6, 674, 1971.
2. **Blank, M. L., Kasama, K., and Snyder, F.,** *J. Lipid Res.,* 13, 390, 1972.
3. **Ackman, R. G., Eaton, C. A., Kinneman, J., and Litchfield, C.,** *Lipids,* 10, 44, 1975.

Table 36
TRIACYLGLYCEROLS (GLC)

Column packing	P1	P2	P1	P2	P1	P2
Temperature program	T1	T1	T2	T2	T3	T3
Gas	N_2	N_2	N_2	N_2	N_2	N_2
Flow rate (mℓ/min)	35	35	35	35	35	35
Column						
Length(cm)	168	53	168	53	168	53
Diameter						
(cm I.D.)	0.3	0.3	0.3	0.3	0.3	0.3
(cm O.D.)	0.65	0.65	0.65	0.65	0.65	0.65
Form	Coil	Coil	Coil	Coil	Coil	Coil
Material	SS	SS	SS	SS	SS	SS
Detector	FI	FI	FI	FI	FI	FI
Reference	1	1	1	1	1	1

Carbon Number	Relative Elution Temperature[a]					
6	0.493	0.493	0.485	0.425	0.485	0.418
9	0.523	0.463	0.533	0.469	0.535	0.482
12	0.576	0.522	0.593	0.534	0.596	0.549
15	0.648	0.599	0.664	0.610	0.668	0.628
18	0.710	0.671	0.724	0.679	0.729	0.691
21	0.763	0.735	0.778	0.744	0.781	0.754
24	0.813	0.796	0.826	0.802	0.825	0.811
30	0.932	0.908	0.932	0.906	0.929	0.914
36	1.000	1.000	1.000	1.000	1.000	1.000
42	1.08	1.08	1.08	1.09	1.07	—
48	1.14	—	1.14	—	—	—
54	1.20	—	—	—	—	—

[a]   Mean of three replicates.

Column packing	P1 =	3% QF-1 on acid-washed silanized Chromosorb W (100-120 mesh)
	P2 =	10% SE-30 on Gas Chrom Q (100-120 mesh)
Temperature program	T1 =	118°C to 322°C at 1.85°C/min
	T2 =	118°C to 322°C at 3.71°C/min
	T3 =	188°C to 322°C at 5.56°C/min

REFERENCE

1. **Watts, R. and Dils, R.,** *J. Lipid Res.,* 9, 40, 1968.

Table 37
1,3-DIOXOLANETRIACYLGLYCEROLS (GLC)

Column packing	P1
Temperature program	T1
Gas	He
Flow rate (mℓ/min)	60
Column	
Length(cm)	65
Diameter	
(cm I.D.)	0.2
(cm O.D.)	0.3
Form	Coil
Material	SS
Detector	FI
Reference	1

Compound Class[a]	RRT × 100
PPS	144
PSS	154
SSS	164
SSA	175
VSP	183
VSS	192
VAS	202
VVP	209
VVS	210
VVA	228
VVV	242

Column packing	P1 =	3% OV-1 on Gas Chrom Q (60-80 mesh)
Temperature program	T1 =	260-374°C at 4°C/min

REFERENCE

1. **Fioriti, J. A., Kanuk, M. J., and Sims, R. J.,** *J. Chromatogr. Sci.,* 7, 448, 1969.

Table 38
ACETYL DIACYLCLYCEROLS (GLC)

Column packing	P1	P2	P2	P3	P3	P2
Temperature program	T1	T2	T3	T4	T5	T6
Gas	N_2	N_2	N_2	N_2	He	N_2
Flow rate (mℓ/min)	150	150	150	150	100	40
Column						
Length(cm)	60	60	60	46	70	61
Diameter						
(cm I.D.)	0.2	0.2	0.2	0.2	0.25	0.2
(cm O.D.)	0.3	0.3	0.3	0.3	0.4	0.3
Form	U-tube	U-tube	U-tube	U-tube	U-tube	Coil
Material	SS	SS	SS	SS	Glass	SS
Detector	FI	FI	FI	FI	FI	FI
Reference	1	2	3	4	5	6

1,2(2,3)-Isomers		RRT × 100				
16	—	40	41	—	—	—
18	—	48	50	—	—	—
20	—	51	60	—	—	—
22	—	68	71	—	—	30
24	—	76	78	—	—	45
26	—	84	85	—	—	61
28	—	92	93	—	61	80
30	100	100	100	100	—	100
32	108	108	108	—	81	120
34	117	116	115	—	—	140
36	124	123	122	167	100	160
38	131	129	128	182	—	175
40	—	—	—	200	—	187
42	—	—	—	233	—	—

Column packing
P1 = 1% OV-17 on Gas Chrom Q (100-120 mesh)
P2 = 3% JXR on Gas Chrom Q (100-120 mesh)
P3 = 1% OV-1 on Gas Chrom Q (100-120 mesh)

Temperature program
T1 = 100-325°C at 4°C/min
T2 = 90-270°C at 4°C/min
T3 = 100-300°C at 4°C/min
T4 = 210-280°C at 4°C/min
T5 = 150-275°C at 5°C/min
T6 = 150-300°C at 4°C/min

REFERENCES

1. **Kuksis, A., Marai, L., and Gornall, D. A.,** *J. Lipid Res.,* 8, 352, 1967.
2. **Breckenridge, W. C. and Kuksis, A.,** *J. Lipid Res.,* 9, 388, 1968.
3. **Marai, L., Breckenridge, W. C., and Kuksis, A.,** *Lipids,* 4, 562, 1969.
4. **Kuksis, A., Breckenridge, W. C., Marai, L., and Stachnyk, O.,** *J. Am. Oil. Chem. Soc.,* 45, 537, 1968.
5. **Wood, R., Baumann, W. J., Snyder, F., and Mangold, H. K.,** *J. Lipid Res.,* 10, 128, 1969.
6. **Parodi, P. W.,** *J. Chromatogr.,* 111, 223, 1975.

Table 39
ACETYL DIACYLGLYCEROLS (GLC)

Column packing	P1	P2	P3	P4	P5	P5
Temperature program	T1	T2	T3	T4	T5	T5
Gas	N_2	N_2	Ar	He	N_2	N_2
Flow rate(mℓ/min)	150	100	150	50	20	20
Column						
Length (cm)	60	50	60	35	150	150
Diameter						
(cm I.D.)	0.2	0.2	0.3	0.3	0.3	0.3
(cm O.D.)	0.3	0.3	0.635	0.635	0.635	0.635
Form	U-tube	U-tube	U-tube	Coil	Coil	Coil
Material	SS	SS	Glass	Glass	Glass	Glass
Detector	FI	FI	Ar	FI	FI	FI
Reference	1	2	3	4	5	5

1,2-Isomers	RRT × 100				RRT × 1000	ECL
Saturates						
28:0	—	—	—	—	589	28.00
30:0	—	—	—	—	766	30.00
32:0	—	70	—	—	1000	32.00
34:0	—	85	—	—	1307	34.00
36:0	—	100	390	—	1724	36.00
38:0	—	114	413	—	2287	38.00
40:0	—	—	—	—	3069	40.00
42:0	—	—	—	—	4147	42.00
Monoenes						
28:1						
30:1	—	—	—	100	—	—
32:1	—	—	—	141	—	—
34:1	85	85	—	155	1525	35.11
36:1	100	100	—	169	2017	37.11
38:1	114	114	—	—	—	—
40:1	130	130	—	—	—	—
30:2	—	—	—	—	—	—
32:2	—	—	—	—	—	—
34:2	83	83	—	157	1815	36.37
36:2 (18:1 18:1)	100	100	384	170	2292	38.01
36:2 (18:0 18:2)	113	100	384	—	2355	38.20
Trienes						
36:3	93	93	—	172	2730	39.21
38:3	113	112	—	—	—	—
40:3	128	128	—	—	—	—
Tetraenes						
34:4	—	—	—	—	—	—
36:4	93	93	—	168	3281	40.43
38:4	113	112	384	191	—	—
40:4	128	127	423	—	—	—
42:4	138	138	—	—	—	—
44:4	—	—	—	—	—	—

Table 39 (continued)
ACETYL DIACYLGLYCEROLS (GLC)

1,2-Isomers	RRT × 100				RRT × 1000	ECL
Pentaenes						
34:5	—	—	—	—	—	—
36:5	93	92	—	—	3999	41.72
38:5	106	105	—	—	—	—
40:5	119	118	—	—	—	—
42:5	133	132	—	—	—	—
44:5	—	—	—	—	—	—
Hexaenes						
34:6						
36:6	93	92	—	—	4903	43.1
38:6	106	105	—	—	—	—
40:6	118	118	—	—	—	—
42:6	134	132	—	—	—	—
44:6	—	—	—	—	—	—

Column packing	P1 =	1% OV-17 on Gas Chrom Q (100-120 mesh)
	P2 =	3% JXR on Gas Chrom Q (100-120 mesh)
	P3 =	3% JXR on Gas Chrom Q (100-120 mesh)
	P4 =	2.5% OV-17 on Shimalite W (60-80 mesh)
	P5 =	3% Silar 10C on Gas Chrom Q (100-120 mesh)
Temperature program	T1 =	195-280°C at 4°C/min
	T2 =	200-280°C at 4°C/min
	T3 =	200-280°C at 5°C/min
	T4 =	180-280°C at 3°C/min
	T5 =	270°C isothermal

REFERENCES

1. **Kuksis, A. and Marai, L.,** *Lipids,* 2, 217, 1967.
2. **Kuksis, A., Breckenridge, W. C., Marai, L., and Stachnyk, O.,** *J. Lipid Res.,* 10, 25, 1969.
3. **Holub, B. J., Breckenridge, W. C., and Kuksis, A.,** *Lipids,* 6, 307, 1971.
4. **Hasegawa, K., and Suzuki, T.,** *Lipids,* 8, 631, 1973.
5. **Itabashi, Y. and Takagi, T.,** *Lipids,* 15, 205, 1980.

Table 40
ACETYL DIACYLGLYCEROLS (GLC)

Column packing	P1	P2	P3	P1
Temperature (°C)	235	250	260	240
Gas	N_2	N_2	N_2	He
Flow rate (mℓ/min)	30	40	40	50
Column				
Length (cm)	180	180	180	183
Diameter				
(cm I.D.)	0.2	0.3	0.3	0.3
(cm O.D.)	0.3	0.635	0.635	0.635
Form	U-tube	U-tube	U-tube	U-tube
Material	Glass	Glass	Glass	Glass
Detector	FI	FI	FI	FID
Reference	1	2	3	4

1,2(2,3)-Isomers	RRT × 100			ECL
10:0 6:0	33	—	—	—
12:0 4:0	37	—	—	—
12:0 6:0	55	—	—	—
14:0 4:0	59	50	50	—
15:0 4:0	76	—	—	—
18:0 2:0	—	—	—	18.000
16:0 4:0	—	—	—	18.975
14:0 6:0	—	—	—	16.667
12:0 8:0	—	—	—	16.475
10:0 10:0	—	—	—	16.466
12:0 8:0	36	—	—	—
14:0 6:0	91	—	99	—
16:0 4:0	100	100	100	—
16:0 4:0	118	112	—	—
17:0 4:0	128	—	—	—
14:0 8:0	142	—	—	—
16:0 6:0	152	—	—	—
18:0 4:0	164	160	160	—
18:1 4:0	188	182	162	—
16:2 4:0	236	—	—	—
16:0 8:0	248	—	—	—
18:0 6:0	265	260	265	—
18:1 6:0	300	280	265	—
10:0 16:0	380	—	—	—
16:0 14:0	45	—	—	—
18:1 8:0	45	—	—	—

Column packing	P1	=	3% EGSS-X on Gas Chrom Q (100-120 mesh)
	P2	=	3% NPGS on Gas Chrom Q (100-120 mesh)
	P3	=	3% CHDMS on Gas Chrom Q (100-120 mesh)
	P4	=	3% Silar 5CP on Gas Chrom Q (100-120 mesh)

REFERENCES

1. **Kuksis, A., Marai, L., and Myher, J. J.,** *J. Am. Oil Chem. Soc.,* 50, 193, 1973.
2. **Kuksis, A.,** unpublished results.
3. **Kuksis, A.,** unpublished results.
4. **Myher, J. J. and Kuksis, A.,** unpublished results.

Table 41
TRIMETHYLSILYL DIACYLGLYCEROLS (GLC)

Column packing	P1	P2	P3	P4	P4
Temperature program	T1	T2	T3	T4	T4
Gas	N_2	N_2	N_2	N_2	N_2
Flow rate (mℓ/min)	150	80	80	45	45
Column					
Length (cm)	60	50	50	360	360
Diameter					
(cm I.D.)	0.2	0.2	0.2	0.3	0.3
(cm O.D.)	0.3	0.3	0.3	0.4	0.4
Form	U-tube	U-tube	U-tube	W-tube	W-tube
Material	SS	SS	SS	Glass	Glass
Detector	FI	FI	FI	FI	FI
Reference	1	2	3	4	5

1,2(2,3)-Isomers	RRT × 100					MU Value
14:0 14:0	—	84	100	—	—	35.12
14:0 16:0	—	90	—	—	—	—
16:0 16:0	—	95	123	—	—	38.94
14:0 18:0	86	100	131	—	—	—
16:0 18:0	100	105	141	—	—	—
16:0 18:1	100	113	149	80	—	—
16:0 18:2	100	117	157	79	—	—
16:0 20:4	111	127	—	—	—	—
16:0 22:6	120	—	—	—	—	—
18:0 18:0	111	—	—	—	—	42.85
18:0 18:1	111	—	—	100	—	42.63
18:1 18:1	111	—	—	—	—	—
18:0 18:2	—	—	—	—	—	—
18:0 20:3	120	—	—	—	—	44.50

1,3-Isomers

14:0 14:0	—	—	—	—	—	35.40
14:0 16:0	—	—	—	—	—	—
16:0 16:0	—	—	—	—	—	39.27
16:0 18:0	—	—	—	—	—	—
16:0 18:1	—	—	—	83	—	—
16:0 18:2	—	—	—	82	—	—
16:0 20:4	—	—	—	—	—	—
16:0 22:6	—	—	—	—	—	—
18:0 18:0	—	—	—	—	—	43.20
18:0 18:1	—	—	—	106	—	42.48
18:1 18:1	—	—	—	—	—	42.79
18:0 18:2	—	—	—	—	—	—
18:0 20:3	—	—	—	—	—	44.87

Column packing P1 = 1% OV-17 on Gas Chrom Q (100-120 mesh)
P2 = 3% OV-1 on Gas Chrom Q (100-120 mesh)
P3 = 3% OV-1 on Gas Chrom Q (100-120 mesh)
P4 = 1% SE-30 on Gas Chrom P (100-120 mesh)

Temperature program T1 = 190-315°C at 4°C/min
T2 = 100-380°C at 4°C/min
T3 = 175-350°C at 4°C/min
T4 = 270-350°C at 2°C/min

Table 41 (continued)
TRIMETHYLSILYL DIACYLGLYCEROLS (GLC)

REFERENCES

1. **Kuksis, A., Stachnyk, O., and Holub, B. J.,** *J. Lipid Res.,* 10, 660, 1969.
2. **Kuksis, A., Marai, L., and Myher, J. J.,** *J. Am. Oil Chem. Soc.,* 50, 193, 1973.
3. **Kuksis, A., Myher, J. J., Marai, L., and Geher, K.,** *J. Chromatogr. Sci.,* 13, 423, 1975.
4. **Horning, M. G., Casparrini, G., and Horning, E. C.,** *J. Chromatogr. Sci.,* 7, 267, 1969.
5. **Casparrini, G., Horning, M. G., and Horning, E. C.,** *Anal. Lett.,* 1, 481, 1968.

Table 42
ISOVALEROYL
MONOACYLGLYCEROLS[a] (GLC)

Column packing	P1	P2	P1
Temperature program	T1	T2	T3
Gas	He	He	He
Flow rate (mℓ/min)	100	30	100
Column			
Length (cm)	56	200	60
Diameter			
(cm I.D.)	0.2	0.3	0.2
(cm O.D.)	0.3	0.635	0.3
Form	U-tube	U-tube	U-tube
Material	SS	Glass	SS
Detector	FI	FI	FI
Reference	1	2	3

Carbon Numbers		RRT × 100	
23	—	47	—
24 (iso-acyl)	16	53	23
24	—	56	24
25	—	61	28
26 (iso-acyl)	24	70	31
26	—	75	38
28	31	100	41
48 (16:0 16:0 16:0)	100	—	100

[a]   *sn*-2-Isomers.

Column packing	P1 =	3% JXR on Gas Chrom Q (100-120 mesh)
	P2 =	1% SE-30 on Chromosorb W (100-120 mesh)
Temperature program	T1 =	160-350°C at 4°C/min
	T2 =	220-290°C at 10°C/min
	T3 =	160-360°C at 4.6°C/min

REFERENCES

1.   **Litchfield, C., Ackman, R. G., Sipos, J. C., and Eaton, C. E.,** *Lipids,* 6, 674, 1971.
2.   **Blomberg, J.,** *Lipids*, 9, 461, 1974.
3.   **Ackman, R. G., Eaton, C. A., Kinneman, J., and Litchfield, C.,** *Lipids*, 10, 44, 1975.

Table 43
TRIACYLGLYCEROLS (GLC)

Column packing	P1	P1	P2	P2	P3
Temperature program	T1	T1	T2	T3	T4
Gas	N_2	N_2	He	N_2	N_2
Flow rate (mℓ/min)	120	120	100	100	80
Column					
Length (cm)	50	50	180	50	45
Diameter					
(cm I.D.)	0.2	0.2	0.25	0.2	0.25
(cm O.D.)	0.3	0.3	0.635	0.3	0.635
Form	Coil	Coil	U-tube	Coil	Coil
Material	SS	SS	Glass	SS	Glass
Detector	FI	FI	FI	FI	FI
Reference	1	2	3	4	5

Carbon Number: Double Bond Number	RRT × 100				
Saturates					
24:0	18	—	—	—	—
26:0	30	—	—	—	—
28:0	44	—	—	—	—
30:0	58	—	—	—	—
32:0	73	—	—	—	—
33:0	79	—	—	—	—
34:0	87	—	—	—	—
35:0	93	—	—	—	—
36:0	100	100	—	100	100
37:0	106	106	—	—	—
38:0	112	112	—	113	—
39:0	—	—	—	—	—
40:0	125	125	—	124	—
41:0	130	130	—	—	—
42:0	136	136	—	135	170
43:0	143	143	—	—	—
44:0	150	150	85	148	—
45:0	156	156	89	—	—
46:0	162	162	93	159	—
47:0	167	167	96	—	—
48:0	172	172	100	170	234
49:0	178	178	103	—	—
50:0	183	183	107	180	—
51:0	188	188	110	—	—
52:0	192	192	114	189	—
53:0	—	—	117	—	—
54:0	202	202	121	200	288
56:0	—	—	127	—	—
Monoenes					
32:1	73	—	—	—	—
34:1	87	—	—	—	—
36:1	100	—	—	—	—
38:1	112	—	—	—	—
40:1	125	125	—	—	—
42:1	136	136	—	134	—
44:1	150	150	85	148	—
46:1	162	162	93	157	—
48:1	—	172	100	168	—
50:1	—	183	38107	181	—

Table 43 (continued)
TRIACYLGLYCEROLS (GLC)

Carbon Number: Double Bond Number		RRT × 100			
Monoenes					
52:1	—	192	114	189	—
54:1	—	202	121	198	—
56:1	—	—	129	207	—
58:1	—	—	—	216	—
Dienes					
32:2	68	—	—	—	—
34:2	80	—	—	—	—
36:2	95	—	—	—	—
38:2	108	—	—	—	—
40:2	120	119	—	—	—
42:2	132	130	—	—	—
44:2	145	143	—	141	—
46:2	157	155	93	152	—
48:2	—	167	100	163	—
50:2	—	177	107	171	—
52:2	—	187	114	181	—
54:2	—	198	121	189	—
56:2	—	—	129	198	—
58:2	—	—	—	206	—
Trienes					
46:3	—	153	91	150	—
48:3	—	164	98	160	—
50:3	—	174	105	170	—
52:3	—	184	112	180	—
54:3	—	195	119	190	298
56:3	—	—	127	200	—
Hexaenes					
54:6	—	—	—	—	312

Column packing	P1 =	2% JXR on Gas Chrom Q (100-120 mesh)
	P2 =	3% JXR on Gas Chrom Q (100-120 mesh)
	P3 =	2% Poly-S-179 on Supelcoport (100–120 mesh)
Temperature program	T1 =	180-305°C at 3°C/min
	T2 =	200-350°C at 4°C/min
	T3 =	180-330°C at 3°C/min
	T4 =	220-335°C at 3°C/min

REFERENCES

1. **Breckenridge, W. C. and Kuksis, A.,** *Lipids*, 3, 291, 1968.
2. **Breckenridge, W. C. and Kuksis, A.,** *Lipids*, 4, 197, 1969.
3. **Culp, T. W., Creger, C. R., Swanson, A. A., Couch, J. R., and Harlow, R. D.,** *Exp. Eye Res.*, 7, 134, 1968.
4. **Breckenridge, W. C., Marai, L., and Kuksis, A.,** *Can. J. Biochem.*, 47, 761, 1969.
5. **Aneja, R., Bhati, A., Hamilton, R. J., Padley, F., and Steven, D. A.,** *J. Chromatogr.*, 173, 392, 1979.

Table 44
TRIACYLGLYCEROLS (GLC)

	P1	P2	P3	P1	P4
Column packing	P1	P2	P3	P1	P4
Temperature (°C)	270	250	245	270	300
Gas	N_2	N_2	N_2	N_2	Ar
Flow rate(mℓ/min)	90	75	120	90	300
Column					
Length (cm)	50	40	50	50	60
Diameter					
(cm I.D.)	0.3	0.3	0.2	0.3	0.3
(cm O.D.)	0.635	0.635	0.3	0.635	0.635
Form	U-tube	U-tube	Coil	U-tube	Coil
Material	Glass	Glass	SS	Glass	SS
Detector	FI	FI	FI	FI	ArI
Reference	1	2	3	1	4

Carbon Number: Double Bond Number	RRT × 1000			ECL	
Saturates					
30:0	—	80	20	—	—
32:0	—	130	30	—	—
34:0	—	170	50	—	—
36:0	188	270	80	36.00	110
38:0	244	400	120	38.00	—
40:0	321	630	190	40.00	—
42:0	423	1000	290	42.00	320
44:0	561	—	440	44.00	—
46:0	742	—	660	46.00	—
48:0	1000	—	1000	48.00	—
50:0	1362	—	—	50.00	—
52:0	1880	—	1550	52.00	—
54:0	2595	—	2500	54.00	2610
56:0	3613	—	—	56.00	—
Monoenes					
36:1	209	—	—	36.81	—
38:1	275	—	—	38.87	—
40:1	357	—	—	40.77	—
42:1	454	—	—	42.50	—
44:1	604	—	—	44.53	—
46:1	804	—	—	46.54	—
48:1	1084	—	—	48.52	—
50:1	1471	—	—	50.49	—
52:1	1982	—	—	52.33	—
54:1	2702	—	—	54.25	—
Dienes					
38:2	336	—	—	40.32	—
40:2	422	—	—	41.98	—
42:2	560	—	—	43.98	—
44:2	716	—	—	45.75	—
46:2	961	—	—	47.74	—
48:2	1214	—	—	49.25	—
50:2	1625	—	—	51.10	—
52:2	2191	—	—	52.96	—
54:2	2959	—	—	54.79	—

Table 44 (continued)
TRIACYLGLYCEROLS (GLC)

Carbon Number: Double Bond Number	RRT × 1000			ECL	
Trienes					
42:3	674	—	—	45.32	—
44:3	852	—	—	46.92	—
46:3	1090	—	—	48.55	—
48:3	1410	—	—	50.21	—
50:3	1858	—	—	51.93	—
52:3	2496	—	—	53.76	—
54:3	3206	—	—	55.28	—
Tetraenes					
48:4	1665	—	—	51.24	—
50:4	2178	—	—	52.79	—
52:4	2852	—	—	54.57	—
54:4	3689	—	—	56.12	—
Hexaenes					
54:6	4841	—	—	57.70	—
Nonaenes					
54:9	7906	—	—	60.32	—

Column packing   P1 = 5% Silar 10C on Gas Chrom Q (100-120 mesh)
                      P2 = 3% JXR on Gas Chrom Q (100-120 mesh)
                      P3 = 2.25% SE-30 on Chromosorb W (60-80 mesh)
                      P4 = 3/4% SE-30 on Gas Chrom P (80-100 mesh)

REFERENCES

1. **Takagi, T. and Itabashi, Y.,** *Lipids*, 12, 1062, 1977.
2. **Kuksis, A.,** *Fette Seifen Anstrichm.*, 73, 332, 1971.
3. **Liegwater, D. C. and Van Gend, H. W.,** *Fette Seifen Anstrichm.*, 67, 1, 1967.
4. **Pelick, N., Supina, W. R., and Rose, A.,** *J. Am. Oil Chem. Soc.*, 38, 506, 1961.

Table 45
TRIACYLGLYCEROLS (GLC)

Column packing	P1
Temperature (°C)	265
Gas	N_2
Flow rate(mℓ/min)	40
Column	
Length (cm)	180
Diameter	
(cm I.D.)	0.2
(cm O.D.)	0.625
Form	U-tube
Material	Glass
Detector	FI
Reference	1

Molecular Species	RRT × 100
16:0 10:0 4:0	31
14:0 14:0 4:0	46
12:0 14:0 6:0	56
16:0 14:0 4:0	84
16:0 14:0 6:0	96
16:0 16:0 4:0	100
16:0 16:1 4:0	106
18:0 16:0 4:0	141
14:0 18:1 6:0	141
14:0 18:2 6:0	151
16:0 18:1 4:0	151
16:0 18:2 4:0	160
16:0 18:1 6:0	210
18:0 18:0 4:0	210
16:0 18:2 6:0	234
18:0 18:1 4:0	234
18:0 18:2 4:0	251

Column packing P1 = 3% EGSS-X on
Gas Chrom Z
(100-120 mesh)

REFERENCE

1. **Kuksis, A., Marai, L., and Myher, J. J.,** *J. Am. Oil Chem. Soc.,* 50, 193, 1973.

Table 46
TRIMETHYLSILYL DIACYLGLYCEROLS (GLC)

Column packing	P1	P2	P3	P4	P5	P6	P7	P1	P7
Temperature (°C)	270	265	298	250	280	280	265	270	270
Gas	He	N₂	Ar	N₂	He	He		He	N₂
Flow rate(mℓ/min)	50	30	75	30	40	40		50	13
Column									
Length (cm)	180	120	240	180	180	180	240	180	200
Diameter									
(cm I.D.)	0.3	0.3	0.3	0.3	0.3	0.3	0.3	0.3	0.3
(cm O.D.)	0.635	0.635	0.635	0.635	0.635	0.635	0.635	0.635	0.635
Form	U-tube	U-tube	U-tube	U-tube	U-tube	U-tube	U-tube	U-tube	Coil
Material	Glass	Glass	Glass	Glass	Glass	Glass	Glass	Glass	Glass
Detector	FI	FI	Ar	FI	FI	FI	FI	FI	FI
Reference	1	2	3	2	4	4	5	1	6

1,2(2,3)-Isomers	RRT × 100							ECL	RRT	ECL
16:0 16:0	1000	14	100	45	45	47	100	32.00	944	31.59
16:0 18:0	1517	65	146	65	65	67	—	34.00	1258	33.60
16:0 18:1	1669	73	—	72	80	71	147	34.46	1415	34.41
16:0 18:2	1918	81	195	80	88	88	172	35.12	1659	35.49
16:0 18:3	2261	—	—	—	—	—	—	35.91	—	—
16:0 20:4	3070	—	175	—	—	—	247	37.40	—	—
18:0 18:0	2301	100	209	100	100	100	—	36.00	1691	35.62
18:0 18:1	2531	111	—	108	105	103	194	36.46	1892	36.38
18:1 18:1	2794	115	—	115	111	116	—	36.93	2127	37.18
18:0 18:2	2908	125	—	124	123	122	219	37.12	2164	37.30
18:1 18:2	3227	131	—	132	137	137	—	37.62	2492	38.25
18:0 18:3	3448	152	—	147	—	137	—	37.94	—	—
18:2 18:2	3721	147	—	150	153	157	—	38.31	2950	39.38
18:0 18:3	3794	160	—	160	—	156	—	38.40	—	—
18:2 18:3	4369	174	—	178	—	173	—	39.08	3543	40.57
18:0 20:4	4660	—	230	—	—	—	319	39.40	—	—
18:3 18:3	5163	195	—	196	—	196	—	39.88	4296	41.80
18:0 20:0	3840	—	—	—	—	—	—	38.00	—	—
18:0 22:6	—	—	296	—	—	—	—	—	—	—

1,3-Isomers

1,3-Isomers								ECL	RRT	ECL
16:0 16:0	1000	—	108	—	—	—	—	32.00	1000	32.00
16:0 18:0	1517	—	158	—	—	—	—	34.00	1332	34.00
16:0 18:1	1669	—	—	—	—	—	—	34.46	1504	34.82
16:0 18:2	1918	—	212	—	—	—	—	35.12	1788	35.95
16:0 18:3	2261	—	—	—	—	—	—	35.91	—	—
16:0 20:4	3070	—	—	—	—	—	—	37.40	—	—
18:0 18:0	2301	—	230	—	—	—	—	36.00	1788	36.00
18:0 18:1	2531	—	—	—	—	—	—	36.46	2021	36.83
18:1 18:1	2794	—	—	—	—	—	—	36.93	2264	37.61
18:0 18:2	2918	—	—	—	—	—	—	37.12	2333	37.81
18:1 18:2	3237	—	—	—	—	—	—	37.62	2676	38.73
18:0 18:3	3458	—	—	—	—	—	—	37.94	—	—
18:2 18:2	3721	—	—	—	—	—	—	38.31	3184	39.88
18:1 18:3	3794	—	—	—	—	—	—	38.40	—	—
18:2 18:3	4379	—	—	—	—	—	—	39.08	3828	41.06
18:0 20:4	4660	—	—	—	—	—	—	39.40	—	—
18:3 18:3	5163	—	—	—	—	—	—	39.88	4629	42.3
18:0 20:0	3840	—	—	—	—	—	—	38.00	—	—

Table 46 (continued)
TRIMETHYLSILYL DIACYLGLYCEROLS (GLC)

Column packing   P1 = 3% Silar 5CP on Gas Chrom Q (100-120 mesh)

P2 = 10% EGSS-X on Gas Chrom P (100-120 mesh)

P3 = 3% OV-1 on Gas Chrom Q (100-120 mesh)

P4 = 3% ECNSS-M on Gas Chrom Q (100-120 mesh)

P5 = 3% NPGS on Gas Chrom Q (60-80 mesh)

P6 = 3% CHDMS on Gas Chrom Q (60-80 mesh)

P7 = 10% Apolar (Silar) 10C on Gas Chrom Q (100-120 mesh)

REFERENCES

1. **Myher, J. J. and Kuksis, A.,** *J. Chromatogr. Sci.,* 13, 138, 1975.
2. **Kuksis, A.,** *Can. J. Biochem.,* 49, 1245, 1971.
3. **O'Brien, J. F. and Klopfenstein, W. E.,** *Chem. Phys. Lipids,* 6, 1, 1971.
4. **Kuksis, A.,** unpublished.
5. **Kuksis, A.,** unpublished.
6. **Itabashi, Y. and Takagi, T.,** *Lipids,* 15, 205, 1980.

Table 47
ACETYL DIACYLGLYCEROLS (GLC)

Column packing	P1	P2	P3	P4	P5	P6
Temperature (°C)	260	300	290	249	290	280
Gas	He	He	He	Ar	N_2	N_2
Flow rate (mℓ/min)	30	40	40	40	80	80
Column						
Length (cm)	180	180	180	43	35	180
Diameter						
(cm I.D.)	0.2	0.2	0.2	0.4	0.3	0.3
(cm O.D.)	0.3	0.3	0.3	0.635	0.635	0.635
Form	U-tube	U-tube	U-tube	U-tube	Coil	U-tube
Material	Glass	Glass	Glass	Glass	Glass	Glass
Detector	FI	FI	FI	Ar	FI	FI
Reference	1	2	2	3	4	5

1,2(2,3)-Isomers			RRT × 100			
12:0 12:0	—	—	—	—	—	20
14:0 14:0	—	—	—	—	—	45
14:0 16:0	—	—	—	40	—	—
16:0 16:0	48	47	42	64	73	100
16:0 18:0	69	48	64	100	—	—
16:0 18:1	76	71	67	—	88	—
16:0 18:2	90	80	72	—	—	—
18:0 18:0	100	100	100	—	—	—
18:0 18:1	110	103	100	—	—	—
18:1 18:1	120	110	104	—	100	200
18:0 18:2	120	110	111	—	—	—
18:1 18:2	139	123	114	—	—	—
18:0 18:3	—	136	124	—	—	—
18:2 18:2	162	136	124	—	—	—
18:1 18:3	—	—	130	—	—	—
18:2 18:3	—	—	141	—	—	—
18:0 20:4	—	—	—	—	—	—
18:3 18:3	—	—	163	—	—	—

Column packing	
	P1 = 3% EGSS-X on Gas Chrom Q (100-120 mesh)
	P2 = 3% NPGS on Gas Chrom Q (60—80 mesh)
	P3 = 3% CHDMS on Gas Chrom P (60-80 mesh)
	P4 = 3% SE-30 on Chromosorb W (60-80 mesh)
	P5 = 3% Dexsil 300 GD on Anachrom BS (70-80 mesh)
	P6 = 3% OV-1 on Gas Chrom Q (100-120 mesh)

REFERENCES

1. **Kuksis, A.**, *Can. J. Biochem.*, 49, 1245, 1971.
2. **Kuksis, A.**, *J. Chromatogr. Sci.*, 10, 53, 1972.
3. **Renkonen, O.**, *Biochim. Biophys. Acta*, 125, 288, 1966.
4. **Hasegawa, K. and Suzuki, T.**, *Lipids*, 10, 667, 1975.
5. **Banschbach, M. W., Geison, R. L., and O'Brien, J. F.**, *Anal Biochem.*, 59, 617, 1974.
6. **Wood, R., Baumann, W. J., Snyder, F., and Mangold, H. K.**, *J. Lipid Res.*, 10, 128, 1969.

Table 48
TRIMETHYLSILYLDIRADYLGLYCEROLS
(GLC)

Column packing	P1	P2	P3
Temperature (°C)	250	250	260
Gas	He	He	He
Flow rate (mℓ/min)	40	40	40
Column			
Length (cm)	180	180	180
Diameter			
(cm I.D.)	0.3	0.3	0.3
(cm O.D.)	0.625	0.625	0.625
Form	U-tube	U-tube	U-tube
Material	Glass	Glass	Glass
Detector	FI	FI	FI
Reference	1	1	1

Molecular Species	RRT × 100		
Alkylacyl			
32:0	100	100	14
32:1	110	112	20
34:0	150	149	65
34:1	170	169	72
34:2	190	185	80
36:0	230	225	100
36:1	250	240	112
36:2	290	285	125
36:4	310	300	170
38:4	460	440	—
Alkenylacyl			
32:0	100	100	14
32:1	110	112	21
34:0	150	149	66
34:1	170	169	73
34:2	190	185	81
36:0	230	225	100
36:1	250	240	111
36:2	290	285	124
36:4	310	300	171
38:4	460	440	—

Column packing P1 = 3% Silar 5CP on Gas Chrom Q (100-120 mesh)

P2 = 3% Silar 10C on Gas Chrom Q (100-120 mesh)

P3 = 10% EGSS-X on Gas Chrom P (100-120 mesh)

REFERENCE

1. **Myher, J. J. and Kuksis, A.,** unpublished results, 1978.

Table 49
ACETYLDIRADYLGLYCEROLS (GLC)

Column packing	P1	P2	P3
Temperature (°C)	255	260	270
Gas	He	He	He
Flow rate (mℓ/min)	40	40	40
Column			
Length (cm)	180	180	180
Diameter			
(cm I.D.)	0.3	0.3	0.3
(cm O.D.)	0.625	0.625	0.625
Form	U-tube	U-tube	U-tube
Material	Glass	Glass	Glass
Detector	FI	FI	FI
Reference	1	1	1

Molecular Species	RRT × 100		
Alkylacyl			
32:0	100	100	50
32:1	109	112	55
34:0	150	148	70
34:1	167	170	75
34:2	192	184	90
36:0	230	226	100
36:1	255	241	110
36:2	292	286	120
36:4	315	310	165
38:4	470	460	—
38:6	—	—	—
40:6	—	—	—
Alkenylacyl			
32:0	100	—	—
32:1	109	—	—
34:0	151	—	—
34:1	168	—	—
34:2	193	—	—
36:0	229	—	—
36:1	253	—	—
36:2	290	—	—
36:4	465	—	—
38:4	—	—	—
38:6	—	—	—
40:6	—	—	—

Column packing P1 = 3% Silar 5CP on Gas
Chrom Q (100-120 mesh)
P2 = 3% Silar 10C on Gas
Chrom Q (100-120 mesh)
P3 = 10% EGSS-X on Gas
Chrom P (100-120 mesh)

REFERENCE

1. **Kuksis, A.,** unpublished results, 1978.

Table 50
ACETYL MONOACYLGLYCEROLS (GLC)

Column packing	P1	P1	P2	P3	P3
Temperature (°C)	248	248	220	240	240
Gas	He	H	N_2	N_2	N_2
Flow rate (mℓ/min)	40	40	40	20	20
Column					
Length (cm)	180	180	180	200	200
Diameter					
(cm I.D.)	0.3	0.3	0.2	0.3	0.3
(cm O.D.)	0.635	0.635	0.6	0.635	0.635
Form	U-tube	U-tube	U-tube	Coil	Coil
Material	Glass	Glass	Glass	Glass	Glass
Detector	FI	FI	FI	FI	FI
Reference	1	1	2	3	3
1(3)-Isomers	**RRT**	**ECL**	**RRT**	**RRt**	**ECL**
4:0	—	—		—	—
6:0	—	—	20	—	—
8:0	—	—	60	—	—
10:0	—	—	120	—	—
12:0	—	—	220	—	—
14:0	393	14.00	—	721	14.00
14:1	440	—	440	—	—
16:0	627	16.00	600	1000	16.00
16:1	699	—	700	1222	17.22
17:0	780	17.00	780	—	—
18:0	1000	18.00	1000	1388	18.00
18:1	1101	18.38	1170	1661	19.12
18:2 (n-6)	1258	18.96	1470	2090	20.56
18:3 (n-3)	1470	19.63	—	2679	22.11
18:4	1565	19.88	—	—	—
20:0	1595	20.00	—	1911	20.00
20:1	1749	20.36	—	2679	22.11
20:4 (n-6)	—	—	—	3622	23.98
20:4 (n-3)	—	—	—	4288	25.03
20:5 (n-3)	—	—	—	4600	25.47
22:0	2544	22.00	—	2630	22.00
22:1	2737	22.29	—	3067	22.95
22:6 (n-3)	4223	24.14	—	7022	28.09
24:0	—	—	—	3630	24.00
2-Isomers					
4:0	—	—	—	—	—
6:0	—	—	—	—	—
8:0	—	—	110	—	—
10:0	—	—	140	—	—
12:0	—	—	230	—	—
14:0	393	14.00	370	—	—
14:1	460	—	460	—	—
16:0	627	16.00	610	—	—
16:1	699	—	720	—	—
18:0	1000	18.00	1000	—	—
18:1	1101	18.38	1160	—	—
18:2	1258	18.96	1460	—	—
18:3	1470	19.63	—	—	—
18:4	1565	19.88	—	—	—

Table 50 (continued)
ACETYL MONOACYLGLYCEROLS (GLC)

1(3)-Isomers	RRT	ECL	RRT	RRt	ECL
20:0	1595	20.00	—	—	—
20:1	1749	20.36	—	—	—
20:5	2445	21.29	—	—	—
22:0	2544	22.00	—	—	—
22:1	2737	22.29	—	—	—
22:6	4223	24.14	—	—	—

Column packing   P1 = 3% Silar 5CP on Gas Chrom Q (100-120 mesh)
                   P2 = 3% EGSS-X on Gas Chrom Q (100-120 mesh)
                   P3 = 5% Silar 10C on Gas Chrom Q (100-120 mesh)

REFERENCES

1. **Myher, J. J. and Kuksis, A.,** *Lipids,* 9, 382, 1974.
2. **Kuksis, A., Marai, L., and Myher, J.,** *J. Am. Chem. Soc.,* 50, 193, 1973.
3. **Itabashi, Y. and Takagi, T.,** *Lipids*, 15, 205, 1980; see also **Søe, J. B.,** *Fette, Seifen, Anstrichm.,* 85, 72, 1983.

Table 51
TRIMETHYLSILYL
MONOACYLGLYCEROLS (GLC)

Column packing	P1	P1	P2	P2
Temperature (°C)	220	220	190	190
Gas	N_2	N_2	N_2	N_2
Flow rate (mℓ/min)	40	40	20	20
Column				
Length (cm)	180	180	200	200
Diameter				
(cm I.D.)	0.3	0.3	0.3	0.3
(cm O.D.)	0.635	0.635	0.635	0.635
Form	U-tube	U-tube	Coil	Coil
Material	Glass	Glass	Glass	Glass
Detector	FI	FI	FI	FI
Reference	1	1	2	2

1(3)-Isomers	**RRT**	**ECL**	**RRT**	**ECL**
14:0	318	14.00	605	14.00
16:0	558	16.00	1000	16.00
16:1	630	16.41	1174	16.62
18:0	1000	18.00	1672	18.00
18:1	1100	18.30	1875	18.47
18:2	1276	18.83	2307	19.26
18:3	1540	19.48	2946	20.22
18:4	1628	19.68	—	—
20:0	1792	20.00	2785	20.00
20:1	1951	20.28	3097	20.41
20:5	2885	21.61	5539	22.63
22:0	3211	22.0	4676	22.00
22:1	3384	22.17	5063	20.38
22:6	5420	23.80	9814	24.79
24:0	—	—	7971	24.00

2-Isomers				
14:0	318	14.00	531	13.50
16:0	558	16.00	881	15.50
16:1	630	16.41	1019	16.07
18:0	1000	18.00	1474	17.51
18:1	1100	18.30	1649	17.95
18:2	1276	18.83	2009	18.72
18:3	1540	19.48	2564	19.68
18:4	1628	19.68	—	—
20:0	1792	20.00	2467	19.53
20:1	1951	20.28	—	—
20:5	2885	21.61	4743	22.03
22:0	—	—	4147	21.54
22:1	3384	22.17	—	—
22:6	5420	23.80	8322	24.16

Column packing   P1 =   3% Silar 5CP on Gas Chrom Q
                         (100-120 mesh)
                 P2 =   5% Silar 10C on Gas Chrom Q
                         (100-120 mesh)

REFERENCES

1. **Myher, J. J. and Kuksis, A.,** *Lipids,* 9, 382, 1974.
2. **Itabashi, Y. and Takagi, T.,** *Lipids,* 15, 205, 1980; see also **Søe, J. B.,***Fette, Seifen, Anstrichm.,* 85, 72, 1983.

Table 52
ACYLGLCEROLS (GLC-OPEN TUBULAR)

Liquid phase	P1	P2	P2	P3
Temperature program	T1	T2	T3	T4
Gas	He	He	He	H_2
Flow rate (mℓ/min)	—	5	5	2 bar
Column				
Length (cm)	2800	3000	3000	2000
Diameter				
(cm I.D.)	0.03	0.05	0.05	0.025
(cm O.D.)	0.08	—	—	0.08
Form	Coil	Coil	Coil	Coil
Material	Glass	Glass	Glass	Glass
Detector	FI	MS	MS	FI
Reference	1	2	2	3

Molecular Species	RRT × 100			
Triacyl				
12:0 12:0 12:0	29	—	—	28
12:0 12:0 14:0	35	—	—	36
12:0 14:0 14:0	45	—	—	48
14:0 14:0 14:0	58	—	—	60
14:0 14:0 16:0	71	—	—	74
14:0 16:0 16:0	80	—	—	88
16:0 16:0 16:0	100	—	—	100
16:0 16:0 18:0	—	—	—	114
16:0 18:0 18:0	—	—	—	130
18:0 18:0 18:0	—	—	—	140
sn-1,2-Diacyl				
16:0 18:2	—	100	—	—
16:0 18:1	—	100	—	—
18:1 18:2	—	139	—	—
18:0 18:2	—	148	—	—
18:1 18:1	—	148	—	—
sn-1(3)-Monoacyl				
16:1	—	—	93	—
16:0	—	—	100	—
17:0	—	—	130	—
18:1	—	—	155	—
18:0	—	—	167	—

Liquid phase	P1 = Dexsil 300 GC
	P2 = OV-1/SILANOX
	P3 = Poly S-179
Temperature program	T1 = 300-335°C at 0.5°C/min
	T2 = 300°C isothermal
	T3 = 230°C isothermal
	T4 = 300-400°C at 4°C/min

REFERENCES

1. **Novotny, M., Sequra, R., and Zlatkis, A.,** *Anal. Chem.,* 44, 9, 1972.
2. **Gaskell, S. J. and Brooks, C. J. W.,** *J. Chromatogr.,* 142, 469, 1977.
3. **Schomburg, G., Dielmann, H., Husmann, H., and Weeke, F.,** *J. Chromatogr.,* 122, 55, 1976.

Table 53
TRIMETHYLSILYL AND TERTIARY BUTYLDIMETHYLSILYL DIACYLGLYCEROLS (GLC-OPEN TUBULAR)

Liquid phase	P1	P1
Temperature (°C)	250	250
Gas	H_2	H_2
Head pressure	5 psi	10 psi
Column		
Length (cm)	1000	1000
Diameter		
(cm I.D.)	0.025	0.025
Form	Coil	Coil
Material	Glass	Glass
Detector	FI	FI
Reference	1	1

Molecular Species	Carbon Number	RRT × 1000	
16:0 16:0	32 (0,0)	630	632
16:0 16:1ω9	32 (0,1)	690	672
16:0 16:1ω7	32 (0,1)	698	686
14:0 18:2	32 (0,2)	781	760
	33 (0,0)	—	763
	33 (0,1)	—	817
	33 (0,2)	—	912
16:0 18:0	34 (0,0)	949	948
16:0 18:1ω9	34 (0,1)	1000	1000
16:0 18:1ω7	34 (0,1)	1025	1018
16:1ω9 18:1ω9	34 (1,1)	1068	1060
16:0 18:2	34 (0,2)	1132	1113
16:0 18:3ω6	34 (0,3)	—	1154
14:0 20:4	34 (0,4)	1184	1156
16:1ω9 18:2	34 (1,2)	1217	1220
16:1ω7 18:2	34 (1,2)	1237	1261
17:0 18:2	35 (0,2)	1344	1361
17:0 18:2	35 (0,1)	1213	1225
16:0 18:3ω3	34 (0,3)	1311	1331
17:0 18:2	35 (0,2)	1366	1353
18:0 18:0	36 (0,0)	—	1436
18:0 18:1ω9	36 (0,1)	1481	1501
18:0 18:1ω7	36 (0,1)	1517	1530
18:1ω9 18:1ω9	36 (1,1)	1586	1582
18:1ω9 18:1ω7	36 (1,1)	—	1614
18:1ω7 18:1ω7	36 (1,1)	—	1646
18:0 18:2	36 (0,2)	1659	1665
16:0 20:4	36 (0,4)	1757	1694
16:0 20:3ω6	36 (0,3)	—	1731
18:1ω9 18:2	36 (1,2)	1791	1760
18:1ω7 18:2	36 (1,2)	1831	1798
16:1ω9 20:4	36 (1,4)	—	1806
18:0 18:3ω3	36 (0,3)	1896	1990
16:0 20:5	36 (0,5)	2010	1933
18:2 18:2	36 (2,2)	2039	1974
18:1ω9 18:3ω3	36 (1,3)	2075	2102
17:0 20:4	37 (0,4)	2109	2042
18:2 18:3ω3	36 (2,3)	2374	2373
16:1ω7 20:4	36 (1,4)	—	1863

Table 53 (continued)
TRIMETHYLSILYL AND TERTIARY BUTYLDIMETHYLSILYL DIACYLGLYCEROLS (GLC-OPEN TUBULAR)

Molecular Species	Carbon Number	RRT × 1000	
18:0 20:4	38 (0,4)	2581	2526
18:0 20:3ω6	38 (0,3)	—	2609
18:1ω9 20:4	38 (1,4)	2769	2681
18:1ω7 20:4	38 (1,4)	2839	2739
18:3ω3 18:3ω3	36 (3,3)	2795	2648
18:0 20:5	38 (0,5)	2920	2825
16:0 22:6	38 (0,6)	3041	2926
18:2 20:4	38 (2,4)	3133	3010
16:0 22:5	38 (0,5)	—	3050
16:1ω9 22:6	38 (1,6)	—	3091
16:1ω7 22:6	38 (1,6)	—	3189
18:1ω9 20:5	38 (1,5)	—	3368
18:1ω7 20:5	38 (1,5)	—	3435
18:0 22:6	40 (0,6)	4398	4275
18:0 22:5	40 (0,5)	—	4400
18:1ω9 22:6	40 (1,6)	4736	4542
18:1ω7 22:6	40 (1,6)	4877	4639

Liquid phase   P1 = SP-2330 wall coated

REFERENCE

1. **Myher, J. J. and Kuksis, A.**, *Can. J. Biochem.*, 60, 638, 1982.

Chapter 13

GLYCEROPHOSPHOLIPIDS

S. S. Radwan

I. INTRODUCTION

The present chapter describes and evaluates chromatographic techniques that can be applied in the analyses of phospholipids, especially the glycerophospholipids with reference to less common compounds, such as the alkoxy phospholipids and the phosphono analogs of the common glycerophospholipids (see Figure 1 for formulas). It is neither attempted to give a review of studies concerned with this subject nor to present full details of a variety of procedures as adopted in various research laboratories. In the interest of the reader, emphasis is placed on presenting each chromatographic technique in possibly one single version that, in fact, has been devised by considering valuable details of several variations of the method. The majority of the $R_f \times 100$ values given in the Tables have been calculated by the author from typical pictures of original chromatograms published in the literature. Since the chromatographic behavior of the different compounds depends on several factors that are difficult to control, it is obvious that the $R_f \times 100$ values should be considered solely as a guide; they should be useful, for example, in predicting the sequential migration of different fractions on a chromatogram.

II. PREPARATION OF SAMPLES

A. Extraction

Phospholipids together with proteins, mainly in a lipoprotein combination, occur as major constituents of biological membranes of living cells and as functional groups associated with several enzyme systems. As a rule, they occur with other lipid classes, and there is no method for their selective extraction. They are usually extracted almost completely by the conventional methods of lipid extraction that are described in Chapter 2 of this volume. The proportions of the phospholipids in the total lipid extracts depend on their concentrations in the tissue extracted as well as on the extracting solvents. Thus phospholipids usually represent minor constituents in lipid extracts from storage tissues such as mature oilseeds, but they are major compounds in lipid extracts from physiologically active cells. Being highly polar compounds, phospholipids are best extracted with polar solvents such as alcohols, chloroform, etc. The presence of an alcohol during extraction is essential because it denatures proteins, and thus helps in liberating lipids from lipoproteins. Less polar solvents such as hexane and diethyl ether are not suitable for extracting phospholipids. Solvent mixtures containing high proportions of acetone are also not suitable, although they are very efficient for extracting glycolipids.

There are specific problems associated with the extraction and storage of phospholipids, which the analyst should consider. One of these problems is the possible decomposition of intact molecules due to the activation of enzymes liberated as a result of tissue and cell destruction. Phospholipase D, which degrades intact phospholipids to phosphatidic acids, can be deactivated by propanol[1] as well as by boiling the material in this solvent or in water for a few minutes before extraction. Obviously, boiling should be done under oxygen-free nitrogen. Another problem is the possible transesterification of phospholipids, due to even minor changes of the pH value, in the presence of alcohols, resulting in the formation of alkyl esters of fatty acids in the extracts. As a rule, mild conditions and a rather short time

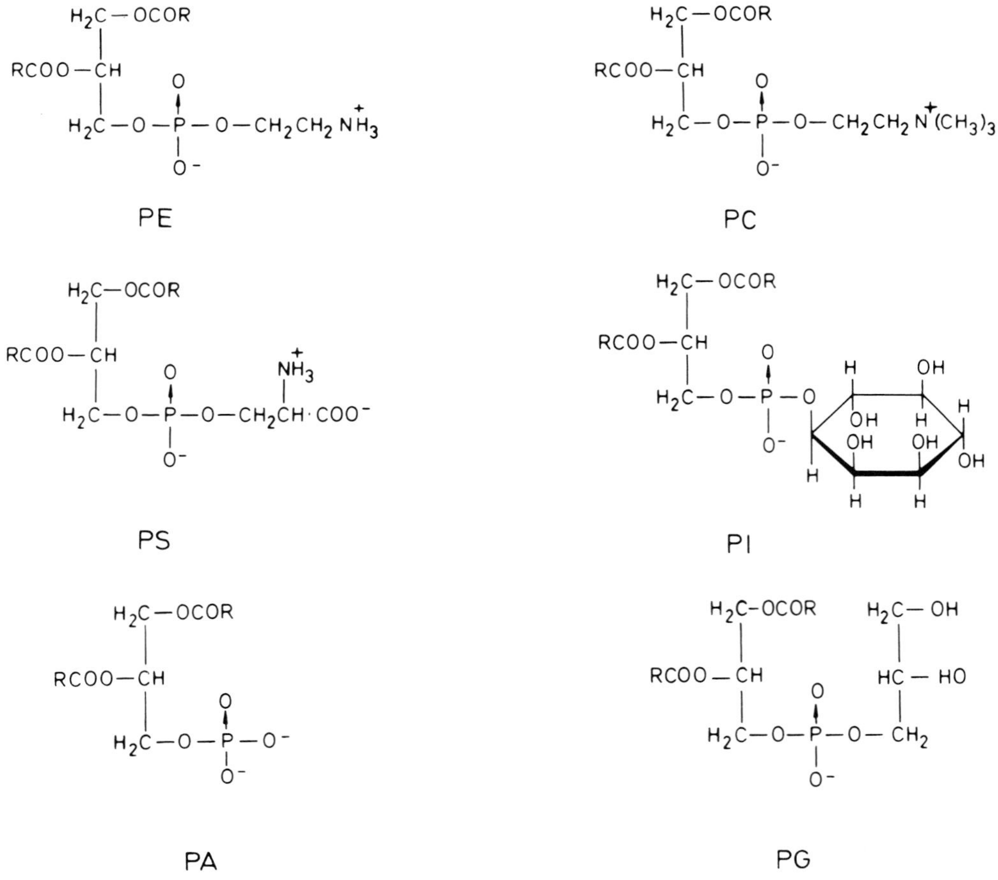

FIGURE 1. The most common phospholipids: PE, Phosphatidylethanolamines (Diacylglycero-phosphoethanolamines), PC, Phosphatidylcholines (Diacylglycerophosphocholines), PS, Phosphatidylserines (Diacylglycerophosphoserines), PI, Phosphatidylinositides, PA, Phosphatidic acids, PG, Phosphatidylglycerols.

are sufficient for hydrolysis and transesterification of phospholipids (see later). Total lipids extracted with, and stored carelessly in solvents containing methanol, usually contain variable proportions of methyl esters of fatty acids. Such esters arise, as artifacts, from phospholipids and, probably, from galactolipids, by methanolysis; this reaction is enhanced by traces of sodium carbonate or sodium bicarbonate coextracted from the tissue.[2] In general, the occurrence of unusually high levels of phosphatidic acids and alkyl esters of fatty acids in the total lipid extracts may indicate that the lipid extracts have been spoiled and that phospholipids probably have been partially decomposed during extraction and storage.

B. Derivatization

Intact phospholipids can be analyzed directly by chromatography; however, for certain purposes, derivatives are prepared and analyzed.

1. Methanolysis

In order to study the patterns of the constituent fatty acids of phospholipids, the fractions resolved by chromatography are subjected to methanolysis, and the methyl esters produced are analyzed by argentation thin-layer chromatography (TLC) and/or gas chromatography (GC). The phospholipids adsorbed may or may not be eluted prior to methanolysis. The

Table 1
REACTIONS OF PHOSPHOLIPIDS TO MILD METHANOLYSIS[a]

No.	Phospholipid	Reaction to mild methanolysis[b]	Products
1.	Diacylglycerophosphocholines and acylglycerophosphocholines	Positive	Methyl esters plus glycerophosphocholine
2.	Diacylglycerophosphoethanolamines and acylglycerophosphoethanolamines	Positive	Methyl esters plus glycerophosphoethanolamine
3.	Diacylglycerophosphoserines and acylglycerophosphoserines	Postive	Methyl esters plus glycerophosphoserine
4.	Diacylglycerophosphoinositols and acylglycerophosphoinositols	Positive	Methyl esters plus glycerophosphoinositol
5.	Diacylglycerophosphoglycerols	Positive	Methyl esters plus diglycerophosphoate
6.	Phosphatidic acid	Positive	Methyl esters plus glycerophosphate
7.	Alkylacylglycerophosphoethanolamines and -cholines, alk-1-enylacyl-glycerophosphoethanolamines and -cholines (choline and ethanolamine plasmalogens)[c]	Partially positive	Methyl esters plus lysoplasmalogens
8.	Lysoplasmologen of choline, ethanolamine and serine phospholipids[c]	Negative	—

[a] According to **Marinetti, G. V.**, in *New Biochemical Separations,* **James, A. T.** and **Morris, L. J.**, Eds., D. van Nostrand, London, 1964, 339.

[b] Mild methanolysis in methanolic sodium hydroxide at 37°C for 15 minutes according to **Dawson, R. M. C.**, *Biochem. J.*, 75, 45, 1960.

[c] The acyl moiety is removed whereas the alk-1-enyl ether moiety is not attacked.

preparation of methyl esters of the constituent fatty acids of phospholipids may be done by acid- or base-catalyzed transesterification. Unlike other classes of acyl lipids, such as the mono-, di-, and triacylglycerols and steryl esters, phospholipids can be methanolyzed completely under relatively mild conditions. Addition of dilute methanolic solutions of sodium hydroxide, sodium methoxide, hydrochloric acid, or sulfuric acid leads to quantitative methanolysis, usually within 5 min.[3,4] Table 1 indicates that practically all the acyl moieties in phospholipid fractions can be obtained as methyl esters by mild methanolysis.

Procedure for the Methanolysis of Phospholipids
Thus the procedure consists essentially of adding 2 to 4 mℓ of 0.5 N sodium methoxide or 1 N sodium hydroxide in absolute methanol to up to 50 mg of the phospholipid sample. After standing for 10 to 15 min at room temperature, 10 mℓ of water is added and the methyl esters of fatty acids are extracted with 5 aliquots, each, of 5 mℓ of hexane. Methanolysis can also be achieved by using 5% (w/v) anhydrous hydrochloric acid in absolute methanol or, even more simply, by using 1% concentrated sulfuric acid in absolute methanol[5] instead of sodium methoxide or sodium hydroxide. Anhydrous hydrochloric acid in methanol is prepared by bubbling HCl gas into dry methanol, or by adding 5 mℓ of acetyl chloride slowly to 50 mℓ of cool dry methanol.[6] It is important that water be rigorously excluded, otherwise some free fatty acids instead of methyl esters may be produced. Prior to their GC analysis the methyl esters are purified by preparative TLC on silica gel using hexane-diethyl ether (90:10 v/v).[7]

Procedure for the Methanolysis of Phospholipids on Chromatoplates
Methyl esters of constituent fatty acids in phospholipids can also be prepared simply by direct methanolysis on TLC plates.[4] The lipid mixture (1 mg) is applied as a spot

Table 2
ENZYMES USED FOR THE HYDROLYSIS OF PHOSPHOLIPIDS

No.	Enzyme	Origin	Site of action	Products of hydrolysis
1.	Pancreatic lipase	Pancreas of pig and other animals	Ester bond at position 1 of the glycerol moiety	Fatty acids plus 2-acyl-glycerophospholipid
2.	Phospholipase A_2	Snake venoms	Ester bond at position 2 of the glycerol moiety	Fatty acids plus 1-acylglycerophospholipid
3.	Phospholipase C	Bacteria	Bond of the glycerol moiety to the phosphate moiety	Diacylglycerols plus phosphate-base derivative
4.	Phospholipase D	Green plants	Bond of the phosphate moiety to the base moiety	Phosphatidic acids plus base

or a short streak at the lower right corner of a thin-layer plate, and another sample is applied as a spot at the lower left corner to serve for identification. After developing with a solvent such as chloroform-methanol-acetic acid (85:20:1 v/v),[8] the phospholipid fractions resolved along the left side of the plate are visualized by iodine vapor, and the parallel fractions on the right side are marked and subjected to methanolysis directly on the plate. Methanolysis is achieved by irrigating the adsorbent areas carrying the fractions with aliquots of 20 μℓ of a 12% solution of sodium hydroxide in absolute methanol, and keeping the fractions wet by pipetting on absolute methanol regularly over a period of 5 min. After applying a standard sample of methyl esters as a spot at the upper right corner, the plate is turned 90° clockwise, and the methyl esters are separated in the second direction using hexane-diethyl ether (90:10 v/v).[7] The standard sample of methyl esters is exposed to iodine vapor, and thus the methyl esters resulting from the individual phospholipid fractions can be detected, eluted by hexane-diethyl ether (1:1 v/v), and analyzed by GC. The author has found recently[63] that monogalactosyldiacylglycerols and digalactosyldiacylglycerols can also be subjected to methanolysis directly on the adsorbent layer.

2. Enzymatic Hydrolysis

In biochemical studies it is often desired to investigate the distribution of fatty acids between the 1 and 2 positions of the phospholipid molecule. This can be accomplished by enzymatic hydrolysis as shown in Table 2.

Procedure for the Hydrolysis of Phospholipids with Pancreatic Lipase
Preparations of pancreatic lipase are available commercially. For hydrolysis,[9,10] 10 mg of the phospholipid, 3 mg of bile salts, and 4.5 mg of bovine serum albumin are dispersed by ultra-sonification in 1 mℓ of 0.1 M borate buffer, pH 8, which is made 5 $\times 10^{-3}$ M with calcium chloride at the time of addition of 10 mg of pancreatic lipase. The mixture is shaken for 6 hr until complete hydrolysis occurs. Fatty acids that have been esterified in position 1 of the glycerol moiety as well as the corresponding lysophospholipid are the end products; they can be extracted with chloroform-methanol (7:3 v/v) and separated by preparative TLC.

Procedure for the Hydrolysis of Phospholipids with Phospholipase A
Phospholipase A of snake venoms, especially from Crotalus adamanteus, C. atrox, and Ophiophagus hannah, catalyzes the hydrolysis of acyl moieties in position 2 of phospholipids. The reaction occurs only in the presence of calcium ions and is stimulated by diethyl ether. For hydrolysis,[11] 2 to 5 mg of venom from O. hannah, which

has been found superior for this purpose to venoms of other snakes, is dissolved in 0.5 mℓ of tris buffer, pH 7.5, containing 4 mM calcium chloride. 5 to 50 mg of phospholipid is dissolved in 2 mℓ of diethyl ether and 100 μℓ of venom solution is added followed by shaking for 1 hr. 10 mℓ of methanol and 20 mℓ of chloroform are added, and the system is dried over anhydrous sodium sulfate. After filtration the solvents are removed under vacuum, and the products are separated by preparative TLC. This method is less efficient with diacylglycerophosphoethanolamines and diacylglycerophosphoinositols than with diacylglycerophosphocholines; however, the addition of 0.25 mℓ of 0.1 M borate buffer,[12,13] pH 7, increases the efficiency considerably. The end products of the hydrolysis reaction are fatty acids and lysophospholipids.

Procedure for the Hydrolysis of Phospholipids with Phospholipase C
Diacylglycerols are sometimes more readily separable into molecular species by chromatographic procedures than are their parent phospholipids. Phospholipase C is the enzyme used for catalyzing the derivatization of diacylglycerols from phospholipids. The enzyme is available commercially; however, very active preparations can be prepared from the liquid media that have supported the growth of Bacillus cereus or, preferably, Clostridium welchii, simply by ammonium sulfate precipitation.[12] For hydrolysis,[13] 1 mg of phospholipase C from C. welchii dissolved in 2 mℓ of 0.5 M tris buffer, pH 7.5, and containing 2 × 10⁻³ M calcium chloride is added to 10 mg of phospholipid (diacylglycerophosphocholines) dissolved in 2 mℓ diethyl ether. After shaking for 3 hr, the diacylglycerols produced are extracted three times with diethyl ether and purified by preparative TLC.

Procedure for the Hydrolysis of Phospholipids with Phospholipase D
Phosphatidic acids can also be derived from phospholipids for the purpose of better resolution into molecular species by chromatographic procedures. Phospholipase D catalyzes the hydrolysis of phospholipids into phosphatidic acids. The enzyme is available commercially, but an active preparation can be obtained easily by homogenizing cabbage with an equal volume of water; after filtration and centrifugation at 13,000 g for 30 min the supernatant solution can be used.[14] For hydrolysis,[14] 1.25 mℓ of 0.1 M sodium acetate, 0.25 mℓ of 1 M calcium chloride, and 2 mℓ of the enzyme preparation are added to 10 mℓ of the phospholipid dissolved in 1 mℓ of diethyl ether. After shaking for 3 hr, glacial acetic acid is added until the pH is brought to 2.5. Phosphatidic acids derived from various phospholipids are extractd three times with chloroform-methanol (2:1 v/v) and purified by preparative TLC.

3. Silylation

Dephosphorylated and deacylated phospholipids are prepared for GC and coupled GC/MS by converting them into the trimethylsilyl (TMS) ether derivatives. Hydroxy compounds as well as amino compounds are silylated[15] by means of different silylating reagents, viz. hexamethyldisilazane, *N*-trimethyl-silylimidazole, bis (trimethylsilyl) acetamide or bis (trimethylsilyl) fluoroacetamide, using solvents such as pyridine and acetonitrile in the presence of trimethyl-chlorosilane or traces of water as catalyst. The reaction is usually carried out at ro room temperature for 24 hr, or at 60°C for only 2 hr.

III. CHROMATOGRAPHY

Phospholipids present in total lipid extracts from animals, plants, or microorganisms can be separated by the conventional chromatographic techniques. In general, column chromatography (CC) is applied when relatively large amounts of lipids are to be fractionated; for the separation and analysis of rather small amounts, paper chromatography and TLC are more suitable. Combining more than one technique facilitates complete analysis. Thus eluates from column chromatograms can be tested for purity by TLC. Also, the coupling of TLC and GC provides a good tool for the quantitative determination of individual fractions and their constituent fatty acids.

In the following discussion the applications of the various chromatographic techniques in fractionation and analysis of phospholipids are considered separately; however, it is expected that the analyst should choose the combination of procedures suitable for his study.

A. Separation of Phospholipids by Column Chromatography (CC)

As mentioned before, CC is a useful preparative technique whenever large amounts of lipid mixtures are to be fractionated. The fractions eluted from the columns are usually analyzed by TLC and/or by chemical means, for example, by phosphorus determinations of the various fractions. For isolating pure compounds, fractions of the eluate that contain more than one compound may be subjected to CC under different conditions such as various packing materials, other solvents for elution, or to preparative TLC. The principles of separation processes in CC involve either adsorption on adsorbents such as silicic acid, alumina, or Florisil (magnesium silicate) or, at least partly, ion-exchange and adsorption chromatography on diethylaminoethyl (DEAE) cellulose.

1. Column Chromatography on Silicic Acid

The adsorbent that finds widest application in column chromatography of lipid extracts is silicic acid. The size of the adsorbent particles is of special importance; it should be small enough so that it may provide adequate surface area for good separation, but it should be large enough to permit reasonable flow rates of the eluting solvents through the column. 200 mesh silicic acid fulfills these two conditions well, and hence, silicic acid with this particle size is commonly used. Also, the degree of hydration of the adsorbent affects the separation. Too much water bound to the particles will reduce the adsorption of lipids to silicic acid and, conversely, in the presence of too little water, the lipids become adsorbed to silicic acid so firmly that they are not eluted easily and, hence, tailing occurs. About 5% should be the optimal water content of silicic acid used in the CC of phospholipids.[5] This low water content is reached either by activating the adsorbent at 110°C for several hours prior to packing, or by dehydration of the adsorbent after packing the column by passing acetone through it.[16]

A slurry of activated silicic acid in chloroform is packed into a glass column provided with a sintered disc and a stopcock. The lipid mixture to be fractionated is dissolved in the least possible amount of chloroform and applied to the column. For each gram of adsorbent, an amount of not more than 30 mg of lipid mixture can be fractionated. The lipid components are eluted successively with solvents of gradually increasing polarity. The flow rate of the solvents through the column should be adjusted to between 1 and 3 mℓ/min. If, under normal conditions, the flow rate is much slower, it is necessary to apply pressure from a nitrogen cylinder to the top of the column. The composition of the fractions eluted is assessed by TLC and/or by chemical means.

In general, it is rather difficult to obtain pure individual compounds from lipid mixtures by this technique alone. Hence, CC should be coupled with preparative TLC in order to accomplish complete separations. Table 3 summarizes the procedure for separating phospholipids in natural mixtures by silicic acid CC. For more details the reader should refer to pertinent reviews.[17-20] As a rule, when chloroform-methanol mixtures are used as eluting solvents, both acidic phospholipids and nonacidic phospholipids are eluted independently in the sequence shown in Table 4. Where the individual components of the two groups overlap along the column depends on the properties of the adsorbent, which in turn can vary according to the manufacturer and to the batch as well as to the cations with which the acidic phospholipids in natural mixtures are associated. Commonly, the diacylglycerophosphoserines are eluted together with the diacylglycerophosphoethanolamines, whereas the

Table 3
SEPARATION OF PHOSPHOLIPIDS BY SILICIC ACID COLUMN CHROMATOGRAPHY

Column material	Fraction number	Eluting solvent	Column volumes of solvent	Composition of fraction
A. TOTAL LIPIDS				
Silicic acid	I	Chloroform	10	Simple lipids
	II	Acetone	40	Glycolipids plus trace of acidic phospholipids
	III	Methanol	10	Phospholipids
B. PHOSPHOLIPIDS (FRACTION III)				
Silicic acid	1	Chloroform-methanol (95:5 v/v)	10	Residual phosphatidic acids and cardiolipins
	2	Chloroform-methanol (80:20 v/v)	20	Diacylglycerophosphoethanolamines and diacylglycerophosphoserines
	3	Chloroform-methanol (50:50 v/v)	20	Diacylglycerophosphoinositols and diacylglycerophosphocholines
	4	Methanol	20	Sphingomyelin and lysophospholipids
C. FRACTION 2				
Silic acid - silicate	i	Chloroform-methanol (80:20 v/v)	8	Diacylglycerophosphoethanolamines
	ii	Methanol	3	Diacylglycerophosphoserines

Table 4
SEQUENCE OF ELUTION OF ACIDIC AND NONACIDIC PHOSPHOLIPIDS ON SILICIC ACID COLUMNS WITH CHLOROFORM-METHANOL MIXTURES

Nonacidic phospholipids	Acidic phospholipids
Acylglycerophosphocholines	
Sphingomyelins	
Acylglycerophosphoethanolamines	
Diacylglycerophosphocholines	
Diacylglycerophosphoethanolamines	
	Diacylglycerophosphoinositols
	Diacylglycerophosphoserines
	Diacylglycerophosphoglycerols
	Phosphatidic acids
	Cardiolipins

diacylglycerophosphoinositols are usually very close to the diacylglycerophosphocholines (Table 4). Combinations of acidic and nonacidic classes of phospholipids can be separated completely[21,22] by further chromatography on silicic acid – silicate column (Table 3), which can be prepared by washing a silicic acid column with one column volume of chloroform-methanol, (4:1 v/v) containing 1% ammonia. Alternatively, such compounds can be separated conveniently by preparative TLC using slightly basic solvents (see later).

2. Column Chromatography on Other Adsorbents

In comparison to silicic acid, other adsorbents, including Florisil (magnesium silicate) and alumina, have found only limited application in CC of phospholipids.

Florisil is a very strong adsorbent, and hence, it is not suitable for the fractionation of phospholipid mixtures. On the other hand, Florisil washed with acid can give separations similar to those obtained with silicic acid.

Alumina is also not a satisfactory adsorbent for the fractionation of phospholipids because, being basic, it frequently leads to partial hydrolysis of these compounds. Neutral alumina, however, is useful in the separation of some fractions such as diacylglycerophosphocholines[23] and diacylglycerophosphoinositols[24] from natural mixtures in a pure form. On neutral alumina, the nonacidic phospholipids are eluted first with chloroform-methanol-water in different proportions; then, the acidic fractions can be separated after adding ammonium salts to the solvent.[24]

3. Column Chromatography on Modified Cellulose

Columns of diethylaminoethyl (DEAE) cellulose, in the acetate form, are as useful as those of silicic acid in the chromatography of phospholipids.

Procedure for the Chromatography of Lipids on DEAE Cellulose
The acetate form of DEAE cellulose is obtained[21] first by washing the commercial preparation in three successive cycles, each consisting of 3 volumes of 1 N HCl, excess of water, 3 volumes of 0.1 N KOH and, finally, excess of water. The washed DEAE cellulose is then kept overnight in 3 volumes of glacial acetic acid. A column is packed[21] with a slurry of DEAE cellulose in acetic acid, and the latter is removed by washing the column successively with 5 volumes of methanol, 3 volumes of chloroform-methanol (1:1 v/v), and 5 volumes of chloroform. Up to 300 mg lipid mixture can be applied safely to a 30 × 2.5 cm column.[5] By eluting with chloroform-methanol mixtures, the nonacidic phospholipids are obtained first; weakly acidic compounds, such as diacylglycerophosphoserines, are then eluted with acetic acid, whereas highly acidic phospholipids can only be eluted with chloroform-methanol mixtures to which ammonia or inorganic salts have been added. Table 5 summarizes procedures for separating phospholipids in lipid extracts from animals and plants by CC on DEAE cellulose acetate.

The identity of the fractions can be confirmed by TLC, and the individual classes of compounds in each fraction of the eluate can be separated completely by preparative TLC.

Columns of DEAE cellulose in the borate form as well as of triethylaminoethyl (TEAE) cellulose are useful for special purposes. Since DEAE cellulose (borate) retains diacylglycerophosphoethanolamines (when elution is carried out with chloroform-methanol mixtures or with pure methanol), the coupling of chromatography on DEAE cellulose acetate with that on DEAE cellulose borate is useful for the separation of diacylglycerophosphoethanolamines from ceramide polyhexosides.[20] The same separation can also be achieved by coupling chromatography on DEAE cellulose acetate with that on TEAE cellulose.

For further details the reader may refer to recent reviews.[20,25]

B. Separation of Phospholipids by Paper Chromatography (PC)

When only small amounts of lipid mixtures are available, separations should be carried out by chromatography either on silicic acid impregnated paper or on thin layers of silicic acid. In fact, TLC techniques are much more common in current use than PC procedures, although the quality of separation is similar in both cases. Workers who prefer analyzing and separating lipids by chromatography on silicic acid impregnated paper attribute several advantages to this technique (see for example Reference 26); however, most of the advantages listed actually hold true also for TLC.

Occasionally, PC of lipids is performed on silicic acid-impregnated glass fiber, phosphate-impregnated paper, formalin-treated paper, as well as on nontreated paper.[27] However, silicic acid impregnated paper is most commonly used.

For preparation and storage of silicic acid-impregnated paper, special techniques are used.[26,28]

Table 5
SEPARATION OF PHOSPHOLIPIDS BY DEAE CELLULOSE COLUMN CHROMATOGRAPHY

Fraction number	Eluting solvent	Column volumes of solvent	Composition of fraction
A. ANIMAL LIPID MIXTURES[a]			
1	Chloroform	10	Simple lipids
2	Chloroform-methanol (9:1 v/v)	10	Diacylglycerophosphocholines, acylglycerophosphocholines, sphingomyelin/cerebrosides
3	Chloroform-methanol (1:1 v/v)	10	Diacylglycerophosphoethanolamines, acylglycerophosphoethanolamines, ceramides, di- and polyhexosides
4	Ethanol	10	—
5	Glacial acetic acid	10	Diacylglycerophosphoserines
6	Ethanol	4	—
7	0.05 M ammonium acetate in chloroform-methanol (4:1 v/v) + 20 mℓ 28% ammonia/ℓ	10	Cardiolipins, phosphatidic acids, diacylglycerophosphoglycerols, diacylglycerophosphoinositols, cerebroside sulfates
8	Methanol	10	—
B. PLANT LIPID MIXTURES[b]			
I	Chloroform	10	Simple lipids
II	Chloroform	10	Monogalactosyldiacylglycerols
III	Chloroform-methanol (95:5 v/v)	10	Diacylglycerophosphocholines, cerebrosides, sterylglycosides
IV	Chloroform-methanol (90:10 v/v)	10	Digalactosyldiacylglycerols
V	Chloroform-methanol (60:40 v/v)	10	Diacylglycerophosphoethanolamines
VI	as in fraction number 7 (above)	10	Cardiolipins, phosphatidic acids, diacylglycerophosphoglycerols, diacylglycerophosphoinositols, sulfolipids
VII	Methanol	10	—

[a] Rouser, G., Kritchevsky, G., and Yamamoto, A., in *Lipid Chromatographic Analysis*, Vol. I, Marinetti, G. V., Ed., Marcel Dekker, New York, 1967, 99.

[b] Nichols, B. W. and James, A. T., *Fette Seifen Anstrichm.*, 66, 1003, 1964.

Procedure for the Paper Chromatography of Phospholipids
Filter paper is first dipped in a solution of sodium silicate and, after draining, it is submersed in an acid bath to produce silicic acid. The sodium silicate solution is prepared by dissolving 310 g of silicic acid in 1 ℓ of 28.8% aqueous sodium hydroxide solution. After cooling to room temperature, the volume is increased to 1.5 ℓ with water. This amount is sufficient for impregnation of about 48 strips, 11 × 45 cm, of paper. Strips of Whatman 3 MM paper are dipped individually in the sodium silicate solution and then left to hang for a few minutes for draining. The paper is then kept in 1 ℓ of 6 N hydrochloric acid (500 mℓ hydrochloric acid in 500 mℓ water) for 30 min, after which it is washed in running tap water for 2 hr, and finally, in several changes of distilled water for another 2 hr. After hanging overnight to dry, the impregnated paper strips are developed with chloroform-methanol (1:1 v/v), in order to free them from hydrochloric acid and/or lipids that may be present. The strips are hung up to dry, and then they are pressed between glass plates, after which they are stored by wrapping them in filter paper sheets and aluminum foil. It is necessary to protect the paper strips against moisture during storage. Moist paper should be dried under vacuum prior to use.

Commercial paper impregnated with silicic acid (SG-81 paper) is also available.

The lipid mixture to be analyzed is dissolved in chloroform, and aliquots containing 1 to 10 μg of lipid phosphorus are applied as spots or narrow streaks to the paper. When phosphorus determination of the fractions separated by chromatography is to be achieved, amounts containing at least 5 μg of lipid phosphorus should be applied. As low as 0.1 μg of lipid phosphorus, however, can be detected qualitatively on the chromatograms.

The chromatograms are developed in the ascending direction, usually at room temperature, in tight glass chambers. For this purpose, the paper carrying the lipid mixture to be fractionated is rolled into a cylinder that is fixed with stainless steel wire clips and inserted into the chamber containing the developing solvent at the bottom. Especially in humid and hot weather, it is necessary to remove moisture from the paper by keeping it under vacuum for 15 min prior to development. A too-high moisture content of the paper leads to much faster migration of the fractions, and in extreme cases, all fractions are squeezed together at the solvent front.

Chromatograms are developed, either by a single solvent system in one dimension or by two solvent systems in two perpendicular dimensions. In the latter case, chromatograms that have been developed with the first solvent should be made completely free of it, preferably under vacuum and oxygen-free nitrogen, before they are developed in the second direction. Unidimensional chromatograms are usually developed within 3 to 4 hr, whereas two-dimensional chromatograms need about 2 to 3 hr in each direction.

As a rule, there is not any single solvent system which alone can achieve complete separation of the phospholipid fractions in natural mixtures. Several solvent systems are known, including neutral, slightly acidic, and slightly basic ones, each of which is suitable for particular separations. For example, slightly basic solvents usually lead to better resolution of acidic phospholipids than slightly acidic or neutral solvents. The development of the chromatogram in two perpendicular directions with two different solvents provides a possibility of obtaining complete separation on a single chromatogram. Table 6 lists the composition of the solvents commonly used in unidimensional chromatography on silicic acid-impregnated paper. The $R_f \times 100$ values of different phospholipid fractions using various developing solvents are presented in Table 7.

It is apparent that certain fractions, for example diacylglycerophosphoinositols and acylglycerophosphoethanolamines, usually migrate close to each other on the chromatograms. This holds true also for certain phospholipid and glycolipid fractions; for example, diacylglycerophosphocholines migrate close to digalactosyldiacylglycerols.

Two-dimensional chromatograms on silicic acid impregnated paper can be developed with any of the pairs of solvent systems given in Table 8. Both solvent systems may be neutral, or one may be neutral or slightly acidic and the other slightly basic. When, in the first direction, developing solvents with relatively low volatility are used, for example, pyridine and diisobutylketone, they must be completely removed before the second development can be started. Traces of such solvents result in modification of the chromatographic behavior of the fractions in the second direction. Solvents are usually removed by evaporation in a stream of oxygen-free nitrogen and/or under vacuum. If traces of solvents are still present, they may be removed by washing the chromatogram with diethyl ether and subsequent drying before the second development is done. For further details on the chromatography of lipids on silicic acid impregnated paper, the reader is referred to several review articles.[26,28,29]

C. Separation of Phospholipids by Thin-Layer Chromatography (TLC)

TLC has become the technique most widely used in the separation of lipids, including phospholipids, especially when only small amounts of lipid mixtures are available. The procedure is simple, sensitive, and quick. The adsorbents used in TLC can be impregnated with certain substances in order to achieve separations which are not possible on nonim-

Table 6
COMPOSITION OF SOLVENTS USED FOR THE SEPARATION OF PHOSPHOLIPIDS AND OTHER POLAR LIPIDS BY UNIDIMENSIONAL PAPER CHROMATOGRAPHY

No.	Solvent	Ratio (v/v)	Ref.
1	Chloroform-methanol-water	80:20:2.5	1
2	Diisobutylketone-acetic acid-water	40:20:3	2
3	Diisobutylketone-acetic acid-water	40:25:5	3
4	Diisobutylketone-acetic acid-saline	40:20:3	2
5	Diisobutylketone-pyridine-water	54:40:6	2
6	Diisobutylketone-n-dibutylether-acetic acid-water	20:20:20:3	2
7	Chloroform-methanol-5 *N* ammonia	60:35:5	4
8	Chloroform-methanol-5 *N* ammonia	64:34:4	4
9	Chloroform-diisobutylketone-acetic acid-methanol-water	45:30:20:15:4	4
10	Chloroform-diisobutylketone-acetic acid-methanol-water	27:38:23:8:4	4
11	Chloroform-diisobutylketone-pyridine-methanol-water	26.5:22:31:15.5:5	4
12	Chloroform-diisobutylketone-pyridine-methanol-water	25:20:28:20:7	4
13	Chloroform-diisobutylketone-pyridine-methanol-acetic acid-formic acid-water	33:10:25:20:8:2:2	4
14	Chloroform-diisobutylketone-pyridine-methanol-acetic acid-formic acid-water	20:30:20:15:10:2:3	4

REFERENCES

1. **Lea, C. H., Rhodes, D. N., and Stoll, R. D.,** *Biochem. J.,* 60, 353, 1955.
2. **Marinetti, G. V.,** in *New Biochemical Separations,* James, A. T. and Morris, L. J., Eds., D. van Nostrand, London, 1964, 339.
3. **Marinetti, G. V. and Stotz, E.,** *Biochim. Biophys. Acta,* 21, 168, 1956.
4. **Wuthier, R. E.,** in *Lipid Chromatographic Analysis,* Vol. I, 2nd ed., Marinetti, G. V., Ed., Marcel Dekker, New York, 1976, 59.

Table 7
PAPER CHROMATOGRAPHY OF PHOSPHOLIPIDS

Serial number in Table 6	Diacyl-glycero-phospho-inositols	Diacyl-glycero-phospho-serines	Diacyl-glycero-phospho-cholines	Diacyl-glycero-phospho-ethanol-amines	Phospha-tidic-acids	Diacyl-glycero-poly-phospho-glycerols	Sphingo-myelins
			R$_f$ × 100				
2[a]		45		50	79-90	57-67	
3[b]	36	54	48	59	95		
7[c]	67	42	64	82	19	90	44
8[c]	30	19	46	65	8	78	28
9[c]	27	48	39	62	68	73	23
10[c]	24	45	39	55	59	66	27
11[c]	45	34	14	54	77	83	9
12[c]	58	55	30	69	74	81	23
13[c]	62	57	31	67	63	79	18
14[c]	53	54	43	65	63	67	32

Note: Chromatograms developed on silicic acid-impregnated paper with solvents specified in Table 6. References a, b, and c correspond to References 1, 2, and 3, respectively, in Table 6.

Table 8
COMPOSITION OF SOLVENTS USED IN TWO-DIMENSIONAL CHROMATOGRAPHY OF PHOSPHOLIPIDS ON SILICIC ACID-IMPREGNATED PAPER

No.	First solvent[a]	Second solvent[a]
1	Chloroform-methanol-water (65:25:4)	Diisobutylketone-acetic acid-water (40:25:5)
2	Chloroform-methanol-water (65:25:4)	Diisobutylketone-pyridine-water (25:25:4)
3	Chloroform-methanol-28% ammonia (65:35:5)	Chloroform-acetone-methanol-acetic acid-water (50:20:10:10:5)
4	Chloroform-methanol-5 *N* ammonia (64:34:4)	Chloroform-diisobutylketone-pyridine-methanol-acetic acid-formic acid-water (23:20:25:20:8:1:3)
5	Chloroform-methanol-5 *N* ammonia (64:28:4)	Chloroform-diisobutylketone-methanol-acetic acid-water (27:28:8:23:4)
6	Chloroform-diisobutylketone-methanol-acetic acid-water (23:45:10:25:4)	Chloroform-diisobutylketone-pyridine-methanol-0.5 *N* ammonia (30:25:35:17.5:6)
7	Chloroform-diisobutylketone-methanol-acetic acid-water (45:30:15:20:4)	Chloroform-diisobutylketone-pyridine-methanol-0.5 *N* ammonia (30:25:35:25:8)
8	Chloroform-diisobutylketone-methanol-acetic acid-water (45:30:15:20:4)	Chloroform-diisobutylketone-pyridine-methanol-water (30:25:35:25:8)
9	Chloroform-diisobutylketone-pyridine-methanol-acetic acid-formic acid-water (33:10:25:20:8:2:2)	Chloroform-diisobutylketone-methanol-acetic acid-water (27:38:8:23:4)

[a] All proportions are v/v.

From Wuthier, R. E., in *Lipid Chromatographic Analysis*, Vol. I, 2nd ed., Marinetti, G. V., Ed., Marcel Dekker, New York, 1976, 59. With permission.

pregnated layers. The advantages of TLC are fully exploited by coupling this technique with GC.

The adsorbent most commonly used in TLC is a very fine grade silica gel; however, other substances such as alumina, cellulose, kieselguhr and others, may be applied for specific purposes. Depending upon the sorbent coated on the plates, separations may be based on adsorption, partition, ion-exchange, or a combination of mechanisms.

For coating glass plates with silica gel, a slurry of 35 g of the adsorbent in 70 mℓ of water is spread at the desired thickness using a suitable spreader on five glass plates, each 20 × 20 cm. Usually, silica gel layers, 0.2 to 0.3 mm thick, are applied in qualitative analyses; however, layers 0.5 mm thick, or even up to 1 mm, are useful in preparative work. After coating, the adsorbent layers are air dried, activated at 110°C for a few hours, and stored in a desiccator until they are used. For coating impregnated layers the same procedure is followed. The slurry of adsorbent is prepared using a solution of the impregnating substance at the desired concentration, instead of water. Precoated silica gel plates are also available from many suppliers. These can be impregnated by spraying, dipping, or irrigation.

For obtaining satisfactory separations, amounts of 10 to 50 μg total lipids, dissolved in chloroform, should be applied to the plates. However, up to 10 mg can be fractionated by preparative TLC on a single 20 × 20 cm plate, 0.5 mm thick. The plates should be developed in tight glass jars lined with filter paper. The developing solvent ascends by capillarity at a rate of 15 to 18 cm within 25 to 40 min. The chromatograms are dried under nitrogen before they are sprayed with detecting reagents. Similar to chromatography on silicic acid-impregnated paper, TLC may be developed in one or two dimensions.

1. Unidimensional TLC

Table 9 summarizes the composition of single solvent systems commonly used in unidimensional TLC of phospholipids on silica gel and silica gel impregnated with different

Table 9
COMPOSITION OF SOLVENTS USED FOR THE SEPARATION OF PHOSPHOLIPIDS AND OTHER POLAR LIPIDS BY UNIDIMENSIONAL THIN-LAYER CHROMATOGRAPHY

No.	Adsorbent	Solvent[a]	Ref.
1.	Boric acid impregnated silica gel	Chloroform-methanol (80:25)	1
2.	Silica gel	Chloroform-methanol-water (40:10:4)	2
3.	Silica gel	Chloroform-methanol-water (65:35:8)	3
4.	Silica gel	Chloroform-methanol-water (25:10:1)	4
5.	Silica gel	Chloroform-methanol-water (65:25:4)	5
6.	Magnesium silicate impregnated silica gel	Chloroform-methanol-water (65:25:4)	6
7.	Silica gel	Chloroform-acetone-water (30:60:2)	7
8.	Silica gel	Acetone-benzene-water (91:30:8)	8
9.	Ammonium sulfate impregnated silica gel	Acetone-benzene-water (91:30:8)	9
10.	Silica gel	Chloroform-methanol-ammonia (60:25:4)	1
11.[b]	Potassium oxalate impregnated silica gel H	Chloroform-methanol-4 *N* ammonia (9:7:2)	10
12.	Silica gel	Chloroform-methanol-ammonia [30% w/v]-water (60:35:5:2.5)	7
13.	Silica gel	Chloroform-methanol-acetic acid (80:25:1)	1
14.	Silica gel	Chloroform-methanol-acetic acid-water (65:15:10:4)	1
15.	Sodium carbonate impregnated silica gel	Chloroform-methanol-acetic acid-water (25:15:4:2)	11
16.	Ammonium sulfate impregnated silica gel H	Chloroform-methanol-acetic acid-water (50:25:8:1)	12
17.	Silica gel	Chloroform-acetone-methanol-acetic acid (73:25:1.5:0.5)	7
18.	Silica gel	Diisobutylketone-acetic acid-water (80:50:7)	1

[a] All proportions are v/v.
[b] For separation of diacylglyceropolyphosphoinositols.

REFERENCES

1. **Nichols, B. W.,** in *New Biochemical Separations,* James, A. T. and Morris, L. J., Eds., D. van Nostrand London, 1964, 321.
2. **Body, D. R. and Gray, G. M.,** *Chem. Phys. Lipids,* 1, 254, 1967.
3. **Fischer, W., Ishizuka, I., Landgraf, H. R., and Herrmann, J.,** *Biochim. Biophys. Acta,* 296, 527, 1973.
4. **Christie, W. W.,** *Lipid Analysis,* Pergamon Press, Oxford, 1976, 191.
5. **Wagner, H., Hörhammer, L., and Wolff, P.,** *Biochem. Z.,* 334, 175, 1961.
6. **Rouser, G., Galli, C., Lieber, E., Blank, M. L., and Privett, O. S.,** *J. Am. Oil Chem. Soc.,* 41, 836, 1964.
7. **Clayton, T. A., MacMurray, T. A., and Morrison, W. R.,** *J. Chromatogr.,* 47, 277, 1970.
8. **Pohl, P., Glasl, H., and Wagner, H.,** *J. Chromatogr.,* 49, 488, 1970.
9. **Khan, M-U. and Williams, J. P.,** *J. Chromatogr.,* 140, 179, 1977.
10. **Gonzalez-Sastre, F. and Folch-Pi, J.,** *J. Lipid Res.,* 9, 532, 1968.
11. **Skipski, V. P., Peterson, R. F., and Barclay, M.,** *Biochem. J.,* 90, 374, 1964.
12. **Kaulen, H. D.,** *Anal. Biochem.,* 45, 664, 1972.

substances. The $R_f \times 100$ values of phospholipid fractions with different solvents are given in Table 10. It is to be noted that the solvent systems which contain large proportions of acetone do not yield good separations of phospholipids, but they give efficient resolution of glycolipids. Phospholipid fractions are separated well by using neutral, slightly acidic, or slightly basic mixtures of chloroform and methanol. In many cases, it has been found that the separation is improved when the adsorbent is impregnated with ammonium sulfate.[30,31]

There is not any single solvent system which alone can give complete separation. Although

Table 10
THIN-LAYER CHROMATOGRAPHY OF PHOSPHOLIPIDS[a]

$R_f \times 100$

Serial number in Table 9	Diacylglycerophosphoinositols	Diacylglycerophosphoserines	Diacylglycerophosphocholines	Diacylglycerophosphoethanolamines	Phosphatidic acids	Diacylglycerophosphoglycerols	Diacylglycerolpolyphosphoglycerols	Sphingomyelins
1				10.1				
4			23.7	65.8				13.2
5			38.9	61.1				
7	5.8		1.9	9.6		13.5		
8			1.1	17.0		60.2		
9	10.4		29.8	41.0		41.0	70.9	
10			17.9	32.1				
12	23.0	25.0	49.0	54.2				
13				27.2		9.6		
14	8.5		25.4	47.7	70.0	47.7	95.4	
15	45.8	55.2	31.3	83.3				17.7
18	9.8		15.7	41.2	64.7	37.3	64.7	

[a] References are given in Table 9.

Table 11
COMPOSITION OF SOLVENTS USED IN TWO-DIMENSIONAL THIN-LAYER CHROMATOGRAPHY OF PHOSPHOLIPIDS

No.	Adsorbent	First solvent[a]	Second solvent[a]	Ref.
1	Silica gel	Chloroform-methanol-water (65:15:2)	Chloroform-methanol-acetone-acetic acid-water (65:10:20:10:3)	1
2	Silica gel	Chloroform-methanol-water (65:25:4)	Chloroform-methanol-ammonia-2-aminopropane (60:30:4:0.5)	2
3	Silica gel H	Chloroform-methanol-water (65:25:4)	Tetrahydrofuran-methylal-methanol-2 M ammonia (10:5:5:1)	3
4	Magnesium silicate impregnated silica gel	Chloroform-methanol-water (65:25:4)	n-Butanol-acetic acid-water (60:20:20)	4
5	Silica gel	Chloroform-methanol-7 N ammonia (65:30:4)	Chloroform-methanol-acetic acid-water (170:25:25:6)	5
6	Magnesium silicate impregnated silica gel	Chloroform-methanol-28% ammonia (65:35:5)	Chloroform-acetone-methanol-acetic acid-water (10:4:2:2:1)	4
7	Silica gel	Chloroform-methanol-7 N ammonia (65:35:5)	Chloroform-methanol-7 N ammonia (35:60:5)	6
8	Sodium borate impregnated silica gel	Chloroform-methanol-7 N ammonia (65:35:5)	Chloroform-methanol-7 N ammonia (35:30:5)	7
9	Ammonium sulfate impregnated silica gel H	Acetone-benzene-water (91:30:8)	Chloroform-methanol-acetic acid-water (85:10:10:2)	8

[a] All proportions are v/v.

REFERENCES

1. **Lepag, M.,** *Lipids,* 2, 244, 1967.
2. **Kleinig, H. and Kopp, C.,** *Planta,* 139, 61, 1978.
3. **Gray, G. M.,** *Biochim. Biophys. Acta,* 144, 519, 1967.
4. **Rouser, G., Kritchevsky, G., and Yamamoto, A.,** in *Lipid Chromatographic Analysis,* Vol. I, Marinetti, G. V., Ed., Marcel Dekker, New York, 1967, 99.
5. **Nichols, B. W.,** in *New Biochemical Separations,* James, A. T. and Morris, L. J., Eds., D. van Nostrand, London, 1964, 321.
6. **Skidmore, W. D. and Entenman, C.,** *J. Lipid Res.,* 3, 471, 1962.
7. **Bunn, C. R., Keele, B. B., Jr., and Elkan, G. H.,** *J. Chromatogr.,* 45, 326, 1969.
8. **Radwan, S. S.,** *J. Chromatogr. Sci.,* 16, 538, 1978.

solvents that contain diisobutyl ketone give good resolutions of phospholipids, they have the drawbacks that they migrate slowly and that they can be removed only with difficulty. In addition, such solvents fail to separate diacylglycerophosphocholines and diacylglycerophosphoinositols. Slightly acidic mixtures of chloroform and methanol are, in general, very efficient in resolving phospholipid fractions. However, the separation of diacylglycerophosphoethanolamines and diacylglycerophosphoglycerols as well as the separation of diacylglycerophosphocholines and digalactosyldiacylglycerols are not satisfactory. Moreover, in such slightly acidic solvents, phosphatidic acids usually migrate as long streaks that may mask some other fractions on the chromatogram. Although slightly basic mixtures of chloroform-methanol generally give poor resolution of phospholipids, they have the advantage of separating diacylglycerophosphocholines and diacylglycerophosphoinositols as well as of resolving phosphatidic acids and polyglycerophospholipids.

2. Two-Dimensional TLC

The merits of two single solvent systems can be combined in two-dimensional TLC, thus affording complete separation in many cases. Table 11 summarizes the composition of the sorbents and of the solvent pairs which are commonly used in two-dimensional TLC. A neutral, slightly acidic, or slightly basic solvent is used in one direction, whereas a contrasting solvent is used in the other direction. When the lipid mixtures are rich in glycolipids, for example lipids extracted from green organs of plants, it is recommended that, in one direction, a solvent system be used containing acetone, which retards the migration of phospholipids. The latter compounds can then be fractionated in the other direction by slightly acidic chloroform-methanol mixtures.

It is possible to separate up to 1 mg of lipid on a single 0.3 mm thick layer, or even up to 3 mg on a single 0.5 mm thick layer by two-dimensional TLC, and to determine the fractions resolved quantitatively (see below).

For more details on the application of TLC in the analysis of lipids, the reader may refer to several reviews.[32-37]

D. Separation of Phospholipids by Gas Chromatography and by Gas Chromatography/Mass Spectrometry (GC/MS)

Although intact phospholipids are not separable by GC, some of their derivatives, especially the trimethylsilyl (TMS) ethers, are rather volatile and may thus be fractionated.

First, the phospholipids should be dephosphorylated, either chemically in phenyl ether at 250°C for 15 min,[38] or enzymatically using phospholipase C (see before). Then the diacyl-glycerols produced from the phospholipids are silylated. Components of a phospholipid mixture can also be prepared for GC/MS by deacylation followed directly by silylation to obtain the TMS ethers, i.e., without dephosphorylation.

Deacylation[39,40] is performed by treating phospholipid mixtures with 0.1 N aqueous sodium hydroxide followed by neutralization with 0.1 N aqueous acetic acid and finally extraction with chloroform-isobutanol-methanol-water (1.8:1:0.2:2 v/v), whereby the sodium salts of the glycerophosphate ester backbone are separated in the aqueous phase. By passing the components of the aqueous phase through a cation exchange resin (Dowex 50 W, H^+ form), the salts are converted to acids, which are eluted, lyophilyzed, and silylated as described before.

Table 12 describes the experimental conditions for the gas chromatographic separation of trimethylsilyl ethers of glycerophospholipids as described by various authors. Experimental conditions for the separation of TMS ethers of sphingolipids are summarized in Table 13.

Although GC is suitable for separating TMS-ethers derived from glycerophospholipid mixtures (Table 12), this technique can, so far, be used only as a qualitative but not as a quantitative method of analysis.[39] Eluates from gas chromatographs can be characterized by transferring them through some connected molecular separator interface to a mass-spectrometer. Readers interested in the various devices as well as in details of the technique may refer to Reference 41.

E. Separation of Molecular Species of Phospholipids

As a rule, fatty acids esterified in position 2 of the various phospholipids are up to 90% unsaturated,[42] So far, satisfactory chromatographic procedures for subfractionation of intact phospholipid species are not available, although, if the phosphate-base moieties are eliminated from the glycerophospholipids, the resulting diacylglycerols can be acetylated and separated more satisfactorily into molecular species.

1. Separation of Intact Phospholipids

Molecular species of intact phospholipids can be resolved partially by argentation-TLC as well as by reversed-phase TLC. As a rule, such separations require skill and experience and, even then, the separations obtained are usually not very satisfactory. The reason lies certainly in the complex nature of natural mixtures as far as the degree of unsaturation of the acyl moieties is concerned.

Table 14 summarizes the experimental conditions for the partial separation of diacylgly-

Table 12
GLC OF TMS ETHERS DERIVED FROM GLYCEROPHOSPHOLIPIDS

No.	Column Material	Column Dimensions	Column temperature	Gas; flow rate	Phospholipid analyzed	Derivative formed	Retention time	Ref.
1	1% SE-30 on Gas Chrom Q	2 ft × 1/4″ glass column	Programming from 180—240°C at 4°/min	Helium; 75 mℓ/min	1,2-Diacyl-*sn*-glycero-3-phosphoinositols	Oktakis (TMS)-*sn*-glycero-3-phosphoinositol	1.00	1
2	1% SE-30 on Gas Chrom Q	2 ft × 1/4″ glass column	Programming from 180—240°C at 4°/min	Helium; 75 mℓ/min	*Bis*(1,2-diacyl-*sn*-glycero-3-phospho)-1′,3′-*sn*-glycerol-3	Heptakis (TMS)-*bis*(*sn*-glycero-3-phospho)-1′,3′-*sn*-glycerol	1.38	1
3	1% SE-30 on Gas Chrom Q	2 ft × 1/4″ glass column	Programming from 180—240°C at 4°/min	Helium; 75 mℓ/min	1,2-Diacyl-*sn*-glycero-3-phosphoinositol monophosphates	Nonakis(TMS)-*sn*-glycero-3-hosphoinositol monophosphate	1.62	1
4	1% SE-30 on Gas Chrom Q	2 ft × 1/4″ glass column	Programming from 180—240°C at 4°/min	Helium; 75 mℓ/min	1,2-Diacyl-*sn*-glycero-3-phosphoinositol diphosphates	Decakis (TMS)-*sn*-glycero-3-phosphoinositol diphosphate	2.12	1
5	1% OV-17 on Supelcoport, 80—100 mesh	6 ft × 2 mm, glass column	Programming from 150—250°C at 5°/min	Helium; 25 mℓ/min	1,2-Diacyl-*sn*-glycero-3-phosphates	Tetrakis(TMS)-*sn*-glycero-3-phosphate Tetrakis(TMS)-*sn*-glycero-2-phosphate	0.10	2
6	1% OV-17 on Supelcoport, 80—100 mesh	6 ft × 2 mm, glass column	Programming from 150—250°C at 5°/min	Helium; 25 mℓ/min	1,2-Diacyl-*sn*-glycero-3-phosphoglycerols	Pentakis(TMS)-*sn*-3-glycerophosphoglycerol	0.41	2
7	1% OV-17 on Supelcoport, 80—100 mesh	6 ft × 2 mm, glass column	Programming from 150—250°C at 5°/min	Helium; 25 mℓ/min	1,2-Diacyl-*sn*-glycero-3-phosphoethanolamines	Tris(TMS)-*N,N-bis*(TMS)-*sn*-glycero-3-phosphoethanolamine	0.49	2
8	1% OV-17 on Supelcoport, 80—100 mesh	6 ft × 2 mm, glass column	Programming from 150—250°C at 5°/min	Helium; 25 mℓ/min	1,2-Diacyl-*sn*-glycerophosphoserines	Tetrakis(TMS)-*N*-TMS-*sn*-glycero-3-phosphoserine	0.52	2

REFERENCES

1. **Cicero, T. J. and Sherman, W. R.,** *Biochem. Biophys. Res. Commun.,* 43, 451, 1971.
2. **Duncan, J. H., Lennarz, W. J., and Fenselau, C. C.,** *Biochemistry,* 10, 927, 1971.

Table 13
GLC OF SPHINGOLIPID DERIVATIVES

No.	Column		Temperature	Lipid analyzed	Lipid derivative	Ref.
	Material	Dimensions				
1	1-2% OV-1 on Gas Chrom Q, 100-120 mesh	1.4 m × 3 mm glass column	Isothermal 320°C	Cerebrosides	TMS-ether	1, 2
2	1% SE-30 on acid-washed and silanized Gas Chrom P, 100-120 mesh	12 ft × 3.5 mm glass column	Programming from 230°C to 350°C at 2°/min	Ceramides with normal fatty acids	TMS-ether	3
3	1% SE-30 on acid-washed and silanized Gas Chrom P, 100-120 mesh	12 ft × 3.5 mm glass column	Programming from 270°C to 350°C at 2°/min	Ceramides with hydroxy fatty acids	TMS-ether	3
4	1-2.5% OV-1 on Gas Chrom Q, 60-80 mesh	1-1.7 m × 3 mm glass column	Isothermal between 275°C to 350°C	Ceramides with sphingosine or phytosphingosine and normal or hydroxy fatty acids	TMS-ether	4-8
5	1% OV-1 on acid-washed Chromosorb W, 68-80 mesh	2 m × 3 mm glass column	Isothermal at 280°C	Ceramide aminoethyl phosphonates	TMS-ether	9
6	2% SE-30 on Chromosorb W, 60-80 mesh	1 m × 3 mm glass column	190°C	2-Amino-1,3-diol alkenes	Di-O-TMS ether N-acetyl di-O-TMS ether	9,10
7	15% DEG polyester on Chromosorb W, 80-100 mesh	2 m × 3 mm steel column	120°C for aldehydes and 140°C for methyl esters	2-Amino-1,3-diol alkenes	Aldehydes, fatty acid methyl esters	9
8	2% SE-30 on Chromosorb W, 60-80 mesh	1 m × 3 mm glass column	205°C	2-Amino-1,3-diol alkenes	N-acetyl poly TMS ether	9, 10

| 9 | 3% SE-30 or 3% OV-1(nonpolar) 3% XE - 60 or 3% OV - 17(polar) on acid-washed Gas Chrom S, 100-120 mesh | 6 ft × 3 mm glass column | 220°C | 2-Amino-1,3-diol alkenes | N-acetyl di-*O*-TMS ether | 11 |
| 10 | 6% Silicone on silanized Gas-Chrom P, 80-100 mesh | 2 m × 3.5 mm glass column | 207°C | 2-Amino-1,3-diol alkenes | Di-*O*-TMS ether | 12 |

REFERENCES

1. **Samuelsson, K. and Samuelsson, B.,** *Biochem. Biophys. Res. Commun.,* 37, 15, 1969.
2. **Hammarström, S.,** *Eur. J. Biochem.,* 21, 388, 1971.
3. **Casparrini, G., Horning, E. C., and Horning, M. G.,** *Chem. Phys. Lipids,* 3, 1, 1969.
4. **Samuelsson, B. and Samuelsson, K.,** *J. Lipid Res.,* 10, 47, 1969.
5. **Samuelsson, B. and Samuelsson, K.,** *Biochim. Biophys. Acta,* 164, 421, 1968.
6. **Samuelsson, B. and Samuelsson, K.,** *J. Lipid Res.,* 10, 41, 1969.
7. **Hammarström, S., Samuelsson, B., and Samuelsson, K.,** *J. Lipid Res.,* 11, 150, 1970.
8. **Hammarström, S.,** *J. Lipid Res.,* 11, 175, 1970.
9. **Matsubara, T. and Hayashi, A.,** *Biochim. Biophys. Acta,* 296, 171, 1973.
10. **Gaver, R. C. and Sweeley, C. C.,** *J. Am. Oil Chem. Soc.,* 42, 294, 1965.

Table 14
SEPARATION OF MOLECULAR SPECIES OF PHOSPHOLIPIDS BY ARGENTATION TLC

No.	Adsorbent (% of silver nitrate)	Solvent (v/v)	Phospholipid (origin)	Species $R_f \times 100$					Ref.
				Monoenoic	Dienoic	Trienoic	Tetraenoic	Hexaenoic	
1	Silica gel H (30%)	Chloroform-methanol-water (60:30:5)	Diacylglycero-phosphocholines (rat liver)	59.1	50.4		21.7	8.7	1
2	Silica gel H (30%)	Chloroform-methanol-water (55:35:7)	Diacylglycero-phosphoethanolamines (rat liver, egg)	66.3	62.0		21.7	10.9	1
3	Silica gel G (48%)	Chloroform-methanol-water (65:30:3.5)	Diacylglycero-phosphoglycerols (spinach leaves)	62.5		40.0			2
4	Silica gel (10%)	Chloroform-methanol-water (65:35:5)	Diacylglycero-phosphoinositols (rat liver)	62.0		53.4	39.7	15.5	3

REFERENCES

1. **Arvidson, G. A. E.,** *Eur. J. Biochem.,* 4, 478, 1968.
2. **Haverkate, F. and van Deenen, L. L. M.,** *Biochim. Biophys. Acta,* 106, 78, 1965.
3. **Holub, B. J. and Kuksis, A.,** *J. Lipid Res.,* 12, 510, 1971.

cerophosphocholines, diacylglycerophosphoethanolamines, diacylglycerophosphoglycerols, and diacylglycerophosphoinositols into simple groups of molecular species according to the number of double bonds using layers of silica gel impregnated with silver nitrate. In Table 14, the origin of each phospholipid fraction is also specified. Corresponding phospholipid fractions from other origins will not necessarily exhibit a chromatographic behavior identical to that described in Table 14 because the composition of their acyl moieties is certainly not identical with that of the fractions given in this Table. Furthermore, the environmental conditions affect the degree of unsaturation of phospholipids of one and the same origin. It is essential that the impregnated layers should be highly active. Thus, for example, in separation of diacylglycerophosphocholines from rat liver, 0.35-mm thick layers of silica gel H impregnated with 30% silver nitrate are activated at 180°C for 24 hr.[43-45]

Fractions of diacylglycerophosphocholines that have been separated by argentation chromatography can be subfractionated by reversed-phase partition on silanized kieselguhr G impregnated with undecane as a stationary phase; methanol-water (90:10 v/v) saturated with undecane to 70% serves as mobile phase.[44] In this manner, each fraction is subfractionated into two bands, one containing the species with acyl moieties having 16 carbon atoms, and the other band containing the species with acyl moieties having 18 carbon atoms.

2. Separation of Diacylglycerols, Phosphatidic Acids, and Deacylation Products Derived from Phospholipids

Diacylglycerols derived from intact phospholipids should be acetylated before they are resolved into molecular species, although the free diacylglycerols can also be separated.[46]

For acetylation,[47] up to 50 mg of diacylglycerols are dissolved in 2 mℓ of acetic anhydride-pyridine (5:1 v/v) and solution is kept at room temperature overnight. After evaporating the solvents, the diacylglycerol acetates can be analyzed directly or after being purified by preparative TLC.

The molecular species of diacylglycerol acetates derived from phospholipids can be resolved partially by argentation TLC and reversed-phase TLC as well as GLC. It has been possible[48] to separate 1-alkenyl-2-acyl-3-acetylglycerols, 1-alkyl-2-acyl-3-acetylglycerols, and 1,2-diacyl-3-acetylglycerols whose parent phospholipids are not separable. Such separations have been achieved on silica gel G using, first, hexane-diethyl ether (1:1 v/v) to a height of 7 cm, followed by toluene to a height of 15 cm. The three compounds can be arranged according to increasing migration rates in the order: diacyl derivatives, alkyl-acyl derivatives, and alkenyl-acyl derivatives. On silver nitrate impregnated silica gel, using the solvent chloroform – absolute ethanol (94:6 v/v), the diacylglycerols can be fractionated partially according to the number of double bonds.

Phosphatidic acid dimethyl esters derived from phospholipids can also be fractionated into groups of molecular species, according to the number and positions of double bonds, by argentation TLC.[49] Dimethyl esters can be prepared by esterifying phosphatidic acids with diazomethane.[50]

It is sometimes recommended that the various phospholipid classes be analyzed by separating their deacylation products, because better resolution is thus obtained. Table 15 gives the $R_f \times 100$ values of deacylation products of different glycerophospholipids in paper chromatography using various solvent systems.

F. Separation of Alkoxy Phospholipids and Phosphonolipids

Methods have been described for the separation of intact alkyl-acyl-glycerophosphocholines from intact diacylglycerophosphocholines from *Tetrahymena pyriformis* and intact 1-alk-1′-enyl-acyl-glycerophosphoethanolamines from intact diacylglycerophosphoethanolamines from sheep brain by ascending chromatography on a silica gel G column.[51] Methods for the separation of ether phospholipids by TLC have been described and re-

Table 15
SEPARATION OF DEACYLATED PHOSPHOLIPIDS BY PAPER CHROMATOGRAPHY

No.	Deacylation product	$R_f \times 100$ with solvents				
		I	II	III	IV	V
1	Glycerophosphocholine	83	88	22	67	41
2	Glycerophosphoethanolamine	62	66	17	47	41
3	Glycerophosphoserine		30		50	41
4	Glycerophosphoinositol	12	20	6	45	26
5	Glycerophosphoglycerol	36	51	11		56

I. Phenol-water (5:2 by weight), **Ferrari, R. A. and Benson, A. A.,** *Arch. Biochem. Biophys.*, 93, 185, 1961.

II. Phenol, saturated with water-acetic acid-ethanol (50:5:6 v/v); **Dawson, R. M. C., Hemington, N., and Davenport, J. B.,** *Biochem. J.*, 84, 497, 1962.

III. n-Butanol-propionic acid-water (151:75:100 v/v); for reference see I.

IV. Methanol-98% formic acid-water (80:13:7 v/v); **Dawson, R. M. C.,** *Biochem. J.*, 75, 45, 1960.

V. t-Butanol-water (62:38 v/v) plus 10% (w/v) trichloroacetic acid; **Wheeldon, L. W.,** *J. Lipid Res.*, 1, 439, 1960.

viewed.[51] Reference has already been made to the thin-layer chromatographic separation of derivatives of dephosphorylated alkoxyphospholipids.

Lipids from marine invertebrates and protozoa include glycerophospholipids that contain phosphonic acid esterified to glycerol, and a carbon-phosphorus bond to the nitrogenous base. The latter bond is remarkably resistant toward acid hydrolysis. The phosphonolipids include compounds analogous to diacylglycerophosphocholines and diacylglycerophosphoethanolamines, and the corresponding analogs have similar polarities; hence, they are not easily separable by chromatographic procedures. Table 16 gives the $R_f \times 100$ values of these compounds when mixtures of chloroform-acetic acid-water, in various proportions, are used for separation on silicic acid layers.

G. Separation of Radioactive Phospholipids

Radioactively labeled phospholipids can be prepared by providing biological systems such as tissue slices or homogenates, tissue cultures, and microorganisms with suitable radioactive precursors, followed by extracting and separating the phospholipids produced. Mixtures of radioactive phospholipids can be fractionated by any of the chromatographic methods described before. The detection and assay of radioactive fractions that have been resolved by TLC can be achieved quite easily. For documentation, autoradiographs can be obtained.

For quantitation of radioactivity in the various fractions, either the gas-flow counting method or the liquid-scintillation counting technique may be used. For gas-flow counting, TLC scanners are available commercially. Radioactivity is assayed through triangulation of the peaks recorded. Liquid-scintillation counting is, however, more accurate.[52] Fractions are removed from the chromatograms into vials and counting can be performed, in the presence of the adsorbent[53] or after its removal; 10 to 15 mℓ of a suitable scintillation solution is added. For example, the following mixture can be used: 5 g of 2.5 diphenyloxazole; 0.3 g of 1,4-bis-2-(5-phenyloxazolyl) benzene, 130 mℓ of methanol; and 100 mℓ of Bio-Solv solubilizer (Beckman) in 1000 mℓ of toluene. Scintillation solutions are also available commercially, for example ''Aquasol'', a xylene-based scintillation-fluor from New England Nuclear Corporation, North Billerica, Mass.

Table 16
TLC OF PHOSPHOLIPIDS AND THEIR PHOSPHONO ANALOGS

No.	Solvent[a]	Diacylglycero-phosphocholines	Diacylglycero-phosphono-cholines	Diacylglycero-phosphoethanol-amines	Diacylglycero-phosphono-ethanolamines
		$R_f \times 100$			
1	Chloroform-acetic acid-water (60:36:4)	20.8	26.3	62.5	76.4
2	Chloroform-acetic acid-water (20:76:4)	14.3	17.1		
3	Chloroform-acetic acid-water (40:45:8)	31.4	38.6	52.9	67.1
4	Chloroform-acetic acid-water (60:34:6)	38.6	47.1		
5	Chloroform-acetic acid-water (50:45:5)	23.9	35.2	60.6	70.4

[a] All proportions are v/v.

Results from **Kapoulas, V. M.**, *Biochim. Biophys. Acta*, 176, 324, 1969.

IV. DETECTION

Phospholipid fractions that have been resolved by PC or TLC can be detected with nonspecific as well as with specific reagents. Nonspecific reagents include iodine vapor. The chromatograms are brought into tight glass chambers containing several grams of solid iodine dispersed on the bottom. The chamber atmosphere becomes saturated with iodine vapor. One may use a 10 mℓ pipet whose narrow end is packed loosely with a small cotton plug, 1 to 2 g of solid iodine, and another small cotton plug in order to direct a stream of iodine vapor to certain areas of the chromatogram without staining the rest of the plate. By blowing at the wide end of the pipet a gentle stream of iodine vapor is produced. All lipid fractions, especially unsaturated compounds, appear as brown spots on a pale-brown background. An advantage of iodine vapor is that it volatilizes completely on exposing the chromatograms for some time to the atmosphere. Highly unsaturated compounds retain iodine for a longer time and lose it with difficulty. Phospholipid fractions that have been visualized with iodine vapor can be removed from the chromatograms and analyzed quantitatively for phosphorus or nitrogen, but not for the patterns of their constituent fatty acids. Gas chromatography of iodine-treated fatty acids reveals the presence of several artifacts.

Another nonspecific method of detection depends on charring the fractions by heating at 180 to 200°C after spraying the chromatograms with 50% aqueous sulfuric acid or chromic-sulfuric acid. Obviously, detection of fractions by charring cannot be achieved using paper chromatograms. The method is destructive and the fractions cannot be applied for any further analyses.

Fractions that are to be analyzed for the patterns of their constituent fatty acids can be detected by spraying the chromatograms with Rhodamine 6G or 2′,7′-dichlorofluorescein and immediately viewing in UV light while the chromatograms are still wet.

Several specific spray reagents can be used in order to detect certain moieties in the phospholipid fractions, and are thus useful for identification.

1. Detection of phosphorus.[54] Two solutions are prepared by dissolving 10 g of sodium molybdate in 100 mℓ of 4 N hydrochloric acid and 1 g of hydrazine hydrochloride in 100 mℓ of water. These two solutions are mixed and heated in a boiling water bath

for 5 min. After cooling the reagent is made up to 1000 mℓ. On spraying the chromatograms with this reagent, the phospholipid zones immediately yield blue colors on a white background. The latter darkens with time.

2.   Detection of amino groups. Free amino groups in diacylglycerophosphoethanolamines and diacylglycerophosphoserines as well as in their lysoderivatives can be detected by spraying the chromatograms with 0.2% ninhydrin in butanol saturated with water. After heating at 100 to 110°C for about 20 min, the zones that contain amino groups are stained red to violet

3.   Detection of choline.[55] Choline in diacylglycerophosphocholines, acylglycerophosphocholines, and sphingomyelins is detected by spraying the chromatograms with Dragendorff reagent. Solutions of 40 g of potassium iodide in 100 mℓ of water (solution A) and 1.7 g of bismuth subnitrate in 100 mℓ of 20% acetic acid (solution B) are prepared. Directly prior to use, 5 mℓ of solution A is mixed with 20 mℓ of solution B, and 75 mℓ of water is added. Fractions containing choline give orange-red colors with this reagent. Gentle warming may be needed to develop the colors. Additional spray reagents are described in Chapter 28.

V. QUANTITATIVE DETERMINATION

Although several methods have been suggested for the quantitation of phospholipid fractions that have been resolved by chromatography, only a few of them are useful.

An earlier procedure consists of charring the fractions on TLC plates and photodensitometrically determining the carbon produced.[56] This method frequently gives nonreproducible results.[57,58] It is certainly not very useful. The same drawback holds also for the fluorometric method[59] which has been developed for the determination of phospholipid fractions on TLC plates.

Phospholipids that have been fractionated by chromatography can be analyzed quantitatively through determination of their phosphorus or through determination of their constituent fatty acids. Analytical methods for phosphorus determination are destructive and thus, in order to achieve further analyses, such as the determination of fatty acid patterns, new fractions are needed. Determining phospholipid fractions through assaying their acyl moieties is a useful technique, because it gives information not only on the quantities of phospholipids but also on the patterns of their constituent fatty acids.

Procedure for the Analysis of Phospholipid Fractions through Determination of Their Phosphorus[60]
Phospholipid fractions that have been resolved by unidimensional or two-dimensional chromatography are detected with iodine vapor, Rhodamine 6G or 2',7'-dichlorofluorescein and the compounds are eluted with chloroform-methanol (1:1 v/v) containing 10% water.[5] If necessary, fractions removed from TLC plates or paper chromatograms can be analyzed directly without removing the adsorbent or the paper. The following method is suitable for analyzing fractions that have been resolved by TLC. A solution is prepared by dissolving 4.4 g of ammonium molybdate in 500 mℓ water; 14 mℓ concentration of sulfuric acid is added and the volume is made up to 1000 mℓ. A reducing reagent is prepared by dissolving 2.5 g sodium bisulfite, 0.5 g sodium sulfite and 0.42 g 1-amino-2-naphthol-4-sulfonic acid in 250 mℓ water and, after it is left in the dark for several hours, it is filtered into a dark bottle. This reagent is stable in the refrigerator for about 1 month. The standard solution for calibration consists of 0.5 m M sodium dihydrogen phosphate in water. The phospholipid sample in a test tube is dried in a stream of nitrogen, then 0.4 mℓ of at least 70% perchloric acid is added and digestion is performed by heating in a sand bath for 20 min whereby the tube is to act as an air condenser. After cooling 2.4 mℓ of the ammonium molybdate reagent

followed by 2.4 mℓ of the reducing reagent are added and the mixture is heated in a boiling water bath for 10 min. After cooling, the absorbance of the solution is measured at 820 nm, together with a blank prepared in exactly the same manner but without phospholipid sample or standard phosphate. The weight of phosphorus in the fraction analyzed is read from a calibration curve constructed by analyzing known weights of the standard solution, following the same procedure. The weight of phospholipids is obtained by multiplying the weight of phosphorus times a factor which can be calculated by considering the molecular weights of the phospholipids and the atomic weight of phosphorus. For diacylglycerophosphocholines and diacylglycerophosphoethanolamines the factors are 25.8 and 24.6, respectively. In calculating the factors, it should be remembered that some phospholipids contain more than one phosphorus atom per molecule and that other phospholipids, namely the lysoderivatives, contain only a single acyl moiety, instead of two, per molecule.

Procedure for the Analysis of Phospholipid Fractions through Determination of Their Constituent Fatty Acids
The phospholipid fractions are subjected to methanolysis, the resulting methyl esters are separated, and after adding the methyl ester of an unusual fatty acids as an internal standard, the amounts of the constituent fatty acids can be determined by GLC. Analyses can be achieved with fractions that have been resolved by unidimensional[61] or two-dimensional[62] TLC. In the first case, amounts of 5 to 10 mg of total lipids can be fractionated, whereas in the second case, only 1 to 3 mg of lipids can be resolved on one plate. After the fractions have been visualized with Rhodamine 6G or 2',7'-dichlorofluorescein in UV light, the adsorbent areas carrying the different compounds are removed into glass tubes and the internal standard is added. Methyl heptadecanoate is suitable as internal standard; however, if the fractions analyzed contain large proportions of heptadecanoic acid, another internal standard should be selected (e.g., C15:0, C20:0). As a rule, 0.02-0.05 mg of the internal standard should be added to each fraction resulting from the separation of 1 mg of total lipids. For methanolysis, 5 mℓ aliquots of 5% hydrochloric acid or 1% sulfuric acid in absolute methanol are added and, after sealing under oxygen-free nitrogen, the mixtures are heated at 90°C for 90 min, although room temperature and shorter duration are often adequate for complete methanolysis. After cooling, aliquots of 10 mℓ water are added and the methyl esters in each fraction are extracted with 5 aliquots each of 5 mℓ hexane. The combined hexane extracts are concentrated to a minimal volume and the methyl esters are purified by TLC before they are analyzed by GC. The weights of methyl esters in each phospholipid fraction can be calculated as follows:

$$\text{Weight of methyl esters} = \frac{\text{Total peak area of methyl esters} \times \text{Weight of methyl heptadecanoate}}{\text{Peak area of methyl heptadecanoate}}$$

The weight of each phospholipid fraction is calculated by multiplying the weight of methyl esters of the constituent fatty acids by the factor 1.371.[61] This method can be applied also in the determination of other acyl lipids.[61,62]

TLC has been used widely for the determination of the lecithin to sphingomyelin ratio in amniotic fluid.[63-69] The ratio of these two classes of phospholipids aids in assessing fetal lung maturity.[63] The various procedures that have been described yield different ratios. In order to establish a reliable range of values for a given method, a large number of samples must be collected at parturition, and the results of their analyses must be correlated with fetal lung maturity.

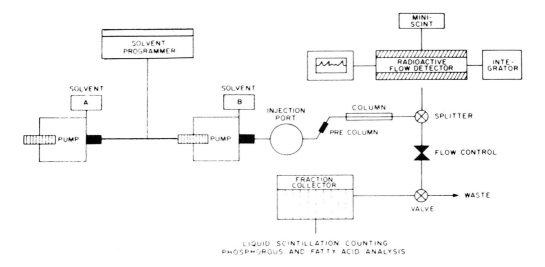

FIGURE 2.   System for the separation of mixtures of radioactively labeled phospholipids by high-performance liquid chromatography.[78] (From Alam, I., Smith, J. B., Siver, M. J., and Ahern, D., *J. Chromatogr.*, 234, 218, 1982. By permission of Elsevier North-Holland Biomedical Press, Amsterdam.)

It appears that a complete analysis of the phospholipid composition of amniotic fluid would be of greater diagnostic value than a determination of the lecithin to sphingomyelin ratio. High performance liquid chromatography will probably become the method of choice for such analyses.[70]

Numerous procedures for the fractionation of phospholipids by high performance liquid chromatography (HPLC) have been described.[70-79] Yet none of these has found wide application so far. In the author's opinion, TLC is still indispensable as a routine method for the qualitative and semiquantitative analysis of phospholipids. There is no doubt, however, that HPLC will become the method of choice for the quantitative analysis of complex mixtures of phospholipids once reliable detection systems have been developed. HPLC has been recommended in conjunction with an automated method of phosphorus analysis but the entire system is certainly rather complex.[74]

Precise analyses of mixtures of radioactively labeled phospholipids are possible.[73,78] As an example, Figure 2 shows a flow chart of a HPLC system equipped with a flow-through radioactivity detector and a stream splitter for diverting part of the effluent from the column to a fraction collector. An application of this system to the fractionation of phospholipids is described in Chapter 14.

Quite recently, an application of the Iatroscan-Chromarod method to the analysis of phospholipid mixtures has been described.[80].

REFERENCES

1. **Kates, M.,** *Can. J. Biochem. Physiol.,* 34, 967, 1956.
2. **Lough, A. K., Felinski, L., and Garton, G. A.,** *J. Lipid Res.,* 3, 478, 1962.
3. **Hübscher, G., Hawthorne, J. N., and Kemp, P.,** *J. Lipid Res.,* 1, 433, 1959.
4. **Kaufmann, H. P., Radwan, S. S., and Ahmad, A.-K. S.,** *Fette Seifen Anstrichm.,* 68, 261, 1966.
5. **Christie, W. W.,** *Lipid Analysis,* Pergamon Press, Oxford, 1982, 53.
6. Applied Science Laboratories, *Gas-Chrom Newsletter,* Vol 11(4), 1970.
7. **Mangold, H. K. and Malins, D. C.,** *J. Am. Oil Chem. Soc.,* 37, 383, 1960.

8. **Nichols, B. W.,** in *New Biochemical Separations,* James, A. T. and Morris, L. J., Eds., Van Nostrand, London, 1964.
9. **Slotboom, A. J., de Haas, G. H., Bonsen, P. P. M., Burbach-Westerhuis, G. J., and van Deenen, L. L. M.,** *Chem. Phys. Lipids,* 4, 15, 1970.
10. **Slotboom, A. J., de Haas, G. H., Burbach-Westerhuis, G. J., and van Deenen, L. L. M.,** *Chem. Phys. Lipids,* 4, 30, 1970.
11. **Robertson, A. F. and Lands, W. E. M.,** *Biochemistry,* 1, 804, 1962.
12. **Ottolenghi, A. C.,** *Methods Enzymol.,* 14, 188, 1969.
13. **Renkonen, O.,** *J. Am. Oil Chem. Soc.,* 42, 298, 1965.
14. **Wurster, C. F. and Corpenhaver, J. H.,** *Lipids,* 1, 422, 1966.
15. *Handbook of Silylation,* Pierce Chemical Company, Rockford, Ill.
16. **Hirsch, J. and Ahrens, E. J., Jr.,** *J. Biol. Chem.,* 233, 311, 1958.
17. **Rouser, G., Kritchevsky, G., and Yamamoto, A.,** *Lipid Chromatographic Analysis,* Vol. 1, Marinetti, G. V., Ed., Marcel Dekker, New York, 1967, 99.
18. **Sweeley, C. C.** *Methods Enzymol.,* 14, 254, 1969.
19. **Christie, W. W.,** *Lipid Analysis,* Pergamon Press, Oxford, 1973, chap. 3.
20. **Rouser, G., Kritchevsky, G., and Yamamoto, A.,** in *Lipid Chromatographic Analysis,* Vol. 3, Marinetti, G. V., Ed., Marcel Dekker, New York, 1976, 713.
21. **Rouser, G., Kritchevsky, G., Heller, D., and Lieber, E.,** *J. Am. Oil Chem. Soc.,* 40, 425, 1963.
22. **Rouser, G., O'Brien, J., and Heller, D.,** *J. Am. Oil Chem. Soc.,* 38, 14, 1961.
23. **Singleton, W. S., Gray, M. S., Brown, M. L., and White, J. L.,** *J. Am. Oil Chem. Soc.,* 42, 53, 1965.
24. **Luthra, M. G. and Sheltawy, A.,** *Biochem. J.,* 126, 251, 1972.
25. **Rouser, G., Kritchevsky, G., Yamamoto, A., Simon, G., Galli, C., and Bauman, A. J.,** *Methods Enzymol.,* 14, 272, 1969.
26. **Wuthier, R. E.,** in *Lipid Chromatographic Analysis,* Vol. 1, Marinetti, G. V., Ed., Marcel Dekker, New York, 1976, 59.
27. **Marinetti, G. V.,** *J. Lipid Res.,* 3, 1, 1962.
28. **Kates, M.,** *Techniques of Lipidology,* North-Holland, Amsterdam, 1972, chap. 5.
29. **Kates, M.,** in *Lipid Chromatographic Analysis,* Vol. 1, Marinetti, G. V., Ed., Marcel Dekker, New York, 1967.
30. **Mangold, H. K. and Kammereck, R.,** *J. Am. Oil Chem. Soc.,* 39, 201, 1962.
31. **Khan, M.-U. and Williams, J.,** *J. Chromatogr.,* 140, 179, 1977.
32. **Mangold, H. K., Schmid, H. H. O., and Stahl, E.,** *Methods Biochem. Anal.,* 12, 393, 1964.
33. **Morris, L. J.,** *J. Lipid Res.,* 7, 717, 1966.
34. **Skipski, V. P. and Barclay, M.,** *Methods Enzymol.,* 14, 530, 1969.
35. **Renkonen, O. and Luukkonen, A.,** in *Lipid Chromatographic Analysis,* Vol. 1, Marinetti, G. V., Ed., Marcel Dekker, New York, 1976, 1.
36. **Mangold, H. K.,** in *Thin-Layer Chromatography,* Stahl, E., Ed., Springer-Verlag, Berlin, 1969, 363.
37. **Stahl, E.,** *Thin-Layer Chromatography,* Springer-Verlag, Berlin, 1969.
38. **Horning, M. G., Murakami, S., and Horning, E. C.,** *Am. J. Clin. Nutr.,* 24, 1086, 1971.
39. **Duncan, J. H., Lennarz, W. J., and Fenselau, C. C.,** *Biochemistry,* 10, 927, 1971.
40. **Cicero, T. J. and Sherma, W. R.,** *Biochem. Biophys. Res. Commun.,* 42, 428, 1971.
41. **Ryhage, R. and Wikström, S.,** in *Mass Spectrometry, Techniques and Application,* Milne, G. W. A., Ed., John Wiley & Sons, New York, 1971, 91.
42. **Hanahan, D. J., Brockerhoff, H., and Barron, E. J.,** *J. Biol. Chem.,* 235, 1917, 1960.
43. **Arvidson, G. A. E.,** *J. Lipid Res.,* 6, 574, 1965.
44. **Arvidson, G. A. E.,** *J. Lipid Res.,* 8, 155, 1967.
45. **Arvidson, G. A. E.,** *Eur. J. Biochem.,* 4, 478, 1968.
46. **van Golde, L. M. G., Pieterson, V. A., and van Deenen, L. L. M.,** *Biochim. Biophys. Acta,* 152, 84, 1968.
47. **Renkonen, O.,** *Biochim. Biophys. Acta,* 125, 288, 1966.
48. **Renkonen, O., Liusvaara, S., and Miettinen, A.,** *Ann. Med. Exp. Fenniae,* 43, 200, 1965.
49. **Renkonen, O.,** *J. Lipid Res.,* 9, 34, 1968.
50. **Renkonen, O.,** *Biochim. Biophys. Acta,* 152, 114, 1968.
51. **Viswanathan, C. V.,** *J. Chromatogr.,* 98, 129, 1974.
52. **Snyder, F.,** in *The Current Status of Liquid Scintillation Radioassay of Thin-Layer Chromatograms,* Bransome, E. D., Jr., Ed., Grune & Stratton, New York, 1970, 248.
53. **Kates, M.,** *Techniques of Lipidology,* North-Holland, Amsterdam, 1972, chap. 6.
54. **Vaskovsky, V. E. and Svetashev, V. I.,** *J. Chromatogr.,* 65, 451, 1972.
55. **Wagner, H., Hörhammer, L., and Wolff, P.,** *Biochem. Z.,* 334, 175, 1961.
56. **Rouser, G., Galli, C., Lieber, E., Blank, M. L., and Privett, O. S.,** *J. Am. Oil Chem. Soc.,* 41, 836, 1964.

57. **Rouser, G., Kritchevsky, G., Galli, C., and Heller, D.,** *J. Am. Oil Chem. Soc.,* 42, 215, 1965.
58. **Mangold, H. K. and Mukherjee, K. D.,** *J. Chromatogr. Sci.,* 13, 398, 1975.
59. **Heyneman, R. A., Bernard, D. M., and Vercauteren, R. E.,** *J. Chromatogr.,* 68, 285, 1972.
60. **Bartlett, G. R.,** *J. Biol. Chem.,* 234, 466, 1959.
61. **Christie, W. W., Noble, R. C., and Moore, J. H.,** *Analyst,* 95, 940, 1970.
62. **Radwan, S. S.,** *J. Chromatogr. Sci.,* 16, 538, 1978.
63. **Gluck, L., Kulovich, M. V., Borer, R. C., Brenner, P. H., Anderson, G. G., and Spellacy, W. N.,** *Am. J. Obstet. Gynecol.,* 109, 440, 1971.
64. **Wilde, C. E. and Oakley, R. E.,** *Ann. Clin. Biochem.,* 12, 83, 1975.
65. **Ng, D. S. and Blass, K. G.,** *J. Chromatogr.,* 163, 37, 1979.
66. **Blass, K. G., Briand, R. L., Ng, D. S., and Harold S.,** *J. Chromatogr.,* 182, 311, 1980.
67. **Blass, K. G. and Ho, C. S.,** *J. Chromatogr.,* 208, 170, 1981.
68. **Kolins, M. D., Epstein, E., Civin, W. H., and Weiner, S.,** *Clin. Chem. (Winston-Salem, N.C.),* 26, 403, 1980.
69. **Larsen, H. F. and Trostmann, A. F.,** *J. Chromatogr.,* 226, 484, 1981.
70. **Briand, R. L., Harold, S., and Blass, K. G.,** *J. Chromatogr.,* 223, 277, 1981.
71. **Hax, W. M. A. and Geurts van Kessel, W. S. M.,** *J. Chromatogr.,* 142, 735, 1977.
72. **Geurts van Kessel, W. S. M., Hax, W. M. A., Demel, R. A., and de Gier, J.,** *Biochim. Biophys. Acta,* 486, 524, 1977.
73. **Blom, C. P., Deierkauf, F. A., and Riemersma, J. C.,** *J. Chromatogr.,* 171, 331, 1979.
74. **Kaitaranta, J. K. and Bessman, S. P.,** *Anal. Chem.,* 53, 1232, 1981.
75. **Yandrasitz, J. R., Berry, G., and Segal, S.,** *J. Chromatogr.,* 225, 319, 1981.
76. **Guerts van Kessel, W. S. M., Tieman, M., and Demel, R. A.,** *Lipids,* 16, 58, 1981.
77. **Chen, S. S.-H. and Kou, A. Y.,** *J. Chromatogr.,* 227, 25, 1982.
78. **Alam, I., Smith, J. B., Silver, M. J., and Ahern, D.,** *J. Chromatogr.,* 234, 218, 1982.
79. **Chen, S. S.-H., Kou, A. Y., and Chen, H.-H. Y.,** *J. Chromatogr.,* 276, 37, 1983.
80. **du Plessis, L. M. and Pretorius, H. E.,** *J. Am. Oil Chem. Soc.,* 60, 1261, 1983.

Chapter 14

ETHER LIPIDS (ALKOXYLIPIDS)

F. Spener

I. INTRODUCTION

In man, animals, and bacteria, the common neutral acylglycerols and phospholipids are accompanied by a group of isosteric molecules having an ether linkage in place of an ester bond. In a chemical sense, these ether lipids, often referred to as "alkoxylipids", are described as derivatives of *sn*-glycerol,[1] where the 1-hydroxy group of the *sn*-glycerol backbone is usually substituted with saturated and monounsaturated C_{16} or C_{18} O-alkyl chains. In addition to alkyl groups, alk-1-enyl groups are also found, rendering these "aldehydogenic" molecules open to acidic attack (see Figure 1).

The proportions of ether lipids found in human and animal tissues, as well as in bacteria, range between 0.5 and 1% of the total lipids; somewhat higher amounts are deposited in muscle and nerve tissues.[1,2] Thus high proportions of neutral ether lipids have only been found together with storage triglycerides in liver oils of sharks and ratfish.[3] Whereas these monoalkyl- and monoalk-1-enyl glycerolipids possess the D-configuration,[2] the dialkylglycerolipids found in halophilic bacteria have the L-configuration (2,3-dialkyl).[4] Dialkylglycerols have also been detected in bovine heart.[5] In recent years, chemically synthesized di- and trialkylglycerols have gained wide recognition as biochemical model substances for isosteric acylglycerolipids.[6,7] Although the biochemistry, including biosynthesis, of naturally occurring ether lipids has been largely elucidated,[8,9] their function remains to be determined.

Further types of ether lipids[1,2] have been detected as follows: methoxyalkylglycerols in marine animals and man,[10,11] hydroxyalkylglycerols in shark and Harderian gland of rabbit,[12,13] alkyl- and alk-1-enyl phosphonolipids in protozoa,[14] alkylacylglyceroglycolipids in brain, nerves and mucus of mammals,[15,16] thioalkylglycerols in mammalian heart muscle,[17] and ether diol lipids in marine and terrestrial animals.[3,18]

In the last two decades, adsorption and partition chromatography have greatly aided in the resolution of complex lipid mixtures into individual classes. However, when analyzing ether lipids, two problems have to be dealt with, i.e., the overwhelming presence of acyl lipids and the rather small difference in polarity of ester, ether, and enol ether linkages. Although it is possible today to fractionate neutral lipids into diacyl-, alkylacyl-, and alk-1-enylacylglycerols and to proceed to the isolation of molecular species, the resolution of phospholipids into these classes is barely attainable, even when masking the polar groups chemically.[9] Because of the pronounced polarity, phosphate and base groups of phospholipids are removed prior to analysis of molecular species. In a number of publications, the chromatographic analysis of ether lipids has been described and reviewed.[1,2,20-27]

In the sections to follow, references will be made to other chapters of this book, where methods also applicable to ether lipids are described in great detail.

II. PREPARATION OF SAMPLES

A. Extraction

Ether lipids are processed together with the bulk acyl lipids, specific extraction procedures for ether lipids are not available. Homogenization of tissues, extraction, isolation, and purification of total lipids[24,28,29] is treated extensively elsewhere in this book (Chapter 2).

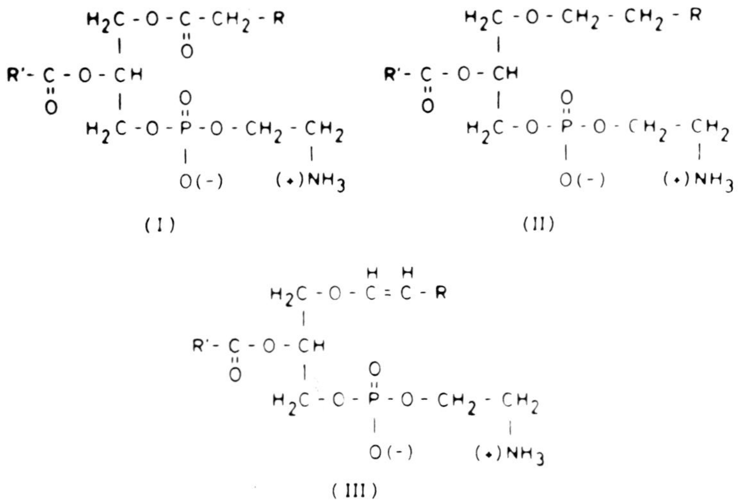

FIGURE 1.   Three forms of ethanolamine phospholipids:
(I) Diacylglycerophosphoethanolamines (Phosphatidylethanolamines)
(II) Alkylacylglycerophosphoethanolamines
(III) Alk-1-enylacylglycerophosphoethanolamines (Ethanolamine plasmalogens)

B. Isolation

1. Isolation of Individual Lipid Classes of General Occurrence

Naturally occurring neutral lipids containing O-alkyl- and O-alk-1-enyl groups can be separated by chromatographic methods from their corresponding acyl species. This is not the case with polar lipids, viz., phospholipids and glycolipids.

Gross fractionation of total lipids is achieved by column chromatography (CC) on silicic acid[24,30] or acid treated Florisil.[24,31] Neutral lipids as a whole are eluted with chloroform, glycolipids with acetone, and phospholipids with increasing amounts of methanol in chloroform. Preparative thin-layer chromatography (TLC) offers more accuracy and speed as compared to CC; however, its use is confined to samples not exceeding 10 mg, in particular when fractionating polar lipids.

Neutral ether lipids and acyl lipids have been separated from each other and from other lipid classes by CC (Tables 1, 2*),[32] by TLC (Tables 3, 4),[33] and partly by gas-liquid chromatography (GLC) (Tables 11, 12).[34] The separation of phospholipid classes is treated in Chapter 13. Some degree of resolution by TLC into alk-1-enylacyl-, alkylacyl-, and diacylphosphatidylethanolamines has been achieved after masking the phosphoryl-base moiety with a methyl and dinitrophenyl group (Table 6).[35]

2. Methoxyalkyl- and Hydroxyalkylglycerols

Neutral and polar ether lipids containing the (2-methoxy)alkyl or (2-hydroxy)alkyl moiety occur as minor constituents or in trace quantities in fish and mammals, including man. Upon saponification of extracted lipids, the unsaponifiable material was chromatographed on silicic acid columns for the removal of hydrocarbons, sterols, and alcohols with petroleum hydrocarbon-ethanol (19:1 v/v) whereas unsubstituted and 2-substituted alkylglycerols were eluted with ethanol. After derivatization to isopropylidenes, the alkyl- and (2-methoxy)alkyl derivatives could be separated by CC on silicic acid with 1 and 5% diethyl ether in petroleum hydrocarbon, respectively.[11] Similarly, 1-(2'-hydroxy)alkyl-2,3-isopropylideneglycerols were isolated by CC and TLC.[11,12]

---

*   All tables follow text.

1-Hydroxyalkyl-2-acyl- and 1-hydroxyalkyl-2,3-diacyl-*sn*-glycerols in rather substantial amounts have been isolated from the pink portion of rabbit Harderian gland.[13,36] The exact position of the hydroxy group has not been established as yet; however, chromatographic and chemical evidence proved the major lipid of this gland to be the totally acylated derivative, i.e., 1-(*O*-acyloxy)alkyl-2,3-diacyl-*sn*-glycerol.[13] Chromatographic data for the separation of substituted alkylglycerols and their derivatives are shown in Tables 7 (TLC) and 20 (GLC).

3. Phosphonolipids

The phosphatidylethanolamine fraction from lipids of *Tetrahymena pyriformis* contains substantial amounts of alkylacyl species having the 2-aminoethylphosphonate group.[37] Treatment of this fraction with ethanolic KOH afforded 1-alkyl-3-phosphonoethanolamine-*sn*-glycerols in the organic phase. 1-Alkyl-*sn*-glycerols for further chromatographic processing were finally obtained after splitting off the 2-aminoethylphosphonate group with 6 *N* HCl at 110°C for 24 hr.[37] A plasmalogen-type lipid containing the 2-aminoethylphosphonate group has been detected in lipids from rumen protozoa.[38] The vinyl ether bond has been cleaved to yield aldehydes by reacting the deacylated 1-alk-1'-enylphosphonolipid with the HgCl$_2$-trichloroacetic acid reagent.[38]

Ascending dry-column chromatography was used to isolate pure 1-alkyl-2-acyl-*sn*-glycero-3-(2'-aminoethylphosphonate) from total lipids of *T. pyriformis*. The column was filled with activated silica gel G and developed with chloroform-acetic acid-methanol-water (75:25:5:1.5 v/v). This technique is applicable for fractionations of lipids up to 6 g.[39,40]

4. Alkylacylglyceroglycolipids

Ether lipids containing galactose, i.e., 1-alkyl-2-acyl-3-galactosylglycerols, have been detected in lipid extracts from mammalian brain[41] and nerve tissue.[15] Lipids were applied to a silicic acid column and neutral lipids eluted with chloroform and chloroform-acetone (19:1 v/v). The crude galactolipid fraction was then recovered with chloroform-acetone (3:1 v/v) and purified by TLC using chloroform-methanol (185:15 v/v) as solvent. This fraction was hydrolyzed with 3 *N* H$_2$SO$_4$ for 90 min at 100°C in sealed tubes to remove galactose and acyl groups. Complete removal was finally achieved by methanolysis with BF$_3$-methanol. The resulting alkylglycerols and methyl esters were separated by TLC, and alkylglycerols were analyzed by GLC as their trimethylsilyl derivatives.[15]

Recently, human gastric secretions and saliva were shown to contain glycolipids composed of alkylacylglycerol and one to eight glucose residues.[16,42] These lipids were extracted from lyophilized secretions and fractionated by CC on silicic acid. After the first elution with chloroform, glucolipids were eluted with acetone and acetone-methanol 9:1. Purification and isolation of alkylacylglucolipids according to carbohydrate content was achieved by preparative TLC on silica gel HR (Table 8). Alkylglycerols were obtained after methanolysis of purified compounds.[16,43]

5. Thioalkylglycerols

In small amounts *S*-alkylglycerols have been isolated after alkaline hydrolysis of neutral lipids from bovine and human heart.[17,44] From the unsaponifiable material *S*-alkylglycerols were recovered together with their isosteric *O*-alkyl counterparts. These two classes of compounds could only be separated, after their conversion to isopropylidene derivatives, by TLC (Table 9) and GLC (Table 21).[44,45]

6. Ether Diol Lipids

Naturally occurring alkyl and alk-1-enyl ethers of ethanediol have been extracted in high quantities from tissues of the starfish *Distolasterias nipon*.[46] After prefractionation of neutral lipids on a column of silica gel with hexane-diethyl ether (95:5 and 80:20 v/v) the less-polar fractions eluted were rechromatographed on layers of grade IV alumina with hexane-diethyl ether (20:1 v/v). Pure alk-1-enylacylethanediols (R_f 0.60) were isolated and treated with 1% $HgCl_2$ in 0.1 M HCl to liberate aldehydes. The more polar fractions obtained from CC were fractionated on layers of silica gel with hexane-diethyl ether (85:15 v/v) Pure alkylacylethanediols (R_f 0.60) were recovered from the adsorbent and subjected to methanolysis and the resulting monoalkylethanediols analyzed directly by temperature-programmed GLC.[46]

Unusual di-*O*-alkyl-1,5-pentanediols in amounts of 0.5 to 1% of total lipids have been isolated from the jaw oil of the porpoise *Phocoena phocoena*.[47] Fractionation of neutral lipids by TLC on silica gel showed a band at R_f 0.62 (Table 10). Chromatographic and chemical evidence revealed the lipids of this zone as dialkylethers of 1,5-pentanediol having predominantly C_{18}-chains.[47]

Alkyl- and alk-1-enyl ethers of ethanediol have been synthesized,[48-50] and model mixtures together with glycerol-derived ether lipids were separated by CC on Sephadex LH 20 (Table 2),[51] by TLC on silica gel (Table 10),[48,50,52] and by GLC after derivatization (Table 23),[49,53] in addition to molecular species (Table 22).[48]

C. Degradation and Derivatization

Lipids are partially degraded and modified by various chemical and enzymic means. When extracts of natural sources are under investigation, ester bonds are usually cleaved, and the bulk of ester lipids is removed. The resulting alkyl- and alk-1-enylglycerols are then transformed to derivatives suitable for chromatographic analysis.

1. Saponification

When extracts are treated with mineral acids, lipids containing enol ether bonds immediately release long-chain aldehydes; thus hydrolysis under alkaline conditions is preferred. Complete hydrolysis of all ester bonds is achieved when total lipids are refluxed for 1 to 2 hr in 0.3 N NaOH in methanol, affording alkyl- and alk-1-enylglycerols as unsaponifiable material.[24] On the other hand, the phosphate ester bond resists saponification when samples are treated with 0.1 N NaOH in methanol for 15 min at 38°C.[54] Further procedures for controlled deacylation can be found in great detail in Chapter 15.

2. Methanolysis

Acid-catalyzed transesterification is carried out in tightly closed glass tubes by reacting 1 to 10 mg of lipid in a 1 to 2 mℓ solution of 5% dry HCl in absolute methanol at 80°C for 2 hr.[55] When a mixture of methanol-benzene-concentrated H_2SO_4 (86:10:4 v/v) is used instead of methanolic HCl, the reaction can be carried out in the presence of silica gel scrapings from TLC plates.[56] Only neutral lipids can be transesterified with 10% BCl_3 or 14% BF_3 in methanol.[57,58] The lipophilic products of acid methanolysis are separated best by preparative TLC on silica gel using a two- to threefold development in hexane-diethyl ether (95:5 v/v). Dimethyl acetals, originating from all types of plasmalogens, migrate ahead of methyl esters, and alkylglycerols remain at the start.[21]

Alkali-catalyzed transesterification of neutral lipids is carried out with 4% NaOH in a mixture of methanol-benzene (60:40 v/v).[24] Methanolysis is completed after 20 min at room temperature, the resulting alkyl- and alk-1-enyl-glycerols are separated from methyl esters by preparative TLC.

3. Acetolysis

The glycerophosphate bond of phospholipids is selectively cleaved by treating the sample (0.5 to 50 mg) with 2 mℓ of acetic acid-acetic anhydride (3:2 v/v) in a sealed vessel for 4 hr at 145°C.[59] The 1-alkyl-2-acyl-3-acetyl-*sn*-glycerols and 1,2-diacyl-3-acetyl-*sn*-glycerols produced are conveniently separated by TLC (Table 3) and GLC (Tables 12, 13). It should be noted, however, that isomerization of 1,2- to 1,3-species occurs to some extent during the reaction. In addition, the enol ether bond in plasmalogens is destroyed under the conditions employed.[27]

4. Grignard Reagent

Ethyl magnesium bromide affects random deacylation of neutral glycerolipids in anhydrous diethyl ether with a minimum of acyl migration.[60] Thus 1-alkyl-3-acyl- and 1-alkyl-2-acylglycerols are obtained for further studies of molecular species.

5. Enzymic Degradation

The use of enzymes permits the facile and selective removal of acyl, phosphate, and phosphoryl-base groups.

a. Pancreatic Lipase

This enzyme deacylates the *sn*-3-position of neutral ether lipids and the *sn*-1- and 3-position of triacylglycerols.[61] When a pancreatic lipase preparation free of phospholipase A_2 is used, *sn*-1-acyl groups from phospholipids are selectively removed, whereas *sn*-1-alkyl and 1-alk-1'-enyl groups are not attacked.[62]

b. Phospholipase A_2

The enzyme from snake venom hydrolyzes the *sn*-2-acyl group from phospholipids of all types; however, some sluggishness was observed towards substrates having alkyl and alk-1-enyl groups at the *sn*-1-position of glycerol.[63]

c. Phosphatases

Glycerol-bound phosphate groups can be cleaved off with alkaline phosphatase from *Escherichia coli,* this enzyme being specific for acyl and alkyl lysophosphatidic acids and acyl- and alkyl dihydroxyacetonephosphate.[64,65] An acid phosphatase from wheat exhibited no restriction towards the phosphate group of phosphatidic and lysophosphatidic acids.[64]

d. Phospholipase C

The release of phosphoryl-base from phospholipids by action of phospholipase C is the preferred method for the preparation of 1-alkyl-2-acyl-, 1-alk-1'-enyl-2-acyl-, and 1,2-diacyl-*sn*-glycerols. However, the source of the enzyme is critical: phospholipase C from *Bacillus cereus* accepts phosphatidylethanolamines as substrates, whereas the hydrolysis of phosphatidylserines and -inositols is catalyzed only in the presence of lysophosphatidylcholines and sphingomyelins.[66] No problems are encountered in the removal of phosphorylcholine from the corresponding phospholipid with phospholipase C from *Clostridium welchii.* In contrast, this enzyme exerts its action upon phosphatidylethanolamines only in the presence of lysophosphophatidylcholines and sphingomyelins.[55]

6. Hydrogenolysis

The method of choice in preparing alkyl and alk-1-enyl glycerols from complex lipid mixtures is reductive hydrogenolysis with $LiAlH_4$[24] or $NaAlH_2(OCH_2OCH_3)_2$ in 70% benzene solution.[68] The latter reagent, known as "Vitride", is said to allow a better recovery of alk-1-enylglycerols.

Procedure for the Hydrogenolysis of Ether Lipids[33]
A procedure for small lipid samples is given: to 10 to 20 mℓ of a concentrated solution of LiAlH₄ in diethyl ether, cautiously add 5 to 10 mg lipid in 1 to 2 mℓ diethyl ether. The mixture is stirred and refluxed for 2 hr. Excess reagent is then destroyed by addition of wet diethyl ether and water. Upon careful neutralization with 2 N H₂SO₄,

long-chain alcohols, alkyl- and alk-1-enylglycerols are extracted with diethyl ether and separated by preparative TLC.[33]

7. Aldehydes

Aldehydes are simply released from all types of plasmalogens and alk-1-enylglycerols by treatment with hydrochloric acid, either in ethereal solution,[68] or directly on a chromatoplate with HCl fumes.[69]

Procedure for the Liberation of Aldehydes from Alk-1-enyl Ethers[68]
One to 5 mg of alk-1-enyl ethers, or silica gel containing such an amount, 3 mℓ of peroxide-free diethyl ether, and 1 mℓ of concentrated HCl are stirred in an atmosphere of nitrogen for 30 min. After addition of 5 mℓ of hexane and 10 mℓ of water, the reaction mixture is shaken and centrifuged to break emulsions. The upper layer, which contains the long-chain aldehydes, is washed with several 10 mℓ portions of water until neutral.

Supposedly, trimerization and polymerization of aldehydes is prevented by storage in CS_2;[70] they are readily analyzed by TLC (Tables 3, 4) and GLC (Table 15). Aldehydes liberated from plasmalogens have also been converted to dioxanes with 1,3-propanediol using *p*-toluenesulfonic acid as catalyst and subsequently analyzed by GLC (Table 16).[71] Acid-catalyzed methanolysis of lipids containing vinyl ether bonds leads to dimethyl acetals, however, GLC of these compounds deserves special precautions.[72]

8. Acetates

Neutral lipids having one or two hydroxy groups can be converted to their corresponding acetates by reacting these compounds with acetic anhydride in the presence of catalytic amounts of pyridine for 7 hr at 80°C.[27,33] Acetates of glycerolipids containing alkyl, alk-1-enyl, and acyl groups are suitable derivatives for separations by TLC (Tables 3, 7) and GLC (Tables 12, 13, 14, 17, 20, 23).

9. Isopropylidenes

The preferred derivatives for TLC (Tables 7, 9) and GLC (Tables 17, 20, 21) analysis of alkylglycerols are their cyclic ketals with acetone.

Procedure for the Preparation of Isopropylidene Derivatives of Alkylglycerols[73]
The ketalization of alkylglycerols with acetone is achieved by dissolving around 5 mg alkylglycerols in 5 mℓ absolute acetone and allowing the solution to stand at room temperature for 0.5 hr in the presence of small amounts (5 μℓ of 72% perchloric acid.[27,33,73]

10. Trimethylsilyl Ethers (TMS)

Hydroxy groups of neutral lipids are protected with the trimethylsilyl group. Alkyl- and alk-1-enylglycerols, as well as mixed ether-ester-glycerols of various forms, are conveniently chromatographed by GLC as their trimethylsilyl ethers (Tables 14, 18, 20, 23). In the presence of acyl groups, care must be taken to prevent isomerization; a reagent containing pyridine-hexamethyldisilazane-trimethylchlorosilane has been recommended.[27,74,75]

11. Alkoxy Acetaldehydes

Glycol cleavage of alkyl and alk-1-enylglycerols with sodium metaperiodate in pyridine leads to alkoxy acetaldehydes.[76] This method has successfully been used for the detection and analysis of biosynthetically produced ether lipids.[65]

12. Reductive Ozonolysis

Reductive ozonolysis cleaves unsaturated alkylglycerols at the location of double bonds, by reducing the ozonides formed, with triphenylphosphine, for example, to aldehyde moieties. With isopropylidene derivatives of alkylglycerols as starting products, the resulting aldehydic derivatives of isopropylidene glycerol and aldehydes can be directly analyzed by GLC.[77-80]

13. Ancillary Derivatives

Occasionally alkyl- and alk-1-enylglycerols have been analyzed as trifluoroacetates (Tables 18, 23).[81]

14. Determination of Lipids Containing Alkyl and Alk-1-enyl Moieties

The proportions of lipids containing alkyl moieties can be determined by GLC of the isopropylidene derivatives of the constituent alkylglycerols in the presence of an internal standard.[82] A method for the simultaneous quantification of the constituent alkyl and alk-1-enyl moieties involves the use of the dimethylacetal of heptadecyl aldehyde and the isopropylidene derivative of heptadecylglycerol as internal standards.[83] Another method is based on the isotopic derivative technique (see Chapter 25). A mixture of ether lipids and ester lipids is reduced with LiAlH$_4$ to yield alkylglycerols, alk-1-enylglycerols, and long-chain alcohols. Acetylation of these compounds with radioactively labeled acetic anhydride leads to a mixture of labeled acetates which are resolved by TLC and then estimated by scanning of the chromatogram or by liquid-scintillation counting of the eluted fractions.

Procedure for the Quantitative Determination of Ether Lipids[84]
Radioactively labeled acetic anhydride, 1 mCi, is diluted with nonactive acetic anhydride and the resulting mixture (^{1}1 mℓ) is dissolved in ten times its volume of anhydrous pyridine to yield 10 mℓ of reagent solution. The lipid mixture, 5 to 10 mg, that is obtained from the complex lipid extract by hydrogenolysis with LiAlH$_4$ is reacted in a capped 5-mℓ vial with 0.5 mℓ of the acetic anhydride-pyridine solution at 80°C for 30 min. After cooling to room temperature, the vial is opened, the reaction mixture diluted with 1 mℓ of water, and the acetylated lipids are extracted with four 1-mℓ portions of hexane. The combined hexane extracts are washed with four 1-mℓ portions of water and dried over anhydrous sodium sulfate. Most of the solvent is evaporated off with a stream of nitrogen and the volume of the solution made up to 0.5 mℓ by the addition of hexane. About 100 to 200 μg of the acetylated lipids are resolved by chromatography on 0.3 mm thick layers of silica gel H, with benzene-chloroform (1:3) as developing solvent. The ratios of the three fractions of ^{14}C-labeled acetates are determined by scanning the chromatograms with a Berthold LB 2760 instrument at a speed of 300 mm/hr. Alternatively, the fractions of ^{14}C- or ^{3}H-labeled acetates are made visible by exposing the chromatograms to iodine vapor and then eluted with water-saturated diethyl ether. The eluates are filtered through sintered-glass funnels, the solvent is evaporated with a stream of nitrogen, and the radioactivity is determined by liquid-scintillation counting.

Procedure for the Simultaneous Gas Chromatographic Analysis of Alkyl and 1-Alkenyl Moieties[88]
A mixture of alkylglycerols and 1-alkenylglycerols is isolated from the products of hydrogenolysis of a total lipid extract by chromatography on a thin layer of silica gel with hexane-diethyl ether (1:4 v/v). The alkylglycerols and 1-alkenylglycerols (1-10 mg) are acetylated in a capped 5 mℓ vial with 1 mℓ of acetic anhydride-pyridine (1.4) at 80°C for 30 min. After cooling to room temperature, the reaction mixture is diluted

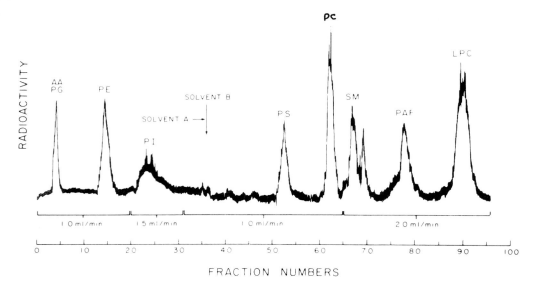

FIGURE 2.   Separation of PAF and other phospholipids by HPLC.[87]
Column: 250/4.6 mm I.D. LiChrosorb Si60 (10 μm particle size) in a Waters Assoc. Liquid Chromatograph equipped with a Model 6000 A solvent delivering system and a radioactive flow detector.
Solvents: (A) Propanol-ethyl acetate-benzene-water (130:80:30:20 v/v); (B) Propanol-toluene-acetic acid-water (93:110:15:15 v/v).
PE, Phosphatidylethanolamines; PI, phosphatidylinositols; PS, phosphatidylserines; PC, phosphatidylcholines; SM, sphingomyelins; PAF, platelet activating factor; LPC, lysophosphatidylcholines. (By permission of Elsevier Biomedical Press, Amsterdam.)

with 1 mℓ of water and the alkyldiacetylglycerols and 1-alkenyldiacetylglycerols are extracted with four portions of hexane, 1 mℓ each. The combined hexane extracts are washed with four portions of water, 1 mℓ each. After evaporation of the solvent, 1.5 mℓ of diethyl ether and 1.5 mℓ of 2 N HCl are added, and the mixture is stirred magnetically at room temperature for 30 min. The alkyldiacetylglycerols and the long-chain aldehydes formed by hydrolysis of the 1-alkenyldiacetylglycerols are extracted with 4 mℓ of hexane. The hexane extract is washed with water until neutral and dried over Na_2SO_4. Most of the solvent is evaporated by a stream of nitrogen, and the volume of the solution is made up to 100 μℓ by the addition of hexane.

The composition of the alkyl and 1-alkenyl moieties is determined by gas chromatography on a column (200 × 0.5 cm) packed with 10% Silar 5 CP on Gas-Chrom Q (80-100 mesh) at 220°C. The aldehydes derived from 1-alkenyl groups are eluted in the first 25 min; they are recorded at a chart speed of 20 mm/min, and the alkyldiacetylglycerols are recorded at a speed of 5 mm/min.

15. Platelet Activating Factor (PAF)

Recent investigations have documented the role of 1-*O*-alkyl-2-*O*-acetyl-*sn*-glycero-3-phosphocholines ("Platelet activating factor", PAF) as a highly active class of compounds mediating a variety of physiological reactions in man and animals.[85,86] PAF can be separated from other phospholipids by TLC on silica gel with the solvent system chloroform-methanol-water, (65:35:6 v/v); it migrates between sphingomyelins and acylglycerophosphocholines (lysolecithins).[85]

High-performance liquid chromatography[87] allows the analysis of phospholipids including PAF in less than 2 hr; experimental conditions are given in Figure 2.

It appears that the quantitative determination of PAF is best carried out by the isotope dilution method[88] in conjunction with TLC or HPLC; ^{3}H-labeled PAF is available.*

III. COMPARISON OF CHROMATOGRAPHIC METHODS

A thorough description and comparison of chromatographic methods pertaining to alkyl- and acylglycerols is given in Chapter 12.

REFERENCES

1. **Snyder, F., Ed.,** *Ether Lipids: Chemistry and Biology,* Academic Press, New York, 1972.
2. **Mangold, H. K. and Paltauf, F.,** *Ether Lipids: Biochemical and Biomedical Aspects,* Academic Press, New York, 1983.
3. **Malins, D. C. and Wekell, J. C.,** in *Progress in the Chemistry of Fats and Other Lipids,* Vol. 10, Holman, R. T., Ed., Pergamon Press, Oxford, 1970, 339.
4. **Kates, M.,** in *Ether Lipids: Chemistry and Biology,* Snyder, F., Ed., Academic Press, New York, 1972, 351.
5. **Pugh, E. L., Kates, M., and Hanahan, D. J.,** *J. Lipid Res.,* 18, 710, 1977.
6. **Spener, F., Paltauf, F., and Holasek, A.,** *Biochim. Biophys. Acta,* 152, 368, 1968.
7. **Mangold, H. K.,** in *Ether Lipids: Chemistry and Biology,* Snyder, F., Ed., Academic Press, New York, 1972, 158.
8. **Snyder, F.,** *Adv. Lipid Res.,* 10, 233, 1972.
9. **Wykle, R. L. and Snyder, F.,** in *The Enzymes of Biological Membranes,* Vol. 2, Martonosi, A., Ed., John Wiley & Sons, London, 1976, 87.
10. **Hallgren, B. and Ställberg, G.,** *Acta Chem. Scand.,* 21B, 1519, 1967.
11. **Hallgren, B., Niklasson, A., Ställberg, G., and Thorin, H.,** *Acta Chem. Scand.,* 28B, 1029, 1974.
12. **Hallgren, B. and Ställberg, G.,** *Acta Chem. Scand.,* 28B, 1074, 1974.
13. **Kasama, K., Rainey, W. T., Jr., and Snyder, F.,** *Arch. Biochem. Biophys.,* 154, 648, 1973.
14. **Thompson, G. A., Jr.,** in *Ether Lipids: Chemistry and Biology,* Snyder, F., Ed., Academic Press, New York, 1972, 321.
15. **Rumsby, M. G. and Rossiter, R. J.,** *J. Neurochem.,* 15, 1473, 1968.
16. **Slomiany, B. L., Slomiany, A., and Glass, G. B. J.,** *Eur. J. Biochem.,* 84, 53, 1978.
17. **Ferrell, W. J.,** *Lipids,* 8, 234, 1973.
18. **Bergelson, L. D.,** in *Progress in the Chemistry of Fats and Other Lipids,* Vol. 10, Holman, R. T., Ed., Pergamon Press, Oxford, 1970, 241.
19. **Renkonen, O.,** *J. Lipid Res.,* 9, 34, 1968.
20. **Viswanathan, C. V.,** *Chromatogr. Rev.,* 10, 18, 1968.
21. **Mangold, H. K.,** in *Thin-Layer Chromatography,* 2nd ed., Stahl, E., Ed., Springer-Verlag, Berlin, 1969, 363.
22. **Snyder, F.,** in *Progress in Thin-Layer Chromatography and Related Methods,* Vol. 2, Niederwieser, A. and Pataki, G., Eds., Ann Arbor Science, Michigan, 1971, 105.
23. **Hanahan, D. J.,** in *Ether Lipids: Chemistry and Biology,* Snyder, F., Ed., Academic Press, New York, 1972, 25.
24. **Kates, M.,** *Techniques of Lipidology: Isolation, Analysis and Identification of Lipids,* Work, T. S. and Work, E., Eds., North-Holland, Amsterdam, 1972.
25. **Schmid, H. H. O., Bandi, P. C., and Su, K. L.,** *J. Chromatogr. Sci.,* 13, 478, 1975.
26. **Snyder, F.,** in *Lipid Chromatographic Analysis,* Vol. 1, 2nd ed., Marinetti, G. V., Ed., Marcel Dekker, New York, 1976, 111.
27. **Myher, J. J.,** in *Handbook of Lipid Research,* Vol. 1, Kuksis, A., Ed., Plenum Press, New York, 1978, 123.
28. **Folch, J. M., Lees, M., and Sloane Stanley, G. H.,** *J. Biol. Chem.,* 226, 497, 1957.
29. **Bligh, E. G. and Dyer, W. J.,** *Can. J. Biochem. Physiol.,* 37, 911, 1959.
30. **Rouser, G., Kritchevsky, G., and Yamamoto, A.,** in *Lipid Chromatographic Analysis,* Vol. 3, 2nd ed., Marinetti, G. V., Ed., Marcel Dekker, New York, 1976, 713.

---

* The Radiochemical Center, Amersham, Bucks., England. New England Nuclear, 549 Albany Street, Boston, MA 02118, U.S.

31. **Carrol, K. K., Cutts, J. H., and Murray, G. D.,** *Can. J. Biochem.,* 46, 899, 1968.
32. **Curstedt, T.,** *Biochim. Biophys. Acta,* 489, 79, 1977.
33. **Schmid, H. H. O. and Mangold, H. K.,** *Biochem. Z.,* 346, 13, 1966.
34. **Wood, R., Baumann, W. J., Snyder, F., and Mangold, H. K.,** *J. Lipid Res.,* 10, 128, 1969.
35. **Renkonen, O.,** in *Progress in Thin-Layer Chromatography and Related Methods,* Vol. 2, Niederweiser, A. and Pataki, G., Eds., Ann Arbor Science, Michigan, 1971, 143.
36. **Rock, C. O. and Snyder, F.,** *Arch. Biochem. Biophys.,* 171, 631, 1975.
37. **Thompson, G. H., Jr.,** *Biochemistry,* 6, 2015, 1967.
38. **Dawson, R. M. G. and Kemp, P.,** *Biochem. J.,* 105, 837, 1967.
39. **Viswanathan, C. V. and Nagabhushanam, A.,** *J. Chromatogr.,* 75, 227, 1973.
40. **Jacini, G. and Fedeli, E.,** in *Adv. Exp. Med. Biol.,* Vol. 4, Holmes, W. L., Carlson, L., and Paoletti, R., Eds., Plenum Press, New York, 1969, 639.
41. **Norton, W. T. and Brotz, M.,** *Biochem. Biophys. Res. Commun.,* 12, 198, 1963.
42. **Slomiany, B. L., Slomiany, A., and Glass, G. B. J.,** *Biochemistry,* 16, 3954, 1977.
43. **Slomiany, A. and Slomiany, B. L.,** *Biochem. Biophys. Res. Commun.,* 76, 115, 1977.
44. **Ferrell, W. J. and Radloff, D. M.,** *Physiol. Chem. Physics,* 2, 551, 1970.
45. **Wood, R., Piantadosi, C., and Snyder, F.,** *J. Lipid Res.,* 10, 370, 1969.
46. **Vaver, V. A., Pisareva, N. A., Rozynov, B. V., Ushakov, A. N., and Bergelson, L. D.,** *Chem. Phys. Lipids,* 7, 75, 1971.
47. **Varanasi, U. and Malins, D. C.,** *Science,* 166, 1158, 1969.
48. **Baumann, W. J., Schmid, H. H. O., Ulshöfer, H. W., and Mangold, H. K.,** *Biochim. Biophys. Acta,* 144, 355, 1967.
49. **Baumann, W. J., Schmid, H. H. O., Kramer, J. K. G., and Mangold, H. K.,** *Z. Physiol. Chem.,* 349, 1677, 1968.
50. **Kramer, J. K. G. and Mangold, H. K.,** *Chem. Phys. Lipids,* 4, 332, 1970.
51. **Calderon, M. and Baumann, W. J.,** *J. Lipid Res.,* 11, 167, 1970.
52. **Kaufmann, H. P., Mangold, H. K., and Mukherjee, K. D.,** *J. Lipid Res.,* 12, 506, 1971.
53. **Schupp, E. and Baumann, W. J.,** *J. Lipid Res.,* 14, 121, 1973.
54. **Pietruszko, R. and Gray, G. M.,** *Biochim. Biophys. Acta,* 56, 232, 1962.
55. **Stoffel, W., Chu, F., and Ahrens, E. H., Jr.,** *Anal. Chem.,* 31, 307, 1959.
56. **Chalvardjian, A.,** *Biochem. J.,* 90, 518, 1964.
57. **Metcalfe, L. D. and Schmitz, A. A.,** *Anal. Chem.,* 33, 363, 1961.
58. **Morrison, W. R. and Smith, L. M.,** *J. Lipid Res.,* 5, 600, 1964.
59. **Renkonen, O.,** *J. Am. Oil Chem. Soc.,* 42, 299, 1965.
60. **Wood, R. and Snyder, F.,** *Arch. Biochem. Biophys.,* 131, 478, 1969.
61. **Snyder, F. and Piantadosi, C.,** *Biochim. Biophys. Acta,* 152, 794, 1968.
62. **De Haas, G. H., Sarda, L., and Roger, J.,** *Biochim. Biophys. Acta,* 106, 638, 1965.
63. **Gottfried, E. L. and Rapport, M. M.,** *J. Biol. Chem.,* 237, 329, 1962.
64. **Blank, M. L. and Snyder, F.,** *Biochemistry,* 9, 5034, 1970.
65. **Snyder, F., Malone, B., and Blank, M. L.,** *J. Biol. Chem.,* 245, 1800, 1970.
66. **Yeung, S. K. F. and Kuksis, A.,** *Can. J. Biochem.,* 52, 830, 1974.
67. **Snyder, F., Blank, M. L., and Wykle, R. L.,** *J. Biol. Chem.,* 216, 3639, 1971.
68. **Bandi, Z. L.,** *Chem. Phys. Lipids,* 3, 409, 1969.
69. **Schmid, H. H. O. and Mangold, H. K.,** *Biochim. Biophys. Acta,* 125, 182, 1966.
70. **Wood, R. and Healy, K.,** *Lipids,* 5, 661, 1970.
71. **Rao, P. V., Ramachandran, S., and Cornwell, D. G.,** *J. Lipid Res.,* 8, 380, 1967.
72. **Gray, G. M.,** in *Lipid Chromatographic Analysis,* Vol. 3, 2nd ed., Marinetti, G. V., Ed., Marcel Dekker, New York, 1976, 897.
73. **Hanahan, D. J., Ekholm, J., and Jackson, C. M.,** *Biochemistry,* 2, 630, 1963.
74. **Myher, J. J. and Kuksis, A.,** *Lipids,* 9, 382, 1974.
75. **Wood, R. and Snyder, F.,** *Lipids,* 2, 161, 1967.
76. **Baumann, W. J., Schmid, H. H. O., and Mangold, H. K.,** *J. Lipid Res.,* 10, 132, 1969.
77. **Ramachandran, S., Sprecher, H. W., and Cornwell, D. G.,** *Lipids,* 3, 511, 1968.
78. **Spener, F. and Mangold, H. K.,** *J. Lipid Res.,* 12, 12, 1971.
79. **Schmid, H. H. O., Bandi, P. C., Mangold, H. K., and Baumann, W. J.,** *Biochim. Biophys. Acta,* 187, 208, 1969.
80. **Snyder, F. and Blank, M. L.,** *Arch. Biochem. Biophys.,* 130, 101, 1969.
81. **Wood, R. and Snyder, F.,** *Lipids,* 1, 62, 1966.
82. **Paltauf, F.,** *Biochim. Biophys. Acta,* 270, 317, 1972.
83. **Su, K. L. and Schmid, H. H. O.,** *Lipids,* 9, 208, 1974.
84. **Totani, N. and Mangold, H. K.,** *Mikrochim. Acta,* 1981 I, 73.

85. **Demopoulos, C. A., Pinckard, R. N., and Hanahan, D. J.,** *J. Biol. Chem.,* 254, 9355, 1979.
86. **Benveniste, J., Tencé, M., Varenne, P., Bidault, J., Boullet, C., and Polonsky, J.,** *C. R. Acad. Sci.,* 289 (D), 1037, 1979.
87. **Alam, I., Smith, J. B., Silver, M. J., and Ahern, D.,** *J. Chromatogr.,* 234, 218, 1982.
88. **Totani, N.,** *Fette Seifen Anstrichm.,* 84, 71, 1982.

IV. TABLES OF CHROMATOGRAPHIC DATA

The tables are presented in a self-explanatory manner; they are arranged in the following order:

Table 1 —   Adsorption column-chromatography (AC)
Table 2 —   Liquid-liquid column chromatography (LLC)
Tables 3 to 10 —   Thin-layer chromatography (TLC)
Tables 11 to 23 — Gas-liquid chromatography (GLC)

Abbreviations used:

Alu,   aluminum
ECL,   equivalent chain-length relative to the saturated straight-chain analog
Fl,   flame ionization
$R_f \times$ 100, retention index times 100
RRT $\times$ 100, relative retention time times 100
SS,   stainless steel
TC,   thermal conductivity
TFA,   trifluoroacetate
TMS,   trimethylsilyl

Table 1
NEUTRAL ETHER LIPIDS (AC)

	A1	A2	A1
Adsorbent	A1	A2	A1
Solvent	S1	S2	S1
Technique	T1	T2	T3
Detection	D1	D1	D1
Reference	1	2	3

Compound Class	% Polar solvent in nonpolar solvent		
1-Alkyl-2,3-diacylglycerols	5	15	—
1-Alkyl-2,3-acylisovaleroylglycerols	10	15	—
1-Alkyl-2,3-isopropylideneglycerols	—	—	1
1-(2'-Methoxy)alkyl-2,3-isopropylideneglycerols	—	—	5

Adsorbent	A1 =	Silicic acid
	A2 =	Unisil
Solvent	S1 =	Diethyl ether in hexane
	S2 =	Chloroform in hexane
	S3 =	Diethyl ether in petroleum hydrocarbon (60-80°C)
Technique	T1 =	20 mg of lipid per g of silicic acid
	T2 =	2 × 30 cm column, 75 mℓ fractions
Detection	D1 =	TLC

REFERENCES

1. **Kasama, K., Uezumi, N., and Itoh, K.,** *Biochim. Biophys. Acta,* 202, 56, 1970.
2. **Blank, M. L., Kasama, K., and Snyder, F.,** *J. Lipid Res.,* 13, 390, 1972.
3. **Hallgren, B., Niklasson, A., Ställberg, G., and Thorin, H.,** *Acta Chem. Scand.,* 28B, 1029, 1974.

Table 2
NEUTRAL ETHER LIPIDS (LLC)

Support	S1	S1	S1	S1	S2
Solvent	S1	S2	S2	S3	S4
Technique	T1	T2	T2	T3	T4
Detection	D1	D1	D1	D1	D2
Reference	1	1	1	1	2

Compound Class	RRT × 100				
1-Alkyl-2-acylglycerols[a]	48	—	—	—	—
32:2, 34:3, 36:4	—	48	—	—	—
33:2	—	54	—	—	—
32:1, 34:2, 32:1	—	62	—	—	—
34:1, 36:2	—	80	—	—	—
36:1	—	100	—	—	—
1-Alk-1'-enyl-2-acylglycerols[a]	74	—	—	—	—
32:2, 34:3, 36:4	—	—	47	—	—
33:2, 35:3	—	—	53	—	—
32:1, 34:2, 36:3	—	—	58	—	—
33:1, 35:2	—	—	67	—	—
32:0, 34:1, 36:2	—	—	77	—	—
35:1	—	—	88	—	—
36:1					
1,2-Diacylglycerols	100	—	—	—	—
1,2-Dialkylglycerols	—	—	—	—	79
1-Alkylethanediols	—	—	—	—	94
Cholesterol	—	—	—	38	100
1-Alkylglycerols	—	—	—	68	—
1-Alk-1'-enylglycerols	—	—	—	100	—

[a] Molecular species (carbon number of long-chain moieties:number of double bonds) as TMS-ethers.

Support	S1	= Lipidex-5000
	S2	= Sephadex LH-20
Solvent	S1	= Heptane-toluene (100:5)
(v/v)	S2	= Acetone-water-heptane-pyridine (87:13:10:1)
	S3	= Hexane-chloroform (4:1)
	S4	= Ethanol
Technique	T1	= 1.6 × 98 cm column, flow rate 34 mℓ/hr, 5 mℓ fractions
	T2	= 0.15 × 130 cm column, flow rate 1.8 mℓ/hr, 0.5-1 mℓ fractions
	T3	= 0.42 × 29 cm column, flow rate 6 mℓ/hr
	T4	= 2.5 × 79 cm column, flow rate 0.5 mℓ/hr fractions
Detection	D1	= By GLC as TMS-derivatives
	D2	= By GLC as acetates

REFERENCES

1. **Curstedt, T.,** *Biochim. Biophys. Acta,* 489, 79, 1977.
2. **Calderon, M. and Baumann, W. J.,** *J. Lipid Res.,* 11, 167, 1970.

Table 3
NEUTRAL ETHER LIPIDS (TLC)

	1	2	3	4	5	6	7	8	9	10	11	12	13
Layer	L1	L1	L1	L2	L3	L1	L3	L1	L1	L2	L3	L3	L3
Solvent	S1	S2	S3	S4	S4	S4	S5	S5	S6	S7	S8	S9	S10
Technique	T1	T1	T1	T2	T1	T1	T1	T1	T1	T3	T1	T1	T1
Detection	D1	D1	D1	D2	D1	D1	D1	D1	D1	D1	D1	D1	D1
Reference	1	1	1	2	3	4	5	6	7	8	9	5	10
Compound Class $R_f \times 100$													
1,2,3-Trialkylglycerols	—	—	—	—	—	—	45	—	89	—	—	—	82
1,2-Dialkyl-3-acylglycerols	—	57	82	—	—	—	35	32	82	—	—	—	69
1-Alk-1'-enyl-2,3-diacylglycerols	37	49	79	59	39	39	30	—	74	—	45	58	—
1-Alkyl-2,3-diacylglycerols	30	—	—	52	29	27	25	24	—	—	35	—	53
1-Alk-1'-enyl-2-acyl-3-acetylglycerols	—	—	—	—	—	—	—	—	—	55	—	—	—
1-Alkyl-2-acyl-3-acetylglycerols	—	—	—	—	—	—	—	—	—	55	—	—	—
1,2-Diacyl-3-acetylglycerols	—	—	—	36	22	18	17	14	49	37	29	50	31
1,2,3-Triacylglycerols	—	—	—	—	—	—	6	—	—	—	—	7	11
1,2-Dialkylglycerols	—	—	—	—	—	—	—	—	—	—	—	12	—
1,2-Diacylglycerols	—	—	—	—	—	—	—	—	—	—	—	—	—
1,3-Diacylglycerols	—	—	—	—	—	—	—	—	—	—	—	—	—
Wax esters	—	—	—	—	75	74	80	78	94	—	61	68	—
Sterol esters	—	—	—	—	—	—	—	84	94	—	—	58	—
Fatty acid methyl esters	48	57	70	—	48	52	54	—	81	—	—	—	18
Long-chain aldehydes	—	—	—	—	—	—	50	—	—	—	15	17	—
Fatty acids	—	—	—	—	—	—	—	—	—	—	—	12	—
Long-chain alcohols	—	—	—	—	—	—	—	—	—	—	—	7	—
Sterols	—	—	—	—	—	—	—	—	—	—	8	1	—
Phospholipids	—	—	—	—	—	—	—	—	—	—	1	—	—

Layer   L1 = Silica gel G, 0.3 mm
         L2 = Silica gel G, 0.5 mm
         L3 = Silica gel G, 0.25 mm

Solvent (v/v)

S1 = Petroleum hydrocarbon-diethyl ether (95:5 [twice])
S2 = Petroleum hydrocarbon-diethyl ether (90:10)
S3 = Petroleum hydrocarbon-diethyl ether (80:20)
S4 = Hexane-diethyl ether (95:5 [twice])
S5 = Hexane-diethyl ether (95:5)
S6 = Hexane-diethyl ether (90:10)
S7 = Hexane-diethyl ether (88:12)
S8 = Benzene
S9 = Hexane-diethyl ether-acetic acid (80:20:1)
S10 = Hexane-diethyl ether-acetic acid (90:10:1)

Technique

T1 = Chambers lined with filter paper
T2 = Preparative TLC
T3 = Samples derived from beef heart phospholipids

Detection

D1 = 50% Sulfuric acid, charring; or alc.sol. of 2,4-dinitro phenylhydrazine
D2 = 0.2% 2',7'-Dichlorofluorescein in ethanol

REFERENCES

1. **Schmid, H. H. O. and Mangold, H. K.,** *Biochim. Biophys. Acta,* 125, 182, 1966.
2. **Schmid, H. H. O. and Mangold, H. K.,** *Biochem. Z.,* 346, 13, 1966.
3. **Schmid, H. H. O., Jones, L. L., and Mangold, H. K.,** *J. Lipid Res.,* 8, 692, 1967.
4. **Spener, F. and Mangold, H. K.,** *J. Lipid Res.,* 10, 609, 1969.
5. **Snyder, F.,** *J. Chromatogr.,* 82, 7, 1973.
6. **Sansone, G. and Hamilton, J. G.,** *Lipids,* 4, 435, 1969.
7. **Wood, R. and Snyder, F.,** *J. Lipid Res.,* 8, 494, 1967.
8. **Viswanathan, C. V., Phillips, F., and Lundberg, W. O.,** *J. Chromatogr.,* 38, 267, 1968.
9. **Snyder, F.,** *Meth. Cancer Res.,* 6, 399, 1971.
10. **Baumann, W. J. and Mangold, H. K.,** *Biochim. Biophys. Acta,* 116, 570, 1966.

Table 4
NEUTRAL ETHER LIPIDS INCLUDING DERIVATIVES OF
DIHYDROXYACETONE (TLC)

Layer	L1	L1	L1	L1	L1	L1
Solvent	S1	S2	S3	S4	S5	S2
Technique	T1	T1	T1	T1	T1	T2
Detection	D1	D1	D1	D1	D1	D2
Reference	1	1	1	1	1	2
Compound Class			$R_f \times 100$			
1-Alkyldihydroxyacetones	41	54	68	35	7	51
1-Acyldihydroxyacetones	—	54	—	29	—	—
1-Alkyldihydroxyacetonephosphates	—	—	—	—	—	6
1-Alkyl-2,3-diacylglycerols	—	86	—	—	—	—
1,2,3-Triacylglycerols	—	86	—	—	—	—
1,2-Dialkylglycerols	—	—	—	49	51	—
1-Alkyl-2-acylglycerols	—	66	—	—	—	—
1-Alkyl-3-acylglycerols	—	75	—	54	51	—
1,2-Diacylglycerols	46	61	—	41	44	—
1,3-Diacylglycerols	52	71	—	49	44	—
1-Alk-1'-enylglycerols	24	26	29	—	—	—
1-Alkylglycerols	19	22	24	14	—	—
1-Acylglycerols	—	17	19	10	—	—
Wax esters	88	—	—	—	—	83
Long-chain aldehydes	52	81	—	—	—	—
Fatty acids	—	54	65	60	2	—
Long-chain alcohols	46	49	59	41	47	43
Sterols	41	44	55	—	—	—
Phospholipids	1	2	2	1	—	—

Layer   L1 = Silica gel G, 0.25 mm
Solvent   S1 = Hexane-diethyl ether-methanol-acetic acid (70:30:5:1)
(v/v)   S2 = Chloroform-methanol-acetic acid (98:2:1)
          S3 = Diethyl ether-acetic acid (100:0.5)
          S4 = Hexane-diethyl ether-acetic acid (40:60:1)
          S5 = Hexane-diethyl ether-conc. ammonia (40:60:1)
Technique T1 = Chambers lined with filter paper
          T2 = Separation of ^{14}C-labeled products
Detection D1 = 50% Sulfuric acid, charring
          D2 = ^{14}C-zonal profile scans

REFERENCES

1. **Snyder, F.**, *J. Chromatogr.*, 82, 7, 1973.
2. **Snyder, F., Blank, M. L., Malone, B., and Wykle, R. L.**, *J. Biol. Chem.*, 245, 1800, 1970.

Table 5
1- AND 2-ALKYLGLYCEROLS (TLC)

Layer	L1	L2	L3	L4	L4	L2
Solvent	S1	S1	S1	S2	S3	S4
Technique	T1	T1	T2	T2	T2	T2
Detection	D1	D1	D1	D1	D1	D1
Reference	1	1	2	3	3	3

Compound Class	$R_f \times 100$					
1-Alkylglycerols	55	44	—	—	—	—
18:0[a]	—	—	72	—	—	—
18:1[a]	—	—	65	—	—	—
18:2[a]	—	—	55	—	—	—
2-Alkylglycerols	30	33	—	—	—	—
3-Phytanyl-*sn*-glycerol	—	—	—	25	32	10
2-Phytanylglycerol	—	—	—	25	32	18
2,3-Diphytanylglycerol	—	—	—	55	90	—

[a] Molecular species (carbon number in aliphatic chain:number of double bonds).

Layer	L1 = Silica gel G + 10% sodium arsenite, 0.25 mm
	L2 = Silica gel G + 5% boric acid, 0.25 mm
	L3 = Silica gel G + 5% silver nitrate, 0.25 mm
	L4 = Silica gel G, 0.25 mm
Solvent (v/v)	S1 = Chloroform-ethanol (90:10)
	S2 = Chloroform-diethyl ether (90:10)
	S3 = Diethyl ether
	S4 = Chloroform-methanol (98:2)
Technique	T1 = Chambers lined with filter paper
	T2 = Analytical TLC
Detection	D1 = 50% Sulfuric acid, charring

REFERENCES

1. **Snyder, F.,** *J. Chromatog.* 82, 7, 1973.
2. **Wood, R. and Snyder, F.,** *Lipids* 1, 62, 1966.
3. **Joo, C. N., Shier, T., and Kates, M.,** *J. Lipid Res.* 9, 782, 1968.

Table 6
PHOSPHOLIPIDS (TLC)

Layer	L1	L1	L1	L1	L1	L1	L1	L1	L1	L1
Solvent	S1	S2	S3	S4	S5	S6	S7	S8	S9	S10
Technique	T1	T1	T1	T1	T1	T1	T1	T2	T1	T1
Detection	D1	D1	D1	D1	D1	D1	D1	D1	D1	D1
Reference	1	1	1	1	2	2	2	3	3	4

| Compound Class | \multicolumn{10}{c}{$R_f \times 100$} |

Compound Class										
Derivatization to dimethylphosphatidates										
Alk-1-enylacyl	—	53	—	40	—	—	—	—	—	—
Alkylacyl	78	53	—	33	—	—	—	—	—	—
Diacyl	70	42	—	26	—	—	—	—	—	—
Derivatization to diradylglyceroacetates										
Alk-1-enylacyl	—	—	47	—	23	78	67	—	—	—
Alkylacyl	—	—	38	—	12	78	67	—	—	—
Diacyl	—	—	24	—	—	61	48	—	—	—
Methylated dinitrophenylphosphatidyl ethanolamines										
Alk-1-enylacyl	—	—	—	—	—	—	—	38	45	—
Alkylacyl	—	—	—	—	—	—	—	—	39	—
Diacyl	—	—	—	—	—	—	—	28	31	—
Specific iodination of phosphatidyl cholines										
Alk-1-enylacyl	—	—	—	—	—	—	—	—	—	62
Diacyl	—	—	—	—	—	—	—	—	—	54

Layer	L1	= Silica gel G, 0.25 mm
Solvent	S1	= Diethyl ether
(v/v)	S2	= Hexane-diethyl ether-acetic acid (20:80:1)
	S3	= 1. Hexane-diethyl ether (80:20) (7 cm), 2. toluene (15 cm)
	S4	= Hexane-chloroform (50:50) (7 times)
	S5	= Toluene
	S6	= Hexane-diethyl ether/acetic acid (70:30:0.5)
	S7	= Hexane-diethyl ether (80:20)
	S8	= Chloroform-methanol (99:1)
	S9	= 1. Hexane-chloroform (40:60) (5 times), 2. toluene-chloroform (40:60) (5 times)
	S10	= Chloroform-methanol-conc. (ammonia 70:37:5)
Technique	T1	= Analytical TLC
	T2	= Preparative TLC
Detection	D1	= 50% Sulfuric acid, charring

REFERENCES

1. **Renkonen, O.,** *Biochim. Biophys. Acta,* 152, 114, 1968.
2. **Renkonen, O.,** *Biochim. Biophys. Acta,* 125, 288, 1966.
3. **Renkonen, O.,** *J. Lipid Res.,* 9, 34, 1968.
4. **Viswanathan, C. V., Hoevet, S., and Lundberg, W. O.,** *Fette Seifen Anstrichm.,* 70, 858, 1968.

Table 7
METHOXY-, HYDROXY-, AND KETOALKYLGLYCEROLS (TLC)

	L1	L2	L2	L2	L2	L2	L2	L2
Layer								
Solvent	S1	S2	S3	S4	S5	S6	S7	S8
Detection	D1	D1	D1	D1	D1	D1	D1	D1
Reference	1	1	2	3	4	4	4	4
Compound Class				$R_f \times 100$				
Wax esters	—	—	—	—	55	—	—	—
Triacylglycerols	—	—	64	—	17	27	—	—
1-(9',10'-Acyloxy)alkyl-2,3-diacylglycerols	—	—	—	—	17	15	—	—
1-(9',10'-Acetoxy)alkyl-2,3-diacylglycerols	—	—	—	—	10	8	—	—
1-Alkyl-2,3-diacylglycerols	—	—	64	—	29	39	—	—
1-Alkyl-2,3-acylisovaleroylglycerols	—	—	—	—	—	25	—	—
1-(2'-Hydroxy)alkyl-2,3-diacylglycerols	—	—	32	—	—	—	—	—
1-(2'-Acetoxy)alkyl-2,3-diacetylglycerols	62	—	—	—	—	—	—	—
1-(9',10'-Acetoxy)alkyl-2,3-isopropylideneglycerols	—	—	—	—	—	—	—	52
1-(2'-Acetoxy)alkyl-2,3-isopropylideneglycerols	86	—	—	—	—	—	—	—
1-(2'-Methoxy)alkyl-2,3-isopropylideneglycerols	—	—	—	30	—	—	—	—
1-Alkyl-2,3-isopropylideneglycerols	—	—	—	—	—	—	—	29
1-(2'-Hydroxy)alkyl-2,3-isopropylideneglycerols	42	—	—	—	—	—	—	—
1-(2'-Keto)alkyl-2,3-isopropylideneglycerols	74	—	—	—	—	—	—	—
1-(9',10'-Hydroxy)alkyl-2-acylglycerols	—	—	24	—	—	—	—	—
1-(9',10'-Acetoxy)alkylglycerols	—	—	—	—	—	—	—	2
1-Alkylglycerols	—	—	—	—	—	—	35	—
1-(9',10'-Hydroxy)alkylglycerols	—	—	—	—	—	—	19	—
1-(2'-Hydroxy)alkylglycerols	0	34	—	—	—	—	5	—
1-(2'-Keto)alkylglycerols	—	37	—	—	—	—	—	—
Sterols	—	—	—	9	2	—	—	—

Layer   L1 = Silica gel H, 0.5 mm
            L2 = Silica gel G, 0.5 mm
Solvent   S1 = Hexane-diethyl ether (50:50)
(v/v)   S2 = Chloroform-methanol-water (80:10:1)
            S3 = Hexane-diethyl ether-acetic acid (60:40:2)
            S4 = Trimethylpentane-ethyl acetate (70:20)
            S5 = Hexane-diethyl ether-acetic acid (90:10:1) (twice)
            S6 = Benzene (twice)
            S7 = Hexane-diethyl ether (100:0.5)
            S8 = Hexane-diethyl ether (60:40)
Detection   D1 = 50% H_2SO_4, charring

REFERENCES

1. **Muramatsu, T. and Schmid, H. H. O.**, *Chem. Phys. Lipids,* 9, 123, 1972.
2. **Rock, C. O. and Snyder, F.**, *Arch. Biochem. Biophys.,* 171, 631, 1975.
3. **Hallgren, B., Niklasson, A., Ställberg, G., and Thorin, H.**, *Acta Chem. Scand.,* 28B, 1029, 1974.
4. **Kasama, K., Rainey, W. T., Jr., and Snyder, F.**, *Arch. Biochem. Biophys.,* 154, 648, 1973.

Table 8
ALKYLACYLGLYCEROGLYCOLIPIDS (TLC)

Compound Class				$R_f \times 100$				
Layer	L1	L1	L1	L1	L2	L3	L3	L4
Solvent	S1	S2	S3	S4	S5	S5	S6	S7
Detection	D1	D1	D1	D1	D2	D3	D3	D3
Reference	1	2	2	2	3	4	4	4
1-Alkyl-2-acyl-3-monoglucosylglycerols	90	—	—	—	—	—	—	—
1-Alkyl-2-acyl-3-diglucosylglycerols	80	—	—	—	—	—	—	—
1-Alkyl-2-acyl-3-triglucosylglycerols	73	—	—	—	—	—	—	—
1-Alkyl-2-acyl-3-tetraglucosylglycerols	62	—	—	16	—	—	—	—
1-Alkyl-2-acyl-3-sulfotriglucosylglycerols	57	22	15	7	—	—	—	—
1-Alkyl-2-acyl-3-hexaglucosylglycerols	32	13	10	3	—	—	—	—
1-Alkyl-2-acyl-3-octaglucosylglycerols	19	—	—	—	—	—	—	—
1-Alkyl-2-acyl-3-sulfogalactosylglycerols	—	—	—	—	59	30	62	43
1,2-Diacyl-3-monogalactosylglycerols	—	—	—	—	90	67	78	48
1,2-Diacyl-3-digalactosylglycerols	—	—	—	—	68	35	44	15
1,2-Diacyl-3-monoglycosylglycerols	—	—	—	—	—	68	—	79
1,2-Diacyl-3-sulfoquinovosylglycerols	—	—	—	—	—	21	49	45
Cerebroside-3-sulfates	—	—	—	—	—	23	62	—
Cerebroside-6-sulfates	—	—	—	—	—	—	—	43

Layer   L1 = Silica gel HR, 0.25 mm
L2 = Silica gel G
L3 = Silica gel + 5% CaSO₄, 0.25 mm
L4 = Silica gel + 2% NaBO₃, 0.25 mm

Solvent   S1 = Chloroform-methanol-water (65:30:8)
(v/v)   S2 = Chloroform-methanol-acetic acid-water (60:20:20:1)
S4 = Chloroform-acetone-methanol-water (50:40:20:5)
S5 = Chloroform-methanol-water (65:25:4)
S6 = Chloroform-methanol-conc. ammonia-water (60:40:5:5)
S7 = Chloroform-methanol-acetic acid-water (50:25:8:4)

Detection D1 = Resorcinol reagent
           D2 = Aniline-diphenylamine reagent
           D3 = Iodine vapor, or anthrone reagent

REFERENCES

1. **Slomiany, B. L., Slomiany, A., and Glass, G. B. J.**, *Eur. J. Biochem.*, 84, 53, 1978.
2. **Slomiany, B. L., Slomiany, A., and Glass, G. B. J.**, *Biochemistry*, 16, 3954, 1977.
3. **Kornblatt, M. J., Schachter, H., and Murray, R. K.**, *Biochem. Biophys. Res. Commun.*, 48, 1489, 1972.
4. **Ishizuka, I., Suzuki, M., and Yamakawa, T.**, *J. Biochem. (Tokyo)*, 73, 77, 1973.

Table 9
THIOALKYLGLYCEROLS (TLC)

Layer		L1	L1	L2
Solvent		S1	S1	S2
Technique		T1	T2	T2
Detection		D1	D2	D2
Reference		1	2	3

Compound Class	$R_f \times 100$		
1-Thioalkyl-2,3-isopropylideneglycerols	39	38	21
1-Alkyl-2,3-isopropylideneglycerols	29	26	16

Layer	L1 = Silica gel G, 0.25 mm
	L2 = Anasil G, 0.25 mm
Solvent	S1 = Hexane-diethyl ether (90:10)
(v/v)	S2 = Hexane-diethyl ether-acetic acid (30:70:1)
Technique	T1 = Preparative TLC
	T2 = Analytical TLC
Detection	D1 = Iodine vapors, Rhodamine 6G
	D2 = 50% H_2SO_4, charring

REFERENCES

1. **Ferrell, W. J. and Radloff, D. M.,** *Physiol. Chem. Phys.,* 2, 551, 1970.
2. **Wood, R., Piantadosi, C., and Snyder, F.,** *J. Lipid Res.,* 10, 370, 1969.
3. **Ferrell, W. J.,** *Lipids,* 8, 234, 1973.

Table 10
ETHER DIOL LIPIDS (TLC)

Layer	L1	L2	L1	L1
Solvent	S1	S2	S3	S4
Technique	T1	T2	T1	T3
Detection	D1	D1	D1	D1
Reference	1	2	3	4

Compound Class	$R_f \times 100$			
1,2-Dialkylethanediols	65	—	—	62
1-*trans*-Alk-1′-enyl-2-acylethanediols	—	—	61	—
1-*cis*-Alk-1′-enyl-2-acylethanediols	—	82	54	—
1-Alkyl-2-acylethanediols	55	—	40	—
1,2-Diacylethanediols	43	—	21	—
1,3-Dialkylpropanediols	65	—	—	—
1-Alkyl-3-acylpropanediols	52	—	—	—
1,3-Diacylpropanediols	43	—	—	—
1,5-Dialkypentanediols	—	—	—	62
1,2,3-Trialkylglycerols	80	53	—	—
1,2-Dialkyl-3-acylglycerols	68	—	—	—
1-Alkyl-2,3-diacylglycerols	52	—	—	—
1,2,3-Triacylglycerols	35	—	—	60

Layer	L1 =	Silica gel G, 0.25 mm
	L2 =	Silica gel KSK (150-200 mesh), 0.5 mm
Solvent	S1 =	Hexane-diethyl ether (90:10)
(v/v)	S2 =	Hexane-diethyl ether (85:15)
	S3 =	Hexane-diethyl ether (95:5)
	S4 =	Hexane-diethyl ether (80:20)
Technique	T1 =	Tanks lined with filter paper
	T2 =	Adsorbent without binder, air-dried
	T3 =	Separation after enrichment of ether diol lipid
Detection	D1 =	50% H_2SO_4, charring

REFERENCES

1. **Baumann, W. J., Schmid, H. H. O., Ulshöfer, H. W., and Mangold, H. K.,** *Biochim. Biophys. Acta,* 144, 355, 1967.
2. **Vaver, V. A., Pisareva, N. A., Rosynov, B. V., Ushakov, A. N., and Bergelson, L. D.,** *Chem. Phys. Lipids,* 7, 75, 1971.
3. **Kramer, J. K. G. and Mangold, H. K.,** *Chem. Phys. Lipids,* 4, 332, 1970.
4. **Varanashi, U. and Malins, D. C.,** *Science,* 166, 1158, 1969.

Table 11
TRIRADYLGLYCEROLS (GLC)

		P1	P2	P3
Column packing		P1	P2	P3
Temperature (°C)		T1	T2	T3
Gas		He	He	N$_2$
Flow rate (mℓ/min)		100	100	120
Column				
Length (cm)		50	70	50
I.D. (cm)		0.15	0.25	0.2
Material		Glass	Glass	SS
Detection		F1	F1	F1
Reference		1	2	3

Compounds	Carbon number[a]	RRT × 100		
1-Alkyl-2,3-diacylglycerols				
	46	55	42	—
	48	65	56	—
	50	77	71	—
	52	88	83	—
	54	100	100	—
	56	112	115	—
	58	124	132	—
	60	136	147	—
	62	147	165	—
Trimyristoylglycerol	42	—	—	39
Trihexadecylglycerol	48	—	—	79
1,2-Dihexadecyl-3-palmitoylglycerol	48	—	—	90
1-Hexadecyl-2,3-dipalmitoylglycerol	48	—	—	100
Tripalmitoylglycerol	48	—	—	108

[a]   Carbon numbers represent the sum of the carbon atoms in long-chain moieties linked to glycerol.

Column packing	P1 = 3% JXR on Gas-Chrom Q (100-120 mesh)
	P2 = 1% OV-1 on Gas-Chrom Q (100-120 mesh)
	P3 = 3% JXR on Gas-Chrom P (100-120 mesh)
Temperature program	T1 = 270-390°C at 6.5°C/min
	T2 = 200-335°C at 5°C/min
	T3 = 200-330°C at 4°C/min

REFERENCES

1. **Wood, R. and Snyder, F.,** *J. Lipid Res.*, 8, 494, 1967.
2. **Wood, R. and Snyder, F.,** *Arch. Biochem. Biophys.*, 131, 478, 1969.
3. **Kuksis, A.,** in *Lipid Chromatographic Analysis*, Vol. 1, 2nd ed., Marinetti, G. V., Ed., Marcel Dekker, New York, 1976, 215.

Table 12
ALK-1-ENYLACYL-, ALKYLACYL-, AND DIACYLGLYCEROLS (GLC)

Column packing	3% SE-30 on Chromosorb W (60-80 mesh)
Temperature (°C)	250
Gas	Ar
Flow rate (mℓ/min)	50
Column	
Length (cm)	100
I.D. (cm)	0.4
Material	Glass
Detection	Fl
Reference	1

Carbon number[a]	ECL			
	Saturates[c]	Monoenes[c]	Dienes[c]	Polyenes[c]
1,2-Diacylglycerols[b]				
30	30.00	—	—	—
32	32.00	31.8	31.8	—
34	34.00	33.8	33.8	—
36	—	35.8	35.8	35.3
38	—	—	37.8	37.3
40	—	—	—	39.4
1-Alk-1'-enyl-2-acylglycerols[b]				
30	28.7	—	—	—
32	30.8	30.5	30.5	—
34	32.7	32.6	32.6	—
36	—	34.6	34.4	34.1
38	—	—	36.4	36.0
40	—	—	—	38.0
1-Alkyl-2-acylglycerols[b]				
30	29.3	—	—	—
32	31.3	31.1	31.0	—
34	33.5	33.2	33.0	—
36	—	35.2	35.0	34.5
38	—	—	37.0	36.4
40	—	—	—	38.5

[a] Carbon numbers represent the sum of the carbon atoms in long-chain moieties at 1- and 2-positions of glycerol.
[b] As acetate derivatives.
[c] Double bonds in aliphatic moieties, except the enol ether bond.

REFERENCE

1. **Renkonen, O.**, *Biochim. Biophys. Acta,* 137, 575, 1967.

Table 13
ALK-1-ENYLACYL- AND
ALKYLACYLGLYCEROLS (GLC)

Column packing	P1	P2	P1	P2
Temperature (°C)	T1	275	T1	275
Gas	He	He	He	He
Flow rate (mℓ/min)	100	—	100	—
Column				
Length (cm)	70	70	70	70
I.D. (cm)	0.25	0.3	0.25	0.3
Material	Glass		Glass	
Detection	F1	F1	F1	F1
Reference	1	2	1	2

Carbon number[a]	RRT × 100			
1-Alk-1'-enyl-2-acylglycerols[b]				
32	50	—	—	
34	72	57	—	—
36	100	100	—	—
38	128	151	—	—
40	156	230	—	—
42	183	—	—	—
1-Alkyl-2-acylglycerols[b]				
30	—	—	21	—
32	—	—	46	45
34	—	—	71	69
36	—	—	100	100
38	—	—	131	151
40	—	—	158	—
42	—	—	183	—

[a]   Carbon numbers represent the sum of the carbon atoms in long-chain moieties at 1- and 2-positions of glycerol.

[b]   As acetates.

Column packing	P1 =	1% OV-1 on Gas-Chrom Q (100-120 mesh)
	P2 =	1% OV-1 on Chromosorb WHP (100-120 mesh)
Temperature program	T1 =	200-335°C at 5°C/min

REFERENCES

1. **Wood, R. and Snyder, F.,** *Arch. Biochem. Biophys.,* 131, 478, 1969.
2. **Baumann, W. J., Takahashi, T., Mangold, H. K., and Schmid, H. H. O.,** *Biochim. Biophys. Acta,* 202, 468, 1970.

Table 14
DIALKYLGLYCEROLS (GLC)

Column packing			P1	P1
Temperature (°C)			T1	T1
Gas			He	He
Flow rate (mℓ/min)			100	100
Column				
Length (cm)			70	70
I.D. (cm)			0.25	0.25
Material			Glass	Glass
Detection			F1	F1
Reference			1	1

Carbon number[a]	1-Position	2-Position	Acetates RRT × 100	TMS RRT × 100
1,2-Dialkylglycerols				
30	18:0	12:0	68	66
32	16:0	16:0	78	76
34	18:0	16:0	89	88
36	18:0	18:0	100	100
38	18:0	20:0	112	113
1,2-Diacylglycerols				
28	14:0	14:0	58	66
32	16:0	16:0	85	88
36	18:0	18:0	109	113

[a] Carbon numbers represent the sum of the carbon atoms in long-chain moieties at 1- and 2-positions of glycerol.

Column packing   P1 = 1% OV-1 on Gas-Chrom Q (100-120 mesh)
Temperature program T1 = 150-275°C at 5°C/min

REFERENCE

1. **Wood, R., Baumann, W. J., Snyder, F., and Mangold, H. K.,** *J. Lipid Res.,* 10, 128, 1969.

Table 15
ANALYSIS OF ALK-1-ENYL MOIETIES: ALDEHYDES AND DIMETHYLACETALS (GLC)

Column packing	P1	P2	P3	P4	P1	P5	P6	P7
Temperature (°C)	192	190	T1	215	192	150	197	100
Gas	He	He	He	He	He	N$_2$	Ar	Ar
Flow rate (mℓ/min)	98		40-60		98	50	112	100
Column								
Length (cm)	300	150	150	180	300	150	180	120
I.D. (cm)	0.4	0.3	0.3	0.2	0.4	0.4	0.3	0.4
Material	SS	Glass	Glass	SS	SS	Glass	Glass	Glass
Detection	TC	Fl	Fl	TC	TC	Fl	Fl	Fl
Reference	1	2	3	1	1	4	5	4

Chain length: Number of double bonds	Aldehydes RRT × 100				Dimethylacetals RRT × 100			
12:0	17	—	—	22	18	10	9	7
13:0	—	—	—	—	—	15	—	11
14:0	33	—	45	37	28	21	18	17
15:0	42	—	—	—	40	31	27	27
16:0	57	54	69	61	55	46	42	42
16:1	—	—	—	—	—	52	36	37
17:0	75	—	—	—	74	68	65	65
18:0	100	100	100	100	100	100	100	100
18:1	118	121	112	—	—	113	87	84
18:2	146	—	—	—	—	135	78	79
18:3	—	—	—	—	—	170	—	79
19:0	134	—	—	—	135	—	—	—
20:0	—	179	—	—	—	—	—	—

Column packing
P1 = 10% EGS on Gas-Chrom P (60-80 mesh)
P2 = 10% EGSS-X on Gas-Chrom P (100-120 mesh)
P3 = 15% EGSS-X on Gas-Chrom P (100-120 mesh)
P4 = 10% Apiezon M on Gas-Chrom P (60-80 mesh)
P5 = 15% EGSS-X on Gas-Chrom CLH (80-100 mesh)
P6 = 16% Apiezon M on Gas-Chrom S AW-DMCS (60-80 mesh)
P7 = 12.5% Apiezon M on 'alkaline' Gas-Chrom P (80-100 mesh)
Temperature program T1 = 130-180°C at 3°C/min

REFERENCES

1. **Venkata Rao, P., Ramachandran, S., and Cornwell, D. G.,** *J. Lipid Res.,* 8, 380, 1967.
2. **Snyder, F.,** *Adv. Exp. Med. Biol.,* 4, 609, 1969.
3. **Wood, R. and Harlow, R. D.,** *J. Lipid Res.,* 10, 463, 1969.
4. **Gray, G. M.,** in *Lipid Chromatographic Analysis,* Vol. 3, 2nd ed., Marinetti, G. V., Ed., Marcel Dekker, New York, 1976, 897.
5. **Gilbertson, J. R., Ferrell, W. J., and Gelman, R. A.,** *J. Lipid Res.,* 8, 38, 1967.

Table 16
ANALYSIS OF ALK-1-ENYL MOIETIES: 2-ALKYL-1,3-DIOXANES AND ALKYLACETATES (GLC)

Column packing	P1	P2	P3	P4	P1	P5	P4
Temperature (°C)	192	180	200	215	192	180	215
Gas	He	He	He	He	He	He	He
Flow rate (mℓ/min)	98	—	—	—	98	—	—
Column							
Length (cm)	300	300	180	180	300	180	180
I.D. (cm)	0.4	0.2	0.4	0.2	0.4	0.3	0.2
Material	SS	SS	Alu	SS	SS	Alu	SS
Detection	TC	Fl	Fl	TC	TC	Fl	TC
Reference	1	1	2	1	1	3	1

Chain length: Number of double bonds	2-Alkyl-1,3-dioxanes RRT × 100				Alkylacetates RRT × 100		
12:0	19	20	—	22	19	—	22
14:0	32	34	—	37	32	30	37
14:1	—	—	—	—	—	36	—
15:0	43	—	—	—	—	41	—
16:0	58	58	60	63	61	56	62
16:1	—	—	—	—	—	66	—
17:0	75	77	77	—	—	74	—
18:0	i00	100	100	100	100	100	100
18:1	115	116	117	—	—	118	—
18:2	139	146	—	—	—	—	—
19:0	135	—	—	—	—	—	—

Column packing P1 = 10% EGS on Gas-Chrom P (60-80 mesh)

P2 = 15% EGS on Gas-Chrom P (60-80 mesh)

P3 = 10% EGSS-X on Gas-Chrom P (100-120 mesh)

P4 = 10% Apiezon M on Gas-Chrom P (60-80 mesh)

P5 = 20% DEGS on Anakrom A (100-110 mesh)

REFERENCES

1. **Venkata Rao, P., Ramachandran, S., and Cornwell, D. G.,** *J. Lipid Res.,* 8, 380, 1967.
2. **Su, K. L. and Schmid, H. H. O.,** *Lipids,* 9, 208, 1974.
3. **Spener, F. and Mangold, H. K.,** *Biochim. Biophys. Acta,* 192, 516, 1969.

Table 17
1-ALKYLGLYCEROLS (GLC)

Column packing	P1	P1	P2	P3	P4	P5	P6	P7	P8
Temperature (°C)	200	200	210	200	219	220	230	160	190
Gas	He	He	He	He	Ar	He	Ar	Ar	Ar
Flow rate (mℓ/min)	—	30	—	30	40	40	60	80	80
Column									
Length (cm)	180	180	190	90	360	180	122	180	180
I.D. (cm)	0.4	0.3	0.4	0.3		0.3	0.5	0.4	0.4
Material	Alu	Glass	Alu	Glass	Glass	Glass	Glass	Glass	Glass
Detection	Fl	Fl	Fl	Fl	Fl	Fl	Fl	Fl	Fl
Reference	1	2	3	2	4	5	6	7	7

Chain length: Number of double bonds	Isopropylidenes RRT × 100				Acetates RRT × 100			Alkoxyacetaldehydes RRT × 100	
12:0	—	19	—	—	—	—	—	13	10
14:0	—	33	32	25	43	32	—	25	22
14:1	—	39	—	—	—	—	—	—	—
16:0	60	61	54	50	61	57	55	50	47
16:1	—	69	—	—	—	—	—	—	—
17:0	72	—	—	—	—	—	—	—	—
18:0	100	100	100	100	100	100	100	100	100
18:1	115	121	—	—	118	—	91	115	89
20:0	—	—	178	—	—	—	—	—	—

Column packing
P1 = 10% EGSS-X on Gas-Chrom P (100-120 mesh)
P2 = 20% EGS on Gas-Chrom E (60-80 mesh)
P3 = 2.5% SE-30 on Aeropak 30 (100-120 mesh)
P4 = 10% EGSS-X
P5 = 3% Silar-5 CP on Gas-Chrom Q (100-120 mesh)
P6 = 3% SE-30 on Chromosorb W AW HMDS (80-100 mesh)
P7 = 15% DEGS on Anakrom A (100-120 mesh)
P8 = 3.5% SE-30 on silanized Gas-Chrom S (80-100 mesh)

REFERENCES

1. **Su, K. L. and Schmid, H. H. O.**, *Lipids*, 9, 208, 1974.
2. **Snyder, F. and Blank, M. L.**, *Arch. Biochem. Biophys.*, 130, 101, 1969.
3. **Schmid, H. H. O., Baumann, W. J., and Mangold, H. K.**, *Biochim. Biophys. Acta*, 144, 344, 1967.
4. **Renkonen, O.**, *Biochim. Biophys. Acta*, 125, 288, 1966.
5. **Myher, J. J. and Kuksis, A.**, *Lipids*, 9, 382, 1974.
6. **Albro, P. W. and Dittmer, J. C.**, *J. Chromatogr.*, 38, 230, 1968.
7. **Gelman, R. A. and Gilbertson, J. R.**, *Anal. Biochem.*, 31, 463, 1969; see also **Wijsman, J. A. and Hilwig, G. N. G.**, *Fette Seifen Anstrichm.*, 83, 275, 1981.

Table 18
1- AND 2-ALKYLGLYCEROLS (GLC)

Column packing	P1	P2	P3	P4	P3	P5	P4	P6
Temperature (°C)	170	190	200	210	200	200	230	240
Gas	He	He	He	He	He	He	He	N$_2$
Flow rate (mℓ/min)	90	80	90	90	90	10	90	28
Column								
Length (cm)	150	50	150	150	150	180	150	180
I.D. (cm)	0.3	0.3	0.3	0.3	0.3	0.2	0.3	0.3
Material	Alu	Glass	SS	SS	SS	SS	SS	Glass
Detection	Fl	Fl	Fl	Fl	Fl	Fl	Fl	Fl
Reference	1	2	1	1	1	3	1	4

Chain length: Number of double bonds	Trifluoroacetates RRT × 100				TMS RRT × 100			n-Butylboronates RRT × 100
1-12:0[a]	—	10	16	14	16	—	17	—
2-12:0[a]	—	—	18	14	16	—	17	—
1-14:0	29	26	29	26	30	—	30	—
2-14:0	36	—	34	26	30	—	30	—
1-16:0	—	52	55	52	55	56	55	—
2-16:0	—	—	63	52	55	56	55	—
1-18:0	100	100	100	100	100	100	100	100
2-18:0	121	—	118	100	100	100	100	118
1-18:1	112	114	—	91	—	108	89	—
2-18:1	134	—	—	91	—	108	89	—
1-18:2	134	—	—	—	—	—	—	—
2-18:2	172	—	—	—	—	—	—	—
1-20:0	—	170	—	—	—	—	—	—

Column packing P1 = 10% EGGS-X on Gas-Chrom P (100-120 mesh)
P2 = 20% EGS on Gas-Chrom E (60-80 mesh)
P3 = 2.5% SE-30 on Aeropak 30 (100-120 mesh)
P4 = 10% EGSS-X
P5 = 3% Silar-5 CP on Gas-Chrom Q (100-120 mesh)
P6 = 3% SE-30 on Chromosorb W AW HMDS (80-100 mesh)
P7 = 15% DEGS on Anakrom A (100-120 mesh)
P8 = 3.5% SE-30 on silanized Gas-Chrom S (80-100 mesh)

REFERENCES

1. **Su, K. L. and Schmid, H. H. O.,** *Lipids,* 9, 208, 1974.
2. **Snyder, F. and Blank, M. L.,** *Arch. Biochem. Biophys.,* 130, 101, 1969.
3. **Schmid, H. H. O., Baumann, W. J., and Mangold, H. K.,** *Biochim. Biophys. Acta,* 144, 344, 1967.
4. **Renkonen, O.,** *Biochim. Biophys. Acta,* 125, 288, 1966.
5. **Myher, J. J. and Kuksis, A.,** *Lipids,* 9, 382, 1974.
6. **Albro, P. W. and Dittmer, J. C.,** *J. Chromatogr.,* 38, 230, 1968.
7. **Gelman, R. A. and Gilbertson, J. R.,** *Anal. Biochem.,* 31, 463, 1969.

Table 19
PHYTANYLGLYCEROLS (GLC)

Column packing	P1	P2
Temperature (°C)	180	220
Gas	Ar	Ar
Flow rate (mℓ/min)		
Column		
Length (cm)	120	50
I.D. (cm)	0.4	0.4
Material	Glass	Glass
Detection	F1	F1
Reference	1	2

Compounds	RRT × 100	
Monoalkylglycerols		
(as di-*O*-methyl ethers)		
1-Hexadecylglycerol	100	—
3-Phytanyl-*sn*-glycerol	145	—
2-Phytanylglycerol	125	—
Dialkylglycerols		
1,2-Dihexadecylglycerol	—	100
2,3-Diphytanyl-*sn*-glycerol	—	210
1,2-Diphytanyl-*sn*-glycerol	—	215

Column packing P1 = 10% BDS on Anakrom
A
P2 = 2% SE-52 on Gas-
Chrom Z (80-100
mesh)

REFERENCES

1. **Joo, C. N., Shier, T., and Kates, M.,** *J. Lipid Res.*, 9, 782, 1968.
2. **Kates, M., Palameta, B., and Yengoyan, L. S.,** *Biochemistry*, 4, 1595, 1965.

Table 20
HYDROXY- AND KETOALKYLGLYCEROLS (GLC)

	P1	P2	P3	P3
Column packing	P1	P2	P3	P3
Temperature (°C)	260	260	200	200
Gas	He	He	He	He
Column				
Length (cm)	190	300	180	180
I.D. (cm)	0.2	0.2	0.3	0.3
Material	Alu	Alu	Alu	Alu
Detection	Fl	Fl	Fl	Fl
Reference	1	1	2	2
Compounds	**ECL**	**ECL**	**RRT × 100**	**RRT × 100**
2,3-Isopropylidenes				
1-Hexadecyl	20,55	20,50		
1-Heptadecyl	21,50	21,50		
1-(2′-Keto)hexadecyl	21,95	24,90		
1-(2′-Keto)heptadecyl	22,90	25,90		
1-(2′-Hydroxy)hexadecyl	22,25	25,45		
1-(2′-Hydroxy)heptadecyl	23,20	26,45		
1-(2′-Acetoxy)hexadecyl	22,70	24,60		
1-(2′-Acetoxy)heptadecyl	23,65	25,60		
1-(9′,10′-Acetoxy)hexadecyl				65
1-(9′,10′-Acetoxy)octadecyl				100
2,3-Diacetylglycerols				
1-(2′-Acetoxy)hexadecyl	24,60			
1-(2′-Acetoxy)heptadecyl	25,55			
2,3-di-TMS ethers of glycerol				
1-(9′,10′-TMS)hexadecyl			56	
1-(9′,10′-TMS)heptadecyl			73	
1-(9′,10′-TMS)octadecyl			100	

Column packing P1 = 6% SE-30 on Anakrom ABS (90-100 mesh)
P2 = 10% SP-1000 on Gas-Chrom P (80-100 mesh)
P3 = 10% EGSS-X on Gas-Chrom P

REFERENCES

1. **Muramatsu, T. and Schmid, H. H. O.,** *Chem. Phys. Lipids,* 9, 123, 1972.
2. **Kasama, K., Rainey, W. T., Jr., and Snyder, F.,** *Arch. Biochem. Biophys.,* 154, 648, 1973.

Table 21
1-THIOALKYLGLYCEROLS (GLC)

Column packing	P1	P2	P3	P4
Temperature (°C)	185	185	175	T1
Gas	N$_2$	He	He	He
Flow rate (mℓ/min)	60	40-60	40-60	40-60
Column				
Length (cm)	180	150	150	150
I.D. (cm)	0.4	0.3	0.3	0.3
Material	Glass	Glass	Glass	SS
Detection	F1	F1	F1	F1
Reference	1	2	2	2

Chain length: Number of double bonds	RRT × 100			
2,3-Isopropylideneglycerols				
10:0	—	31	28	29
12:0	—	43	41	41
13:0	12	—	—	—
14:0	23	58	56	56
15:0	35	—	—	—
16:0	50	77	76	76
16:1	63	—	—	—
18:0	100	100	100	100
18:1	122	—	—	—

Column packing   P1 = 14.5% EGSS-X on Gas-Chrom P (100-120 mesh)

P2 = 15% EGSS-X on Gas-Chrom P (100-120 mesh)

P3 = 15% EGS on D-dusted Gas-Pack WAB (80-100 mesh)

P4 = 5% SE-30 on Chromosorb W (60-80 mesh)

Temperature program   T1 = 180-225°C at 4°C/min

REFERENCES

1. **Ferrell, W. J. and Radloff, D. M.,** *Physiol. Chem. Phys.,* 2, 551, 1970.
2. **Wood, R., Piantadosi, C., and Snyder, F.,** *J. Lipid Res.,* 10, 370, 1969.

Table 22
DIRADYLETHANEDIOLS (GLC)

Column packing	P1	P1
Temperature (°C)	T1	T2
Gas	He	He
Flow rate (mℓ/min)	70	70
Column		
Length (cm)	70	70
I.D. (cm)	0.25	0.25
Material	Glass	Glass
Detection	F1	F1
Reference	1	1

Chain length[a]		RRT × 100	
1-Position	**2-Position**		
cis/trans-alk-1'-enyl	acyl		
cis-16:0	14:0	72	—
trans-16:0	14:0	76	—
cis-18:0	14:0	80	—
trans-18:0	14:0	84	—
cis-16:0	18:0	88	—
trans-16:0	18:0	92	—
cis-18:0	18:0	100	—
trans-18:0	18:0	104	—
cis-16:0	16:0	—	80
cis-18:0	18:0	—	100
alkyl	acyl		
16:0	16:0	—	83
18:0	18:0	—	102
acyl	acyl		
16:0	16:0	—	87
18:0	18:0	—	105

[a] Enol ether double bond not counted.

Column packing P1 = 1% OV-1 on Chromosorb W (WP) (100-200 mesh)
Temperature program T1 = 150 - 270°C at 5°C/min
T2 = 150-300°C at 5°C/min

REFERENCE

1. **Kramer, J. K. G. and Mangold, H. K.,** *Chem. Phys. Lipids,* 4, 332, 1970.

Table 23
ALK-1-ENYL- AND ALKYLETHANEDIOLS (GLC)

Column packing	P1	P2	P3	P2	P3	P2	P3
Temperature (°C)	200	195	T1	T2	T1	T3	T1
Gas	He	He	He	He	He	He	He
Flow rate (mℓ/min)	—	50	50	50	50	50	50
Column							
Length (cm)	180	152	150	152	150	152	150
I.D. (cm)	0.4	0.175	0.3	0.175	0.3	0.175	0.3
Material	Glass	Glass	SS	Glass	SS	Glass	SS
Detection	F1	F1	F1	F1	F1	F1	F1
Reference	1	2	2	2	2	2	2

Chain length: Number of Double Bonds	Acetates RRT × 100		Trifluoroacetates RRT × 100		TMS RRT × 100		
Alk-1-enylethanediols							
16:0	58	—	—	—	—	—	—
18:0	100	—	—	—	—	—	—
18:1	115	—	—	—	—	—	—
Alkylethanediols							
12:0	—	24	39	36	27	31	34
14:0	—	39	57	54	45	48	52
16:0	—	61	77	75	70	71	75
18:0	—	100	100	100	100	100	100
18:1	—	120	96	108	95	106	96
20:0	—	162	127	131	132	136	128

Column packing	P1	=	18% HiEFF on Gas-Chrom P (60-100 mesh)
	P2	=	15% EGSS-X on Gas-Chrom P (100-120 mesh)
	P3	=	5% SE-30 on Chromosorb W (60-80 mesh)
Temperature program	T1	=	150-235°C at 4°C/min
	T2	=	125-190°C at 3°C/min
	T3	=	125-185°C at 3°C/min

REFERENCES

1. **Baumann, W. J., Schmid, H. H. O., Kramer, J. K. G., and Mangold, H. K.,** *Z. Physiol. Chem.,* 249, 1677, 1968.
2. **Wood, R. and Baumann, W. J.,** *J. Lipid Res.,* 9, 733, 1968.

Chapter 15

PRODUCTS OF PARTIAL DEGRADATION OF PHOSPHOLIPIDS

R. M. C. Dawson

I. INTRODUCTION

Since the very early days of research into the phospholipids contained in biological samples, total hydrolysis by strong alkali or acid has been used to identify and estimate the molecular entities in these complex lipids. By using milder degradation procedures it is possible to limit the destruction to specific bonds and from the nature of the products obtained, deduce much more detailed information concerning the nature of the parent phospholipid. This is often of great use in helping to determine the structure of an unknown phospholipid, to decide between known structures where ambiguity exists, or to assist in analyzing the components of a complex mixture.

By far the most widely used degradative procedure has been the schism of *O*-acyl bonds using alcoholic alkali. Such bonds in the common diacylated-phosphoglycerides or their lyso counterparts are very rapidly broken even at room temperature by treatment with dilute alcoholic solutions of strong alkalis that can produce a pH of above 12 (e.g., KOH or tetramethylammonium hydroxide). Nevertheless, there are differences in the rate at which deacylation of *O*-acyl groups occur, e.g., phosphatidylcholine is deacylated much more rapidly than either phosphatidylethanolamine or phosphatidylserine,[1] while the deacylation of the *O*-acyl group in plasmalogens and alkylacyl phospholipids (ether phospholipids) is quite slow.[2,3] Consequently the concentration and type of alkali used and the length and temperature of treatment needs to be a compromise if a mixture of phospholipids is being examined. An inadequate treatment can lead to plasmalogens being left intact, while a more vigorous alkaline degradation can lead to secondary decomposition, e.g., cyclic-1,2-gly-cerophosphate can easily form during the alkaline alcoholysis of phosphatidylcholine[3] and inositol monophosphate from phosphatidylinositol,[4] as well as the expected glycerylphos-phorylcholine and glycerylphosphorylinositol derivatives. A method of avoiding this secondary decomposition during deacylation is to use a recently introduced reagent containing monomethylamine.[14] Complete transacylation of *O*-acyl groups on the phospholipids, including those on plasmalogens, to the amino group can be brought about by heating the mixture. Afterwards the excess volatile base can be disposed of by evaporation, thus avoiding any necessity for removing alkali by ion exchange columns or other means.

If an investigation is being carried out on an unknown phospholipid, additional information on its structure can be gained by carrying out a brief acid hydrolysis of the primary water-soluble phosphorus-containing deacylation product of the alkaline treatment. Thus the primary deacylation product of *N*(1-carboxyethyl) phosphatidylethanolamine, *N*(1-carboxyethyl) glycerylphosphorylethanolamine is split into glycerophosphate and hydroxyethylalanine by 1 *N* HCl hydrolysis (15 min 100°C).[5]

After complete deacylation of the *O*-acyl groups has been carried out, a number of phospholipids or their partial degradation products still distribute in the solvent phase of a water-lipid solvent mixture. For convenience these will subsequently be referred to as "alkali-stable". These include the alkenyl and alkyl derivatives of the glycerol phosphoryl "bases" derived from the plasmalogens and alkylacyl phospholipids, respectively, and the *N*-acylated phospholipids such as the sphingolipids or *N*-(acyl) glycerylphosphoryl-ethanolamine (-serine) derived from the *N*-acylated forms of phosphatidylethanolamine or phosphatidylser-ine, respectively. Such mixtures can be investigated further by treatment with mercuric ions to hydrolyze catalytically the alkenyl group of the plasmalogens thus leaving the water-soluble glycerylphosphoryl base residues of the plasmalogens for examination. Finally a treatment with methanolic HCl at 100°C will split off phosphoryl base units from sphingomyelin or ceramide phosphorylethanolamine. If it is suspected that phosphonolipids may be present in the original lipid, then any of the water-soluble phosphate ester produced, particularly ''glycerylphosphorylethanolamine'' or ''phosphorylethanolamine'' can be examined for the presence of its phosphono analog by strong acid hydrolysis; the C–P bond is totally resistant to such treatment, whereas the C–O–P bond will always be split.

II. PREPARATION OF SAMPLES

A. Extraction

Often unknown phospholipids in a relatively pure form are simply recovered by elution from preparative thin-layer chromatograms or chromatographic columns.

Procedure for the Extraction of Lipids from Animal Tissue
Mammalian tissues are extracted by homogenization in 20 volumes of chloroform-methanol (2:1 v/v) using a Waring-type blender. After standing 20 min, the supernatant is collected by centrifugation and the pellet washed twice with 5 volumes of the same solvent. The combined extracts are washed once with 0.2 volumes of 0.9% NaCl to remove contaminating water-soluble phosphorus compounds, and the lower chloroform-rich phase is recovered by centrifuging. It is evaporated to dryness in a rotary dryer to decompose proteolipids, and the lipids are extracted from the residue with chloroform. Triphosphoinositide and diphosphoinositide remain in the chloroform-methanol extracted pellet and can be removed by several washes with acidified chloroform-methanol (chloroform-methanol [2:1 v/v] containing 0.25% by volume concentrated HCl). This solution is washed with 0.9% NaCl as above and then the lower phase several times with 0.9% NaCl-methanol-CHCl$_3$ (47-45-3 v/v) to remove HCl. At each wash the proteinaceous interface is preserved, and finally the interface and lower phase are taken to dryness. The residue is boiled several times for 10 min with a little acetone or ethanol, which is taken to dryness after each treatment. The residue is finally extracted with water-saturated chloroform to solubilize the polyphosphoinositides. Unfortunately, this acid-solvent extraction of polyphosphoinositides cannot be applied directly to the original biological sample to extract all the phospholipids because extensive hydrolysis of the acid-labile plasmalogens occurs.

Procedure for the Extraction of Lipids from Microorganisms
Microorganisms are refluxed for 30 min with ethanol-water (4:1 v/v) and then extracted with chloroform-methanol as above. The combined extracts are taken to dryness in a rotary dryer and taken up in chloroform-methanol (2:1 v/v) for washing to remove contaminating water-soluble components as described above for mammalian tissues.

B. Degradation of Phospholipids

Theoretically, the strong alkali used to bring about the alcoholysis of phospholipids is only acting as a catalyst so that minimal quantities will suffice. In practice, excess neutral triglyceride (e.g., fatty livers) can interfere and have to be removed before the alkali treatment is used or more alkali should be added so that the pH remains strongly alkaline.[6] If the sample contains only di-O-acylated phosphoglycerides or their lyso analogs, or if it is only desired to examine this group of phospholipids, a very mild alkaline treatment will suffice, just sufficient to deacylate and cause little additional breakdown of the water-soluble phosphate esters. Such a procedure has been described by Brockerhoff[7] (A, Table 1).

Procedure for the Alkaline Alcoholysis of Lipids[7]
The sample (100 to 500 μg P of a phospholipid mixture or less if individual
lipids are used) in chloroform (0.2 mℓ) is treated with 0.8 mℓ of 0.125 M
LiOH·H₂O in methanol. After 15 min, 3 mℓ H₂O, 2 mℓ of ethanol, and 4 mℓ of
chloroform are added in this sequence. After shaking and centrifuging, the upper
aqueous layer is carefully removed and passed through a short (1 × 4 cm)
column of cation exchange resin. (Dowex 50, pyridinium form; Amberlite IRC
50, H⁺ form) to remove alkali. The transfer is completed by washing the lower
phase and column with methanol-H₂O (1:1 v/v). After adding a drop of concen-
trated ammonia to the eluate plus washings, the solution is concentrated in vacuo
for subsequent chromatography. If subsequent examination of the more alkali-
stable phospholipids is also desired, a somewhat stronger alkaline alcoholysis is
used. The sample of lipid is taken as above to dryness and then 0.8 mℓ of carbon
tetrachloride, 7.5 mℓ of ethanol, 0.65 mℓ of H₂O and 0.25 mℓ of aqueous N
NaOH are added.[8] The sample is incubated for 20 to 60 min (see below) at 37°C,
and the reaction is stopped by adding 0.4 mℓ of ethyl formate that neutralizes
the excess alkali by the reaction:

$$ethyl\ formate\ +\ NaOH\ =\ Na\ formate\ +\ ethanol$$

without causing acid hydrolysis of the aldehydogenic group in the plasmalogens.

The ethyl formate neutralized reaction mixture is taken to dryness, and 1 mℓ of
H₂O and 2 mℓ of isobutanol-CHCl₃ (1:2 v/v) are added to the residue. After
solubilization, the mixture is centrifuged and the phases separated; persistent
emulsions are unusual, but can be broken by freezing in solid carbon dioxide
and centrifuging after thawing. If the alkaline alcoholysis is carried out for 20
min, it still allows the water-soluble decomposition products of the O-acylated
phosphoglycerides to be examined, although some slight secondary decomposi-
tions of the phosphate esters may have occurred. The O-acyl grouping in the
plasmalogens is not entirely decomposed. If the alcoholysis is prolonged for 60
min, some further decomposition of the water-soluble phosphate ester occurs,
but the plasmalogens are completely O-deacylated. This allows for a better ex-
amination of the alkali-stable phospholipids. If, therefore, the amount of mate-
rial permits, it is preferable to carry out a limited alkaline alcoholysis on one
portion for examining the water-soluble phosphate esters poduced from the O-
acylated phosphoglycerides and a more vigorous alcoholysis (60 min) for sub-
sequent examination of the ''alkali = stable'' phospholipids.

Procedure for Alkaline O→N Transacylation [(A) Table 1]
This procedure removes all O-acyl groups with little secondary decomposi-
tion.Monomethylamine gas from a cyclinder is passed into 20 mℓ of a mixture
containing methanol-water n-butanol (4:3:1) v/v using an anti-suckback device
until the volume has increased to 32.5 mℓ. Alternatively the commercial solu-
tions of monomethylamine solutions in alcohol diluted with water can be used as
the reagent although the reaction is slower. The phospholipid solution (250 μg
of lipid P) freed from Ca²⁺ and Mg²⁺ by washing is taken to dryness in a 10 mℓ
tube, the residue treated with 3 mℓ of the methylamine reagent, and heated at

Table 1

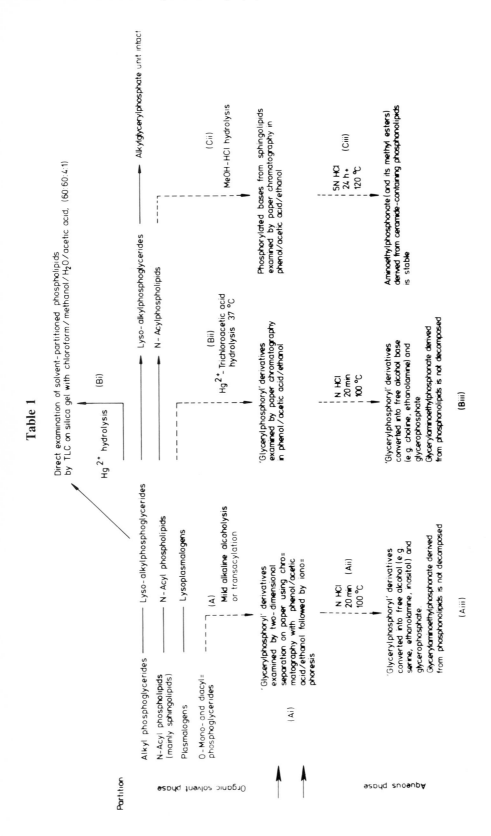

53°C for one hour with a spring-loaded stopper on the tube (30 min is sufficient to deacylate diacylated phospholipids such as phosphatidylcholine).
The reactants are cooled in ice, and mixed with a half volume of n-propanol. The mixture is carefully evaporated in vacuo using swirling and a splash head and finally taken to dryness at 50°C. To the residue is added 1 mℓ H_2O and 1.2 mℓ n-butanol-petroleum hydrocarbon-ethyl formate (20:4:1 v/v). (The latter ester neutralizes any trace of monomethylamine remaining.) After shaking and centrifuging the upper solvent phase containing lysophospholipids with alkenyl or alkyl groups is removed, leaving a lower aqueous phase with the water soluble phosphate ester decomposition products.

Procedure for the Separation of the Water-Soluble P-Containing Compounds formed by Alkali Treatment (Table 1 Ai)
The water-soluble phosphate-esters formed by deacylation of the phospholipids with alkali are separated on a two-dimensional paper system. An aliquot of the aqueous phase formed by either method described above is spotted onto paper. If the alkali has been removed with cation exchange resin, all the aliquot can be used if necessary. If ethyl formate has been used to neutralize the alkali, only 0.2 mℓ aliquots can be safely used because of the distorting effects of sodium formate on the separation. The phosphate esters are separated in the first dimension with phenol saturated with water-acetic acid-ethanol (100:10:12 v/v) (overnight, descending). After air drying at low temperatures (diethyl ether washing of the chromatogram can be used to facilitate the removal of phenol once the water has evaporated) the phosphate esters are separated in the second dimension by ionophoresis at pH 3.6 with a volatile buffer (pyridine-acetic acid-H_2O [1:10:89 v/v]). The dried paper supported on glass rods or a hydrophobic notwettable plastic mesh framework is wetted with the pyridine-acetic acid buffer from a hand-operated bulb pipet. The buffer is finally allowed to flow by capillary action into the theoretical line of phosphate esters separated by the phenol solvent, so this is wetted symmetrically from both sides. Ionophoresis is carried out under water-cooled high flash point white spirit (kerosene, Varsol) in an apparatus described previously[9] and which is commercially available in similar design. A period of 1 to 1.5 hr at 40 v/cm is usually sufficient, but this can be varied depending on the temperature of the tank, type of paper, etc. and the separation desired.
The dried paper is sprayed with 0.25% (w/v) ninhydrin in acetone followed by heating the paper at 80°C for 5 min. The phosphate esters are then located with a molybdate spray. (5 mℓ of 60% perchloric acid + 10 mℓ /N HCl + 25 mℓ of 4% [w/v] ammonium molybdate made to 100 mℓ with water). After drying at room temperature (higher temperatures can lead to a blue background diminishing resolution), the blue colors of the phosphate-esters are developed by irradiation under a UV light. A two-dimensional map of the relative positions of the various phosphate esters that can be derived from phospholipids is given in Figure 1A with the parent phospholipids identified in the legend.

1. Acid Hydrolysis of Phosphate Esters Produced by Alkali Treatment (see Aii, Table 1)
If any of the phosphate-esters after the above two-dimensional separation cannot be identified, or confirmation of identity is required, spots can be located on the paper by nondestructive methods (e.g., autoradiography after ^{32}P labeling, dipping in 0.025% quinine sulfate, $2H_2O$, in ethanol and examination under UV light[10,11]), eluted, and decomposed by a brief acid hydrolysis. The eluted phosphate ester is mixed with 0.1 volume of 10 *N* HCl

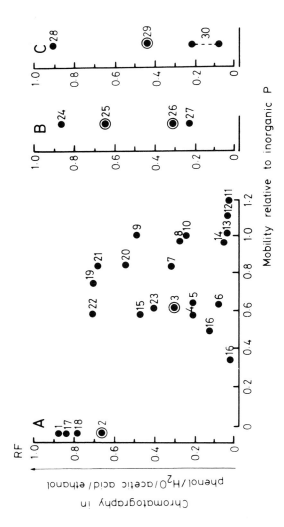

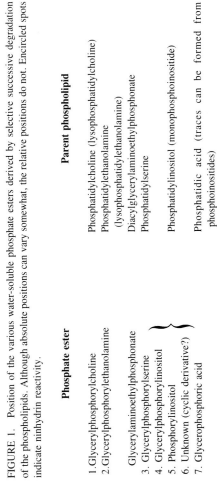

FIGURE 1.   Position of the various water-soluble phosphate esters derived by selective successive degradation of the phospholipids. Although absolute positions can vary somewhat, the relative positions do not. Encircled spots indicate ninhydrin reactivity.

Phosphate ester

1. Glycerylphosphorylcholine
2. Glycerylphosphorylethanolamine

   Glycerylaminoethylphosphonate
3. Glycerylphosphorylserine
4. Glycerylphosphorylinositol
5. Phosphorylinositol
6. Unknown (cyclic derivative?)
7. Glycerophosphoric acid

Parent phospholipid

Phosphatidylcholine (lysophosphatidylcholine)
Phosphatidylethanolamine
(lysophosphatidylethanolamine)
Diacylglycerylaminoethylphosphonate
Phosphatidylserine

Phosphatidylinositol (monophosphoinositide)

Phosphatidic acid (traces can be formed from phosphoinositides)

8. Bis(glycerylphosphoryl) glycerol	Bis(phosphatidyl)-glycerol (cardiolipin)
9. Cyclic glycerophosphoric acid	Phosphatidylcholine, phosphatidylinositol
10. Inorganic phosphate	Acyldihydroxyacetone phosphate. Traces often tenaciously accompany higher phosphoinositides of brain
11. Inositol trisphosphate	Triphosphoinositide
12. Glycerylphosphorylinositolbisphosphate	
13. Inositol bisphosphate	
14. Glycerylphosphorylinositolmonophosphate	Diphosphoinositide
15. Bis(glyceryl)phosphate	Phosphatidylglycerol
	Acylphosphatidylglycerol
	Amino acid esters of phosphatidylglycerol
	Phosphatidylinositotrimannoside
16. Deacylation products of phosphatidylinositotrimannoside (the slowest running is predominant)	
17. Glycerylphosphoryl-N-dimethylethanolamine	Phosphatidyl-N-dimethylethanolamine
18. Glycerylphosphoryl-N-monomethylethanolamine	Phosphatidyl-N-monomethylethanolamine
19. Glycerylphosphoryl-n-propanol	Phosphatidyl-n-propanol*
20. Glycerylphosphorylmethanol	Phosphatidylmethanol*
21. Glycerylphosphorylethanol	Phosphatidylethanol*
22. Glycerylphosphoryl-N-(1-carboxyethyl) ethanolamine	N(1-carboxyethyl)-phosphatidyl-ethanolamine
23. Glycerylphosphoryl-galactoglycerol	Glycerylphosphoryl-galactodiglyceride
24. Glycerylphosphorylcholine	Choline plasmalogen
25. Glycerylphosphorylethanolamine	Ethanolamine plasmalogen
Glycerylaminoethylphosphonate	Aminoethylphosphonate plasmalogen
26. Glycerylphosphorylserine	Serine plasmalogen
27. Glycerylphosphorylinositol	Inositol plasmalogen
28. Phosphorylcholine + sphingosylphosphorylcholine	Sphingomyelin
29. Phosphorylethanolamine	Ceramide phosphorylethanolamine
Aminoethylphosphonate	Ceramide aminoethylphosphonate
30. Inorganic P	Derived by secondary decomposition of 28 - 30, e.g., when higher plant tissues are extracted with aqueous alcohol.

* Formed by transfer of phosphatidyl unit to alcohol catalyzed by phospholipase D, e.g., when higher plant tissues are extracted with aqueous alcohol.

and heated in a boiling water bath for 15 to 20 min. The hydrolyzate is taken to dryness *in vacuo* and the components examined by conventional chromatographic methods to identify the free base, amino acid, or sugar originally attached to the acid-stable glycerophosphate moiety.

2. Hydrolysis of the Alkali-Stable Phospholipids

The alkali-stable phospholipids in the chloroform-isobutanol phase isolated above can be examined (1) directly by thin-layer chromatography (TLC), with and without catalytic decomposition of the plasmalogens, or (2) by determining the water-soluble phosphate esters split off by a mild acid-Hg^{2+} hydrolysis of the residual phospholipids.

Procedures for the Hydrolysis of the Alkali-Stable Phospholipids (1) (Bi, Table 1).
This method can be used after a 20 min alkaline hydrolysis although if sufficient material is available, a 60 min hydrolysis is better since all the plasmalogens will have been O-deacylated (see section on alkali treatment). The chloroform-isobutanol lower phase is split into two 0.8 mℓ portions, one of which is reserved for direct TLC, and the other is hydrolyzed with a $HgCl_2$-containing reagent followed by a similar TLC of the stable phospholipids. The 0.8 mℓ aliquot is mixed with 0.05 mℓ of a reagent containing 0.27 g of $HgCl_2$ in 10 mℓ of 10% water in methanol, and the mixture is incubated for 20 min at 37°C. To this mixture is added 0.03 mℓ of 1 N methanolic NaOH, and the incubation is continued for a further 20 min. The solution is neutralized by mixing with ethyl formate (0.05 mℓ), and it is then shaken with 1 mℓ of water. The lower phase is recovered by centrifuging. Both this lower phase and the original unhydrolyzed 0.8 mℓ aliquot are taken to dryness, using ethanol (3 to 4 volumes) to assist in the volatilization of the isobutanol. The lipid residues are chromatographed on silica gel plates (Merck F-254) using chloroform-methanol-acetic acid-H_2O, (60:60:1:4 v/v). The plates are sprayed to detect P,[12] giving spots that can be identified as shown in Figure 2. It is clear that by measuring the P in the spots by any suitable method available, one can obtain a quantitative assessment of the ethanolamine- or choline-containing alkyl acyl or alkenyl acyl phosphoglycerides and sphingomyelin present in a given tissue.

Procedure for the Hydrolysis of the Alkali-Stable Phospholipids (2) (Bii,Ci Table 1)
This method of examining the alkali-stable phospholipids again gives better results if it can be used after a complete O-deacylation of the plasmalogens (i.e., a 60 min alkali treatment — see previous section), although a 20 min alkali treatment can be used if necessary. A 1.6 mℓ portion of the chloroform-isobutanol phase containing the alkali-stable phospholipids is shaken vigorously at 37°C for 30 min with 0.8 mℓ of 1% (w/v) trichloroacetic acid containing $HgCl_2$ (5 mM). After cooling, 2 mℓ of diethyl ether is added, and after shaking, the upper ethereal layer is removed and collected. The lower aqueous phase is then extracted twice with 2 mℓ chloroform-diethyl ether-isobutanol (1:1:2 v/v) and once more with diethyl ether (2 mℓ). The solvent extracts are combined for methanolic-HCl hydrolysis (below). The aqueous phase is made slightly alkaline by holding a small piece of filter paper soaked in strong ammonia in the tube, which is then stoppered and shaken. A sample (0.3 mℓ) is subjected to descending paper chromatography (18 hr) in phenol-saturated with water acetic acid-ethanol (100:10:12 v/v). After drying at 70°C, the papers are sprayed with 0.25% (w/v) ninhydrin in acetone to locate glycerylphosphorylethanolamine and glyceryl-

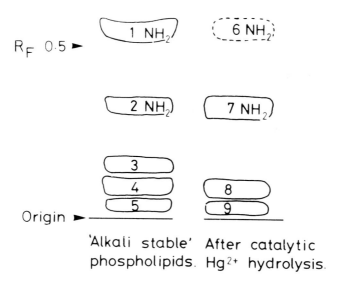

FIGURE 2.   TLC of alkali-stable P-containing lipids produced by the alkaline treatment of a mammalian tissue phospholipid mixture. Adsorbent: Merck Silica gel 60 F 254; Solvent: Chloroform-methanol-acetic acid-water, 60:60:1:4; Spray reagent: see Reference 12. Spots marked NH_2 can be revealed by a preliminary spray of 0.25 % ninhydrin in acetone and warming at 80°C.

1. Ethanolamine plasmalogens (seen after 20 min alkaline alcoholysis but virtually disappear after 60 min alkaline alcoholysis).
2. Alkenyl plus alkyl glycero phosphorylethanolamines.
3. Choline plasmalogens (see note for 1. above).
4. Sphingomyelins
5. Alkenyl plus alkyl glycerophosphorylcholines
6. Dialkyl glycerophosphorylethanolamines
7. Alkyl glycerophosphorylethanolamines
8. Sphingomyelins
9. Alkyl glycerophosphorylcholines

phosphorylserine (if present) and then with a reagent to detect phosphorus (see above). The positions of various phosphorus-containing spots arising from the plasmalogens are indicated in Figure 1B. The combined solvent fractions obtained from the Hg^{2+}-trichloroacetic acid hydrolysis are evaporated to dryness and heated in a sealed tube for 4 hr with 1.25 mℓ of dry 2 N methanolic HCl at 100°C (Ci, Table 1). The mixture is evaporated to dryness in vacuo and 2 mℓ of diethyl ether and 0.5 mℓ of H_2O are added. After centrifuging, the aqueous layer is reextracted with a further 2 mℓ of diethyl ether. The ether extracts contain the total phosphorus of the combined alkyl phosphoglycerides. An aliquot of the aqueous layer (0.2 mℓ) is chromatographed on paper (descending, 18 hr: with water phenol saturated acetic acid-ethanol (100:10:12 v/v) and then sprayed with ninhydrin followed by P-detecting reagent as above (Table 1). The position of the phosphorylated bases derived from the sphingolipids is illustrated in Figure 1C.

C. Degradation of Phosphonolipids

These phospholipids have been found in appreciable amounts only in the Protozoa Coelenterata and Mollusca, but traces may be present in the tissues of higher animals which ingest these organisms in their diet and also in bacteria. The phosphonolipids occur almost exclusively as 1.2 diacyl-*sn*-glycerol aminoethylphosphonate (a plasmalogen analog has been described) or as ceramide aminoethylphosphonate or its *N*-methylated derivatives. These phosphonolipids can be distinguished from their analogs (phosphatidylethanolamine and ceramide phosphorylethanolamine) by taking advantage of the great stability of the C–P bond in the water-soluble phosphorus-containing compounds derived from them to acid hydrolysis.[13] If phosphonolipids are suspected, the water-soluble fraction prepared by alkaline treatment (see above) can be examined both before and after acid hydrolysis (Figure 1).

Procedure for the Detection of Phosphonolipids
An aliquot of the aqueous layer is made acid with 0.1 volume of 10 N HCl and heated at 100°C for 20 min. The hydrolysate is taken to dryness in vacuo and the phosphate esters reexamined in the two-dimensional chromatographic-iono-phoresis system (Ai, Table 1). 1,2-Diacyl-sn-glyceryl aminoethyl phosphonate is deacylated by the alkali giving glycerylaminoethylphosphonate, which is not sep-arated from the glycerylphosphorylethanolamine spot derived from phosphati-dylethanolamine. However, glycerylphosphorylethanolamine is completely hydrolyzed by the acid treatment, whereas the glycerylaminoethylphosphonate remains totally stable. Thus any part of this spot remaining after acid hydrolysis represents the phosphonolipid. The glycerylphosphorylethanolamine contained in the water-soluble fraction split off from the plasmalogen fraction by Hg²⁺ hydrolysis can be subjected to the same type of acid treatment (1 N HCl, 100°C, 20 min) to check for the presence of the phosphonolipid analog of ethanolamine plasmalogen (Biii, Table 1).

On methanolic-HCl hydrolysis of the alkali and Hg^{2+}-acid stable phospholipids any cer-amide aminoethylphosphonate liberates water-soluble aminoethylphosphonate which co-chromatographs with the ethanolamine phosphate derived from ceramide phosphorylethan-olamine. However, the two can be distinguished by making the water-phase derived from the methanolic-HCl hydrolysis 5 *N* with respect to HCl, sealing it in a thick glass tube, and heating it in an oven at 120°C for at least 24 hr (Ciii, Table 1). The aminoethylphosphonate is stable while the phosphorylethanolamine is completely hydrolyzed to inorganic P and ethanolamine. If, therefore, a ninhydrin-reacting P-containing spot remains in the phos-phorylethanolamine position, it represents ceramide aminoethylphosphonate in the original lipid fraction. At the same time, if any of the phosphorylcholine spot derived from sphin-gomyelin remains stable, this might indicate the presence of ceramide (*N*-trimethyl) aminoethylphosphonate.

REFERENCES

1. **Pries, C.,** *Proefschrift,* Rijksuniversitat, Leiden, Netherlands, 1965.
2. **Ansell, G. B. and Spanner, S.,** *J. Neurochem.,* 10, 941, 1963.
3. **Maruo, B. and Benson, A. A.,** *J. Biol. Chem.,* 234, 254, 1959.
4. **Dawson, R. M. C.,** *Biochem. J.,* 75, 45, 1960.
5. **Kemp, P. and Dawson, R. M. C.,** *Biochem. J.,* 113, 555, 1969.
6. **Prottey, C. and Hawthorne, J. N.,** *Biochem. J.,* 101, 191, 1966.
7. **Brockerhoff, H.,** *J. Lipid Res.,* 4, 96, 1963.
8. **Dawson, R. M. C., Hemington, N., and Davenport, J. B.,** Biochem. J., 84, 497, 1962.
9. **Dawson, R. M. C.,** in *Lipid Chromatographic Analysis,* Vol. 1, Marinetti, G. V., Ed., Marcel Dekker, New York, 1978, 149.
10. **Dawson, R. M. C. and Dittmer, J. C.,** *Biochem. J.,* 81, 540, 1961.
11. **Rorem, E. S.,** *Nature (London),* 183, 1739, 1959.
12. **Vaskovsky, V. E. and Kostetsky, E. Y.,** *J. Lipid Res.,* 9, 396, 1968.
13. **Dawson, R. M. C. and Kemp, P.,** *Biochem. J.,* 105, 837, 1967.
14. **Clarke, N. G. and Dawson, R. M. C.,** *Biochem. J.,* 195, 301, 1981.

Chapter 16

GLYCOGLYCEROLIPIDS*

W. Fischer

I. INTRODUCTION

Glycoglycerolipids are common constituents of photosynthetic membranes in higher plants and blue-green algae, are widespread among cytoplasmic membranes of Gram-positive bacteria and some other microorganisms, and have recently been discovered in the secretions of the human digestive tract. The higher homologs found in these secretions are on a borderline with lipoglycans, but are included in this chapter because they behave essentially as lipids during extraction and chromatography. Included also are sulfo- and phosphoglycoglycerolipids. Derivatives of glycoglycerolipids with polymeric hydrophilic chains such as the lipoteichoic acids of Gram-positive bacteria or the lipoglycans found in some microorganisms have been omitted because they are extracted and purified by different procedures. The structures of some representative compounds are shown in Schemes 1 to 4.

II. PREPARATION OF SAMPLES

A. Extraction

Most extraction procedures go back to the classical work of Folch et al.[1] in which chloroform was used as lipid solvent and methanol for breaking lipid-protein complexes. Animal tissues are first minced, leaves and plant roots are torn or cut into pieces, and then the material is homogenized in an adequate blender during extraction. Microorganisms are best extracted as wet cells; extraction of freeze-dried cells results in a lower yield.[65] Secretions from the digestive tract are dialyzed against water, freeze-dried, and then extracted.

All solvents should be freshly distilled and peroxide-free before use. With lipids containing highly unsaturated fatty acids, the solvents also should be deaerated by bubbling nitrogen through them, and all operations should be carried out under nitrogen.

The presence of degradative enzymes which often become active on homogenization must be considered. Plant leaves, for example, contain glycolipases,[2] phospholipases,[2] and transacylases that convert glycolipids into lyso-compounds and derivatives with three fatty acids.[3,4] Immediate addition of alcohol-containing solvents inactivates most of them, but dipping leaves into hot water is necessary to inactivate the stable phospholipase D.[2] With bacteria the formation of lyso-compounds and other artifacts can be avoided by rapidly cooling down the culture to 2°C, harvesting the cells by centrifugation at 2°C, and immediate extraction of the wet cell paste.[5] Phenol at a concentration of 1% (v/v) prevents, in addition, conversion of phosphatidyl glycerols to bisphosphatidyl glycerols which may occur in Gram-positive bacteria during harvesting.[65]

Procedure for the Extraction of Lipids[1]
Biological material is homogenized for a few minutes and/or stirred up to 24 hr in 20 volumes of chloroform-methanol (2:1 v/v), and then the homogenate is

---

* Symbols and shorthand designations: The structures of glycoglycerolipids and their derivatives are given by using the symbols recommended by the IUPAC-IUB Commission on Biochemical Nomenclature (see Chapter 1, Appendix). Unless otherwise stated, symbols for hexoses and their derivatives designate pyranose form. In the heads of the Tables the following shorthand designations for common solvents are used: A, acetone; C, chloroform; M, methanol.

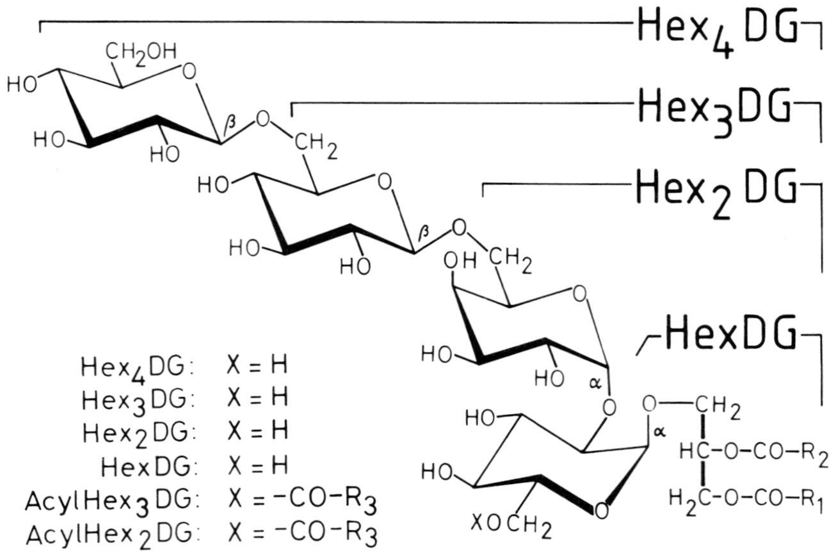

SCHEME 1. Glycoglycerolipids.

Note: As an example, the glycoglycerolipids which have been extracted from *Lactobacillus casei*[5] are shown. A two-dimensional thin-layer chromatogram of these lipids is depicted in Figure 3. DG, 1,2-di-*O*-acylglycerol; R, hydrocarbon chain (C_{16}-C_{18}).

filtered through a sintered glass funnel with suction. To remove water-soluble nonlipid contaminants, the crude extract is mixed with exactly two tenths of its volume of either water or an appropriate salt solution (0.7% NaCl, 0.9% KCl, 0.05% CaCl₂, or 0.04% MgCl₂), and the mixture is allowed to separate into two phases by standing or centrifugation. The upper phase is siphoned off as completely as possible, and remaining contaminants are removed by rinsing the walls of the vessel and the interface three times with small amounts of pure solvent upper phase in such a way as not to disturb the lower phase. The lower phase is taken to dryness in a rotary evaporator (30 to 35°C bath temperature), and the lipids are immediately dissolved in chloroform-methanol (2:1). The combined upper phases may be dialyzed or freeze-dried, and the residue extracted with chloroform-methanol (2:1 or 1:1) and tested for lipids by thin-layer chromatography (TLC).

Modifications — In many cases the extraction with chloroform-methanol has been repeated twice or three times. (For recovery, see References 1, 6, 7.) Since more polar neutral glycoglycerolipids are thought to escape into the water phase,[8] the phase partition was occasionally omitted and the nonlipid contaminants were removed by partition chromatography on Sephadex[9,10] of the crude lipid extract. However, the recovery of polar glycoglycerolipids has not been studied. It should be noted that the use of salts of mono- instead of divalent cations for phase partition causes a large loss of glycerophosphoglycolipids into the water phase.[65]

Procedure for the Extraction of Lipids[11-13]
To 100 g of fresh material, 300 mℓ of chloroform-methanol (1:2 v/v) is added. Animal tissues or plant leaves are homogenized for 2 min at room temperature; cell organelles or microorganisms are magnetically stirred for 30 min. After standing for a few hours with occasional mixing, extract is separated from the

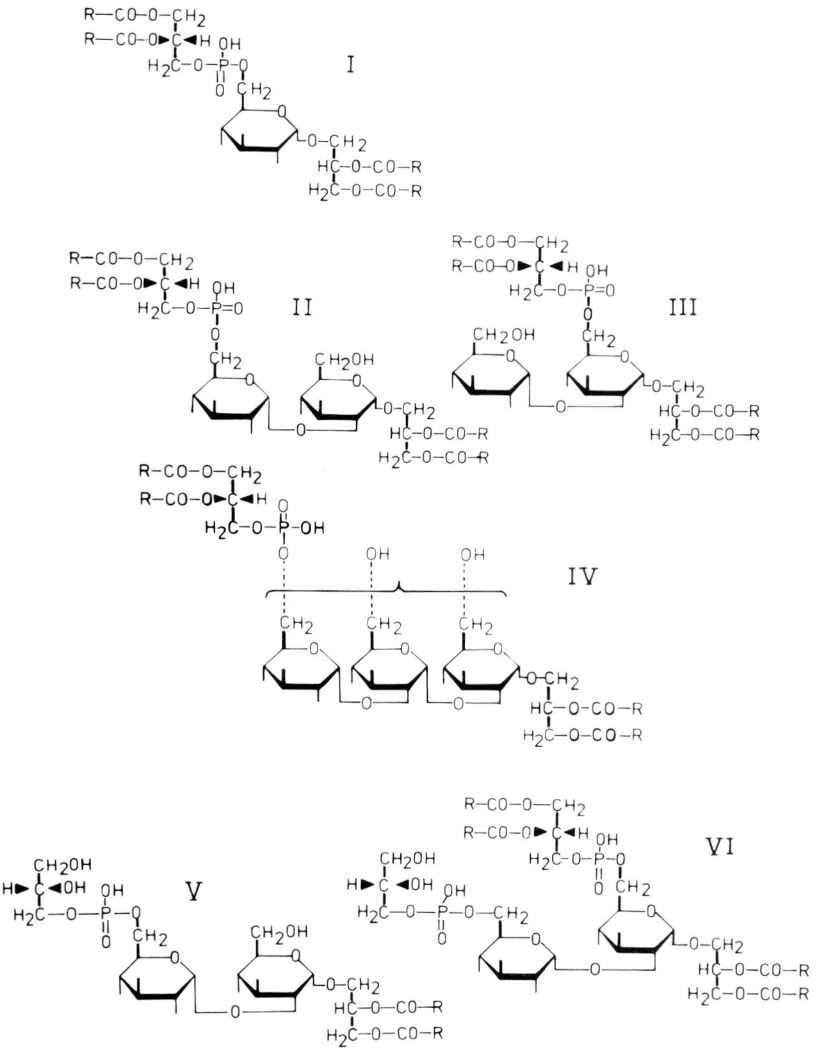

SCHEME 2. Phosphatidyl and glycerophosphoglycolipids from Streptococci.
Note: (I, Ptd-6Glc(α1-3)acyl$_2$Gro; II, Ptd-6Glc(α1-2)Glc(α1-3)acyl$_2$Gro; III, Glc(α1-2),Ptd-6Glc(α1-3)acyl$_2$Gro; IV, Ptd-[Glc(α1-2)Glc(α1-2)Glc(α1-3)acyl$_2$Gro]; V, GroP-6Glc(α1-2)Glc(α1-3)acyl$_2$Gro; VI, GroP-6Glc(α1-2),Ptd-6Glc(α1-3)acyl$_2$Gro. For structural analysis and occurrence of these compounds and their possible metabolic relationship to lipoteichoic acids, see References 42 and 42a.

cell residue by centrifugation or, in the case of leaves, by filtration, and transferred to a separatory funnel. The extraction of the residue is repeated with 380 mℓ of chloroform-methanol-water (1:2:0.8) and finally with 150 mℓ of chloroform-methanol (1:2). To the combined extracts, 250 mℓ of chloroform and 290 mℓ of water are added and, after shaking, the phases are allowed to separate. With smaller amounts of starting material, phase separation is performed by centrifugation. The lower chloroform phase is withdrawn, diluted with benzene, and brought to dryness in a rotary evaporator (30 to 35°C). For complete recovery, the water-methanol layer may be washed with the initial volume of a theoretical lower phase. The lipids are immediately dissolved in chloroform-methanol (2:1). The upper phase may be tested for lipids as described above.

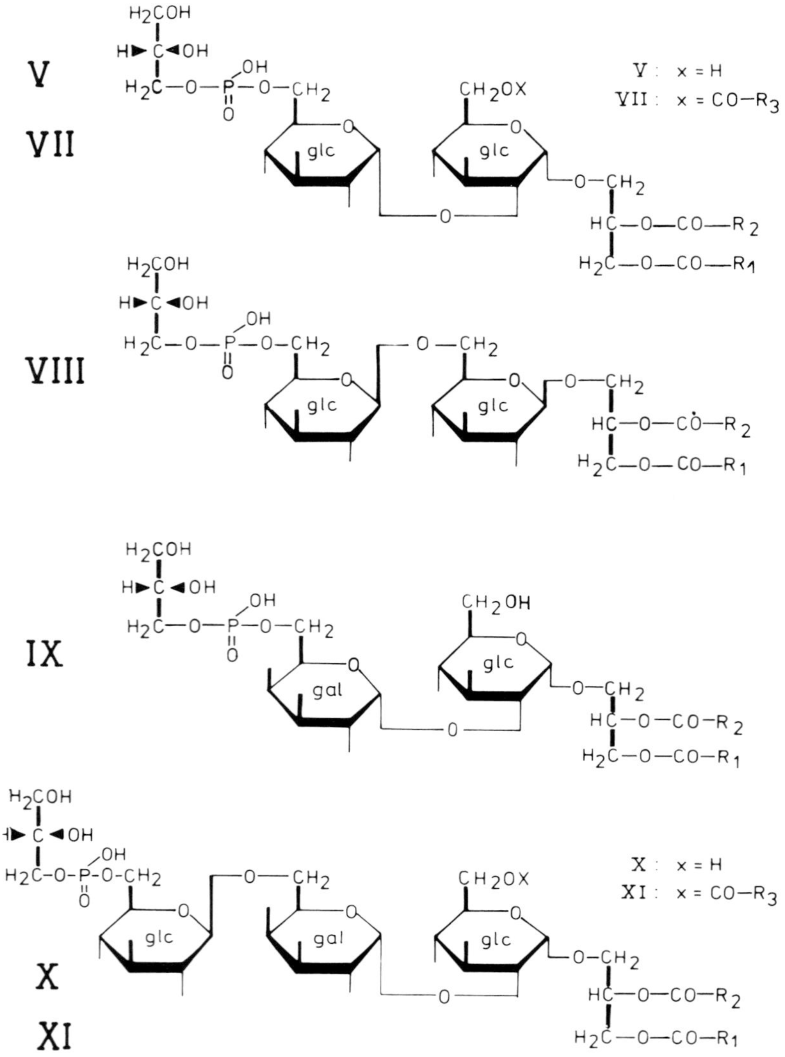

V

VII

$V: x = H$
$VII: x = CO—R_3$

VIII

IX

$X: x = H$
$XI: x = CO—R_3$

X

XI

SCHEME 3. Structural variations in bacterial glycerophosphoglycolipids.

Note: Compound V was isolated from *Streptococcus faecalis, S. hemolyticus, S. lactis, and S. group B;* VII from *S. lactis;* VIII from *Staphylococcus aureus, Bacillus licheniformis,* and *B. subtilis;* IX from *Lactobacillus plantarum;* X and XI from *L. casei.*[16,42a] Glycerophosphoglycolipids are derived from membrane glycoglycerolipids, which are characteristic for the bacteria indicated.

Modifications — Minor modifications for various purposes have been described in detail.[13] When the lipid extract contains polar glycophospholipids[14] or glycerophosphoglycolipids,[15-17] 0.2% $BaCl_2$ has been used for phase separation instead of water. For complete recovery of aminoacylphosphatidylglycerols, bacterial cells are extracted at pH 4.7 by addition of acetate buffer.[17a] It should be noted that tetrahexosyldiacylglycerol[5] and such polar compounds as di(glycerophospho)-glycolipids[17] or trigalactosylglycerophosphoglycolipids[65] are almost completely recovered in the chloroform layer. Lipoteichoic acid of Gram-positive bacterial membranes is not extracted by the foregoing procedure. It can be extracted from the defatted cell residues by phenol water.[18] Complete extraction of biophosphatidylglycerol from Gram-positive bacteria requires wall-depleted or disrupted cells.[18a,18b] Since during

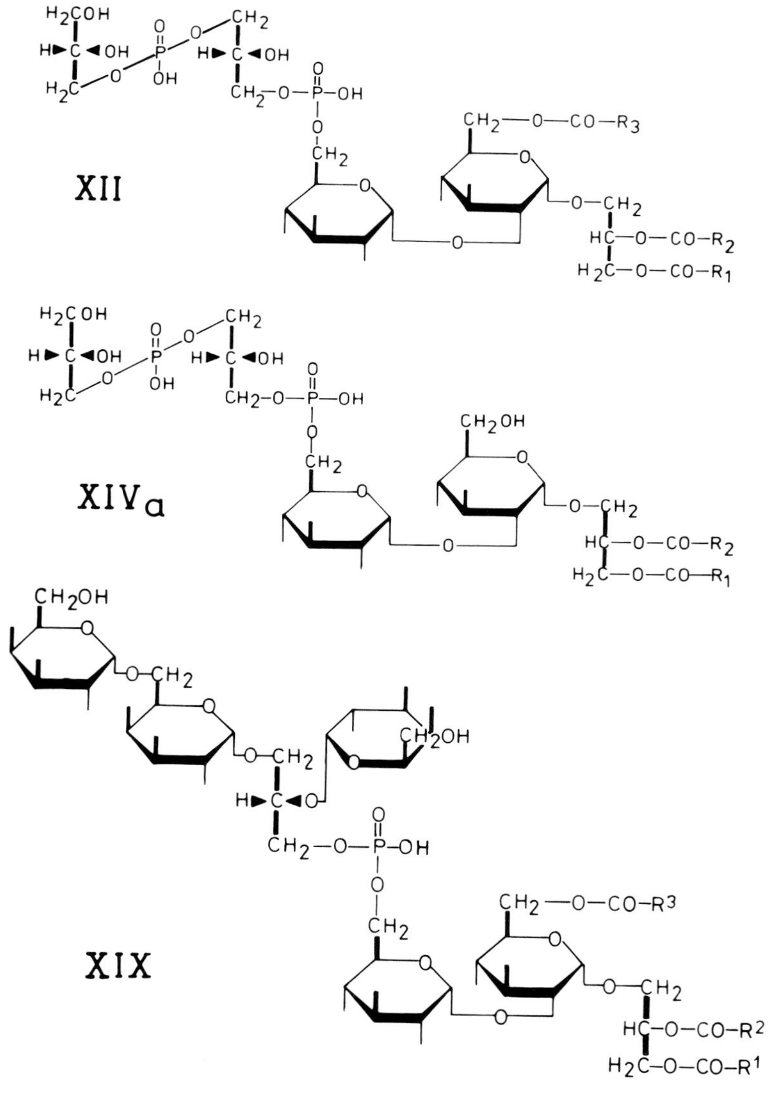

XII

XIV_a

XIX

SCHEME 4. Unusually polar glycerophosphoglycolipids.

Note: Thus far, these compounds have been detected in group N Streptococci.[17,42a] Isomeric mono- and digalactosyl derivatives of compound XIX have also been described.[66]

treatment with lysozyme phosphatidyglycerol may be converted into biphosphatidygly-cerol,[18a,65] mechanical disruption at low pH in the presence of 1% (w/v) phenol is preferable.[65]

Procedure for the Extraction of Lipids from Photosynthetic Tissue[13,19]
For leaves that contain rather stable degradative enzymes, the following proce-dure is recommended: 100 g of fresh leaves are cut and added portionwise to 300 mℓ hot isopropanol. After homogenization for 1 to 2 min, the hot homoge-nate is filtered with suction and the filter residue is washed with hot isopropanol (200 mℓ). The filter cake is then blended with 200 mℓ chloroform-isopropanol (1:1 v/v), the homogenate is filtered, and the filter residue is washed with chlo-roform-isopropanol (1:1) and finally with chloroform. The combined filtrates are concentrated in vacuo, the lipids dissolved in chloroform (200 mℓ), and the

solution is washed several times with 100 mℓ portions of water (or 1% NaCl solution). The chloroform layer is then diluted with benzene and concentrated to dryness in vacuo (30 to 35°C); the lipids are immediately dissolved in chloroform-methanol (2:1).

B. Solvent Fractionation

This method is based on the insolubility in cold acetone of most phospholipids, and on the fact that glycerides, pigments, and other neutral lipids are soluble in cold acetone. Some applications to the purification of glycoglycerolipids can be found in References 20 to 23. The procedure, however, is of limited value for this purpose since monohexosyldiacylglycerol is soluble in acetone, and dihexosyldiacylglycerol is recovered in both fractions,[20] whereas sulfoglycolipids[22] and oligohexosyldiacylglycerols[23] are precipitated together with phospholipids.

C. Countercurrent Distribution

Countercurrent distribution has been used occasionally in crude fractionation of lipid extracts from plant material (for a review, see Reference 12). This procedure permits superior fractionation of the nonpolar lipids, but this advantage is not sufficient to recommend its use generally prior to column chromatography.

III. CHROMATOGRAPHY

A. Column Chromatography (CC)

Column chromatography of lipids is performed as adsorption chromatography or as combined adsorption and ion exchange chromatography. For detailed reviews, see References 24 to 26.

A good adsorbent is magnesium silicate (Florisil), when used in the absolute absence of water; otherwise, adsorption chromatography changes partly into partition chromatography.[25] It has proved to be useful in the separation of glycolipids and sulfolipids from pigments and phospholipids in plant extracts (Table LC-1*). A more versatile adsorbent is silicic acid, which has been of great value in the purification of glycolipids when methanol-chloroform mixtures were replaced by acetone in chloroform.[27,28] As shown in Tables LC-2 and LC-3, glycoglycerolipids are eluted in the order of increasing polarity, whereas phospholipids are not eluted with chloroform-acetone and acetone.[28] The disadvantage in fractionating crude lipid extracts lies in the fact that huge volumes of acetone are necessary to elute plant sulfolipids (Table LC-1) or tri- and tetrahexosyldiacylglycerols (Tables LC-3, LC-4). However, the higher homologs are readily purified with a gradient of methanol in chloroform (Table LC-3), provided that glyco- and phospholipids have previously been separated by CC on DEAE-cellulose (see below). In the separation and purification of certain phosphoglycolipids, NaHCO$_3$-treated silicic acid proved to be superior to nontreated silicic acid.[7,29] Chromatography on totally porous silica spheres (Iatrobeads), which was successfully used in the purification of glycosphingolipids[30] and sulfolipids,[31] has thus far not been applied to glycoglycerolipids.

Chromatography on DEAE-cellulose (acetate form) combines adsorption chromatography, probably by hydrogen bonding, and anion exchange chromatography.[24-26] An example by Landgraf[7] is shown in detail in Figure 1; other examples are given in Tables LC-1 to LC-4. The least polar lipids (sterols, sterol esters, glycerides, hydrocarbons, pigments, etc.) are eluted with chloroform. Next, glycoglycerolipids up to tetra- and pentahexosyldiacylglycerols (Tables LC-3, LC-4) are eluted in the order of increasing polarity by a stepwise gradient of

---

*   All tables appear following the text.

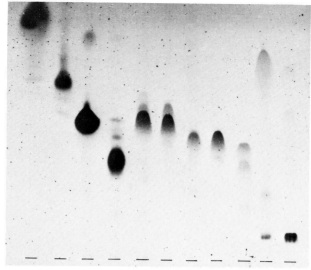

Fraction Nr	1	2	3	4	5	6	7	8	9	10	11
C/M (v/v)	1/0	97/3	9/1	3/1	2/1	2/1	2/1	2/1	2/1	2/1	2/1
NH_4OAc (g/ℓ)	–	–	–	–	0.1	0.1	0.5	0.5	1	2	4

FIGURE 1. Lipid composition of fractions from column chromatography on DEAE-cellulose, acetate form, of crude lipids from *Streptococcus hemolyticus* D-58. For lipid load and column dimensions see Table LC-2, column 4; fractions of 300 mℓ each were collected. The lipids of each fraction (1 to 11) are given below from top to bottom, with the main compound marked by an asterisk, and the lipid weight of the total fraction added in brackets: 1, Neutral lipids and pigments (700 mg); 2, GlcNAc(α1-3)acyl$_2$Gro (not stained) migrating with Glc(α1-3)acyl$_2$Gro* (40 mg); 3, Glc(α1-2)Glc(α1-3)acyl$_2$Gro* (350 mg); 4, Glc(α1-2)Glc(α1-2)Glc(α1-3) acyl$_2$Gro* (35 mg); 5, acylphosphatidylglycerol (trace), Ptd-6Glc(α1-3)acyl$_2$Gro, Ptd-6Glc(α1-2)Glc(α1-3)acyl$_2$Gro* (53 mg); 6, Ptd-6Glc(α1-3)acyl$_2$Gro, Ptd-6Glc(α1-2)Glc(α1-3)acyl$_2$Gro* (24 mg); 7, Glc(α1-2),Ptd-6Glc(α1-3)acyl$_2$Gro (7 mg); 8, Glc(α1-2),Ptd-6Glc(α1-3)acyl$_2$Gro*, phosphatidylglycerol (54 mg); 9, Glc(α1-2), Ptd-6Glc(α1-3)acyl$_2$Gro, Ptd-6Glc(α1-2)Glc(α1-2)Glc(α1-3)acyl$_2$Gro (24 mg); 10, bisphosphatidylglycerol*, GroP-6Glc(α1-2)Glc(α1-3)acyl$_2$Gro* (110 mg); 11, GroP-6Glc(α1-2)Glc(α1-3)acyl$_2$Gro (32 mg). Precoated silica gel plate (Woelm), developed in chloroform-methanol-water, (65:25:4). Visualization: α-naphthol/H_2SO_4 (Chapter 28, Reagent No. 22). (Courtesy of Landgraf, H. R., Thesis, University of Erlangen-Nuremberg, 1976. With permission.)

methanol in chloroform; larger glycolipids have been successfully separated by addition of small amounts of water.[32] If present, zwitterionic[12,25] and cationic phospholipids[14,33] are eluted between, or along with, glycolipids. On further elution with a continuous or stepwise gradient of ammonium acetate in chloroform-methanol (2:1), anionic lipids appear according to the number of negative charges[14,34,35] (Table LC-2) or, if equally charged as is the case with many phosphoglycolipids, according to their relative polarity (Figure 1, Tables LC-2, LC-3). These resolution properties together with the high load capacity make chromatography on DEAE-cellulose the most efficient procedure for separation of complex lipid mixtures. Some additional advantages have been recently reported for DEAE-Sephadex.[36]

To get pure individual lipids, more than one column is usually necessary. Possible combinations, used in purification of plant glycoglycerolipids, are found in Table LC-1. In our experience, the best combination is fractionation of the crude lipid extract on DEAE-cellulose or DEAE-Sephadex, and further purification on columns of silicic acid (Tables LC-2, LC-3). Minor components, enriched by chromatography on DEAE-cellulose, are more rapidly purified by preparative TLC.[7,16,17]

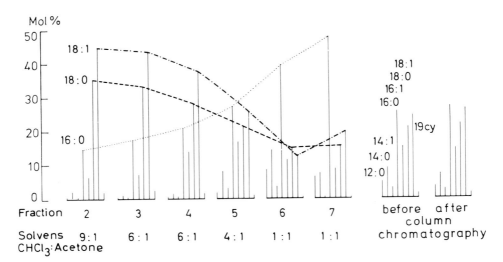

FIGURE 2.   Fractionation owing to different acyl chains of monoglucosyldiacylglycerols in silicic acid column chromatography. The solvent gradient used served to remove neutral lipids and pigments; 35 and 40% of the lipid was eluted in fraction 4 and 5, respectively. (Reproduced from Fischer, W., *Biochim. Biophys. Acta*, 487, 89, 1977. With permission.)

If the fatty acid composition is of interest, the individual lipids must be isolated as completely as possible. As shown in Figure 2 and Table LC-1, molecular species with longer hydrocarbon chains elute from silicic acid and DEAE-cellulose columns ahead of those with shorter chains. If, therefore, fractions with overlapping composition are discarded, the final fatty acid composition can be quite different from the original one.

Procedure for the Isolation of Glycoglycerolipids by Column Chromatography on DEAE-Cellulose[25,26]

The washing of DEAE-cellulose is performed conveniently in centrifuge beakers (250 mℓ) by gently stirring the slurry with a glass rod, and subsequent centrifugation and decantation. DEAE-cellulose (60 g) is washed first with 1 M HCl (800 mℓ) followed by water to neutral pH, and then 0.1 M KOH (800 mℓ) followed by washing with water. When the pH is neutral, the material is converted into the acetate form by stirring it in five volumes of acetic acid. Excess acid is removed by washing with methanol, and the material is then dried in vacuo over KOH.

For packing columns, the dried preparation (15 to 20 g) is left overnight in acetic acid, and remaining aggregates are very gently broken up with a glass rod. A glass tube (2.5 × 30 cm) equipped with a Teflon stopcock and a female joint at the top to take up a solvent reservoir (1 ℓ) is used. A small plug of glass wool is positioned at the bottom to retain the adsorbent. The slurry is poured in portions into the column and homogenously packed under a slight pressure of nitrogen. Then the bed is washed in sequence with 3 to 5 volumes each of methanol, chloroform-methanol (1:1), and chloroform. Uniformity of packing is tested with a few milligrams of azobenzene or azulene dissolved in chloroform, which should move down the column in an even band and elute between 70 and 90 mℓ of the effluent. If the performance is insufficient, repacking is best done in chloroform-methanol (1:1), followed by washing with chloroform. The lipid is dissolved in chloroform, washed into the column with this solvent, and the elution is performed at a flow rate of 2.5 to 3 mℓ/min. Examples for lipid load and elution schemes are given in Tables LC-1 to LC-4. In stepwise elution, 4 to 6

*bed volumes of each solvent should be used and collected in two fractions. Be-
fore changing the solvent, 0.5 mℓ of the effluent should be collected separately,*
concentrated, and its lipid composition tested by TLC on microslides. Fractions
containing no salt can be concentrated directly. Ammonium acetate is removed
previously by Folch partition[7,14] or dialysis,[35] or, if polar phosphoglycolipids
are present, by passage through a column of cation exchange resin (Merck S
1080), H+ form, and subsequent evaporation with several additions of carbon
tetrachloride.[16] The resultant acidic lipids are dissolved in chloroform-methanol
(2:1), excess ammonia is added, and the mixture is brought to dryness in a rotary
evaporator. Before storage, the ammonium salts are converted into the more
stable barium salts.[14]*

*If not allowed to run dry, a column can be used many times. Before reuse it is
washed in the following sequence with three bed volumes each of methanol,
acetic acid, methanol, chloroform-methanol (1:1), and chloroform.*

*Procedure for the Isolation of Glycoglycerolipids by Column Chromatography
on Silicic Acid*

*Silicic acid (Mallinckrodt, 100 mesh) is washed several times with water to re-
move fine particles. Then the adsorbent is washed twice with methanol and dried
at room temperature. Before use, it is activated at 110°C for several hours.
Other silicic acids, such as Unisil (100 to 200 or 200 to 300 mesh, Clarkson
Chemical Co.), Bio-Sil BH (100 to 200 or 200 to 325 mesh, Biorad Laboratories), or Absorbosil CAB (100 to 140 mesh, Applied Science Laboratories), may
be used without additional washing and other preliminary treatment except heat
activation if the adsorbent has been exposed to moisture.[13,25] A chromatography
tube (2.5 cm I.D.), equipped as described above, is used. A bed 5 cm high is
prepared by pouring a slurry (about 15 g silicic acid) in chloroform into the
tube, and the bed is washed with three column volumes of chloroform. The lipid
(100 to 200 mg) is usually applied in chloroform. Elution is accomplished at a
flow rate of 3 mℓ/min. Elution schemes can be adapted from the examples given
in Tables LC-1 to LC-4. In a stepwise procedure, eight to ten column volumes
of each solvent may be used and collected in two or three fractions. Control of
the effluent before changing the solvent by TLC on microslides is recommended.
With a continuous gradient smaller fractions are collected automatically.*

B. Paper Chromatography and Thin-Layer Chromatography

1. Native Lipids

Chromatography of native lipids on silicic acid-impregnated paper has been reviewed in
detail.[38] Resolution is, however, generally inferior to that obtained by TLC that has, therefore,
become the method of choice. The most important adsorbent used in the TLC of native
lipids is silicic acid or silica gel. Commercially available precoated glass plates or plastic
sheets, usually give better separations than plates coated by hand. There are, however, great
differences in resolution, R_f values, and spot forms on plates from various producers.

TLC is used for qualitative and quantitative analysis of crude lipid extracts (Figure 3),
monitoring column chromatography, preparative purposes, and testing the homogeneity of
isolated lipids. Another application, on silver nitrate-impregnated layers, concerns the separation of molecular species of isolated individual lipids according to the number of double
bonds in their long aliphatic chains.

Identification of glycoglycerolipids by TLC is only tentative. It must not be based on R_f

---

*   If the effluent contains ammonium acetate, 0.5 mℓ of chloroform-methanol (2:1) and 0.25 mℓ of 0.2% BaCl₂
in water are added. After mixing, the upper layer is discarded and the remainder is concentrated for TLC.

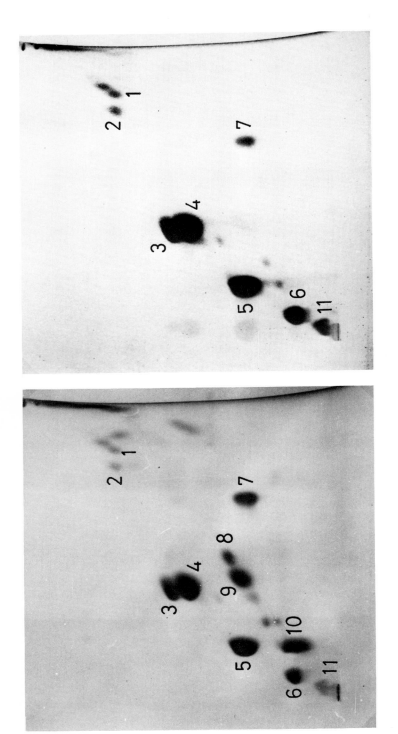

FIGURE 3.   Two-dimensional TLC of crude lipid extract from *Lactobacillus casei* on a precoated plate (silica gel 60, Merck). The first dimension (upward) was developed in chloroform-methanol-water (65:25:4), and the second dimension in chloroform-acetone-methanol-acetic acid-water (50:20:10:10:5). All lipids were stained with iodine vapor (left), then the glycolipids with α-naphthol/H$_2$SO$_4$ (right). 1, Glc(α1-3)acyl$_2$Gro; 2, Gal(α1-2),acyl-6Glc(α1-3)acyl$_2$Gro; 3, Glc(β1-6)Gal(α1-2),acyl-6Glc(α1-3)acyl$_2$Gro; 4, Gal(α1-2)Glc(α1-2)Glc(α1-2)Gal(α1-2)Glc(β1-6)Glc(β1-6)Gal(α1-2) Glc(α1-3)acyl$_2$Gro; 7, polar pigment; 8, bisphosphatidylglycerol; 9, phosphatidylglycerol; 10, lysylphosphatidylglycerol; 11, glycerophosphoglycolipids.

values, which may even vary on the same kind of plate in the same laboratory. Furthermore, glycoglycerolipids with the same structure differ in their R_f values, if their long chains are of different length.[16] Reference compounds are needed for identification, but even then isomeric glycolipids are difficult to differentiate and more than one solvent system (cf. TLC-1) or chromatography of the deacylation products (PC,TLC and GC-2) may be necessary. Isomers with glucosyl- and galactosyl-residues are readily resolved on borate-impregnated silica gel layers (TLC-1, column 4).

Procedure for the Two-Dimensional Thin-Layer Chromatography of Polar Lipids
Crude lipid extract (0.5 to 1 mg), dissolved in chloroform-methanol (2:1 v/v), is spotted with a microsyringe as a 5 to 8 mm long band at a corner of each of three chromatoplates (20 × 20 cm). HPTLC plates (Merck, Kieselgel 60, 10 × 10 cm) need less lipid extract, are developed in shorter time, and give better resolution. Reference compounds, chromatographed on the free edges of the plates in each dimension, facilitate identification.[7] Appropriate solvent combinations used for the development in the first and second dimension have been described for plant[25,39] and bacterial lipids.[7,14,17,29,40-42] The lipids on the three plates are first visualized with iodine vapor. After evaporation of iodine, one plate is stained for glycolipids with α-naphthol/H_2SO_4 reagent, the second plate is stained in sequence for amino- and phospholipids with ninhydrin and Dittmer Lester-reagent, respectively, and the third plate for 1,2-glycol group-containing lipids with the periodate Schiff reagent. Phosphoglycolipids may be identified by staining first with α-naphthol/H_2SO_4 and, after cooling, by the Dittmer Lester-reagent on the same plate.
Each staining is documented by photostatic copy or photography.

For comparative studies on the occurrence of glyco- and phosphoglycolipids in crude extracts from bacteria, one-dimensional TLC in an appropriate solvent system is satisfactory in combination with a staining procedure specific for carbohydrates.[5,15,29,43] Glyceroglycolipids from secretions of the digestive tract can be analyzed in the same way.[44]

Thin-layer chromatography can be carried out on microslides.[13] Commercially available plastic sheets precoated with silicic acid (Merck, Kieselgel 60) are cut into pieces of 2 × 9 cm. Lipids (5 nmol each) from column fractions and reference compounds are spotted as 3 mm bands. A small tank, as used for staining histological specimens, is satisfactory for development with a suitable solvent (about 10 min).

Procedure for the Quantitative Analysis of Complex Lipid Mixtures[25]
Each sample is spotted on four plates for two-dimensional TLC as described above. After development in an appropriate solvent combination, the lipids are stained faintly by iodine vapor, and the spots are encircled with a pencil. Then iodine is removed in a stream of air, the spots are wetted with water using a microsyringe, and they are scraped off into reaction vials. Lipids of larger spots are measured individually, minor components from two or four plates are combined. Acid hydrolysis or color reactions are conveniently performed in the presence of silica gel, which is afterwards removed by centrifugation.

Glycolipids can be measured directly by an anthrone procedure[45,46] or, after acid hydrolysis (2 M HCl, 100°C, 2.5 hr) and subsequent neutralization, with enzymes specific for one or more of the constituent monosaccharides.[14,15,47] Phosphoglycolipids like phospholipids have usually been measured[14,15] by a molybdenum blue procedure.[48]

On chromatoplates, quantitative analysis can be performed by charring the lipids and photodensitometric measurement.[25,49]

Procedure for the Isolation of Glycoglycerolipids by Thin-Layer Chromatography[13,14]

Dependent on the amount of lipid to be separated, normal (0.25 mm, Merck, Kieselgel 60) or preparative layers (2 mm, Merck, Kieselgel 60) are used. Before use, the plates are washed by chromatography in methanol-diethyl ether (75:25 v/v). The sample is applied as a 16 cm band with a microsyringe or a Pasteur pipet (tip drawn out and melted round). After development in a suitable solvent, excess solvent is removed in a stream of cold air or by flushing with nitrogen in a tank. If possible, the lipid bands are visualized by spraying with water. Otherwise, the central part of the plate is covered with a clean glass plate, and the edges of the bands are stained with iodine vapor. After marking the bands, iodine is removed in a stream of air, and the plate is wetted with water. If the fatty acid pattern is of interest, the iodine-stained edges are discarded. The damp adsorbent of identical bands is scraped into 30 mℓ screw-capped centrifuge tubes (Teflon lined cap), 25 mℓ of chloroform-methanol-water (1:2:0.8) is added, the mixture vortexed for 1 to 2 min, centrifuged, and the supernatant decanted into a 200 mℓ centrifuge beaker. The extraction is repeated twice more. To the combined extracts, 20 mℓ each of chloroform and water is added, and the extract is mixed and centrifuged. The chloroform layer is withdrawn, diluted with benzene, and brought to dryness in vacuo.

More polar glycerophosphoglycolipids have been successfully eluted as follows:[14,16,17] in a small chromatography column (1 to 1.5 cm I.D.) equipped with a Teflon stopcock, a plug of washed cotton wool is positioned on the bottom and overlaid with washed silicic acid (2 to 3 cm high) in chloroform. The area of adsorbent containing the spot is scraped into the column and lipids are eluted in sequence with chloroform-methanol-acetic acid (2:1:0.03), (1:1:0.02), and chloroform-methanol-water-acetic acid (65:25:4:1). The combined effluents are diluted with benzene and carbon tetrachloride and brought to dryness in a rotary evaporator. Anionic lipids are successively converted into ammonium and barium salts. After taking up in chloroform-methanol (1:1), residual silica gel is removed by centrifugation.

In every case, plates should be immediately developed after application of the sample, and the lipids eluted after development.

A procedure for the chromatography on thin layers of boric acid impregnated silica gel[50] is useful for separating all polyhydroxy compounds.

Molecular species of plant glycoglycerolipids have been separated on silica gel layers containing 10 to 20% silver nitrate. Plates may be prepared by hand, suspending silica gel H or G (with or without binder) in an aqueous solution of silver nitrate,[51,52] or brushing precoated silica gel plates with a 10% solution of silver nitrate in acetonitrile.[53] Before use, layers are activated at 110°C for 1 to 18 hr. For suitable solvents see TLC-3. Development at 7°C improves separation.[53]

2. Products of Deacylation

Partition chromatography of deacylated glycoglycerolipids and their derivatives is very efficient for identification since minor differences in the core structure become more apparent than on adsorption chromatography of the native lipids. We prefer one-dimensional TLC on silica gel layers, using two different solvent systems, which permits one to differentiate almost all naturally occurring glycosylglycerols (PC,TLC). Furthermore, on silica gel the compounds can be stained with reagents specific for carbohydrates, which cannot be used on cellulose. Paper chromatography, formerly employed for identification, is still applicable for separations on a preparative scale.

Procedure for the Deacylation of Lipids[54]

Lipids (10 to 20 μmol) are dissolved in 2 mℓ of chloroform, 0.5 mℓ of 0.4 M NaOH in methanol (prepared from 4 M NaOH in water) is added and, after vortexing, the mixture is allowed to stand at room temperature for 5 min. Then 4 mℓ of water is added, followed by portions of cation exchange resin (Merck S 1080), H⁺ form, until the water phase is slightly acidic, or strongly acidic in the case of glyco- and phosphoglycolipids, respectively. The upper layer is withdrawn, the organic layer washed twice with water, and the combined water phases, if acidic, are neutralized with ammonia and brought to dryness in a rotary evaporator. Glycosylglycerols may be stored in a dry form, and phosphodiester-containing compounds in water-methanol (1:1 v/v) at −20°C.

Procedure for the Deacylation of Lipids[55]

Lipid (up to 20 μmol) is dissolved in a screw-capped centrifuge tube in 0.2 mℓ of chloroform. Methanol (0.3 mℓ) and 0.5 mℓ 0.2 M KOH in methanol are added; after vortexing, the mixture is kept at room temperature for 15 min. Next, 0.2 mℓ of methanol, 0.8 mℓ chloroform and 0.9 mℓ of water are added and, after mixing, phase separation is done by centrifugation. The upper layer is transferred into another screw-capped centrifuge tube that contains about 0.5 mℓ of cation exchange resin (Merck, S 1080), NH₄⁺-form. After mixing, removal of sodium ions is tested with indicator paper, and judged as complete when the blue-green color disappears rapidly. The supernatant is withdrawn from the resin. The chloroform-layer in the first tube is twice washed with 1 mℓ of methanol-water (10:9), and the supernatants are used to wash the resin. The combined methanol-water mixtures are brought to dryness, and the deacylation products are stored as above. The chloroform layer contains the fatty acids as methyl esters, which can be directly used for analysis by GLC. This procedure gives incomplete deacylation of phosphodiester-containing compounds, if these are present as salts of divalent cations.[14]

The following reagents (see Chapter 28) are used for the detection of lipids on thin-layer chromatograms: chromic-sulfuric acid solution (Reagent No. 38) (charring of all organic compounds), iodine vapors[56] (Reagent No. 18) (all unsaturated lipids), α-naphthol-sulfuric acid[57] (Reagent No. 22) (glycolipids including phosphoglycolipids), ninhydrin[58] (Reagent No. 24) (lipids containing an amino group), molybdenum blue reagent[59] (Reagent No. 1) (phospholipids including glycoglycerolipids containing phosphate groups), and periodate-Schiff reagent[60] (Reagent No. 33) (compounds with 1.2-glycol groups, such as phosphatidylglycerol, glycerophosphoglycolipids, and deacylated glyceroglycolipids).

C. Gas Chromatography (GC)

1. Native Lipids

GC of individual glycoglycerolipids has been used to identify molecular species differing in the chain length of their constituent fatty acids (GC-1). Previous to analysis double bonds were hydrogenated using PtO_2 as a catalyst.

Procedure for the Preparation of Trimethylsilyl Ethers of Glycoglycero-lipids[61,62]

Hydrogenated galactolipid or sulfolipid methylester (1 mg) is dissolved in pyridine (50 μℓ) and N-trimethylsilylimidazole (25 μℓ) is added. After 30 min the mixture is diluted with ethyl benzene (50 μℓ), and portions of the solution are directly injected for GC.

Procedure for the Preparation of Methyl Esters of Sulfolipids[62]
*Before trimethylsilylation, hydrogenated sulfoquinovosyldiacylglycerol (100 mg)
is dissolved in chloroform-methanol (65:35) and slowly passed through a column
(1 × 10 cm) of Dowex 50 W × 4, H⁺ form, equilibrated with the same solvent
mixture. The eluate is treated with diazomethane, and complete esterification is
checked by TLC on silica gel in chloroform-methanol (9:1).*

2. Products of Deacylation*

GC in combination with mass spectral analysis proved to be useful in the identification
of desulfated and deacylated ether-containing seminolipid (GC-1).

Like TLC, GC analysis of deacylation products is an efficient procedure for differentiating
isomeric glycoglycerolipids. Trimethylsilyl and trifluoroacetyl derivatives have been used,
the former giving a superior resolution with monohexosyl-, and the latter with trihexosyl-
glycerols. Members of a homologous series when present in a mixture are readily quantitated
as trifluoroacetates at a programmed rise of temperature[63] (GC-2, last column).

Procedure for the Preparation of Trimethylsilyl Ethers of Glycosylgylcerols[14]
*Deacylation products (100 nmol) are dried over P_2O_5 in vacuo. Pyridine, 5 μℓ,
and trimethylchlorosilane, 5 μℓ, and 20 μℓ of N,O-bis(trimethylsilyl)-trifluo-
roacetamide are added. After standing at room temperature for a few minutes,
the mixture is directly injected.*

*Procedure for the Preparation of Trifluoroacetyl Derivatives of Glycosyl-
glycerols*[64]
*Deacylation product (100 nmol) is dried as above in a microvial fitted with a
screw-cap. Trifluoroacetic anhydride, 60 μℓ, containing 5 μℓ pyridine per mil-
liliter is added, and the mixture allowed to stand at room temperature overnight
before GC analysis.*

ACKNOWLEDGMENT

The comparative study by the author on the chromatographic behavior of native and
deacylated glycoglycerolipids became possible through the kind gift of plant glycolipids by
Professor E. Heinz (Köln); of synthetic Gal(α1-6)Gal(α1-6)Gal(β1-3)acyl₂Gro by Dr. R.
Gigg (London); and the isolation of bacterial glycolipids and related compounds by Drs. H.
U. Koch, R. A. Laine, H. R. Landgraf, M. Nakano, and D. Schuster in this laboratory.
This work was supported by the Deutsche Forschungsgemeinschaft.

REFERENCES

1. **Folch, J., Lees, M., and Sloane-Stanley, G. H.,** *J. Biol. Chem.,* 226, 497, 1957.
2. **Kates, M.,** *Adv. Lipid Res.,* 8, 225, 1970.
3. **Critchley, C. and Heinz, E.,** *Biochim. Biophys. Acta,* 326, 184, 1973.
4. **Heinz, E., Bertrams, M., Joyard, J., and Douce, R.,** *Z. Pflanzenphysiol.,* 87, 325, 1978.
5. **Nakano, M. and Fischer, W.,** *Hoppe Seyler's Z. Physiol. Chem.,* 358, 1439, 1977.
6. **Prottey, C. and Ballou, C. E.,** *J. Biol. Chem.,* 243, 6196, 1968.
7. **Landgraf, H. R.,** Thesis, University of Erlangen-Nuremberg, 1976.
8. **Shaw, N.,** *Bacteriol. Rev.,* 34, 365, 1970.

---

* For deacylation, see procedures in Chapter 15.

9. **Wells, M. A. and Dittmer, J. C.,** *Biochemistry,* 2, 1259, 1963.
10. **Siakotos, A. N. and Rouser, G.,** *J. Am. Oil Chem. Soc.,* 42, 913, 1965.
11. **Bligh, E. G. and Dyer, W. J.,** *Can. J. Biochem. Physiol.,* 37, 911, 1959.
12. **Allen, C. F., Good, P., Davis, H. F., Chisum, P., and Fowler, S. D.,** *J. Am. Oil Chem., Soc.,* 43, 223, 1966.
13. **Kates, M.,** *Techniques of Lipidology — Laboratory Techniques in Biochemistry and Molecular Biology,* Vol. 3, (Part 2), Work, T. S. and Work, E., Eds., North-Holland, Amsterdam, 1972.
14. **Fischer, W.,** *Biochem. Biophys. Acta,* 487, 74, 1977.
15. **Fischer, W., Nakano, M., Laine, R. A., and Bohrer, W.,** *Biochim. Biophys. Acta,* 528, 288, 1978.
16. **Fischer, W., Laine, R. A., and Nakano, M.,** *Biochim. Biophys. Acta,* 528, 298, 1978.
17. **Laine, R. A. and Fischer, W.,** *Biochim. Biophys. Acta,* 529, 250, 1978.
17a. **Houtsmiller, U. M. T. and Van Deenen, L. L. M.,** *Biochim. Biophys. Acta,* 106, 564, 1965.
18. **Nakano, M. and Fischer, W.,** *Hoppe Seyler's Z. Physiol. Chem.,* 359, 1, 1978.
18a. **Bertsch, L. L., Bonsen, P. P. M., and Kornberg, A.,** *J. Bacteriol.,* 98, 75, 1969.
18b. **Filgueiras, M. H. and Op den Kamp, J. A. F.,** *Biochim. Biophys. Acta,* 620, 332, 1980.
19. **Kates, M. and Eberhardt, F. M.,** *Can. J. Bot.,* 35, 895, 1957.
20. **Sastry, P. S. and Kates, M.,** *Biochemistry,* 3, 1271, 1964.
21. **Weenink, R. O.,** *Biochem. J.,* 93, 606, 1964.
22. **Ishizuka, I., Suzuki, M., and Yamakawa, T.,** *J. Biochem.,* 73, 77, 1973.
23. **Oshima, M. and Yamakawa, T.,** *Biochemistry,* 13, 1140, 1974.
24. **Rouser, G., Kritchevsky, G., Heller, D., and Lieber, E.,** *J. Am. Oil Chem. Soc.,* 40, 425, 1963.
25. **Rouser, G., Kritchevsky, G., and Yamamoto, A.,** *Lipid Chromatographic Analysis,* Vol. 1, Marinetti, G. V., Ed., Marcel Dekker, New York, 1967, 99.
26. **Rouser, G., Kritchevsky, G., Yamamoto, A., Simon, G., Galli, C., and Baumann, A. J.,** *Methods Enzymol.,* 14, 272, 1969.
27. **Smith, L. M. and Freeman, N. K.,** *J. Dairy Sci.,* 42, 150, 1959.
28. **Vorbeck, M. L. and Marinetti, G. V.,** *J. Lipid Res.,* 6, 3, 1965.
29. **Fischer, W. and Landgraf, H. R.,** *Biochim. Biophys. Acta,* 380, 227, 1975.
30. **Ando, S., Isobe, M., and Nagai, Y.,** *Biochim. Biophys. Acta,* 424, 98, 1976.
31. **Ishizuka, I., Inomata, M., Ueno, K., and Yamakawa, T.,** *J. Biol. Chem.,* 253, 898, 1978.
32. **Stettner, K., Watanabe, K., and Hakomori, S. I.,** *Biochemistry,* 12, 656, 1973.
33. **Schmid, R. D.,** Thesis, University of Erlangen-Nuremberg, 1971.
34. **Hendrickson, H. S. and Ballou, C. E.,** *J. Biol. Chem.,* 239, 1369, 1964.
35. **Olson, R. W. and Ballou, C. E.,** *J. Biol. Chem.,* 246, 3305, 1971.
36. **Momoi, T., Ando, S., and Nagai, Y.,** *Biochim. Biophys. Acta,* 441, 488, 1976.
37. **Fischer, W.,** *Biochim. Biophys. Acta,* 487, 89, 1977.
38. **Kates, M.,** *Lipid Chromatographic Analysis,* Vol. 1, Marinetti, G. V., Ed., Marcel Dekker, New York, 1967, 1.
39. **Nichols, B. W.,** *New Biochemical Separations,* James, A. T. and Morris, L. J., Eds., Van Nostrand, New York, 1969, 334.
40. **Komaratat, P. and Kates, M.,** *Biochim. Biophys. Acta,* 398, 464, 1975.
41. **Minnikin, D. E. and Abdolrahimazadeh, H.,** *J. Bacteriol.,* 120, 999, 1974.
42. **Fischer, W.,** *Lipids,* Vol. 1, Paoletti, R., Porcellati, G., and Jacini, G., Eds., Raven Press, New York, 1976, 255.
42a. **Fischer, W.,** *Chemistry and Biological Activities of Bacterial Surface Amphiphiles,* Shockman, G. D. and Wicken, A. J., Eds., Academic Press, New York, 1981, 209.
43. **Fischer, W., Laine, R. A., Nakano, M., Schuster, D., and Egge, H.,** *Chem. Phys. Lipids,* 21, 103, 1978.
44. **Slomiany, B. L., Slomiany, A., and Glass, G. B. J.,** *Eur. J. Biochem.,* 84, 53, 1978.
45. **Jatzkewitz, H.,** *Hoppe Seyler's Z. Physiol. Chem.,* 336, 25, 1965.
46. **Heinz, E.,** *Biochim. Biophys. Acta,* 144, 333, 1967.
47. **Ueno, K., Ishizuka, I., and Yamakawa, T.,** *Biochim. Biophys. Acta,* 487, 61, 1977.
48. **Rouser, G., Fleischer, S., and Yamamoto, A.,** *Lipids,* 5, 494, 1970.
49. **Renkonen, O. and Varo, P.,** *Lipid Chromatographic Analysis,* Vol. 1, Marinetti, G. V., Ed., Marcel Dekker, New York, 1967, 42.
50. **Kean, E. L.,** *J. Lipid Res.,* 7, 449, 1966.
51. **Nichols, B. W. and Moorhouse, R.,** *Lipids,* 4, 311, 1969.
52. **Heinz, E. and Haarwood, J. L.,** *Hoppe Seyler's Z. Physiol. Chem.,* 358, 897, 1977.
53. **Siebertz, H. P. and Heinz, E.,** *Z. Naturforsch.,* 32c, 193, 1977.
54. **Ballou, C. E., Vilkas, E., and Lederer, E.,** *J. Biol. Chem.,* 238, 69, 1963.
55. **Marshall, M. O. and Kates, M.,** *Biochim. Biophys. Acta,* 260, 558, 1972.
56. **Mangold, H. K. and Malins, D. C.,** *J. Am. Oil Chem. Soc.,* 37, 383, 1960.

57. **Jacin, H. and Mischkin, A. R.,** *J. Chromatogr.,* 18, 170, 1965.
58. **Marinetti, G. V.,** *New Biochemical Separations,* James, A. T. and Morris, L. J., Eds., Van Nostrand, New York, 1969, 339.
59. **Dittmer, J. C. and Lester, R. L.,** *J. Lipid Res.,* 5, 126, 1964.
60. **Shaw, N.,** *Biochim. Biophys. Acta,* 164, 435, 1968.
61. **Auling, G., Heinz, E., and Tulloch, A. P.,** *Hoppe Seyler's Z. Physiol. Chem.,* 352, 905, 1971.
62. **Tulloch, A. P., Heinz, E., and Fischer, W.,** *Hoppe Seyler's Z. Physiol. Chem.,* 354, 879, 1973.
63. **Koch, H. U. and Fischer, W.,** *Biochemistry,* 17, 5275, 1978.
64. **Shapira, J.,** *Nature (London),* 222, 792, 1969.
65. **Fischer, W.,** unpublished observations.
66. **Fischer, W., Schuster, D., and Laine, R. A.,** *Biochim. Biophys. Acta,* 575, 319, 1979.

GENERAL REFERENCES AND BOOKS

Kates, M., Plant phospholipids and glycolipids, *Adv. Lipid Res.,* 8, 225, 1970.

Shaw, N., Bacterial glycolipids, *Bacteriol. Rev.,* 34, 365, 1970.

Sastry, P. S., Glycosyl glycerides, *Adv. Lipid Res.,* 12, 251, 1974.

Hakomori, S.-I. and Ishizuka,I., *Glycolipids — Handbook of Biochemistry and Molecular Biology,* 3rd ed., Fasman, G. D., Ed., *Lipids, Carbohydrates and Steroids,* CRC Press, Boca Raton, Fla., 1975, 416.

Shaw, N., Bacterial glycolipids and glycophospholipids, *Adv. Microbial Physiol.,* 12, 141, 1975.

Fischer, W., Glycerophosphoglycolipids, presumptive biosynthetic precursors of lipoteichoic acids, in *Chemistry and Biological Activities of Bacterial Surface Amphiphiles,* Shockman, G. D. and Wicken, A. J., Eds., Academic Press, New York, 1981, 209.

Marinetti, G. V., Ed., *Lipid Chromatographic Analysis,* Vol. 1, Marcel Dekker, New York, 1967; 2nd ed., 1976.

Lowenstein, J. M., Ed., *Lipids — Methods in Enzymology,* Vol. 14, Colowick, S. P. and Kaplan, N. O., Eds., Academic Press, New York, 1969.

Kates, M., *Techniques of Lipidology — Laboratory Techniques in Biochemistry and Molecular Biology,* Vol. 3 (Part 2), Work, T. S. and Work, E., Eds., North-Holland, Amsterdam, 1972.

Table 1
FRACTIONATION OF INDIVIDUAL LIPIDS BY COLUMN CHROMATOGRAPHY ACCORDING TO THE CHAIN LENGTH OF THE CONSTITUENT FATTY ACIDS

	12:0	14:0	14:1	16:0	16:1	18:0	18:1	19-Cyclopropane-carboxylic acid
Diglucosyldiacylglycerols								
Fraction 1 (71 mg)	1.6	8.6	1.2	33.5	17.5	6.0	30.2	1.3
Fraction 2 (64 mg)	9.9	17.3	n.d.	38.9	17.5	3.0	13.0	n.d.
Phosphatidylglycerols								
Fraction 1 (238 mg)	—	1.0	—	21.1	13.0	24.7	31.4	8.7
Fraction 2 (78 mg)	1.2	5.8	1.6	32.0	18.1	16.9	18.5	5.8

Note: Diglucosyldiacylglycerols from *Streptococcus* B 20789 were eluted from a column of DEAE-cellulose, acetate form, in two fractions with chloroform-methanol (9:1 v/v). Phosphatidylglycerols (barium salt) from *Streptococcus* B 090 were eluted from a silica gel column with chloroform-methanol (6:1 and 4:1, v/v). The lipids were proven pure by TLC. Values of fatty acids are given as mol percent. n.d., not determined.

From Fischer, W., *Biochim. Biophys. Acta,* 487, 89, 1977. With permission.

Table 2
GLYCEROGLYCOLIPIDS (PLANTS) (LC-1)

	P1	P2	P3	P2	P4	P2	P5	P5
Packing								
Weight (g)	—	—	—	15	15	—	—	—
Column								
Length (cm)	15	25	15	18	5	20	—	—
Diameter (cm)	4.5	4.5	2.5	2.5	2.5	2.5	—	—
Lipid								
Load (g)	1.0	—	—	0.5	0.1	—	—	—
Pretreatment	T1	T2	T3	T1	T1	T4	T1	T1
Elution	E1	—	—	E2	E3	E4	E3	E3
Detection	D1	D1	D1	D2	D3	—	—	—
References	1	1	1	2	3	4	5	6

Compound	C:M	C:M:NH$_4$OH[a]	C:M	C:M:NH$_4$OAc[b]	C:A	C:M	C:A	C:A
Gal (β1-3) acyl$_2$Gro	2:1	2:1:0	9:1	98:2:0	1:1	98:2	75:25	—
Gal (α1-6) Gal (β1-3) acyl$_2$Gro	2:1	2:1:0	4:1	9:1:0	0:1	9:1	—	—
HO$_3$S-6Qui (α1-3) acyl$_2$Gro	2:1	4:1:0.2	—	2:1:<5	0:1	—	—	—
Acyl-6Gal (β1-3) acyl$_2$Gro	—	—	—	—	—	4:1/NH$_4$-salt[c]	9:1	—
Acyl-6Gal (β1-3) acylGro	—	—	—	—	—	—	75:25	—
Acyl-6Gal (α1-6) Gal (β1-3) acyl$_2$Gro	—	—	—	—	—	—	—	From 55:45 to 2:8

[a] NH$_4$OH, concentrated.

[b] NH$_4$OAc, g/ℓ.

[c] NH$_4$-salt, 0.01 or 0.05 *M* ammonium or potassium acetate and 20 mℓ 28% (w/w) aqueous ammonia per liter.

Packing		
P1	=	Florisil (60-100 mesh)
P2	=	DEAE cellulose (Standard Selectacel), acetate form
P3	=	Silicic acid (Mallinckrodt, 120 mesh)
P4	=	Unisil silicic acid (100-200 mesh)
P5	=	Silicic acid

Pretreatment		
T1	=	Crude lipid extract
T2	=	Separated from neutral lipids, chlorophylls and phospholipids on Florisil (Column 1)
T3	=	Separated from sulfolipid on DEAE cellulose (Column 2)
T4	=	Separated from neutral lipids, chlorophylls and phospholipids on silicic acid

Elution (v/v)   E1 = Chloroform-methanol, containing 5% 2,2-dimethoxypropane, stepwise
   E2 = Chloroform-methanol, stepwise; ammonium acetate in chloroform-methanol (2:1) gradient from 0 to 5 g/ℓ
   E3 = Chloroform; chloroform-acetone, stepwise
   E4 = Chloroform-methanol-NH$_3$-salt, stepwise

Detection   D1 = By PC
   D2 = By TLC
   D3 = By weight

REFERENCES

1. **O'Brien, J. S. and Benson, A. A.,** *J. Lipid Res.,* 5, 432, 1964.
2. **Allen, C. F., Good, P., Davis, H. F., and Chisum, P.,** *J. Am. Oil Chem. Soc.,* 43, 223, 1966.
3. **Rouser, G., Kritchevsky, G., Simon, G., and Nelson, G. J.,** *Lipids,* 2, 37, 1967.
4. **Rouser, G., Kritchevsky, G., and Yamamoto, A.,** in *Lipid Chromatographic Analysis,* Vol. 1, Marinetti, G. V., Ed., Van Nostrand, New York, 1967, 99.
5. **Critchley, C. and Heinz, E.,** *Biochim. Biophys. Acta,* 326, 184, 1973.
6. **Heinz, E., Rullkötter, J., and Budzikiewicz, H.,** *Hoppe Seyler's Z. Physiol. Chem.,* 335, 612, 1974.

Table 3
GLYCOGLYCEROLIPIDS (STREPTOCOCCI) (LC-2)

	P1	P2	P3	P4	P4
Packing					
Weight (g)	10	—	—	15	15
Column					
Length (cm)	12			19	18
Diameter (cm)	1.5			2.5	2.5
Lipid					
Load	500 μmol P			1.5 g	100 mg
Pretreatment	T1	T1	T2	T1	T3
Elution	E1	E1	E2	E3	E4
Detection	D1	D1	D2	D3	D3
References	1	2	3	4	4

Compound	C:A	C:A	C:A	C:M:NH$_4$OAc[a]	C:M:NH$_4$OAc[a]
Glc(α1-3)acyl$_2$Gro	1:1	1:1	3:7	97:3:0	—
GlcNAc(α1-3)acyl$_2$Gro	—	1:1	0:1	97:3:0	—
Glc(α1-2)Glc(α1-3)acyl$_2$Gro	0:1	0:1[b]	—	9:1:0	—
Glc(α1-2)Glc(α1-2)Glc(α1-3)acyl$_2$Gro	—	0:1[b]	—	3:1:0	—
Ptd-6Glc(α1-3)acyl$_2$Gro	—	—	—	2:1:0.1	98:2:1[c]
Ptd-6Glc(α1-2)Glc(α)1-3)acyl$_2$Gro	—	—	—	2:1:0.1	98:2:1[c]
Glc(α1-2)-,Ptd-6Glc(α1-3)acyl$_2$Gro	—	—	—	2:1:0.5	—
Ptd-6Glc(α1-2)Glc(α1-2)Glc(α1-3)acyl$_2$Gro	—	—	—	2:1:1	—
GroP-6Glc(α1-2)Glc(α1-3)acyl$_2$Gro	—	—	—	2:1:2-4	—
(GroP)$_2$-6Glc(α1-2),acyl-6Glc(α1-3) acyl$_2$Gro	—	—	—	2:1:10	—
(GroP)$_2$-6Glc(α1-2)Glc(α1-3)acyl$_2$Gro	—	—	—	2:1:20	—

[a] NH$_4$OAc, g/ℓ.
[b] Elute separately.
[c] Elute separately (column volumes 21-25, 26-30, respectively).

Packing P1 = Unisil silicic acid, 100-200 mesh
         P2 = Silicic acid (Mallinckrodt), 100 mesh
         P3 = Kanto-Kagaku silicic acid
         P4 = DEAE-cellulose (Merck), acetate form

Pretreatment   T1 = Crude lipid extract
               T2 = Fraction from chromatography on silicic acid, Mallinckrodt (lane 2)
               T3 = Fraction from chromatography on DEAE cellulose, Merck, acetate form (lane 4)

Elution   E1 = Chloroform; chloroform-acetone, stepwise
               E2 = Chloroform-acetone, stepwise
               E3 = Chloroform; chloroform-methanol-ammonium acetate, stepwise.
               E4 = Chloroform -methanol-ammonium acetate, without gradient

Detection   D1 = Anthrone sulfuric acid procedure, TLC
               D2 = Anthrone sulfuric acid procedure and quantitation of hexosamine
               D3 = By TLC

REFERENCES

1. **Vorbeck, M. L. and Marinetti, G. V.**, *J. Lipid Res.* 6, 3, 1965; **Fischer, W. and Seyferth, W.**, *Hoppe Seyler's Z. Physiol. Chem.*, 349, 1662, 1968.
2. **Ishizuka, I. and Yamakawa, T.**, *J. Biochem.*, 64, 13, 1968.
3. **Ishizuka, I. and Yamakawa, T.**, *Jpn. J. Exp. Med.*, 39, 321, 1969.
4. **Landgraf, H. R.**, Thesis, University of Erlangen-Nuernberg, 1976; **Laine, R. A. and Fischer, W.**, *Biochem. Biophys. Acta*, 529, 250, 1978.

Table 4
GLYCOGLYCEROLIPIDS (LACTOBACILLI) (LC-3)

	P1	P2	P2	P2	P2
Packing					
Weight (g)	25	25	30	25	50
Column					
Length (cm)	32	8	10	8	17
Diameter (cm)	2.5	2.5	2.5	2.5	2.5
Lipid					
Load	1.5 g	162 mg	145 mg	127 mg	300 mg
Pretreatment	T1	T2	T2	T2	T2
Elution	E1	E2	E3	E2	E2
Detection	D1	D1	D1	D1	D1
Reference	1, 2	1	1	1	1

Compound	**C:M:NH$_4$OAc**[a]	**C:A**	**C:M**	**C:A**	**C:A**
Glc(α1-3)acyl$_2$Gro	97:3:0	—	—	3:1	—
Gal(α1-2)-,acyl-6Glc(α1-3)acyl$_2$Gro	97:3:0	—	—	1:1	—
Glc(β1-6)Gal(α1-2)-,acyl-6Glc(α1-3)acyl$_2$Gro	9:1:0	—	—	—	1:2
Gal(α1-2)Glc(α1-3)acyl$_2$Gro	9:1:0	0:1[b]	9:1	—	0:1
Glc(β1-6)Gal(α1-2)Glc(α1-3)acyl$_2$Gro	3:1:0	0:1[b]	5:1	—	—
Glc(β1-6)Glc(β1-6)Gal(α1-2)Glc(α1-3)acyl$_2$Gro	2:1:0	0:1[b]	4:1	—	—
GroP-6Glc(β1-6)Gal(α1-2)-,acyl-6Glc(α1-3)acyl$_2$Gro	2:1:4	—	—	—	—
GroP-6Glc(β1-6)Gal(α1-2)Glc(α1-3)acyl$_2$Gro	2:1:10	—	—	—	—

[a]   NH$_4$OAc, g/ℓ.

[b]   Elute separately, but for the elution of Hex$_4$acyl$_2$Gro about 55 column volumes (1600 mℓ) are necessary.

Packing	P1 =	DEAE-cellulose (Merck), acetate form
	P2 =	Silicic acid (Mallinckrodt), 100 mesh, washed
Pretreatment	T1 =	Crude lipid extract.
	T2 =	Fractions from DEAE-cellulose (column 1)
Elution	E1 =	Chloroform; chloroform-methanol-ammonium acetate, stepwise
	E2 =	Chloroform-acetone, stepwise
	E3 =	Chloroform-methanol, stepwise
Detection	D1 =	By TLC

REFERENCES

1. **Nakano, M. and Fischer, W.,** *Hoppe Seyler's Z. Physiol. Chem.,* 358, 1439, 1977, and unpublished data.
2. **Fischer, W., Laine, R. A., and Nakano, M.,** *Biochim. Biophys. Acta,* 528, 298, 1978, and unpublished data.

Table 5
GLYCOGLYCEROLIPIDS (VARIOUS MICROORGANISMS) (LC-4)

Packing	P1	P2	P1	P3	P4	P5
Column						
Length (cm)	—	4.5	18	—	—	22
Diameter (cm)	—	1.2	1.8	—	—	2.5
Lipid						
Load (mg)	336	1.6	250	—	—	—
Pretreatment	T1	T2	T3	T4	T5	T6
Elution	E1	—	E2	E3	E4	—
Detection	D1	D1	D1	D1	D1	—
Reference	1	2	3	4	4	5

Compound	C:M	C:M:NH$_4$OAc[a]	C:A	C:M	C:M	C:M
Glc(α1-3)acyl$_2$Gro	97:3	7:3:0	—	—	—	—
GlcUA(α1-3)acyl$_2$Gro	97:3	7:3:6	—	—	—	—
Ptd-6Glc(α1-3)acyl$_2$Gro	97:3	—	—	—	—	—
Glc(β1-4)GlcUA(α1-3)acyl$_4$Gro	9:1	—	—	—	—	—
Acyl$_2$Gal-acyl$_2$Gro, acylGal-acyl$_2$Gro, Gal-acyl$_2$Gro	—	—	3:1	—	—	—
AcylGal$_2$-acyl$_2$Gro, Gal$_2$-acyl$_2$Gro, Gal-acylGro	—	—	1:1	—	—	—
Gal$_2$-acyl$_2$Gro	—	—	1:2	—	—	—
Gal$_2$-acylGro, Gal$_3$-acyl$_2$Gro	—	—	0:1	—	—	—
Galf(β1-2)Gal(α1-6)GlcNAcyl-(β1-2)Glc(α1-3)acyl$_2$Gro	—	—	—	From 72:28 to 64:36[b]	7:3[c]	—
Gal(α1-2)Gal(α1-3)mannoheptosyl-(β1-3)Glc(α1-2)Glc(α1-3)acyl$_2$Gro	—	—	—	—	—	7:3[c]

[a]   NH$_4$OAc, g/ℓ.
[b]   Not elutable from silicic acid with acetone.
[c]   Separation from contaminating phospholipids.

Packing	P1	=	Silicic acid
	P2	=	DEAE-cellulose (Whatman D 23), acetate form
	P3	=	Silicic acid (Konto-Kagaku), 100 mesh
	P4	=	DEAE-cellulose (Seikagaku-kogyo), acetate form
	P5	=	DEAE-cellulose, acetate form
Pretreatment	T1	=	Fraction of crude lipids, soluble in ethereal chloroform
	T2	=	Mixture of purified lipids
	T3	=	Glycolipid fraction from chromatography on silicic acid
	T4	=	Fraction of crude lipids, insoluble in acetone
	T5	=	Fraction from chromatography on silicic acid (column 4)
	T6	=	Polar lipid fraction from chromatography on silicic acid
Elution (v/v)	E1	=	Chloroform; chloroform-methanol, stepwise
	E2	=	Chloroform-acetone, stepwise
	E3	=	Chloroform-methanol, stepwise, followed by a gradient from 7:3 to 6:4
	E4	=	Chloroform-methanol, stepwise
Detection	D1	=	By TLC

REFERENCES

1. **Wilkinson, S. G.,** *Biochim. Biophys. Acta,* 187, 492, 1969.
2. **Wilkinson, S. G.,** *Biochim. Biophys. Acta,* 164, 148, 1968.
3. **Exterkate, F. A. and Veerkamp, J. H.,** *Biochim. Biophys. Acta,* 176, 65, 1969.
4. **Oshima, M. and Yamakawa, T.,** *Biochemistry,* 13, 1140, 1974.
5. **Mayberry, W. R., Langworthy, T. A., and Smith, P. F.,** *Biochim. Biophys. Acta,* 441, 115, 1976.

Table 6
GLYCOGLYCEROLIPIDS (MAMMALIAN ORIGIN) (LC-5)

Packing							
	P1	P2	P3	P4	P5	P6	P5
Weight (g)	200	—	—	—	—	—	—
Column							
Length (cm)	—	—	—	82	40	35	30
Diameter (cm)	—	—	—	0.7	2.5	1.2	1.2
Lipid							
Load	5 g	25 mg/g[a]	25 mg/g[a]	—	—	630 mg	—
Pretreatment	T1	T2	T3	T4	T5	T5	T6
Elution	E1	E1	E2	E3	E4	E5	E6
Detection	D1	D2	D2	D3	D1	D1	D1
References	1	2	2	3	4	5	5

Compound	C:M	C:M	C:A	C:M:W	A	C:M:W	A:M
Gal(β1-3)alkyl, acylGro	97:3:96:4	90:10	7:3	~81:18:1.3	—	—	—
HO₃S-3Gal(β1-3)alkyl, acylGro	—	85:18	3:7	—	—	—	—
HO₃S-3Gal(β1-3)acyl₂Gro	—	—	—	~81:18:1.3	—	—	—
HO₃S-3Gal(β1-3)alkylGro	—	80:20	2:8	—	—	—	—
HO₃S-6Glc(α1-6)Glc α1-6)Glc(α1-3)alkyl, acylGro	—	—	—	—	1	—	—
Glc(α1-6)Glc(α1-6)Glc(α1-6)Glc(α1-6)Glc-(α1-6)	—	—	—	—	—	30:60:8	9:1
Glc(α1-3)alkyl, acylGro	—	—	—	—	—	—	—
Glc(α1-6)Glc(α1-6)Glc(α1-6)Glc(α1-6)Glc-(α1-6)	—	—	—	—	—	30:60:8	9:1
Glc(α1-6)Glc(α1-6)Glc(α1-3)alkyl, acylGro	—	—	—	—	—	—	—

[a] Lipid (mg) per g adsorbent.

Packing		
P1	=	Unisil silicic acid
P2	=	Florisil with 1% water
P3	=	Silicic acid AR CC-7 (Mallinckrodt), 200-325 mesh
P4	=	Iatrobeads
P5	=	Silicic acid (Bio-Rad), 200-400 mesh
P6	=	DEAE-Sephadex

Pretreatment	T1	=	Crude lipid extract from brain
	T2	=	Fraction of lipid extract from testis, insoluble in acetone and diethyl ether, soluble in pyridine
	T3	=	Crude lipid extract from testis or treated as given under T2
	T4	=	Sulfolipid fraction from brain, partially purified on columns of Florisil and DEAE-Sephadex
	T5	=	Lipid extract from freeze-dried human gastric secretion
	T6	=	Neutral lipids from DEAE-Sephadex (column 6)
Elution (v/v)	E1	=	Chloroform; chloroform-methanol, stepwise
	E2	=	Chloroform; chloroform-acetone, stepwise
	E3	=	Chloroform-methanol-water; linear gradient from (93:7:0.5) to (69:29:2)
	E4	=	Chloroform acetone
	E5	=	Chloroform-methanol-water (30:60:8) continuously
	E6	=	Chloroform; acetone; acetone-methanol (9:1)
Detection	D1	=	By TLC
	D2	=	Anthrone-sulfuric acid method; phosphorus determination; TLC
	D3	=	^{35}S-label, scintillation counter

REFERENCES

1. **Norton, W. T. and Brotz, M.**, *Biochem. Biophys. Res. Commun.*, 12, 198, 1963.
2. **Ishizuka, I., Suzuki, M. and Yamakawa, T.**, *J. Biochem.*, 73, 77, 1973.
3. **Ishizuka, I., Inomata, M., Ueno, K., and Yamakawa, T.**, *J. Biol. Chem.*, 253, 898, 1978.
4. **Slomiany, B. L., Slomiany, A., and Glass, B. J.**, *Eur. J. Biochem.*, 78, 33, 1977.
5. **Slomiany B. L., Slomiany, A., and Glass, B. J.**, *Biochemistry*, 16, 3954, 1977.

Table 7
GLYCOGLYCEROLIPIDS (TLC-1)

	L1	L1	L1	L2	L3		L4	L4	L4	L4	L4
Layer	S1	S2	S3	S3	S4		S5	S6	S7	S8	S9
Solvent											
Detection	D1	D2	D2	D2	D3		D2	D2	D2	D2	D2
References	1	2	2	2	3		4	4	4	4	4
Compound						$R_f \times 100$					
Acyl-6Gal(β1-3)acyl$_2$Gro	—	—	—	—	—		~100	90	93	99	90
Gal(β1-3)acyl$_2$Gro	—	67	78	49	—		~100	70	66	81	65
Glc(α1-3)acyl$_2$Gro	—	68	—	79	—		~100	69	71	83	66
Glc(β1-3)acyl$_2$Gro	—	—	—	—	—		~100	69	69	80	64
GlcNAc(α1-3)acyl$_2$Gro	—	—	—	—	—		~100	71	75	83	—
Gal(β1-3)alkyl,acylGro	—	67	78	48	—		—	—	—	—	—
Glc(α1-3)alkyl>acylGro	—	—	—	—	89		—	—	—	—	—
HO$_3$S-3Gal(β1-3)alkyl,acylGro	—	30	62	43	—		—	45	—	—	—
HO$_3$S-6Qui(α1-3)acyl$_2$Gro	—	21	49	45	—		—	32	34	27	6
Acyl-6Gal(α1-6)Gal(β1-3)acyl$_2$Gro	—	—	—	—	—		~100	65	58	71	42
Gal(α1-2)-,acyl-6Glc(α1-3)acyl$_2$Gro	—	—	—	—	—		~100	69	64	76	47
Glc(α1-2)-,acyl-6Glc(α1-3)acyl$_2$Gro	—	—	—	—	—		~100	70	67	78	51
Gal(α1-6)Gal(β1-3)acyl$_2$Gro	—	35	44	15	—		66	42	31	30	—
Gal(α1-2)Glc(α1-3)acyl$_2$Gro	—	—	—	—	—		75	47	39	35	—
Glc(α1-2)Glc(α1-3)acyl$_2$Gro	89	—	—	—	—		86	48	39	43	16
Glc(β1-6)Glc(β1-3)acyl$_2$Gro	—	—	—	—	—		75	43	34	34	—
Man(α1-3)Man(α1-3)acyl$_2$Gro	—	—	—	—	—		85	47	38	45	—
Glc(α1-6)Glc(α1-3)alkyl,acylGro	—	—	—	—	71, 79[a]		—	—	—	—	—
Glc(β1-6)Gal(α1-2)-,acyl-6Glc(α1-3)acyl$_2$Gro	—	—	—	—	—		88	51	39	35	—
Gal(α1-6)Gal(α1-6)Gal(β1-3)acyl$_2$Go	—	—	—	—	—		43	26	16	14	—
Glc(β1-6)Gal(α1-2)Glc(α1-3)acyl$_2$Gro	—	—	—	—	—		50	34	25	17	—
Glc(α1-2)Glc(α1-2)Glc(α1-3)acyl$_2$Gro	70	—	—	—	—		51	32	23	19	4
Glc(β1-6)Glc(β1-6)Glc(β1-3)acyl$_2$Gro	—	—	—	—	—		45	27	18	14	—
Glc(α1-6)Glc(α1-6)Glc(α1-3)alkyl,acylGro	—	—	—	—	60		—	—	—	—	—
HO$_3$S-6Glc(α1-6)Glc(α1-6)Glc(α1-3)alkyl,acylGro	—	—	—	—	56		—	—	—	—	—
Glc(β1-6)Glc(β1-6)Gal(α1-2)Glc(α1-3)acyl$_2$Gro	—	—	—	—	—		33	22	9	7	—
Galf(β1-2)Gal(α1-6)GlcNacyl(β1-3)Glc(α1-3)acyl$_2$Gro	79	—	—	—	—		—	—	—	—	—
Glc(α1-6)Glc(α1-6)Glc(α1-6)Glc(α1-6)Glc-(α1-6) Glc(α1-3)alkyl,acylGro	—	—	—	—	31		—	—	—	—	—
Glc(α1-6)Glc(α1-6)Glc(α1-6)Glc(α1-6)Glc-(α1-6) Glc(α1-6)Glc(α1-6)Glc(α1-3)alkyl, acylGro	—	—	—	—	20		—	—	—	—	—

[a]   Double spot, possibly due to different chains.

Layer	L1	=	Silica gel
	L2	=	Silica gel, impregnated with borate, 2%
	L3	=	Silica gel HR 250 μm
	L4	=	Precoated silica gel plates (Merck, silica gel 60)
Solvent (v/v)	S1	=	Chloroform-methanol-water (65:35:8)
	S2	=	Chloroform-methanol-water (65:25:4)
	S3	=	Chloroform-methanol-25% ammonia-water (60:40:5:5)
	S4	=	Chloroform-methanol-water (65:30:8)
	S5	=	Chloroform-methanol-acetic acid-water (65:25:10:5)
	S6	=	Chloroform-methanol-water (65:25:5)
	S7	=	Chloroform-methanol-25% ammonia-water (65:30:3:3)
	S8	=	Chloroform-acetone-methanol-acetic acid-water (50:20:10:10:5)
	S9	=	Chloroform-acetone-methanol-acetic acid-water (80:20:10:10:4)
Detection	D1	=	Anthrone/H$_2$SO$_4$
	D2	=	Iodine vapor, followed by α-naphthol/H$_2$SO$_4$
	D3	=	Orcinol reagent (see Chapter 28)

Table 7 (continued)
GLYCOGLYCEROLIPIDS (TLC-1)

REFERENCES

1. **Oshima, M. and Yamakawa, T.,** *Biochem. Biophys. Res. Commun.,* 49, 185, 1972.
2. **Ishizuka, I., Suzuki, M., and Yamakawa, T.,** *J. Biochem.,* 73, 77, 1973.
3. **Slomiany, B. L., Slomiany, A., and Glass, G. B. J.,** *Eur. J. Biochem.,* 84, 53, 1978: R_F-values personal communication.
4. **Fischer, W.,** unpublished data.

Table 8
GLYCOGLYCEROLIPIDS (TLC-2)

Layer	L1	L1	L1
Solvent	S1	S2	S3
Detection	D1,2	D1,2	D1,2
References	1	2,3	3,4
Compound		$R_f \times 100$	
Ptd-6Glc(α1-3)acyl$_2$Gro	65	—	—
Ptd-6Glc(α1-2)Glc(α1-3)acyl$_2$Gro	47	—	—
Glc(α1-2)-,Ptd-6Glc(α1-3)acyl$_2$Gro	34	—	—
Ptd-Glc (α1-2)Glc(α1-2)Glc(α1-3)acyl$_2$Gro	25	—	—
Gro*P*-6Glc(α1-2)Glc(α1-3)acyl$_2$Gro	5	30	21
Gro*P*-6Glc(α1-2)-,Ptd-6Glc(α1-3)acyl$_2$Gro	7	—	—
Gro*P*-6Glc(α1-2)-,acyl-6Glc(α1-3)acyl$_2$Gro	—	41	36
Gro*P*-6Glc(β1-6)Glc(β1-3)acyl$_2$Gro	—	23	—
Gro*P*-6Glc(β1-6)Gal(α1-2)Glc(α1-3)acyl$_2$Gro	—	21	—
Gro*P*-6Glc(β1-6)Gal(α1-2)-,acyl-6Glc(α1-3)acyl$_2$Gro	—	32	—
(Gro*P*)$_2$-6Glc(α1-2)Glc(α1-3)acyl$_2$Gro	—	—	5
(Gro*P*)$_2$-6Glc(α1-2)-,acyl-6Glc(α1-3)acyl$_2$Gro	—	—	13
Gal$_3$Gro*P*-6Glc(α1-2)-,acyl-6Glc(α1-3)acyl$_2$Gro	—	7	—
Glc(α1-2)Glc(α1-3)acyl$_2$Gro[a]	39	60	53
Glc(α1-2)Glc(α1-2)Glc(α1-3)acyl$_2$Gro[a]	19	32	24

[a] Given for comparison.

Layer	L1 =	Precooled plates, silica gel 60, Merck
Solvent (v/v)	S1 =	Chloroform-acetone-methanol-acetic acid-water (50:20:10:10:5)
	S2 =	Chloroform-methanol-25% ammonia-water (65:35:4:4) developed twice
	S3 =	Chloroform-methanol-25% ammonia-water (65:35:3:3) developed twice
Detection	D1 =	Chapter 28, Reagent No. 22
	D2 =	Chapter 28, Reagent No. 1

REFERENCES

1. **Fischer, W.,** *Biochim. Biophys. Acta,*487, 74, 1977.
2. **Fischer, W., Laine, R. A., and Nakano, M.,** *Biochim. Biophys. Acta,* 528, 298, 1978.
3. **Fischer, W., Nakano, M., Laine, R. A., and Bohrer, W.,** *Biochim. Biophys. Acta,* 528, 288, 1978.
4. **Laine, R. A. and Fischer, W.,** *Biochim. Biophys. Acta,* 529, 250, 1978.

Table 9
GLYCOGLYCEROLIPIDS, MOLECULAR SPECIES (TLC-3)

		L1	L2	L3
Layer		L1	L2	L3
Solvent		S1	S1	S2
Detection		D1,D2	D3,D4	D5
References		1	2	3

Compound	Fatty acid combinations	R_f	R_x^a	R_x
Gal(β1-3)acyl$_2$Gro[b]	16:0/18:1	0.66	—	—
	16:1/18:1	0.59	—	—
	16:1/18:2 ⎫ 16:2/18:1 ⎭	0.50	—	—
	16:2/18:2	0.41	—	—
	16:2/18:3	0.26	—	—
	Hexaenoic	0.16	—	—
Gal(β1-3)acyl$_2$Gro	16:0/16:0	—	1.00	—
	18:1/18:1	—	0.87	—
	18:2/18:2	—	0.68	—
	18:0/18:3	—	0.63	—
	18:3/18:3	—	0.29	—
	18:3/16:3	—	0.21	—
HO$_3$S-6Qui(α1-3)acyl$_2$Gro[b]	16:0/16:0	—	—	1.00
	16:0/18:1	—	—	0.88
	16:0/18:2 ⎫ 18:1/18:1 ⎭	—	—	0.83
	16:0/18:3	—	—	0.75
	18:2/18:2 ⎫ 18:1/18:3 ⎭	—	—	0.66
	18:2/18:3	—	—	0.54
	18:3/18:3	—	—	0.27

[a] R_X, Mobility relative to that of the compound with the value 1.00.
[b]   Fatty acid combinations derived by the compiler from the given fatty acid compositions.

Layer	L1 = Silica gel (without binder), containing 10% (w/w) silver nitrate
	L2 = Silica gel H (Merck), containing 10% (w/w) silver nitrate
	L3 = Silica gel G - silver nitrate (4:1) (w/w)
Solvent (v/v)	S1 = Chloroform-methanol-water (60:21:4) containing 0.005% butylated hydroxy toluene (for preparative TLC)
	S2 = Chloroform-methanol-water (65:25:4)
Detection	D1 = Chlorsulfonic acid and charring (analytical)
	D2 = Dichlorofluorescein (preparative)
	D3 = 50% H_2SO_4 and charring (analytical)
	D4 = Aniline naphthalenesulfonate (preparative)
	D5 = Rhodamine 6G (see Chapter 28)

REFERENCES

1. **Nichols, B. W. and Moorhouse, R.,** *Lipids,* 4, 311, 1969.
2. **Siebertz, M. and Heinz, E.,** *Hoppe Seyler's Z. Physiol. Chem.,* 358, 27, 1977. R_X-values taken from **Siebertz, H. P. and Heinz, E.,** *Z. Naturforsch.,* 32c, 193, 1977.
3. **Heinz, E. and Haarwood, J. L.,** *Hoppe Seyler's Z. Physiol. Chem.,* 358, 897, 1977.

Table 10
GLYCOGLYCEROLIPIDS, DEACYLATED (PC, TLC)

Technique	PC1	PC2	PC2	PC2	TLC	TLC
Solvent	S1	S2	S2	S3	S4	S5
Detection	—	D1	D2	D2	D3	D3
References	1	2	3	4	5	5
Compound[a]	R_f	R_X	R_X	R_X	R_X	R_X
Gal (reference)	—	1.00	—	—	—	—
Glc (reference)	—	—	1.00	—	1.25	1.52
Gal(α1-3)Gro[b]	—	—	0.86	—	1.40	1.53
Gal(α1-2)Gro[b]	—	—	—	—	1.46	1.69
Gal(β1-3)Gro	0.62	0.89	0.85	—	1.34	1.43
Galf(β1-3)Gro	—	1.61	—	—	—	—
Glc (α1-3)Gro	—	—	1.00	1.00	1.46	1.71
Glc(α1-2)Gro[b]	—	—	—	—	1.54	1.83
Glc(α1-1)Gro[c]	—	—	1.00	—	—	—
Glc(β1-3)Gro	—	—	1.02	—	1.51	1.74
Glc(β1-1)Gro[c]	—	—	1.00	—	—	—
GlcNAc(α1-3)Gro	—	—	—	—	1.61	1.87
GlcUA(α1-3)Gro	—	—	—	0.41	—	—
HO$_3$S-6Qui(α1-3)Gro	0.18	—	—	—	—	—
Gal(α1-6)Gal(α1-3)Gro[b]	—	—	—	—	0.80	0.74
Gal(α1-6)Gal(β1-3)Gro	0.46	0.20	—	—	0.78	0.69
Gal(α1-2)Glc(α1-3)Gro	—	—	0.43	—	0.89	0.91
Glc(α1-2)Glc(α1-3)Gro	—	—	0.52	—	1.00	1.00
Glc(β1-2)Glc(α1-3)Gro[c]	—	—	0.57	—	—	—
Glc(β1-2)Glc(β1-3)Gro[c]	—	—	0.65	—	—	—
Glc(β1-3)Glc(α,β1-3)Gro[c]	—	—	0.79	—	—	—
Glc(β1-4)Glc(β1-3)Gro[c]	—	—	0.43	—	—	—
Glc(β1-6)Glc(β1-3)Gro	—	—	0.49	—	1.22	1.03
Man(α1-3)Man(α1-3)Gro	—	—	0.51	—	1.23	1.18
Glc(β1-4)GlcUA(α1-3)Gro	—	—	—	0.29	—	—
Gal(α1-6)Gal(α1-6)Gal(β1-3)Gro	0.38	—	—	—	0.28	0.28
Gal(α1-6)Gal(α1-3)Gro(2←1α)Gal[b]	—	—	—	—	0.38	0.32
Glc(α1-2) Glc(α1-2)Glc(α1-3)Gro	—	—	—	—	0.60	0.63
Glc(β1-6)Gal(α1-2)Glc(α1-3)Gro	—	—	—	—	0.61	0.52
Glc(β1-6)Glc(β1-6)Glc(β1-3)Gro	—	—	—	—	0.74	0.51
Glc(β1-6)Glc(β1-6)Gal(α1-2)Glc(β1-3)Gro	—	—	—	—	0.29	0.23

[a] R_X, Mobility relative to that of the compound with the value 1.00.
[b]   Prepared from lipoteichoic acids and glucosylated bisphosphatidylglycerol, respectively.
[c]   Chemically synthesized.

Technique	PC1 =	Whatman No. 4 paper
	PC2 =	Whatman No. 1 paper
		TLC, precoated plates, silica gel 60 (Merck)
Solvent (v/v)	S1 =	Phenol-water (5:2)
	S2 =	Butanol-pyridine-water (6:4:3)
	S3 =	1-Butan-l-ol - ethanol-water-ammonia (ρ = 0.88), (40:10:49:1) upper phase
	S4 =	1-Propanol-pyridine-water (7:4:2) developed once (13 cm)
	S5 =	1-Propanol-ethyl acetate-water (7:2:2) developed twice (12 cm each)
Detection	D1 =	Reagent No. 35
	D2 =	Reagents No. 33 or 35
	D3 =	α-Naphthol/H$_2$SO$_4$ or alkaline permanganate (see Chapter 28)

Table 10 (continued)
GLYCOGLYCEROLIPIDS, DEACYLATED (PC, TLC)

REFERENCES

1. **Ferrari, R. A. and Benson, A. A.,** *Arch. Biochem. Biophys.,* 93, 185, 1961.
2. **Plackett, P.,** *Biochemistry,* 6, 2746, 1967.
3. **Brundish, D. E. and Baddiley, J.,** *Carbohyd. Res.,* 8, 308, 1968.
4. **Wilkinson, S. G.,** *Biochim. Biophys. Acta,* 187, 492, 1969.
5. **Fischer, W.,** unpublished experiments.

Table 11
GLYCOGLYCEROLIPIDS, MOLECULAR SPECIES (GC-1)

Column Packing	P1		P2	P3
Temperature	265-340		240-320	220-300
Range (°C)				
Program (°C/min)	2		2	3
Gas	He		He	—
Flow rate (mℓ/min)	45		60	—
Column				
Length (cm)	60		100	100
Diameter (cm)	0.3		0.3	0.3
Form	—		—	—
Material	SS		SS	Glass
Detector	FID		FID	FID
References	1		2	3

Compound	Fatty acid combinations		Retention time (min)[a]	
Tetrakis(trimethylsilyl)-	16:16	10.7	—	—
Gal(β1-3)acyl$_2$Gro,	16:18	13.6	—	—
hydrogenated	18:18	16.5	—	—
Tris(trimethylsilyl)-	14:14	—	8.7	—
MeO$_3$S-6Qui(α1-3)acyl$_2$Gro,	14:16	—	12.5	—
hydrogenated	16:16, 18:14	—	16.1	—
	18:16, 20:14	—	20.6	—
	16:20, 18:18, 22:14	—	24.9	—
	18:20, 24:14	—	29.3	—
	24:16	—	33.5	—

Compound	Hydrocarbon chain		Retention time (min)[a]	
Hexakis(trimethylsilyl)-				
Gal(β1-3)Gro[b]	None	—	—	6.4
Pentakis(trimethylsilyl)-	14:0	—	—	25.6
Gal(β1-3)alkylGro[b]	16:0	—	—	29.2
	18:0, 18:1	—	—	32.3

[a] Calculated from published GC tracings.
[b] Obtained by solvolysis and deacylation of HO$_3$S-3Gal(β1-3)acyl$_2$- and alkyl,acylGro (seminolipid).

Column packing P1 = 2% SE 30 on AW-DMCS Chromosorb W (60-80 mesh)
P2 = 1.5% Dexsil 300 on AW-DMCS Chromosorb W (80-100 mesh)
P3 = 4% OV-101 on Celite 545 HMDS (80—100 mesh)

REFERENCES

1. **Auling, G., Heinz, E., and Tulloch, A. P.,** *Hoppe Seyler's Z. Physiol. Chem.,* 352, 905, 1971.
2. **Tulloch, A. P., Heinz, E., and Fischer, W.,** *Hoppe Seyler's Z. Physiol. Chem.,* 354, 879, 1973.
3. **Ishizuka, I., Inomata, M., Ueno, K., and Yamakawa, T.,** *J. Biol. Chem.,* 253, 898, 1978.

Table 12
GLYCOGLYCEROLIPIDS, DEACYLATED (GC-2)

Column packing	P1	P2	P1	P2	P2	P2
Temperature (°C)	185	210	255	180	210	130-260
Program (°C/min)	—	—	—	—	—	4
Gas	N_2	N_2	N_2	N_2	N_2	N_2
Pressure (kg/cm^2)	1.06	1.2	1.06	1.2	1.2	1.2
Column						
Length (cm)	150	210	150	210	210	210
Diameter (cm I.D.)	—	0.2	—	0.2	0.2	0.2
Form	—	Coiled	—	Coiled	Coiled	Coiled
Material	—	Glass	—	Glass	Glass	Glass
Detector	—	FID	—	FID	FID	FID
Derivatives of compounds	D1	D1	D1	D2	D2	D2
References	1	2	1	2	2	2

Compound	Retention time (min)					
Gal(α1-2)Gro[a]	—	7.5	—	—	—	6.6
Gal(α1-3)Gro[a,b]	16.5	8.9	—	—	—	6.8
Gal(β1-3)Gro	18.4	9.8	—	—	—	7.8
Glc(α1-1)Gro[b]	17.0	—	—	—	—	—
Glc(α1-2)Gro	—	8.0	—	—	—	—
Glc(α1-3)Gro	17.0	9.2	—	—	—	6.7
Glc(β1-1)Gro[b]	23.2	—	—	—	—	—
Glc(β1-3)Gro	23.2	11.5	—	—	—	—
GlcNAc(α1-3)Gro	—	18.3	—	—	—	—
Gal(α1-6)Gal(α1-3)Gro[a]	—	—	—	9.3	—	15.5
Gal(α1-6)Gal(β1-3)Gro	—	—	—	14.2	—	17.5
Gal(α1-2)Glc(α1-3) Gro	—	—	13.2	9.1	—	15.2
Glc(α1-2)Glc(α1-3)Gro	—	—	14.0	11.2	—	16.3
Glc(β1-2)Glc(α1-3)Gro[b]	—	—	14.0	—	—	—
Glc(β1-2)Glc(β1-3)Gro[b]	—	—	12.3	—	—	—
Glc(β1-3)Glc(α,β1-3)Gro[b]	—	—	12.0	—	—	—
Glc(β1-4)Glc(β1-3)Gro[b]	—	—	14.6	—	—	—
Glc(β1-6)Glc(β1-3)Gro	—	—	17.6	15.8	—	18.1
Man(α1-3)Man(α1-3)Gro	—	—	10.1	9.1	—	15.2
Gal(α1-6)Gal (α1-6)Gal(β1-3)Gro	—	—	—	—	8.9	23.0
Gal(α1-6)Gal(α1-3)Gro(2←1α)Gal[a]	—	—	—	—	7.6	22.2
Glc(α1-2)Glc(α1-2)Glc(α1-3)Gro	—	—	—	—	6.4	21.4
Glc(β1-6)Gal(α1-2)Glc(α1-3)Gro	—	—	—	—	10.1	23.8
Glc(β1-6)Glc(β1-6)Glc(β1-3)Gro	—	—	—	—	12.7	25.0
Glc(β1-6)Glc(β1-6)Gal(α1-2)Glc(α1-3)Gro	—	—	—	—	38.3	30.2

[a] Prepared from lipoteichoic acids and glucosylated bisphosphatidylglycerol, respectively.
[b] Chemically synthesized.

Column packing	P1 =	3% SE-52 on Celite (80-100 mesh)
	P2 =	2% SE-30 on Chromosorb G-AW-DMCS (80-100 mesh)
Derivatives of compounds	D1 =	Trimethylsilyl derivatives
	D2 =	Trifluoroacetyl derivatives

REFERENCES

1. **Brundish, D. E. and Baddiley, J.,** *Carbohyd. Res.,* 8, 308, 1968.
2. **Fischer, W.,** unpublished data. For separation of monohexosylglycerols see also: **Ishizuka, I. and Yamakawa, T.,** *Jpn. J. Exp. Med.,* 39, 321, 1969; **Ishizuka, I., Inomata, M., Ueno, K., and Yamakawa, T.,** *J. Biol. Chem.,* 253, 898, 1978.

INDEX

A

AC, see Adsorption chromatography
Acer griseum, 327
Acetates, 514
Acetolysis of alkoxylipids, 512—513
Acetoxyarachidic acid, 303
Acetoxybehenic acid, 303
Acetoxylignoceric acid, 303
Acetoxystearic acid 303
Acetylacylglycerols,422
Acetylation, 501
Acetyl diacylglycerols, 434, 458—461, 472
    GLC of, 474
    silver nitrate-TLC, of 435
Acetyl monoacylglycerols, 475—476
Acid-catalyzed esterification, 41
Acid-catalyzed transesterification, 41
Acid hydrolysis on phosphate esters, 549—552
Acids, see specific acid
Acyl-1-(l-alkenyl)-*sn*-glycerol-2-phosphate, 22
Acylglycerols, 381—480
    AC of, 412
    chemical properties, 381—382
    chromatography, 385—392
    fractionation, 53
    GLC of, 397—403, 478
    LLC of, 392—397, 437—438
    multidimensional, 404—405
    physical properties, 381—382
    sample preparation, 382—385
    tabulated data, 410—480
    TLC of, 420
Adjusted retention time, 97
Adsorption chromatography (AC)
    acylglycerols, 412
    epoxytriacylglycerols, 414
    estolide acylglycerols, 417
    fractionation by, 148
    hydroxytriacylglycerols, 416
    ketotriacylglycerols, 415
Mercuric acetate adducts of triacylglycerols, 418
    neutral ether lipids, 520
    silver nitrate, 419
    triacylglycerols, 413, 419
Agave americana, 343
Alcohols, see also specific types
    chromatography, 75—76
    fatty, 16
    fractionation, 53
    long-chain, 15, 77
    preparation of samples, 74
    primary, 73, 348, 360, 362, 372, 377
    secondary, 73, 348, 360, 372
Aldehyde esters, 90—91
Aldehydes, 514
    alkoxy, 90—91
    chromatography, 75—76

dimethyl acetals, 85—87
    esters, 90—91
    fatty, 82—84, 88—89
    fractionation, 53
    GLC of, 536
    preparation of samples, 74—75
    surface structure of, 73
    surface lipids, 348, 360, 362, 372
Alkaline hydrolysis (saponification)
    alkoxylipids, 512
    amide bonds, 38
    ester bonds, 37—38
Alkali-stable phospholipids, 552—553
Alk-l-enylacylglycerols, 533, 534
Alk-1-enyldiacylglycerols, 53
Alk-l-enylethanediols, 544
Alkoxy acetaldehydes, 514
    GC of, 90—91
Alkoxyaldehydes, 90—91
Alkoxylipids, see also Ether lipids
    acetolysis of, 512—513
    degradation and derivatization of, 512—517
    enzymatic degradation of, 513
    extraction of, 509
    isolation of, 510—512
    methanolysis of, 512
    saponification of, 512
Alkoxy phospholipids, 501—502
Alkyl acetates, 78—81, 537
Alkylacylglyceroglycolipids, 511, 528—529
Alkylacylglycerols, 533, 534
Alkyldiacylglycerols, 53
2-Alkyl-1,3-dioxanes, 537
Alkylethanediols, 544
1-Alkylglycerols, 53, 525, 538, 539
2-Alkylglycerols, 525, 539
Alkylhydroxypropanedioic acid, 362
Amides, 38, 54
Amines, 54
Amino groups, 504
Amyrin ether, 372
Animals, surface lipids, 347—380
    CC, 366—369
    classes, 348—349
    esters, 350—351
    GC, 369—371, 373—374
    preparation of samples, 353—356
    quantitative analysis, 374—376
    reference compounds, 377
    TLC, 357—359, 362—366
Anteiso-alkanes, 58
Apple (*Malus pumila*), 321, 343
Argentation chromatography, 148
Artemisia absinthium, 300
Autoxidation
    food fats, 6
    samples, 36

B

Bacillus
   licheniformis, 558
   sp., 241
   subtilis, 558
Bacterial glycerophosphoglycolipids, 558
Base-catalyzed transesterification, 40—41
BDS, see 1,4-Butanediol succinate
Beeswax as reference compound, 377
Betula pendula (silver birch), 327
BHT, see Butylated hydroxy toluene
Biological aspects of lipids, 8—9
Birds, 355
Bloor's classification, 3
Bonds
   amide, 38
   double, 42—44
   ester, 37—38
Botrytis
   cinerea, 281
   fabae, 281
Bottle cork (*Quercus suber*), 344
Branched-chain fatty acids, 241—275
   nonvolatile, 244—246
   tabulated data, 247—275
   volatile medium-length, 243—244
   volatile short-length, 242—243
Brassica species, 377
Broad bean (*Vicia faba*), 321
Butanal, 82, 84, 90
1,4-Butanediol succinate (BDS), 102
Butylated hydroxy toluene (BHT), 47

C

Candelilla wax as reference compound, 377
Carbocyclic fatty acids
   chromatography, 282—283
   preparation of samples, 277—281
   tabulated data, 284—288
Carbon numbers (CN), see also Chain lengths, 98—
      101, 398—399
β-Carotene, 58
Catalytic hydrogenation, 42
CBN, see Commission on Biochemical
      Nomenclature
CC, see Column chromatography
Cellulose, 488
Cerebroside, defined, 25
Chain lengths
   equivalent, defined, 98—101
   hydrocarbons, 68
   modified equivalent, 100
Chemical synthesis of lipids, 9—10
Cholesterol, 54
Cholesteryl, 54
Choline, 504
Chromatography, see specific types
Chysobalanus icaco, 300

Citrus limon (lemon), 343
Classification of lipids, 2—5
Cleavage, oxidative, 43
CN, see Carbon numbers
Column chromatography (CC), 49, 50, 485—488
   cutins, 332—334
   fatty acids, 298—299
   glycoglycerolipids, 560—563, 571
   hydroxy acids, 303
   modified cellulose, 488
   silicic acid, 486—487
   suberins, 332—334
   surface lipids, 366—369, 375
   triacylglycerols, 385—387, 392—395
Columns
   open tubular, 401—403
   packed, 397—401
   variables in, 101—102
Commission on Biochemical Nomenclature (CBN),
      14
Complex mixture fractionation, 47—55
Compound lipids, 3
Constituent fatty acids, 505
Contamination, 36, 353
Cork (*Quercus suber*), 321, 327, 344
Countercurrent distribution of glycoglycerolipids,
      560
Crotalus
   adamanteus, 39
   atrox, 39
Cutin, 321—345
   CC of, 332—334
   classes of, 322—324
   depolymerization of, 333, 336, 339, 341
   GC of, 338—342
   preparation of, 325—331
   TLC of, 333—338
Cyanosilicone liquid phases, 102—103
Cyclic acetals of aldehydes, 86—87
ω-Cyclohexyl fatty acids, 280—281
ω-Cyclopentenyl fatty acids, 280—281
Cyclopropane fatty acids, 277—278
Cyclopropene fatty acids, 279—280

D

Deacetylation of Grignard reagent, 513
Deacylation, 496, 501
   glycoglycerolipids, 566—568
DEAE, see Diethylaminoethyl
Decanal, 82 84, 88, 90
Decanoic acid, 237, 238
Decanolide, 316
Dedecanal, 84
DEGS, see Diethylene glycol succinate
Demethyloctadecanedionate, 340
Depolymerization
   cutin, 333, 336, 339, 341
   reagents for, 329
   samples, 328—331

suberin, 333, 336, 340, 341
Derived lipids, 3
Deuterohydrazine reduction, 42—43
Dexsil profile, 374
Diacylglycerols, 501, 541
  fractionation, 53
  GLC of, 533
  resolution of, 405
  silver nitrate-TLC of, 433
Dialdehydes, 90—91
Dialkylglycerols, 53, 535
Diazomethane, 41—42
Diepoxystearate, 315
Diepoxystearic acid, 301
Diesters, 348, 360, 362
Diethylaminoethyl (DEAE) cellulose, 488
Diethylene glycol succinate (DEGS), 102
Dihexadecylglycerol, 340
Dihydroxyacetone, 524
Dihydroxybehenate, 305
Dihydroxybehenic acid, 306
Dihydroxy fatty acids
  GC of, 311—313
  TLC of, 305, 306
10,16-Dihydroxyhexadecanoic acid, 329
Dihydroxyoleic acid, 301
Dihydroxypalmitic acid, 313
Dihydroxystearate, 305
Dihydroxystearic acid
  elution volume for, 303
  equivalent chain lengths for, 312
  relative retention times for, 301, 306, 313
Diketones, 348, 360
β-Diketones, 377
Dimethyl acetals
  GC of, 85—89
  GLC of, 536
Dimethylborate (DMB) diacylglycerols, 449—450
Dimethylbutanoate, 249
Dimethylbutanoic acid, 247
Dimethyldecanoate, 249
Dimethyldihydroxyoctadecanedioate, 340
Dimethyldocosanedioate, 340
Dimethyldodecanoate, 260
Dimethylepoxyoctadecanedioate, 340
Dimethyltetradecanoate, 261
Dimethylhexadecanedioate, 340
Dimethylhexadecanoate, 261, 262
Dimethylhydroxyhexadecane, 339
Dimethylhydroxymethoxyoctadecanedioate, 340
Dimethylicosanedioate, 340
Dimethylnonaoate, 249
Dimethyloctadecanoate, 262
Dimethyloctanoate, 249
Dimethylpentadecanoate, 261
Dimethyltetradecanoate, 261
Dimethyltridecanoate, 260, 261
Dimethylundecanoate, 260
Dinitrophenylhydrazones, 90—91
Diols, 348, 360
1,3-Dioxolanetriacylglycerols, 457

Diphytanylglycerol, 340
Diradylethanediols, 543
Direct gas chromatography, 62
Distolasterias nipon, 511
DMB, see Dimethylborate
Docosadexaenyl acetate, 81
Docosadiene, 67
Docosahexaene, 67
Docosahexenyl acetate, 79
Docosanediol, 341
Docosanol, 340, 341
Docosanyl acetate, 81
Docosapentaene, 67
Docosapentaenyl acetate, 79, 81
Docosatetraenyl acetate, 79, 81
Docosatriene, 67
Docosene, 67
Docosenol, 77
Docosenyl acetate, 79, 81
Dodecanal, 82, 85, 86, 88, 90
Dodecanoic acid, 237, 238
Dotriacontanoic acid, 372
Dotriacontanol, 372
Dotriacontranal, 372
Double bonds
  fatty acids, 42—44
  stereomutation of, 43—44

E

ECL, see Equivalent chain lengths
EGS, see Ethylene succinate
Enzymatic degradation of alkoxylipids, 513
Enzymatic hydrolysis, 38—39, 484—485
Epicuticular lipids, 372, 377
Epoxidation, 43
Epoxy acids
  chromatography, 298—300
  preparation of samples, 295—298
  tabulated data, 301, 308, 315
Epoxybehenic acid, 308
Epoxyoctadecadienoate, 315
Epoxyoctadecenoate, 315
Epoxyoleate, 315
Epoxyoleic acid, 301
Epoxystearate, 315
Epoxystearic acid, 301, 308
Epoxy triacylglycerols, 414, 426
Equivalent chain lengths (ECL), 98—101
Ester bonds
  alkaline hydrolysis (saponification) of, 37—38
  hydrogenolysis of, 38
Esterification
  acid-catalyzed, 41
  free fatty acids, 41—42
  lipids, 39—42
Estolide triacylglycerols
  AC of, 417
  LLC of, 444
  structure of, 381

TLC of, 428
Ethanal, 82, 84, 90
Ether diol lipids, 511—512, 531
Ether lipids, 509—544
   preparation of samples, 509—517
   tabulated data, 520—544
Ethylbutanoate, 249
Ethylbutanoic acid, 247
Ethyldimethyloctadecanoate, 263
Ethyldimethyltetradecanoate, 263
Ethyldimethylundecanoate, 263
Ethylene succinate (EGS), 102
Ethylmethyldecanoate, 249
Ethylmethylhexadecanoate, 262
Ethylmethyloctadecanoate, 262
Ethylmethyloctanoate, 249
Ethyloctadecanoate, 257
Ethyltridecanoate, 251
Exocarpus cupressiformis, 281
Extraction, 35—36
   hydrocarbon, 57—59
   lipid, 33—37
   solvent, 34—35

F

FAO, see Food and Agricultural Organization
Fats, 6, 15
Fatty acids, see also specific types, 15
   branched chain, see Branched-chain fatty acids
   carbocyclic, see Carbocyclic fatty acids
   CC of, 298—299
   constituent, 505
   ω-cyclohexyl, 280—281
   ω-cyclopentenyl, 280—281
   cyclopropane, 277—278
   cyclopropene, 279—280
   defined, 14
   derivatives of, 296—298
   dihydroxy, see Dihydroxy fatty acids
   double bonds of, 42—44
   epoxy, see Epoxy acids
   free, 41—42
   furanoid, see Furanoid fatty acids
   GC of, 242—246, 300
   hydroxy, see Hydroxy acids
   keto, see Keto acids
   monohydroxy, see Monohydroxy fatty acids
   monomeric, 322—324
   PC of, 298
   straight chain, see Straight-chain fatty acids
   in surface lipids of plants and animals, 348, 361, 362, 372
   tetrahydroxy, 306, 307
   TLC of, 243, 246, 296, 299—300
   trihydroxy, 307, 313
Fatty alcohols, 16
Fatty aldehydes, 82—84, 88—89
Flame-TLC, 391
Food and Agricultural Organization (FAO), 9

Food fat autoxidation, 6
Fractionation
   adsorption chromatography, 148
   argentation chromatography, 148
   complex mixtures, 47—55
   gas-liquid partition chromatography, 148
   reversed-phase partition chromatography, 148
   solvents for, 53—54, 560
Free fatty acid esterification, 41—42
Fucolipids, defined, 25
Fungal spores, 356—357
Furanoid fatty acids,
   chromatography, 282—283
   preparation of samples, 281—282
   tabulated data, 289—293

G

Gangliosides, 31
Gas chromatography (GC), 95, 96, 482, 485, 486
   aldehyde esters, 90—91
   alkoxy acetaldehydes, 90—91
   alkyl acetates, 78—81
   cutin, 338—342
   cyclic acetals, 86—87
   dialdehydes, 90—91
   dihydroxy fatty acids, 311—313
   dimethyl acetals, 85—89
   dinitrophenylhydrazones, 90—91
   direct, 62
   fatty acids, 242—246, 300, 311—313
   fatty aldehydes, 82—84, 88—89
   glycoglycerolipids, 567—568, 586
   hydroxy fatty acids, 314
   keto acids, 316, 317—318
   long-chain alcohols, 77
   mass spectrometry coupled with, 63, 338, 496
   methyl ketones, 92
   monohydroxy fatty acids, 310—313
   oximes, 90—91
   phospholipids, 496
   plant epicuticular lipids, 372
   suberin, 338—342
   surface lipids, 369—376
   TLC coupled with, 62
   trihydroxy fatty acids, 313
Gas-liquid chromatography (GLC), 7, 42, 61—62, 97, 501
   acetyl diacylglycerols, 458—461, 472, 474
   acetyl monoacylglycerols, 475—476
   acylglycerols, 478
   aldehydes of, 536
   alk-l-enylacylglycerols, 533, 534
   alk-l-enylethanediols, 544
   alkylacetates, 537
   alkylacylglycerols, 533, 534
   2-alkyl-1,3-dioxanes, 537
   alkylethanediols, 544
   alkylglycerols, 538, 539

diacylglycerols, 455, 458—463, 470—472, 533, 535
dimethylacetals, 536
1,3-dioxolanetriacylglycerols, 457
diradylethanediols, 543
hydroxyalkylglycerols, 541
isovaleroyl diacylglycerols, 455
isovaleroyl monoacylglycerols, 464
ketoalkylglycerols, 541
monoacylglycerols, 464, 475—477
phytanylglycerols, 540
sphingolipid derivatives, 498—499
l-thioalkylglycerols, 542
TMS ether derivatives, 497
triacylglycerols, 451—454, 456, 465—469
trimethylsilyl diacylglycerols, 462—463, 470—471, 473
trimethylsilyl monoacylglycerols, 477
triradylglycerols, 532
Gas-liquid partition chromatography, 148
Gasteria planifolia, 321
GC, see Gas chromatography
Gel, 364
Gel permeation chromatography, 63
GLC, see Gas-liquid chromatography
Glycerides, see Acylglycerols
Glyceroglycolipids, 572—573
Glycerol derivatives, 18—19
Glycerolphosphoglycolipids, 557—559
Glycerophospholipids, 20, 22, 481—506
    chromatography, 485—502
    detection, 503—504
    preparation of samples, 481—485
    quantitative determination, 504—506
Glycoglycerolipids, 555—587
    CC of, 560—563, 571
    countercurrent distribution of, 560
    deacylation of, 566—568
    defined, 23
    distribution of, 560
    extraction of, 555—560
    fractionation, 571
    GC of, 567—568, 586—587
    lactobacilli, 576
    LC of, 574—579
    mammals, 578—579
    microorganisms, 574—577
    PC of, 563—567, 584—585
    plants, 572—573
    preparation of samples, 555—560
    solvent fractionation of, 560
    streptococci, 574—575
    TLC of, 563—567, 580—585
Glycolipids, 23—27
Glycosphingolipids, 23
Glycosylceramide, 25
Glycosylsphingoid, 25
Gooseberry (*Ribes grossularia*), 343
Grignard reagent deacetylation, 513
Group N streptococci, 559

H

Hentriacontane, 372
Hentriacontanol, 372
Heptacosane, 372
Heptadecane, 67, 68
Heptadecanol, 77
Heptadecanyl acetate, 78, 80
Heptadecenyl acetate, 78
Heptanal, 84, 88
Hexacosanal, 372
Hexacosanoic acid, 372
Hexacosanol, 340, 341, 372
Hexadecanal, 82, 85, 86, 88, 90
Hexadecane, 67, 68
Hexadecanediol, 341
Hexadecanetriol, 341
Hexadecanoic acid, 237, 238, 372
Hexadecanol, 77, 341
Hexadecanyl acetate, 78, 80
Hexadecenal, 85, 86, 88
Hexadecylglycerol, 340
Hexadedecenyl acetate, 78, 80
Hexahydroxystearic acid, 301
Hexanal, 82, 84, 88, 90
Hexanoic acid, 237
High performance liquid chromatography (HPLC), 8, 506
    triacylglycerols, 393
HPLC, see High performance liquid chromatography
Hydrazine reduction, 42—43
Hydrocarbons, 57—71
    chain lengths of, 68
    chromatography, 59—66
    derivatives of, 59
    extraction of, 57—59
    fractionation, 53
    polyunsaturated, 58
    retention times for, 67
    sample preparation, 57, 59
    separation of, 63—64
    in surface lipids, 348, 360, 362, 372
Hydrogenation, 42—43
Hydrogen flame-TLC, 391
Hydrogenolysis
    ester bonds, 38
    phospholipase C, 513—514
Hydrolysis, 333
    acid, 549—552
    alkaline, 37—38
    alkali-stable phospholipids, 552—553
    enzymatic, 38—39, 484—485
    lipids, 37—39
    pancreatic lipase and, 484
    phospholipase and, 484, 485
    phospholipids, 484, 485
    triacylglycerols, 38
Hydroxy acids
    chromatography, 298—300
    fractionation, 54

preparation of samples, 295—298
surface lipids, 348
tabulated data, 301, 303—307, 310—314
Hydroxyalkanoic acid, 362
Hydroxyalkylglycerols
GLC of, 541
isolation of, 510—511
TLC of, 527
Hydroxyarachidate, 310
Hydroxyarachidic acid, 313, 314
Hydroxybehenate, 310
Hydroxybehenic acid, 313, 314
Hydroxybutyric acid, 304
Hydroxycaprate, 305, 310
Hydroxycapronic acid, 304
Hydroxycaprylate, 310
Hydroxycerotate, 310
Hydroxydecadienoic acid, 314
Hydroxy β-diketones, 348, 360
Hydroxydodecendioic acid, 314
Hydroxyelaidic acid, 304, 311
Hydroxyenanthic acid, 314
Hydroxyethylamides, 54
Hydroxyheptanoic acid, 304
Hydroxyhexacosenoate, 310
Hydroxyicosadienoic acid, 312
Hydroxyisobutyric acid, 304
Hydroxyisocaproic acid, 304
Hydroxyisovaleric acid, 304
Hydroxylation, 43
Hydroxylaurate, 305, 310
Hydroxylignocerate, 310
Hydroxylignoseric acid, 301
Hydroxylinoleate, 310
Hydroxylinolenate, 310
Hydroxymyristate, 305, 310
Hydroxymyristic acid, 301
Hydroxynonacosan, 372
Hydroxyoctadecadienoic acid, 304, 305, 312
Hydroxyoctadecenoic acid, 311
Hydroxyoleate, 305, 310
Hydroxyoleic acid, 301, 304, 311, 313
Hydroxypalmitate, 310
Hydroxypalmitic acid, 301
Hydroxypentacosanoate, 310
Hydroxypentacosenoate, 310
Hydroxypentadecanoate, 310
Hydroxypropionic acid, 304
Hydroxystearate, 305, 310
Hydroxystearic acid, 301, 304, 311, 313, 314
Hydroxytetracosenoate, 310
Hydroxytriacylglycerols, 381, 416, 427
Hydroxytricosanoate, 310
Hydroxyundecanoate, 310
Hydroxyvaleric acid, 304

I

Icosadetraceneyl acetate, 78, 79
Icosadiene, 67
Icosadienyl acetate, 78, 80
Icosane, 68
Icosanediol, 341
Icosanoic acid, 237, 238
Icosanol, 340, 341
Icosanyl acetate, 80
Icosapentaene, 67
Icosapentaenyl acetate, 79, 81
Icosatetraene, 67
Icosatetraenyl acetate, 81
Icosatrienyl acetate, 80
Icosene, 67
Icosenol, 77
Icosenyl acetate, 78
Infrared (IR) spectroscopy, 63, 96
Insect surface lipids, 355—356
Intact phospholipids, 496—501
Internal standard, 376
International Union of Biochemistry, 10
International Union of Pure and Applied Chemistry, 10
Ion-exchange chromatography, 63
IR, see Infrared
Isolation, 5—8
Isopropylideneglycerols, 542
Isopropylidenes, 514, 541
Isovaleric acid, 247
Isovaleroyl acylglycerols, 424
Isovaleroyl diacylglycerols, 455
Isovaleroyl monoacylglycerols, 464

K

Keto acids
chromatography, 298—300
preparation of samples, 295—298
tabulated data, 302, 309, 316—318
Ketoalkylglycerols, 527, 541
Ketobutyric acid, 302, 309, 316
Keotcaproic acid, 302, 309
Ketocaprylic acid, 309
Keto fatty acids, 309
Ketoheptadecyl, 541
Ketoheptanoic acid, 309
Ketohexadecyl, 541
Ketoisocaproic acid, 302, 309, 316
Ketoisovaleric acid, 302, 309, 316
Ketols, 348, 360, 372
Ketomethylbutyric acid, 316
Ketomethylthiobutyric acid, 302, 316
Ketomethylvaleric acid, 302, 316
Ketones
chromatography, 75—76
methyl, 92
preparation of samples, 75
in surface lipids, 348, 360, 362, 372

Ketopelargonic acid, 309
Ketopropionic acid, 309
Ketotriacylglycerols, 415
Ketovaleric acid, 302, 309, 316

L

Lactobacilli, 576
Lactobacillus
    casei, 558
LC, see Liquid chromatography
Lemon (*Citrus limon*), 343
Licanic acid, 295
Lipidology, 1—32
    analysis, 5—8
    biological aspects, 8—9
    chemical synthesis, 9—10
    classification, 2—5
    definitions, 2—5
    historical aspects of research, 1—2
    isolation, 5—8
    nomenclature, 10, 14—31
Liquid chromatography (LC)
    glyceroglycolipids, 572—573
    glycoglycerolipids, 574—579
    high performance, see High performance liquid
        chromatography
    silicic acid, 241
Liquid-liquid chromatography (LLC)
    acylglycerols, 437—438
    diacylglycerols, 446
    dimethylborate (DMB) diacylglycerols, 449—450
    estolide triacylglycerols, 444
    neutral ether lipids, 521
    tertiarybutyldimethylsilyl diacylglycerols, 446
    triacylglycerols, 439—442, 444, 446—450
    trimethylsilyl diacylglycerols, 446
Liquid phases, 102—103
Liquid-solid chromatography, 59—60
LLC, see Liquid-liquid chromatography
Long-chain alcohols, 15, 77
Lycopersicon esculentum (tomato), 321, 343
Lysoplasmenic acid, 21, 22

M

Malus pumila (apple), 321, 343
Mammals
    isolation of surface lipids of, 354—355
    LC-5 of glycoglycerolipids from, 578—579
Mass spectrometry (MS), 96, 485
    GC combined with 63, 338, 496
    phospholipids, 496
MECL, see Modified equivalent chain length
Medical aspects of lipids, 8—9
Membrane isolation, 325
Mercuric acetate adducts, of triacylglycerols, 418
Mercuric acetate derivatives, 44
Methanolysis

alkoxylipids, 512
    phospholipids, 482—484
Methoxyalkylglycerols, 510—511, 527
Methylalkane structure, 58
Methyl arsolate ether, 372
Methylbutanoic acid, 247
Methyl decanoate, 185, 249
Methyl decohexaenoate, 188
Methyl diepoxystearate, 308
Methyl dihydroxyhexadecanote, 339
Methyl dimethyleneoctadecanoate, 287
Methyl docosadexaenoate, 119, 122, 127
Methyl docosadienoate, 110
    equivalent chain lengths for, 133, 140, 143, 181
    relative retention times for, 116, 119, 121, 124,
        153, 157
Methyl docosahexaenoate, 106, 111, 129
    equivalent chain lengths for, 131, 135, 138, 141,
        143, 150, 159, 162, 165, 167, 170, 174
    relative retention times for, 112, 116, 124, 147,
        153, 155, 172
Methyl docosanoate, 110, 129
    equivalent chain lengths for, 131, 133, 135, 137,
        140, 143, 150, 159, 162, 164, 167, 170,
        174, 181, 183, 191
    relative retention times for, 112, 116, 119, 121,
        122, 124, 127, 145, 147, 153, 155, 157,
        172, 188
    retention index for, 340
Methyl docosapentaenoate, 111
    equivalent chain lengths for, 131, 133, 140, 141,
        143, 150, 159, 164, 236
    relative retention times for, 112, 122, 145, 147,
        153, 155, 158
Methyl docosatetraenoate, 110, 129
    equivalent chain lengths for, 131, 135, 138, 150,
        162, 164, 167, 170, 181
    relative retention times for, 116, 121, 124, 127,
        145, 147, 155, 188
Methyl docosatrienoate, 179
Methyl docosenoate, 110, 129
    equivalent chain lengths for, 131, 133, 135, 138,
        140, 143, 150, 159, 164, 167, 174
    relative retention times for, 112, 114, 119, 121,
        122, 124, 127, 145, 147, 155, 172
Methyl dodecanoate, 185, 251, 255
Methyl dodecenoate, 227, 265
Methyl eicosanoate, 253
Methylenedodecanoate, 265
Methylenetrimethylhexadecanoate, 272
Methyl epoxybehenate, 308
Methyl epoxyoleate, 308
Methyl epoxystearate, 308
Methyl hemicosanoate, 116, 155
Methyl hemicosapentaenoate, 127
Methyl henicosanoate, 129
    equivalent chain lengths for, 131, 133, 135, 137,
        140, 143, 150, 159, 162, 170, 174
    relative retention times for, 119, 124, 145, 153,
        172, 188
Methyl henicosapentaenoate, 110

equivalent chain lengths for, 131, 137, 140, 170, 174
relative retention times for, 112, 116, 119, 147, 155
Methyl hentriacontanoate, 258
Methyl heptacosanoate, 258
Methyl heptacosatrienoate, 177
Methyl heptadecanoate, 105, 110, 129
    equivalent chain lengths for, 131, 133, 135, 137, 139, 142, 149, 159, 161, 164, 166, 169, 174, 190, 252, 253, 257
    relative retention times for, 112, 116, 118, 120, 122, 124, 126, 145, 147, 152, 154, 157, 172, 187
Methyl heptadecenoate, 110, 129
    equivalent chain lengths for, 131, 135, 137, 139, 142, 149, 161, 164, 166, 169
    relative retention times for, 112, 116, 118, 120, 122, 124, 126, 147, 152, 154, 172
Methyl heptadenoate, 174
Methyl heptanoate, 249
Methyl hetradecanoate, 252
Methyl hexacosadienoate, 177
Methyl hexacosahexaenoate, 177
Methyl hexacosanoate, 138, 177, 258, 340
Methyl hexacosapentaenoate, 177
Methyl hexacosatrienoate, 177
Methyl hexadecadienoate, 110
    equivalent chain lengths for, 133, 149
    relative retention times for, 116, 118, 126, 147, 154
Methyl hexadecanoate, 104, 105, 110, 129
    equivalent chain lengths for, 131, 133, 135, 137, 139, 142, 149, 159, 161, 164, 166, 169, 174, 180, 182, 190, 252, 256, 262
    relative retention times for, 112, 114, 116, 118, 120, 122, 124, 126, 145, 147, 152, 154, 157, 172, 185, 187
    retention index for, 339
Methyl hexadecatetraenoate
    equivalent chain lengths for, 131, 133
    relative retention times for, 116, 118, 124, 126, 145, 147, 152
Methyl hexadecatrienoate, 110
    equivalent chain lengths for, 135, 137, 139, 149, 161, 169
    relative retention times, for, 120, 145, 147, 152
Methyl hexadecenoate
    equivalent chain lengths for, 137, 139, 142, 149, 166, 266
    relative retention times for, 120, 145, 147, 154, 157, 186, 272
Methyl hexanoate, 249
Methyl hexicosanoate, 157
Methyl hydroxyarachidate, 301
Methyl hydroxydocosanoate, 340
Methyl hydroxyheptadecanoate, 301
Methyl hydroxyhexadecanoate, 273, 339, 340
Methyl hydroxycosanoate, 340
Methyl hydroxymyristate, 301
Methyl hydroxynonadecanoate, 301

Methyl hydroxyoctadecenoate, 339, 340
Methyl hydroxypalmitate, 301
Methyl hydroxystearate, 301
Methyl hydroxytetracosanoate, 340
Methyl hydroxytetradecanoate, 273
Methyl icosatrienoate, 159
Methyl icosanoate, 122
Methyl icosadienoate, 110
    equivalent chain lengths for, 131, 159, 164, 167, 181, 183, 191
    relative retention times for, 112, 116, 157, 172, 188
Methyl icosanoate, 104, 105, 107, 108, 110, 129
    equivalent chain lengths for, 131, 133, 135, 137, 140, 143, 150, 159, 162, 164, 166, 169, 174, 181, 183, 191
    relative retention times for, 112, 114, 116, 119, 121, 124, 127, 145, 147, 153, 155, 157, 172, 188
    retention index for, 340
Methyl icosapentaenoate, 110, 129
    equivalent chain lengths for, 131, 133, 137, 140, 143, 150, 162, 167, 170, 174
    relative retention times for, 116, 119, 121, 122, 124, 145, 153, 155
Methyl icosatetraenoate, 110, 129
    equivalent chain lengths for, 133, 135, 143, 162, 164, 167, 170, 181, 183, 236, 268
    relative retention times for, 112, 116, 153, 157, 172
Methyl icosatrienoate, 110
    equivalent chain lengths for, 137, 140, 150, 162, 174, 179, 268
    relative retention times for, 119, 121, 124, 127, 147, 153, 155
Methyl icosaynoate, 235
Methyl icosabatrienoate, 133
Methyl icosenoate, 110, 129
    equivalent chain lengths for, 131, 133, 135, 143, 166, 170
    relative retention times for, 119, 121, 122, 124, 127, 145, 155
Methyl icosopentaenoate, 106
Methyl ketocaprate, 316
Methyl ketolaurate, 316
Methyl ketones, 53, 73, 92
Methyl methoxyoctadecanoate, 340
Methyl methylenedocosanoate, 285
Methyl methylenedodecanoate, 284
Methyl methyleneheptadecanoate, 284
Methyl methylenehexadecanoate, 284
Methyl methyleneicosanoate, 285
Methyl methyleneoctadecanoate, 284—286
Methyl methylenetetradecanoate, 284
Methyl methyleneundecanoate, 284
Methyl monanoate, 249
Methyl nonacosanoate, 258
Methyl nonadecanoate, 105, 110, 129
    equivalent chain lengths for, 131, 133, 135, 137, 140, 142, 149, 159, 162, 164, 166, 169

relative retention times for, 112, 116, 119, 121, 122, 124, 127, 145, 147, 153, 155, 157, 172, 188
Methyl nonadecenoate, 129
  equivalent chain lengths for, 131, 135, 137, 142, 150, 162, 166
  relative retention times for, 112, 116, 119, 121, 124, 127, 147, 153, 155
Methyl nonynoate, 225
Methyl octacosanoate, 258
Methyl octacosatrienoate, 177
Methyl octadecadienote, 104, 105, 106, 107, 108, 109, 110
  equivalent chain lengths for, 140, 149, 162, 164, 166, 169, 199, 201, 203, 205, 207—211, 213, 214, 216, 220, 222, 230, 232
  relative retention times for, 122, 152, 187
Methyl octadecanoate, 104, 105, 107—110, 129
  equivalent chain lengths for, 131, 133, 135, 137, 139, 142, 149, 159, 161, 164, 166, 169, 174, 176, 180, 182, 190, 191, 236, 253, 257, 262
  relative retention times for, 112, 114, 116, 118, 120, 122, 124, 126, 145, 147, 152, 154, 157, 172, 186, 187
Methyl octadecatetraenoate
  equivalent chain lengths for, 131, 133, 140, 142, 149, 159, 164, 166, 169, 180, 183
  relative retention times for, 112, 119, 121, 122, 124, 127, 145, 152
Methyl octadecatrienoate, 104—110
  equivalent chain lengths for, 135, 137, 169, 174, 217, 219, 224, 235
  relative retention times for, 116, 118, 120, 155, 157
Methyl octadecenoate, 104—108, 129
  equivalent chain lengths for, 131, 133, 137, 139, 142, 161, 191, 194, 196, 197, 228, 233, 234, 266
  relative retention times for, 118, 120, 124, 126, 147, 154, 186
Methyl octanoate, 249
Methyl oleanolate ether, 372
Methyl oxohexadecanoate, 339
Methyl pentaocsadienoate, 177
Methyl pentacosanoate, 258
Methyl pentadecanoate, 105, 110, 129
  equivalent chain lengths for, 131, 135, 137, 139, 142, 149, 159, 161, 164, 166, 174, 182, 252, 256
  relative retention times for, 112, 116, 118, 120, 122, 124, 126, 145, 147, 152, 154, 157, 172, 187
Methyl pentadecenoate
  equivalent chain lengths for, 137, 139, 142, 161, 164, 174, 190, 266
  relative retention times for, 112, 118, 120, 124, 147, 152, 154, 172
Methyl pentanoate, 249
Methylpentanoic acid, 247
Methylpropanoic acid, 247

Methyl tetracosadienoate, 176
Methyl tetracosahexaenoate, 177
Methyl tetracosanoate, 111, 129
  equivalent chain lengths for, 131, 133, 135, 141, 143, 150, 159, 162, 165, 167, 170, 174, 176, 181, 258
  relative retention times for, 116, 119, 121, 122, 124, 127, 145, 153, 155, 172, 188
  retention index for, 340
Methyl tetracosapentaenoate, 176
Methyl tetracosatrienoate, 176, 179
Methyl tetracosenoate, 129
  equivalent chain lengths for, 131, 133, 135, 141, 143, 150, 159, 162, 165, 170, 174, 181
  relative retention times for, 121, 122, 124, 127, 153, 155, 172, 188
Methyl tetradecanoate, 104, 105, 110, 129
  equivalent chain lengths for, 131, 135, 137, 139, 142, 149, 159, 161, 164, 166, 180, 182, 190, 251, 256
  relative retention times for, 112, 114, 116, 118, 120, 122, 124, 126, 145, 147, 152, 154, 157, 172, 185, 187
Methyl tetradecenoate, 129
  equivalent chain lengths for, 131, 135, 137, 139, 142, 149, 161, 166, 252, 265
  relative retention times for, 112, 118, 120, 124, 126, 145, 147, 152, 154, 172, 272
Methyl tetramethyleneicosanoate, 287
Methyl triacontapentaenoate, 177
Methyl triacontatrienoate, 177
Methyl tricontanoate, 258
Methyl tricosanoate, 258
Methyl tridecanoate, 251, 255, 256
Methyl trihydroxyoctadecanoate, 340
Methyl trihydroxyoctadecenoate, 339
Methyl trimethyleneoctadecanoate, 287
Methyl undecanoate, 249, 251, 255
Methyl undecenoate, 193, 226
Microorganisms, see also specific types
  LC-4 of glycoglycerolipids from, 57
Modified cellulose, 488
Modified equivalent chain length (MECL), 100
Modified layers, 337
Molecular species of phospholipids, 496—501
Monoacylglycerols, 405
Monoalkylglycerols, 436
Monoesters, 360, 362
Monohydroxy fatty acids
  GC of, 310—313
  TLC of, 304, 305
Monomeric fatty acids, 322—324
MS, see Mass spectrometry

N

Neuraminic acid, 26, 27
Neutral ether lipids, 520—523
Neutral fats, defined, 15
Nitriles, 54

NMR, see Nuclear magnetic resonance
Nomanal, 82
Nomenclature, 10, 14—32
Nonacosane, 372
Nonacosanol, 372
Nonacosanone, 372
Nonanal, 84, 88, 90
Nonpolar solvents, 34
Norphytene structure, 58
Nuclear magnetic resonance (NMR) spectroscopy, 63, 96

O

Octacosonal, 372
Octacosanoic acid, 372
Octacosanol, 372
Octadecadienal, 82, 85, 86, 88
Octadecadiene, 68
Octadecadienol, 77
Octadecadienyl acetate, 78, 80
Octadecanal, 82, 85, 86, 88, 90
Octadecane, 67, 68
Octadecanediol, 341
Octadecanetetraol, 341
Octadecanetriol, 341
Octadecanoic acid, 237, 238, 372
Octadecanol, 77, 341
Octadecantrienal, 88
Octadecanyl acetate, 78, 80
Octadecatetraenyl acetate, 80
Octadecatrienal, 85, 86
Octadecatriene, 68
Octadecatrienyl acetate, 78, 80
Octadecenal, 82, 85, 86, 88, 90
Octadecene, 341
Octadecenetetraol, 341
Octanal, 82, 88, 90
Octanoic acid, 237
Oils, 295—296
Open tubular columns for triacylglycerols, 401—403
Ophiophagus hannah, 39
Oxidation, 43
Oxidative cleavage, 43
Oxidized triacyglycerols, 430
Oximes, 90—91
Oxygenated seed oil extraction, 295—296
Ozonolysis, 515

P

Packed columns for triacylglycerols, 397—401
PAF, see Platelet activating factor
Palmitoyl-2-stearoyl-*sn*-glycero-3-phosphoethano-lamine, 22
Pancreatic lipase, 38, 484, 513
Paper chromatography (PC), 485, 488—490, 503
    epoxy acids, 301

fatty acids, 298
glycoglycerolipids, 563—567, 584—585
hydroxy acids, 301
keto acids, 302
methyl esters, 301
phospholipids, 489, 491
triacyglycerols, 395—397
Parinarium lauricum, 300
Parsnip (*Pastinaca sativa*), 321
Partition chromatography, 148
Pastinaca sativa (parsnip), 321
PC, see Paper chromatography
Pentacyclic triterpenoids, 348, 349
Pentadecane, 67, 68
Pentadecanol, 77
Pentadecanyl acetate, 78, 80
Pentanal, 82, 84, 90
Phosphatases, 513
Phosphate esters, 549—552
Phosphatidic acids, 20, 21, 482, 501
Phosphatidylcholines, 482
Phosphatidylethanolamines, 482
Phosphatidylglycerols, 482
Phosphatidylinositides, 482
Phosphatidylserines, 482
Phospholipase, 39, ss481
    degradation of, 513
    hydrogenolysis of, 513—514
    hydrolysis of phospholipids with, 484, 485
    reductive ozonolysis of, 515
Phospholipids
    alkali-stable, 552—553
    chromatography of, 485—502
    defined, 20
    degradation of, 546—553
    extraction of, 546
    GC of, 496
    hydrolysis of, 484, 485
    intact, 496—501
    methanolysis of, 482—484
    molecular species of, 496—501
    MS of, 496
    PC of, 489, 491
    radioactive, 502
    TLC of, 494, 526
Phosphonolipids, 501—502
    degradation of, 553—554
    isolation of, 511
Phosphorus, 503, 504
Phytanoylacylglycerols, 425
Phytanylglycerol, 340, 540
Plants, see also specific plants
    epicuticular lipids of, 372
    glyceroglycolipids from, 572—573
    surface lipids, 347—380
        CC, 366—369
        classes, 348—349
        esters, 350—351
        GC, 369—374
        preparation of samples, 353—354, 356—357
        quantitative analysis, 374—376

reference compounds, 377
    TLC, 357—361, 364—366
Plasmalogen, 21
Plasmanic acid, 22
Plasmenic acid, 21, 22
Platelet activating factor (PAF), 516—517
PLC, see Preparation layer chromatography
PMR, see Proton magnetic resonance
Polar glycerophosphoglycolipids, 559
Polarity and separation of triacylglycerols, 386—388
Polar solvents, 35
Polyesters, 349
Polymeric plant lipids, see Cutins; Suberins
Polyunsaturated hydrocarbons, 58
Potato (*Solanum tuberosum*), 321
Preparation layer chromatography (PLC), 337—338
Preparative thin-layer chromatography (Prep-TLC) of surface lipids, 375
Prep-TLC, see Preparative thin-layer chromatography
Primary alcohols, 73, 348, 360, 362, 372, 377
Pristane, 58
Propanal, 82, 84, 90
Proton magnetic resonance (PMR) spectroscopy, 278
Psychosine, defined, 23

Q

Quantitative chromatography of whole surface lipids, 374—376
Quercus suber (cork), 321, 327, 344

R

Radioactive phospholipids, 502
Reduction, 42—43
Reductive ozonolysis of phospholipase C, 515
Relative response factor (RRF), defined, 375
Resolution
    diacylglycerols, 405
    monoacylglycerols, 405
    triacylglycerols, 404
Reversed-phase partition chromatography, 148
Reversed-phase TLC, 8, 496, 501
Ribes grossularia (gooseberry), 343
Ricinoleic acid, 295
RRF, see Relative response factor

S

Sanseviera trifasciata, 343
Saponification, see Alkaline hydrolysis
Secondary alcohols, 73, 348, 360, 372
Seed oils, 295—296
Sialic acid, defined, 27
Sialoglycosphingolipids, 26

Silica gel, 364
Silicic acid, 241, 486—487
Silver birch (*Betula pendula*), 327
Silver nitrate-AC of triacylglycerols, 419
Silver nitrate-TLC
    acetyldiacylglycerols, 434
    acetyldiradylglycerols, 435
    diacylglycerols, 433
    monoalkylglycerols, 436
    triacylglycerols, 431—433
Simple lipids, 3
Solanum tuberosum (potato), 321
Solvents
    extraction of, 34—35
    fractionation, 53—54, 560
    nonpolar, 34
    polar, 35
Spectroscopy, see also specific types
    infrared (IR), 63, 96
    mass, see Mass spectrometry
    nuclear magnetic resonance (NMR), 63, 96
    proton magnetic resonance (PMR), 278
    ultraviolet (UV), 63, 96
Spermaceti as reference compounds, 377
Sphinganine, 16
Sphingoid, 15, 16
Sphingoid base, 15
Sphingolipid derivatives, 498—499
Spinach (*Spinacia oleracea*), 321
Spinacia oleracea (spinach), 321
Spores, 356—357
Squalene, 58
Standards, 376
Stenacherium sp., 300
Stereomutation of double bonds, 43—44
Sterols, 348, 360, 362
Storage of tissues, 33—34
Straight-chain fatty acids, 95—240
    chromatography, 97—103
    preparation of samples, 97
    tabulated data, 104—240
Streptococci, 557
    group N, 559
    LC of glycoglycerolipids from, 574—575
Streptococcus
    faecalis, 558
    group B, 558
    hemolyticus, 558, 561
    lactis, 558
Suberin, 321—345
    CC of, 332—334
    classes of, 322—324
    depolymerization of, 333, 336, 340, 341
    GC of, 338—342
    preparation of samples, 327—331
    TLC, of, 333—338, 340
Surface lipids, in animals and plants, see also Animals; Plants
    in animals, 362—363
    in birds, 355
    CC of, 366—369, 375

in fungal spores, 356—357
GC of, 368—376
in insects, 355—356
isolation of, 354—356
in mammals, 354—355
in plants, 356, 360—361
prep-TLC of, 375
TLC of, 357—366, 375
whole, 374—376
Synthesis of chemicals, 9—10

T

Tertiarybutyldimethylsilyl diacylglycerols, 446
Tetracosanediol, 341
Tetracosanol, 340, 341, 372
Tetracosene, 67
Tetracosenyl acetate, 79, 81
Tetradecanal, 82, 85, 86, 88, 90
Tetradecane, 67, 68
Tetradecanoic acid, 237, 238
Tetradecanol, 77
Tetradecanyl acetate, 78, 80
Tetrahydroxy fatty acids, 306, 307
Tetrahydroxystearate, 307
Tetrahydroxystearic acid, 301, 303, 306
Tetrahymena pyriformis, 511
Tetramethyldecanoate, 263
Tetramethyldodecanoate, 263
Tetramethylheptadecanoate, 269, 272
Tetramethylhexadecanoate, 269, 272
Tetramethylhexadecenoate, 271, 272
Tetramethyloctadecanoate, 263, 269, 272
Tetramethylpentadecanoate, 269, 272
Tetramethyltridecanoate, 263
Tetramethylundecanoate, 263
Thin-layer chromatography (TLC), 7, 52—54
    acetylacylglycerols, 422
    acylglycerols, 420
    alcohols, 75
    aldehydes, 75
    alkylacylglyceroglycolipids, 528—529
    alkylglycerols, 525
    cutin, 333—338
    dihydroxyacetone, 524
    dihydroxy fatty acids, 305, 306
    dimethylborate (DMB) diacylglycerols, 449—450
    epoxy fatty acids, 308
    epoxy triacyglycerols, 426
    estolide triacylglycerols, 428
    ether diol lipids, 531
    fatty acids, 243, 246, 296, 299—300, 304, 305,
        307—309
    flame, 391
    GC coupled with, 62
    glycerophospholipids, 482, 485, 486, 488, 490—
        496, 501, 503, 505, 506
    glycoglycerolipids, 563—567, 580—585
    hydrocarbons, 60—61
    hydroxyalkylglycerols, 527

    hydroxytriacylglycerols, 427
    isovaleroylacylglycerols, 424
    ketoalkylglycerols, 527
    keto fatty acids, 309
    ketones, 75
    methoxyalkylglycerols, 527
    methyl esters, 308
    modified layers, 337
    monohydroxy fatty acids, 304, 305
    neutral ether lipids, 522—523
    oxidized triacylglycerols, 430
    phospholipids, 494, 526
    phytanoylacylglycerols, 425
    preparative, 375
    reversed-phase, 8, 496, 501
    silica gel, 364
    silver nitrate-, see Silver nitrate-TLC
    suberin, 333—338, 340
    surface lipids, 357—366, 375
    tetrahydroxy fatty acids, 306, 307
    thioalkylglycerols, 530
    triacylglycerols, 387—392, 395—397, 421, 426,
        428, 430, 449—450
    tri-hydroxy fatty acids, 307
    two-dimensional, 51, 495—496
    unidimensional, 492—495
Thioalkylglycerols, 511, 530, 542
Tissue
    animal, 546
    storage of, 33—34
TLC, see Thin-layer chromatography
TMS, see Trimethylsilyl
Tomato (*Lycopersicon esculentum*), 321, 343
Transesterification, 37, 333
    acid-catalyzed, 41
    base-catalyzed, 40—41
    lipids, 39—42
Triacontanoic acid, 372
Triacontanol, 372
Triacylglycerols
    AC of, 413, 418
    CC of, 385—387, 392—395
    derivatization of, 384—385
    detection of, 387, 391—392, 394—395, 397, 403
    extraction of, 383
    fractionation, 53
    GLC of, 451—454, 456, 465—469
    HPLC of, 393
    hydrolysis of, 38
    isolation of, 383—384
    LLC of, 439—442, 447—450
    open tubular columns for, 401—403
    oxidized, 430
    packed columns for, 397—401
    PC of, 395—397
    resolution of, 404
    separation of, 386—391, 398—400
    silver nitrate-AC of, 419
    silver nitrate-TLC of, 431—433
    structure of, 381
    in surface lipids, 360, 362

TLC of, 387—392, 395—397, 421—433, 449—450
Trialkylglycerols, 53
Triepoxystearate, 315
Triesters, 348, 362
Trihydroxy fatty acids, 307, 313
Trihydroxystearate, 307
Trihydroxystearic acid, 313,314
Trimethylacetic acid, 248
Trimethylhexadecanoate, 263
Trimethyloctadecanoate, 263
Trimethylpentadecanoate, 263
Trimethylsilyl diacylglycerols, 473
   GLC of, 462—463, 470—471
   LLC of, 446
Trimethylsilyl (TMS) ethers, 485, 497, 514
Trimethylsilyl monoacylglycerols, 477
Trimethyltridecanoate, 263, 269
Triradylglycerols, 532
Triterpenoid acids, 361, 377
Triterpenoids, 348, 349, 372
Triterpenols, 360
Tubular columns, 401—403
Two-dimensional TLC, 51, 495—496

U

Ultraviolet (UV) spectroscopy, 63, 96
Undecanal, 82, 84, 85, 88, 90

Unidimensional TLC, 492—495
Unsaturation and separation of triacylglycerols,
   386—391, 399—400
UV, see Ultraviolet

V

Variables in columns, 101—102
Vernolic acid, 295
Vernonia anthelmintica, 300
Vicia faba (broad bean), 321
Vitamin A, 54

W

Wax, 53, 377
WHO, see World Health Organization
Whole surface lipids, 374—376
World Health Organization (WHO), 9